T0371618

Unconventional Reservoir Geomechanics

Since the beginning of the US shale gas revolution in 2005, the development of unconventional oil and gas resources has gathered tremendous pace around the world. This book provides a comprehensive overview of the key geologic, geophysical, and engineering principles that govern the development of unconventional reservoirs. The book begins with a detailed characterization of unconventional reservoir rocks: their composition and microstructure, mechanical properties, and the processes controlling fault slip and fluid flow. A discussion of geomechanical principles follows, including the state of stress, pore pressure, and the importance of fractures and faults. After reviewing the fundamentals of horizontal drilling, multi-stage hydraulic fracturing and stimulation of slip on pre-existing faults, the key factors impacting hydrocarbon production are explored. The final chapters cover environmental impacts and how to mitigate hazards associated with induced seismicity. This text provides an essential overview for students, researchers and industry professionals interested in unconventional reservoirs.

Dr. Mark D. Zoback is the Benjamin M. Page Professor of Geophysics at Stanford University. He conducts research on *in situ* stress, fault mechanics, and reservoir geomechanics with an emphasis on shale gas, tight gas and tight oil production. His first book, *Reservoir Geomechanics*, published by Cambridge University Press in 2007 is now in its 15th printing. His online course in reservoir geomechanics has been completed by approximately 10,000 students around the world. Dr. Zoback has received a number of awards and honors, including election to the US National Academy of Engineering.

Dr. Arjun H. Kohli is a Research Scientist and Lecturer in the Department of Geophysics at Stanford University. He conducts research on earthquake physics with an emphasis on plate boundary faults and induced seismicity in geologic reservoirs. He co-developed two massive open online courses on reservoir geomechanics, which feature interactive exercises designed for students ranging from high school to industry professionals.

Unconventional Reservoir Geomechanics

Shale Gas, Tight Oil, and Induced Seismicity

MARK D. ZOBACK AND ARJUN H. KOHLI

Stanford University, California

CAMBRIDGE
UNIVERSITY PRESS

CAMBRIDGE
UNIVERSITY PRESS

University Printing House, Cambridge CB2 8BS, United Kingdom

One Liberty Plaza, 20th Floor, New York, NY 10006, USA

477 Williamstown Road, Port Melbourne, VIC 3207, Australia

314–321, 3rd Floor, Plot 3, Splendor Forum, Jasola District Centre, New Delhi – 110025, India

79 Anson Road, #06–04/06, Singapore 079906

Cambridge University Press is part of the University of Cambridge.

It furthers the University's mission by disseminating knowledge in the pursuit of education, learning, and research at the highest international levels of excellence.

www.cambridge.org
Information on this title: www.cambridge.org/9781107087071
DOI: 10.1017/9781316091869

© Mark D. Zoback and Arjun H. Kohli 2019

This publication is in copyright. Subject to statutory exception
and to the provisions of relevant collective licensing agreements,
no reproduction of any part may take place without the written
permission of Cambridge University Press.

First published 2019

Printed and bound in Great Britain by Clays Ltd, Elcograf S.p.A. 2019

A catalog record for this publication is available from the British Library.

Library of Congress Cataloging-in-Publication Data
Names: Zoback, Mark D., author. | Kohli, Arjun H., author.
Title: Unconventional reservoir geomechanics : shale gas, tight oil and induced seismicity / Mark
D. Zoback, Stanford University, California, Arjun H. Kohli, SLAC National Accelerator Laboratory.
Description: Cambridge, United Kingdom ; New York, NY, USA : Cambridge University Press, 2019. |
Includes bibliographical references and index.
Identifiers: LCCN 2018044699 | ISBN 9781107087071
Subjects: LCSH: Hydrocarbon reservoirs – Technological innovations. | Energy development –
Environmental aspects.
Classification: LCC TN870.57.Z64 2019 | DDC 553.2/8–dc23
LC record available at https://lccn.loc.gov/2018044699

ISBN 978-1-107-08707-1 Hardback

Cambridge University Press has no responsibility for the persistence or accuracy of
URLs for external or third-party internet websites referred to in this publication
and does not guarantee that any content on such websites is, or will remain,
accurate or appropriate.

Contents

	Preface	*page* ix
Part I	**Physical Properties of Unconventional Reservoirs**	1
1	**Introduction**	3
	Unconventional Resources	6
	Types of Unconventional Reservoirs	9
	Recovery Factors and Production Rates	17
	Horizontal Drilling and Multi-Stage Hydraulic Fracturing	21
2	**Composition, Fabric, Elastic Properties and Anisotropy**	31
	Composition and Fabric	31
	Elastic Properties	41
	Elastic Anisotropy	50
	Poroelasticity	53
	Estimating Elastic Properties from Geophysical Data	56
3	**Strength and Ductility**	65
	Rock Strength	65
	Time-Dependent Deformation (Creep)	69
	Stress and Strain Partitioning	73
	Modeling Time-Dependent Deformation	76
	Estimating *In Situ* Differential Stress from Viscoelastic Properties	80
	Brittleness and Stress Magnitudes	84
4	**Frictional Properties**	91
	Fault Strength and Stress Magnitudes	92
	Rock Friction	95
	Frictional Strength and Stability	101
	Implications for Induced Shear Slip during Hydraulic Stimulation	111

5	**Pore Networks and Pore Fluids**	115
	Matrix Porosity	115
	Matrix Pore Networks	132
	In Situ Pore Fluids	138

6	**Flow and Sorption**	149
	Matrix Flow	150
	Permeability	154
	Sorption	168

7	**Stress, Pore Pressure, Fractures and Faults**	181
	State of Stress in US Unconventional Reservoirs	182
	Measuring Stress Orientation and Magnitude	191
	Pore Pressure in Unconventional Reservoirs	204
	Fractures and Faults in Unconventional Reservoirs	211
	Utilizing 3D Seismic Data to Map Fault zones and Fractures	225

Part II	**Stimulating Production from Unconventional Reservoirs**	231

8	**Horizontal Drilling and Multi-Stage Hydraulic Fracturing**	233
	Horizontal Drilling	233
	Multi-Stage Hydraulic Fracturing	238
	Fracturing Fluids and Proppants	254

9	**Reservoir Seismology**	263
	Microseismic Monitoring during Reservoir Stimulation	263
	Seismic Wave Radiation	270
	Earthquake Source Parameters and Scaling Relationships	282
	Earthquake Statistics	287
	Locating Microearthquakes	289

10	**Induced Shear Slip during Hydraulic Fracturing**	301
	Shear Stimulation and Production	301
	Coulomb Faulting and Slip on Poorly Oriented Fracture and Fault Planes	304
	Shear Slip and Permeability	315

11	**Geomechanics and Stimulation Optimization**	322
	Landing Zones	323
	Optimizing Completions I: Field Tests and Reservoir Simulation	328
	Vertical Hydraulic Fracture Growth	332

	Optimizing Completions II: Reservoir Simulation and 3D Geomechanics	334
	Viscoplastic Stress Relaxation and Varying Stress Magnitudes with Depth	339
	Targeting Geomechanical Sweet Spots: Fractures, Faults and Pore Pressure	344

12	**Production and Depletion**	345
	Production Decline Curves and One-Dimensional Flow	346
	Using Microseismicity to Estimate Total Fracture Area	349
	Evolution of a Shear Fracture Network	353
	Matrix Damage and Permeability Enhancement	358
	Seismic and Aseismic Fault Slip	359
	Depletion of Ultra-Low Permeability Formations with High Permeability Fractures	360
	The Frac Hit Phenomenon and Well-to-Well Communication	362
	Modeling Poroelastic Stress Changes	370

Part III Environmental Impacts and Induced Seismicity

375

13	**Environmental Impacts and Induced Seismicity**	377
	Overview of Environmental Issues	378
	Induced Seismicity	389

14	**Managing the Risk of Injection-Induced Seismicity**	406
	Avoiding Injection Near Potentially Active Faults	406
	Estimation of Fault Slip Potential in the Permian and Fort Worth Basins	415
	Risk Management and Traffic Light Systems	422
	Utilizing Seismogenic Index Models to Manage Produced Water Injection	426
	Site Characterization Risk Frameworks	434

| | *References* | 442 |
| | *Index* | 479 |

Preface

For the past eleven years, the students and researchers in the Stress and Reservoir Geomechanics group at Stanford University have been working to improve our understanding of a wide range of multi-scale processes related to unconventional reservoir development and induced seismicity. Research topics have ranged from imaging pores at the nanometer scale to studying the state of stress in basins hundreds of km across. In addition to studying fundamental processes in the laboratory and carrying out a number of theoretical studies, we have carried out over a dozen comprehensive case studies in collaboration with a number of oil and gas companies that have allowed us to examine basic research questions in a variety of field datasets. Examples from these studies are used throughout the book. While much remains to be learned about these highly complex reservoirs, in this book we attempt to address a number of fundamental issues that affect successful exploitation of these resources while minimizing the environmental impacts of production, in particular the occurrence of induced seismicity.

We would like to acknowledge the many Stanford researchers who made important contributions to the content of this book. These include, in alphabetic order: Gader Alalli, Maytham Al-Ismael, Richard Alt, Indrajit Das, Noha Farghal, Yves Gensterblum, Alex Hakso, Rob Heller, Sander Hol, Owen Hurd, Lei Jin, Madhur Johri, Wenhuan Kuang, Cornelius Langenbruch, Jens-Erik Lund Snee, Xiaodong Ma, Fatemeh Rassouli, Julia Reece, Ankush Singh, Hiroki Sone, John Vermylen, Rall Walsh, Randi Walters, Matt Weingarten, Wei Wu and Shaochuan Xu. Special thanks go to Jens, Kate Matney and Fatemeh for their help preparing a number of figures in the book. We also thank Usman Ahmed and Francesco Mele for providing several figures.

We would also thank colleagues at universities and private companies that provided us with images, data and helpful comments on earlier drafts of the book chapters. These include Cal Cooper, Leo Eisner, Bo Guo, Paul Hagin, Peter Hennings, Tony Kovscek, Lin Ma, Shawn Maxwell, Julian Mecklenburgh, Mike Ming, Kris Nygaard, Cindy Ross, Vik Sen, Julie Shemeta, Ed Steele, Richard Sullivan and especially Steve Willson. We appreciate their generosity as we have benefited greatly from their insights. We also thank Mark McClure for allowing us to use the ResFrac software package for some of the calculations shown in Chapters 8 and 11.

MDZ was able to prepare material for this book while on sabbatical at Victoria University in Wellington, New Zealand and ETH in Zurich, Switzerland. Professors John Townend and Domenico Giardini are thanked for generously hosting these sabbaticals. Financial support for preparation of this book came in part from the Stanford Natural Gas Initiative and the Stanford Center for Induced and Triggered Seismicity.

Part I

Physical Properties of Unconventional Reservoirs

1 Introduction

The goal of this book is to address a range of topics that affect the recovery of hydrocarbons from extremely low-permeability unconventional oil and gas reservoirs. While there are various definitions of *unconventional reservoirs*, in this book we consider oil- and gas-bearing formations with permeabilities so low that economically meaningful production can only be realized through horizontal drilling and multi-stage hydraulic fracturing. These reservoirs have permeabilities measured in nanodarcies, not millidarcies – in other words, a million times lower than conventional reservoirs. Despite their ultra-low permeability, there is no question about the scale and impact of production from unconventional oil and reservoirs in the US and Canada over the past decade.

The topics addressed in this book are presented from the perspectives of first principles. To establish these fundamentals, we draw on extensive laboratory investigations of core samples (from nanometer to cm scale) as well as studies in the fields of geology, geophysics, earthquake seismology, rock mechanics and reservoir engineering. We will discuss first principles in the context of multi-disciplinary case studies, mostly from unconventional reservoirs of various types in North America. In the chapters that follow, we integrate and synthesize information about the response of unconventional reservoir to stimulation (as indicated by hydraulic fracturing, microseismic and production data) with available geologic and geophysical data and geomechanics (knowledge of the state stress, natural fractures and faults and pore pressure) in both the producing formations and those surrounding it.

Part I of this book (Chapters 1–7) addresses the physical properties of unconventional reservoirs. In seven chapters, we discuss a number of the formations exploited to date in North America in terms of their composition, microstructure, mechanical and flow properties, state of stress and pore pressure and pre-existing fractures and faults. To some degree, Chapters 2–7 consider topics that progressively increase in scale, starting with laboratory studies on core samples that focus on the physical properties from the cm- to nanometer scale (the rocks matter) and concluding by discussing basin-scale stress fields and fracture and fault systems (which control hydraulic fracture propagation and the efficiency of reservoir stimulation). In fact, an interesting aspect of unconventional reservoir development is that the factors and processes that affect development span an extraordinary wide range of scales – from hydrocarbon flow through nanometer-scale pores to variations of stress orientation and magnitude that can occur at scales ranging from tens of meters to hundreds of kilometers.

Chapters 2–5 focus on the composition, fabric, physical properties and pore networks of different types of unconventional reservoir rocks, which despite their common feature of ultra-low permeability, are currently being economically produced. Chapter 6 addresses the physics of flow in the ultra-low permeability matrix and its sensitivity to pressure depletion. Although there is always a great deal of discussion related to optimizing stimulation in an unconventional play from operational perspectives (well spacing and hydraulic fracturing operations), the fundamental challenge is that hydrocarbon flow must be stimulated from very small scale (micrometer- to nanometer-scale) pores in an economically viable manner.

Chapter 7 concludes Part I with a discussion of stress fields, pore pressure and natural fracture and fault systems. As mentioned above, the physical properties of the reservoir rocks matter but so does their geomechanical setting. The physical state of the formations *in situ* affects both hydraulic stimulation and the production of hydrocarbons. This includes the state of stress in the formations (both the producing formations and those surrounding it), temperature and thermal history, pore pressure and the characteristics of pre-existing fractures and faults. These topics were discussed at some length in the context of conventional reservoirs by Zoback (2007).

Perhaps the most challenging aspect of optimizing production from unconventional reservoirs is related to the linkages among what appear to be intrinsic properties of the rocks (their composition, fabric, the degree of diagenesis, kerogen content and maturity, etc.) and apparently extrinsic attributes (the state of stress, temperature, pore pressure and the presence of fractures and faults, etc.). We attempt to clarify many of these linkages throughout the book. For example, in Chapter 6 we discuss the ultra-low matrix permeability of unconventional reservoir rocks which, in Chapter 12, we use to develop a conceptual framework for understanding the relationship between depletion and production. This, in turn, helps the reader understand the applicability of microseismic data for defining the stimulated rock volume, or SRV (discussed in Chapters 9, 10 and 12). Another example of linkages is encapsulated in the mechanical property known as *brittleness*, which is frequently used in the industry as a metric identifying desirable intervals for drilling and stimulation in unconventional reservoirs. As discussed in Chapter 3, brittleness is conventionally defined in terms of elastic stiffness and the nature of rock failure in compression. In Chapters 10 and 11 we discuss the relationships between factors affecting variations of brittleness and stress magnitude, which has a first-order effect on vertical hydraulic fracture growth and proppant placement. Still another example is how detailed knowledge of stress state and pore pressure (discussed in Chapter 7) is of first-order importance when considering stimulation (Chapter 10) and the effects of depletion (Chapter 12).

Part II of the book addresses the process of stimulating production from unconventional reservoirs using horizontal drilling and multi-stage hydraulic fracturing. While the importance of horizontal drilling and multi-stage hydraulic fracturing are well understood, the third key technological development that led to successful exploitation of unconventional reservoirs was the utilization of low viscosity hydraulic fracturing fluids (often referred to as *slickwater*). Chapter 8 reviews several important engineering aspects of horizontal drilling and multi-stage hydraulic fracturing as well as a few engineering and geologic issues that affect both vertical and lateral hydraulic fracture propagation. Chapter 9 covers the basics of microseismic monitoring, which is currently the best tool available for monitoring the spatial and temporal characteristics of the stimulation process. The technologies associated with microseismic monitoring are largely based on principles developed over many decades to study natural earthquakes. In Chapter 10 we discuss the importance of interactions among the state of stress, pre-existing fractures and faults, and hydraulic fractures, which are critical to understanding hydraulic stimulation. In Chapter 11 we discuss hydraulic fracture propagation and proppant placement in the context of operational parameters, geomechanical implications for exploiting stacked pay and what it means to stay *in zone* for some formations. Chapter 12 presents a unified overview of flow from nano-scale pores to hydraulic fractures via the stimulated fracture network. We attempt to show how the matrix composition, fabric, permeability and mechanical properties of the producing formation (as well as the surrounding formations) dictate how hydraulic stimulation should be performed.

Part III of the book addresses the environmental impacts of unconventional development, in particular the occurrence of induced seismicity. Chapter 13 provides an approach for understanding how to identify (and minimize) the environmental impacts of development, while balancing the benefits of producing enormous supplies of natural gas and oil from domestic sources. One of the important benefits of the recent abundance of natural gas in the US is related to the greenhouse gas emission reductions resulting from switching from coal to natural gas for producing electricity. Figure 1.1 illustrates the dramatic increase in the availability of natural gas in the United States as production began to increase from the Barnett formation in the Fort Worth Basin of northeast Texas from 2005 to 2006. Note the marked reduction in CO_2 emissions resulting from fuel-switching from coal to natural gas from electrical power generation (red and blue curves) that began at about that time. In about a decade, coal went from providing 48.2% of electricity in the US to 33.4%, as the use of gas for electrical power generation increased from 21.4% to 32.5%. Consequently, CO_2 emissions dropped about 15% to levels not seen for over 25 years, as did emissions of particulate matter and of a number of other air pollutants associated with coal use. Another point worth recognizing is that at the time the shale gas revolution began, natural gas was in short supply, gas prices were quite high and a number of liquified natural gas (LNG) import terminals were under construction in the US. Because of the increases in the production from unconventional gas reservoirs, gas production after 2008 continued to increase, despite historically low prices. Fortunately, none of the LNG import terminals were completed, and by 2015 they began to be converted to LNG export terminals.

That said, there are a number of environmental and social impacts associated with unconventional reservoir development, which are exacerbated by the very large number of wells being drilled. Although it is beyond the scope of the book to discuss this topic comprehensively, at the beginning of Chapter 13 we address the potential for contamination of aquifers and methane leakage related to improper well construction, as well as whether hydraulic fracturing operations might compromise well integrity. A somewhat

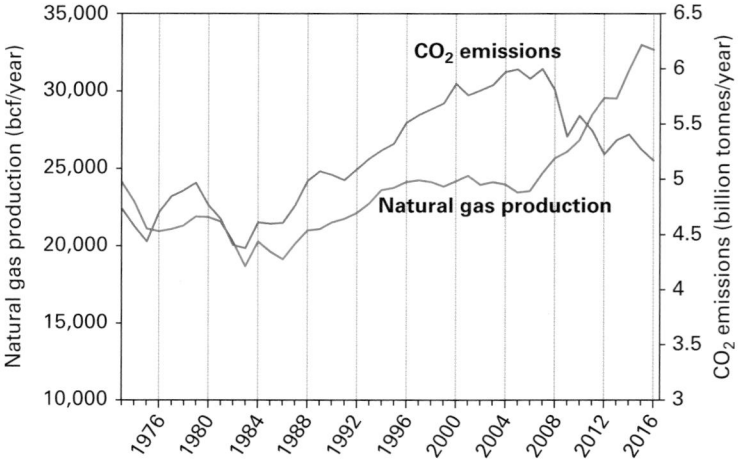

Figure 1.1 Annual production of natural gas in the United States over the past 43 years (red curve) and CO_2 emissions from the electrical power sector (blue curve). Data from the Energy Information Agency (2017).

unexpected environmental issue that has arisen in the US and Canada has been the increase in seismicity in a number of areas where unconventional reservoir development has been occurring. Chapter 13 discusses seismicity induced by hydraulic fracturing, the disposal of flowback water after hydraulic fracturing and the disposal of produced water. Chapter 14 presents strategies for managing the risks associated with induced seismicity. Some of these are also applicable to the exploitation of geothermal reservoirs as well as CO_2 injection and storage.

In the remainder of this introductory chapter, we will establish a framework for understanding a number of the topics discussed later in the book. This includes a brief discussion of the different types of unconventional formations being produced and the fundamental challenge of low recovery factors which is discussed at length in Chapter 12. We also provide a brief overview of how horizontal drilling and multi-stage hydraulic fracturing are carried out in unconventional reservoirs. The chapters that follow will address many of the issues introduced here in much more detail.

Unconventional Resources

It is incontrovertible that an enormous quantity of both natural gas and oil have been produced from unconventional reservoirs in just over a decade. Figure 1.2 is a summary of cumulative production of natural gas (Fig. 1.2a) and oil (Fig. 1.2b) from ~100,000 horizontal wells drilled in the major unconventional plays in the US. The plots show a three-fold increase in unconventional gas production and five-fold increase in unconventional oil production. So much natural gas is being produced in the US that the LNG

Introduction 7

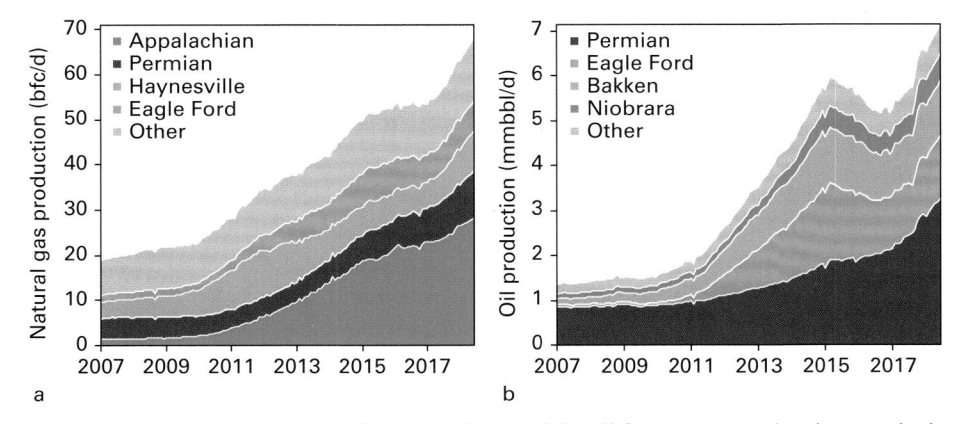

Figure 1.2 Cumulative production of (a) natural gas and (b) oil from unconventional reservoirs in the United States. Data from US Energy Information Agency (2017).

import terminals that were under construction in the early 2000s were being converted to LNG export terminals a little more than a decade later. While the initial production of natural gas from the Barnett shale occurred at a time of unusually high gas prices, the steady increase in production of shale gas in the US since 2008 occurred despite a steady *decrease* in gas prices. As discussed below, this was achieved through remarkable improvements in operational efficiency. The situation in Canada is similar. Thus, both private citizens and companies using large amounts of energy throughout North America (Mexico has started to import significant quantities of natural gas from the US and Canada) have benefited from the increasingly abundant supplies of inexpensive natural gas while experiencing decreasing rates of air pollution and greenhouse gas emissions.

Similarly, production from ultra-low-permeability oil reservoirs in the US has increased markedly, dramatically reducing the volume of imported oil (Fig. 1.2b). By 2017, the US was importing less than half as much oil as it did a decade earlier, resulting in a dramatic reduction of the US trade deficit and its reliance on imported oil as well as growth of the US economy. As oil is used primarily for transportation in the US, improved efficiency of cars and trucks has also contributed to the decrease in imported oil, but so has production of millions of barrels of oil per day from unconventional reservoirs.

The great majority of unconventional production had been from five gas producing regions and four oil producing regions. Some regions which experienced extensive development over the past decade will decline in the face of oil prices decreasing from about $100/BBL prior to 2015 to about $60/BBL in 2018. One such area is the Williston Basin, which hosts the Bakken shale. However, other areas like the Permian Basin are currently seeing a marked increase in the amount of drilling and production because it has been recognized that *stacked pay* (multiple producing formations at the same location) can be economically produced due to continued improvements in operational efficiency.

Past and Future Unconventional Development in the US – As impressive as the production figures shown in Fig. 1.2 are, according to Bureau of Economic Geology at the University of Texas as many as 1.6 million new wells could be drilled in these plays in the coming decades (Fig. 1.3). It should be noted that this estimate is based on estimates of technically recoverable reserves, a realistic density of wells in any given area and the presumption that drilling will occur in areas where drilling was allowed in 2017. Also shown on this map are important plays like the Niobrara and Mancos shales of Colorado and New Mexico, respectively, the SCPOOP, STACK and Merge plays in the Anadarko Basin of Oklahoma, which will likely involve tens of thousands of new wells. Not shown are the Utica formation (principally natural gas in Ohio) and important plays in Canada such as the Montney and Duvernay as well as other potentially significant plays that have not yet been identified (or announced).

The need for significant continued improvements in recovery factors from unconventional reservoirs is introduced later in this chapter and revisited throughout the book. Suffice it to say at this point that the importance of improving recovery factors goes well beyond its impact for any one company, any one play or any one country.

Global Unconventional Resources – Figure 1.4 summarizes the potential for unconventional oil and gas production in many parts of the world. While most unconventional production to date has been in North America, unconventional oil and gas development in other parts of the world could have a tremendous impact assuming that it is done in a manner that builds upon both the operational experience and knowledge gained over

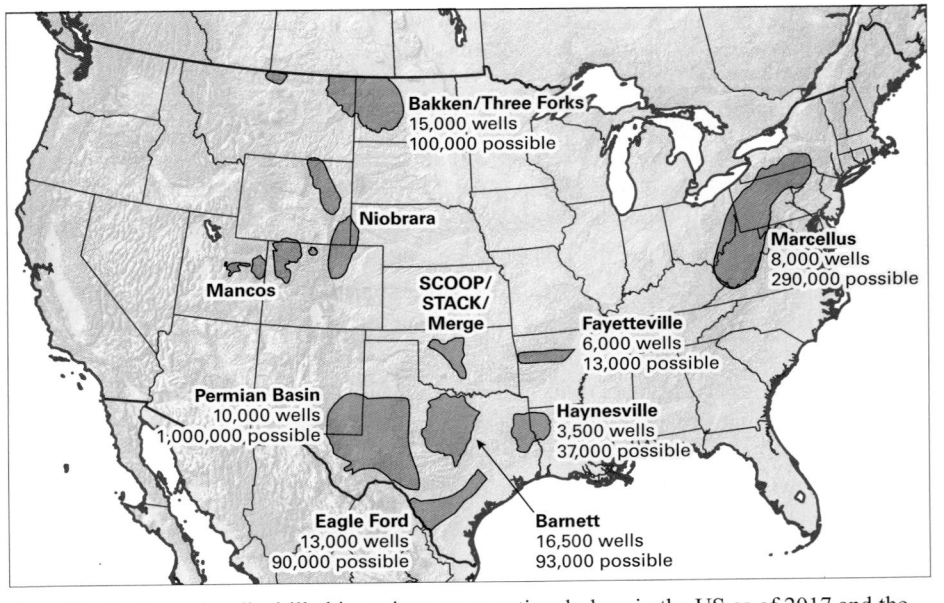

Figure 1.3 Horizontal wells drilled in major unconventional plays in the US as of 2017 and the numbers of wells that could be drilled in the future based on technically recoverable reserves. From Svetlana Ikonnikova, Bureau of Economic Geology, 2017.

Introduction 9

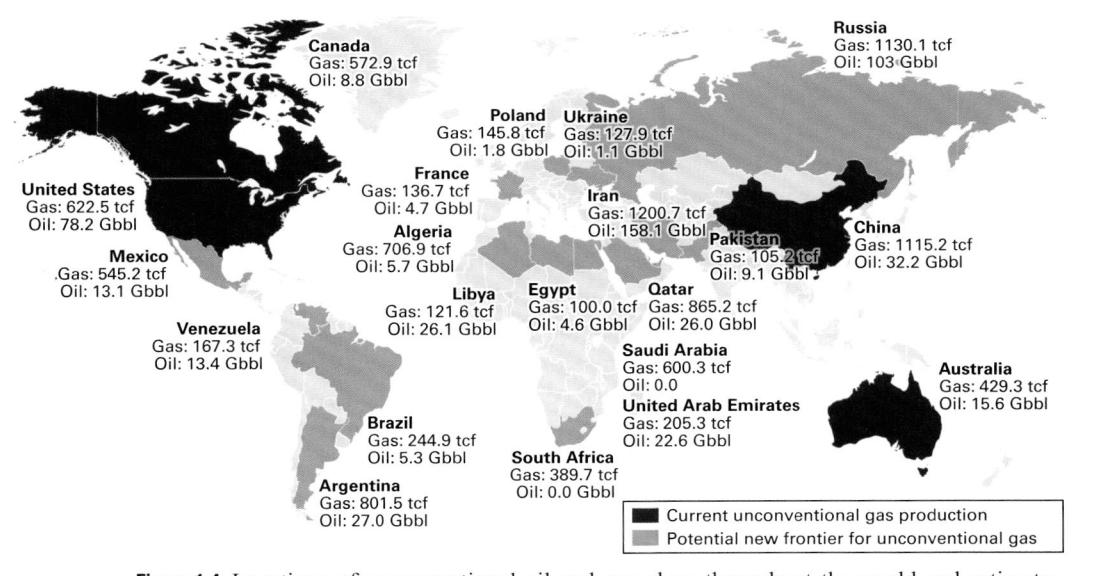

Figure 1.4 Locations of unconventional oil and gas plays throughout the world and estimates of technically recoverable reserves. World Energy Council, 2016; US Energy Information Agency.

the past decade of development in North America. Note that globally, unconventional production could increase available *wet* natural gas (natural gas and natural gas liquids) by almost 50%. Due to the vast quantities of proven conventional reserves in the Middle East, the cumulative impact on oil is lower (11% globally) but the distributed nature of these resources is extremely important.

It is abundantly clear that as the global energy system progressively decarbonizes over the next half century, it is important to recognize that there are more than adequate supplies of unconventional oil and gas from widely distributed sources to meet global needs, even accounting for the marked increase in energy use from the rapidly expanding economies of the developing world. Thus, the critical questions concern how to develop the use of these resources in a manner that produces the maximum economic benefit with the least environmental, social and climate impacts.

Types of Unconventional Reservoirs

The unconventional reservoirs that are contributing to significant oil and gas production (those listed in Fig. 1.2 and others) are geologically different in significant ways, thus necessitating different approaches for how they should be optimally exploited. While these differences will be highlighted throughout this book, it is important to recognize that there are generally three basic types of unconventional reservoirs successfully being exploited with horizontal drilling and multi-stage hydraulic fracturing.

Organic-Rich Source Rocks and Maturation – The first basic type of unconventional reservoirs are organic-rich source rocks, sometimes called *resource* plays. These are exemplified by the Barnett shale, the first unconventional formation to be extensively developed with horizontal drilling and multi-stage hydraulic fracturing. Conventional oil and gas are produced from relatively permeable reservoirs into which oil and gas have migrated from organic-rich source rocks. In contrast, resource plays refer to producing zones that are organic-rich (and often clay-rich) source rocks themselves, which are characterized, as mentioned above, by ultra-low permeability. While hydrocarbon source rocks have been recognized and studied for over a century, the potential for economic production from source rocks was considered impossible until successful development of the Barnett shale around 2004, by Mitchell Energy. Figure 1.5 (from Loucks & Ruppel, 2007) illustrates the depositional environment of the Barnett shale, where various types of organic matter were deposited in an anaerobic, relatively deepwater coastal environment, leading to preservation of the organic matter in clay-rich sediments during Mississippian time (~325 Ma).

As is the case in most unconventional reservoirs, different lithofacies comprise the Barnett shale which are defined on the basis of mineralogy, fabric, biota and texture. An example is shown in Fig. 1.6 (Sone & Zoback, 2014a) in which three distinct Barnett lithofacies are defined on the basis of mineralogy and a gamma ray log. Most production has come from the lower "hot" part of the Barnett. Notice that even within the lower Barnett, sub-units have been defined and even within the sub-units, there are individual bedding units or lithofacies with distinct composition at relatively fine scale. This multiscale heterogeneity is discussed in detail later in this chapter.

Peters et al. (2005) discuss the organic content of hydrocarbon source rocks and maturation in great detail. It is obvious that the amount of organic matter that is present

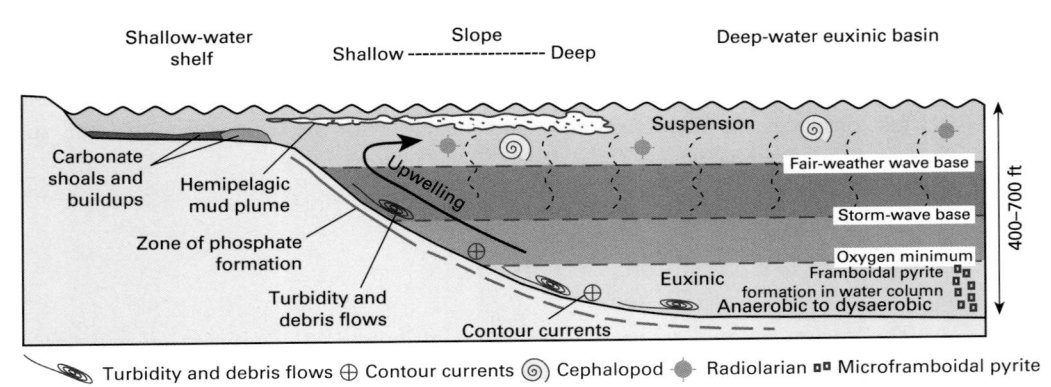

Figure 1.5 Schematic illustration of the manner in which various types of organic material were deposited and preserved in a deep water euxinic basin to produce the Barnett shale. From Loucks and Ruppel (2007). *Reprinted by permission of the AAPG whose permission is required for further use.*

Introduction   11

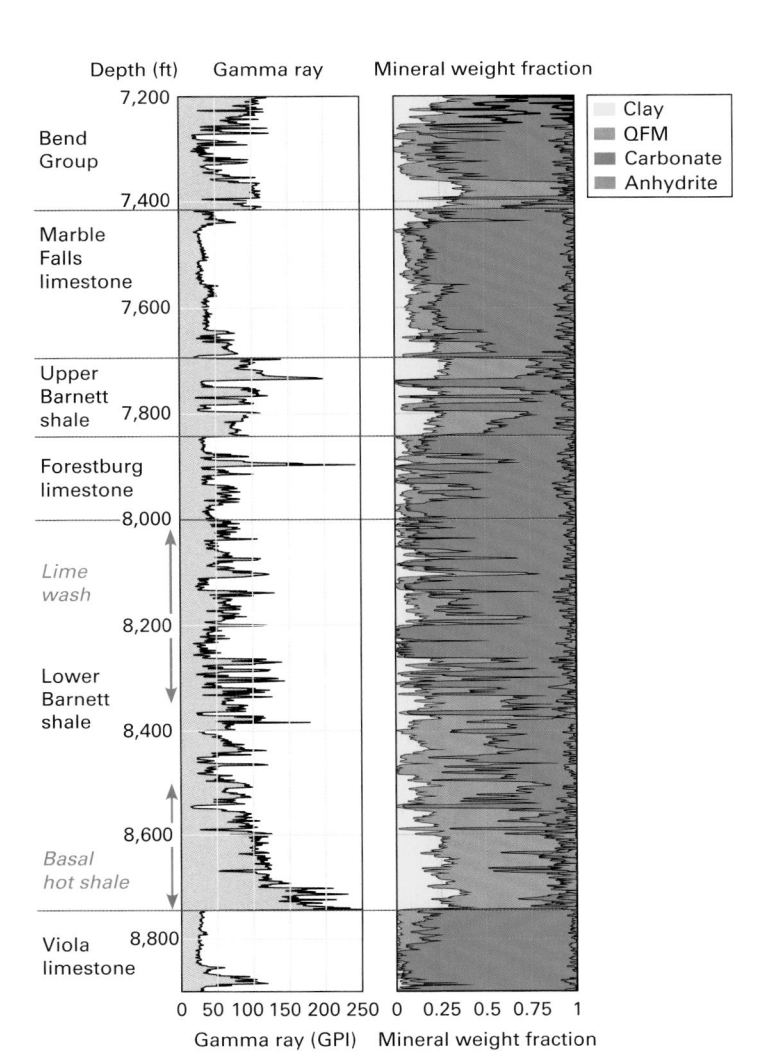

Figure 1.6 Stratigraphy of the Barnett and adjacent formations on the basis of lithology and physical properties. From Sone & Zoback (2014). *Reprinted by permission of Elsevier whose permission is required for further use.*

in a resource play is important for production, as is the degree of maturation of the organic matter. Figure 1.7 (from Peters et al., 2005) schematically shows the approximate depth and temperatures at which kerogen is converted to dry gas, wet gas (natural gas plus natural gas liquids) and oil. It also indicates the degree of vitrinite reflectance, R_o, a commonly used metric for maturation. The temperatures indicated correspond to average geothermal gradients of about 25 °C/km. At shallow depth, bacteria in groundwater consume coal and other organic matter in sediments and generate biogenic gas. This is typically the origin of coal bed methane resources. Only the Antrim shale in Michigan appears to be associated with biogenic gas.

Unconventional Reservoir Geomechanics

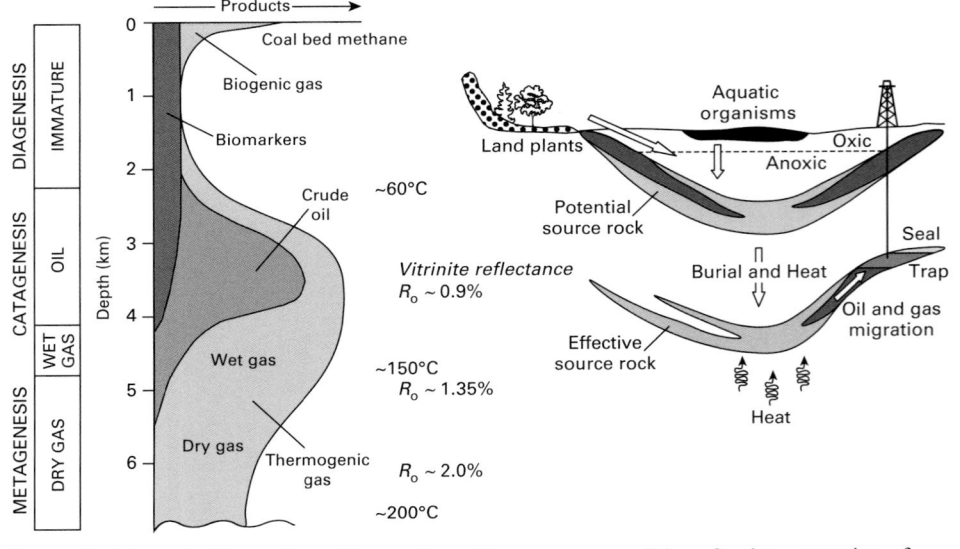

Figure 1.7 Schematic illustration of the depth and temperature conditions for the maturation of organic material in sediments into oil, wet gas (natural gas and condensates) and dry gas. Modified from Peters et al. 2005.

Thermogenic oil and gas production occurs over a wide range of temperatures, in part due to the type of kerogen involved (and other factors). Peak oil generation occurs at ~100 °C (R_o ~ 1.0) as well as generation of wet gas natural gas and natural gas liquids. At temperatures exceeding ~150 °C (R_o > 1.4), thermal maturation of kerogen produces principally dry gas.

The red dots in Fig. 1.8a show that the concentration of Barnett shale gas wells in the Fort Worth Basin (from Pollastro et al., 2007) is correlated with expected values of vitrinite reflectance values of 1.3–1.7 (Fig. 1.8b, from Montgomery et al., 2005). Correspondingly, other indicators of kerogen maturation, such as the hydrogen index (HI) and HI-based kerogen transformation ratio (Figs. 1.8c,d from Jarvie et al., 2007) also indicate that the region of maximum gas production is correlated with kerogen maturation. The hydrogen index (HI) parameterizes the quantity of petroleum generated during pyrolysis (thermal decomposition) relative to the total organic carbon in the sample. It is expressed in units of milligrams of hydrocarbons per gram of TOC (Peters et al., 2005). Hydrogen index is an indication of source rock quality; higher hydrogen indices indicate that the kerogen is more oil prone whereas lower HI indicates the kerogen is more gas prone.

Other factors are also clearly important in defining the area of maximum Barnett productivity as the zone of optimal maturation extends well to the southeast beyond the producing area. The maps in Figs. 1.8c,d also suggest potential gas production farther to the west in the Fort Worth Basin than do the vitrinite reflectance data (e.g., Montgomery et al., 2005). Note that Barnett oil production (green dots in Fig. 1.8a) extends to the northwest of the core area of gas production and is associated with lower values of vitrinite reflectance, as expected.

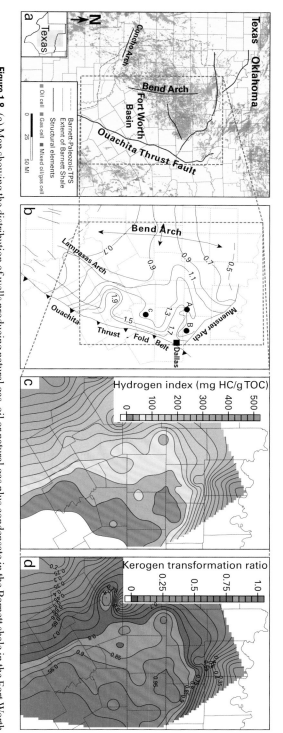

Figure 1.8 (a) Map showing the distribution of wells producing natural gas, oil or natural gas plus condensate in the Barnett shale in the Fort Worth Basin. From Pollastro et al. (2007). The dashed rectangle is shown in Fig. 7.21d. (b) Vitrinite reflectance data in the productive parts of the Barnett shale correspond well with hydrocarbon production. From Montgomery et al. (2005). (c) Barnett kerogen conversion based on HI and (d) HI-based transformation ratio. From Jarvie et al. (2007). *Reprinted by permission of the AAPG whose permission is required for further use.*

The general properties of a number of organic-rich shale formations in the US and Canada are summarized in Table 1.1. Note that mass volume of kerogen in unconventional reservoirs is generally quite modest, and average TOC by weight is never more than about 7%, although individual samples can be as much as twice that. As this organic material is the source of hydrocarbons, it is obviously critically important for production (as are other important factors, such as maturation, which is discussed below). As discussed in Chapters 3–6, the presence of organic material has an important impact on a wide variety of shale properties. These include mechanical properties, texture and anisotropy (Chapters 2 and 3), friction (Chapter 4), pore structure (Chapter 5) and permeability (Chapter 6). There are two principal reasons why a small mass percentage or organic content has an important effect on many of these key properties of organic-rich shales. First, because of the low density of kerogen, its volume percentage in a shale is generally about a factor of 2 higher than the weight percentage. In other words, a shale with 7% weight of kerogen has roughly 14% by volume. Second, in many basins there is a direct correlation between TOC and clay content (as discussed in Chapter 2). While the nature of this correlation is variable, it is often observed that TOC content increases with increasing clay content. As we will discuss in the following chapters, the combined effects of organic matter and clay in the rock matrix significantly impact mechanical and flow properties.

In general, most resource plays are currently at a shallower depth and lower temperatures than those associated with maximum burial. Evidence for this is the fact that appreciable diagenesis of the shales has occurred. As discussed in Chapters 2 and 3, most unconventional basins are composed of strong, stiff rocks. Also, as shown in Chapter 2, most of the clays in unconventional play have passed through the smectite to illite transition (at ~100 °C), even though current reservoir temperatures might be lower than 100 °C. Hence, one might assume that hydrocarbon maturation may have occurred in the geologic past. However, the presence of exploitable hydrocarbons in resource plays is likely associated with ongoing, or geologically recent, maturation – depending on the complex interaction between the amount and type of organic material that is present, its thermal/maturation history and whether the rate of hydrocarbon generation is faster than the rate at which the hydrocarbons can diffuse out of the source rock. In fact, ongoing or recent maturation is the likely source of excess pore pressure that is observed in many resource plays, which can be critically important for successful production from these ultra-low permeability reservoirs (Engelder et al., 2009). The degree of overpressure in various unconventional reservoirs and its relation to stress magnitudes and faulting is described in more detail Chapter 7. Its importance for production is discussed in Chapter 12.

Figure 1.9 (from Loucks et al., 2012) provides a useful framework for thinking about diagenesis (compaction and cementation), kerogen maturation, clay conversion and other aspects of source rock evolution, several of which will be explored further in subsequent chapters. For example, the evolution of pore space development in the kerogen is important for fluid transport (Chapter 5) and processes such as cementation and the conversion of smectite to illite have an important effect on the physical properties of the shales (Chapter 3). As indicated in Fig. 1.9, these processes are controlled to first order by the time, temperature and pressure path of basin evolution, i.e. when did burial occur, how deep and how hot did the formation get, etc.

Table 1.1 General properties of organic-rich shale formations.

Formation	Barnett	Duvernay	Eagle Ford	Haynesville	Wolfcamp	Marcellus	Horn River	Woodford
Depth of formation top (km)	1.4–2.6#	1.8–3.6#	3.3–3.5#		1.2–2.6#	2.42#		2.0–3.9#
Formation temp (°C)			150#		60–80#	Avg 180 wells 129#		
Porosity (%)	4.5–6.5#		8–12#		5–7#	Avg 118 wells 3–8#		6#
TOC range/Average (wt %)	0–14.0/7.1* 2–7#	4–11^	1.9–5.7/3.5* 0–8#	1.6–3.1/2.5*	2.3*	2–10#	1–5/3.6#	Avg 200 samples 0–13/5.3#
Number of samples for TOC	16*	17*	13*		8*			193#
Average depth for TOC (km)	2.6* 2.0#	3.9* 2.7#	3.5* 3.4#		2.0*	2.5#	2.4#	2.9#

*Samples used in Stanford laboratory experiments. ^Samples reviewed by Khosrokhavar et al. (2014), #Studies reviewed by Edwards and Celia (2018), blank fields indicate data not reported in original reference

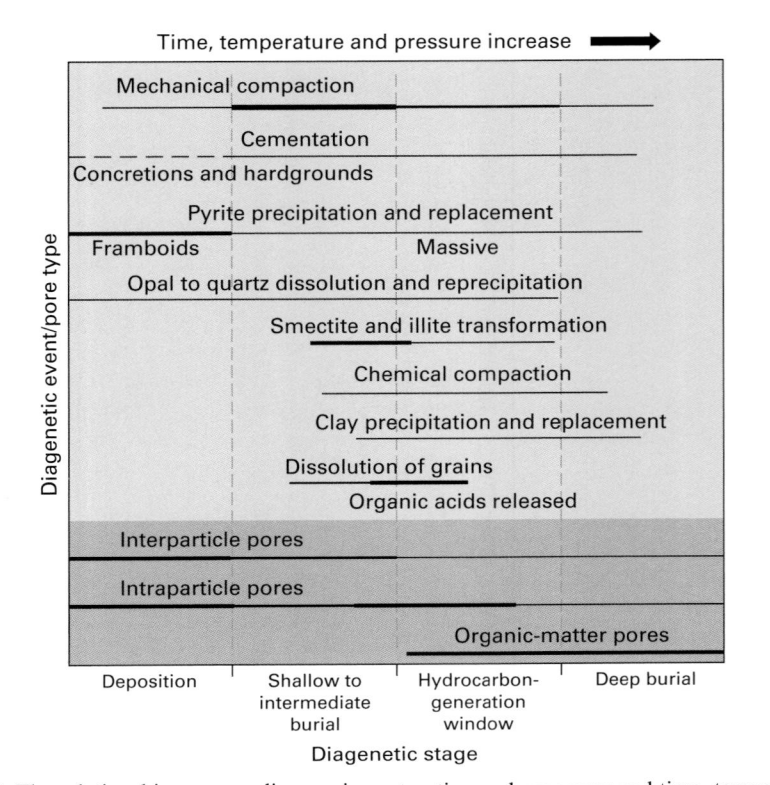

Figure 1.9 The relationships among diagenesis, maturation and processes and time, temperature and depth are of organic-rich source rocks. From Loucks et al. (2012). *Reprinted by permission of the AAPG whose permission is required for further use.*

Conventional Unconventionals and Hybrid Plays –

Tight oil reservoirs, the second type of unconventional reservoir could be equally termed conventional unconventionals because hydrocarbons have migrated into the reservoirs from conventional source rocks. What makes these reservoirs unconventional is that their ultra-low permeability requires horizontal wells with multiple hydraulic fractures to achieve economically viable production. The Bakken formation is a good example of this. As illustrated in Fig. 1.10 (from Miller et al., 2008) the producing zone is the low permeability Middle Bakken sandstone member. Hydrocarbons have migrated into the Middle Bakken from source rocks of both the Upper and Lower Bakken, which are immediately above and below the reservoir. Although there has been *hit and miss* production from the Bakken since the early 1960s, sustained production was not economically viable until horizontal drilling and multi-stage hydraulic fracturing began to be widely utilized around 2008 (Fig. 1.2b).

The third general type of unconventional reservoir are "hybrid" plays in which production occurs from both source rocks and low permeability conventional reservoirs. The Monterey formation of western California is an excellent example of this, although it was mostly developed prior to utilization of horizontal drilling and multi-stage hydraulic

Introduction   17

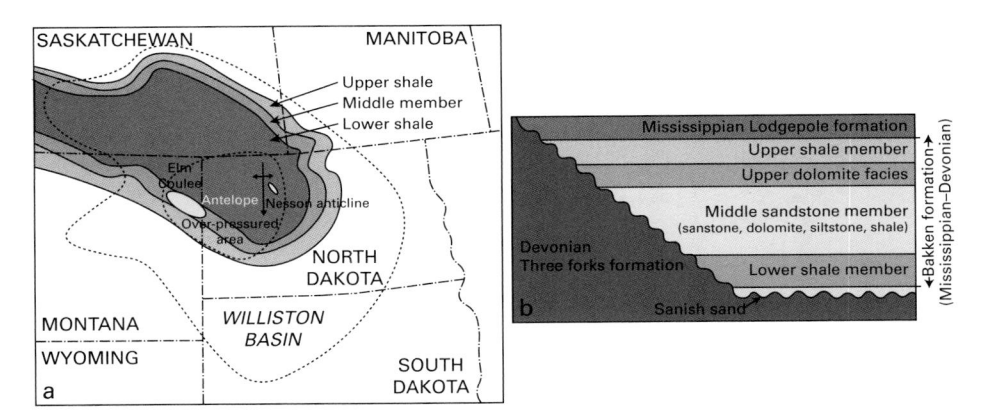

Figure 1.10 (a) Location of the Bakken formation in the Williston Basin of the north-central US and south-central Canada. After Miller et al. (2008). (b) Schematic stratigraphic column illustrating that the very low permeability Middle sandstone member of the Bakken formation is the producing interval, while hydrocarbons are sourced from both the lower and upper organic-rich Bakken shale members.

fracturing. The Monterey is a very low permeability siliceous shale that is highly fractured in many areas (Graham & Williams, 1985). Successful wells are often those that intersect pre-existing fractures and faults which increases access to the low permeability matrix. Thus, individual intervals could represent production from both source rock intervals or low-permeability reservoirs into which hydrocarbons have migrated.

In this regard, the Eagle Ford play in southeastern Texas, is similar. Figure 1.11a (US Energy Information Agency, 2013) shows the location of the play and the areas that tend to produce, oil, gas or wet gas and condensate (depending on the degree of maturation). As illustrated in Fig. 1.7, the oil window would have been at shallowest depth at the time of maturation, the gas window at greatest depth and the gas plus condensate window at intermediate depth. Figure 1.11b (from Jweda et al., 2017) shows the high degree of layering in both the Upper and Lower Eagle Ford. As described by Jweda et al., it is typical to produce from the Eagle Ford with one horizontal well in the Upper Eagle Ford and two in the thicker Lower Eagle Ford that alternate with one near the top and one near the bottom, thus producing a "W" pattern in cross-section (this will be discussed further in Chapter 11). As the hydraulic fractures generated in each horizontal well are expected to span the thickness of the upper or lower unit, it is obvious that production could be coming from both organic-rich source rocks and reservoir rocks into which the hydrocarbons have migrated. For similar reasons, a number of the other tight oil plays could be considered as hybrid plays.

Recovery Factors and Production Rates

Despite the success and impact of production from unconventional oil and gas development in the US and Canada over the past decade, there are three outstanding issues that point to the

18   Unconventional Reservoir Geomechanics

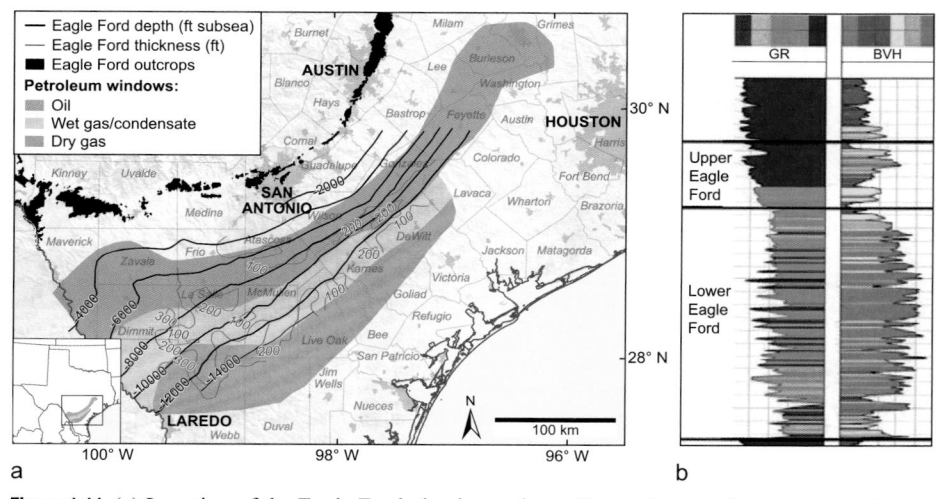

Figure 1.11 (a) Location of the Eagle Ford play in southeast Texas showing the regions producing oil, gas or gas condensate (from US Energy Information Agency, 2013). (b) Schematic stratigraphic column that illustrates the fine layering of the organic-rich shale. GR represents gamma ray (increasing to the left) and BVH represent bulk volume of hydrocarbons (increasing to the right). From Jweda et al. (2017).

critical need to improve our understanding of unconventional oil and gas resources to improve hydrocarbon recovery. The first is the rapid decrease of production rates seen in the first two to three years of production from both shale gas and tight oil wells.

Rate and Cumulative Production with Time – Figure 1.12 (from Hakso & Zoback, 2019) shows the average monthly oil and gas production and cumulative production from wells in the Eagle Ford (oil production only), Marcellus (gas production only; one mcf is a thousand cubic feet under standard pressure and temperature conditions), Bakken (oil production only) and Barnett (gas production only) in successive two-year periods. Production rates are shown starting in the third month after production and only for years in which data are available from more than 200 wells. In general, improved development practices over time (longer laterals, more hydraulic fracturing stages, improved well and hydraulic fracturing spacing, more effective fracturing procedures, etc.) result in increased production rate and (increased cumulative production) as operational practices improved from year to year. A notable exception to this is the Eagle Ford where there has been little change in initial production rates after 2010. Common to each region is the rapid decrease in production rates in the first years of production.

Analysis of a smaller dataset from the Barnett shale (Patzek et al., 2013) shows a similar pattern of exponential production rate decline that follows a simple $t^{-1/2}$ form which is discussed in detail in Chapter 12. Rapidly declining shale gas production rates are not limited to the four areas shown in Fig. 1.12. Sandrea & Sandrea (2014) point out that initial production rates in the Haynesville are quite high (probably due to very high pore pressures at depth), but so are depletion rates.

Key to economics of development, of course, is the cumulative production per well over time. If we consider cumulative production over three years as a reasonable metric

Introduction   19

(a)

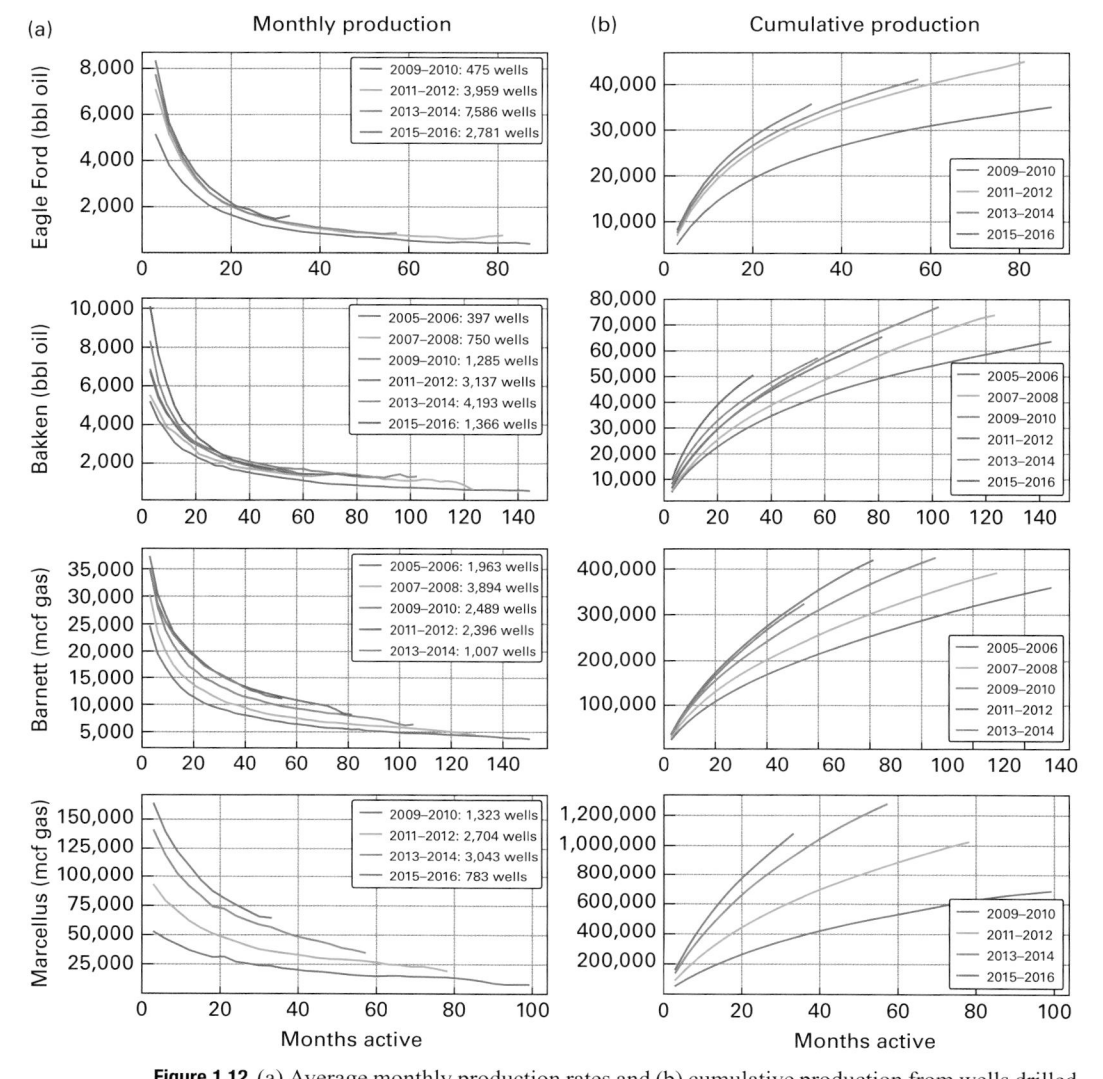

Figure 1.12 (a) Average monthly production rates and (b) cumulative production from wells drilled in the Eagle Ford (oil production only), Marcellus (gas production only), Bakken (oil production only) and Barnett (gas production only) in successive two-year periods. Data courtesy of Drillinginfo. The number of wells in each monthly average is shown. Only years in which data is available for more than 200 wells are shown.

of whether an average well will be profitable, we see about a doubling of cumulative 3-year production in the Marcellus as development has proceeded but a much lower increase (approximately 50%) in the other plays. If we consider just the Bakken and Barnett for which production data are available for the longest periods, the first four years of production accounts for 62% of the long-term cumulative production in the Bakken and 73% of the long-term cumulative production in the Barnett.

Table 1.2 Best 3-months average gas rate, mscf/day (from Baihly et al., 2010).

Basin	Well count	P10	P25	P50	P75	P90
Barnett	1,143	737	1,118	1,569	2,121	2,927
Fayetteville	1,037	976	1,429	2,020	2,657	3,289
Woodford	417	821	1,389	2,153	3,336	4,959
Haynesville	625	5,578	7,179	8,970	11,370	14,111
Eagle Ford	389	2,151	2,883	4,107	5,442	6,696
Marcellus	446	1,926	2,927	4,757	6,366	7,679

Of course, the other keys to economics of the developments are the resource price and the cost per well. Detailed consideration of these issues is beyond the scope of this book. It is important to recognize, however, that while the costs associated with drilling and hydraulic fracturing have decreased markedly in recent years due to remarkable improvements in operational efficiency, the level of effort needed to extract hydrocarbons also has increased. In other words, whether it is more fracturing stages per well (discussed in Chapter 8), closer well spacing and/or longer laterals, more fracturing fluid and proppant, etc. much of the increase in cumulative production per well seen in Fig. 1.12 results from the increase in effort and resources (e.g., lateral length, number of hydraulic fracturing stages, amount of water, sand, etc.) expended to extract hydrocarbons. Chapters 10 and 12 cover the necessity and application of these procedures in various settings.

Recovery Factors – The second issue related to recovery that is important to discuss is the overall low recovery factors (the ratio of produced hydrocarbon to the estimated total in place) for both individual shale gas and tight oil wells and for unconventional plays overall. While there is little published data on this topic, there seems to be a consensus that recovery from dry gas reservoirs (such as the Barnett) is about 25% after the first few years of production. Obviously, such numbers are uncertain as it is difficult to estimate gas in place and total recovery is not known until a well is eventually plugged and abandoned. The situation is even worse for tight oil as most estimates place recovery factors at 2–10% of the oil in place. In other words, the vast majority of hydrocarbons are left in the ground following primary production from unconventional plays.

Despite the difficulty estimating recovery factors accurately, the numbers cited above are clearly quite low when compared with conventional reservoirs. Moreover, when looking at any play in detail, the recovery factors vary considerably from the best parts of the play between the "fairway," or "sweet spot," to the more peripheral areas of lower production and recovery factors can sometimes varies considerably from well to well. This point is emphasized in Table 1.2 (from Baihly et al., 2010) which shows that for six unconventional plays, the upper quartile of wells (P75 and P90) produce considerably more than the lower three quartiles combined.

Unproductive Wells – The third issue related to overall hydrocarbon production is the very large number of economically unsuccessful wells that are drilled in unconventional plays. As pointed out by King (2014), "about a third of all shale wells drilled in the United

States have not been economic . . . another one-third are economic only at moderately high gas prices . . . [but] the wells in the top one-third perform so well and have such favorable economics that they may be able to carry the rest of the wells in a development project." Along with significantly improving cumulative recovery and prolonging the producing lifetime of unconventional wells, drilling fewer uneconomic wells and shortening the learning curve for improving the effectiveness of exploitation are high priorities for optimal production from unconventional reservoirs. At the time of this writing, there have been a series of experiments utilizing injection of methane or CO_2 in tight oil reservoirs to enhance production. It is not yet clear what the impact of technologies such as these will have on ultimate production efficiency.

Horizontal Drilling and Multi-Stage Hydraulic Fracturing

The operational processes associated with horizontal drilling and multi-stage hydraulic fracturing are generally well known. We discuss some of the specific engineering aspects of these processes at some length in Chapter 8. At this point we only want to introduce some basic concepts to establish a context for a number of the topics discussed in later chapters. First, the *lateral* (the horizontal section of the well) usually targets a single producing formation. As illustrated in Fig. 1.13, the laterals typically extend 1.3 to 3 km in length (Table 1.3), depending on drilling and geologic conditions, as well as other factors such as lease holdings, etc. Each lateral is hydraulically fractured in *stages*, which refers to an interval that is pressurized simultaneously, starting at the *toe* of the well and proceeding toward the *heel*. In a cased and cemented well, there might be five, or more, perforated sections within a stage, such that multiple hydraulic fractures are propagated simultaneously (see Chapter 8). It is not uncommon for there to be more than 40 or more hydraulic fracture stages in a single lateral. If the lateral is cased and

Table 1.3 General attributes of horizontal wells in different unconventional basins (from Kennedy et al. (2016) and other sources).

Formation	Depth range (m)	Thickness range (m)	Lateral lengths (m)
Bakken	2,920–3,200	12–22	2,650–3,050
Barnett	2,000–2,600	30–180	12,00–1,325
Duvernay	2,500–4,000	20–70	1,830–2,150
Eagle Ford	2,100–3,700	30–145	1,500–2,135
Fayetteville	300–2,150	6–61	1,430–1,680
Haynesville	3,200–4,100	61–91	1,340–1,430
Horn River	2,000–2,750	38–137	1,524–2,000
Marcellus	1,200–2,600	15–61	1,280–1,500
Montney	1,500–3,500	46–305	1,430–1,740
Niobrara	900–4,300	15–91	1,230–1,550
Utica	600–4,300	21–230	1,430–1,890
Wolfcamp	1,676–3,350	457–795	1,390–2,050

cemented (the most common type of well completion), each stage is perforated multiple times with the intent of initiating a fracture from each set of perforations. Hence, if there are 40 frac stages and 4 perfs per stage, the expectation is that 160 hydraulic fractures will be propagated away from a single wellbore.

Table 1.3 (from Kennedy et al., 2016 and other sources) summarizes the attributes of a number of unconventional producing formations in North America. By modern standards for conventional oil and gas development, most unconventional wells are relatively shallow, ranging from about 1,500 m to 3,500 m. Producing formation thicknesses vary from as little as 15 m to almost 300 m, although the most productive intervals could be much smaller. In Chapter 11 we discuss the issue of lateral placement from the perspective of formation properties and vertical hydraulic fracture propagation. Laterals range in length from 800 m (when limited by geologic factors) to as much as 3000 m in the Bakken and other formations. Although 1,500 m is a typical lateral length in many of formations being exploited, longer laterals are currently being drilled in a number of unconventional plays. For example, some recent wells drilled in the Marcellus formation exceed 5,000 m in length. Lateral length is based on practical issues such as land leases (often related to square land sections in the United States that measure 1 mile on each side) as well as geologic and technical limitations in very long laterals, specifically the capability to effectively hydraulically fracture the toe of wells and adequately cement the casing. Moreover, as costs for multi-stage hydraulic fracturing are typically more than those for drilling, the cost for drilling two wells may be acceptable if it results in more effective stimulation of the distal section of the well.

Pad Drilling and Stacked Pay – Figure 1.13 also illustrates pad drilling, where numerous wells are drilled from a single drill site. There are many advantages to doing this. It is operationally efficient to drill multiple wells from a single drilling pad. Once a drill rig and the many pieces of equipment needed to drill have moved to a pad, they are able to drill multiple wells by simply moving the rig a short distance (as little as 10 m) to drill another well. Some operators utilize drill rigs that can "walk" (move themselves without disassembly) the short distance from well head to well head. Pad drilling is one of several technologies developed over the past decade that have led to significant improvements in operational efficiency of exploiting unconventional oil and gas reservoirs that are nothing short of remarkable. As noted above, the unconventional gas boom started during a period of unusually high gas prices. However, due to increased efficiency, shale gas production steadily increased (Fig. 1.1) even after the sharp drop in gas prices that started with the global financial crisis of 2008. Pad drilling is also advantageous because it significantly limits the environmental impact of drilling operations and their impact on local residents (as discussed in Chapter 13) as there is one drill pad constructed, one road used for access and one pipeline needed to transport the oil and/or gas from multiple wells from a single site.

Note in Fig. 1.13 the complex trajectories that result when multiple wells are drilled from a single pad as they must avoid collisions and achieve the correct locations at depth in the producing formation. As discussed in Chapter 11, the spacing of the wells is intended to stimulate as much of the reservoir as possible. Obviously, if the wells are too far apart, there will be unstimulated parts of the producing formation between the wells

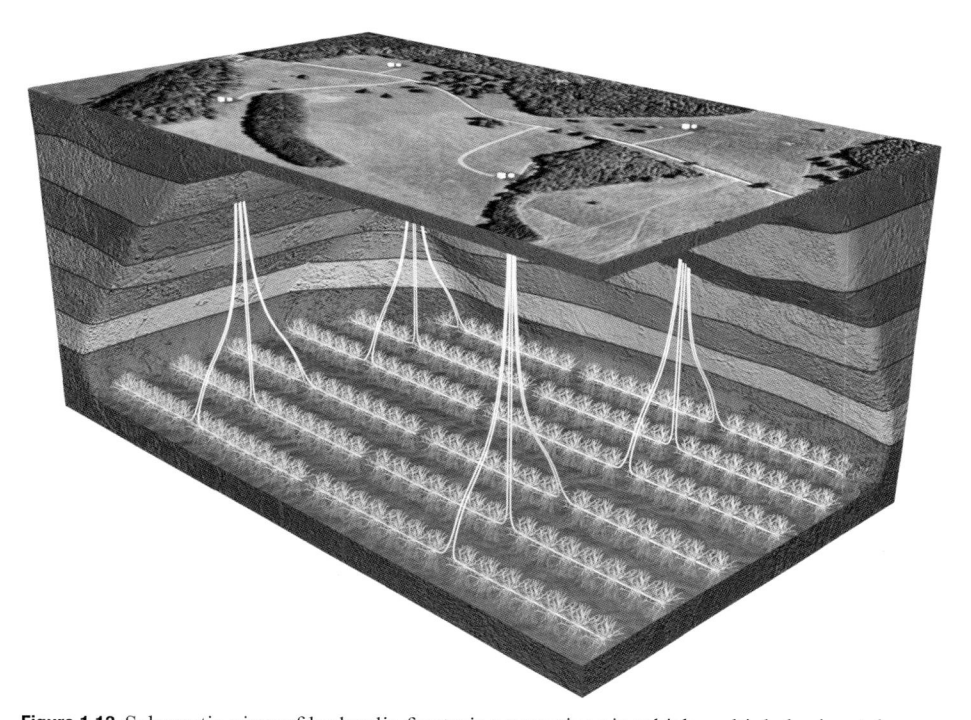

Figure 1.13 Schematic view of hydraulic fracturing operations in which multiple horizontal wells are drilled from each of four pads, and each well produces multiple hydraulic fractures. As shown, it is common for the wells to be drilled at the same azimuth, but in opposite directions. Multiple production zones can be exploited by wells drilled at different depths in approximately the same spatial position. Figure courtesy of Statoil.

but if the wells are too close together, the stimulation zones from multiple wells could interfere and produce from the same parts of the reservoirs.

If there are multiple producing horizons in a given area (*stacked pay*), multiple wells will be drilled at different depths to access the hydrocarbons in each (sometimes called *cube* drilling). Obviously, if one could accurately control the vertical propagation of hydraulic fractures, it might be possible to produce from multiple horizons with the same lateral. However, variability in rock properties from layer to layer, and especially variations of the magnitude of the minimum horizontal principal stress, S_{hmin}, either promotes or impedes vertical hydraulic fracture propagation. How, and why, the magnitude of the least principal stress varies with depth through stratigraphic sequences (and controls hydraulic fracture propagation) is discussed in Chapter 11. There are several reasons why stress magnitudes might vary from layer to layer, including variations of the viscoplastic (time-dependent) deformational properties of the producing formation with respect to those above and below it (as discussed in Chapter 3). This phenomenon illustrates why extremely precise horizontal trajectories might be needed to keep a well in exactly the right lithofacies of a producing horizon, which might be only 5–10 m thick to produce optimal results. An example of this is presented later in this chapter and is also discussed further in Chapter 11.

While the success (and dramatically increased efficiency) of horizontal drilling and multi-stage hydraulic fracturing are frequently cited as the key drivers of production from unconventional reservoirs, the importance of using slickwater should not be overlooked. Slickwater is a low viscosity fluid composed of over 98% water with limited chemical additives. The trade-off of using low viscosity fluid is that it has a greatly reduced capacity for transporting proppant. Initially, the trend in industry prior to "unlocking" the keys to success in developing the Barnett shale was to use highly viscous gels to make individual hydraulic fractures as large, and as filled with as much proppant, as possible. The reason why slickwater works during hydraulic stimulation is that its low viscosity promotes leak-off of the fluid pressure from the hydraulic fractures into fractures in the surrounding rocks, which stimulates shear slip on pre-existing fractures and faults. This, in turn, creates a permeable fracture network (discussed at length in Chapters 10 and 11) adjacent to the hydraulic fracture planes creating what is sometimes called the stimulated reservoir volume, or SRV. This process dramatically increases the overall contact area between the ultra-low permeability matrix and the much higher-permeability fracture network (as discussed in Chapter 12).

Factors that influence the design of a given well and how it should be hydraulically fractured (e.g., the type of hydraulic fracturing equipment to be used, how many/how long the individual hydraulic fracturing stages should be, how many perforations per stage are needed, what type of fracturing fluid and proppant should be used, etc.) are discussed in Chapter 8. Issues related to well spacing are discussed in Chapters 11 and 12 as well as how reservoir depletion from a well or group of wells affects stimulation when infill wells are drilled. This is sometimes referred to as parent well and child well issues. It is also related to the *frac hit* phenomenon – where a hydraulically fracture propagating from a well intersects a previously drilled well. This likely occurs because of the poroelastic stress change due to depletion in the area surrounding the older well. This is also addressed in Chapter 12.

The azimuth at which the horizontal wells are drilled is controlled by the orientation of the least principal horizontal stress, S_{hmin}. The state of stress in unconventional reservoirs is briefly introduced below but is discussed comprehensively in Chapter 7. In the general context of reservoir geomechanics, the reader is referred to Zoback (2007), which discusses the state of stress in the Earth's crust and its significance for issues related to the exploitation of oil and gas. At this point it is important to note that because hydraulic fractures propagate in a plane perpendicular to the least principal stress, usually S_{hmin} (Hubbert & Willis, 1957), the overall geometry with which development is carried out is relatively simple. As shown schematically in Fig. 1.14a, wells are drilled in the direction of S_{hmin} so that hydraulic fractures propagate perpendicular to the well path and the spacing of stages and perforations, as well as the spacing between adjacent wells, is designed to access hydrocarbons from a contiguous volume of the reservoir. In typical operations with cased and cemented wells, each stage pressurizes multiple perforations such that a hydraulic fracture propagates from each. The schematic in Fig. 1.14a illustrates a case in which the stages and perforations are accessing most of the producing zone along the length of the lateral, but because the wells are too far apart, there are portions of the reservoir between the wells that are not being produced.

Introduction   25

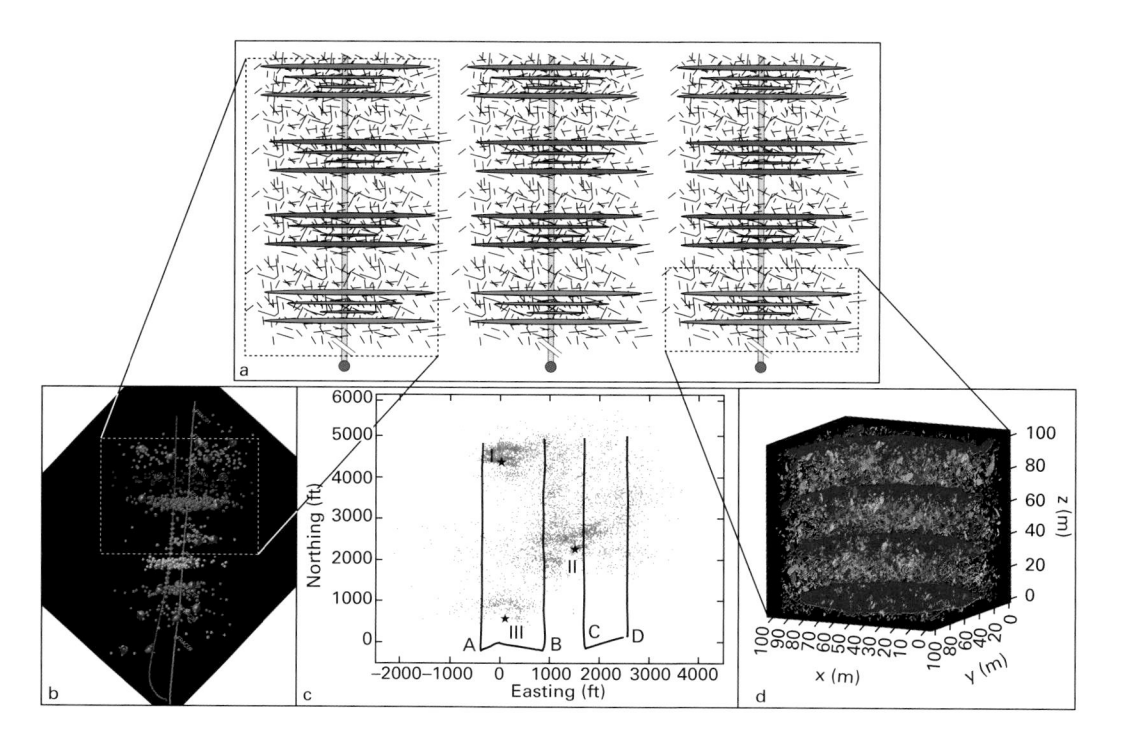

Figure 1.14 (a) Schematic illustration of a well pad with parallel horizontal wells (light gray) drilled in the direction of S_{hmin} that are hydraulically fractured in multiple stages (colored zones). (b) Recorded microseismic events associated with multiple horizontal fractures in a Devonian shale in West Virginia. From Moos et al. (2011). (c) Four wells (black lines) exploiting stacked pay (the Woodford shale and Mississippi lime) in Oklahoma. From Ma & Zoback (2017b). The microseismic events were located using seismic recording arrays in the three vertical wells (stars) labeled I, II and III. (d) A theoretical fracture network model derived from microseismic data obtained during stimulation of a well in the Barnett shale. This model is discussed in more detail in Chapters 10 and 12. From Hakso & Zoback (2019).

One of the most important technological developments associated with hydraulic fracturing surrounds the discovery that literally thousands of microseismic events (extremely small earthquakes) accompany hydraulic fracturing. This is illustrated schematically in Fig. 1.14a and shown in two actual datasets in Figs. 1.14b,c. Surrounding each hydraulic fracture, seismic slip is induced on pre-existing fractures forming a stimulated network of shear faults surrounding the hydraulic fractures. The radiation of seismic waves from these events clearly show that they do not emanate from the hydraulic fractures themselves but, like naturally occurring earthquakes, they result from shear slip on pre-existing planes (Rutledge et al., 2004; Maxwell & Cipolla, 2011; Warpinski et al., 2013).This is discussed at some length in Chapter 9. The typical magnitude of these microseismic events is about M −2, which implies about 0.1 mm of slip on faults roughly 1 m in size. Most importantly, the microseismic events document the fact that pressure leak-off from the hydraulic fractures permeates an interconnected shear fracture network. This network of relatively permeable planes (discussed in Chapter 10) significantly increases the contact area with the very low-permeability matrix and thus contributes substantially to production (Chapter 12). As mentioned above, the success of slickwater hydraulic fracturing is likely due to the fact that its low viscosity enhances pressure diffusion and leak-off out of the hydraulic fractures and into pre-existing fractures and bedding planes.

Because the creation of the stimulated zone surrounding hydraulic fractures is so central to the process of successfully producing hydrocarbons from such low permeability formations, elements of this process are discussed throughout this book. Chapter 7 covers stress, fractures and reservoir-scale faults in more detail. Chapter 9 reviews selected topics in reservoir seismology to provide a context for understanding what we know about the events (their location, size and the geometry of slip) and how they are recorded using downhole (and surface) arrays of seismometers. The events shown in Fig. 1.14b (Moos et al., 2011) were recorded by an array of seismometers in a horizontal well directly beneath the well that was being stimulated. Note the variable number of microseismic events from stage to stage. The microseismic events shown in Fig. 1.14c (Ma & Zoback, 2017b) were recorded in the three vertical wells indicated by stars and labeled I, II and III. Despite the similarity of the stimulation procedures in each well, there was a highly non-uniform response, with numerous microseismic events in some places, but nearly none in others. There are clearly large areas where there were very few microseismic events and some of the microseismic events seen are associated with relatively large-scale faults cutting across the pad (these data are discussed in more detail in Chapter 7 and by Ma & Zoback, 2017b). Thus, while the spatial distribution of microseismic events gives a sense of the volume of the zone of stimulated slip surrounding hydraulic fractures, a variety of factors discussed throughout this book can affect the actual distribution of events. These factors range from physical properties such as the degree of viscoplastic stress relaxation in the lithofacies of interest (Chapters 3 and 11), the frictional behavior of faults in these lithofacies (Chapter 4), the presence of pre-existing fractures and faults (Chapters 7) and the exact well path as it passes through various lithofacies along a lateral. Chapter 9 discusses the degree to which microseismic event detectability and the uncertainty in the event locations affects interpretation.

Chapter 10 focuses on the geomechanical properties that dictate how shear events are triggered during hydraulic fracturing. This chapter also discusses the possibility that there may also be aseismic shear deformation (i.e., slow slip on pre-existing faults) occurring in the reservoir during stimulation, in addition to the slip recorded as microseismic events. Chapter 7 discusses how "pad-scale" faults influence the hydraulic fracturing process in a number of ways. As will be discussed in both Chapters 4 and 10, the composition of the rock and the detailed manner in which shear is stimulated will determine whether or not shear slip results in seismic radiation. In other words, a fracture may shear slowly during stimulation, not generating seismic waves, but may still contribute to creation of a network of permeable fractures surrounding the hydraulic fractures. Chapter 12 considers flow from the nanometer-scale pores in the matrix to hydraulic fractures which illustrates the importance of the shear fracture network. Figure 1.14d shows a theoretical fracture network developed from a microseismic dataset in the Barnett shale that is discussed further in Chapters 10 and 12.

Bedding, Lithofacies and Staying in Zone – In addition to the challenges posed by the ultra-low permeability of unconventional oil and gas reservoirs, they are heterogeneous at a variety of scales and in a variety of ways. One systematic form of heterogeneity is bedding and layering. As discussed at greater length in Chapter 2, at a scale of centimeters and below, these formations are intrinsically anisotropic – they are far more compliant in a direction perpendicular to the bedding than parallel to it. However, they are also anisotropic at larger scales due to bedding (at a scale of 1 m or more), or due to variations of composition within a given bed. In other words, even within a given hydrocarbon formation, there can be different lithofacies, literally meaning *a mappable subdivision of a designated stratigraphic unit, distinguished from adjacent subdivisions on the basis of composition or particular lithologic features*. In the case of exploiting unconventional formations, it is clear that variations of composition and rock properties associated with bedding or variations of lithofacies within a given bed will influence which is the optimal unit for drilling and hydraulic fracturing. Hydraulic fractures must propagate vertically across many bed boundaries in order to access a sufficient thickness of the hydrocarbon bearing formation to account for production.

Distinct bedding in both the upper and lower Eagle Ford is shown by compositional and geophysical data in Fig. 1.11b. Figures 1.15a,b emphasize the thinly layered stratigraphy of the Eagle Ford in road cut photos. Figure 1.15a shows the highly layered nature of a ~20 m thick section of the Boquillas formation (lower Eagle Ford) near Comstock, Texas. The image on the left side of Fig. 1.15b is a close-up view of part of the roadcut seen to the left in Fig. 1.15a. Note that the individual layers range in thickness from 10 cm to 2 m. Obviously, at this site the lower Eagle Ford is both flat laying and relatively undeformed. The photo on the right side of Fig. 1.15b, however, shows another view of lower Eagle Ford stratigraphy near Langtry, Texas. In this area, modest folding can be seen, as well as many near-vertical fractures, often restricted to individual lithologic units. It is also clear from the variability in color that there are significant variations in the composition of the of the individual beds.

In Fig. 1.15c, we show photomicrographs of Eagle Ford samples with their corresponding weight percent of clay and calcite (clay content increases and calcite content decreases from left to right). Note that the fabric induced by fine-scale bedding becomes more pronounced as clay content increases. In addition, as the wavelengths of compressional waves at the typical frequencies used in 3D seismic surveys are tens of meters to 100 m in length (much greater than layer thickness), the layered nature of the bedding creates anisotropy that affects wave propagation and imaging.

The result of intrinsic anisotropy and sub-horizontal layering seen at the multiple scales shown in Fig. 1.15 produce Vertical Transverse Anisotropy, or VTI (Thomsen, 1986). In simple terms, VTI means that compressional waves will travel more slowly in the vertical direction (normal to the layering) than in the horizontal direction (parallel to the layering). However, if there are pervasive near-vertical fractures with a similar trend, there can be Horizontal Transverse Anisotropy (HTI), may result in which horizontally propagating seismic waves travel faster in the directions parallel to the fractures than those normal to them (see Chapter 7). Horizontal stress anisotropy (the difference

Figure 1.15 a) Road cut photo of a ~20 m thick section of the Boquillas (lower Eagle Ford) formation near Comstock, Texas. (b) Left: Closer views of the lower Eagle Ford show the scale of the layering (~10 cm to 2 m) in the road cut in (a). Right: Gentle folding, alternating lithofacies and the occurrence of near vertical fractures in a different outcrop near Langtry, Texas. Photos courtesy of Peter Hennings, University of Texas (c) Photomicrographs of the samples with different clay/calcite contents. Note that the fine-scale bedding (and resultant anisotropy) increases with clay content.

between the magnitudes of the horizontal principal stresses, S_{Hmax} and S_{hmin}), can also produce HTI, as horizontally traveling P-waves are expected to travel faster in the S_{Hmax} direction than the S_{hmin} direction. Thus, when HTI is observed, it may not be clear whether it is caused by aligned fractures or by horizontal stress anisotropy (Chapter 7).

As noted above, the effect of variations of stress magnitude with depth is discussed in some detail in Chapter 11, as well as its impacts on hydraulic fracture initiation and growth (both vertically and horizontally) and identifying the right landing zone for laterals. Two examples of the importance of vertical variations of composition (and corresponding variations of stress magnitudes) are briefly illustrated in Fig. 1.16. Alalli & Zoback (2018) studied a series of wells drilled in the Marcellus formation in West Virginia. When the trajectory of the well was in the Cherry Valley unit of the Marcellus (shown in gray in Fig. 1.16a), vertical hydraulic fractures were produced, as expected. The measured values of the least principal stress were less than the overburden stress and microseismic events propagated vertically up into the Upper Marcellus (UMRC). In contrast, hydraulic fracturing stages in the clay- and organic-rich Lower Marcellus (LMRC) produced horizontal hydraulic fractures, which was indicated both by the fact that measured magnitudes of the least principal stress were equivalent to the overburden stress (S_V) and the induced microseismic events were restricted to the LMRC.

Figure 1.16b shows a similar, but more subtle case where vertical variations of composition played an important role in affecting stimulation, this time for a well drilled

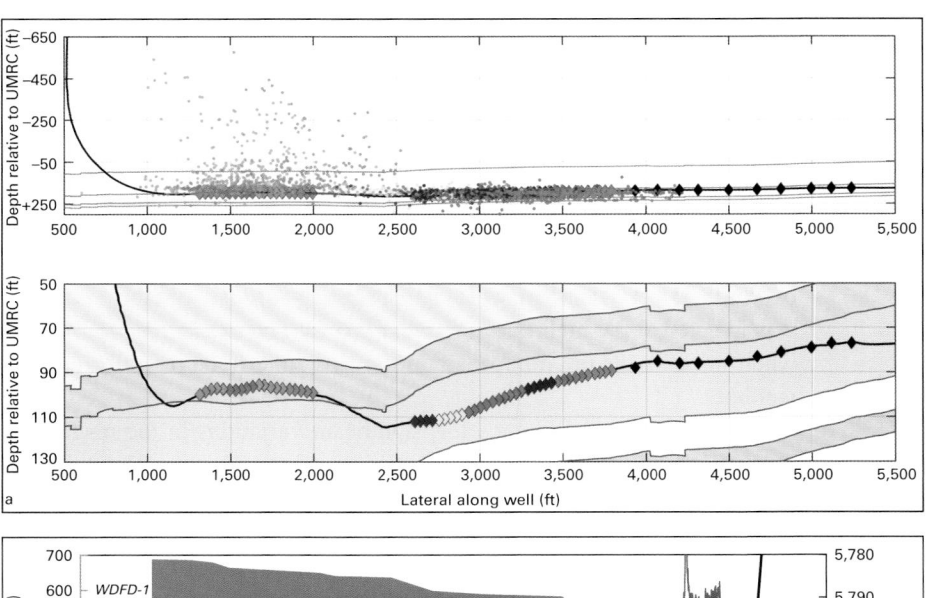

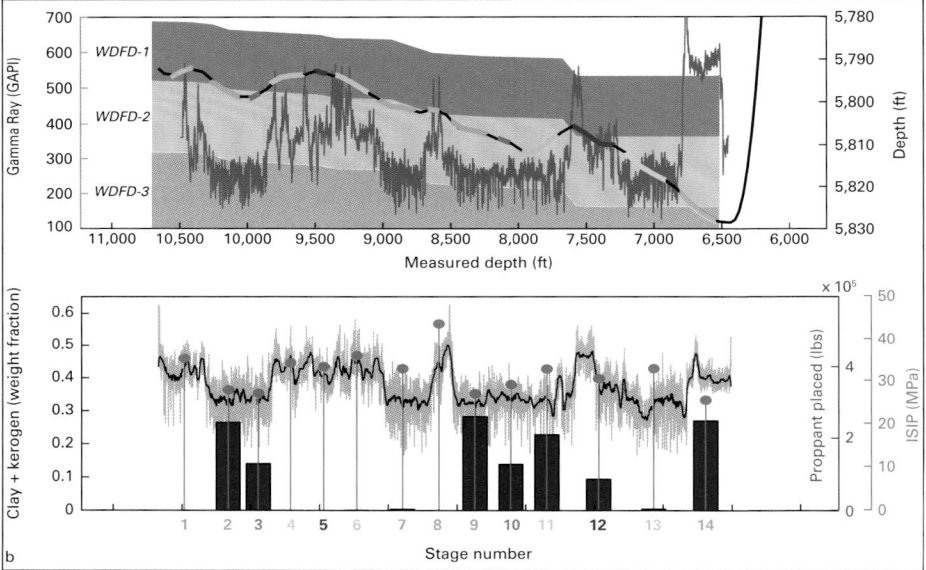

Figure 1.16 (a) A horizontal well drilled into both the Cherry Valley limestone (shown in gray) and Lower Marcellus formation (shown in light green) in West Virginia. From Alalli & Zoback (2018). Horizontal hydraulic fractures were induced during stimulation of the Lower Marcellus. No microseismic data were recorded during stimulation of the first ten stages near the toe of the well. (b) Trajectory of a horizontal well in the Woodford formation in Oklahoma encountered three distinct lithofacies referred to as WDFD 1, 2 and 3. The least principal stress values are shown by the red dots (approximately the overburden stress in WDFD-2). The amount of proppant injected is shown by the blue bars. From Ma & Zoback (2017b).

into the Woodford formation in Oklahoma (from Ma & Zoback, 2017b). In this case the well trajectory stayed, as intended, within the Woodford formation with a "toe up" trajectory. The well path seems quite undulatory due to the vertical exaggeration of the depth scale, but there is actually only 40 ft of elevation difference between the toe and the heel. Nevertheless, the trajectory of the well intersected sections of three distinct lithofacies, each 10–20 ft thick. The hydraulic fracturing stages in the sections of the well that drifted up into the WDFD-1 lithofacies (which is more clay- and organic-rich), encountered such high values of the least principal stress (essentially equivalent to the overburden stress) that the wells were not successfully completed because no proppant was injected at the high injection pressures for fear of a *screen-out*. Screen-outs occur when injection of the proppant into the hydraulic fractures fails and instead fills the wellbore.

The existence, and reasons for, significant variations in the magnitude of the least principal stress from layer to layer is an important topic throughout this book. As noted earlier, this represents an important linkage between intrinsic properties of the lithofacies being hydraulically fractured (as well as the rocks above and below it) and the respective stress magnitudes in these lithofacies. It is frequently argued that some intervals are better for hydraulic fracturing because of different degrees of *brittleness*, which generically describes the relative stiffness of a rock and its tendency for ductile deformation associated with deformation. Over the years, brittleness has been defined in a multitude of ways based on empirical relationships. In Chapter 3 we make the case for *rethinking brittleness* in terms of how the aspects of a rock's composition and microstructure that promote non-brittle, or *viscoplastic* deformation (time-dependent plastic deformation), also affect the state of stress in the rock. This is explored in the context of vertical hydraulic fracture growth and hydraulic fracture initiation in Chapter 11. The example in Fig. 1.16b implies that precise control of the well trajectory in the Woodford (keeping wells in the ~20 ft thick middle lithofacies) is key for successful hydraulic fracturing in that area. These examples are discussed more thoroughly in Chapter 11.

Nearly all of the issues briefly introduced in this chapter will prove to be of importance in subsequent chapters of this book. The primary goal of this chapter was to introduce the idea that the interactions among different geologic and geomechanical factors are key to understanding how to improve recovery from unconventional reservoirs. Essentially, the overall purpose of this book is to establish the fundamental principles related to the multiplicity of attributes of unconventional reservoirs that are critical for optimizing stimulation and production.

2 Composition, Fabric, Elastic Properties and Anisotropy

In this chapter, we explore the composition and fabric of unconventional reservoir rocks in order to understand the variations in elastic properties and anisotropy. First, we survey the range of compositions in unconventional basins and develop a simple classification for the various lithofacies. We then discuss each of the constituents of the rock matrix and describe their role in forming the rock fabric at various scales. The next section covers laboratory measurements of elastic properties and anisotropy, beginning with a brief review of methods for obtaining static, dynamic, anisotropic elastic properties. We discuss the variations of elastic properties with composition and fabric in the context of theoretical bounds from simple rock physics models of layered media to develop a physical understanding of microstructural controls on the stiffness of the rock matrix. The final section covers how elastic properties are estimated from geophysical well logs and reservoir-scale seismic studies. We compare field- and lab-derived elastic properties and discuss their applications for understanding the *in situ* physical properties of unconventional reservoir rocks.

Composition and Fabric

The efficient and intelligent development of unconventional resources depends on optimizing hydraulic stimulation in the context of reservoir properties. In Chapter 1, we discussed how unconventional reservoir rocks are originally deposited as sedimentary sequences, giving rise to variations in composition, organic content and rock fabric. Here, we examine these variations in detail, building up the rock matrix from its constituents. For the purposes of this book, we will mainly focus on samples from the Stanford Stress and Crustal Mechanics Group collections, which were obtained through research partnerships with various energy companies. These samples cover most major unconventional plays in North America, comprising a small, but representative fraction of the reservoir rocks from each basin.

In the scientific and engineering literature, most unconventional reservoir rocks are described by the term "shale." This convention is derived from studies of conventional oil and gas reservoirs, in which "shale" is attributed to fine-grained, ultra-low permeability units or lithofacies acting as source- or cap-rocks (Ulmer-Scholle et al., 2015). For the purposes of this book, rather than attempting to classify unconventional reservoir rocks using geologic terminology (e.g., mudstone, limestone, chert), we adopt a simple ternary classification, which presents mineralogy in terms of QFP (quartz, feldspar, pyrite), carbonates (calcite, dolomite, ankerite, siderite), and the sum of clay minerals and total organic carbon (TOC) (Fig. 2.1a). This classification is particularly useful for comparing rocks with different depositional histories for two reasons. First, it encapsulates ~90% or more of the mineralogy of rocks from the major US unconventional basins (Table 2.1: see pages 61–64). Remaining accessory minerals may include apatite, celestite, gypsum and barite, but these seldom exceed 5 wt%. Also, the three compositional endmembers are grouped according to their relative mechanical and chemical behaviors (Loucks et al., 2012). Clay minerals and organic matter are relatively chemically reactive, mechanically compliant and contribute to anisotropy. QFP are relatively chemically inactive and mechanically stiff. Carbonate minerals are relatively chemical reactive and mechanically stiff. In the ternary classification in Fig. 2.1a, rocks with >50 wt% QFP are termed "siliceous," rocks with >50 wt% carbonate are termed "calcareous" and rocks with >30 wt% clay + TOC are termed "clay-rich." The threshold for clay-rich rocks is slightly lower than the others because 30–40 vol% clay has been shown to be transitional point in terms of microstructural (Revil et al., 2002), mechanical (Kohli & Zoback, 2013) and flow properties (Crawford et al., 2008). We will discuss how variations in clay and organic content impact these physical properties in their respective chapters. It is important to note that in the literature (and studies referenced in this book), both wt% and vol% are widely used. As discussed in Chapter 1, the density of kerogen (1.3 g/cc) is nearly a factor of two less than the measured bulk density of core samples (Sone & Zoback, 2013a), so converting from wt% to vol% increases TOC by nearly a factor of two. Clays, silicates and carbonate minerals are slightly more dense than bulk samples, so their fractions decrease when converting from wt% to vol%. In the following sections, we will examine each of the constituents on the rock matrix in detail.

Clays – Clay contents in unconventional reservoir rocks range from ~0 to 60 wt% (Table 2.1: see pages 61–64). The presence of clay minerals is a key factor determining a range of physical properties, including strength and anisotropy (Chapter 2), ductility (Chapter 3), friction (Chapter 4) and flow (Chapter 6). In most unconventional basins, clay content is positively correlated with TOC, but the individual trend varies between reservoirs due to the specific depositional conditions (Fig. 2.2). The clay mineralogy in most basins is dominated by illite and mixed layer illite–smectite in addition to minor amounts of kaolinite and chlorite (Passey et al., 2010; Milliken & Day-Stirrat, 2013). This is expected since most unconventional basins have experienced pressure–temperature conditions sufficient to convert the majority of clays to illite (Curtis, 2002; Passey et al., 2010; Loucks et al., 2012). Between 70 °C and 100 °C, smectite and mixed layer illite–smectite begins to convert to illite (Ho et al., 1999), and above 130 °C kaolinite is also converted to illite (Bjørlykke, 1998). The illitization process also leads to a marked increase in crystallographic alignment during burial, as the dissolved smectite layers are replaced by illite forming under depositional stresses (Aplin & Macquaker, 2011). In addition, the smectite to illite reaction generates excess silica, which can be trapped within clay layers as microcrystalline quartz, increasing density and rock stiffness during burial (Peltonen et al., 2009).

Composition, Fabric, Elastic Properties and Anisotropy

33

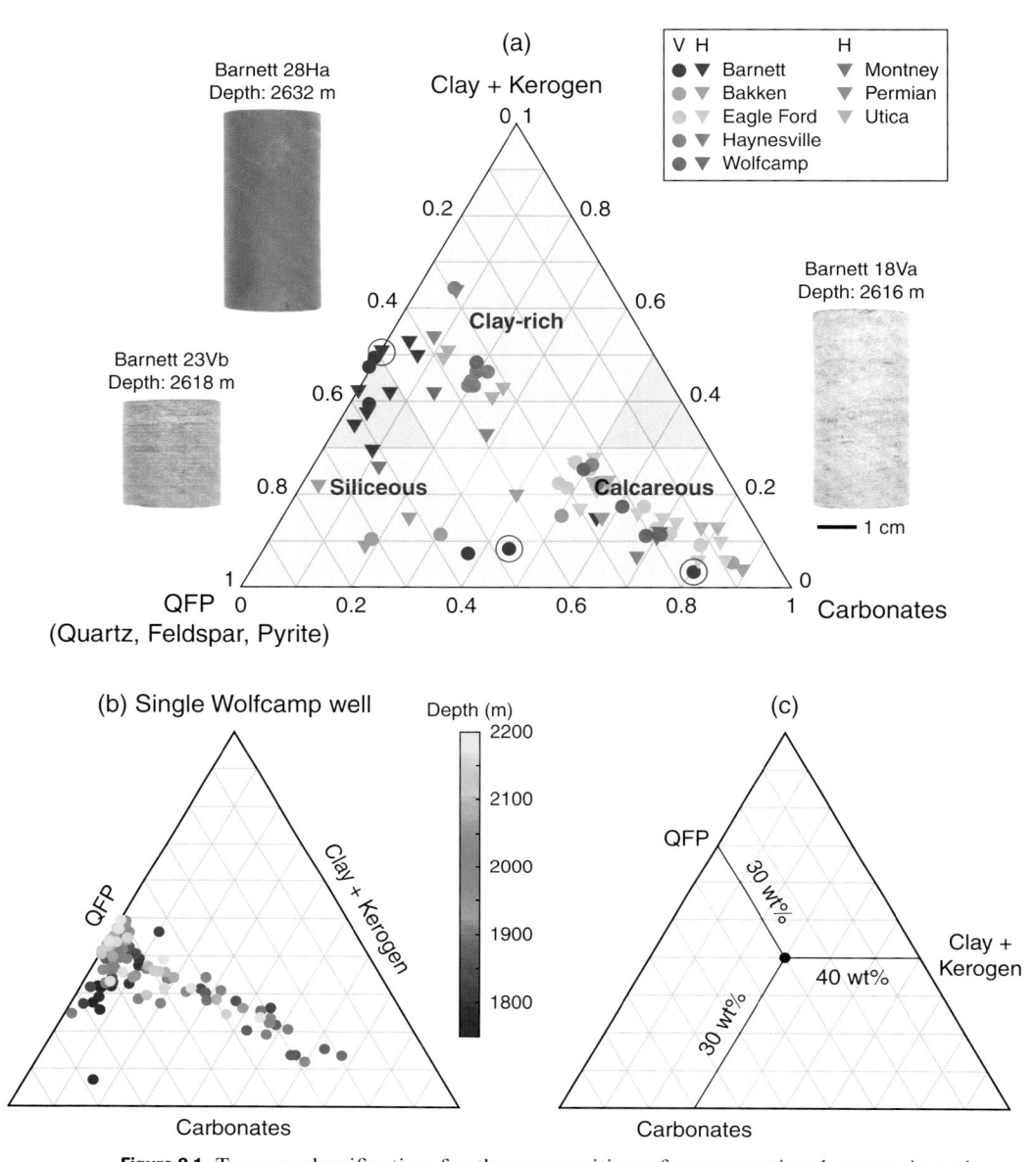

Figure 2.1 Ternary classification for the composition of unconventional reservoir rocks. All values in weight fraction. (a) Select shale samples from major US unconventional basins from the Stanford Stress and Crustal Mechanics Group collection. Photographs of circled core samples from a single Barnett shale well highlight meter-scale variations in composition. Sone (Unpublished). (b) Samples from a single Wolfcamp shale well show systematic variations in composition with depth. (c) Guide for reading values from ternary diagrams.

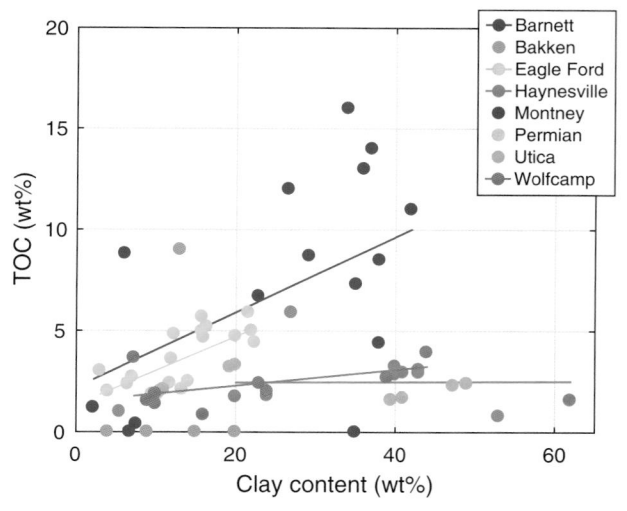

Figure 2.2 Correlation of clay and total organic carbon (TOC) for several major unconventional basins.

This process may be important for creating additional interparticle porosity within clays (Fig. 2.3b, Fig. 5.4). The contribution of clay minerals to matrix porosity is discussed in detail in Chapter 5. The prevalence of illite in unconventional reservoir rocks is also significant because illite is a non-expanding clay, meaning the interlayer cations prevent water from entering its structure and causing expansion (Fig. 2.3a). In contrast, mixed layer illite–smectite clays can expand up to 25% in volume (Williams & Hervig, 2005). We will discuss the causes and consequences of clay swelling further in Chapter 5.

Clay minerals are a major source of mechanical weakness and anisotropy in unconventional reservoir rocks because of their intrinsic (crystallographic) anisotropy (Fig 2.3a) and tendency to align in the bedding plane during deposition and diagenesis (Fig. 2.3b) (Sayers, 1994). Clay minerals manifest in the rock matrix as fine-grained, platy aggregates (~10–500 nm) (Fig. 2.3b; Sondergeld et al., 2010a) that comprise numerous, nm-scale fundamental particles (Fig. 2.3a). The degree of alignment (preferred orientation) of clay aggregates in the rock matrix is dependent on many factors including depositional conditions, clay content and thermal maturity (Revil et al., 2002; Wenk et al., 2008; Kanitpanyacharoen et al., 2011, 2012). In unconventional reservoir rocks, electron microscopy shows clays form local preferred orientations on the μm-scale, which often rotate as layers of clay aggregates appear to "flow" around larger clastic grains (Curtis et al., 2012b). On the mm- to cm-scale, petrographic microscopy and X-ray goniometry indicates clay aggregates form sub-horizontal preferred orientations, which, along with other matrix components, create an anisotropic fabric that defines the bedding plane (Fig. 2.4; Wenk et al., 2008; Kanitpanyacharoen et al., 2011, 2012; Sone & Zoback, 2013a). Sone & Zoback (2013a) observe loss of clay preferred orientation at clay contents less than 30 vol %, which is generally consistent with estimated thresholds for forming an interconnected matrix of clay (Curtis, 1980; Revil et al., 2002). Below this value, clastic grains form a load-bearing framework that disrupts the continuity of clay fabrics from the mm-

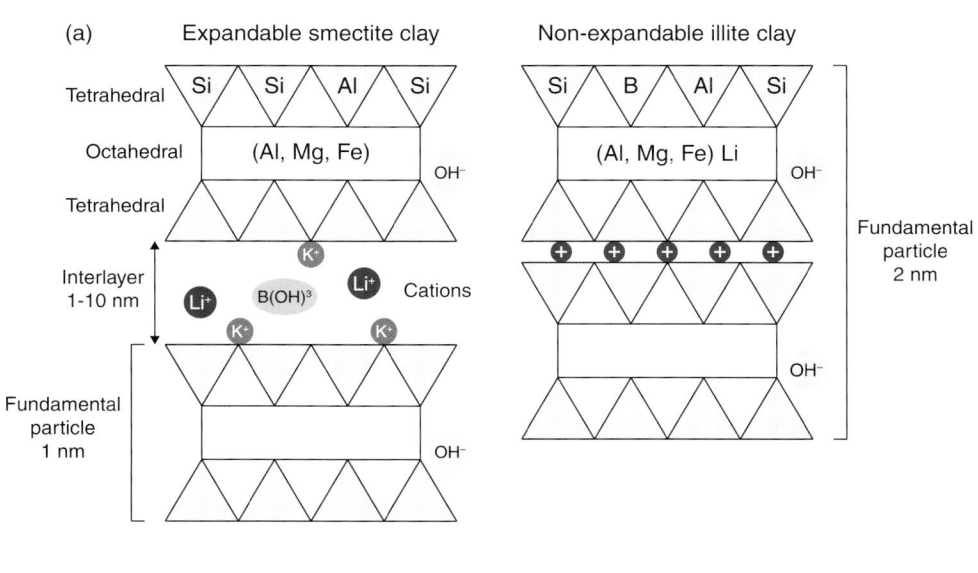

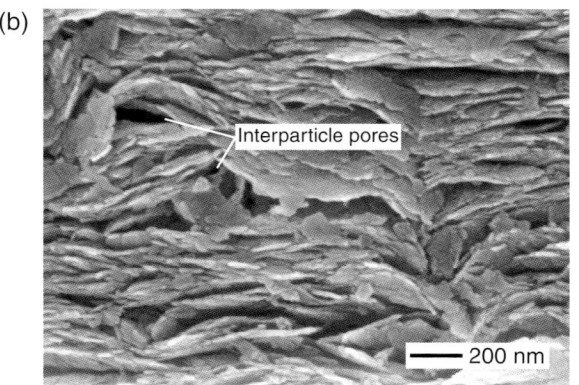

Figure 2.3 (a) Crystallographic structure of smectite and illite clay minerals. Adapted from Williams & Hervig (2005). (b) Clay minerals aggregates (made up of numerous fundamental particles) form strong local preferred orientations and nano-scale interparticle porosity. Haynesville 1–5, secondary electron image. Kohli (Unpublished).

to cm-scale (Fig. 2.8). The formation of clay fabrics also promotes the development of interparticle silt-shaped pores elongated parallel to bedding (Fig. 2.3b; Daigle & Dugan, 2011; Loucks et al., 2012; Leu et al., 2016). The shape preferred orientation of these pores represents another possible source of anisotropy in the mechanical (Chapters 2, 3) and flow (Chapter 5, 6) properties.

Carbonates – Carbonate minerals (principally calcite and dolomite) are important components of the rock matrix. Carbonate contents range from ~0 to 80 wt% in unconventional reservoir rocks (Fig. 2.5). Variations in carbonate content represent one of the major compositional trends between and within individual basins (e.g., Figs. 2.1a, c). The primary source of carbonates in unconventional reservoir rocks is

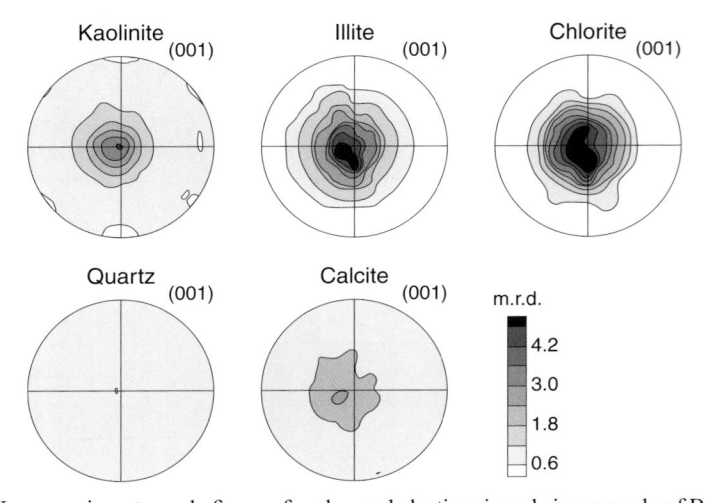

Figure 2.4 X-ray goniometry pole figures for clay and clastic minerals in a sample of Benken shale (France). The strength of mineral preferred orientation is quantified by multiples of a random distribution (m.r.d.). From Wenk et al. (2008). *Reprinted by permission of the Clay Minerals Society whose permission is required for further use.*

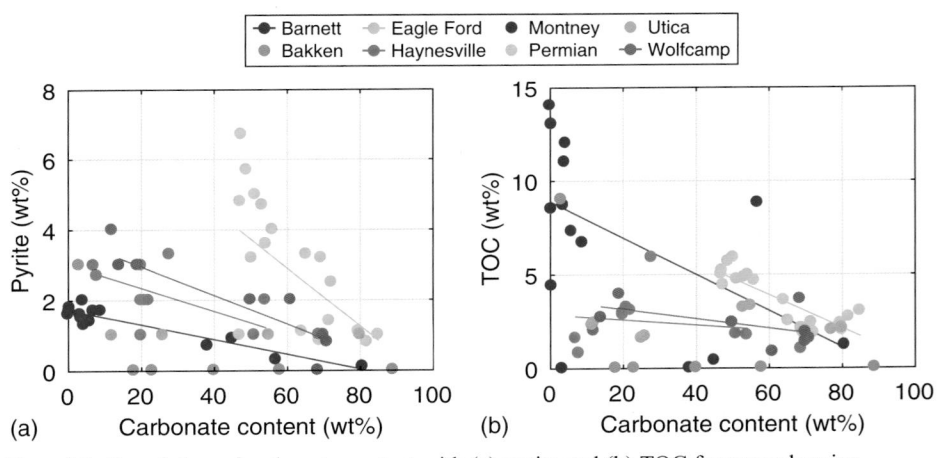

Figure 2.5 Correlation of carbonate content with (a) pyrite and (b) TOC for several major unconventional basins.

fossil marine organisms, including foraminifera (Fig. 2.6c, Fig. 2.8b), calcispheres and bivalves (Loucks & Ruppel, 2007). The biotic origin of carbonates and variations in depositional conditions result in wide variations in grain shapes and sizes, from sub-μm massive fill in fractures and between grains (Fig 2.8a) to μm-scale fossil grains (Fig. 2.8b, Haynesville-1, Eagle Ford-1) to mm-scale angular grains (Fig. 2.8b, Barnett-1). Carbonate content negatively correlates with pyrite content and TOC (Fig. 2.5) because sulfidic and anoxic conditions during deposition favor the preservation of organic matter, while limiting presence of calcareous marine organisms (Loucks

& Ruppel, 2007). Similar to the relationship between clay and TOC, the trend varies between individual reservoirs due to the specific depositional conditions.

Organic Matter – Organic matter is a persistent feature of oil- and gas-bearing unconventional reservoir rocks (Passey et al., 2010). The term kerogen is used to describe a mixture of different organic compounds, but the specific chemical composition can vary greatly between and within individual basins. Kerogen is quantified by chemical measurement of evolved total organic carbon (TOC). Most unconventional source rocks are relatively organic-rich compared to typical sedimentary rocks with TOC ranging from 2 to 15 wt% (Fig. 2.5b; Table 2.1). Organic matter presents a wide range of morphologies in the rock matrix, including μm-scale, bedding parallel lenses (Fig. 2.6a; Fig. 2.8, Eagle Ford-1; Fig. 5.6), mm-scale filled microfractures (Fig. 2.6, Fig. 5.3), isolated deposits within microfossils (Fig. 2.6c, Fig. 5.5). In fact, many different types of organic matter structures can coexist within the same lithofacies (Milliken et al., 2013). In organic-rich rocks, the presence of filled microfractures and lenses of kerogen elongated in bedding plane represents a significant source of elastic and flow anisotropy as kerogen is much more compliant and generally more porous than the other matrix components (Vanorio et al., 2008; Al Ismail & Zoback, 2016).

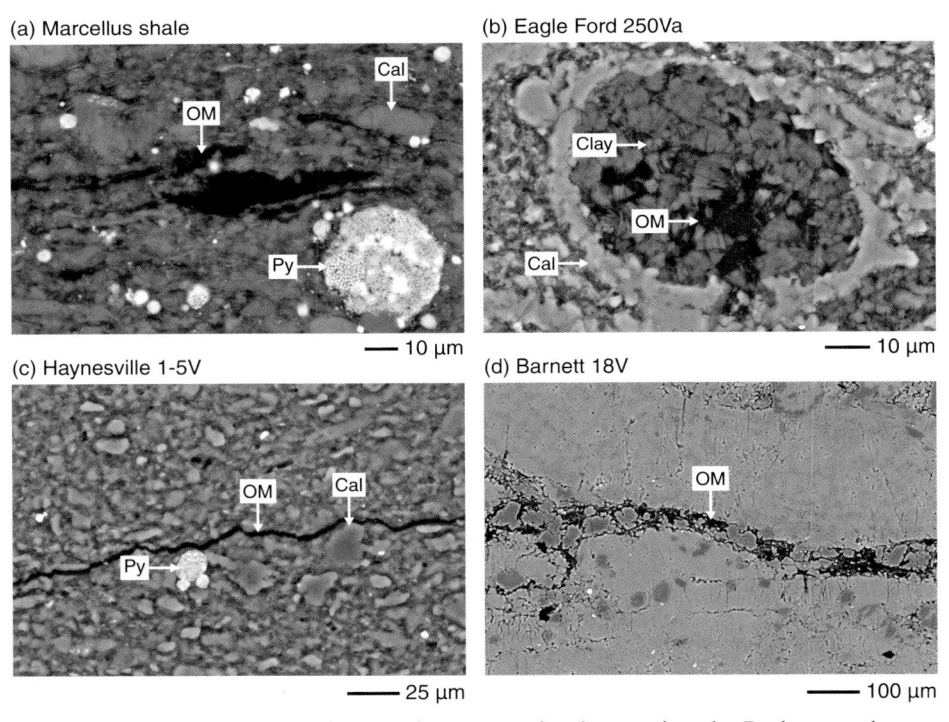

Figure 2.6 Morphologies of organic matter in unconventional reservoir rocks. Backscatter electron images. (a) Lens-shaped deposits 50–100 μm, elongated parallel to bedding, Marcellus shale. (b) Bedding parallel microfracture filled with organic matter, ~1–3 μm thick, Haynesville 1–5. (c) Fossil foraminifera filled with organic matter, Eagle Ford 250Va. (d) Interconnected network of organic matter, Barnett 18Vb. Kohli (Unpublished).

Porosity – Porosity in unconventional reservoir rocks ranges from ~1 to 15 wt% (Fig. 2.7; Chalmers et al., 2012a) and comprises a wide variety of pore types from the sub-nm to sub-μm-scale (Curtis et al., 2012a; Ma et al., 2016). There are three main pore types common in the rock matrix: sub-nm to μm-scale pores in organic matter, nm to μm-scale pores in the mineral matrix (interparticle and intraparticle pores) and μm-scale microfractures mostly found within the bedding plane (Loucks et al., 2012). In Chapter 5, we will explore the characteristics of matrix pore networks in great detail. Here, we consider how porosity relates to compositional trends in terms of the main pore types.

Porosity in major US shale basins is positively correlated with both clay and kerogen (Fig. 2.7). Clay minerals can introduce significant nm-sale interparticle porosity between clay aggregates (Fig. 2.3b) and clay–clastic contacts where fabric continuity is disturbed (Fig. 5.4). Swelling clay minerals such as illite-smectite may also contain smaller intraparticle pores in interlayer spaces (Fig. 2.3a). Due to the tendency of platy clay aggregates to form preferred alignments, clay interparticle porosity is often composed of silt-shaped pores aligned with the local fabric or bedding plane. Organic matter contains sub-nm to sub-μm-scale pores and can range in porosity from ~0–40% within individual deposits (Curtis et al., 2012b). Porosity and pore types within organic matter are dependent on a range of factors including chemical composition, maturity and the mineralogy of the surrounding rock matrix (Curtis et al., 2011, 2012b; Milliken et al., 2013; Anovitz et al., 2015). Porous organic matter often exhibits a spongy texture composed of approximately spherical pore shapes with no apparent preferred orientation (Fig. 5.6; Curtis et al., 2012a; Milliken et al., 2013), but may also contain aligned, silt-shaped pores (Loucks et al., 2012). Organic matter may also form larger interparticle pores up to several μm at contacts with clay or clastic minerals (Fig. 5.6).

Granular silicate, carbonate, phosphate and oxide minerals also contribute interparticle and intraparticle porosity to the rock matrix. Interparticle pores form between clastic

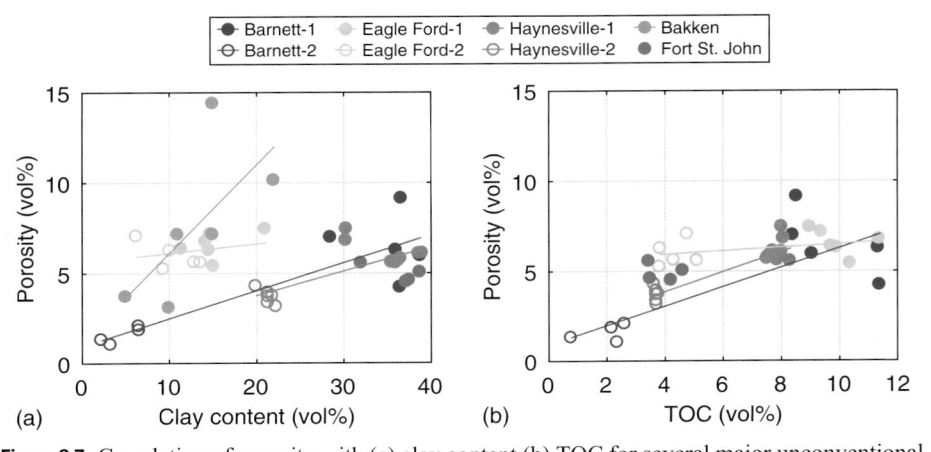

Figure 2.7 Correlation of porosity with (a) clay content (b) TOC for several major unconventional basins. From Sone & Zoback (2013a).

contents and clay–clastic contacts. Pores between clastic contacts are generally larger due to the larger grain size of clastic minerals compared to clays. Carbonate minerals often show significant intragranular porosity, particularly within relic fossils (Fig. 5.5), massive fracture fill (Loucks et al., 2012; Vega et al., 2015) and large grains with dissolution-precipitation features (Rassouli & Zoback, 2018). Pyrite framboids (Fig. 2.6a) also contain significant intragranular porosity between individual euhedral crystals (Fig. 5.5). In most cases, granular matrix porosity shows a wide variety of pore shapes from equant to irregular to silt-shaped and does not exhibit any preferred orientation from the mm- to µm-scale (Loucks et al., 2012). Microfractures in the rock matrix span the cm- to mm-scale in length and range in thickness (aperture) from ~1–100 µm (Fig. 5.3; Vega et al., 2015; Zhang et al., 2017). In most cases, microfractures are filled by either organic matter (Fig. 5.3d) or mineral cement (Loucks et al., 2009, 2012; Vega et al., 2015). Larger fractures may also form on the bedding plane due to mechanical unloading during core recovery, storage or sample preparation (e.g., grinding, epoxy impregnation for thin sections) (Fig. 5.3a).

Considering the variations in porosity and pore types with respect to compositional variations in unconventional reservoir rocks, clay and organic-rich rocks are likely to contain mostly organic matter porosity and clay interparticle porosity oriented in the bedding plane, while clastic rocks are likely to contain a greater fraction of intragranular pores and interparticle pores with no preferred orientation. In clay and organic-rich rocks, microfractures are likely filled with organic matter or may not form at all due to increased ductility relative to clastic rocks (Chapter 3). In siliceous and calcareous lithologies, microfractures may also be filled with organic matter or mineral phases that precipitate during diagenetic fluid flow.

Fabric – In Chapter 1, we discussed the various scales of rock fabric in unconventional basins from outcrop-scale sedimentary layering to cm-scale compositional variations (Fig. 1.15). From the m- to µm-scale, rock fabric (structural anisotropy) is defined by various features, depending on rock matrix composition. Clay minerals form bedding-parallel preferred orientations during diagenesis, compaction and phase transformations. Within these clay fabrics, layering of platy clay aggregates results in the development of bedding-parallel silt-shaped pores, which can contribute to the formation of a fabric in the matrix pore structure. Organic matter distributed in bedding-parallel lenses and veins (microfractures) represents another potential source of anisotropy (Fig. 2.6a,b). Again, the development of porosity in these bedding-parallel features can form a fabric in the pore network. In some cases, organic matter may actually fill nm-scale interparticle spaces between clay aggregates (Kuila et al., 2014), reinforcing the strength of the existing clay fabric. As discussed earlier in this chapter, the morphology, porosity and distribution of organic matter is strongly dependent on the source, depositional conditions and the surrounding rock matrix. Although the major control on fabric strength in unconventional reservoir rocks is the abundance and distribution of clay and organic matter, in some lithofacies, clastic minerals (quartz and carbonates) also exhibit preferred orientations of grain shapes parallel to bedding (Fig. 2.8b, Eagle Ford 176Ha; Klaver et al., 2012). Bedding-parallel microfractures, where present, represent another potential source of anisotropy not necessarily associated with clay and organic matter.

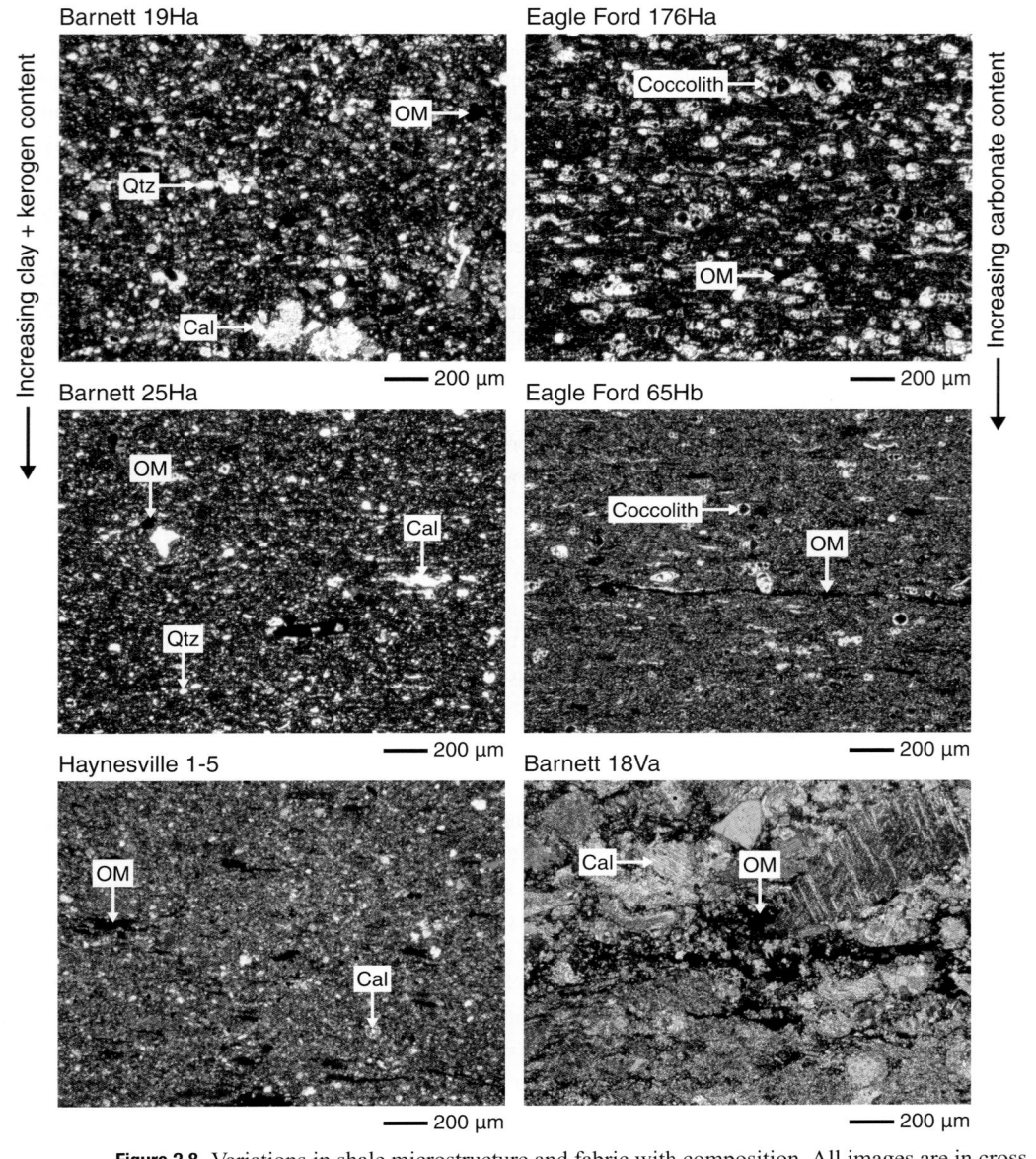

Figure 2.8 Variations in shale microstructure and fabric with composition. All images are in cross polarized light on thin sections cut perpendicular to the bedding plane. (a) Clay-rich shales show increasing fabric strength and decreasing grain size with increasing clay content. (b) Calcareous shales show decreasing clay fabric strength with increasing carbonate content. OM – Organic matter, Qtz – Quartz, Cal – Calcite. Adapted from Sone and Zoback (2013a). See Table 2.1 for sample compositions.

Due to the numerous factors impacting fabric development in unconventional reservoirs, the relationship between anisotropy and matrix composition is not necessarily straightforward. Figure 2.8 shows petrographic images of siliceous rocks ordered by increasing clay and TOC, and calcareous rocks ordered by increasing carbonate content. In the siliceous rocks, increasing the abundance of clay clearly increases the strength of the bedding-parallel fabric; however, whether increasing TOC strengthens the fabric depends on its morphology and distribution (e.g., Fig. 2a Haynesville 1–5 and Barnett 25 Ha). In the calcareous rocks, bedding-parallel preferred orientation of carbonate minerals is present to various degrees at all carbonate contents. As carbonate content increases, any other component of matrix may decrease according to the specific lithological trends in a given reservoir. For example, in Fig 2.8b, carbonate increases mostly at the expense of clay and QFP, while TOC is relatively constant, providing a consistent source of anisotropy in each sample. The complex relationship between anisotropy (fabric) and composition underscores the importance of laboratory testing for understanding the physical properties of different lithofacies.

Scales of Heterogeneity – As we begin to discuss a range of laboratory measurements performed on recovered core samples in Chapters 2–6, it is important to note that, in unconventional reservoir rocks, composition and microstructure may vary significantly from the cm- to the mm-scale. An example of this is seen in Fig. 2.9 in photomicrographs of a horizontal core sample from the Mancos shale. Even at the scale of a 1 inch core sample – the US standard for testing rock properties – there are mm-scale laminations representing significantly different lithofacies. Some mm-scale facies show continuous, bedding-parallel veins of kerogen within a strong clay fabric, while others show a granular structure of quartz with weak clay fabric and only isolated deposits of kerogen. When considering measurements of rock properties at the core-scale, it is critical to remember that any observed behaviors reflect the integrated effects of various microstructures. In Chapter 5, we will discuss efforts to quantify the scale of the representative elementary volume (REV) of the rock matrix.

Elastic Properties

Vertical Transverse Isotropy – The m- to cm-scale fabric of unconventional reservoir rocks shows that the primary variations in composition and microstructure are in the vertical direction, while individual facies are relatively continuous in the horizontal plane (Fig. 1.15; Sayers, 1994; Slatt & Abouseleiman, 2011). Core-scale samples also show horizontal laminations, creating mm-scale compositional variations in the vertical direction (Fig. 2.9). The prevalence of this fabric at various scales suggests shales are transversely isotropic with a vertical axis of rotational symmetry or vertical transverse isotropic (VTI) (Fig. 2.10; Vernik & Nur, 1992). In the following section, we will explore how the VTI model is used to quantify the elastic properties and anisotropy of unconventional reservoir rocks.

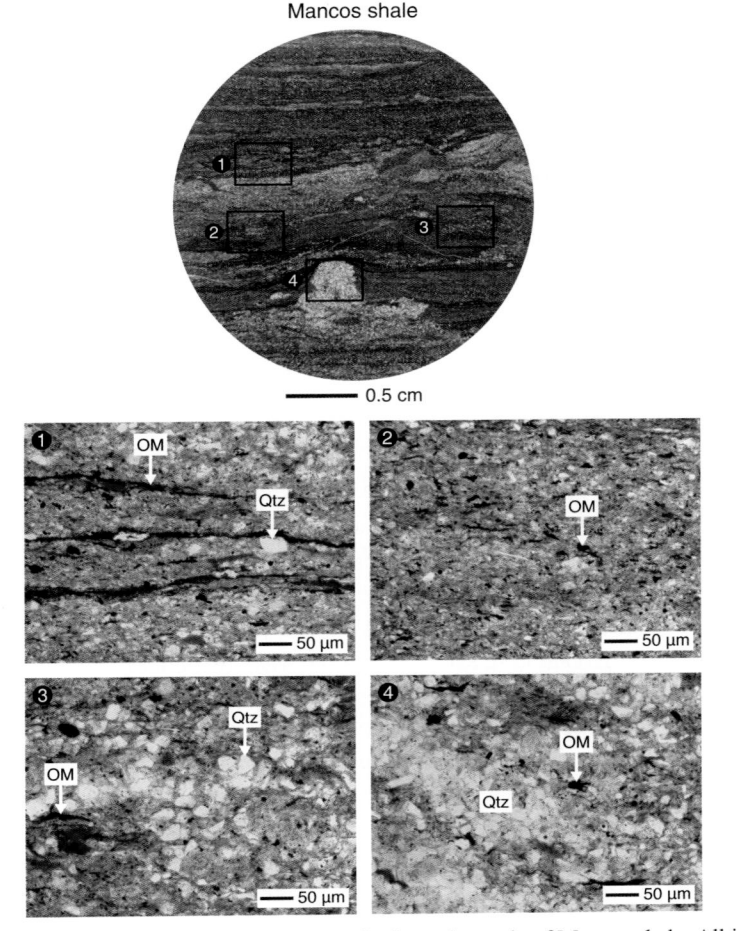

Figure 2.9 Core plug-scale heterogeneity within a horizontal sample of Mancos shale. All images in plane polarized light on a thin section cut parallel to the bedding plane. Kohli (Unpublished).

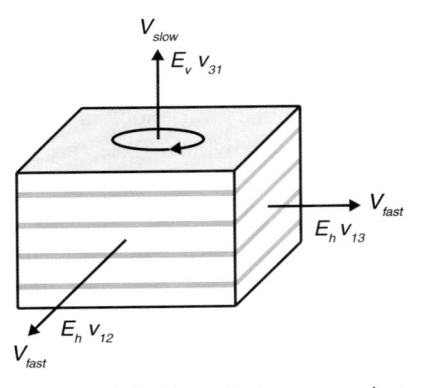

Figure 2.10 Schematic of layered material with vertical transverse isotropy. The vertical direction, x_3, is the axis of symmetry. The fast wave polarizations lie within the horizontal plane (bedding), while the slow wave polarization is vertical, perpendicular to layering.

Composition, Fabric, Elastic Properties and Anisotropy   43

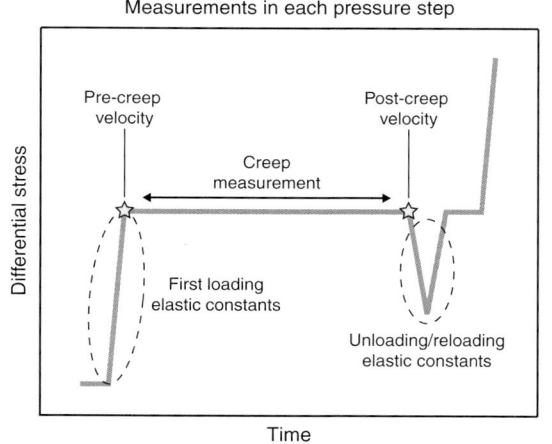

Figure 2.11 Experimental procedure for characterizing static and dynamic elastic properties. Static moduli are determined by the stress–strain relationship upon first loading and after several hours of constant stress application (creep). Sonic velocities are measured before and after the creep step. The full procedure and creep measurements are discussed in detail in Chapter 3 (Fig. 3.1). From Sone & Zoback (2013a).

Static and Dynamic Moduli – Sone & Zoback (2013a) developed a comprehensive procedure for measuring elastic moduli and time-dependent deformation (creep) at various stress levels (Fig 2.11). We will discuss measurements of time-dependent deformation in Chapter 3. Static moduli are determined from the stress–strain response during each loading step. Young's modulus, E, is calculated from the ratio axial stress to axial strain. Poisson's ratio, v, is calculated from the ratio of axial and lateral strains. Dynamic moduli are determined by wave velocity measurements at each stress level before and after application of constant stress (creep measurement). The sample is subsequently partially unloaded and reloaded to the same stress level to re-measure static moduli. This procedure is intended to reveal any hysteresis between first loading and unloading/reloading measurements, which can theoretically provide insights into the former *in situ* stress conditions of the sample. As we will discuss later in the chapter, this is not necessarily straightforward due to the variety of factors that may influence rock stiffness during core recovery and storage.

For VTI materials, only five independent wave-velocity measurements are necessary to calculate the dynamic moduli and quantify anisotropy (Mavko et al., 2009). The relationship between stress and strain (Hooke's law) for a VTI material is:

$$\begin{bmatrix} \sigma_{11} \\ \sigma_{22} \\ \sigma_{33} \\ \sigma_{23} \\ \sigma_{31} \\ \sigma_{12} \end{bmatrix} = \begin{bmatrix} C_{11} & C_{12} & C_{13} & 0 & 0 & 0 \\ C_{12} & C_{11} & C_{13} & 0 & 0 & 0 \\ C_{13} & C_{13} & C_{33} & 0 & 0 & 0 \\ 0 & 0 & 0 & C_{44} & 0 & 0 \\ 0 & 0 & 0 & 0 & C_{44} & 0 \\ 0 & 0 & 0 & 0 & 0 & C_{66} \end{bmatrix} \begin{bmatrix} \varepsilon_{11} \\ \varepsilon_{22} \\ \varepsilon_{33} \\ 2\varepsilon_{23} \\ 2\varepsilon_{31} \\ 2\varepsilon_{12} \end{bmatrix} \qquad (2.1)$$

where σ_{ii} are normal stresses, σ_{ij} are shear stresses, C_{ij} are elastic stiffness coefficients, ε_{ii} are normal strains and ε_{ij} are shear strains. The stiffness coefficients are calculated from wave velocity measurements at various orientations with respect to bedding:

$$C_{11} = \rho V_{P0}{}^2 \tag{2.2}$$
$$C_{33} = \rho V_{P90}{}^2$$
$$C_{44} = \rho V_{S90}{}^2$$
$$C_{66} = \rho V_{S0}{}^2$$
$$C_{13} = -C_{44} + \sqrt{4\rho^2 V_{P45}{}^4 - 2\rho V_{P45}{}^2(C_{11} + C_{33} + 2C_{44}) + (C_{11} + C_{44})(C_{33} + C_{44})}$$

where ρ is the density, V_{P0} and V_{P90} are velocities of the compressional waves traveling parallel and perpendicular to bedding. V_{S90} and V_{S0} are the velocities of the vertically and horizontally propagating shear waves. V_{P45} is the velocity of the quasi-compressional wave, traveling at 45° to bedding (Fig. 2.12).

Wang (2002a,b) developed a procedure for measuring all five velocities of a VTI material on a single horizontal core sample (x_3 parallel to bedding). This method employs a set of P-wave transducers and two S-wave transducers in the vertical direction, as well as horizontal P-wave transducers orientated perpendicular and at 45° to bedding. For a VTI medium, the variation of P-wave and (fast) S-wave velocities as function of incidence angle can be used to obtain the set of five independent stiffness coefficients (Mavko et al., 2009):

$$V_p = \sqrt{\frac{C_{11}\sin^2\theta + C_{11}\sin^2\theta + C_{44} + \sqrt{M}}{2\rho}} \tag{2.3}$$

where $M = \left[(C_{11} - C_{44})\sin^2\theta + (C_{33} - C_{44})\cos^2\theta\right]^2 + (C_{13} - C_{44})^2$

$$V_{s0} = \sqrt{\frac{C_{66}\sin^2\theta + C_{44}\sin^2\theta}{2\rho}}. \tag{2.4}$$

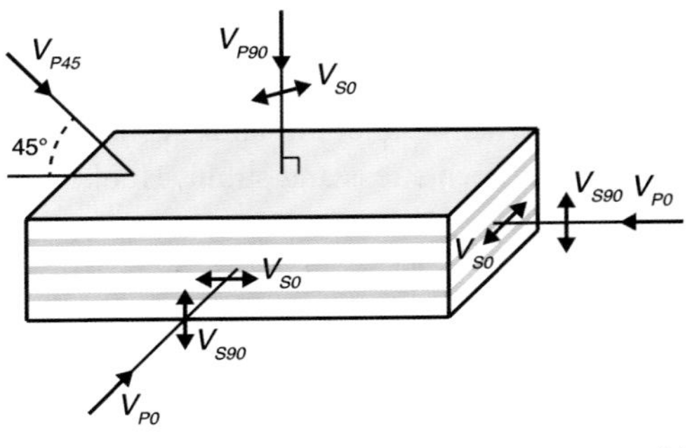

Figure 2.12 Nomenclature for core-scale wave velocity measurements with respect to layering. Adapted from Johnston & Christensen (1995).

Composition, Fabric, Elastic Properties and Anisotropy 45

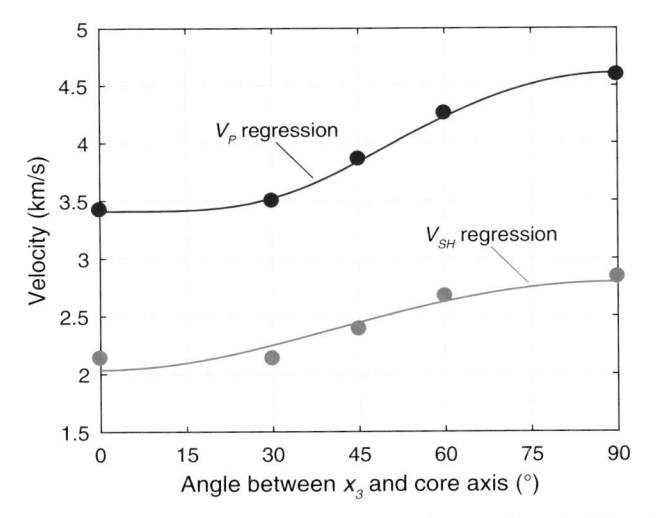

Figure 2.13 Example of velocity measurements made at various angles to bedding for quantifying core-scale anisotropy, Haynesville shale. From Sone & Zoback (2013a).

In the absence of horizontal transducers, vertical velocity measurements can be made on a series of core samples oriented from 0 to 90° with respect to bedding and fitted to Eqns. (2.3) and (2.4) to determine the elastic constants (e.g., Fig. 2.13).

Compositional and Fabric Controls – Sone and Zoback (2013a) measured the static and dynamic elastic properties of unconventional reservoir rocks from several major US basins spanning a range of clay + TOC values from ~0–50 vol%. Figure 2.14 shows the P-wave moduli, S-wave moduli, first-loading Young's moduli and Poisson's ratio as a function of clay + TOC. The data are selected at an axial pressure of 50 MPa, above which only slight stiffening with increasing pressure is observed. As discussed by Sone & Zoback, this ensures the measurements are free from pressure-stiffening effects due to the closing cracks that were induced during core recovery or sample preparation.

All moduli decrease significantly with increasing clay + TOC (Fig. 2.14a–c). Within each reservoir, samples from the relatively low clay + TOC group are consistently stiffer than their counterparts, which reflects the increased proportion of relatively stiff components of the matrix (QFP and carbonates). Samples from all reservoirs also show significant elastic anisotropy. Horizontal samples (x_3 parallel to bedding) are consistently stiffer than vertical samples (x_3 perpendicular to bedding). The difference between the horizontal and vertical stiffness values increases with increasing clay + TOC, reflecting the contributions of clay fabric and organic matter to elastic anisotropy (Fig. 2.8a). The static Poisson's ratio (Fig. 2.14d) shows no clear dependence on composition, but reflects a similar pattern of anisotropy. Within each sample group, the horizontal Poisson's ratio, v_{13}, is larger than the vertical Poisson's ratio, v_{31}.

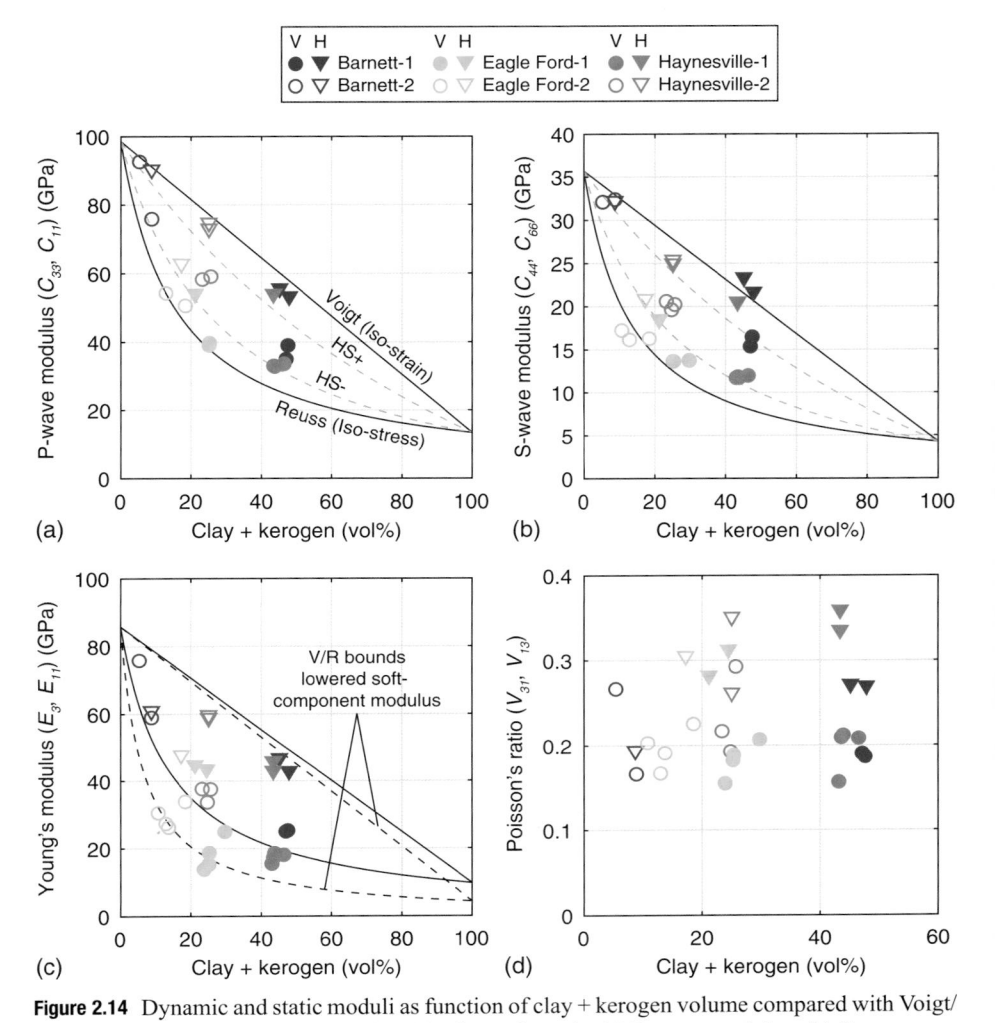

Figure 2.14 Dynamic and static moduli as function of clay + kerogen volume compared with Voigt/ Reuss bounds and upper/lower Hashin–Shtrikman bounds. (a) P-wave modulus. (b) S-wave modulus. (c) Static Young's modulus. (d) Static Poisson's ratio. The Voigt and Reuss bounds are calculated using the values in Table 2.2. In (c), the dashed lines represent the bounds calculated by reducing the Young's modulus of the soft component to half of the value given in Table 2.2. From Sone & Zoback (2013a).

Theoretical Bounds on Elastic Properties – To understand the variations of elastic properties with composition, Sone & Zoback (2013a) compare the laboratory data with theoretical bounds from simple physical models of layered media (Fig. 2.14a–c). Since the major variations in elastic properties arise from variations in the relatively soft components (clay and kerogen), Sone & Zoback simply treat the matrix as a binary mixture of stiff (quartz and calcite) and soft components. The elastic properties of the endmembers are calculated from Hill's (arithmetic) average of the isotropic elastic

Composition, Fabric, Elastic Properties and Anisotropy

Table 2.2 Elastic properties used to calculate the bounds in Fig. 2.14. Stiff and soft components are calculated from Hill's (arithmetic) average of quartz/calcite and clay/kerogen elastic properties. K – bulk modulus, μ – shear (S-wave) modulus, M – compressional P-wave modulus (C_{11} or C_{33}). From Sone & Zoback (2013a).

	K (GPa)	μ (GPa)	M (GPa)	E (GPa)	Source
Quartz	37	44	–	–	Mavko et al. (2009)
Calcite	70.2	29	–	–	Mavko et al. (2009)
Stiff component	51	35.7	98.6	86.9	
Clay	12	6	–	–	Vanorio et al. (2003)
Kerogen	5	3	–	–	Bandyopadhyay (2009)
Soft component	7.8	4.3	13.4	10.8	

properties of the stiff and soft components (Table 2.2). Note that this analysis ignores the effect of porosity because it is a relatively minor component of the matrix ~1–10 vol% (Table 2.1; Fig. 2.4) and is likely already accounted for in elastic properties of the soft components. This is plausible because porosity in the rock matrix is often concentrated in clay and organic matter (Fig. 2.7), so measurements of the pure component elastic properties should capture the effects of interparticle and intraparticle pores (Fig. 2.3b, Fig. 5.6).

Considering the rock matrix in terms of stiff and soft components enables the application of theoretical bounds based on simple physical models of layered media. The Voigt/Reuss bounds represent arithmetic and geometric weighted averages of the elastic properties of stiff and soft components:

Voigt (Iso-stress) Fig. 2.15a

$$E_v = f_{soft}E_{soft} + f_{stiff}E_{stiff} \tag{2.5}$$

Reuss (Iso-strain) Fig. 2.15b

$$E_R^{-1} = f_{soft}E_{soft}^{-1} + f_{stiff}E_{stiff}^{-1} \tag{2.6}$$

where f_i refers to the volume or weight fraction of a given component and E_i refers to the Young's modulus for a given component. The Reuss model represents an iso-stress condition, in which the application of the maximum stress is normal to stiff-soft bedding and strain is concentrated in the soft components (Fig. 2.15a). The Voigt model represents an iso-strain condition in which the application of maximum stress is parallel to bedding and stress is concentrated in the stiff components (Fig. 2.15b).

Figures 2.14a–c show that laboratory measurements of elastic moduli generally lie between the Voigt/Reuss bounds; however, measurements of vertical Young's modulus are slightly lower than the Reuss (iso-stress) bounds (Fig. 2.14c). As discussed by Sone & Zoback (2013a), this discrepancy may arise from differences in measurements of dynamic and static elastic properties of the soft components. The values used in Table 2.2 are derived from laboratory wave velocity (dynamic) measurements. During

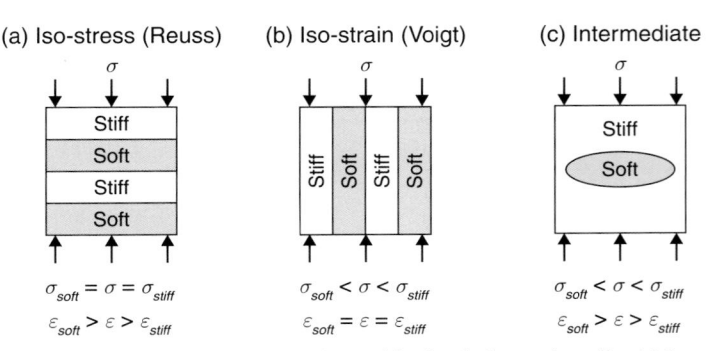

Figure 2.15 Physical models of stress and strain partitioning in layered media. (a) Iso-stress (Voigt). (b) Iso-strain. (c) Intermediate condition with soft inclusions in stiff matrix. From Sone & Zoback (2013b).

static measurements, the soft components may appear more compliant due to increased inelastic deformation at relatively large strains. Lowering the soft component Young's modulus by half is necessary to encapsulate the data for the most compliant shales from the Eagle Ford and Haynesville (Fig. 2.14c). The anomalously low moduli at intermediate clay and kerogen content could be explained by several factors. First, these simple models do not consider possible variations in endmember compositions between different reservoirs. For example, low clay and kerogen samples are more calcareous in the Eagle Ford and more siliceous in the Barnett. Second, the model does not account for any interparticle or intraparticle porosity in the stiff (clastic) components, which could cause increased compliance independent of clay and kerogen content. This may be particularly relevant to relatively calcareous basins such as the Eagle Ford, which exhibits samples with significant intraparticle porosity in calcite and dolomite (e.g., Fig. 5.5, Rassouli & Zoback, 2018). Third, anomalously low moduli could indicate strong fabric anisotropy. The Voigt and Reuss bounds physically represent endmember cases of completely layered media (Fig. 2.15), so data points on or below the Reuss bound may indicate samples with strong fabric anisotropy. This is in fact consistent with the microstructure of the Eagle Ford and Haynesville samples, which show strong fabric anisotropy in petrographic images, due in part to lenses of organic matter elongated in the bedding plane (Fig. 2.8).

Relating Static and Dynamic Elastic Properties – In considering how to use elasticity measurements in the context of *in situ* reservoir deformation, it is important to understand the relationship between static and dynamic elastic properties. Static moduli are often measured at two stages: during first loading and during subsequent unloading/reloading cycles (e.g., Fig. 2.11). Differences between the first loading and reloading elastic properties are commonly observed in rock mechanics and are often attributed to non-recoverable inelastic deformation during first loading above the maximum stress level previously experienced by the rock. Figure 2.16a shows Young's moduli measured during first loading are consistently ~20% lower than during reloading, indicating a significant component of initial inelastic deformation. This effect is observed

Composition, Fabric, Elastic Properties and Anisotropy 49

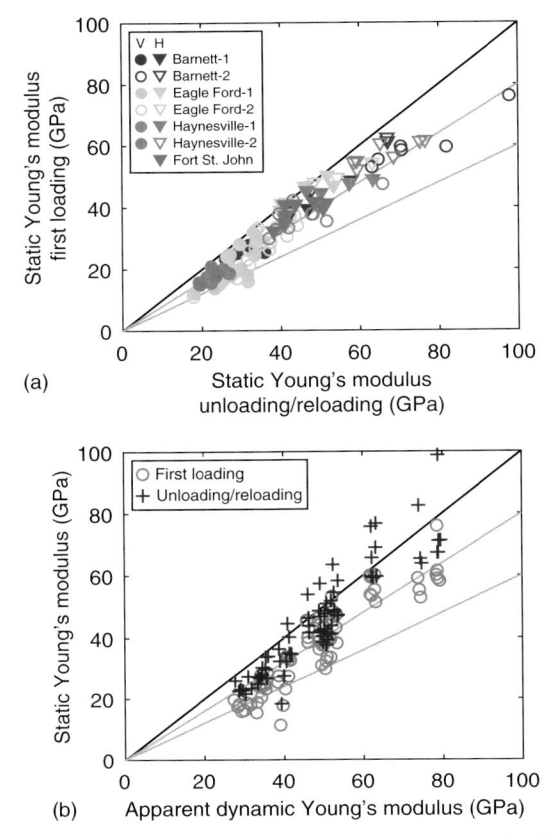

(a)

(b)

Figure 2.16 Comparison between static and dynamic Young's modulus. The black line represents a 1:1 ratio, gray lines indicate 20% and 40% differences. (a) First-loading vs. unloading/reloading modulus. (b) Static vs. apparent dynamic modulus. From Sone & Zoback (2013a).

even below the estimated *in situ* stress, which suggests the core samples may have lost their mechanical "memory" during the recovery process (Sone & Zoback, 2013a). Such decompaction may be caused by various mechanisms including pore expansion by over-pressured gas and/or slow anelastic recovery of the rock frame (Warpinski & Teufel, 1986). Samples from overpressured environments may also exhibit significant hysteresis at low differential stress, as their "memory" reflects very low *in situ* effective stress.

The differences between first loading and reloading static moduli at all stress levels reflects the difficulty of recovering *in situ* elastic properties in laboratory experiments on core samples. Although the reloading measurements do not encapsulate a fully re-compacted state due to the limited timescale of experiments, the first loading measurements capture additional inelastic deformation that is likely not representative of *in situ* behavior. Sone & Zoback (2013a) illustrate this discrepancy by comparing the first loading and unloading/reloading moduli with the apparent dynamic moduli for an isotropic medium:

$$E_1^{apparent} = \frac{C_{66}(3C_{11} - 4C_{66})}{C_{11} - C_{66}}$$

$$E_3^{apparent} = \frac{C_{44}(3C_{44} - 4C_{44})}{C_{33} - C_{44}}.$$

(2.7)

Figure 2.16b shows that the first loading Young's modulus is consistently lower than the apparent dynamic modulus, similar to the static–dynamic relationship observed in many other sedimentary rocks (Mavko et al., 2009). The unloading/reloading Young's modulus is much closer to the dynamic modulus, which implies most of the inelastic deformation in the first loading measurement is permanent (non-recoverable). The variations in static–dynamic relations highlight the importance of understanding the conditions of the static measurement when considering laboratory elastic properties in the context of *in situ* reservoir deformations. Later in this chapter, we will examine the relationship between laboratory elasticity measurements and wellbore measurements performed *in situ*.

Elastic Anisotropy

Elastic anisotropy is quantified in terms of Thomsen's parameters ε, γ and δ (Thomsen, 1986), which are defined by various combinations of the elastic stiffness coefficients (Eqns. 2.8–2.10). In a VTI material, ε or P-wave anisotropy represents the difference between the vertical and horizontal P-wave velocities (V_{P0}, V_{P90}) and γ or S-wave anisotropy, which represents the difference between the vertical and horizontal S-wave velocities (V_{S0}, V_{S90}). δ represents differences in P-wave velocities near the axis of symmetry (x_3), but is somewhat difficult to interpret in physical terms.

$$\varepsilon = \frac{C_{11} - C_{33}}{2C_{33}}$$

(2.8)

$$\gamma = \frac{C_{66} - C_{44}}{2C_{44}}$$

(2.9)

$$\delta = \frac{(C_{13} + C_{44})^2 - (C_{33} - C_{44})^2}{2C_{33}(C_{33} - C_{44})}.$$

(2.10)

Static elastic anisotropy is defined as the ratio of horizontal to vertical Young's modulus, E_h/E_v. Figure 2.17 shows that the degree of anisotropy for each parameter increases with clay and kerogen content, consistent with previous studies that attribute anisotropy in shales to the preferential alignment of clay minerals and the distribution of organic matter (Vernik & Nur, 1992; Sayers, 1994; Vernik & Liu, 1997). For any one sample, it is difficult to distinguish the independent effects of clay and kerogen on anisotropy because their abundance is positively correlated in these samples (Fig. 2.2). Within each basin, samples in the lower clay + TOC group (e.g., Barnett-2) show consistently lower anisotropy. However, between

Composition, Fabric, Elastic Properties and Anisotropy

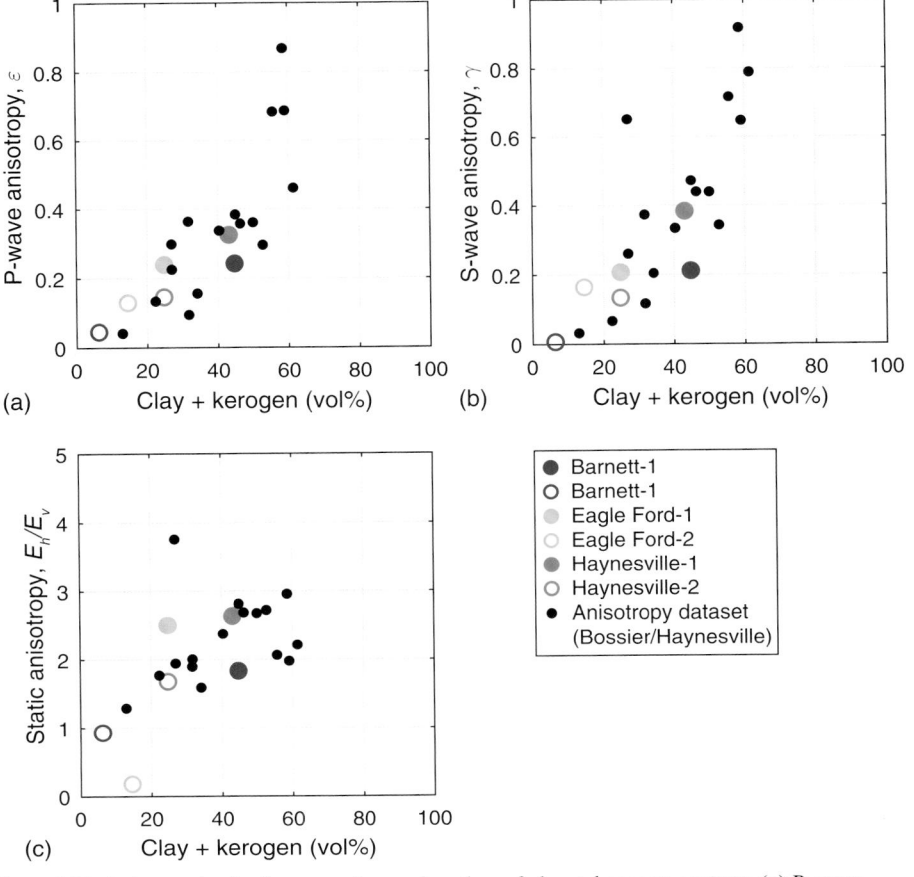

Figure 2.17 Anisotropic elastic properties as function of clay + kerogen content. (a) P-wave anisotropy, ε. (b) S-wave anisotropy, γ. (c) Static anisotropy or ratio of horizontal to vertical Young's modulus. From Sone & Zoback (2013a).

different basins, samples with similar clay + TOC exhibit varying anisotropy (e.g., Haynesville-1 and Barnett-1), which likely reflects differences in the morphology and distribution of the various anisotropic components (clay, kerogen, microfractures) within the rock matrix (Fig. 2.8).

Figure 2.18 shows that the degree of anisotropy for each parameter is negatively correlated with vertical wave velocities and static moduli. Similar trends have been observed in previous studies of shales (Bandyopadhyay, 2009; Mavko et al., 2009) and are attributed to the correlation of clay and kerogen content with both porosity (e.g., Fig. 2.7) and porosity anisotropy (Leu et al., 2016). As discussed in the first half of this chapter, clay and organic matter may impact anisotropy and elastic stiffness in various ways. The intrinsic anisotropy of clay minerals, their relative compliance compared to clastic phases (Table 2.2) and their tendency to align in the bedding plane result in increasing anisotropy and

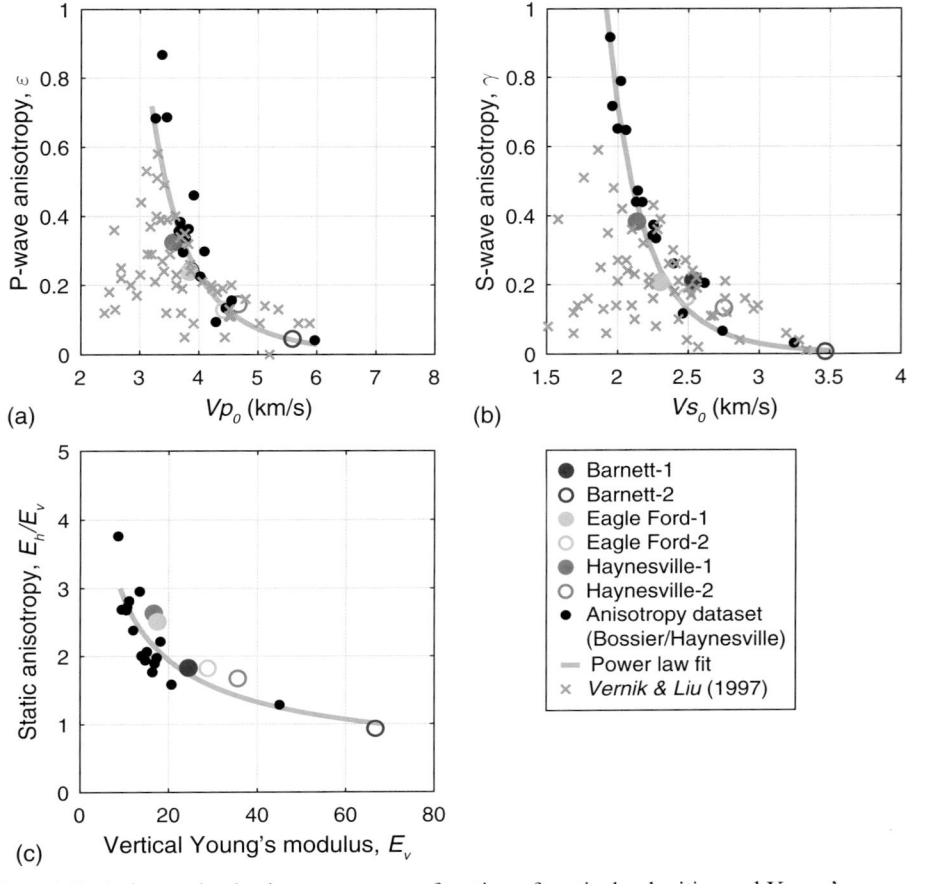

Figure 2.18 Anisotropic elastic parameters as function of vertical velocities and Young's modulus. (a) P-wave anisotropy, ε, vs. vertical P-wave velocity. (b) S-wave anisotropy, γ, vs. vertical S-wave velocity. (c) Static anisotropy vs. vertical Young's modulus. From Sone & Zoback (2013a).

decreasing stiffness with increasing clay content. The porous nature of organic matter, its very low compliance and its tendency to distribute in bedding-parallel features (e.g., Fig. 2.6, Fig. 5.3d, Fig. 5.6c) may also contribute to increasing anisotropy and decreasing stiffness. The correlations between composition, stiffness and anisotropy in Figs. 2.17 and 2.18 appear relatively consistent for samples from the major US unconventional basins, which suggests the proportion of the soft components (clay and kerogen) is a primary control on anisotropy; however, this relationship may vary significantly between basins with significantly different depositional conditions. For example, Vernik & Liu (1997) studied much more organic-rich samples (up to 42 wt%) compared to Sone & Zoback (up to 12 wt %), but found much broader range of anisotropy values for relatively compliant lithologies. This is somewhat counterintuitive, but may be explained by the fact that Vernik & Liu studied samples spanning a much broader range of maturities, while Sone & Zoback focused primarily on mature to post-mature samples from the gas window. Studies of anisotropy in organic-rich shales as function of maturity find that anisotropy is maximized at peak maturity due to the distribution of kerogen in bedding-parallel features (Vanorio et al., 2008; Ahmadov, 2011; Kanitpanyacharoen et al., 2012). In immature shales, kerogen may be distributed in amorphous shapes more closely related to its source (Curtis et al., 2012b), while in post-mature shales it may be dispersed along transport pathways in the matrix (Ahmadov, 2011), which may explain the range of anisotropy values for compliant, organic-rich shales obtained by Vernik & Liu.

Poroelasticity

Most studies of the elastic properties of unconventional reservoir rocks are conducted at dry conditions (no pore fluid pressure) on core samples "as received" or after drying at low temperatures (Sone & Zoback, 2013a; Bonnelye et al., 2016a; Meléndez-Martínez & Schmitt, 2016; Geng et al., 2017). This approach is useful in order to study the fundamental controls on elastic stiffness, such as the effects of matrix composition and microstructure. However, this is not necessarily representative of *in situ* reservoir conditions, in which the rock matrix is likely saturated with brine and/or hydrocarbons. In Chapter 5, we will discuss the range of *in situ* pore fluids in unconventional reservoirs. Here, we review laboratory observations of elastic deformation in unconventional reservoir rocks under fluid-saturated conditions.

In a saturated porous medium, the stress acting on the matrix is a simple effective stress, which is the difference of the confining pressure and the pore pressure, $\sigma = P_c - P_p$. This relationship was developed by Terzaghi (1923) to describe the equal effects of confining and pore pressure on the deformation of unconsolidated soils. In a consolidated rock matrix, the grain structure reduces the area subject to pore pressure, which in turn reduces the relative effect of pore pressure as the rock matrix becomes stiffer. For consolidated porous media, Nur & Byerlee (1971) developed an exact form of the effective stress law:

$$\sigma_{eff} = P_c - \alpha P_p \tag{2.11}$$

where α is the Biot coefficient (Biot, 1941), which describes the relative effect of pore pressure on effective stress, σ_{eff}. Skempton (1960) proposed a form for α in the case of volumetric deformation:

$$\alpha = 1 - \frac{K_{bulk}}{K_{grain}} \tag{2.12}$$

where K_{bulk} is the bulk modulus of the aggregate (rock matrix and pore network) and K_{grain} is the bulk modulus of the matrix components. K_{bulk} is readily measured in the laboratory and may vary with pore and confining pressures; however, for unconventional reservoirs rocks, K_{grain} may be difficult to estimate due to the complex mineralogy and fine grain size of the rock matrix. The difficulty of estimating elastic properties solely on composition is illustrated by the variations in elastic properties between shales with similar matrix compositions from different reservoirs (e.g., Fig. 2.14). K_{grain} may also vary with effective pressure, particularly for matrix constituents with significant intraparticle porosity (clays, organic matter, carbonates). Recognizing that the value of α depends on pore and confining pressure, Todd and Simmons (1972) developed a general expression for α that compares measurements at constant pore pressure and constant effective stress. Ma & Zoback (2017a) adapt this expression to describe volumetric deformation:

$$\alpha = 1 - \frac{(\partial\sigma/\partial\varepsilon_v|P_p)}{(\partial P_p/\partial\varepsilon_v|\sigma)} \tag{2.13}$$

in which the change in simple effective stress with volumetric strain at constant pore pressure, $(\partial\sigma/\partial\varepsilon_v|P_p)$, represents the contribution of effective stress to deformation, while the change in pore pressure with volumetric strain at constant effective stress, $(\partial P_p/\partial\varepsilon_v|\sigma)$, represents the contribution of confining pressure to deformation.

Figure 2.19 shows an example of the relationship confining pressure-volumetric strain for injection and depletion sequences on a sample of Bakken shale (Ma & Zoback, 2017a). Volumetric strain increases with increasing confining pressure and decreases with increasing pore pressure at a given effective stress. The slope of lines of constant pore pressure (solid), which represent K_{bulk}, is consistently lower than the slope of lines of constant simple effective stress (dashed), indicating the value of α is less than 1.

Figure 2.20 shows the evolution of the Biot coefficient as a function of simple effective stress for the depletion and injection sequences in Fig. 2.19. In general, α decreases with increasing simple effective stress (for a given pore pressure) and increases with increasing pore pressure (for a given simple effective stress). The increase in α with pore pressure implies that at low effective stress, pore pressure resists the closure of pores by confining pressure. The measured values of α are generally higher during injection than those during depletion for a given effective stress. Differences in poroelastic behavior between injection and depletion may be attributed to various factors including unloading–reloading hysteresis (Suarez-Rivera & Fjær, 2013), lack of pore pressure equilibrium and/or inelastic damage to the pore network. See Ma & Zoback (2017a) for a detailed discussion of the experimental considerations related to the measurement of poroelastic properties in low porosity, low permeability rocks.

Ma & Zoback studied poroelastic deformation in samples of Bakken shale spanning a range of compositions from calcareous to siliceous. Figure 2.20 shows the variation of the Biot coefficient as function of porosity, clay + kerogen and bulk modulus. Within the calcareous sub-group (filled symbols), α correlates with porosity (Fig. 2.20a). In general, higher porosity rocks are expected to be more strongly influenced by pore pressure due to the increased pore surface area over which pore

Composition, Fabric, Elastic Properties and Anisotropy   55

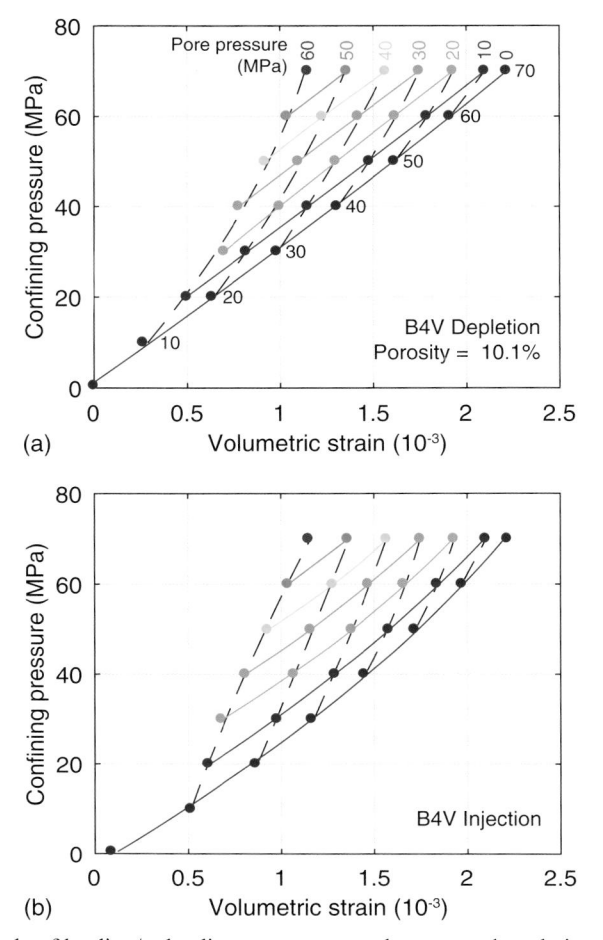

Figure 2.19 Example of loading/unloading sequences used to assess the relationships between effective stress, pore pressure and volumetric strain. Ar pore fluid. Constant pore pressure sequences (solid lines) are denoted by different colored symbols. Constant effective stress sequences are denoted by dashed lines. (a) Depletion (b) Injection. From Ma & Zoback (2017a).

pressure can act. Variability in the Biot coefficient for each lithofacies is also expected to depend on the characteristics of the pore network. The variations in α with effective stress likely result from pressure-dependent deformation of the rock matrix, which may alter fluid access within the pore network. In this case, pore compaction has a greater impact on the poroelastic response of rocks with low porosity (connectivity) and/or compliant pores, which may explain why the highest porosity, stiffest samples show the least variability in α (Figs. 2.21a,c). α also decreases as a function of stiffness (bulk modulus), independent of composition (Fig. 2.21b). Ma & Zoback argue that bulk modulus is more strongly correlated with poroelasticity than composition or porosity for two reasons. First, the bulk modulus integrates all the microstructural

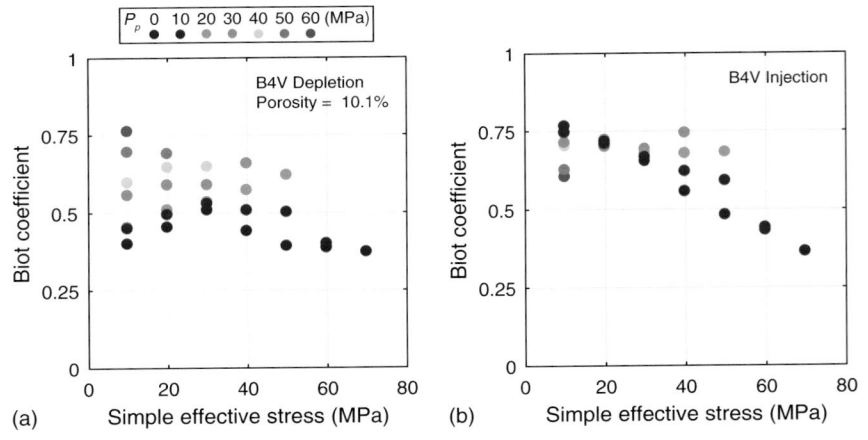

Figure 2.20 Biot coefficient as a function of simple effective stress. Colored points correspond to pore pressures in Fig. 2.19. (a) Depletion (b) Injection. From Ma & Zoback (2017a).

characteristics of the rock matrix, including both composition and porosity. Second, porosity is often measured at ambient pressure, which does not encapsulate the effects of pressure-dependent deformation on poroelasticity. Given the correlation of poroelasticity with stiffness, the variations of bulk modulus with pore pressure and confining pressure may provide insights into the variations of α with pore pressure and confining pressure.

Measurements of poroelasticity at the core-scale provide important constraints on the deformation of the rock matrix under different loading conditions (injection, depletion). However, core samples do not encapsulate all the features of the reservoir that are subject to poroelastic deformation, including cm- to m-scale fractures and faults (Chapter 7). These features will strongly impact the poroelastic response of the reservoir, so, in order to determine reservoir-scale stress changes, it may be necessary to consider a dual-porosity model that isolates the effect of fractures from rock matrix. In Chapter 10, we will explore how fractures and faults respond to changes in effective stress, and in Chapter 12, we will discuss approaches for modeling poroelastic stress changes due to depletion.

Estimating Elastic Properties from Geophysical Data

Detailed laboratory measurements of elastic properties on core samples require a tremendous amount of effort, resources and time. In instances where core recovery is not possible, geophysical measurements (well logs and/or 3D seismic surveys) can be used to estimate the *in situ* elastic properties of cm- to m-scale lithofacies. Elemental capture spectroscopy logs provide estimates of mineralogy to delineate different lithofacies. Dipole sonic logs provide measurements of sonic velocities, which are combined with density logs to determine elastic parameters (Eqn. 2.2).

Composition, Fabric, Elastic Properties and Anisotropy 57

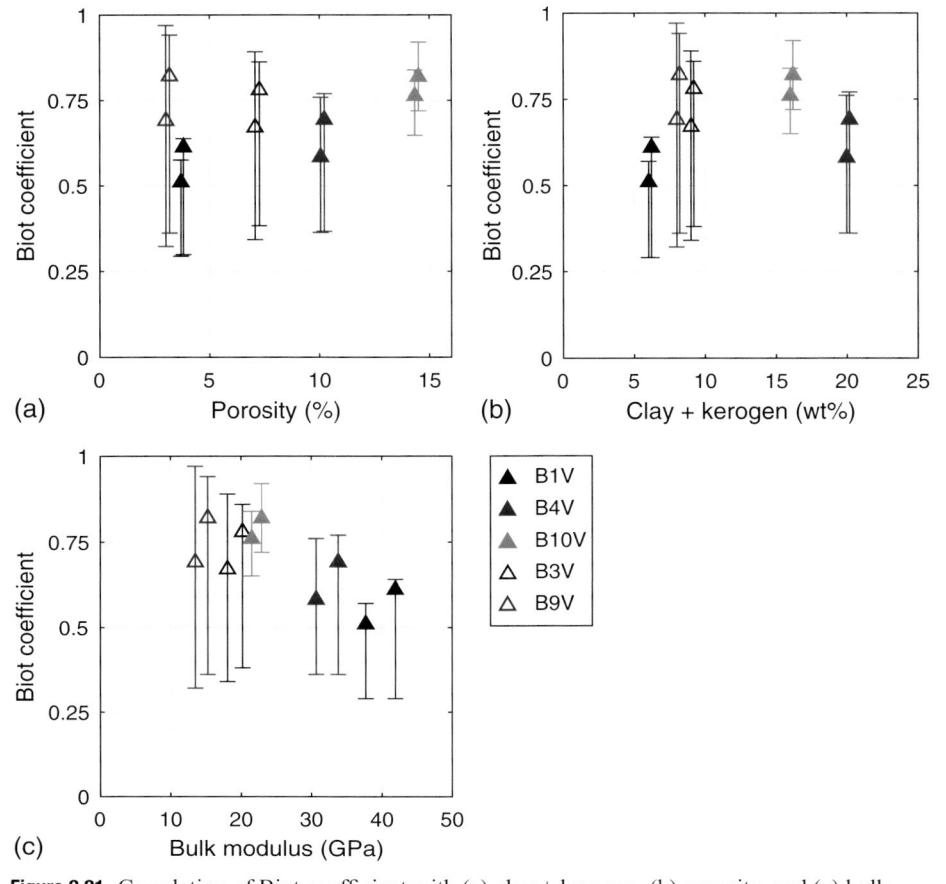

Figure 2.21 Correlation of Biot coefficient with (a) clay + kerogen, (b) porosity, and (c) bulk modulus for samples from the Bakken formation. Points represent data at $P_c = 60$ MPa, $P_p = 30$ MPa. Open and filled symbols represent siliceous and calcareous sub-groups, respectively. Error bars include both injection and depletion measurements. From Ma & Zoback (2017a).

Figure 2.22 shows an example of well log-derived elastic properties for a stratigraphic sequence including the Barnett shale (Perez Altamar & Marfurt, 2014). The data are colored by formation and fall into two distinct groups: relatively compliant rocks with a large range of Poisson's ratios, corresponding to the Barnett shale unit, and relatively stiff rocks with a narrow range of Poisson's ratios, corresponding to the over- and under-lying calcareous (limestone) units. As a thought experiment, we overlay the laboratory data on Barnett shale samples from Sone & Zoback (2013a). Vertical samples from the relatively high clay + kerogen group (Barnett-1) plot within the region for the Barnett shale, while vertical samples from calcareous lithologies (Barnett-2) plot within the region for the limestone units. Horizontal samples appear to define a very different trend, but we might expect this discrepancy as elastic properties derived from log data are

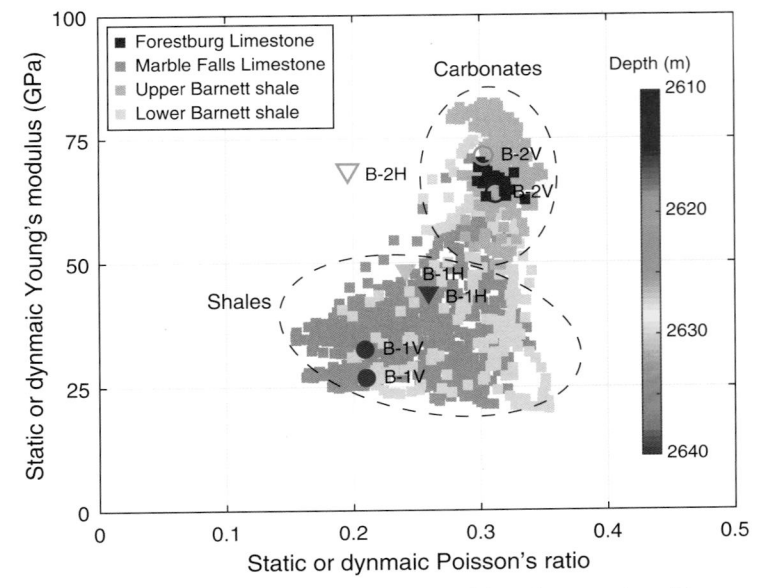

Figure 2.22 Young's modulus vs. Poisson's ratio from well logs for a stratigraphic sequence including the Barnett shale. Data are colored by formation. Adapted from Perez Altamar & Marfurt (2014). Laboratory measurements from Sone & Zoback (2013a) are overlain and colored by depth.

based on measurements of vertical velocities (Perez Altamar & Marfurt, 2014). Coloring the samples by depth illustrates that the Barnett-1 and Barnett-2 groups correspond to a narrow range in the Lower Barnett shale (Sone & Zoback, 2014b), which illustrates fine-scale compositional variations that are not necessarily emphasized in the log data. Although the laboratory samples come from a different location in the basin and therefore cannot be strictly interpreted in the context of these log data, this type of exercise can provide insights into how laboratory measurements relate to *in situ* elastic properties.

Figure 2.23 plots the data for the Barnett shale units in Fig. 2.22 in terms of the product of density and Lamé parameters λ and μ (homogeneous isotropic elastic moduli). These $\mu\rho$–$\lambda\rho$ crossplots are commonly used in geophysical studies to illustrate variations in elastic properties in terms of different lithofacies (Goodway et al., 1997, 2010; Perez Altamar & Marfurt, 2014). It also is a convenient space to compare elastic measurements made at different scales by different methods. In fact, Goodway et al. (1997) developed a methodology for estimating $\mu\rho$ and $\lambda\rho$ values by inverting 3D surface seismic data without density measurements. In Fig. 2.23, Perez Altamar & Marfurt (2014) developed a ternary scheme to understand variations in $\mu\rho$–$\lambda\rho$ values in the context of the elastic properties of major constituent minerals (Table 2.2). We modify this scheme to include vertices for a 50–50 clay–kerogen mixture and pure kerogen, which appear to more completely encapsulate the data. It is important to note that this analysis assumes no porosity, fractures, or fluid saturation. Increasing porosity and/or fracture density would

Composition, Fabric, Elastic Properties and Anisotropy   59

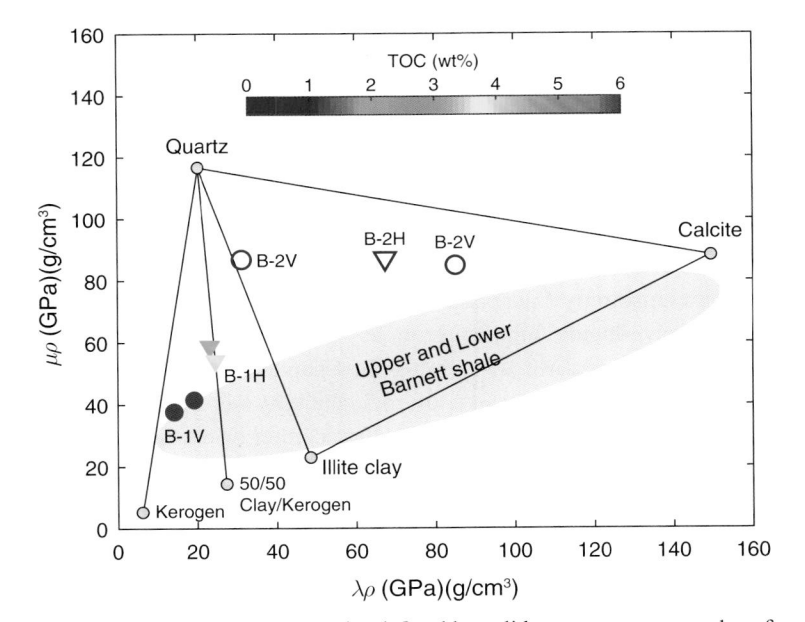

Figure 2.23 Ternary space in $\mu\rho$–$\lambda\rho$ crossplot defined by solid pure component values for quartz, calcite and clay + kerogen (Table 2.2). Square symbols represent data for the Barnett shale from Fig. 2.22. Circles and triangles are laboratory data from Sone & Zoback (2013a). Data are colored by wt% TOC. Adapted from Perez Altamar & Marfurt (2014).

tend to reduce $\mu\rho$–$\lambda\rho$ values. The reduction in density and stiffness (for a given value of porosity) depends on the compliance (composition) and geometry of the specific porosity features. Increasing fluid content lowers compressional stiffness, λ, but does not affect shear stiffness, μ. $\mu\rho$–$\lambda\rho$ crossplots can also be cast in terms of any geophysical well log to assess properties between different lithofacies. For example, contouring the data in Fig. 2.23 by TOC reveals that increasing organic content corresponds with increasing shear stiffness and decreasing compressional stiffness. In the context of the ternary space, this implies high TOC is attributed to relatively siliceous (quartz-rich) lithofacies, which is consistent with well log and core data for the area in the Barnett studied by Perez Altamar & Marfurt.

Similar to Fig. 2.22, we overlay the elastic data from Sone & Zoback (2013a) on Fig. 2.23. For this plot, $\mu\rho$–$\lambda\rho$ values were calculated from static Young's modulus and Poisson's ratio measurements using the relationships between the Lamé parameters for a homogeneous isotropic medium. Samples from the clay and kerogen-rich sub-group (Barnett-1) plot in the region of the organic-rich log data. Samples from the calcareous sub-group (Barnett-2) plot at much higher stiffness away from the Barnett shale trend, which is not surprising considering these samples more closely resemble the adjacent limestone units (Fig. 2.22). Again, the discrepancy in location between the laboratory measurements and log data limits any meaningful comparisons, but this framework has the potential to illustrate characteristic differences between *in situ* and laboratory measurements of elastic properties. At present, geophysical studies cast $\mu\rho$–$\lambda\rho$ crossplots in the context of compositional information to delineate brittle and ductile lithologies in order to predict the effectiveness of hydraulic fracturing (Alzate, 2012; Perez Altamar & Marfurt, 2015). While these methods may be useful in certain contexts, it is also possible to relate geophysical data to direct measurements of ductility in the laboratory. In Chapter 3, we will examine laboratory observations of viscoplastic (time-dependent) deformation in unconventional reservoir rocks to establish a relationship between elastic stiffness and ductility. Increasing viscoplastic deformation decreases stress anisotropy (increases the magnitude of the least principal stress), which causes relatively ductile lithofacies to act as barriers to hydraulic fracturing (Fig. 3.22; Chapter 11). As we will discuss in Chapter 3, quantifying viscoplastic deformation provides a direct measure of rock ductility (inverse of brittleness), in contrast to empirical relationships based on composition or elastic properties (Table 2.2). Future studies of unconventional reservoirs that have access to seismic surveys, well logs and core samples may consider using laboratory-derived relationships between elastic and viscoplastic properties to understand variations in ductility in the context of geophysical classifications of lithofacies (e.g., Fig. 2.23).

Table 2.1 Compositional data for unconventional reservoir core samples from the Stanford Stress and Crustal Mechanics Group collection used in laboratory testing of physical properties (Chapters 2–6). ID refers to the sample name used in this book. Label refers to the sample group (the suffix –1 represents relatively clay + TOC-rich rocks, whereas –2 represents relatively clay + TOC-poor rocks). The symbology in this table will be used in the figures throughout the book to refer to specific samples. Symbol color denotes reservoir, filled/unfilled denotes high/low clay + TOC (–1 or –2), symbol shape denotes sample orientation.

Basin	ID	Label	Orientation	Depth (m)	XRD and Pyrolysis data (wt%)							Laboratory data	Reference	Symbol
					Quartz	Feldspar	Pyrite	Carbonate	Clay	TOC	Other			
Barnett	22Ha	Barnett-1	H	2615	53.00	3.80	1.50	3.60	29.20	8.70	0.20	Stress relaxation	Sone & Zoback (2013a,b)	▼
Barnett	31Ve	Barnett-1	V	2632	42.00	4.10	1.80	0.40	36.00	13.00	2.70	Sone & Zoback, Strength, Creep	Sone & Zoback (2013a,b)	●
Barnett	31Vd	Barnett-1	V	2632	48.00	2.50	1.70	0.30	38.00	8.50	1.00	Elasticity, Strength, Creep	Sone & Zoback (2013a,b)	●
Barnett	28Ha	Barnett-1	H	2632	43.00	4.10	1.60	0.00	37.00	14.00	0.30	Elasticity, Strength, Creep	Sone & Zoback (2013a,b)	▼
Barnett	25Ha	Barnett-1	H	2625	39.00	1.80	2.00	4.00	42.00	11.00	0.20	Elasticity, Strength, Creep	Sone & Zoback (2013a,b)	▼
Barnett	19Ha	Barnett-1	H	2609	46.00	4.00	1.40	5.90	35.10	7.30	0.30	Stress relaxation	Sone & Zoback (2013a,b)	▼
Barnett	18Va	Barnett-2	V	2616	15.00	0.50	0.10	80.70	2.20	1.20	0.30	Elasticity, Strength, Creep	Sone & Zoback (2013a,b)	○
Barnett	18Ha	Barnett-2	H	2616	27.00	0.70	0.30	56.90	6.10	8.80	0.20	Elasticity, Strength, Creep	Sone & Zoback (2013a,b)	▽
Barnett	19Va	Barnett-2	V	2609	51.00	2.20	0.70	38.20	6.80	0.00	1.10	Elasticity, Strength, Creep	Sone & Zoback (2013a,b)	○
Barnett	18Vb	Barnett-2	V	2616	52.00	3.50	1.30	4.30	26.60	12.00	0.30	Elasticity, Strength, Creep	Sone & Zoback (2013a,b)	○
Barnett	23Vb	Barnett-2	V	2618	44.00	2.10	0.90	45.00	7.60	0.40	0.00	Friction	Kohli & Zoback (2013)	○
Barnett	21Va	Barnett-1	V	2632	56.00	4.10	1.60	3.30	35.00	0.00	0.00	Friction	Kohli & Zoback (2013)	●
Barnett	31Ha	Barnett-1	H	2615	52.00	3.50	1.70	0.40	38.00	4.40	0.00	Friction	Kohli & Zoback (2013)	▼
Barnett	25Ha	Barnett-1	H	2625	39.00	1.80	2.00	4.00	42.00	11.00	0.20	Friction	Kohli & Zoback (2013)	▼
Barnett	18Va	Barnett-2	V	2616	15.00	0.50	0.10	80.70	2.20	1.20	0.30	Friction	Kohli & Zoback (2013)	○
Barnett	21Ha	Barnett-1	H	2632	37.00	4.10	1.70	6.90	34.00	16.00	0.30	Permeability, Sorption	Vermylen (2011)	▼
Barnett	27Ha	Barnett-1	H	2615	55.00	3.70	1.70	8.90	22.90	6.70	1.10	Permeability, Sorption	Vermylen (2011)	▼
Barnett	31Ha	Barnett-1	H	2632	52.00	3.00	1.70	0.40	38.00	4.40	0.50	Permeability, Sorption	Vermylen (2011)	▼
Barnett	31Ha	Barnett-1	H	2632	52.00	3.00	1.70	0.40	38.00	4.40	0.50	Permeability, Sorption	Heller et al. (2014)	▼
Barnett	27Ha	Barnett-1	H	2615	55.00	3.70	1.70	8.90	22.90	6.70	1.10	Permeability, Sorption	Heller et al. (2014)	▼
Barnett	18Vb	Barnett-2	V	2616	52.00	3.50	1.30	4.30	26.60	12.00	0.30	Permeability, Sorption	Heller et al. (2014)	○
Barnett	30Vd	Barnett-1	H	2609	—	—	—	—	—	—	—	SEM, µCT	Kohli (Unpublished)	▼
Barnett	18Vb	Barnett-2	V	2616	52.00	3.50	1.30	4.30	26.60	12.00	0.30	SEM, µCT	Kohli (Unpublished)	●
Bakken	1V	Bakken-2	V	3022	—	—	—	87.00	5.00	—	—	Poroelasticity	Ma & Zoback (2017a)	▼
Bakken	3V	Bakken-2	V	3038	—	—	—	31.00	11.00	—	—	Poroelasticity	Ma & Zoback (2017a)	▼
Bakken	3H	Bakken-2	H	3038	—	—	—	23.00	15.00	—	—	Poroelasticity	Ma & Zoback (2017a)	▼
Bakken	4V	Bakken-1	V	3065	—	—	—	47.00	22.00	—	—	Poroelasticity	Ma & Zoback (2017a)	●
Bakken	9V	Bakken-2	V	3069	—	—	—	19.00	10.00	—	—	Poroelasticity	Ma & Zoback (2017a)	●
Bakken	10V	Bakken-2	V	3124	—	—	—	51.00	15.00	—	—	Poroelasticity	Ma & Zoback (2017a)	●
Bakken	001 H3	Lodgepole-1	H	—	5.00	2.00	0.00	89.00	4.00	0.00	0.00	Elasticity, Creep	Yang & Zoback (2016)	▲

Table 2.1 (cont.)

Basin	ID	Label	Orientation	Depth (m)	Quartz	Feldspar	Pyrite	Carbonate	Clay	TOC	Other	Laboratory data	Reference	Symbol
Bakken	001 V3	Lodgepole-1	V	–	6.00	2.00	0.00	87.00	5.00	0.00	0.00	Elasticity, Creep	Yang & Zoback (2016)	
Bakken	003 H2	Lodgepole-2	H	–	42.00	12.00	0.00	23.00	15.00	0.00	8.00	Elasticity, Creep	Yang & Zoback (2016)	
Bakken	003 V3	Lodgepole-2	V	–	40.00	13.00	0.00	31.00	11.00	0.00	5.00	Elasticity, Creep	Yang & Zoback (2016)	
Bakken	004 H2	Middle Bakken-1	H	–	12.00	12.00	0.00	58.00	15.00	0.00	3.00	Elasticity, Creep	Yang & Zoback (2016)	
Bakken	004 V3	Middle Bakken-1	V	–	14.00	14.00	0.00	47.00	22.00	0.00	3.00	Elasticity, Creep	Yang & Zoback (2016)	
Bakken	007 H1	Lower Bakken-1	H	–	61.00	9.00	3.00	3.00	13.00	9.00	2.00	Elasticity, Creep	Yang & Zoback (2016)	
Bakken	009 H1	Middle Bakken-2	H	–	59.00	14.00	0.00	18.00	9.00	0.00	0.00	Elasticity, Creep	Yang & Zoback (2016)	
Bakken	009 V2	Middle Bakken-2	V	–	56.00	14.00	0.00	19.00	10.00	0.00	1.00	Elasticity, Creep	Yang & Zoback (2016)	
Bakken	010 H2	Three Forks-1	H	–	23.00	16.00	0.00	40.00	20.00	0.00	1.00	Elasticity, Creep	Yang & Zoback (2016)	
Bakken	010 V2	Three Forks-1	V	–	16.00	15.00	0.00	51.00	15.00	0.00	3.00	Elasticity, Creep	Yang & Zoback (2016)	
Eagle Ford	176Ha	Eagle Ford-1	H	3916	21.20	0.00	3.60	54.20	15.80	4.97	0.23	Elasticity, Strength, Creep	Sone & Zoback (2013a,b)	
Eagle Ford	121Ha	Eagle Ford-1	H	3890	22.70	1.20	4.70	53.20	12.30	4.83	1.07	Elasticity, Strength, Creep	Sone & Zoback (2013a,b)	
Eagle Ford	288Va	Eagle Ford-1	V	3917	22.10	1.20	5.70	48.90	15.80	5.69	0.61	Elasticity, Strength, Creep	Sone & Zoback (2013a,b)	
Eagle Ford	250Va	Eagle Ford-1	V	3890	24.30	1.90	4.80	47.20	16.40	5.20	0.20	Elasticity, Strength, Creep	Sone & Zoback (2013a,b)	
Eagle Ford	285Va	Eagle Ford-1	V	3915	16.40	1.90	6.70	47.50	22.40	4.42	0.68	Elasticity, Strength, Creep	Sone & Zoback (2013a,b)	
Eagle Ford	246Va	Eagle Ford-1	V	3887	17.90	1.00	4.00	56.00	16.00	4.66	0.44	Elasticity, Strength, Creep	Sone & Zoback (2013a,b)	
Eagle Ford	65Hb	Eagle Ford-2	H	3863	12.30	2.50	0.85	68.90	13.30	2.10	0.05	Elasticity, Strength, Creep	Sone & Zoback (2013a,b)	
Eagle Ford	208Va	Eagle Ford-2	V	3863	13.00	1.50	3.30	65.30	14.10	2.49	0.31	Elasticity, Strength, Creep	Sone & Zoback (2013a,b)	
Eagle Ford	206Va	Eagle Ford-2	V	3861	11.00	2.50	2.50	72.20	9.60	1.86	0.34	Elasticity, Strength, Creep	Sone & Zoback (2013a,b)	
Eagle Ford	254Va	Eagle Ford-2	V	3893	9.70	0.00	1.10	79.60	6.50	2.35	0.75	Elasticity, Strength, Creep	Sone & Zoback (2013a,b)	
Eagle Ford	210Va	Eagle Ford-2	V	3864	13.10	1.80	3.20	69.40	10.30	1.86	0.34	Elasticity, Strength, Creep	Sone & Zoback (2013a,b)	
Eagle Ford	AUK-7	Eagle Ford-2	H	3868	8.00	0.00	1.00	84.99	3.00	3.01	0.00	Creep	Rassouli (2017)	
Eagle Ford	AUK-8	Eagle Ford-2	V	3873	16.00	3.00	5.00	51.27	20.00	4.73	0.00	Creep	Rassouli (2017)	
Eagle Ford	65Hb	Eagle Ford-1	H	3863	12.30	2.50	0.85	68.90	13.30	2.10	0.05	Friction	Kohli & Zoback (2013)	
Eagle Ford	174Ha	Eagle Ford-1	H	3915	16.00	6.00	1.00	47.00	22.00	5.00	3.00	Friction	Kohli & Zoback (2013)	
Eagle Ford	246Va	Eagle Ford-1	V	3887	17.90	1.00	4.00	56.00	16.00	4.66	0.44	Friction	Kohli & Zoback (2013)	
Eagle Ford	254Va	Eagle Ford-2	V	3893	9.70	0.00	1.10	79.60	6.50	2.35	0.75	Friction	Kohli & Zoback (2013)	
Eagle Ford	AUK-8	Eagle Ford-1	V	3873	16.00	3.00	5.00	51.27	20.00	4.73	0.00	Friction	Kohli & Zoback (2013)	
Eagle Ford	AUK-7	Eagle Ford-2	H	3868	8.00	0.00	1.00	84.99	3.00	3.01	0.00	Friction	Kohli & Zoback (2013)	
Eagle Ford	127Ha	Eagle Ford-2	H	3893	7.00	4.00	1.00	80.00	4.00	2.00	2.00	Permeability	Heller et al. (2014)	
Eagle Ford	174Ha	Eagle Ford-1	H	3915	16.00	6.00	1.00	47.00	22.00	5.00	3.00	Permeability	Heller et al. (2014)	
Eagle Ford	MR1	Eagle Ford-2	H	3296	–	–	–	64.10	12.00	3.60	–	Permeability, Sorption, MIP	Al Allali (2018)	
Eagle Ford	MR2	Eagle Ford-2	H	3302	–	–	1.40	71.60	11.80	2.40	–	Permeability, Sorption, MIP	Al Allali (2018)	
Eagle Ford	MR3	Eagle Ford-2	H	3287	–	–	0.80	81.90	7.10	2.70	–	Permeability, Sorption, MIP	Al Allali (2018)	

Formation	Sample	Sub-sample	Orient.	Depth								Measurements	Reference
Eagle Ford	MR4	Eagle Ford-1	H	3298	—	—	3.20	50.30	21.60	5.90	—	Permeability, Sorption, MIP	Al Allai (2018)
Eagle Ford	174Ha	Eagle Ford-1	H	3915	16.00	1.00	—	47.00	22.00	5.00	3.00	SEM, μCT	Kohli (Unpublished)
Eagle Ford	65Hb	Eagle Ford-2	H	3863	12.30	2.50	0.85	68.90	13.30	2.10	0.05	SEM, μCT	Kohli (Unpublished)
Fort St. John	594-3	Fort St John-1	H	—	—	—	—	—	—	—	—	Stress Relaxation	Sone & Zoback (2013a,b)
Fort St. John	611-3	Fort St John-1	H	—	—	—	—	—	—	—	—	Elasticity, Strength, Creep	Sone & Zoback (2013a,b)
Fort St. John	461-1	Fort St John-1	H	—	—	—	—	—	—	—	—	Elasticity, Strength, Creep	Sone & Zoback (2013a,b)
Fort St. John	611-1	Fort St John-1	H	—	6.00	2.00	1.00	20.00	41.00	2.94	0.06	Elasticity, Strength, Creep	Sone & Zoback (2013a,b)
Haynesville	BWH 1-1	Haynesville-1	V	3439	25.00	10.00	1.00	20.00	41.00	2.94	0.06	Stress relaxation	Sone & Zoback (2013a,b)
Haynesville	BWH 1-2	Haynesville-1	V	3439	23.00	6.00	2.00	22.00	43.00	3.08	0.92	Elasticity, Strength, Creep	Sone & Zoback (2013a,b)
Haynesville	BWH 1-3	Haynesville-1	V	3439	23.00	11.00	3.00	20.00	40.00	2.86	0.14	Elasticity, Strength, Creep	Sone & Zoback (2013a,b)
Haynesville	BWH 1-4	Haynesville-1	V	3439	25.00	7.00	2.00	20.00	43.00	2.93	0.07	Elasticity, Strength, Creep	Sone & Zoback (2013a,b)
Haynesville	BWH 1-5	Haynesville-1	V	3439	23.00	11.00	3.00	20.00	40.00	2.84	0.16	Elasticity, Strength, Creep	Sone & Zoback (2013a,b)
Haynesville	BWH1-7	Haynesville-1	V	3439	—	—	—	—	—	—	—	Elasticity, Strength, Creep	Sone & Zoback (2013a,b)
Haynesville	BWH1-8	Haynesville-1	V	3439	—	—	—	—	—	—	—	Stress relaxation	Sone & Zoback (2013a,b)
Haynesville	BWH1-9	Haynesville-1	V	3439	25.00	8.00	2.00	21.00	40.00	3.24	0.76	Elasticity, Strength, Creep	Sone & Zoback (2013a,b)
Haynesville	BWH1-10	Haynesville-1	H	3439	—	—	—	—	—	—	—	Elasticity, Strength, Creep	Sone & Zoback (2013a,b)
Haynesville	Stan25	Haynesville-1	H	3439	—	—	—	—	—	—	—	Elasticity, Strength, Creep	Sone & Zoback (2013a,b)
Haynesville	Stan27	Haynesville-1	H	3439	—	—	—	—	—	—	—	Elasticity, Strength, Creep	Sone & Zoback (2013a,b)
Haynesville	BWH 2-1	Haynesville-2	V	3481	16.00	6.00	1.00	51.00	24.00	1.80	0.20	Elasticity, Strength, Creep	Sone & Zoback (2013a,b)
Haynesville	BWH 2-2	Haynesville-2	V	3481	17.00	5.00	2.00	54.00	20.00	1.74	0.26	Elasticity, Strength, Creep	Sone & Zoback (2013a,b)
Haynesville	BWH 2-3	Haynesville-2	V	3481	—	—	—	—	—	—	—	Elasticity, Strength, Creep	Sone & Zoback (2013a,b)
Haynesville	BWH2-4	Haynesville-2	V	3481	—	—	—	—	—	—	—	Elasticity, Strength, Creep	Sone & Zoback (2013a,b)
Haynesville	Stan29	Haynesville-2	H	3482	—	—	—	—	—	—	—	Elasticity, Strength, Creep	Sone & Zoback (2013a,b)
Haynesville	Stan30	Haynesville-2	H	3482	—	—	—	—	—	—	—	Elasticity, Strength, Creep	Sone & Zoback (2013a,b)
Haynesville	BWH2-6	Haynesville-2	V	3481	20.00	1.00	3.00	7.00	62.00	1.60	5.40	Creep	Sone & Zoback (2013a,b)
Haynesville	Stan 35	Haynesville-1	V	3384	20.00	3.00	1.00	7.00	24.00	1.80	0.20	Friction	Kohli & Zoback (2013)
Haynesville	Stan 37	Haynesville-1	H	3384	—	—	—	51.00	—	2.84	0.16	Friction	Kohli & Zoback (2013)
Permian	BWH 2-2	Permian-2	H	3481	16.00	6.00	3.00	20.00	40.00	1.80	0.20	Friction	Kohli & Zoback (2013)
Haynesville	BWH 1-5	Haynesville-1	V	3439	23.00	11.00	3.00	20.00	40.00	2.84	0.16	Permeability, Sorption	Rassouli & Zoback (2018)
Haynesville	BWH 1-2	Haynesville-1	V	3439	23.00	6.00	2.00	22.00	43.00	3.08	0.92	Permeability, Sorption	Heller et al. (2014)
Monntey		Monntey	H	—	41.00	11.00	4.00	12.00	27.00	2.00	6.00	Permeability, Sorption	Kohli & Zoback (2013)
Monntey		Monntey	H	—	23.00	6.00	2.00	22.00	43.00	3.08	0.92	Permeability, Sorption	Al Ismail & Zoback (2016)
Permian	P1	Permian-1	H	—	35.40	6.50	3.30	27.80	27.00	5.90	—	Permeability, Sorption	Al Ismail & Zoback (2016)
Permian	P2	Permian-1	H	—	17.50	7.20	1.00	68.80	5.50	1.00	—	Permeability, Sorption	Al Ismail & Zoback (2016)
Permian	P4	Permian-2	H	—	25.30	11.10	2.70	7.90	53.00	0.80	—	Permeability, Sorption	Al Ismail & Zoback (2016)
Utica	U1	Utica-1	H	—	28.00	10.00	1.00	12.00	49.00	2.40	—	Permeability, Sorption	Al Ismail & Zoback (2016)
Utica	U2	Utica-1	H	—	26.00	6.00	1.00	26.00	41.00	1.70	—	Permeability, Sorption	Al Ismail & Zoback (2016)
Utica	U3	Utica-2	H	—	19.00	5.00	1.00	55.00	20.00	3.30	—	Permeability, Sorption	Al Ismail & Zoback (2016)
Utica	U4	Utica-2	H	—	7.00	1.00	1.00	80.00	11.00	2.10	—	Permeability, Sorption	Al Ismail & Zoback (2016)

Table 2.1 (cont.)

Basin	ID	Label	Orientation	Depth (m)	XRD and Pyrolysis data (wt%)							Laboratory data	Reference	Symbol
					Quartz	Feldspar	Pyrite	Carbonate	Clay	TOC	Other			
Utica	U1-1	Utica-1	H	1914	–	–	–	11.60	47.30	2.30	–	Permeability, Sorption, MIP	Al Allali (2018)	▲
Utica	U1-2	Utica-1	H	1964	–	–	–	25.10	39.50	1.60	–	Permeability, Sorption, MIP	Al Allali (2018)	▲
Utica	U1-3	Utica-2	H	1978	–	–	–	53.00	19.30	3.20	–	Permeability, Sorption, MIP	Al Allali (2018)	△
Utica	U1-4	Utica-2	H	1990	–	–	–	77.10	10.60	2.00	–	Permeability, Sorption, MIP	Al Allali (2018)	△
Wolfcamp	Cr1	Wolfcamp-2	H	6614	15.00	0.00	1.00	70.00	10.00	1.39	2.61	Creep	Rassouli (2018)	▲
Wolfcamp	Cr2	Wolfcamp-1	H	6355	31.00	2.00	3.00	14.00	39.00	2.69	8.31	Creep	Rassouli (2018)	▲
Wolfcamp	Cr3	Wolfcamp-2	V	6356	17.00	1.00	2.00	50.00	23.00	2.40	4.60	Creep	Rassouli (2018)	○
Wolfcamp	Cr4	Wolfcamp-2	V	6465	18.30	0.00	0.00	68.50	7.30	3.65	2.25	Creep	Rassouli (2018)	○
Wolfcamp	Cr5	Wolfcamp-2	H	6465	15.00	0.00	1.00	70.00	10.00	1.87	2.13	Creep	Rassouli (2018)	△
Wolfcamp	Cr6	Wolfcamp-1	V	6466	27.00	1.00	3.00	19.00	44.00	3.94	2.06	Creep	Rassouli (2018)	●
Wolfcamp	Cr7	Wolfcamp-2	V	6614	14.50	0.00	0.80	71.00	9.00	1.54	3.16	Creep	Rassouli (2018)	○
Wolfcamp	Cr8	Wolfcamp-2	V	6760	18.00	1.00	2.00	61.00	16.00	0.85	1.15	Creep	Rassouli (2018)	○

3 Strength and Ductility

In this chapter, we continue to explore the mechanical properties of unconventional reservoir rocks by considering deformation mechanisms active at various stress and strain conditions. Specifically, we will focus on rock strength – the stress required for brittle failure of intact rock – and ductility – the time-dependent (viscous) strain response as a function of stress.

In the first section, we review laboratory methods for measuring rock strength parameters. We consider the same set of samples from Chapter 2 to assess compositional and fabric controls on strength, as well as the relationship between strength and elastic properties. The next section focuses on how to quantify ductility in terms of time-dependent deformation (creep). We discuss variations in creep with composition and fabric in the context of a simple stress partitioning model, which provides a physical basis for relating elastic stiffness and ductility.

In the following section, we discuss how the observed time-dependent deformation in unconventional reservoir rocks can be modeled using linear viscoelastic theory, specifically a power law function of time. We compare the power law model parameters to elasticity measurements to understand their physical meaning and effects on the deformation response. Finally, we examine the relationship between rock ductility and *in situ* differential stress by considering the impact of viscoelastic stress relaxation in response to tectonic loading over geologic time.

The final section addresses the concept of 'brittleness' and its various definitions in terms of mechanical, compositional and well log measurements. Such brittleness indices are derived from reservoir-specific correlations, not different physical models of brittle behavior. Based on laboratory observations of creep deformation, we describe a framework for rethinking brittleness in terms of the impact of ductile deformation on *in situ* stress magnitudes. The chapter concludes with several examples of how stress relaxation can reduce horizontal stress anisotropy in relatively ductile formations, causing them to act as barriers to vertical hydraulic fracture growth.

Rock Strength

In rock mechanics, strength is quantified by various metrics, depending on specific loading condition. For this chapter, we will focus on conventional triaxial loading, in which differential stress is created by an axial load (P_a) applied to a cylindrical sample subjected to confining pressure (P_c). Sone & Zoback (2013a,b) developed

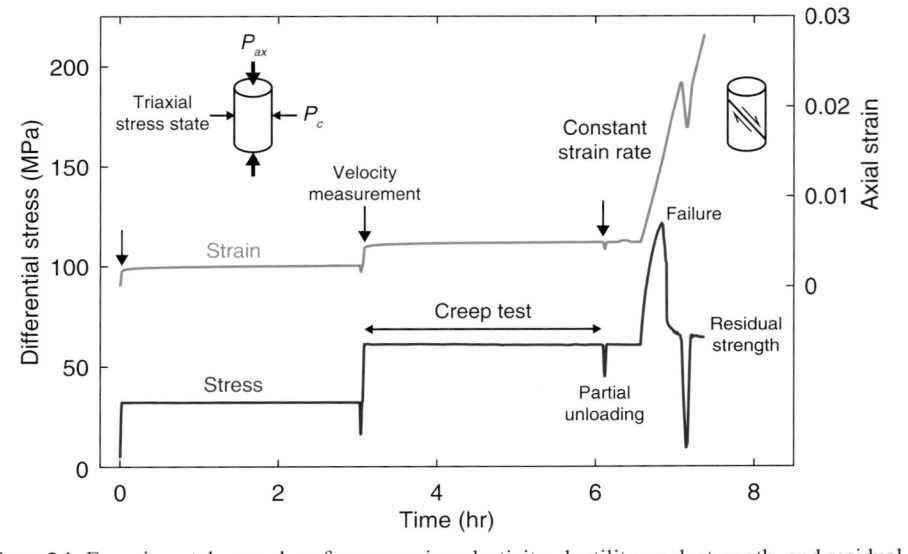

Figure 3.1 Experimental procedure for measuring elasticity, ductility, rock strength, and residual strength after brittle failure (friction). Adapted from Sone & Zoback (2013a).

a comprehensive procedure for measuring rock strength and ductility under triaxial loading (Fig. 3.1). In the first part of the procedure, samples are subjected to a series of steps in differential stress. Each stress step is held for constant for 3 hours and time-dependent strain response is observed. We will discuss these observations in detail later in this chapter. At the beginning and end of each stress step, P-wave and S-wave velocities are measured to quantify the effects of creep deformation on dynamic stiffness. Following the stress steps, the loading condition is changed from constant stress to constant strain rate. Samples are subjected to constant axial strain rate (10^{-5} s^{-1}) until complete brittle failure and the onset of shear deformation on the resultant failure plane. Compressive strength is defined by the maximum axial stress while frictional strength is defined by the ratio of shear to normal stress resolved on the failure plane.

To encapsulate the pressure dependence of maximum axial stress, strength is commonly represented by the uniaxial compressive strength (UCS) and coefficient of internal friction, μ_i. In the Mohr–Coulomb failure criterion, UCS can be inferred from the y-intercept of the trend of maximum axial stress vs. confining pressure (Fig. 3.2a), and the coefficient of internal frictional is calculated from the slope, m:

$$\mu_i = \frac{m - 1}{2\sqrt{m}}. \tag{3.1}$$

The coefficient of sliding friction, μ, is given by the ratio of shear to normal stress resolved on the failure plane (Fig. 3.2b). Inducing brittle failure in this geometry produces a rough failure plane, upon which frictional sliding occurs. This technique is appropriate for estimating relative differences in frictional strength on intact core samples, but because

Strength and Ductility   67

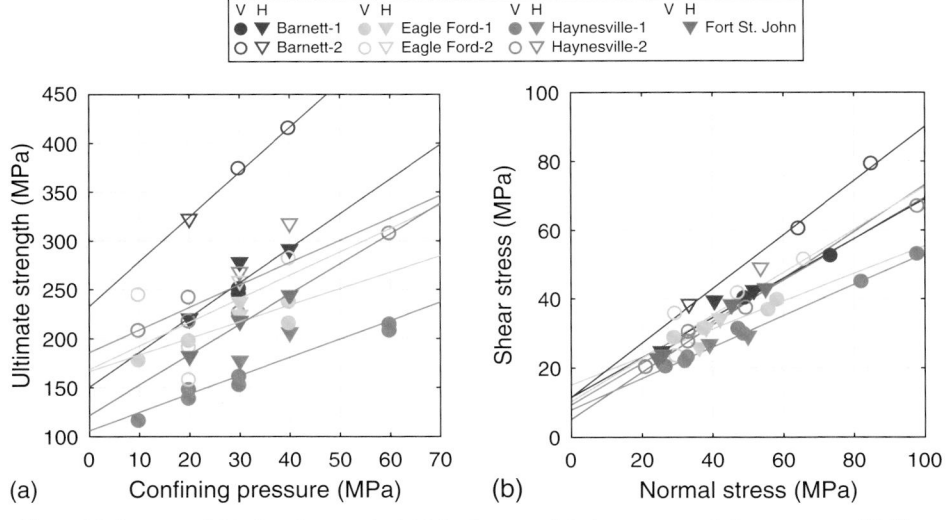

Figure 3.2 Intact and frictional strength. (a) Maximum axial stress vs. confining pressure. (b) Shear vs. normal stress during frictional slip on the failure plane. Adapted from Sone & Zoback (2013b).

the resultant failure planes are geometrically complex, the effects of fault roughness become somewhat convolved with the rock frictional properties (Marone & Cox, 1994). In Chapter 4, we will examine experiments on pre-cut faults in order to understand the controls on frictional strength and the potential for seismic or aseismic slip.

Compositional and Fabric Controls – Unconventional reservoir rocks exhibit a wide range of peak strength values, ranging from ~100 MPa to 400 MPa (Fig. 3.2a). Horizontal samples plot slightly above the trend for each sample group, which implies increased strength in the direction parallel to the bedding plane. Although peak strength has generally been regarded as orientation-independent in rocks with a single well-defined bedding plane (Paterson & Wong, 2005), this result is consistent with recent studies of shales that demonstrate strength is maximized parallel to bedding and minimized when bedding is oriented at 45° to the maximum compressive stress (Lisjak et al. (2014) – Opalinus clay; Chandler et al. (2016) – Mancos shale; Bonnelye et al. (2016a) – Tournemire shale). Bonnelye et al. attribute this orientation dependence to the presence of bedding-parallel microfractures and interparticle pores. At 0° (vertical sample), the maximum compressive stress is orientated perpendicular to long axes of bedding-parallel features, which decreases the critical stress intensity factor (fracture toughness) relative to that at 90° (horizontal sample) (Ashby & Sammis, 1990). At 45°, the shear stress resolved on the bedding plane is maximized, which further reduces fracture toughness and may enable additional, weaker failure mechanisms such as bedding plane delamination (Lisjak et al., 2014; Bonnelye et al., 2016a). Although macroscopic brittle failure is observed in all studies of unconventional reservoir rock strength, Bonnelye et al. note that stress drop and slip accompanying failure were slow compared to typical sedimentary rocks. In addition, no acoustic emissions were recorded during the failure process, which suggests that deformation on both the micro- and

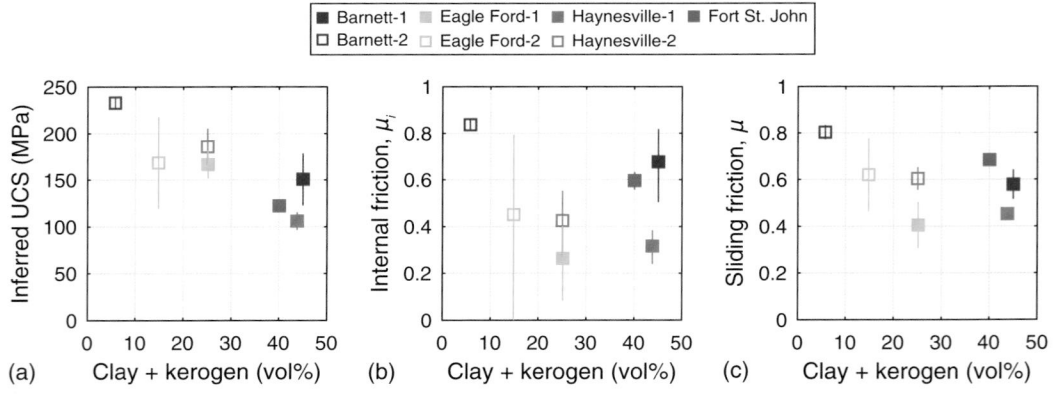

Figure 3.3 Rock strength parameters as function of clay and kerogen content. (a) Inferred UCS (y-intercept in Fig. 3.2a). (b) Coefficient of internal friction. (c) Coefficient of sliding friction. From Sone & Zoback (2013b).

macroscopic scale was mostly aseismic. Geng et al. (2017) studied similar samples from the Tournemire shale and found that, during brittle failure at moderate temperature (75 °C), elastic anisotropy decreases and eventually reverses direction. This reversal in the rock fabric is attributed to stress-induced rotation of microfractures and/or reorientation of clay minerals. If these processes are active on faults in unconventional reservoirs, it is possible that the accumulation of damage due to slip could increase the capacity for elastic and/or inelastic deformation in the fault damage zone, resulting in local stress variations (e.g., Faulkner et al., 2006).

The coefficient of sliding friction varies considerably from ~0.4 to 0.8, but shows no significant difference between vertical and horizontal samples (Fig. 3.2b). This is expected since frictional strength will depend on the characteristics of the failure plane (usually inclined 30–60° to P_d), which may not differ significantly between horizontal and vertical orientations. See Bonnelye et al. (2016b) for a detailed analysis of the structural relationships between brittle failure and bedding. In Chapter 4, we will examine the relationship between composition and frictional properties in detail.

Similar to elastic properties and anisotropy, rock strength is strongly dependent on composition, in particular the presence of relatively compliant, porous phases: clay minerals and organic matter. Figure 3.3 shows the variations in the rock strength parameters as function of clay + TOC. UCS and frictional strength all generally decrease with increasing clay + TOC, although error bars indicate some sample groups contain more variability than others (e.g., Eagle Ford-2). See Table 2.1 for complete sample compositions.

Relating Strength and Elasticity – The similar dependence of strength and elasticity on clay and kerogen content is reflected by the positive correlation of strength parameters with Young's modulus (Fig. 3.4). This relationship is somewhat expected, yet is informative because of the differences in strain magnitude and deformation mechanisms between the measurements of strength and elastic stiffness

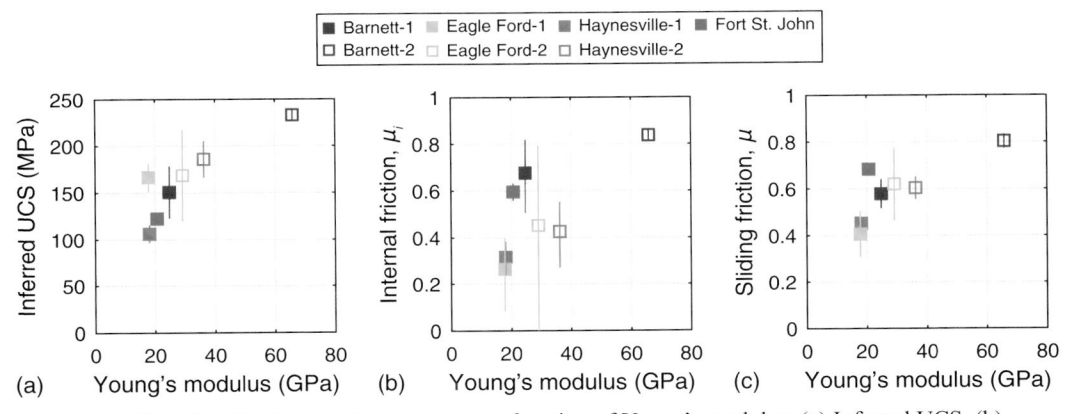

Figure 3.4 Rock strength parameters as function of Young's modulus. (a) Inferred UCS. (b) Coefficient of internal friction. (c) Coefficient of sliding friction. From Sone & Zoback (2013b).

(Chang et al., 2006; Sone & Zoback, 2013b). Brittle failure in sedimentary rocks results from the growth of pre-existing flaws (microcracks and/or pores) at stress-concentrations, which ultimately coalesce into a macroscopic failure plane (Lockner et al., 1992; Bonnelye et al., 2016a). Elastic deformation (static and dynamic) involves much smaller strains over the entire rock matrix, and is also concentrated within the most compliant components – microcracks and pores contained within the most compliant phases, clay minerals and organic matter (Chapter 2). Although crack growth and elastic compression are very different deformation mechanisms, their shared dependence on the presence of weak and compliant phases indicates that matrix composition and fabric are the primary controls on both strength and elastic stiffness. Later in this chapter, we will discuss how quantifying stress and strain partitioning within soft and stiff matrix components can be used to understand the relationships between deformational properties and the state of stress (Sone & Zoback, 2013b).

Time-Dependent Deformation (Creep)

Time-dependent deformation or creep is quantified by the strain response at constant differential stress. At each stress step in Fig. 3.1, Sone & Zoback (2013b) record axial and lateral strains over a 3 hr period and isolate the elastic (instantaneous) and creep (time-dependent) components (Fig. 3.5). Axial strain is much greater than lateral strain, which indicates that most of the ductile response occurs as compaction in the direction of the applied differential stress (parallel to the core axis).

Stress Dependence – To compare creep tests on samples at different stress conditions, the cumulative axial creep strain over 3 hrs is plotted as a function of differential stress (Fig. 3.6). All samples show an approximately linear trend over a large range of stress values. Sone & Zoback (2013b) define the slope for each sample as the "3 hr creep

70   **Unconventional Reservoir Geomechanics**

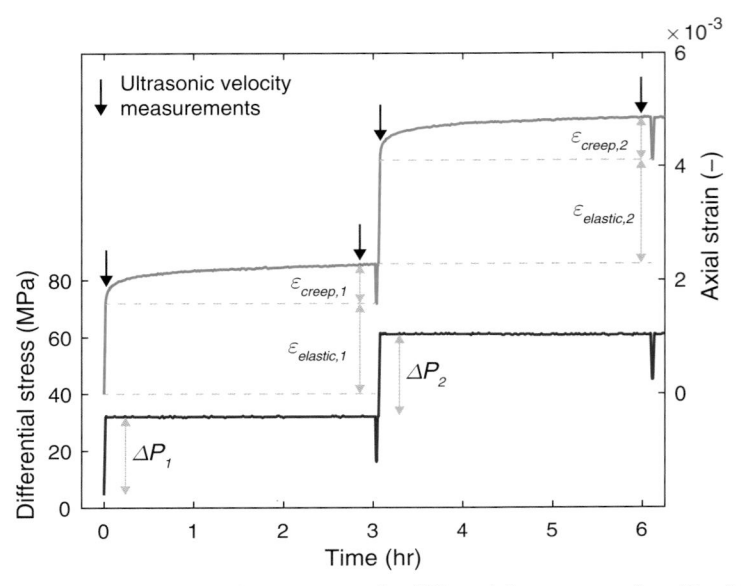

Figure 3.5 Detailed view of the strain response to the differential stress steps from Fig. 3.1. Elastic strain is used to compute Young's modulus and creep strain is used to quantify ductility. From Sone & Zoback (2013b).

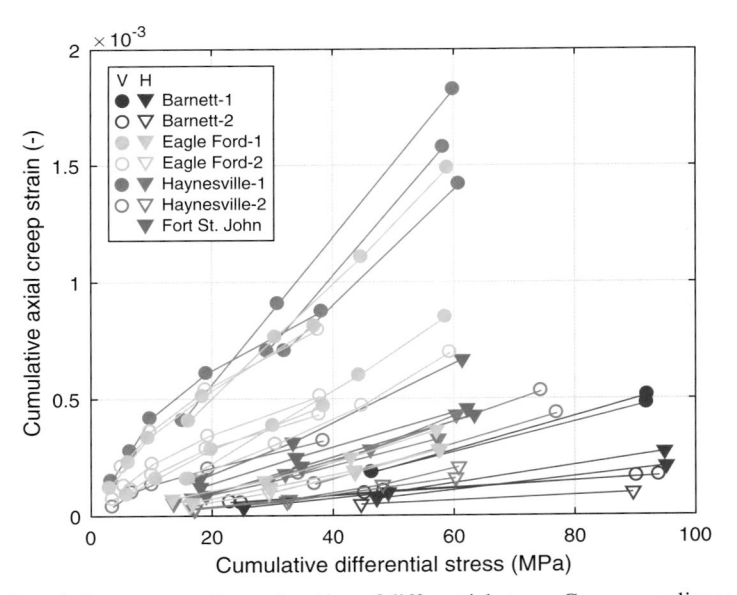

Figure 3.6 Cumulative creep strain as a function of differential stress. Creep compliance (S_{creep}) is calculated from the slope of each sample line. From Sone & Zoback (2013b).

compliance" (S_{creep}), which is a useful metric for describing the creep response (ductility). Sone & Zoback determine S_{creep} values ranging from 10^{-6} to 3×10^{-5} MPa^{-1}, approximately an order of magnitude less than similar studies on unconsolidated sands and shales (Hagin & Zoback, 2004a; Chang & Zoback, 2009).

Strength and Ductility   71

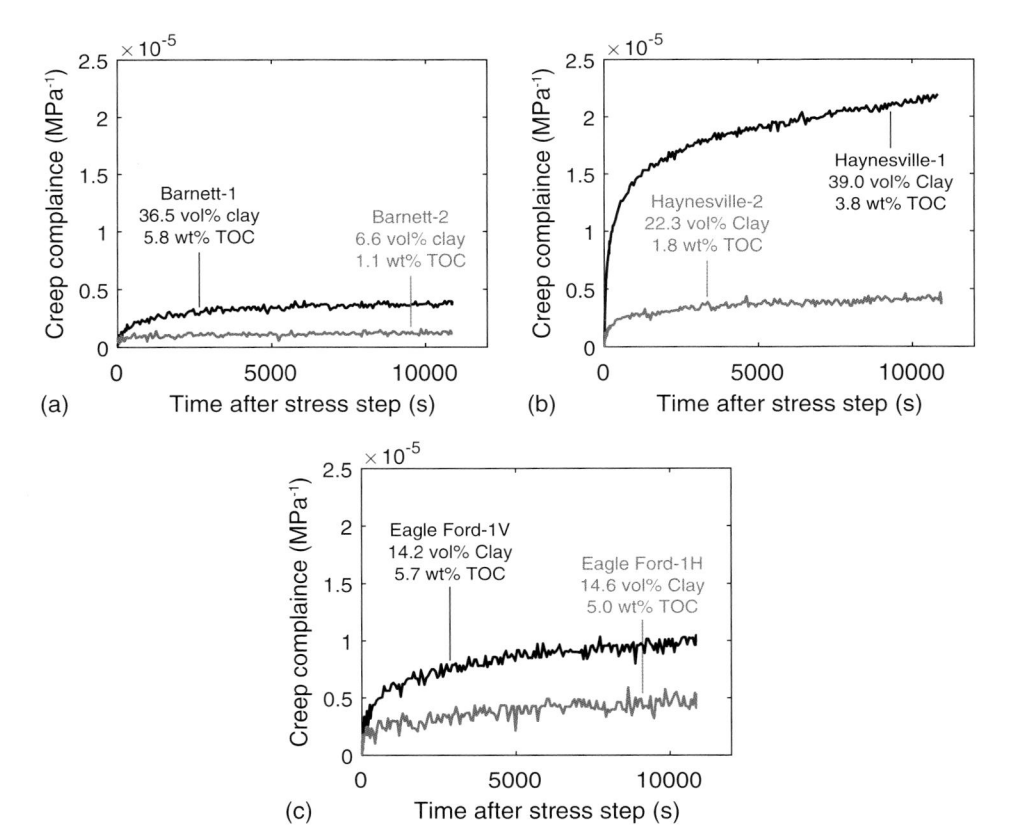

Figure 3.7 Creep strain normalized by differential stress (creep compliance) as a function of time. (a) Barnett clay-rich and -poor endmembers. (b) Haynesville clay-rich and clay-poor endmembers. (c) Eagle Ford-1 vertical and horizontal samples. From Sone & Zoback (2014a). *Reprinted with permission from Elsevier whose permission is required for further use.*

Compositional and Fabric Controls – Figure 3.7a,b shows characteristic creep responses for clay + TOC-rich and TOC-poor endmembers. To compare the strain responses of samples tested at different stress levels, the creep strain is normalized by the differential stress (creep compliance). Again, similar to the trends in elastic stiffness and strength properties, clay + TOC-rich samples show increased creep compliance within samples from the same reservoir (Fig. 3.7a,b). Samples from different reservoirs with similar clay + TOC (e.g., Barnett-1 and Haynesville-1) exhibit significantly different creep compliance. In this case, Haynesville-2 actually shows similar creep compliance to Barnett-1 at approximately half the clay and kerogen content. This signifies that while the presence of weak phases is an important control on ductile deformation, other characteristics of the rock microstructure also impact to ductility, in particular, rock fabric. Figure 3.7c shows creep responses from vertical and horizontal samples from the Eagle Ford shale. Horizontal samples consistently show less creep deformation than vertical samples. Since deformation in these experiments is concentrated in the axial direction, creep compliance is greater in the direction perpendicular to bedding. This is consistent with analogous trends in elasticity (Fig. 2.14) and strength (Fig. 3.2a) and reflects the VTI fabric defined by the horizontal bedding plane (Fig. 2.10).

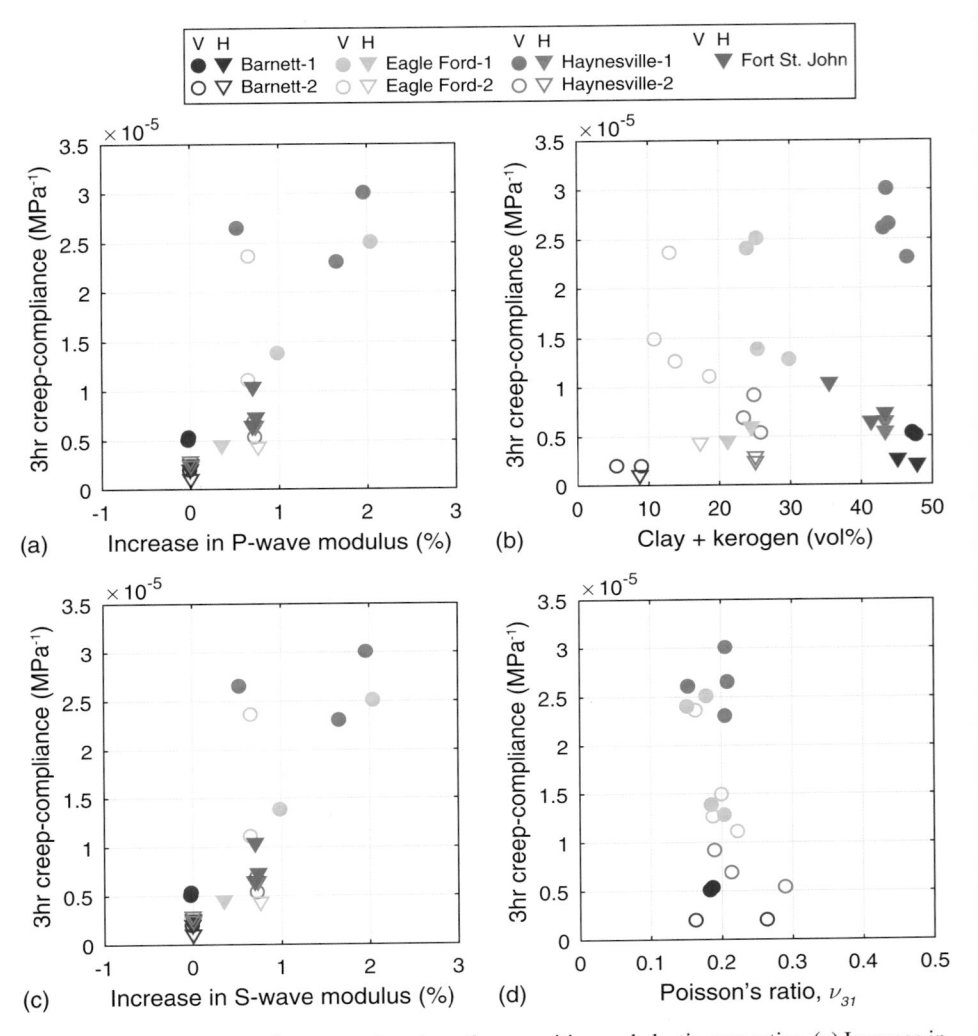

Figure 3.8 3 hr creep compliance as a function of composition and elastic properties. (a) Increase in P-wave modulus from 3 hr creep. (b) Clay + TOC. (c) Increase in S-wave modulus from 3 hr creep. (d) Poisson's ratio. From Sone & Zoback (2013b).

Relating Ductility and Elasticity –

The relationship between elasticity and ductility provides insights into the physical mechanisms responsible for creep. Figure 3.8 shows how creep compliance varies with elastic properties and composition. During each creep step, vertical P-wave moduli (velocities) increase slightly, up to ~2% in most compliant samples (Fig. 3.8a). In similar experiments with continuous velocity recording, Geng et al. (2017) loaded samples close to failure and actually documented an initial decrease in velocities, followed by a larger gradual increase with increasing deformation at constant stress. This suggests that the applied stress was sufficient to exceed fracture toughness and trigger crack growth, but the resultant increased softening was counteracted over time by creep compaction. Geng et al. also find that the greatest change in velocities is in the vertical direction (90° to bedding), indicating again that creep deformation is accommodated by vertical pore compaction. Given that, in unconventional reservoir rocks, most of the porosity is contained within clay minerals and organic matter (Fig. 2.7), it is reasonable to attribute creep deformation and the resultant elastic stiffening to compaction of relatively soft pores within the matrix.

Figure 3.8b shows creep compliance as a function of clay and organic content. Creep compliance is positively correlated with soft component volume within each reservoir sample group, but no overall trend is present. For example, Eagle Ford vertical samples show much greater creep compliance than Barnett vertical samples with more than double the clay and organic content. Again, this signifies that other characteristics of the rock microstructure contribute to ductility such as fabric, thermal maturity (distribution and morphology of organic matter), and porosity within clastic mineral phases (Rassouli & Zoback, 2018). Rock fabric is particularly important, since, similar to the trends in elasticity, all samples show more creep deformation in the direction perpendicular to bedding (vertical). This is evident in Eagle Ford and Haynesville vertical samples, which show both anomalously high creep and elastic compliance (Fig. 2.14). Recall that these samples plot on the iso-stress bound in terms of physical models of layered media (Fig. 2.15), which, along with petrographic observations (Fig. 2.8), implies a strong VTI fabric. Thus, structural anisotropy may have a similar effect on both ductility and elasticity due to their shared dependence on pore compliance.

Figure 3.8c,d shows creep compliance as a function of static elastic properties – Young's modulus and Poisson's ratio. Creep compliances is negatively correlated with Young's modulus for samples from all reservoirs, in contrast to the individual reservoir trends related to composition (Fig. 3.8b). This is expected since the same microstructural factors causing variable creep deformation at similar compositions also impact elastic stiffness. In the next section, we will examine how stress and strain are partitioned in the matrix in order to develop an analytical relationship between creep and elasticity.

Stress and Strain Partitioning

To understand the anisotropy of creep deformation, Sone & Zoback (2013b) return to simple models of rock microstructure, in which the matrix is composed of alternating soft and stiff layers (Fig. 2.15). Soft layers are composed of clay and organic matter, while stiff layers are composed of the dominant clastic minerals, quartz and calcite. See Table 2.2 for the elastic properties of the individual components and soft/stiff endmembers.

In order to determine the stress in the soft and stiff layers, Sone & Zoback (2013b) reduce the problem to one-dimension, treating stress, strain and elastic stiffness as scalar quantities. When the maximum stress is applied perpendicular to layering, the stresses in each layer are equal and the stiffness of rock is equal to the geometric (Reuss) average of the stiffness of each layer (Fig. 2.15a). When the maximum stress is applied parallel to layering, the strains in each layer are equal and the stiffness of rock is equal to the arithmetic (Voigt) average of the stiffness of each layer (Fig. 2.15b). In this case, the stress is greater in the stiff components than in the soft components because the ratio of stress to stiffness in each layer is equal (i.e., $\frac{\sigma_{soft}}{C_{soft}} = \frac{\sigma_{stiff}}{C_{stiff}}$). Assuming that creep deformation occurs in the soft layers and the amount of creep per unit time scales linearly with stress (Fig. 3.6), this model predicts lower creep compliance in horizontal samples because σ_{soft} is less than σ_{stiff} when the applied stress is parallel to layering. For the same reason, the model predicts lower elastic compliance in horizontal samples, which is consistent with the fact that all horizontal samples plot closer to the Voigt average bound (Fig. 2.14).

From this simple model of stress partitioning, it is clear that both elastic stiffness and creep compliance are similarly affected by the relationship between layering and the applied stress. However, the Voigt and Reuss models both represent extreme cases of anisotropy that do not fully describe the variations in elastic moduli with composition (Fig. 2.14). In order to quantify stress and strain partitioning, Sone & Zoback (2013b) extend the one-dimensional model above to account for intermediate microstructures (Fig. 2.15c). Following the approach of Hill (1963), the stress in each component is related to the applied stress on the whole rock volume by the stress partitioning factor, P, where

$$\sigma_i = P_i\sigma(i : soft, stiff). \tag{3.2}$$

Given that the matrix is a binary mixture of soft and stiff components, the average stress and strain in each component is a function of the relative fraction of that component (x_i):

$$\sigma = x_{soft}\sigma_{soft} + x_{stiff}\sigma_{stiff} \tag{3.3}$$

$$\varepsilon = x_{soft}\varepsilon_{soft} + x_{stiff}\varepsilon_{stiff}. \tag{3.4}$$

Sone & Zoback (2013b) combine Eqns. (3.2)–(3.4) to obtain an expression for the average elastic compliance of the whole rock:

$$\frac{\varepsilon}{\sigma} = \frac{1}{C} = \frac{r_{soft}}{C_{soft}} + \frac{1 - r_{soft}}{C_{stiff}} \tag{3.5}$$

in which $r_i = x_iP_i$. In this model, intermediate states represent modifications of the Voigt and Reuss averages through the weighting of the component fractions (x_i) by the stress partitioning factor (P_i). Using Eqn. (3.5) to calculate stress partitioning may be difficult because it requires some knowledge of the rock microstructure, so Sone & Zoback (2013b) solve for P_i to obtain an expression that depends only on the average elastic stiffness and the relative fractions of each component:

$$P_i = \frac{1}{x_i} \frac{C_i}{C} \frac{\Delta C - |C - C_i|}{\Delta C} \tag{3.6}$$

in which $\Delta C = C_{stiff} - C_{soft}$. This expression is used to determine the stresses carried by the soft and stiff components for samples in between the Voigt and Reuss bounds. See Sone & Zoback (2013b) for further details.

This one-dimensional stress partitioning model can also be applied to develop an analytical relationship between elasticity and ductility (creep). Analogous to Eqn. (3.4), creep strain can be decomposed into soft and stiff contributions:

$$\varepsilon_{creep} = x_{soft}\sigma_{soft}S_{soft} + x_{stiff}\sigma_{stiff}S_{stiff}. \qquad (3.7)$$

The expression for average creep compliance, S_{creep}, can then be obtained by replacing the stresses with stress partitioning factors (Eqn. 3.2):

$$S_{creep} = r_{soft}S_{soft} + (1 - r_{soft})S_{stiff}. \qquad (3.8)$$

Combining Eqns. (3.6) and (3.8) yields a relationship between the creep compliance and average elastic stiffness:

$$S_{creep} = \frac{1}{C}\frac{C_{stiff}C_{soft}}{\Delta C}(S_{soft} - S_{stiff}) + \frac{C_{stiff}S_{stiff} - C_{soft}S_{soft}}{\Delta C} \qquad (3.9)$$

in which creep compliance is inversely proportional to elastic stiffness, similar to the relationship between elastic stiffness and elastic compliance.

In order to compare Eqn. (3.9) to laboratory data, the elastic and creep properties of the stiff and soft components must be known. See Table 2.2 for the elastic properties. Since creep compliance (S_i) is not well defined for the individual components, Sone & Zoback

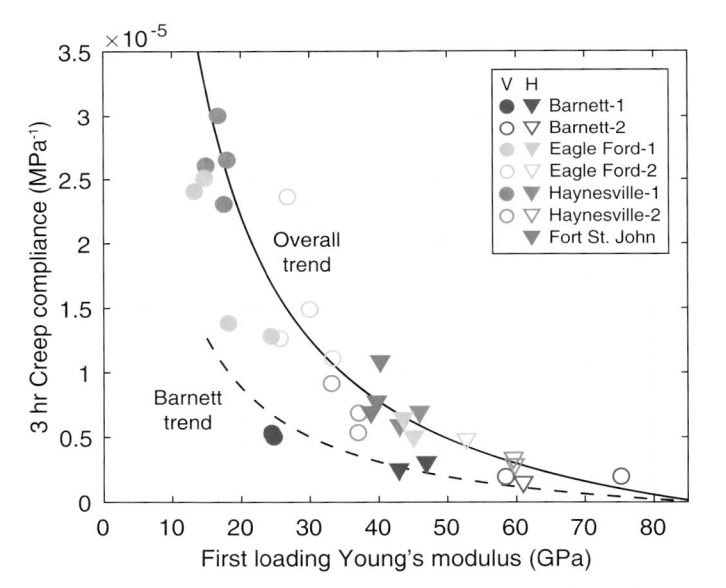

Figure 3.9 Creep compliance as a function of Young's modulus. The solid black line represents the analytical relationship between creep and elasticity (Eqn. 3.9). The dashed black line represents the specific trend for the Barnett shale. Adapted from Sone & Zoback (2014b). *Reprinted with permission from Elsevier whose permission is required for further use.*

(2013b) infer the values by considering stress and strain partitioning in samples near the Voigt and Reuss bounds. The value of S_{soft} is several orders of magnitude higher than S_{stiff}, which reinforces the idea that creep deformation is localized in the soft components. Figure 3.9 compares Eqn. (3.9) to the relationship between creep compliance and Young's modulus. While the overall trend is well characterized by Eqn. (3.9) using S_i values estimated from the full dataset, Sone & Zoback (2014b) also attempt to fit the data from specific reservoirs by manipulating the value of the soft component creep compliance.

The relationship between creep compliance and Young's modulus is not only important for validating stress partitioning as an explanation for the correlation of ductility and elasticity, but can also be used to find a relationship between model parameters of ductile constitutive laws, which is discussed in the following section.

Modeling Time-Dependent Deformation

In this section, we focus on how time-dependent deformation can be described by constitutive laws relating stress, strain and time. Although Sone & Zoback (2014a) document significant irrecoverable (plastic) deformation during loading and unloading cycles, simple monotonic loading yields a linear relationship between creep strain and differential stress (Fig. 3.6), which suggests the observed time-dependent deformation may be described by linear viscoelastic theory. A similar approach is taken to model the deformation of weak, unconsolidated sands (Hagin & Zoback, 2004a,b).

Linear Viscoelasticity – In linear viscoelastic materials, the stress/strain response scales linearly with the stress/strain input and follows the principle of linear superposition. The relationship between stress and strain (or vice versa) is represented by:

$$\varepsilon[c\sigma(t)] = \varepsilon c[\sigma(t)] \tag{3.10}$$

where σ is the stress input, ε is the strain output, and c is a constant. For any arbitrary series of stress/strain inputs, the total output is the sum of the individual responses from each input:

$$\varepsilon(t) = \int_0^t J(t-\tau)\frac{d\sigma(\tau)}{dt}d\tau \tag{3.11}$$

$$\sigma(t) = \int_0^t E(t-\tau)\frac{d\varepsilon(\tau)}{dt}d\tau \tag{3.12}$$

where $J(t)$ and $E(t)$ are the creep compliance function and relaxation modulus function. $J(t)$ describes the time-dependent strain response to a step change in stress and $E(t)$ describes the time-dependent stress response to a step change in strain. This relationship dictates that the time-dependent stress or strain responses can be used to determine the response to any arbitrary input loading history by convolving $J(t)$ or $E(t)$ with the derivative of stress or strain. Taking the Laplace transforms of Eqns. (3.11) and (3.12) yields an expression for the relationship between $J(t)$ and $E(t)$ in terms of dummy variable s:

$$E(s)J(s) = \frac{1}{s^2}. \tag{3.13}$$

As long as either $J(t)$ or $E(t)$ is known, the other function can be obtained through Laplace transforms. This will be particularly important for calculating stress relaxation from laboratory creep data.

Constitutive Laws – To model the laboratory creep data, Sone & Zoback (2014a) use the strain response at each differential stress step normalized by the differential stress (Fig 3.6). The strain data do not exactly represent the creep compliance function, $J(t)$, because each stress step is applied over a finite duration (~60 s), which results in a lag in the strain data compared to a true impulse response. To account for this lag, it is customary to discard the initial portion of the strain data and use the remaining data to determine the creep compliance function (Lakes, 1999). Sone & Zoback demonstrate that this approximate approach yields similar results to the more exact approach, which involves deconvolving the strain response with the stress history via Eqn. (3.5).

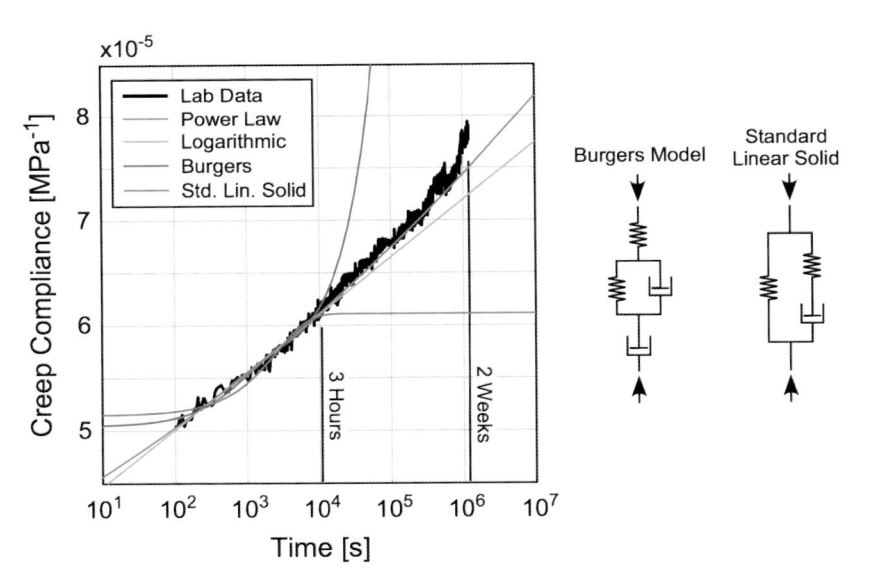

Figure 3.10 2 week creep data fit with various viscoelastic constitutive laws. All constitutive laws fit the data after only 3 hrs, but only the power law model captures the non-asymptotic behavior after 2 weeks. Adapted from Sone & Zoback (2014a). *Reprinted with permission from Elsevier whose permission is required for further use.*

To find the form of $J(t)$, Sone & Zoback consider creep compliance data from relatively short- (3 hr) and long-term (2 weeks) experiments (Fig. 3.10). The creep compliance data show increasing strain with time in log t space. Strain increases with the same trend between 3 hrs and 2 weeks and does not reach an asymptote or constant rate. The continuous, smooth relation between creep compliance and log t suggests there are no characteristic time scales for deformation, i.e., the time dependence of creep is self-similar.

In Fig. 3.10, Sone & Zoback compare various viscoelastic constitutive laws to the creep compliance data to evaluate the best fit model. While all models capture the initial trend up to three hours, typical viscoelastic models composed of springs and dashpots do not fit the data because of their tendency to reach a strain asymptote (e.g., Standard Linear Solid) or stable strain rate (e.g., Burgers) at long times. The power law and logarithmic models both fit the self-similar character of creep compliance, but the logarithmic model does not capture the trend of increasing strain rate at $>10^5$ s. Therefore, Sone & Zoback select the power law model to quantitively characterize creep behavior.

Viscoelastic Power Law – The creep compliance function for the viscoelastic power law is

$$J(t) = Bt^n \tag{3.14}$$

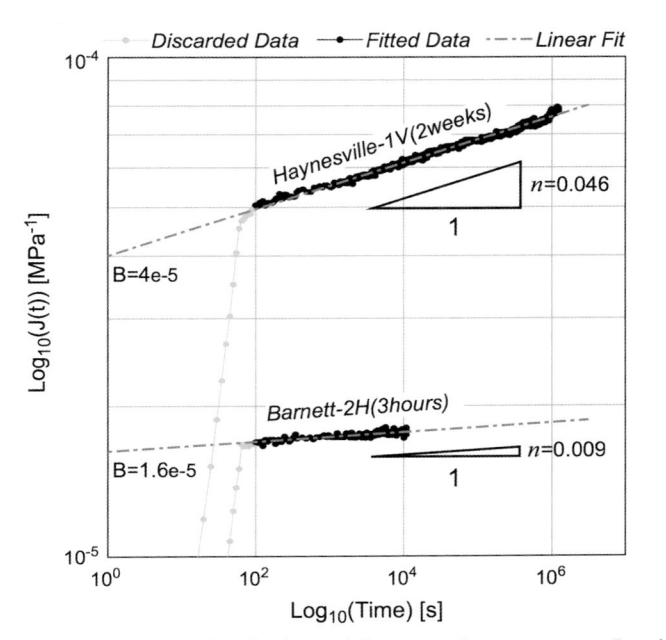

Figure 3.11 Example of linear regression for determining power law parameters B and n from creep compliance data. Adapted from Sone & Zoback (2014a). *Reprinted with permission from Elsevier whose permission is required for further use.*

where B and n are constants. For this 1D form, the boundary conditions are constant vertical stress and variable horizontal stress (or vice versa). To determine model parameters from laboratory data, Sone & Zoback perform a linear regression in log(t)–log(J) space (Fig. 3.11). The slope of the linear fit represents the value of n and the y-intercept represents log(B).

As expected, the variations in B and n values closely resemble the variations in time-dependent deformation in terms of composition, orientation and reservoir locality. With each reservoir group, samples from the clay and organic-rich sub-group show consistently greater values of both constitutive parameters. Within each sub-group, vertical samples show slightly greater values of both parameters than horizontal samples, consistent with the observed anisotropy of creep compliance (Fig. 3.8). Between reservoir groups, B and n values again vary independently of composition, again emphasizing the importance of the rock microstructure in controlling the deformation behavior.

Relating Viscoelastic and Elastic Properties – The simple form of the viscoelastic power law enables basic physical interpretations of the constitutive parameters. Given the form of Eqn. (3.14), constitutive parameter B represents the amount of strain per unit stress after 1 s. Although this does not strictly represent a mathematically instantaneous response, the form of the creep compliance function (Fig. 3.11) suggests that B reflects the elastic compliance. This interpretation is reinforced by the approximately 1:1 correlation of Young's modulus and $1/B$ (Fig. 3.12). The constitutive parameter n can

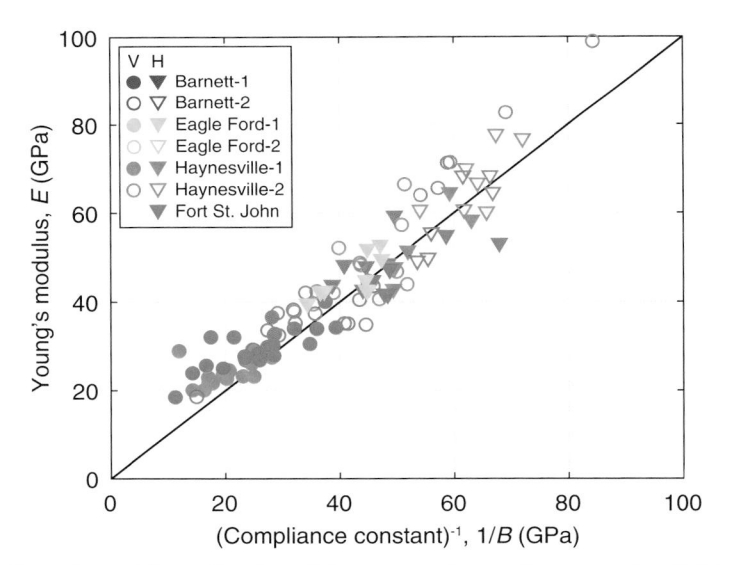

Figure 3.12 Young's modulus as function of the inverse of power law parameter B. Adapted from Sone & Zoback (2014a). *Reprinted with permission from Elsevier whose permission is required for further use.*

be thought of as the capacity of the rock to undergo time-dependent deformation. After the initial elastic response at 1 s, the magnitude of any additional time-dependent deformation depends on the value of n. If $n = 0$, no time-dependent deformation occurs and the rock is purely elastic.

Stress Relaxation – Assuming linear viscoelasticity, Sone & Zoback (2014a) utilize the principle of linear superposition to obtain the relaxation modulus function $E(t)$ from the power law form of the creep compliance function (Eqn. 3.14) using the relationships in Eqns. (3.2–3.4).

$$E(t) \approx \frac{1}{B} t^{-n} \ (n \ll 1) \tag{3.15}$$

Like the creep compliance function, the form of the relaxation modulus is fairly simple and provides the same physical interpretations of the constitutive parameters. At $t \leq 1$, the instantaneous elastic stress response is described by $1/B$, which essentially represents Young's modulus (Fig. 3.12). Time-dependent stress relaxation is described by the term t^{-n}, so n determines the fraction of the elastic stress that is relaxed at a given time. By using constitutive parameters derived from the creep experiments, Sone & Zoback demonstrate that Eqn. (3.15) fits laboratory stress relaxation data, providing additional support for the application of linear viscoelasticity to time-dependent deformation.

Estimating *In Situ* Differential Stress from Viscoelastic Properties

To study the effects of viscoelastic deformation on *in situ* stresses, Sone & Zoback (2014a) consider a simple tectonic history of constant strain rate loading. Combining Eqns. (3.12) and (3.15) yields an expression for time-dependent stress relaxation:

$$\sigma(t) = \dot{\varepsilon} \frac{1}{B(1 - n)} t^{1 - n} \tag{3.16}$$

where $\dot{\varepsilon}$ is the tectonic strain rate. The manner in which stress relaxation affects stress magnitudes is addressed in more detail at the end of this chapter. In Chapter 11 we examine the principle of viscoelastic stress relaxation as the mechanism responsible for variations in stress magnitudes in adjacent lithofacies within a given formation. The result of this are variations in frac gradient from layer to layer which impact the vertical growth of hydraulic fractures.

There are four important points about using Eqn. (3.16) to describe the relationship between stress and strain as function of time. First, in brittle rocks (low B and low n), stress relaxation resulting from viscoelastic deformation is much slower than stress accumulation due to geologic processes. As discussed in Chapters 4 and 7, active geologic processes and the frictional strength of faults control the overall stress state and magnitude of the principal stresses in the Earth's crust. That is, both the relative magnitudes of the principal stresses

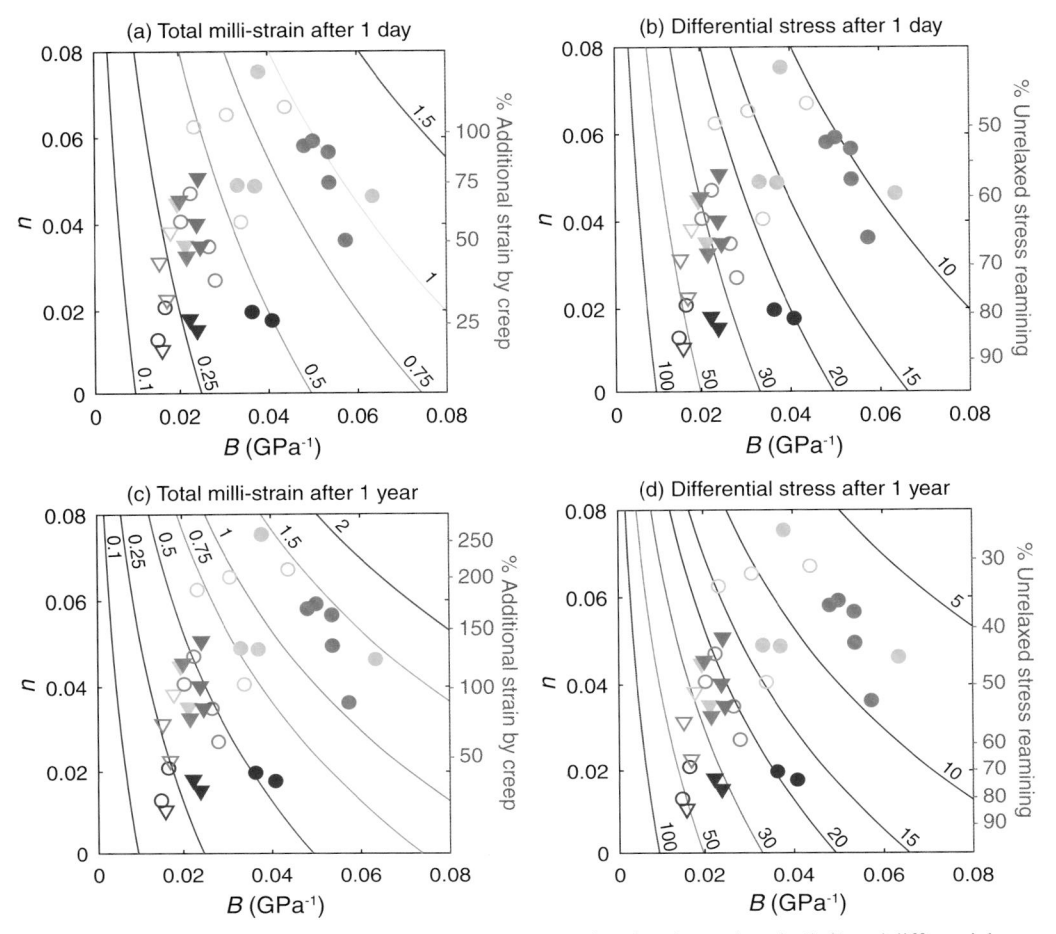

Figure 3.13 A hypothetical laboratory experiment showing the total strain (left) and differential stress (right) in a sample with an initial differential stress of 40 MPa after 1 day and 1 year following an instantaneous application of an incremental stress of 10 MPa. Colored symbols represent power law constitutive parameters derived from laboratory creep data. From Sone & Zoback (2014a). *Reprinted with permission from Elsevier whose permission is required for further use.*

(whether an area is characterized by normal, normal/strike-slip, strike-slip, strike-slip/reverse or reverse faulting) (Fig. 7.1) and the maximum differences among principal stresses (Eqns. 7.3–7.5). In ductile rocks (high B and high n), stress relaxation occurs much more quickly than stress accumulation. This is illustrated in Fig. 3.13, which shows the relative amounts of viscoelastic strain and stress as a function of the power law constitutive parameters (Sone & Zoback, 2014a). Brittle rocks are in the lower left corner of these diagrams and ductile rocks are shown in the upper right corner. Figure 3.14 shows a hypothetical laboratory experiment in which an increment of 10 MPa stress is instantaneously applied to a sample with an initial differential stress of 40 MPa. It is important to

82   Unconventional Reservoir Geomechanics

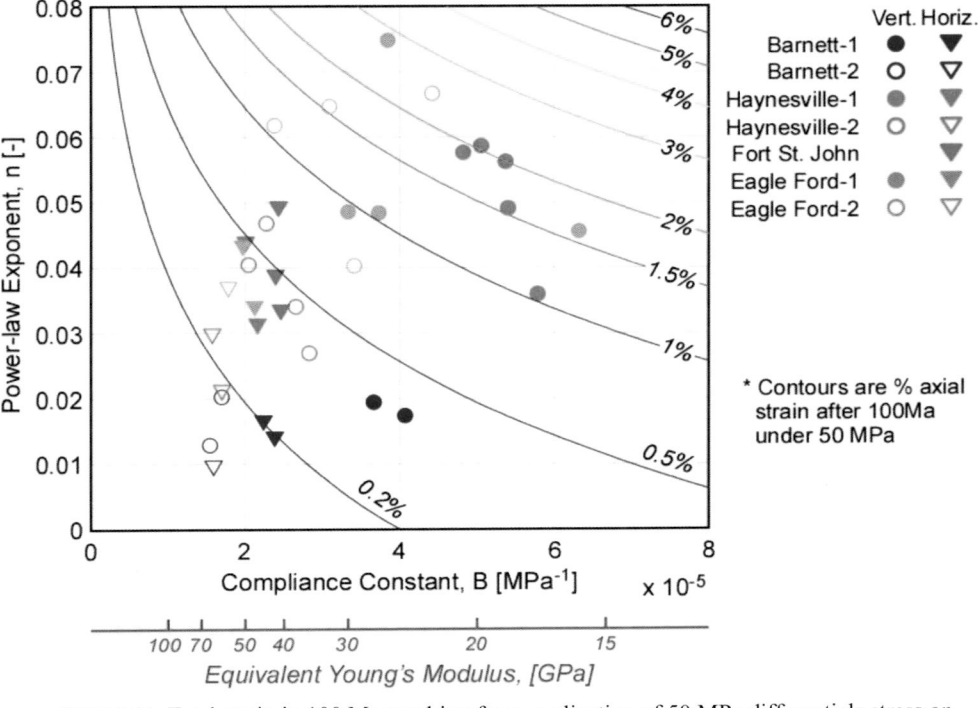

Figure 3.14 Total strain in 100 Ma resulting from application of 50 MPa differentials stress on samples with different values of power law parameters B and n. From Sone (2012).

note that it would take many millions of years for 10 MPa of stress to accumulate in intraplate environments (low tectonic strain rate). The total amount of elastic and viscoelastic strain incurred after 1 day and 1 year after is similar (~0.2–0.25 millistrain for brittle rocks and ~1–1.5 millistrain for ductile rocks) (Fig. 3.13a,c). The same is true of the stresses remaining after viscoelastic stress relaxation. There is essentially no stress relaxation in the relatively brittle rocks (50 MPa remains after 1 day), whereas in ductile rocks, only about 7.5–10 MPa remains (after 1 year and 1 day, respectively) (Fig. 3.13b,d). In other words, in brittle rocks, 90% of the incremental stress remains, whereas in ductile rocks, about half of the incremental stress has relaxed after 1 day, and almost all of the stress (the initial and incremental stress) has relaxed after 1 year.

Second, although the power law model creeps forever, it creeps at rapidly decreasing rates. Hence, the total amount of viscoelastic strain (and total amount of stress relaxation) will be finite. For example, in Fig. 3.13a,c, note that the value of creep compliance, $J(t)$, which is proportional to the total amount of viscoelastic strain (Eqn. (3.11)) increases by about 50% over time periods spanning over 3 orders of magnitude (about 1 day to about 1 year). Applying this principle to geologic time periods, one would expect to see about 100% more strain (~1 millistrain) over a period of a million years.

The third point to note is that the total amount of viscoelastic strain expected to occur in the Earth based on the B and n parameters measured in the laboratory is quite reasonable. Figure 3.14 shows the total amount of viscoelastic (axial) strain that could occur in the most ductile rocks is at most 2–3% after 100 Ma (Sone, 2012). Given that creep deformation is thought to be accommodated primarily by pore compaction (Sone & Zoback, 2013b; Geng et al., 2017), a sedimentary rock experiencing a ~10% reduction in porosity over geologic time is quite reasonable.

Finally, at face value, applying Eqn. (3.16) appears difficult (even when B and n are known) because the tectonic strain rate and the duration of time over which creep has occurred would be essentially impossible to determine geologically and strain rates may vary over time. However, as shown in Fig. 3.15, the pathway by which strain accumulates in a formation is irrelevant, only the total amount of strain (the product of strain rate and time in Eqn. 3.16) affects stress magnitudes. If we assume a constant strain rate of 10^{-19} s^{-1} for 150 Ma (Fig. 3.15b), we see essentially the same stress magnitudes as if the same amount of strain was applied instantaneously 150 Ma ago (Fig. 3.15a) or in the past

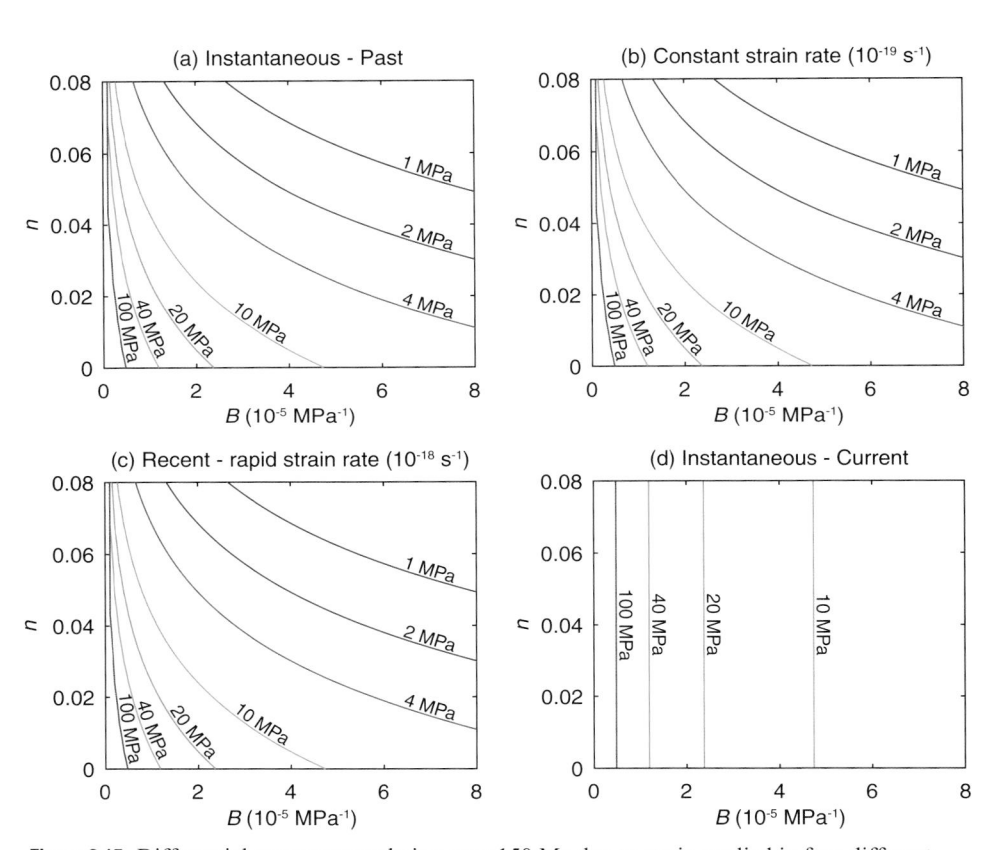

Figure 3.15 Differential stress accumulation over 150 Ma due to strain applied in four different scenarios. (a) Instantaneous strain applied 150 Ma ago. (b) Constant strain rate, 10^{-19} s^{-1}. (c) Rapid strain applied several Ma ago. (d) Instantaneous strain applied in present. Adapted from Sone & Zoback (2014a). *Reprinted with permission from Elsevier whose permission is required for further use.*

few million years (Fig. 3.15c). Only the case of instantaneously applied strain in the present (Fig. 3.15d) produces unreasonable results (no variation in stress magnitudes with creep compliance).

Brittleness and Stress Magnitudes

There is little question that the identification of brittle formations is quite important to optimize the hydraulic fracturing process in unconventional formations. If we consider the schematic representation of hydraulic stimulation of horizontal wells in unconventional reservoirs shown in Fig. 1.15, it is safe to assume that relatively brittle rocks are likely to have more pre-existing fractures and faults available for reactivation in shear by fluid pressurization (Chapters 7 and 10). It is also implicitly assumed that stresses are more anisotropic in brittle rocks (due to their ability to support higher differential stress, e.g., Fig. 3.2), making them better suited for hydraulic fracturing and shear stimulation. In the sections below, we first consider traditional definitions of brittleness, then address *what gets left out*, and ultimately redefine brittleness in terms of the relationship between stress magnitudes and ductility.

Definitions and Empirical Relations – In material science, brittle behavior is defined by a lack of plastic deformation accompanying failure. In rock mechanics, brittle behavior is attributed to rocks that have the tendency to fracture rather than flow, undergo little to no plastic deformation before/after failure, and may fail rapidly

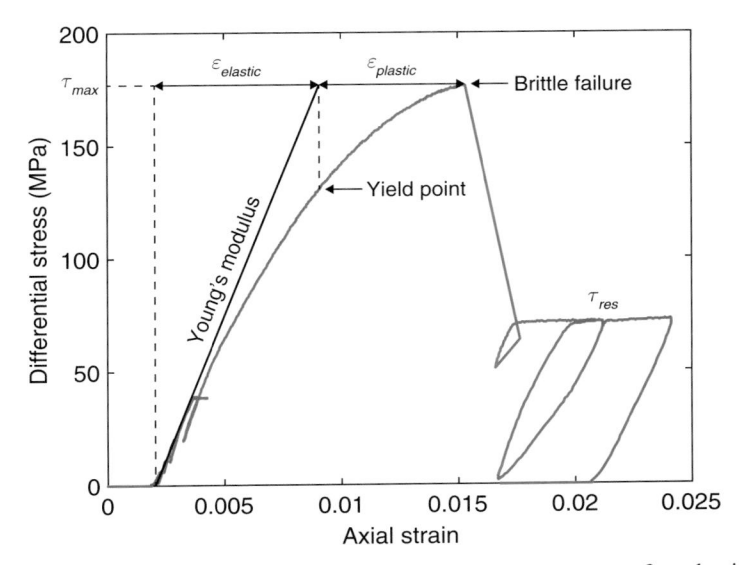

Figure 3.16 Example of an experimental procedure for determining a range of mechanical properties used in brittleness indices. Adapted from Yang et al. (2013).

Table 3.1 Summary of brittleness indices used in the literature. Compiled from Yang et al. (2013) and Zhang et al. (2016).

Index	Reference
Laboratory deformation properties	
$B_1 = \dfrac{\varepsilon_{el}}{\varepsilon_{total}}$	Coates & Parsons (1966)
ε_{el}: elastic strain, ε_{total}: total strain	
$B_2 = \dfrac{W_{el}}{W_{total}}$	Baron et al. (1962)
W_{el}: elastic energy, W_{total}: total strain energy	
$B_3 = \dfrac{C_0 - T_0}{C_0 + T_0}$	Hucka & Das (1974)
C_0: compressive strength, T_0: tensile strength	
$B_4 = \sin\varphi$	Hucka & Das (1974)
φ: friction angle	
$B_5 = \dfrac{\tau_{max} - \tau_{res}}{\tau_{max}}$	Bishop (1967)
τ_{max}: peak shear strength, τ_{res}: residual shear strength	
$B_6 = \left\| \dfrac{\varepsilon_f^p - \varepsilon_c^p}{\varepsilon_c^p} \right\|$	Hajiabdolmajid & Kaiser (2003)
ε_f^p: plastic strain at failure, ε_c^p: specific strain beyond failure	
$B_7 = \dfrac{\lambda}{\lambda + 2\mu}$	Guo et al. (2012)
λ: Lame parameter, μ: shear modulus	
$B_8 = \dfrac{M - E}{M}$	Tarasov & Potvin (2013)
M: post peak strength elastic modulus, E: unloading elastic modulus	
Geophysical / well log properties	
$B_9 = \left(\dfrac{\sigma_{v,max}}{\sigma_v} \right)^b$	Ingram & Urai (1999)
$\sigma_{v,max}$: max previously experienced effective vertical stress, σ_v: vertical stress during measurement, b: empirical value ~0.89	
$B_{10} = \dfrac{1}{2}\left(\dfrac{E_{dyn}(0.8 - \phi)}{8 - 1} + \dfrac{v_{dyn} - 0.4}{0.15 - 0.4} \right) \cdot 100$	Rickman et al. (2008)
E_{dyn}: dynamic Young's modulus, v_{dyn}: dynamic Poisson's ratio, ϕ: porosity	
$B_{11} = -1.4956\phi + 1.0104$ ϕ: neutron porosity	Jin et al. (2014b) – Single Barnett shale well

Table 3.1 (cont.)

Lithology	
$$B_Q = \frac{f_{qtz}}{f_{qtz} + f_{carb} + f_{clay}}$$	Jarvie et al. (2007)
f_{qtz}: wt% quartz, f_{carb}: wt% carbonate, f_{clay}: wt% clay	
$$B_{Q2} = \frac{f_{qtz}}{f_{qtz} + f_{carb} + f_{clay} + TOC}$$	Modified from Jarvie et al. (2007)
TOC: wt% organic matter	
$$B_{Q3} = \frac{f_{qtz} + f_{dol}}{f_{qtz} + f_{cal} + f_{dol} + f_{clay} + TOC}$$	Wang & Gale (2009)
f_{cal}: wt% calcite, f_{dol}: wt% dolomite	
$$B_{Q4} = \frac{f_{QFM} + f_{cal} + f_{dol}}{100}$$	Jin et al. (2014a)
f_{QFM}: wt% quartz, feldspar, mica	

(seismically) (Hucka & Das, 1974). In the context of unconventional reservoirs, the term brittleness is used to describe rocks that are interpreted to exhibit brittle behavior based on certain observations of mechanical properties, microstructure and/or composition. Table 3.1, compiled from Yang et al. (2013) and Zhang et al. (2016), lists a number brittleness indices used in the literature. The first section lists indices derived from laboratory measurements of mechanical properties. Figure 3.16 shows an example of a loading curve to illustrate the meaning of the some of the mechanical parameters. The important thing to note is the variety of different mechanical properties (deformation mechanisms) that are invoked to calculate the indices. Some rely on elastic properties, some on plastic deformation, some on intact strength and some on residual strength. This variation does not result from different mechanistic definitions for brittleness, but rather reflects different correlations between mechanical properties in specific reservoir datasets. Yang et al. (2013) find that, even using mechanical data on the same samples, these indices do not correlate with each other, rock strength or elastic properties. See Yang et al. (2013) for crossplots of the brittleness indices listed in Table 3.1.

The second section of Table 3.1 lists brittleness indices that may be determined by geophysical or well log data, including *in situ* stress, sonic velocity and porosity. While these may be useful for lithological classification in certain contexts, it clear that these parameters do not directly address the mechanisms of brittle behavior. Similar to the laboratory-based indices, these were developed based on correlations in specific datasets rather than any physical model for brittle behavior. The final section of the table lists brittleness indices based on lithology (composition). These may be determined by either well log or core-scale methods. All of the indices are in some way based on the relative fraction of stiff components (quartz and carbonates) in the rock matrix. This approach is somewhat reasonable and generally consistent with the simple binary model of stiff and soft components presented earlier in the chapter; however, as demonstrated by the stress partitioning analysis, brittle or ductile behavior not only depends on composition, but also on how the individual components of the matrix are distributed. Also, given the range of compositional variations within individual reservoirs (Chapter 2), these indices may again only serve as a classification tool in specific datasets.

Although the intended use of brittleness indices is to quantify the effectiveness of hydraulic fracturing in different lithofacies, it clear from our discussion of ductility that the deformation response of a rock cannot be easily packaged into a single parameter. In actuality, hydraulic fracturing and shear stimulation are controlled by a range of factors including the stress state (Chapter 10), the presence of natural fractures (Chapter 7), operational parameters (Chapter 8) and rock mechanical properties (Chapters 2–4). So, rather than using brittleness to predict the effectiveness of hydraulic fracturing, in the next section, we will describe a framework for rethinking brittleness in terms of impact of rock deformation on *in situ* stress.

Rethinking "Brittleness" – While brittleness has many varied definitions in the literature, an important factor to keep in mind is that stress magnitudes become more isotropic in relatively ductile formations. In other words, the response of different formations to hydraulic stimulation is affected both by the mechanical properties of the rock (as defined

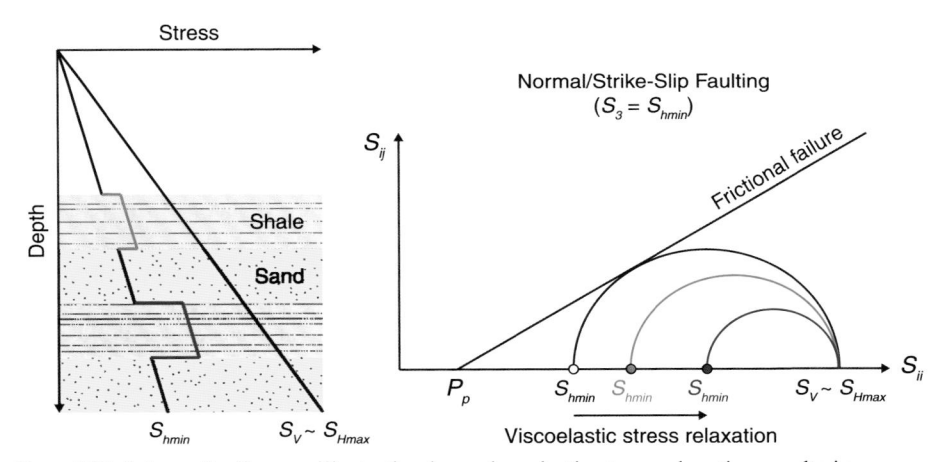

Figure 3.17 Schematic diagram illustrating how viscoelastic stress relaxation results in decreasing stress anisotropy due to increasing the magnitude of the least principal stress. The cartoon on the left shows a moderate increase in the magnitude of the least principal stress above the upper sand resulting from a minor amount of stress relaxation whereas a greater amount of stress relaxation in the shale below the sand creates a larger stress difference and thus a more effective barrier to vertical fracture growth.

by brittleness) and by differences in the magnitudes of principal stress magnitudes. This is illustrated schematically for a layered sedimentary sequence of alternating sands (brittle) and shales (ductile) in a normal faulting environment in Fig. 3.17. Both the stress profile and Mohr–Coulomb diagram show how varying degrees of viscoelastic deformation in the shales causes the least principal stress to increase with respect to the brittle sand units, in which the least principal stress is defined by the frictional strength of normal faults (Chapter 7). The concept of stress relaxation associated with viscoelastic behavior helps us understand why, in hydraulic fracturing of conventional reservoirs, shales are often considered 'frac barriers', as a decrease in stress anisotropy results in an increase in the magnitude of the least horizontal principal stress, S_{hmin}.

As pointed out by Xu et al. (2017), Eqn. (3.16) can be simplified in a normal faulting environments to

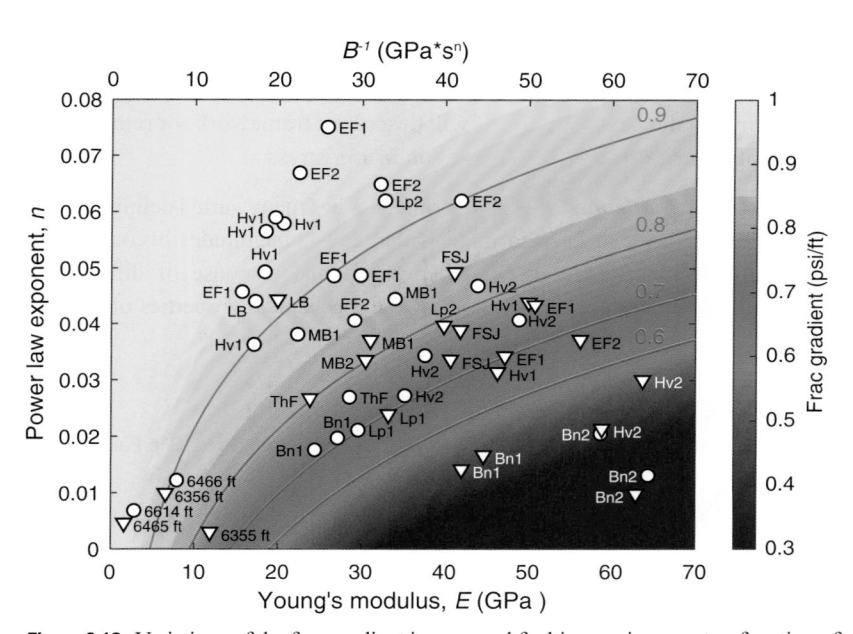

Figure 3.18 Variations of the frac gradient in a normal faulting environment as function of viscoelastic parameters B ($1/E$) and n. Bn: Barnett, Hv: Haynesville, Ef: Eagle Ford, FSJ: Fort St. John, Lp: Lodgepole, MB: Middle Bakken, LB: Lower Bakken, ThF: Three Forks, RV: Reedsville, UU: Upper Utica, BU: Basal Utica, PP: Point Pleasant, LX: Lexington, TR: Trenton. Circles are samples deformed normal to bedding and triangles are samples deformed parallel to the bedding. Adapted from Xu et al. (2017). Original data from Sone & Zoback (2014), Yang & Zoback (2016), and Rassouli & Zoback (2018).

$$S_V - S_{hmin} = \varepsilon_0 t^n \frac{E}{(1-n)} \tag{3.17}$$

Strength and Ductility   89

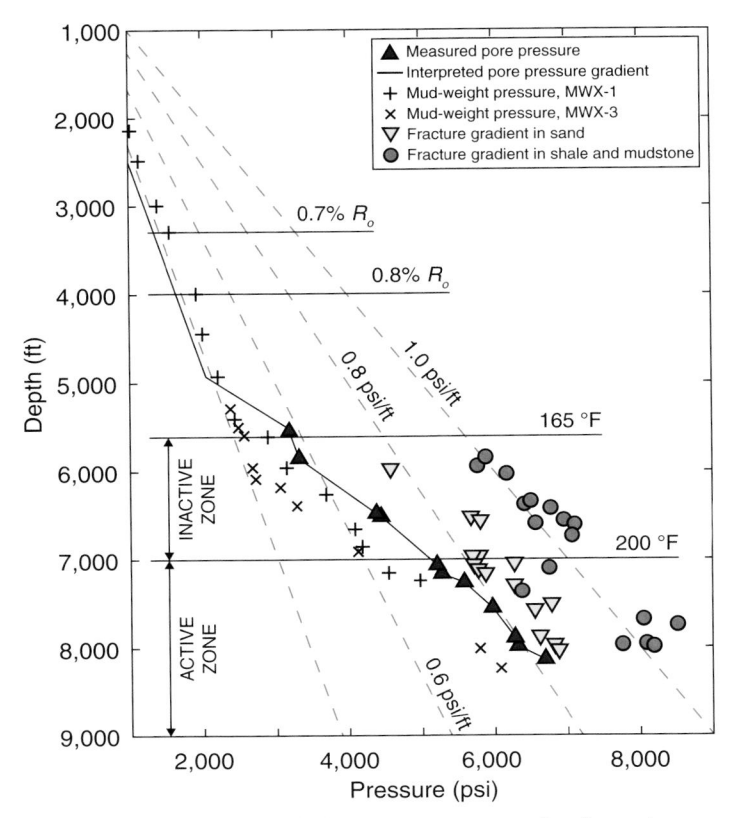

Figure 3.19 Pore pressure and least principal stress measurements in a layered sequence of sands and shales in western Colorado from the Multiwell Experiment (MWX). Elevated pore pressure (blue triangles) was observed at depth in three wells and attributed to hydrocarbon maturation. The least principal stress, S_3, in sands (inverted yellow triangles) increases as the pore pressure, P_p, increases whereas stress measurements in shales (brown dots) are mostly close to the vertical stress, S_V (~1 psi/ft), indicating near complete stress relaxation. Adapted from Nelson (2003).

where in practice, the term $\varepsilon_0 t^n$ is a fitting parameter (as neither the total strain nor time period are known). To illustrate how the viscoelastic power law parameters B (~$1/E$) and n affect the degree of stress relaxation for rocks from different unconventional formations, Fig. 3.18 shows the predicted frac gradient S_{hmin}/depth for a normal faulting environment with hydrostatic pore pressure. A variable amount of stress relaxation occurs, depending weakly on Young's modulus, but strongly on the viscoelastic power law parameter n. Rocks in which we anticipate low fracture gradients (approaching 0.6) might be considered brittle, regardless of relatively low values of Young's modulus. Conversely, Eagle Ford and Haynesville samples with high values of n exhibit frac gradients close to 1.0 and would likely be considered ductile (despite variations in Young's modulus of over a factor of 2). A frac gradient of 1.0 implies near complete stress relaxation (as illustrated in

Figs. 3.14, 3.15). Note that as predicted by Eqn. (7.3), frac gradients less than 0.6 are physically unreasonable for formations with faults exhibiting a coefficient of friction of 0.6. Also note that due to the anisotropy in creep and elastic compliance (Chapters 2 and 3), samples deformed perpendicular to bedding (vertical) generally show greater stress relaxation (frac gradient) than samples deformed parallel to bedding (horizontal).

An example of the effect of stress relaxation in shales is illustrated by stress measurements carried out in the 1980s in the Multiwell Experiment near the town of Rulison, Colorado (a normal faulting area) (Nelson, 2003). Figure 3.19 shows how the measured values of the magnitude of the least principal stress indicates stress magnitudes in sands increase with depth as pore pressure increases (as predicted by Eqn. (7.3)), whereas the stress measurements in the shales are generally very close to the magnitude of the vertical stress, implying nearly total stress relaxation. In Chapter 11, we will examine the issue of variations in stress magnitudes in adjacent formations in detail in order to understand the controls on the vertical growth of hydraulic fractures.

4 Frictional Properties

Production from unconventional reservoirs requires hydraulic fracturing and stimulation of pre-existing faults in order to access more reservoir surface area. Diffusion of fluid pressure from hydraulic fractures induces shear slip on faults by lowering the effective normal stress (Chapter 10). Induced fault slip increases formation permeability through inelastic damage in the surrounding rock and creates a network of relative permeability flow paths that increase access to the ultra-low permeability rock matrix. Slip on pre-existing faults is documented as microseismic events that cluster around hydraulic fractures and are thought to define the stimulated rock volume from which hydrocarbons are produced (Chapter 12). While this paradigm is widely accepted, multiple lines of evidence indicate that the deformation associated with microseismicity can only account for a fraction of production. To understand the relationship between hydraulic stimulation and production, it is important to consider under what conditions faults will slip and whether or not fault slip will cause microseismic events.

In this chapter, we examine the frictional properties of unconventional reservoir rocks in order to understand the controls on frictional strength and frictional stability – the tendency for faults to fail via seismic or aseismic slip. In the first section, we cover Coulomb faulting theory and the role of frictional strength in determining stress magnitudes and the orientation of potentially active (critically stressed) faults in the Earth's crust. We then review the concept of elastodynamic fault friction in order to establish a framework for understanding frictional stability. The next section focuses on how frictional stability can be quantified using rate-state friction theory, which is commonly used to model the dynamics of fault slip.

Using these theoretical foundations, we will discuss observations of frictional strength and stability, and their variations as function of lithology, temperature and fluid pressure. We will consider the grain-scale mechanisms of shear slip to develop a physical basis for the variations in the frictional properties of clay-rich and calcareous facies at reservoir conditions. We will also review recent experiments that focus on the relationship between frictional stability and pore fluid pressure in order to understand under what conditions hydraulic stimulation will induce seismic or aseismic slip.

In the final section, we discuss implications of the physical controls on frictional properties for the occurrence of induced shear slip in unconventional reservoirs. We consider the stress state on faults during hydraulic stimulation in terms of the critical length scales for seismic slip and the effects of lithological variations in frictional strength on the magnitude of the least principal stress.

Fault Strength and Stress Magnitudes

As the focus of this chapter is the frictional properties of pre-existing faults, it is important to review two fundamental concepts related to fault friction in the Earth's crust. First, the frictional strength of pre-existing faults limits the maximum magnitude of principal stress differences in the crust. Second, a wide range of observations support the applicability of laboratory-derived frictional parameters to faults *in situ*.

There is an extensive literature on the application of Mohr–Coulomb theory in rock mechanics. We particularly recommend Jaeger & Cook (1971) for those unfamiliar with Coulomb theory or Mohr circles. Throughout this book, σ will refer to effective stress, $S - P_p$. Figure 4.1a illustrates the basic principles of Mohr–Coulomb theory in terms of a simple laboratory test for measuring frictional sliding on a fault. A cylindrical sample with a natural or pre-cut fault, inclined at β, is jacketed and subjected to both an external confining pressure, σ_3, and an internal pore pressure, P_p. Frictional slip is driven by the application of an axial stress, σ_1, which increases the shear stress, τ, relative to the normal stress, σ_n, on the fault plane.

Disregarding cohesion, S_0, which is likely to be negligible for a rock volume with pre-existing faults, the Coulomb criterion is defined by Eqn. (4.1), which is sometimes referred to as Amonton's law. Frictional sliding will occur when the shear stress is sufficient to overcome the frictional resistance to sliding: the product of the coefficient of friction, μ, and the effective normal stress, which is the difference between the normal stress and pore pressure.

$$\tau = \mu \sigma_n. \tag{4.1}$$

The proximity to frictional failure on a given fault is often referred to as the Coulomb Failure Function, *CFF*, which is simply defined by

$$CFF \equiv \tau - \mu \sigma_n - S_0. \tag{4.2}$$

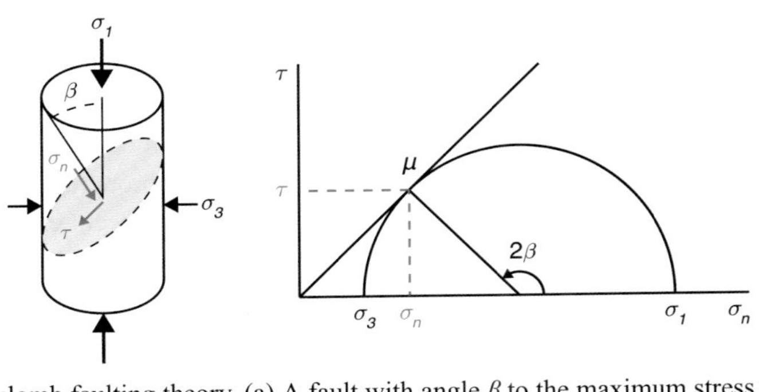

Figure 4.1 Coulomb faulting theory. (a) A fault with angle β to the maximum stress σ_1 will slip at a ratio of shear stress (τ) to normal stress (σ_n) defined by the coefficient of friction μ. (b) A 2D Mohr circle depicts the shear and normal stress values at which frictional failure occurs.

When the *CFF* approaches 0, the condition for frictional failure is satisfied. In the Earth's crust, frictional failure can result from increasing tectonic stresses over time or increasing pore fluid pressure, which reduces effective normal stress on faults.

The Mohr circle is a graphical tool that depicts the shear and normal stress condition at which frictional failure occurs (Fig. 4.1b). Shear and normal stress vary as function of the applied stresses and the angle of fault plane to the axial direction.

$$\tau = \frac{1}{2}(\sigma_1 - \sigma_3)\sin2\beta \tag{4.3}$$

$$\sigma_n = \frac{1}{2}\left[(\sigma_1 + \sigma_3) + (\sigma_1 - \sigma_3)\cos2\beta\right]. \tag{4.4}$$

Frictional sliding is expected to occur when the ratio of shear stress to effective normal stress reaches the coefficient of friction. The angle β of the fault plane to maximum principal stress is given by:

$$\beta = \frac{\pi}{4} + \frac{1}{2}\tan^{-1}\mu. \tag{4.5}$$

In a rock volume with faults at many different orientations, the frictional strength of the faults that are well oriented for sliding controls the maximum difference between the maximum and minimum effective stress. This is illustrated in Fig. 4.2 (again in 2D for simplicity) by faults of three different orientations. Faults with the optimal orientation for sliding (as defined by Eqn. 4.5), are known as "critically stressed" and are represented by fault set 1 for a coefficient of friction of 0.6. As indicated in the schematic map view and the Mohr circle in Fig. 4.2c, fault set 2 has a fault normal at too low an angle with respect to σ_1, and thus has too much normal stress to undergo frictional failure. As fault set 3 is approximately parallel to σ_1, it has less normal stress than fault sets 1 or 2, but has insufficient shear stress to overcome frictional resistance.

Implications for Differential Stress Magnitudes and Orientation – The maximum ratio between the maximum and least principal effective stresses at depth is defined by the frictional strength of well-oriented (*critically stressed*) faults (Jaeger & Cook, 1971).

$$\frac{\sigma_1}{\sigma_3} = \frac{S_1 - P_p}{S_3 - P_p} = \left(\sqrt{\mu^2 + 1} + \mu\right)^2. \tag{4.6}$$

In Chapter 7 (Eqns. 7.3–7.5) we modify Eqn. (4.6) for specific cases of normal, strike-slip and reverse faulting environments. In Chapter 10, we discuss how increased pore pressure during hydraulic stimulation may induce shear slip on faults that are not well oriented in the current stress field. In other words, increasing pore pressure reduces the effective normal stress, allowing Eqn. (4.1) to be satisfied, even though a pre-existing fracture or fault is essentially *dead*, or inactive, from a tectonic perspective.

As mentioned above, Zoback (2007) presents a number of examples in which the magnitudes of *in situ* stresses are consistent with the predictions of Eqn. (4.6). To offer

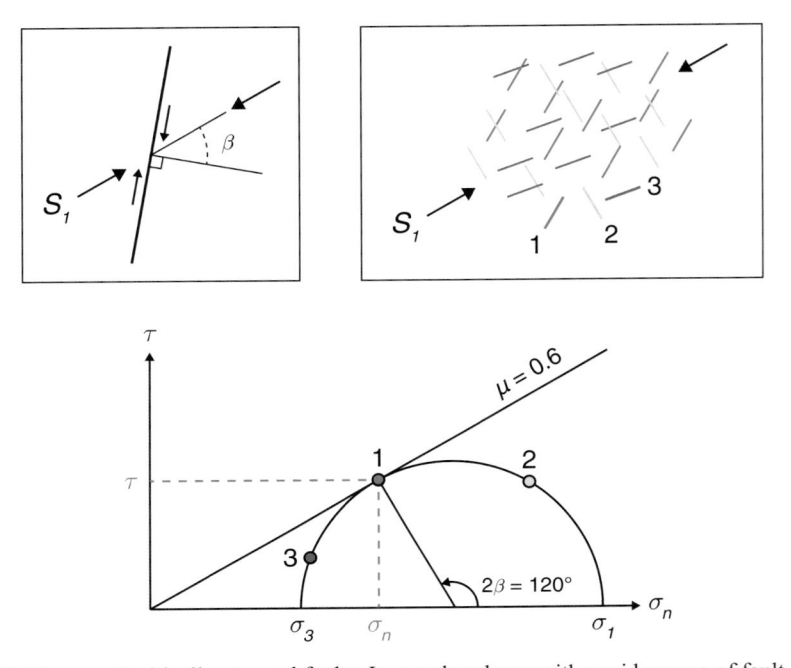

Figure 4.2 Theory of critically stressed faults. In a rock volume with a wide range of fault orientations (illustrated by fault sets 1, 2 and 3) only those that are well oriented with respect to the current stress field will undergo fault slip. Adapted from Zoback (2007).

one example of the application of Coulomb faulting theory in unconventional reservoirs, Fig. 4.3 shows data from the Haynesville shale formation. The estimated vertical stress, pore pressure and least principal stress measurements, S_{hmin}, are shown as a function of depth. Note that the theoretical prediction of Eqn. (4.6) modified for normal faulting (Eqn. 7.3) does a reasonably good job of fitting the measured values of S_{hmin} for a coefficient of friction of 0.6, especially considering that an average value of pore pressure was used to determine the effective stress at each depth.

As Coulomb faulting theory also predicts the orientation of active faults in a known stress field via Eqn. (4.5), another way to demonstrate its applicability is by comparing the orientation of active faults to the current stress field when both are known with some degree of accuracy. This can be done in north-central Oklahoma where appreciable seismicity has been triggered by widespread injection of produced water (Walsh & Zoback, 2015). This is discussed at greater length in Chapters 13 and 14. Relevant to the current discussion, however, is the map in Fig. 4.4a, which shows precisely determined lineations of seismicity in the crystalline basement beneath the injection interval (Schoenball et al., 2018). The orientation of stress throughout this region was determined by Alt & Zoback (2017) through analysis of wellbore stress indicators and inversion of earthquake focal plane mechanisms (Chapter 7). As illustrated by the rose diagram in Fig. 4.4b, the orientation of the fault planes indicated by the seismicity is ±30° to the direction of maximum horizontal stress, exactly as predicted by Coulomb faulting theory for a coefficient of friction of 0.6.

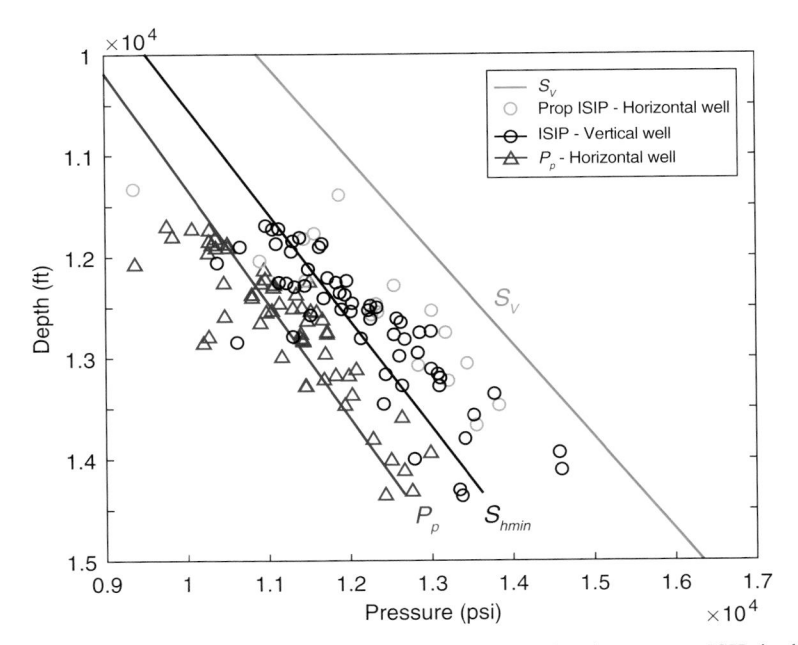

Figure 4.3 Measured values of pore pressure and instantaneous shut-in pressure, ISIP, in the Haynesville shale. ISIP represents the value of the least principal stress, S_{hmin}. The vertical stress, S_V, is determined by integrating density logs as function of depth. Using average values of pore pressure as a function of depth (blue line), the measured values of the least principal stress are relatively well characterized by Coulomb faulting theory for normal faults (black line; Eqn. 7.3).

Rock Friction

Static and Dynamic Friction – A simple, but useful model for rock friction is a spring–slider system (Fig. 4.5a), in which a mass is pulled along a frictional interface by a spring (Byerlee, 1978). Point A represents the position of the mass and point B represents the position of the spring (load point). Figure 4.5b shows the characteristic frictional response in terms of force (stress) vs. displacement for the application of a constant load point velocity, V. Similar to intact rock, initial loading behavior is characterized by a linear elastic response. At point C, the displacement begins to increase more for a given increment of force, which signifies that the mass has overcome static friction and is moving relative to the load point. Force (frictional resistance) continues to increase until point D (peak friction), beyond which several deformation responses are possible. In some cases, the mass may slip suddenly relative to the load point (solid line), causing a drop in force in the spring (stress drop). At constant velocity, force will increase again until enough stored elastic energy is built up for another slip event. These cycles of slip and reloading are termed stick-slip events and are a laboratory analog for earthquakes. In contrast, some materials show a gradual decline in force from peak friction, followed by stable, continuous slip at a constant force (dashed line), which is known as residual or sliding friction.

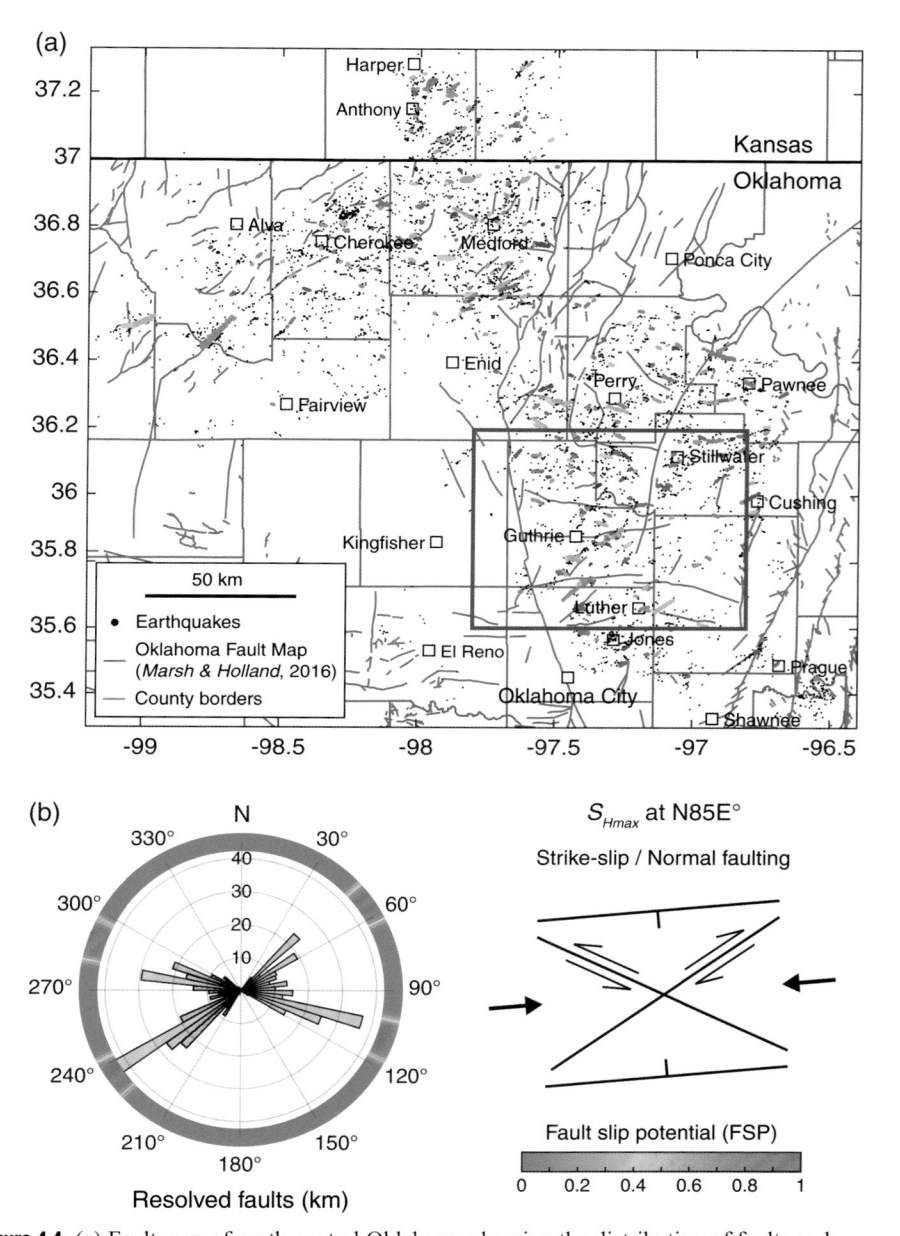

Figure 4.4 (a) Fault map of north-central Oklahoma showing the distribution of faults and seismicity. From Alt & Zoback (2017). (b) Rose diagram of fault orientations, colored on the circumference by fault slip potential, i.e., the value of the Coulomb failure function, *CFF*. The orientation of the maximum principal stress as determined by inversion of earthquake focal mechanisms is N85°E, which is approximately ±30° to the strike of the active fault planes. From Schoenball et al. (2018).

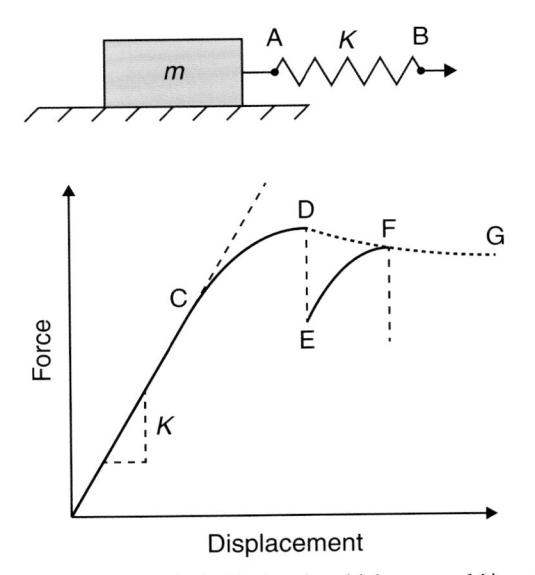

Figure 4.5 (a) Spring–slider model for fault friction, in which a mass M is pulled on a frictional interface by a spring with elastic constant K at loadpoint velocity V. (b) Characteristic frictional response. Loading results in a linear elastic response until point C (static friction), where the mass begins to be displaced relative to the loadpoint (slip). D represents the maximum (peak) frictional resistance. At this point, the mass may slip suddenly (stick-slip), resulting in drop in force in the spring (stress drop), or undergo stable sliding (dashed line), in which displacement of the mass occurs smoothly at relatively constant force (sliding friction). From Byerlee (1978).

Elastodynamic View of Fault Friction – The surfaces of geological faults are not simply planar features that are in completely in contact. This is evidenced by the fact that active faults often act as conduits for fluid flow (Townend & Zoback, 2000), as well as observations of frictional interfaces in laboratory experiments. In nature, fault surfaces have roughness at various scales (Candela et al., 2012) and may consist of fine-grained wear products from past slip events (fault gouge). When two rough surfaces are placed in contact under load (stress), the "real" area of contact, A_r, is the sum of the contact areas of all of the microscopic points of contact, which are known as asperities (Fig. 4.6a). For any rough interface, the real area of contact is less than the nominal area, A, so the stress acting on the asperities for a given applied force is magnified. The magnification of stress on the asperities enables inelastic deformation at relatively modest pressure-temperature conditions. For example, if A_r/A is 1%, an applied remote stress of 100 MPa is felt as 10 GPa, which is on the order of the elastic stiffness for many crustal minerals (Table 2.2). While the asperities may experience inelastic deformation, the area of the fault surface not in contact will only feel the remote stress and will likely deform elastically.

Teufel & Logan (1978) cleverly used thermally activated dyes to determine the asperity contact area in frictional sliding experiments on sandstone at 50 MPa confining pressure. Even at this relatively high pressure (for unconventional reservoirs), the real area of contact is only ~5–15% of the nominal area. Dieterich & Kilgore (1994) actually directly measured the contact area of asperities by optical interferometry on transparent,

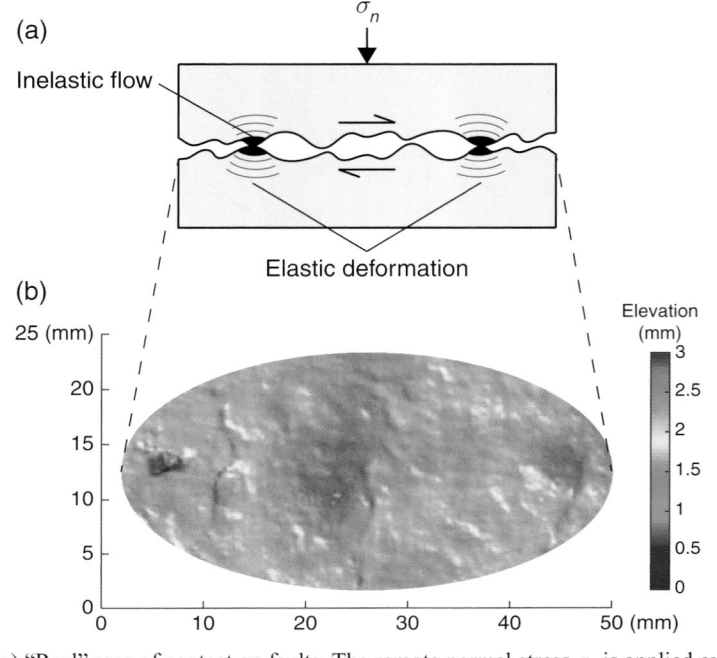

Figure 4.6 (a) "Real" area of contact on faults. The remote normal stress σ_n is applied over the nominal fault area A. Microscopic view of the fault reveals a rough interface with dispersed points of contact known as asperities. The asperity area is the real area of contact A_r. For faults in the crust, $A_r \ll A$, which magnifies effects of remote stress on the asperity contacts, enabling inelastic flow at moderate pressure–temperature conditions. Adapted from Teufel & Logan (1978). (b) Topography of an experimental fault surface in a clay-rich sample of Eagle Ford shale (Wu et al., 2017b). *Reprinted with permission from Elsevier whose permission is required for further use.*

pre-roughened frictional interfaces, and found that the real area of contact ranges from ~0.1% to 3% at similar normal stresses. Both studies also demonstrate that asperity (real) contact area decreases linearly with the logarithm of displacement (slip) rate. Given that the stress on the asperities is a function of real contact area, this implies that, for these surfaces, increasing slip rate decreases frictional resistance. Asperity contact area is also observed to increase with applied normal stress, thereby decreasing the stress felt by the asperities. The frictional properties of asperities are also time-dependent. Teufel & Logan performed hold tests during frictional sliding experiments and observed that the frictional resistance needed to resume sliding (static friction) increases with the logarithm of the hold time. Dieterich (1978) also demonstrated that asperity contact area increases with the logarithm of time, which essentially represents creep or time-dependent deformation at constant stress (Chapter 3). This is consistent with the idea that the magnified stress on asperities is capable of producing inelastic deformation.

Observations of time- and velocity-dependent behavior indicate that friction is controlled by the interactions of two competing processes: (1) frictional resistance increases

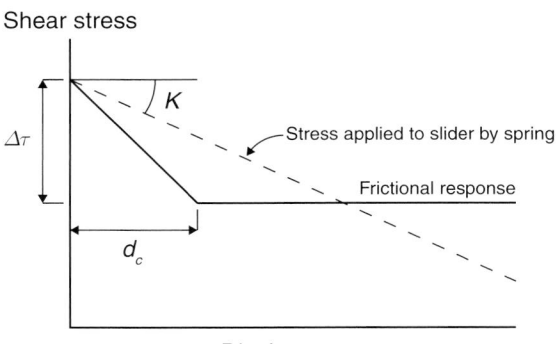

Figure 4.7 Slip weakening during frictional sliding. The spring unloads with slope K, which represents the elastic stiffness of the rock and loading frame. Stable slip occurs if the drop in shear stress, $\Delta\tau$, over a characteristic displacement, d_c, is less than K. Unstable slip occurs when the slope of the unloading curve of the slider (fault) is greater than K. Adapted from Dieterich (1979).

with logarithm of contact time (age), and (2) slip displacement destroys existing contacts and forms new, weaker contacts over a displacement scale related to contact spacing (roughness or grain size). Therefore, slip weakening (by either stick-slip or stable sliding), results when the average lifetime (size) of contacts decreases as a function of slip. The simplest model that encapsulates these effects is linear slip weakening (Fig. 4.7). Again, considering the simple spring–slider system (Fig. 4.5a), the spring, or in the case of laboratory experiments, the sample and loading system, unloads with a slope K (stiffness). After static friction is overcome, the dynamic friction of the slider (fault) is time- and velocity-dependent, and the stress drop of the fault, $\Delta\tau$, is assumed to be linear over the characteristic displacement, d_c. If the slope of the slider unloading curve ($\Delta\tau/d_c$) is greater than that of the spring (K), the slider will jerk forward relative to the load point (unstable slip). If $\Delta\tau/d_c < K$, the slider will slip stably at the imposed displacement rate. Therefore, the critical stiffness for friction instability is $K_c = \Delta\tau/d_c = \sigma_n \Delta\mu/d_c$. In the next section, we will discuss how this relationship can be used to quantify frictional stability in laboratory experiments.

Rate-State Friction – To describe the frictional behavior of faults, Dieterich (1979) and Ruina (1983) developed an expression to address the time- and velocity-dependence of friction:

$$\mu = \mu_0 + a\ln\left(\frac{V}{V_0}\right) + b\ln\left(\frac{V_0\theta}{d_c}\right) \tag{4.7}$$

in which a and b are the rate-state constitutive parameters, μ_0 and V_0 are the steady-state coefficients of sliding friction and the slip velocity respectively, d_c is the characteristic (or critical) slip displacement and θ is a state variable that encapsulates the time-dependence. When slip velocity is increased from V_0 to V, friction increases instantaneously by an amount proportional to the first material constant, a. This is known as the direct effect. Friction then evolves over time to a new steady state over a displacement, d_c, by an amount proportional to b. This is known as the evolution effect. The state variable is used to describe the time-dependence of sliding contacts (asperities) and also evolves over d_c:

$$\frac{d\theta}{dt} = -\frac{V\theta}{d_c} \ln\left(\frac{V_0\theta}{d_c}\right). \tag{4.8}$$

For this particular state evolution law, known as the slip law (Ruina, 1983), at steady-state, the value of the state variable is d_c/V, which is commonly interpreted to represent the average lifetime of sliding contacts. Substituting the steady state value into Eqn. (4.7) yields an expression for the change in the steady-state coefficient of sliding friction, $\mu_{ss} = \mu_0 - \mu$ in terms of the rate-state constitutive parameters:

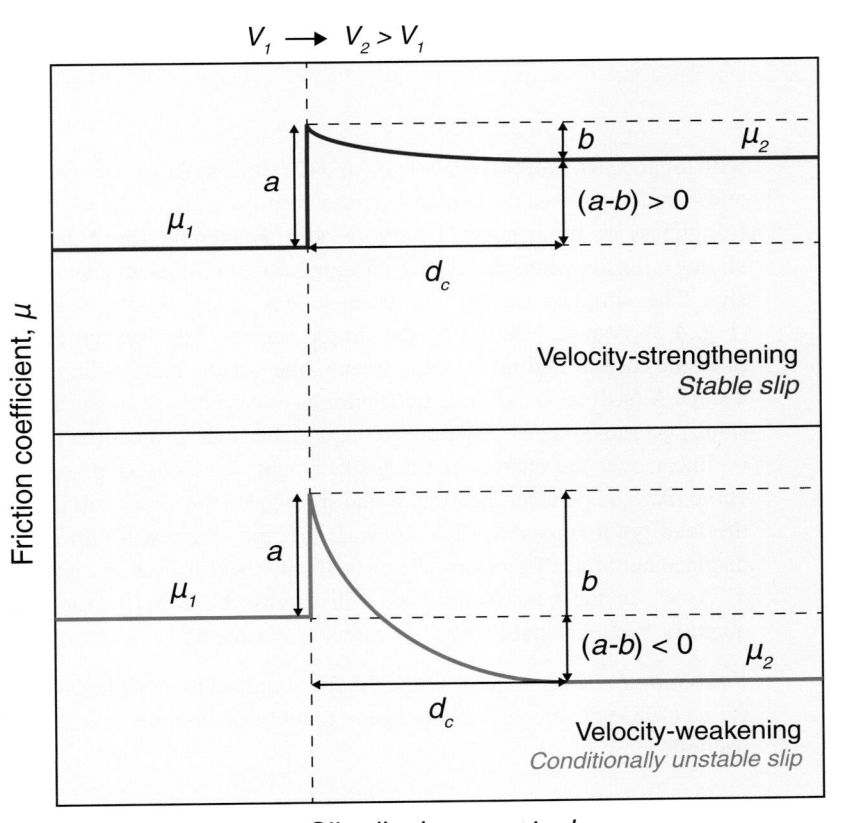

Figure 4.8 Frictional stability in rate-state friction theory. A step velocity increase applied during a frictional sliding experiment causes friction to increase instantaneously and then evolve to a new steady state over critical displacement d_c. Velocity-strengthening (stable) behavior results when instantaneous, velocity-dependent term, a, is greater than the time- and slip-dependent term, b. Velocity-weakening (conditionally unstable) behavior results when $(a - b) < 0$. From Kohli & Zoback (2019).

$$(a - b) = \frac{\Delta \mu_{ss}}{\ln(V/V_0)}. \tag{4.9}$$

If frictional resistance increases with increasing velocity $(a - b) > 0$. This is known as steady-state velocity-strengthening behavior and implies stable (aseismic) slip. If frictional resistance decreases with increasing velocity $(a - b) < 0$. This is known as steady-state velocity-weakening behavior and implies conditionally unstable (seismic) slip. Figure 4.8 illustrates velocity-strengthening and velocity-weakening responses in terms of the rate-state constitutive parameters.

The condition for unstable slip in rate-state friction is simply the critical stiffness expressed in terms of the rate of slip weakening:

$$K_c = \frac{\Delta \tau}{d_c} = \frac{\sigma_n \Delta \mu_{ss}}{d_c} = \frac{\sigma_n (b - a)}{d_c}. \tag{4.10}$$

In the following section, we will discuss measurements and observations of frictional stability using the rate-state friction theory.

Frictional Strength and Stability

Measurement – Friction sliding can be measured in a variety of experimental geometries, including direct shear (biaxial), rotary shear and conventional triaxial. Figure 4.9 illustrates the experimental configuration for friction measurements in a conventional triaxial geometry (e.g., Fig. 4.1a). Friction sliding occurs on either a pre-

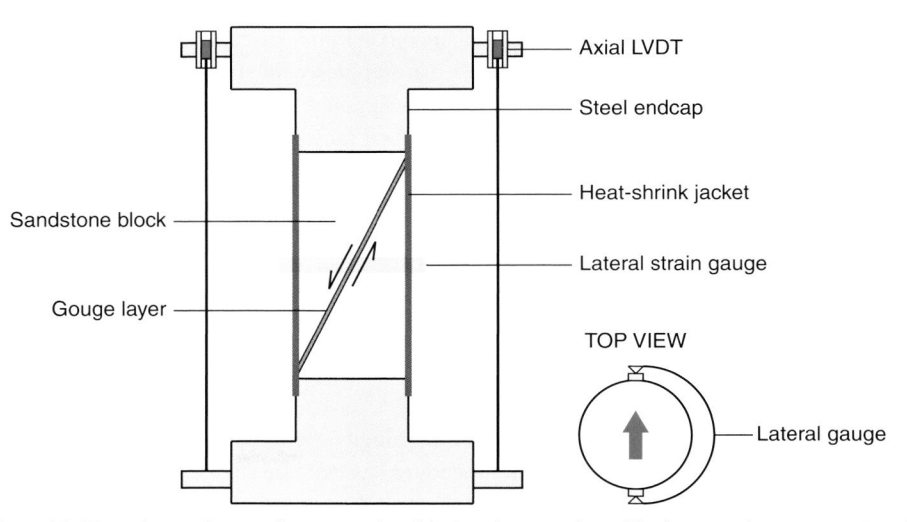

Figure 4.9 Experimental setup for measuring frictional properties of fault gouge in a conventional triaxial apparatus. In the top view, the arrow denotes the slip direction. Adapted from Kohli & Zoback (2013).

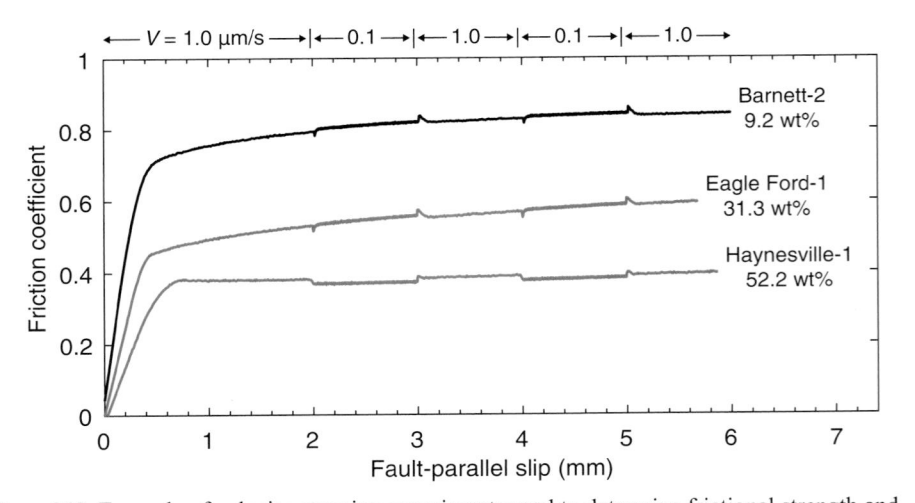

Figure 4.10 Example of velocity-stepping experiments used to determine frictional strength and stability, Barnett shale. The numbers below the sample names denote clay + TOC in wt%. Frictional strength is measured by the steady-state value reached after initial loading behavior (e.g., Fig. 4.5b). Frictional stability is measured by examining the response to step-wise changes in slip velocity. See Table 2.1 for sample compositions. Adapted from Kohli & Zoback (2013).

cut surface or a rough fault formed by compressive failure. In Chapter 3, we briefly review observations of frictional strength on rough faults. Here, we focus on experiments on pre-cut surfaces that are intended to eliminate any confounding effects of roughness or wear products. In addition, our discussion of the controls on frictional strength and stability will focus on experiments that utilize fault gouge (crushed material) on the pre-cut surfaces. Since the elastic properties of unconventional reservoir rocks (Chapters 2 and 3) vary significantly with microstructure (in samples with similar composition) it is important to control for these effects. By using fault gouge that is sieved to a certain particle size, the microstructural details of the intact rock (anisotropy, grain structure and size) are mostly lost, so the frictional response will reflect the composition of sliding contacts. In fault gouge friction experiments, it is common to use pre-cut forcing blocks of a relatively strong, brittle material (e.g., sandstone, granite, or metal). In the conventional triaxial geometry, displacement is measured by axial sensors and is resolved on the fault plane to determine slip. A lateral strain gauge oriented parallel to the direction of slip can be used to resolve changes in thickness of the gouge layer (compaction and dilation). The sample is subjected to uniform confining pressure and frictional sliding is driven by application of axial stress (Fig. 4.1a).

Figure 4.10 shows examples of velocity-stepping friction experiments on samples of Barnett shale. This type of procedure has become standard for measuring both frictional strength and stability. Frictional strength is measured by the steady-state value reached after initial loading behavior. For most rocks, this occurs after only several millimeters of slip displacement. Frictional stability is measured by examining the velocity- and

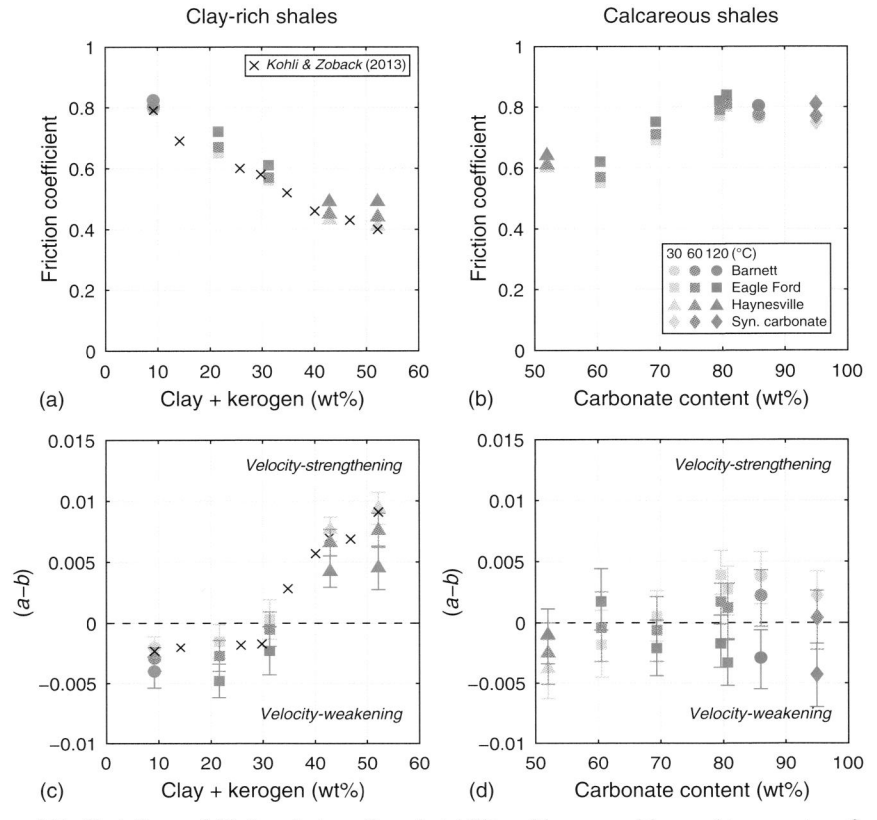

Figure 4.11 Variations of frictional strength and stability with composition and temperature for clay + TOC-rich and calcareous shales. Errors bars for frictional strength are on the order of symbol size. Symbols for frictional stability represent the average value of $(a - b)$ from multiple velocity steps, error bars represent the range of values in a given experiment. See Table 2.1 for sample compositions. Adapted from Kohli & Zoback (2019).

time-dependent response to step-wise changes in imposed slip velocity (e.g., Fig. 4.8). There are several possible ways to quantify frictional stability. In order to obtain the independent values of the constitutive parameters, a, b and d_c, it is necessary to solve Eqn. (4.7) coupled with an expression that accounts for the elastic interactions between the sample and the loading frame (Marone et al., 1990). Considering independent variations of a and b can provide additional insights into the micromechanical details of the frictional response. Alternatively, $(a - b)$ and d_c may simply be obtained.

Compositional and Thermal Controls – To assess compositional controls on friction properties it is important to distinguish the major lithologies. As discussed in Chapter 2, unconventional reservoir rocks span a wide range of lithologies from siliceous to calcareous to clay-rich (Fig. 2.1). Since clay and organic matter are often correlated (Fig. 2.2) and represent the most compliant components of the matrix (Table 2.1), it is convenient to group them together to study mechanical properties. Carbonate contents vary from ~0% to 90% and represent the most significant source of compositional variation in many reservoirs (Fig. 2.1). Clay and carbonate minerals are also known to exhibit distinct frictional properties that depart from Byerlee's law, which states that the coefficient of friction is nearly independent of rock type (Byerlee, 1978). Therefore, to study compositional variations in frictional properties, we distinguish between clay and organic-rich and calcareous shales. In this framework, calcareous shales have 50 wt% or greater carbonate content. Most clay and organic-rich shales have relatively low carbonate content, so the variations in clay and organic matter are mostly compensated by siliciclastic minerals (quartz and feldspar).

Figure 4.11a,b shows the variation in frictional strength as function of composition for clay and organic-rich and calcareous shales. Similar to the results on rough faults (Fig. 3.3c), the coefficient of sliding friction decreases with increasing clay and organic content from ~0.8 to slightly less than 0.4. For calcareous shales, increasing carbonate content from ~50% to 80% corresponds to an increase in friction from ~0.6 to 0.8. Cross-referencing these two datasets (i.e., plotting the clay and organic-rich shales in terms of carbonate content and vice versa), shows general agreement in terms of the variation of frictional strength with the relative proportions of soft and stiff components, which suggests that this framework is appropriate for characterizing unconventional reservoir rocks (Kohli & Zoback, 2019). In fact, some samples can be used to illustrate trends for both calcareous and clay and organic-rich lithologies (e.g., Haynesville BWH2-2, Eagle Ford 65Hb). For all samples, frictional strength increases slightly with increasing temperature up to 120 °C. This effect appears to increase with increasing clay + TOC in clay and organic-rich shales, with the weakest sample showing nearly a ~25% increase in strength. In calcareous shales, frictional strength increases with temperature by a similar amount at all carbonate contents. The observed strengthening with increasing temperature is generally consistent with the results of similar experiments on pure illite (Kubo & Katayama, 2015) and calcite (Verberne et al., 2013) gouges.

Figure 4.11c,d shows that the variation in rate-state frictional stability, $(a - b)$, as function of composition for clay and organic-rich and calcareous shales. Clay and organic-rich samples (>30 wt%) show velocity-strengthening behavior $(a - b > 0)$, which implies stable slip. Below this apparent threshold, clay and organic-poor samples show velocity-weakening behavior $(a - b < 0)$, which implies conditionally unstable slip. In the context of variations in frictional strength (Fig. 4.11a), this signifies that relatively strong rocks $(\mu > 0.6)$ may undergo unstable slip while relatively weak rocks are only capable of stable sliding. In contrast, for calcareous shales, relatively strong samples $(\mu \sim 0.8)$ with >70 wt% carbonate exhibit velocity-strengthening behavior. For all samples, increasing temperature results in lower (or more negative) $(a - b)$ values, which signifies an increased tendency for velocity-weakening behavior. This effect is relatively minor for clay and organic-rich shales, pushing the transition from velocity-weakening to velocity-strengthening behavior to slightly higher clay and TOC. In contrast, the effect of temperature is quite significant in calcareous shales, such that all the carbonate-rich samples (<70 wt%) that are

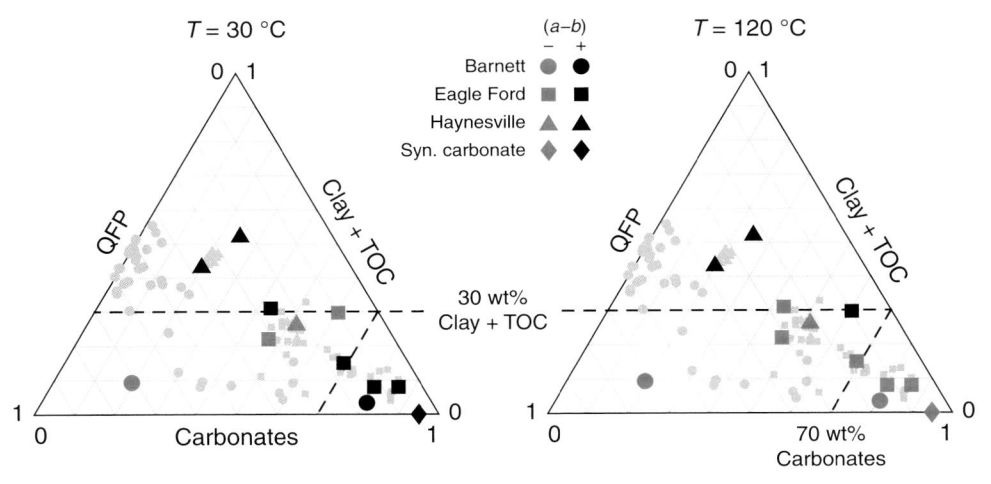

Figure 4.12 Variation of frictional stability with composition in the ternary space QFP-Carbonates-Clay + TOC. Smaller open symbols represent the full range of compositions for the Barnett, Eagle Ford and Haynesville shales. At 30 °C, samples with less than ~30 wt% clay + TOC and ~70 wt% show velocity-weakening responses. At 120 °C, the threshold for stable sliding moves to slightly higher values of clay + TOC and carbonate-rich samples transition from velocity-strengthening to velocity-weakening behavior. From Kohli & Zoback (2019).

velocity-strengthening at ambient temperature are velocity-weakening at 120 °C. These observations of decreasing $(a - b)$ with increasing temperature are again generally consistent with the results of similar experiments on pure illite (Kubo & Katayama, 2015) and calcite (Verberne et al., 2013) gouges, which suggests that temperature dependence of frictional stability may be controlled by the behavior of illite and calcite. Figure 4.12 summarizes the results of Fig. 4.11 within the ternary space for composition of unconventional reservoir rocks. In the following section, we will explore the micromechanical processes responsible for the compositional and thermal controls on frictional properties.

Micromechanical Processes – It is well known that the addition of clays, phyllosilicates and organic matter reduces the coefficient of sliding friction in rocks (Lupini et al., 1981; Takahashi et al., 2007; Crawford et al., 2008; Tembe et al., 2010). Clay and phyllosilicate minerals are weak in shear due to their tendency to align under stress and accommodate slip on the basal crystallographic plane (Kronenberg et al., 1990). The coefficient of pure illite clay, which is the predominant clay mineral in most basins, is ~0.4 at ambient temperature (Tembe et al., 2010; Kubo & Katayama, 2015). Organic matter is even weaker in shear because it is composed predominantly of carbon and hydrogen and lacks a crystal structure. The coefficient of friction of pure graphite, which represents an analog for organic matter, is ~0.1 (Rutter et al., 2013). Given that the coefficients of friction for quartz and calcite range from ~0.6 to 0.8, increasing the abundance of weaker phases obviously results in reduced frictional strength.

In the friction experiments described earlier, the distribution of weak phases is essentially randomized through the use of fault gouge. However, in experiments on synthetic

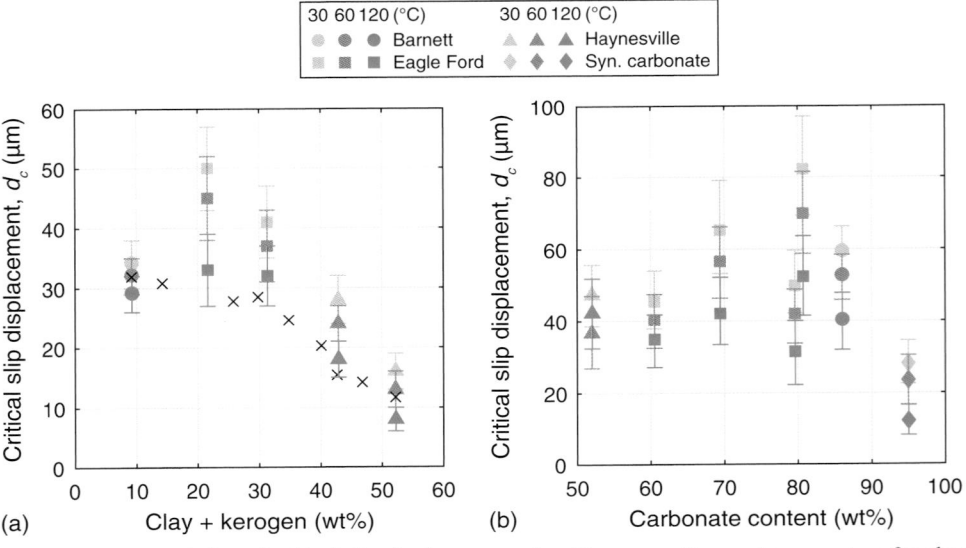

Figure 4.13 Variation of critical slip displacement, d_c, with composition and temperature for clay and organic-rich and calcareous shales. Symbols represent the average value of d_c from multiple velocity steps and error bars represent the range of values in a given experiment. From Kohli & Zoback (2019).

mixtures of a 2:1 ratio of kaolinite:quartz and graphite, Rutter et al. (2013) document a larger than expected impact of graphite on frictional strength at relatively low fractions. This is due to the tendency of organic matter to smear under shear and occupy a larger than representative area of the primary slip surfaces. In the context of unconventional reservoir rocks, this suggests that organic matter may localize on surfaces of active faults from repeated shear events, significantly reducing frictional strength and promoting stable sliding. Careful study of the slip surfaces of natural faults in unconventional basins is needed to assess the role of organic matter on fault frictional properties.

In order to understand the variations in frictional stability (Fig. 4.11c,d), it is important to consider the displacement scale over which frictional evolution occurs. Figure 4.13 shows the variation in critical slip displacement, d_c, as function of composition for clay and organic-rich and calcareous shales. d_c values decrease from ~50 to 5 μm with increasing clay and organic content. In contrast, for calcareous shales, d_c is variable with increasing carbonate content, ranging from ~30 to 100 μm (excluding the synthetic carbonate sample). Recall that the physical interpretation of d_c is the displacement required to renew a population of sliding contacts, or, in other words, the average spacing between sliding contacts. Within a gouge layer, the density of sliding contacts is determined by grain size of the load-bearing framework (Marone & Kilgore, 1993). Therefore, the decrease of d_c with increasing clay and TOC may actually reflect a decrease in average spacing of sliding contacts as the relatively coarse-grained clastic framework is replaced by relatively fine-grained clay and organic matter (Fig. 2.8a). Similarly, the variability in d_c with increasing carbonate content may reflects the variability in the grain sizes (and shapes) of the clastic framework (Fig. 2.8b). See Kohli & Zoback (2019) for a more comprehensive discussion of these results.

For clay and organic-rich shales, the decrease in d_c with clay and organic content also offers some insights into the observed frictional stability (Fig. 4.11a). At these relatively low pressure-temperature conditions, frictional stability is governed by two competing mechanisms: grain or contact-scale crystal plasticity, and cataclastic flow (fracture and grain boundary sliding). Mineral crystal plasticity is stress-, temperature- and time-dependent, and is also inherently velocity-strengthening (increasing strain rate increases stress) (Rutter, 1974). Therefore, it is often invoked to represent the direct effect, a, or the instantaneous, velocity-dependent response. In addition, crystal plasticity is also thought to control the time-dependent growth of sliding contacts under stress. Cataclastic or granular flow is relatively time-insensitive, but depends on slip displacement and the strength and size of sliding contacts. Therefore, the evolution effect, b, represents the competition between the time-dependent growth of contacts, which decreases frictional resistance, and cataclastic flow, which may either increase or decrease frictional resistance depending on the specific grain-scale mechanisms.

In response to an imposed velocity increase (i.e., Fig. 4.8), the stress on the sliding contacts increases, then decays according to the relative contributions of crystal plasticity and cataclastic flow. As the average size of sliding contacts (d_c) decreases with decreasing grain size, for a given velocity, the time for sliding contacts to grow decreases as well, which limits the decay of the direct effect and produces a velocity-strengthening response. For a clay-supported framework, this implies that renewing new sliding contacts by grain boundary sliding is easier than sliding accommodated by crystal plasticity. This is certainly plausible given the observations of fabric development at high clay contents (Chapter 2) and the frictional weakness of slip on the basal plane (Mitchell & Soga, 2005). For a clast-supported framework, the longer characteristic displacement means more time for contact growth; however, the magnitude of crystal plastic deformation is strongly dependent on the mineralogy of the contacts. Quartz has a relatively high crystal plastic strength at low temperatures (<200 °C) while clay and carbonate minerals may deform by dislocation glide on internal slip systems and form crystallographic preferred orientations (Rutter, 1974; Poirier, 1985; Mitchell & Soga, 2005). For this reason, we observe a distinct difference in the frictional stability of siliceous and calcareous shales (e.g., Barnett 23Vb and Eagle Ford 254Va). In siliceous lithologies, high stresses at sliding contacts may lead to grain-scale fracture and comminution, which increase contact area and therefore decrease frictional resistance, leading to velocity-weakening behavior. In calcareous lithologies, high stresses may be dissipated by crystal plastic deformation, but at ambient temperature, the strain rates are too low to completely counteract the effects of grain boundary sliding, resulting in velocity-strengthening behavior.

As temperature increases, crystal plasticity becomes more effective (increased strain rate), which explains the decreased rate dependence for both clay and organic-rich and calcareous shales. In fact, Verberne et al. (2013) estimate that for calcite, the strain rate of dislocation glide increases by nearly a billion times from 30 °C to 120 °C. For a given slip velocity, increasing the contribution of crystal plasticity increases the time-dependent growth of sliding contacts, which further reduces frictional resistance during slip evolution.

For clay and organic-rich shales this effect is relatively minor, whereas for calcareous shales (>70 wt%), the increase in crystal plasticity is sufficient to cause a transition from velocity-strengthening to velocity-weakening between 60 and 120 °C. Verbene et al. (2013) observe a similar transition in frictional stability at ~80 °C in synthetic calcite gouges and document a crystallographic preferred orientation consistent with dislocation glide. This suggests that the temperature dependence of friction stability for calcareous shales is controlled by the behavior of calcite, consistent with the idea that the load-bearing contact structure defines the frictional properties. For a more comprehensive discussion of the micromechanical process responsible for frictional stability, see Kohli & Zoback (2019).

Pore Fluid Effects – Rate-state constitutive parameters represent the frictional behavior of the material on the fault surface (how a fault will slip) whereas the relationship with the elastic stiffness of the fault determines the forces on the fault surface (when a fault will slip). To understand induced slip during hydraulic stimulation, it is necessary to consider how the critical fault stiffness, K_c, is affected by pore pressure. Unstable slip will occur when K_c exceeds the stiffness of the loading system (surrounding rock volume). Replacing σ_n in Eqn. (4.10) gives an expression for K_c in terms of pore fluid pressure:

$$K_c = \frac{(S_n - P_p)(b - a)}{d_c}. \tag{4.11}$$

There are two important points to understand about this relationship. First, increasing pore pressure reduces the fault stiffness, which results in an increased tendency for stable slip. During hydraulic stimulation, the orientation of a fault with respect to the current stress field determines the increase in the pore pressure needed to overcome frictional resistance (Chapter 10). Therefore, we might expect poorly oriented faults to fail at lower effective normal stress via stable slip and well-oriented faults to fail seismically with only minor pore pressure perturbation. Second, in this relationship, unstable (seismic) slip is only possible

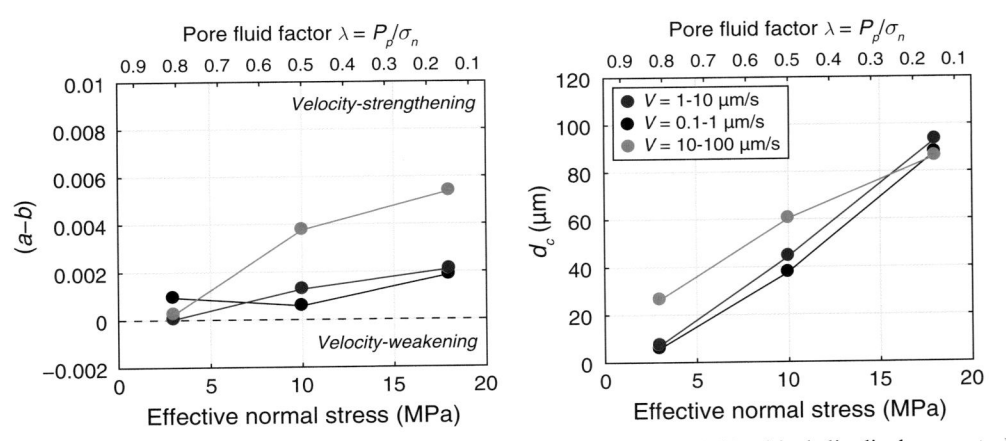

Figure 4.14 Variation of (a) friction rate dependence, $(a - b)$, and (b) critical slip displacement, d_c, with effective normal stress (pore pressure) in a natural limestone gouge. From Scuderi & Colletini (2016). *Reprinted with permission from Elsevier whose permission is required for further use.*

when the fault surface is velocity-weakening or $(a - b) < 0$. Modification of rate-state friction to include a dependence on normal stress has only a minor effect on fault stability (Linker & Dieterich, 1992); however, recent experiments on natural crustal fault gouge demonstrate that the rate-state material properties may actually vary with pore pressure. Scuderi & Collettini (2016) performed velocity-stepping friction experiments on natural carbonate gouges and observed a systematic decrease in friction rate dependence, $(a - b)$, and critical slip displacement, d_c, with increasing pore pressure (Fig. 4.14). At relatively low effective normal stress (<5 MPa), this effect is significant enough to cause a transition from velocity-strengthening to velocity-weakening behavior. In carbonate rocks, the decrease in $(a - b)$ and d_c with increasing pore pressure may reflect an increase in contact-scale crystal

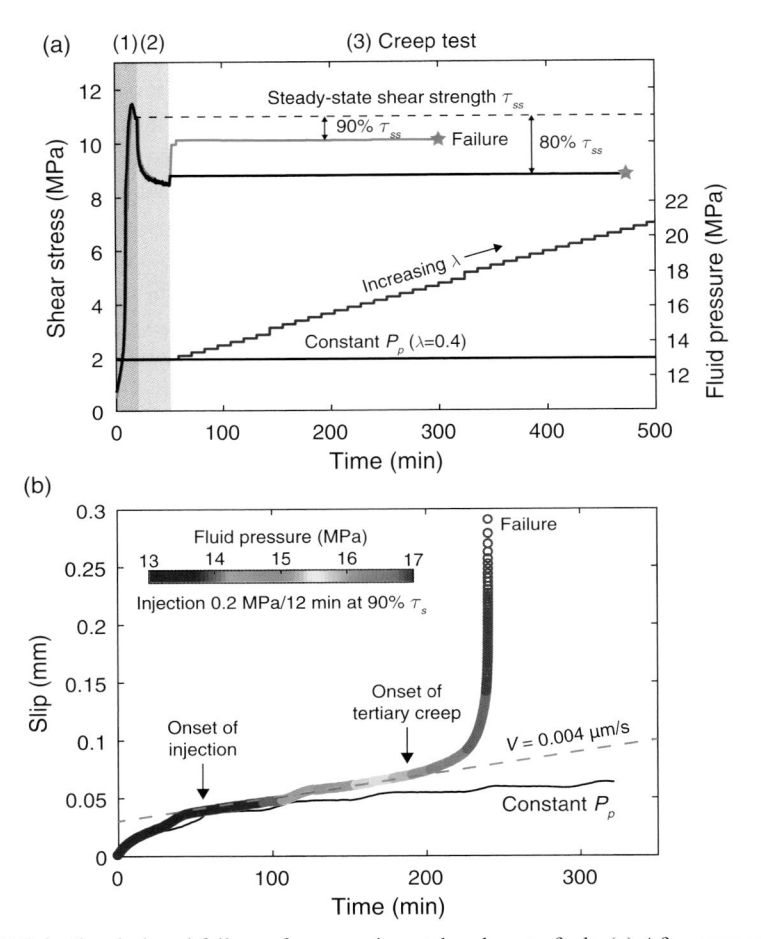

Figure 4.15 Injection-induced failure of an experimental carbonate fault. (a) After measured peak friction and residual friction are reached at hydrostatic pore pressure ($\lambda = P_p/\sigma_n = 0.4$), the fault is subject to a constant shear stress that is ~90% of peak friction (creep test). During the creep test, pore fluid pressure is increased until the fault fails dynamically (unstable slip). (b) Fault slip as a function of time colored by pore fluid pressure. Only a small change in fluid pressure is required to initiate tertiary creep and dynamic failure. Adapted from Scuderi et al. (2017a).

plasticity due to the increased activity of fluid-assisted processes such as a grain boundary diffusion or pressure solution creep. Lower values of $(a - b)$ and d_c both promote unstable fault slip by increasing critical fault stiffness, which counteracts the effect of reduced effective normal stress. This suggests that increasing pore pressure may actually encourage seismic slip, contrary to the idea that poorly orientated faults can only undergo stable slip at low effective normal stress.

Scuderi et al. (2017a) further explored pore fluid effects on frictional stability by injecting fluid into laboratory faults close to failure (Fig. 4.15). The first part of the experiment is constant velocity frictional sliding at hydrostatic pore pressure ($\lambda = P_p/\sigma_n = 0.4$) and yields values for both peak and residual frictional strength. In the second part of the experiment, the fault is subjected to a constant shear stress that is 80 or 90% of the peak strength, which is essentially a fault creep test. The pore pressure is then increased at a constant rate until the fault reaches failure. Figure 4.15b compares the slip response of injection and constant pore pressure for the experiment conducted at 90% of the peak strength. At constant pore pressure, the fault creeps at a slow rate and never reaches failure. In response to injection, the slip rate begins to increase gradually until ~15.5 MPa, at which point slip accelerates rapidly and the fault undergoes dynamic failure. Although this material is velocity-strengthening at hydrostatic conditions, the occurrence of dynamic failure suggests that the variation of rate-state parameters with pore pressure is sufficient to produce velocity-weakening behavior. The accelerating creep due to injection is also correlated with increase in the permeability of the fault, which suggests that the observed behavior represents a competition between dilatant strengthening and slip weakening. Interestingly, Scuderi et al. (2017b) also perform this procedure on a weak ($\mu = 0.28$), less permeable clay-rich shale and observe a slightly different response. Upon injection, fault creep accelerates (up to ~200 μm/s) in periodic cycles, but never reaches dynamic failure. In this case, Scuderi et al. actually document an increase in permeability associated with compaction, which suggests dilatant strengthening may not always be the mechanism limiting instability at low effective normal stress. Regardless, these injection experiments demonstrate that the coupling between pore pressure and frictional properties (rate-state parameters) is an important control on induced fault slip.

In the context of unconventional reservoirs, further study of pore fluid effects over the range of observed lithologies is necessary to understand how faults respond to hydraulic stimulation. In addition, during hydraulic stimulation of pre-existing faults, factors such as injection rate and pore fluid composition may also affect fault stability and may vary with distance from the injection point. For example, faults in the near-field will experience a large, rapid increase in pressure, whereas those further away from the injection point will experience a more gradual increase in pressure. It is important to note that distance from the injection point may also influence the type or composition of the pore fluid. In the near-field, high injection pressures may drive water-based fracturing fluids into fractures and faults, whereas in the far-field, only the pressure signal will be communicated, so *in situ* pore fluid will be present on the fault surfaces. For hydrocarbon reservoirs, pore pressure on far-field faults will be imparted by natural gas and oil, which may have different influences on the frictional properties and need to be considered in future investigations.

Implications for Induced Shear Slip during Hydraulic Stimulation

Frictional Strength and Stress State – Given the compositional layering present in unconventional reservoirs (Fig. 1.15), it is important to understand how variations in reservoir properties may impact stress magnitudes. There are three mechanisms to consider. First, the compartmentalization of elevated pore pressures in ultra-low permeability shale units may result in a near lithostatic stress state. At the end of Chapter 3, we discussed the results of the Multiwell Experiment, which indicate that the least principal stress in shale units remains close to the vertical stress, regardless of changes in the pore pressure gradient. In contrast, in the interbedded sand units, the least principal stress increases with the pore pressure gradient. This seems to indicate that elevated pore pressure may not necessarily be responsible for stress variations due to composition layering. Second, in Chapter 3, we discussed how viscoelastic stress relaxation can increase the relative magnitude of the minimum principal stress and thereby decrease stress anisotropy. Variations in ductility (creep) between different formations can result in variations in stress, causing more ductile units to act as potential barriers to hydraulic fracture growth (Fig. 3.17). In Chapter 11, we will revisit this issue to try to understand variations in the least principal stress in horizontal wells that "porpoise" between different lithofacies (Ma & Zoback, 2017b).

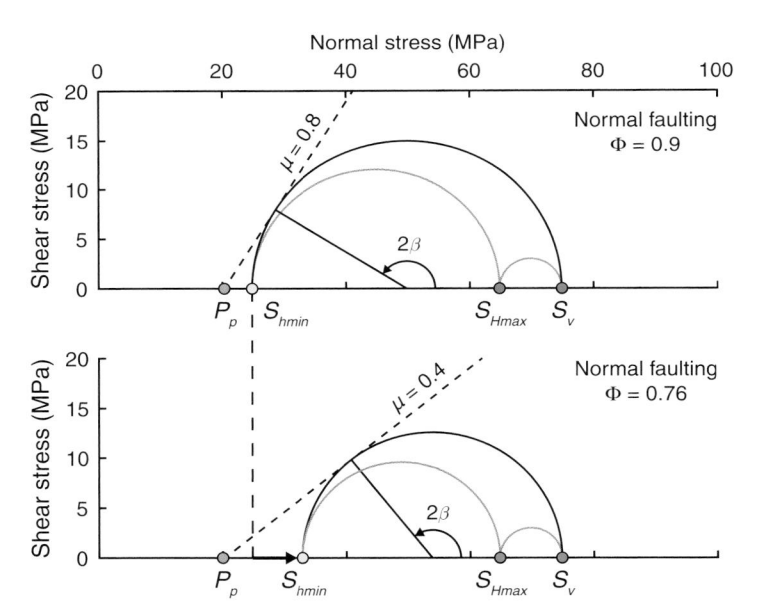

Figure 4.16 Effects of frictional strength on the stress state in a strike-slip/normal faulting environment. The stress state represents the Barnett shale ($S_V = 75$ MPa, $S_{Hmax} = 64$ MPa, $P_p = 20.3$ MPa) (Sone & Zoback, 2014b). For each case, the value of S_{hmin} is determined by the strength of faults in frictional equilibrium. Decreasing frictional strength increases the magnitude of S_{hmin} and decreases the failure angle for critically stressed faults, β.

The final mechanism to consider is the variation of frictional strength with rock composition. If the strength of faults defines the stress state of the crust (Eqn. 4.6), then it is plausible that variations in frictional properties result in variations in stress between different lithofacies. Figure 4.16 shows Mohr circles constructed based on the estimated normal/strike-slip stress state of the Barnett shale (Sone & Zoback, 2014b) in frictional equilibrium with relatively strong ($\mu = 0.8$) and ($\mu = 0.4$) weak lithofacies. Given constant values of S_V, S_{Hmax}, and P_p, reducing the strength of faults in frictional equilibrium causes the minimum principal stress, S_{hmin}, to increase. As a result, the value of Φ, which describes the relationship between the principal stresses, $\Phi = \dfrac{S_2 - S_3}{S_1 - S_3}$, decreases from 0.9 to 0.76, and the angle between S_1 and critically stressed faults, β, increases from 64° to 74°.

Ma & Zoback (2017b) attempted to use the variation in frictional strength with clay + TOC (Fig. 4.13a) to predict the variation of instantaneous shut-in pressure (ISIP), which is a measurement of the least principal stress (Fig. 11.15). They found that the magnitude of the stress variations calculated from the friction-composition trend (Fig. 4.11a) is much smaller than the observed variations in ISIP. This implies that variations in frictional strength alone are not sufficient to explain relatively high values of the least principal stress in clay and organic-rich lithologies. Instead, it may be reasonable to consider the variations in stress in terms of the spectrum of deformation properties. Variations in the frictional strength of faults may be partially responsible for variations in the least principal stress, in addition to viscoplastic stress relaxation on those faults. In fact, experimental faults show a similar type of creep response to intact rock (Fig. 4.15b), which suggests that the viscoplastic properties of faults could similarly contribute to stress evolution. If this is the case, it may be difficult to predict observed stress variations solely based on well log measurements of composition and elastic properties, which sample cm- to m-scale rock volumes and thus, may not represent the material on fault surfaces.

Critical Length Scale for Active Faults – Measuring the rate-state frictional properties of unconventional reservoir rocks enables estimation of the critical length scale for active faults. Dieterich (1978b) proposed that the effective stiffness of a fault in an elastic medium is simply the ratio of the shear modulus, G, and the length, l:

$$K = \frac{G}{l}. \tag{4.12}$$

Combining this expression with Eqn. (4.10) yields an expression for the critical length scale for active faults, l_c:

$$l_c = \frac{G d_c}{\sigma_n (b - a)}. \tag{4.13}$$

Fault patches smaller than l_c are too stiff ($K > K_c$) and will slide stably, while fault patches larger than l_c are sufficiently compliant ($K < K_c$) to fail dynamically. Thus, l_c represents an estimate for the minimum fault patch size needed for unstable (seismic) slip.

Table 4.1 Parameters used for calculation of the critical length scale for active faults for relatively clay + TOC-rich and TOC-poor lithofacies.

	High clay + TOC (30 wt%)	Low clay + TOC (<10 wt%)	Reference
Shear modulus, G (GPa)	12	32	Sone & Zoback (2013a)
$(b - a)$	0.0005	0.005	Kohli & Zoback (2013)
d_c (μm)	20	50	Kohli & Zoback (2013)
μ	0.4	0.8	Kohli & Zoback (2013)
Angle to σ_1 (β)	64°	74°	Fig. 4.16
S_{hmin} (MPa)	32.9	24.8	Fig. 4.16
σ_n (MPa)	19.7	8.3	Fig. 4.16
l_c (m)	24	40	Eqn. (4.13)

In Table 4.1, we calculate values of l_c for high and low clay + TOC samples within the velocity-weakening regime. The shear moduli are obtained from laboratory measurements of dynamic elastic properties on the same samples (Chapter 2; Sone & Zoback, 2013a). The effective normal stress for each case is obtained from the Mohr-circle analysis in Fig. 4.16. Based on these values, relatively weak, high clay + TOC facies show a minimum patch size of ~24 m, about a factor of 2 less than for strong, low clay + TOC facies. In Chapter 9, we discuss the relationship between slip, patch size and moment magnitude in the context of seismic observations of stress drops. Based on these scaling relations, magnitude −2 microearthquakes, which are typical of hydraulic stimulation in unconventional reservoirs, have a fault patch size of ~1 m and a slip of 0.1 mm (Fig. 9.20). There are several possibilities for explaining the order of magnitude difference between value of l_c and patch size estimate based on earthquake scaling relations. First, if effective normal stress is much higher than estimated, the value of l_c will decrease; however, this effect may be counteracted if the rate-state friction properties are dependent on normal stress (e.g., Fig. 4.14). Second, the elastic stiffness of the volume of rock containing a fault may be less than the dynamic elastic stiffness measured on intact core samples. Given that dynamic fault slip creates off-fault damage through inelastic deformation (Johri et al., 2014a), it is plausible that the stiffness of the surrounding is significantly diminished, which would reduce the value of l_c. Integrating observations of natural fractures and faults (Johri et al., 2014b) into m-scale elastic models may help to quantify this effect.

Seismic and Aseismic Slip – In addition to induced seismic slip (microearthquakes), there may also be a significant aseismic slip occurring during hydraulic stimulation. Although initial observations of slow slip on faults ultimately proved to be signals from distant, regional earthquakes (Das & Zoback, 2013a; Caffagni et al., 2015; Zecevic et al., 2016), several recent studies have identified relatively low frequency, emergent signals during hydraulic stimulation in unconventional basins (Hu et al., 2017; Kumar et al., 2017, 2018). These have been termed "long-period long-duration" (LPLD) events and show similar waveforms to volcanic tremor, which results from the propagation of fluid-filled, tensile cracks (Aki et al., 1977; McNutt, 1992). LPLD events show distinct differences from tremor-like events known to be caused by slow shear slip, and are therefore interpreted as originating from tensile fractures that propagate suddenly under high fluid pressures.

Although there is no direct evidence for induced slow slip on faults during hydraulic stimulation, it is plausible that the surface seismic arrays used to detect LPLD are not sensitive to low enough frequencies to capture slow slip. The fact that production is observed to correlate with the density of natural fractures and not the number of microseismic events (Moos et al., 2011), suggests that there is additional deformation on faults that contributes to enhancing formation permeability, but is not captured by seismic instruments. This idea is reinforced by laboratory observations of elevated fault permeability accompanying slow slip (Scuderi et al., 2017a,b; Chapter 10). Therefore, it is worth reviewing the mechanisms that may control the occurrence of slow slip in unconventional reservoirs. First, clay and organic-rich lithologies (>30 wt%) consistently exhibit velocity-strengthening behavior, which signifies that faults in these facies would only fail via stable sliding. Second, under high fluid pressures (low effective normal stress), unstable slip may be limited by dilatant strengthening, which can cause velocity-weakening materials to fail via stable sliding (Segall, 1995; Segall et al., 2010). Third, as we will discuss in Chapter 12, if a fault is poorly oriented in the current stress field and will only slip under high fluid pressures, the rate of migration of fluid along the fault may modulate the slip rate (Zoback et al., 2012). Further study of laboratory faults and induced slip during hydraulic stimulation is necessary to evaluate if any of these mechanisms actually cause slow shear slip on faults. Ultimately, determining how the energy of stimulation is dissipated by the spectrum of deformation associated with pre-existing fractures and faults will contribute to a more complete understanding of the physical processes responsible for production.

5 Pore Networks and Pore Fluids

The pore networks of the rock matrix, and the pore fluids contained within it, determine the flow properties of unconventional reservoir rocks. Before examining the mechanisms and timescales of flow in Chapter 6, we will explore the characteristics of matrix pore networks, as well as the occurrence of *in situ* pore fluids and the flow properties of multiphase systems.

We first review the length scales relevant to pore networks and pore fluids in unconventional reservoirs, and discuss the sources of porosity in the rock matrix. We then address the issue of how to characterize and quantify matrix porosity and pore characteristics (size, shape and orientation). Through a detailed review of characterization methods, we explore how different methods may be validated by each other and/or combined for more complete coverage of length scales. The following section addresses how the characteristics of pore networks vary with matrix composition and fabric. This includes observations of connectivity and tortuosity, which indicate how different pore types link to form flow pathways. Finally, we tackle the issue of upscaling: how to use representative micro- to nano-scale observations of porosity and microstructure to predict the properties of pore networks at the scale of well-logs and beyond.

In order to understand flow behavior within pore networks, it is also essential to understand the distribution and composition of *in situ* pore fluids, as well as the impacts of multiphase effects. We will discuss the occurrence of various pore fluids (natural gas, condensates, oil and brine) in major unconventional basins. The following section focuses on observations of wettability and capillary pressure to establish a physical basis for understanding the relative permeability of water and hydrocarbons in unconventional reservoir rocks. Finally, we will discuss the implications of multiphase effects on the flow properties of pore networks during hydraulic stimulation and production.

Matrix Porosity

Figure 5.1 shows a conceptual model for the matrix of unconventional reservoir rocks in terms of the rock frame and pore space. As discussed in Chapter 2, the rock frame comprises non-clay (clastic) minerals, mainly quartz and carbonates, organic matter and clay minerals. The total pore space, contained within the various solid components, may be filled by water and/or hydrocarbons depending on the specific reservoir conditions.

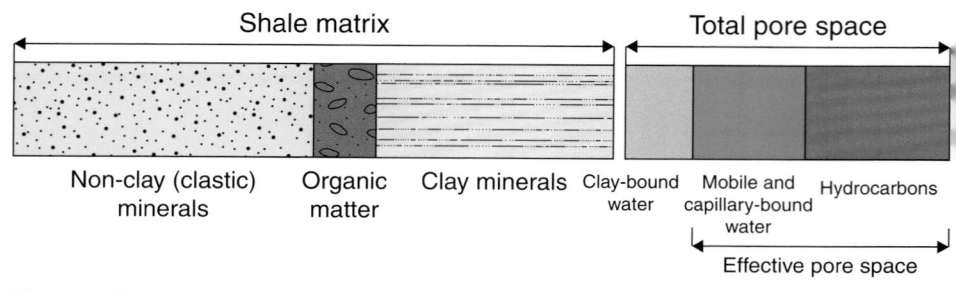

Figure 5.1 Conceptual model for the matrix and pore space of unconventional reservoir rocks. From Passey et al. (2010).

Some fraction of the water may be bound within clay intraparticle spaces or by capillary pressure in very small pores.

Length Scales – To understand how the different components of the matrix contribute to the formation of pore networks, it is critical to place them in the context of scale. Figure 5.2 provides a comprehensive view of the pore fluids, microstructural features and relevant characterization techniques from the Å- to cm-scale. Above the cm-scale, the continuity of the rock matrix may be disrupted by lithological variations (Chapters 1 and 2) and/or the presence of natural fractures (Chapter 7).

For a typical magnitude −2 microearthquake, the fault patch size is approximately ~1 m and the slip is ~0.1 mm (Fig. 9.20). This represents a reasonable upper bound in scale for discussing the characteristics of the surrounding rock matrix. From the cm- to mm-scale, the rock matrix is composed of sedimentary layers with variable composition (lithofacies), which define the bedding plane and represent the coarsest source of anisotropy (see Chapters 1 and 2 for details). Most recovered core samples and core plugs used in laboratory testing (Chapters 2–6) fall within this scale range, enabling a broad range of physical and chemical measurements by well logs, X-ray computed tomography (CT) and X-ray florescence (XRF). Although the resolution of these techniques is far coarser than the scale of the pore networks, characterization of the variations in lithofacies and scales of heterogeneity is important for understanding how different types of pore networks are distributed on the mm- to cm-scale.

From the mm- to μm-scale, the rock matrix also exhibits compositional layering (microfacies; e.g., Fig. 2.9), as well as bedding-parallel (sub-horizontal) microfractures, which may be open or filled with organic matter or mineral cement. This essentially represents the "grain-scale" for unconventional reservoir rocks since clastic grain sizes typically range from ~10 to 100 μm, organic matter deposits range from ~0.1 to 100 μm and clay particle sizes range from ~0.1 to 4 μm. As discussed in Chapter 2, clay in the rock matrix often manifests as aggregates of multiple clay grains (Fig. 2.3). The preferred alignment of clay aggregates, organic matter and microfractures represent the major sources of anisotropy at the grain-scale. A broad range of techniques are available to characterize physical and chemical properties within this range. Physical measurements such as nano-indentation (Ahmadov, 2011) and scratch tests (Akono &

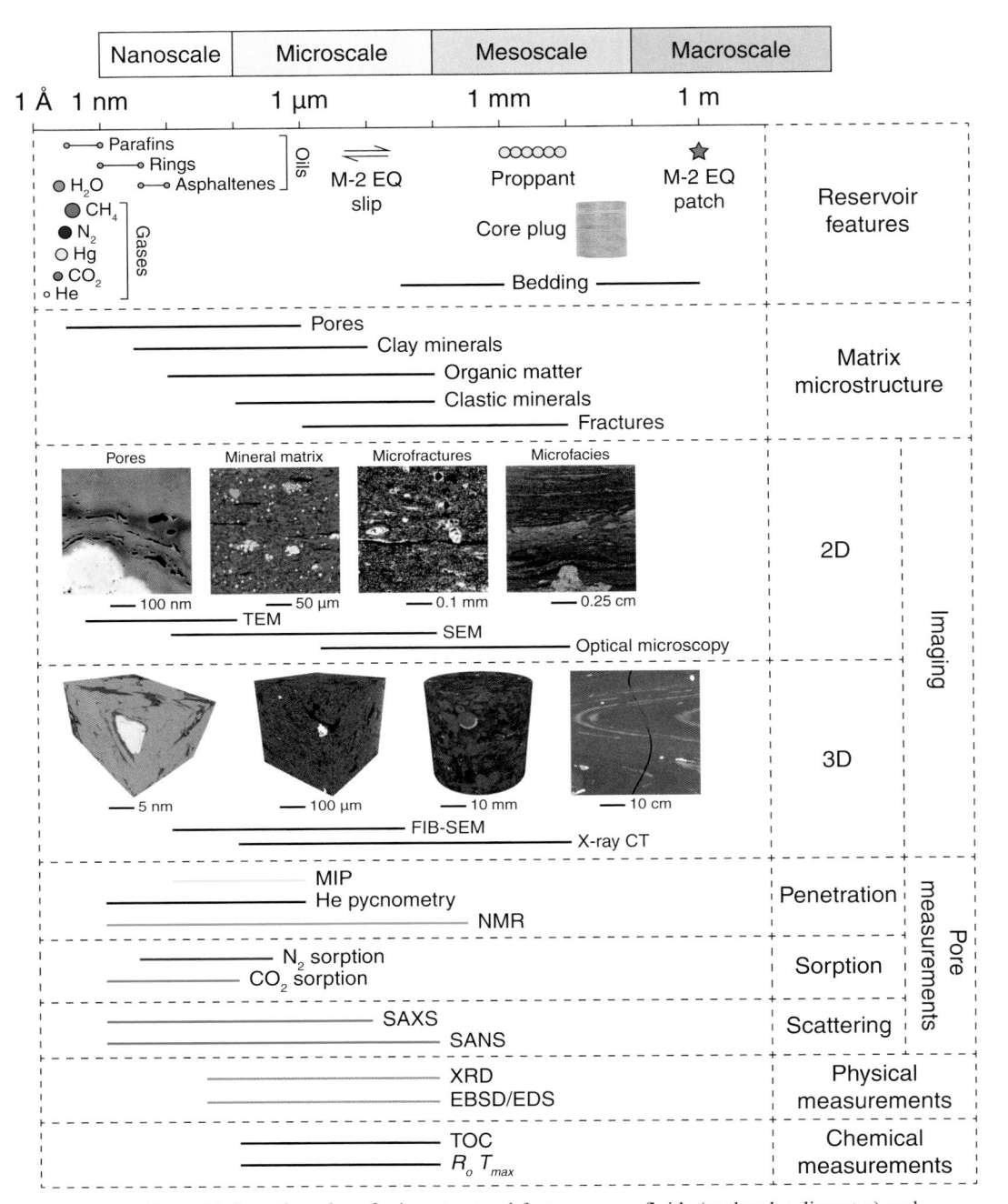

Figure 5.2 Length scales of microstructural features, pore fluids (molecular diameter) and characterization techniques (application range, see Table 5.1 on page 130 for details). Adapted from Nelson (2009) and Ma et al. (2017).

Kabir, 2016) provide micro-scale mechanical properties. Chemical measurements such as X-ray diffraction (XRD) and total organic content (TOC) quantify matrix mineralogy and organic content. Imaging techniques such as optical petrography, scanning electron microscopy (SEM), X-ray micro-computed tomography (μCT) and focused ion beam tomography (FIB-SEM) can visualize matrix microstructure and mineralogy (Table 5.1). It is important to understand that for these imaging techniques, the ratio of sample size to resolution is ~1,000:1; for example, μCT performed on 1 mm samples may yield a resolution of 1 μm, while FIB-SEM performed on 10 μm regions may yield a resolution of ~10 nm. Of course, this depends on the specific instrument/technique, but this heuristic relationship is useful for understanding how imaging techniques may be combined for multi-scale analyses (Ma et al., 2017).

At the μm- to nm-scale are pores in the matrix. As we will discuss in detail below, porosity can be distributed throughout interparticle and/or intraparticle spaces, organic matter and microfractures, with the relative proportions depending on matrix mineralogy, maturity and microstructure. The sizes of larger liquid hydrocarbons, e.g., asphaltenes and ring structures, overlap with the scale of pores in the matrix which underscores the potential for trapping with the rock matrix itself and explains why many unconventional formations are both hydrocarbon sources and reservoirs. Smaller molecules such as methane, water and CO_2 fall between the nm- and Å-scale. Also note the relative molecular sizes of N_2, Hg and He, which are often used for characterizing pore networks and flow properties. From the μm- to sub nm-scale, pores can be quantified by techniques relying on imaging, scattering and flow and adsorption. We will discuss the pro/cons and specific applications of these techniques later in the chapter.

Having surveyed the length scales of the matrix, the production pathway for a fluid or gas molecule begins in nm-scale pores and ends on a m-scale stimulated fault, which represents nearly 10 orders of magnitude in scale.

Sources of Porosity – As discussed above, the μm- to nm-scale porosity in the matrix is contained in three major pore types: microfractures (microcracks), interparticle and intraparticle pores and pores within organic matter. For a comprehensive review of the spectrum of pores types in shale gas reservoir rocks, see Loucks et al. (2012).

At the scale of core plug samples, various types of fractures are visible, so it is important to distinguish *in situ* structures from those induced during core recovery and sample preparation. Figure 5.3 shows an example of induced and *in situ* microfractures in a sample of Eagle Ford shale. In Fig. 5.3a we see a fracture trending sub-parallel to bedding (horizontal) near the top of the frame. Figure 5.3c shows the same type of feature visualized in 3D using μCT. This feature extends the full length of the cm-scale sample and appears filled with epoxy in plane polarized light. For these reasons, we interpret this fracture as induced and do not consider it part of the matrix porosity (although it may inevitably contribute in laboratory measurements of mechanical and flow properties). In the same frame, examining the grain-scale of the matrix reveals bedding-parallel microfractures that are filled with organic matter (Fig. 5.3b). Along with being fully contained in the matrix, the presence of organic matter suggests that it was formed *in situ*. The formation of these microfractures has been attributed to excess

Pore Networks and Pore Fluids   119

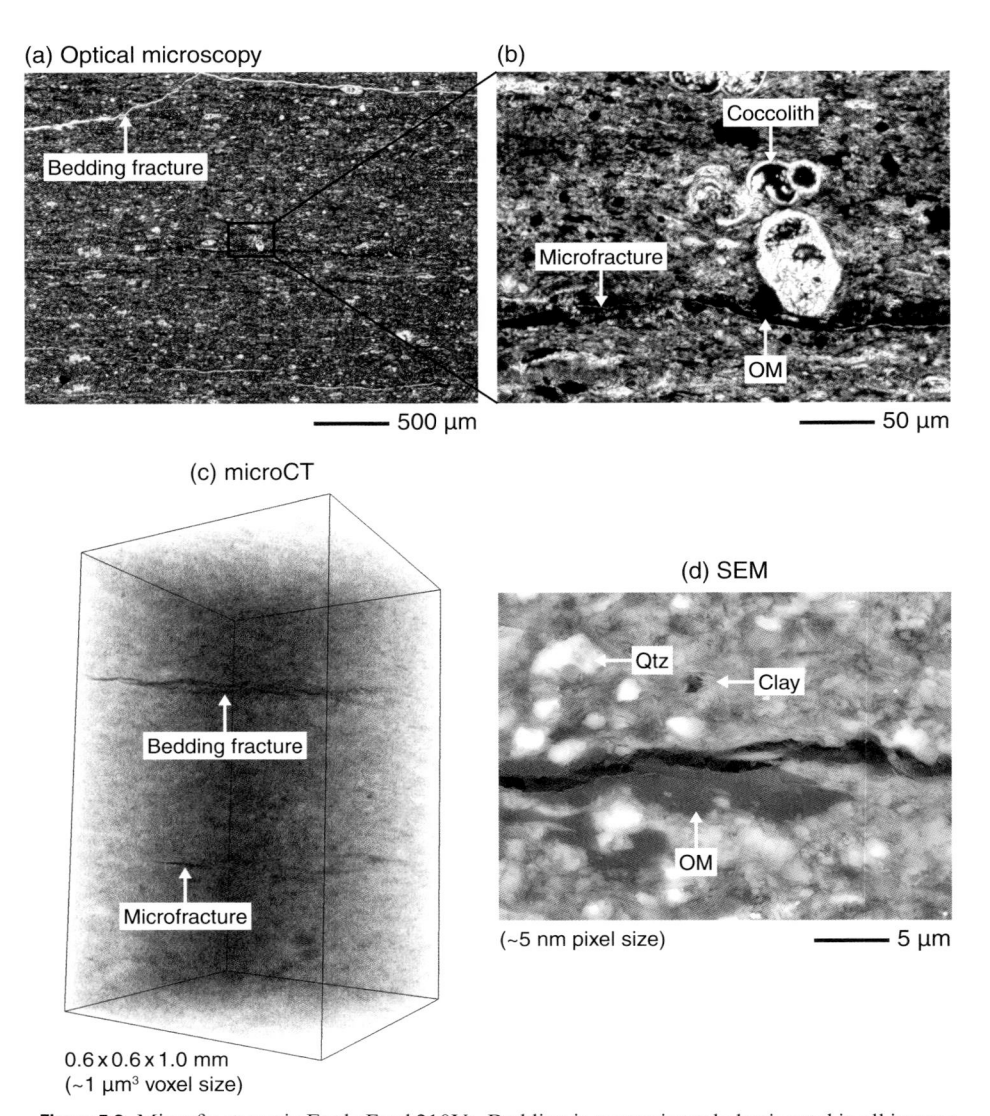

Figure 5.3 Microfractures in Eagle Ford 210Va. Bedding is approximately horizontal in all images. (a,b) Petrographic imaging reveals mm-scale induced fractures and μm-scale microfractures filled with organic matter (OM). Plane polarized light. (c) Segmented porosity from μCT imaging of a mm-scale core shows microfractures span ~10–100 μm while induced fractures are relatively continuous on the mm-scale. (d) A backscattered SEM image shows that microfractures can be composed of interlinked *en echelon* features bordered with organic matter and packets of clay minerals aligned parallel to the fracture. Kohli (Unpublished).

pore pressure generated by the transformation of hydrocarbons during maturation, which essentially hydraulically fractures the matrix (Vernik, 1994; Yang & Mavko, 2018). Figure 5.3d shows that the microfractures may be composed of individual *en enchelon* segments bordered by bedding-parallel clay aggregates and deposits of organic

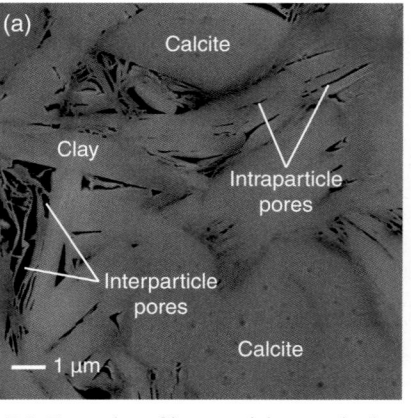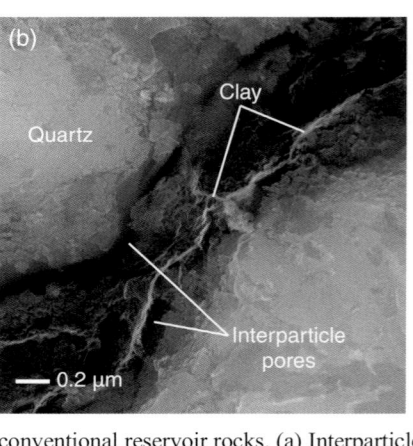

Figure 5.4 Examples of interparticle porosity in unconventional reservoir rocks. (a) Interparticle pores between clay aggregates and adjacent to larger calcite grains, Haynesville shale (Klaver et al., 2015). *Reprinted with permission from Elsevier whose permission is required for further use.* (b) Interparticle pores at the contacts of clay and quartz grains, Barnett 25 Ha. Kohli (Unpublished).

matter. In most cases, microfractures appear to localize within the soft components of matrix and travel around larger clastic grains (Fig. 5.3b,d). This is expected given the inherent low strength of organic matter and clay compared to clastic phases (Chapter 3).

Interparticle and intraparticle pores are distributed throughout the matrix between and within soft and stiff components. Interparticle pores form by sedimentation and compaction, which results in a wide variety of pore shapes and sizes depending on the local mineralogy and microstructure (Fig. 5.4). Contacts between relatively coarse-grained clastic components tend to form angular micropores (up to several μm) with no preferred orientation, while contacts between clay aggregates often form smaller (~100–500 nm) silt-shaped pores preferentially elongated parallel to the local clay fabric (Fig. 2.3b). Clay–clastic contacts may also form elongated (silt-shaped pores), particularly when the basal plane of the clay aggregates is parallel to surface of the adjacent grains (e.g., Fig. 5.4b). This is also commonly observed at the boundaries of large deposits of organic matter (Loucks et al., 2012). Primary interparticle porosity formed during diagenesis may be reduced by hydrocarbon migration and mineral cementation or may be increased by mineral dissolution. Thermal maturation mobilizes hydrocarbons, which may migrate and become trapped in existing interparticle spaces in the matrix. Mineral cementation results from dissolution at stress-concentrated grain contacts and precipitation in interparticle spaces during compaction. The relative proportions of cementation and dissolution will be determined by the specific depositional and fluid chemistry conditions.

Intraparticle pores may form during diagenesis by various mechanisms including crystallization, mineral phase transformations and dissolution/precipitation processes. Figure 5.5 shows examples of intraparticle pores types from several major basins. Intercrystalline pores within pyrite framboids are quite common, but are often filled with clays or organic matter (Fig. 5.5a). Clay minerals and micas form elongated and parallel intraparticle nano-pores (10–100 nm) between aggregates (cleavage-plane pores), in particular within illite–smectite interlayer structures (Fig. 2.3a). Carbonate minerals and

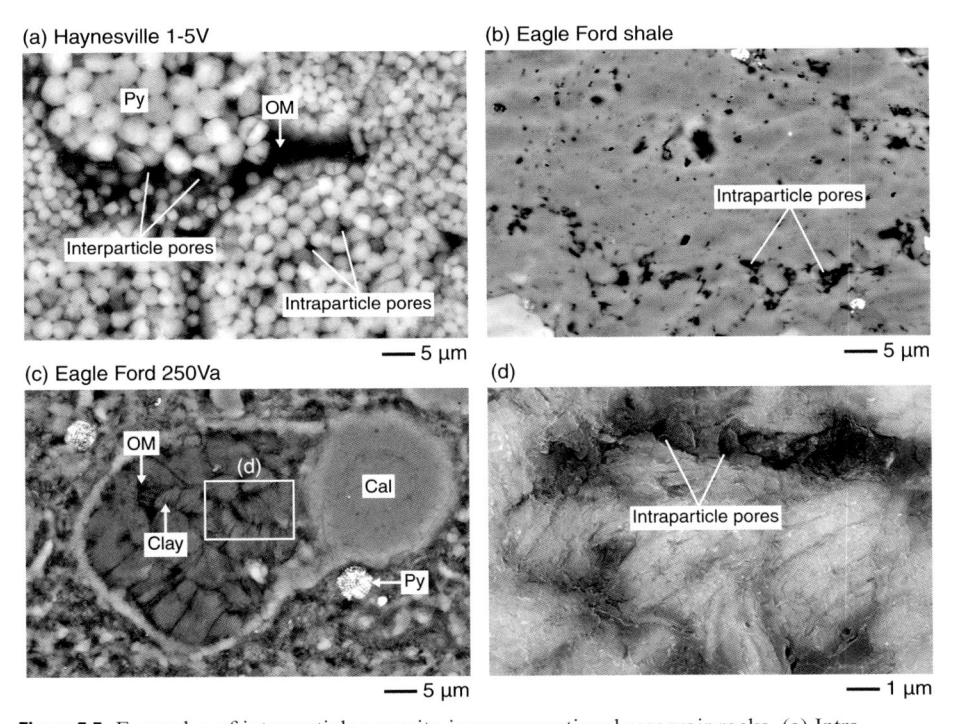

Figure 5.5 Examples of intraparticle porosity in unconventional reservoir rocks. (a) Intra-crystalline pores within pyrite (Py) framboids and organic matter (OM) infilling interparticle spaces, Haynesville 1–5. (b) Intraparticle pitting in dolomite, Eagle Ford shale (Rassouli & Zoback, 2018). (c) Intraparticle pores in clay and organic matter within a multi-chambered foraminifera (globigerinid), Eagle Ford 250Va.

fossils show intraparticle pores in many forms including pitting from the loss of mineral fluid inclusions (Fig. 5.5b), body cavity pores in fossil grains (Fig. 5.5c) and moldic pores from dissolution of crystal or fossil grains (Schieber, 2010). Intraparticle pores may be more resistant to compaction than interparticle pores due to their generally smaller size and tendency to occur in relatively stiff phases, but are also likely to be affected by hydrocarbon migration and mineral cementation.

Organic matter pores are intraparticle pores found within thermally mature ($R_o > 0.6\%$) organic matter, which form via the generation and migration of relatively mobile hydrocarbons (i.e., bitumen, oil, gas) (Dow, 1977). Organic matter pores are generally the smallest pore type in the matrix, ranging from ~0.5 nm to several µm (Fig. 5.6; Loucks et al., 2009; Sondergeld et al., 2010b; Curtis et al., 2012a; King et al., 2015). Depending on chemical composition and maturity, individual deposits of organic matter can range in porosity from ~0–50% (Loucks et al., 2009; Curtis et al., 2011; Curtis et al., 2012b; Loucks et al., 2012; Milliken et al., 2013). Porous organic matter often exhbits a spongy texture with elliptical pores of varying sizes and no apparent preferred orientation (Fig. 5.6a), but may also exhibit a preferred orientation when distributed in low aspect ratio lenses or lamellae (Fig. 5.6b). In some cases, organic matter may also form elongated pores orientated parallel to the boundaries of contacts with clastic grains or clays (shrinkage pores) (Fig. 5.6c). Organic porosity may also vary significantly within the same microfacies. Using FIB-SEM imaging, Curtis et al. (2012b) observed adjacent deposits of porous and non-porous organic matter in the Woodford shale, which may point to the role of chemical composition in determining pore structure. Loucks & Reed (2014) explored criteria for differentiating primary and migrated organic matter and found that migrated organic matter, composed of once mobile hydrocarbons (i.e., bitumen, pyrobitumen), follows the distribution of the pre-existing interparticle pores, creating a more continuous, anisotropic organic pore network than isolated, primary deposits (e.g., Fig. 5.6a,b). The characteristics of organic matter porosity may also vary significantly between different lithofacies within the same formation. Milliken et al. (2013) observe a strong negative correlation of porosity and average pore size with increasing TOC in both mature and post-mature facies in the Marcellus shale. This may reflect the tendency for more complete hydrocarbon expulsion (leading to pore collapse) in lithofacies with interconnected networks of organic matter, which also results

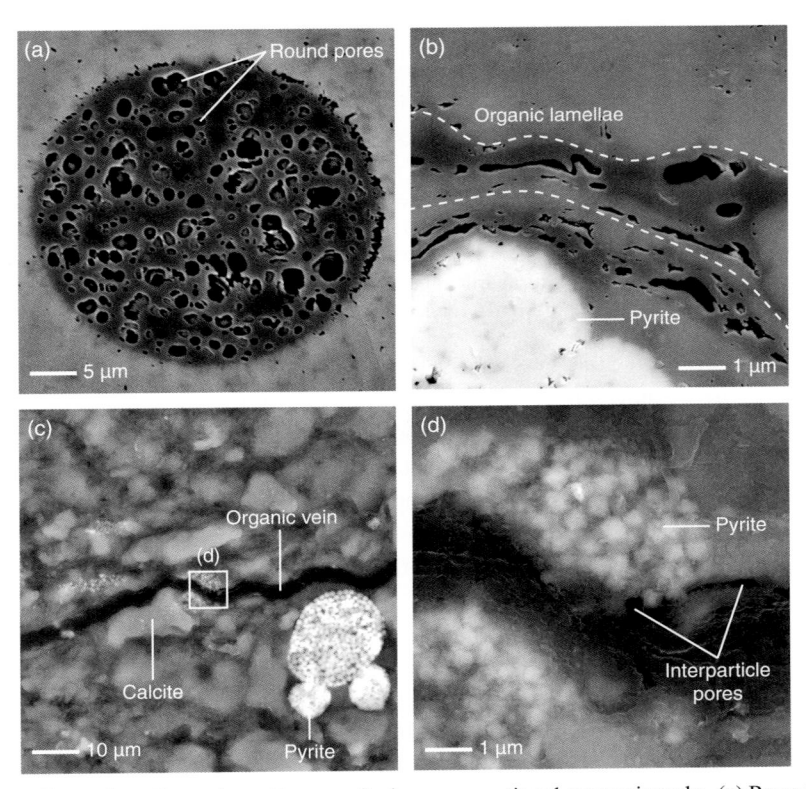

Figure 5.6 Examples of organic matter porosity in unconventional reservoir rocks. (a) Round pores up to ~3 μm in diameter within an isolated deposit of organic matter, Haynesville shale (Klaver et al., 2015). *Reprinted with permission from Elsevier whose permission is required for further use.* (b) Oriented, elongate pores within a thin lens of organic matter, Haynesville shale (Klaver et al., 2015). *Reprinted with permission from Elsevier.* (c) Pores at the boundaries of a bedding parallel vein of organic matter ~2–3 μm thick, Haynesville 1–5. Kohli (Unpublished).

in a more compliant, anisotropic matrix (Chapter 2). In contrast, organic-poor lithofacies with rigid matrix supported by clastic phases may resist compaction and hydrocarbon expulsion, preserving larger intraparticle organic pores.

Quantifying Matrix Porosity and Pore Size – Having covered the relevant length scales and pore types in unconventional reservoir rocks, we are ready to examine the methods for quantitatively characterizing porosity and pore sizes in Fig. 5.2. The inherently challenging nature of the rock matrix (µm- to nm-scale pores, low permeability, complex wettability) demands a battery of different characterization techniques that fall into three main groups based on the mechanism of measurement: imaging, scattering and flow and adsorption. For a comprehensive review of these techniques, see Anovitz & Cole (2015). As we discuss results from different methods, it is important not to focus on which value of porosity and pore size is correct (accurate), but rather what physical properties of the pore network are captured by each method and how different methods can be combined to address specific questions related to the rock matrix. Table 5.1 provides an overview of the methods for quantifying porosity and pore characteristics.

As seen in Fig. 5.2, a wide range of imaging techniques can be applied to characterize the rock matrix from the cm- to sub-nm scale. For a comprehensive review of imaging techniques, see Ma et al. (2017). From the cm- to µm-scale, optical petrography provides a 2D view of mineral structures and rock fabric (Figs. 2.8, 5.3a,b), and X-ray CT provides a 3D view of the distribution of different density matrix components (Fig. 5.3c). Although X-ray CT is capable of quantifying the distribution of porosity through the use of high contrast fluids (Vega et al., 2014; Aljamaan et al., 2017; Peng & Xiao, 2017), quantifying matrix pore sizes via imaging requires the use of higher resolution electron microscopy. In 2D, SEM combined with broad ion beam polishing enables characterization of porosity and mineralogy over cm-scale areas at the nm-scale (Klaver et al., 2015). Image analysis applied to SEM images can quantify porosity, pore size/shape, as well as the relative fractions of porosity in each component (L.Ma et al., 2018). 2D images of ion milled surface will sample a combination of pore bodies and throats. Although beyond the scope of this book, it is important to note that pore characteristics determined from electron microscope and CT images are sensitive to the details of thresholding and segmentation. Serial milling and imaging techniques such as FIB-SEM provide a 3D view of pore networks at the nm-scale, enabling quantification of pore volumes, geometry and connectivity (Curtis et al., 2012a, Ma et al., 2016). Due to the time intensive nature of ion milling at this scale, FIB-SEM imaging has been limited to volumes ~10 × 10 × 10 µm. Recently, the application of Xe ion plasma FIB permits mass removal rates at least 60 times greater than conventional Ga ion FIB systems with comparable or less damage, which extends the image volume by approximately 20 times (Ma et al., 2018). Figure 5.7 shows an example of a FIB-SEM dataset from the Horn River shale (Curtis et al., 2012a). The reconstructed sample volume reveals an interconnected network of organic matter, which contains the majority of the porosity (Fig. 5.7a). By fitting a sphere-filling model to the segmented porosity, Curtis et al. calculate the pore size and pore volume distribution (Fig. 5.7b,c). While pores <10 nm dominate in number, larger pores ~100 nm provide the greatest contribution to the pore volume. The pore network shows connectivity through

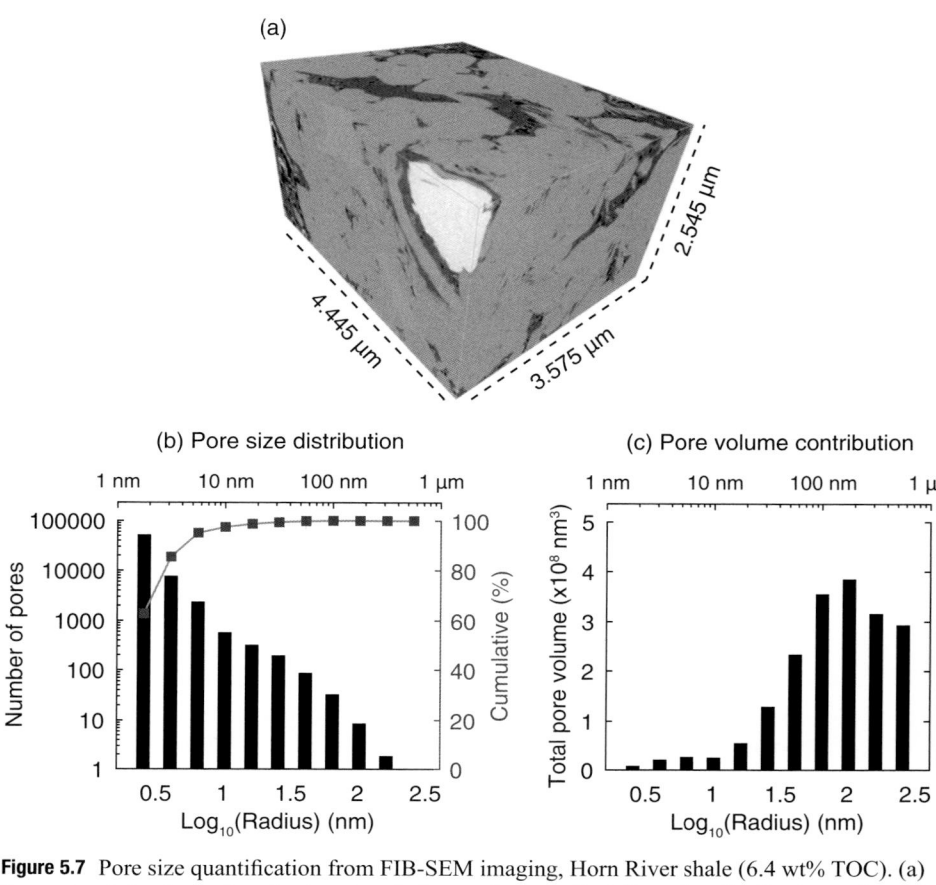

Figure 5.7 Pore size quantification from FIB-SEM imaging, Horn River shale (6.4 wt% TOC). (a) Reconstructed volume from backscattered electron images. (b,c) Pore size distribution estimated from imaging indicate pores <10 nm dominate in number, while larger pores ~100 nm represent most of the pore volume. Adapted from Curtis et al. (2012a). *Reprinted by permission of the American Association of Petroleum Geologists whose permission is required for further use.*

organic matter within the FIB-SEM volume, but it is important to note that the TOC and porosity of volume are 15.6% and 2%, respectively, while those values for the whole sample (measured by helium pycnometry and pyrolysis) are 6.4% and 5.9%. This highlights two of the potential issues with quantifying matrix porosity using FIB-SEM. First, the μm-scale image volume is likely not completely representative of the rock on the mm- to cm-scale due to compositional heterogeneity (Fig. 2.9). In this case, the presence of an interconnected organic network may be overestimated, while the presence of porosity in other features (interparticle and intraparticle pores, microfractures) may be underestimated. Second, the resolution of SEM in these applications is usually ~5–10 nm, which limits quantification of sub-nm pores that are captured by higher resolution imaging techniques (TEM, HIM), as well as scattering and flow and adsorption techniques. While it is unlikely that these sub-resolution pores contribute significant pore volume, they may be important for connectivity within the pore network (Ma et al., 2016).

Small angle X-ray scattering (SAXS) and small angle neutron scattering (SANS), represent another class of techniques for quantifying matrix porosity. Scattering measurements are performed on cm- to mm-scale thin sections (~10–500 μm thick), which allow for sufficient X-ray and/or neutron transmission. The strong contrast in density between the matrix components and air (or fluids) in the pore network produces a characteristic scattering pattern that enables quantification of pore surface area. Similar to FIB-SEM data, sphere-filling models are often applied to extract porosity and pore size distribution (Leu et al., 2016). Depending on the details of the source and detector, SAXS and SANS are sensitive pore sizes ranging from several μm to several Å, allowing for full quantification of matrix porosity and pore size distribution (Anovitz & Cole, 2015). Due to the low flux and limited scattering of neutron beams, SANS is often performed using a relatively large beam of several square mm. In contrast, SAXS may be performed using a more focused X-ray beam, which can be rastered across the sample to create 2D maps of relative porosity (Leu et al., 2016). Scattering techniques also allow for quantification and mapping of pore orientation. In studies of the Opalinus clay, Leu et al. document significant porosity anisotropy defined by elongated pores in the bedding plane. Because scattering is sensitive to pore-matrix density contrast, SAXS and SANS sample pore throats and bodies within both connected and isolated porosity. This is important for comparing results from scattering to those from flow and adsorption, which only encapsulate accessible, connected pores. Figure 5.8 shows pore size distributions for the Opalinus clay from SAXS, SANS and N_2 adsorption plotted in terms of cumulative volume (Leu et al., 2016). In this case, the SAXS measurements from multiple μm-scale sub-volumes show a range of pore distributions, but the average is consistent with results from SANS and N_2 adsorption on mm-scale samples. This suggests that the majority of the total porosity is accessed by nitrogen during the adsorption measurement. Clarkson et al. (2013) obtained similar results on samples from several North American shale gas reservoirs, emphasizing the utility of comparing

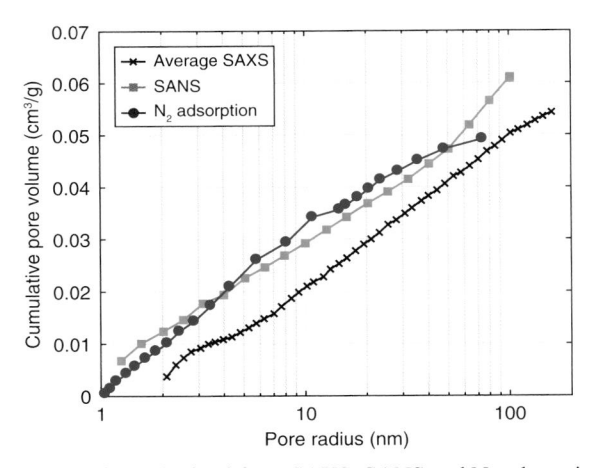

Figure 5.8 Pore size distributions obtained from SAXS, SANS and N_2 adsorption, Opalinus clay (19–5–70 wt% QFP–Carb–Clay + TOC). Adapted from Leu et al. (2016).

scattering and adsorption measurements to distinguish accessible and inaccessible porosity. As discussed by Leu et al., combining FIB-SEM with this comparative analysis also highlights the sources of porosity and their μm-scale connectivity within the matrix.

Methods for quantifying porosity via flow and adsorption include high pressure imbibition, low pressure adsorption and permeability measurements (Chapter 6). We also group nuclear magnetic resonance (NMR) with this class of measurements because although the mechanism of measurement is quite different, it often involves imbibition of fluids into the rock matrix (e.g., Tinni et al., 2017; Reynolds et al., 2018). All methods for quantifying porosity via adsorption are subject to similar experimental challenges. First, the accessibility of the particular fluid to the pore network determines the size range and types of pores that are measured. In addition, the sample size and particle size affects the amount of pore space that is available to access. Larger particles (cores and chips >2 mm) used in mercury intrusion porosimetry (MIP), permeability measurements and NMR have less accessible pore volume than finer particles (~10–100 μm) used in low pressure adsorption, in which internal pore throats may be destroyed, creating additional surface area. Second, the saturation state of the pore space can strongly affect each type of measurement. Many experimental studies use specific procedures for either heating the samples at low temperature to drive off residual water or "cleaning" the samples of residual hydrocarbons with organic solvents such as toluene (Reynolds et al., 2018). Again, the important point here is not to favor any one method, but rather to understand the controls on pore accessibility in a specific sample for a specific application. Lastly, in order to extract information on pore size from each method it is necessary to apply some sort of physical model that assumes pore geometry or distribution. For example, for N_2 adsorption data, there is some debate on the use of the density function theory (DFT) vs. Barrett–Joyner–Halenda (BJH) models. The former assumes pressure-dependent absorption in spherical pore bodies, while the later assumes cylindrical pore throats. Understanding what is being measured (and modeled) is critical for evaluating data from the literature and comparing results from other methods.

It generally accepted that He pycnometry provides the most complete estimate of total accessible porosity (in terms of flow and adsorption methods) because the He molecule has a small kinetic diameter (0.2 nm; Fig. 5.1) and is relatively inert (McPhee et al., 2015). Low pressure adsorption using N_2, CO_2 and CH_4 is commonly used to quantify accessible porosity and pore size distribution in the range ~0.5–500 nm. Results from different gases may be combined to cover a larger range of pore sizes (e.g., Fig. 5.9), but it is important to note that CO_2 and CH_4 adsorption is shown to cause volumetric swelling in clays and organic matter (Chapter 6), which may complicate interpretations of pore accessibility compared to measurements with inert gases. As discussed in detail by Busch et al. (2017), low pressure adsorption measurements record both pore bodies and pore throats. Although a specific model (pore geometry) is applied to determine pore size distribution, similar to scattering methods, the mechanism of measurement is sensitive to all pore surface area. In contrast, pore sizes estimated by MIP reflect a capillary pressure model which depends on interfacial tension and contact angle, and thus represent only pore throats. Depending on the injection pressure, MIP can cover pore sizes ~2 nm–2 μm. During the initial stage of mercury injection, the initial volume uptake in the sample is attributed to filling of surface topography (rugosity) before the capillary pressure required to enter the largest internal pores is reached. Not accounting for the threshold entry pressure may result in an apparent bimodal distribution of nm- and μm-scale pores (e.g., Chalmers et al., 2012b), with the latter likely representing surface features rather than fracture porosity contained within the matrix. Due to differences in sample type (chips vs. crushed), pore accessibility and the assumed physical models, it is clear that MIP is not directly comparable with low pressure adsorption. This is evident in Fig. 5.9, which shows that the magnitude and distribution of pore throats from MIP is consistently different than equivalent low pressure adsorption data representing pore throats and bodies.

For flow and adsorption methods in which pore accessibility determines the measurement, it is critical to understand the "starting point" of the rock matrix in terms of residual water or hydrocarbon saturations. Several recent studies have demonstrated the value of using NMR measurements on different states to assess the presence of bound

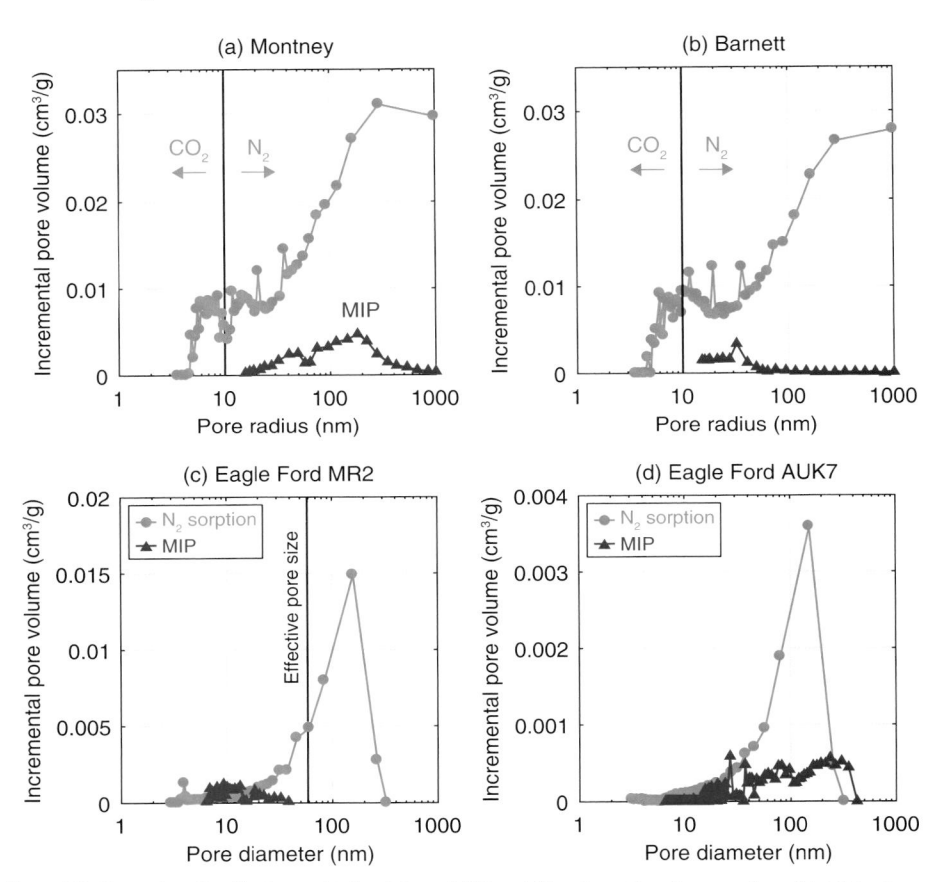

Figure 5.9 Pore size distributions obtained from MIP and N_2 adsorption for samples of (a) Montney shale 37–1–60–1.28 wt% (QFP–Carb–Clay–TOC). (b) Barnett shale, 78–3–15–4.11 wt%. (Clarkson et al., 2013). *Reprinted with permission from Elsevier whose permission is required for further use.* (c) Eagle Ford MR2, 12–72–12–2.4 wt%. The black line represents the effective pore size estimate from a permeability measurement. (d) Eagle Ford AUK7 9–85–3–3 wt%. Adapted from Al Alalli (2018).

and mobile fluids, as well as the characteristics of different wettability pore systems (Tinni et al., 2017; Reynolds et al., 2018). NMR quantifies the distribution of relaxation times related to the response of hydrogen protons within a weak, oscillating magnetic field. In saturated porous media, the T_2 (transverse) relaxation is attributed to surface relaxivity of the pore-fluid interface, and is considered be directly proportional to porosity (Anovitz & Cole, 2015). Short T_2 times indicate small pores with a large surface area to volume ratio and long T_2 times indicate larger pores with a smaller surface area to volume ratio. The conversion of T_2 distributions to pore sizes involves the assumption of a physical model of pore shape geometry. Given that T_2 times are sensitive to surface relaxivity, the calculated pore sizes represent both pore throats and bodies. In order to

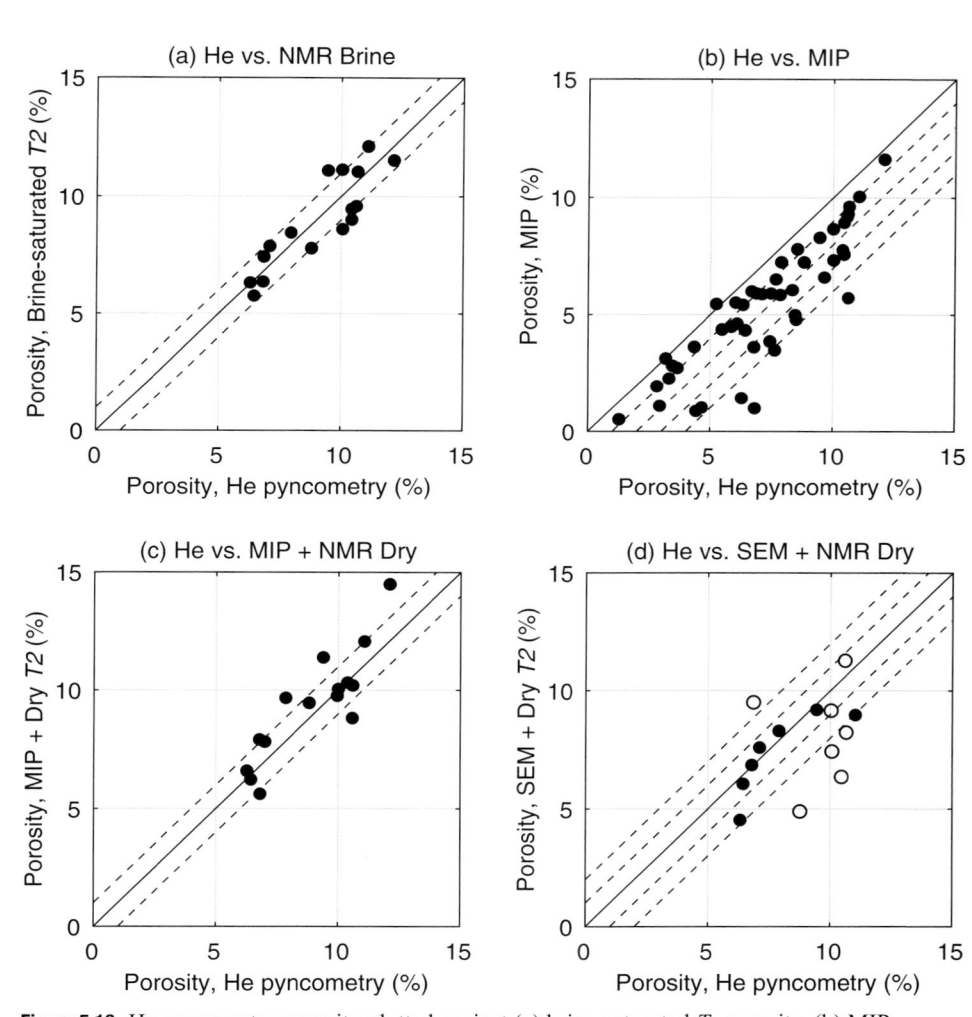

Figure 5.10 He pycnometry porosity plotted against (a) brine-saturated T_2 porosity, (b) MIP porosity, (c) dry T_2 porosity + MIP porosity, (d) dry T_2 porosity + SEM porosity. Unspecified West Texas shales. Black line represents 1:1, dashed lines represent $+/-$ 1% porosity difference. From Reynolds et al. (2018).

establish the "starting point" of the pore network, recent studies perform sequential NMR measurements on "as received," dried, water-saturated and hydrocarbon-saturated states (Tinni et al., 2017; Reynolds et al., 2018). Figure 5.10 shows comparison between NMR derived porosity with other methodologies (Reynolds et al., 2018). In this case, the samples were first cleaned with an organic solvent to remove any residual hydrocarbons. The porosity from He pycnometry agrees with NMR derived porosity on a brine saturated sample, which indicates the fluid accessed the entire pore network (Fig. 5.10a). Porosity from MIP alone is up to 4% less (Fig. 5.10b), but adding the incremental porosity from NMR after drying (to drive off residual water) appears to make up the difference (Fig. 5.10c). This highlights the potential role of residual fluids in limiting access to the pore network during injection or adsorption based measurements. A similar relationship is observed for SEM derived porosity (Fig. 5.10d), which suggests the incremental porosity from drying may correspond to clay intraparticle spaces that were not visible in the "as received" sample. Given the agreement with He pycnometry and the sensitivity to pores not captured by MIP or SEM on "as received" samples, it appears that NMR can provide a comprehensive assessment of the porosity and pore size distribution over the entire range relevant to unconventional reservoir rocks (Reynolds et al., 2018). In addition, NMR uses intact core plugs, which theoretically enables direct comparison with mechanical/flow tests and allows for repeated measurements under different *in situ* conditions (saturation states). At the end of the chapter, we will discuss how NMR can be used to understand the *in situ* saturation state of laboratory samples and the presence of different wettability pore networks.

Combining and Comparing Methodologies – Through our discussion on quantifying porosity and pore characteristics, we have already covered several issues related to combining and comparing results from different techniques, including resolution (accessibility) and pore geometry (pore throats vs. bodies). Table 5.1 summarizes the use case for each method in terms of the applicable range and what pore structures are measured. Ultimately, the goal is not to favor one method over another, but rather to understand how to apply each in the context of the physical mechanism of the specific measurement. For example, if porosity is being measured for the purpose of understanding deformation properties of intact rock, it makes sense to use NMR and/or He pycnometry to measure total porosity and to use imaging to relate pore size to specific components in the matrix. Scattering methods may also be useful for quantifying porosity anisotropy. If porosity is being measured for the purpose of understanding flow properties, it makes sense to use measurements that sample accessible, connected porosity, such as low pressure adsorption and permeability (Chapter 6). Again, imaging methods are useful for visualizing how the flow pathways are distributed within the matrix. NMR may also be used in both cases to assess presence of residual or bound fluids, which may introduce poroelastic effects or capillary blocking.

It is also important to understand to what degree variations in resolution (accessibility) affect the measurement of total porosity. Figure 5.11 shows a comparison of methods for quantifying porosity as a function of resolution for studies on the Opalinus

130 Unconventional Reservoir Geomechanics

Table 5.1 Methods for quantifying matrix porosity and/or pore size distribution.

Method	Range	Structures	Limitations	Use case
Imaging				
Electron microscopy (SEM, TEM)[i]	0.1 nm–1 mm	Pore bodies and throats	2D, no volumetric information, polishing artifacts	Relate pore sizes and shapes to matrix components
FIB-SEM tomography[ii]	5 nm–10 μm	Pore bodies and throats	μm-scale max sample dimension, image artifacts	Pore size distribution, connectivity of microfacies
TEM tomography[iii]	0.1 nm–200 nm	Pore bodies and throats	nm-scale max sample dimension (up to 200 nm thick), image artifacts,	Nanopores within organic matter or between clay minerals
Scattering				
SAXS[iv]	1 nm–1 μm (upper limit depends on sample to detector distance)	Connected and isolated porosity	Pore model dependent, limited availability	Total pore volume, pore size distribution of mm-scale samples
SANS[v]	1 nm–10 μm (upper limit depends on sample to detector distance)	Connected and isolated porosity	Pore model dependent, limited availability	Total pore volume, pore size distribution of cm-scale samples
Flow and adsorption				
He pycnometry [vi]	>0.2 nm (He kinetic diameter)	Accessible porosity	Bulk measurement, crushed sample	Total accessible pore volume
Low pressure adsorption (N_2, CO_2, CH_4)[vii]	1–300 nm	Accessible pore bodies (including surface rugosity) and throats	Particle size and model dependent, crushed sample	Nanoporosity pore size distribution
MIP (thresholded)[viii]	1–300 nm	Accessible pore throats	Particle size and model dependent, mm-scale chips	Assess capillary blocking by residual fluids
MIP[ix]	50 nm–1 μm	Accessible pore bodies (including surface rugosity) and throats	Particle size and model dependent, mm-scale chips	Total sample volume of 2–4 mm particles
NMR[x]	2 nm–2 μm	Accessible porosity	Model dependent, intact cores	Total accessible pore volume and pore size distribution, estimate residual fluids, connectivity and distinguish pore network wettability

Table 5.1 (cont.)

Permeability measurement[xi]	Depends on fluid type and flow pathways	Accessible, permeable pore network	Bulk measurement, model dependent, mm- to cm-scale samples	Estimate of characteristic size of flow paths

[i] Curtis et al. (2012a), Chalmers et al. (2012b), Ma et al. (2016); [ii] Curtis et al. (2012a), Kelly et al. (2016), Ma et al. (2016), Keller & Holzer (2018); [iii] Ma et al. (2018); [iv] Clarkson et al. (2013), Lee et al. (2014), Leu et al. (2016); [v] Anovitz et al. (2015), Leu et al. (2016), Busch et al. (2017); [vi] Chalmers et al. (2012b), Busch et al. (2017); [vii] Chalmers et al. (2012b), Clarkson et al. (2013), Rassouli et al. (2016), Bertier et al. (2016), Busch et al. (2017), Al Alalli (2018); [viii] Rassouli et al. (2016); Ghanbarian & Javadpour (2017), Al Alalli (2018); [ix] Chalmers et al. (2012b); [x] Anovitz & Cole (2015), Tinni et al. (2017);[xi] Heller et al. (2014), Mckernan et al. (2017), Al Alalli (2018).

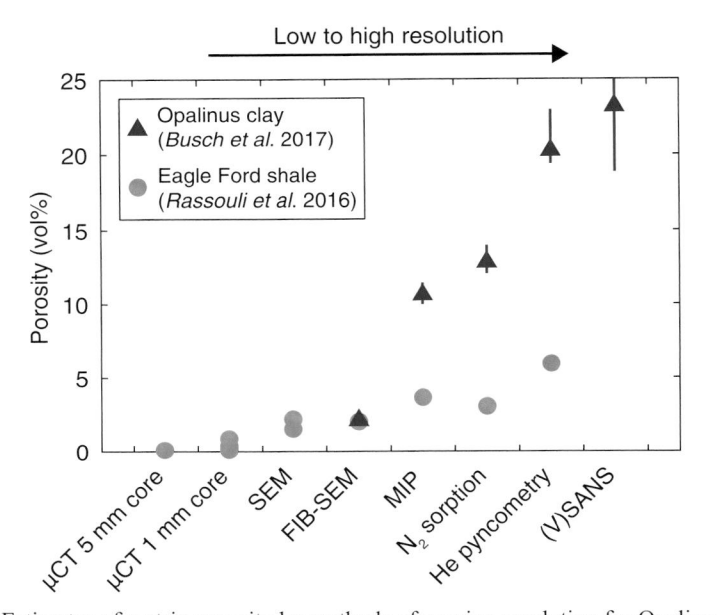

Figure 5.11 Estimates of matrix porosity by methods of varying resolution for Opalinus clay (Busch et al., 2017) and Eagle Ford shale (Rassouli et al., 2016).

clay (Busch et al., 2017) and Eagle Ford shale (Rassouli et al., 2016). Given these data and Fig. 5.10, it is clear that He pycnometry, scattering methods and NMR provide the most comprehensive measurement of total porosity. Low pressure adsorption and MIP are relatively limited in accessibility due to molecular size constraints and/or capillary blocking, and imaging methods, both 2D and 3D, are relatively limited by resolution and field of view. As demonstrated by Rassouli et al. (2016) and imilar multi-scale imaging studies, X-ray CT is not suitable for quantitative characterization of pore size distribution, as the maximum achievable resolution is ~50–100 nm (Ma et al., 2016; Backeberg et al., 2017); however, as we will discuss in the following section, X-ray CT is particularly useful for visualizing the distribution of porosity and porous components, as well as for determining representative elementary volumes for upscaling of pore network properties.

Matrix Pore Networks

Compositional and Fabric Controls – Given the fine-scale heterogeneity (compositional layering) in many unconventional reservoirs (e.g., Fig. 2.9), it is important to understand how the properties of matrix pore networks vary with composition. Figure 5.12 shows the distribution of porosity visualized by μCT in mm-scale samples that represent relatively high and low TOC facies of the Barnett and Eagle Ford shales. It is important to note that since the resolution here is ~1 μm, the segmented porosity likely represents volumes that include both the pores and porous components, and thus, provides only a qualitative view of the distribution of porosity. Although these samples are sourced from core plugs only separated by tens of meters in depth, the variations in composition manifest in completely different pore networks. Organic-rich sample Barnett 18Vb shows a "spongy" texture, in which the pore networks mostly comprise deposits of organic matter hundreds of μm across. In relatively clay-rich and organic-poor sample Barnett 31 Ha, the pore network is dominated by interparticle pores and microfractures, which form an apparent bedding-parallel fabric. The pore network of organic-rich Eagle Ford 174 Ha also shows a bedding-parallel fabric, but is dominated by organic porosity in elongated lenses and within foraminifera fossils (e.g., Fig. 5.5c). Organic-poor sample Eagle Ford 210Va, which also has the lowest clay content, shows a similar pore network to Barnett 31 Ha in terms of pore type (mostly non-organic, interparticle), but differs in terms of fabric. The interparticle pores are not as well oriented, rather structural anisotropy is defined by the presence of a sub-horizontal microfracture network (Fig. 5.3). The variations in pore networks between these samples are obviously not characteristic of the entirety of the Barnett and Eagle Ford, but do illustrate some of the microstructural controls discussed earlier in this chapter and in Chapter 2, as well as highlighting the presence both of inter- and intra-reservoir heterogeneity in unconventional reservoir rocks.

To quantify the distribution of porosity using μCT, several recent studies employ injection or imbibition of high X-ray contrast fluids and gases to assess accessible porosity and/or storage capacity (Vega et al., 2014; Fogden et al., 2015; Aljamaan et al., 2017; Peng & Xiao, 2017). Subtraction of saturated and vacuumed images yields 3D maps of the distribution of the high contrast fluid or gas at voxel (volume pixel) scale resolution. Aljamaan et al. (2017) registered CT maps of porosity with SEM and EDS maps in order to understand how various matrix components and porosity features contribute to core-scale gas storage. An example of this approach is shown in Fig. 5.13. In this sample of Barnett shale, regions with bedding-parallel microfractures show increased gas storage capacity, while high-angle calcite-filled fractures appear relatively inaccessible (Fig. 5.13b). When CO_2 is used, gas storage appears even greater in the relatively accessible regions, which reflects the tendency of CO_2 to adsorb the components in the rock matrix (Chapter 6). Integrated studies such as this have the potential to deconvolve the effects of matrix composition and microstructure on the distribution and accessibility of porosity, which is often not possible when using characterization methods independently.

Pore Networks and Pore Fluids   133

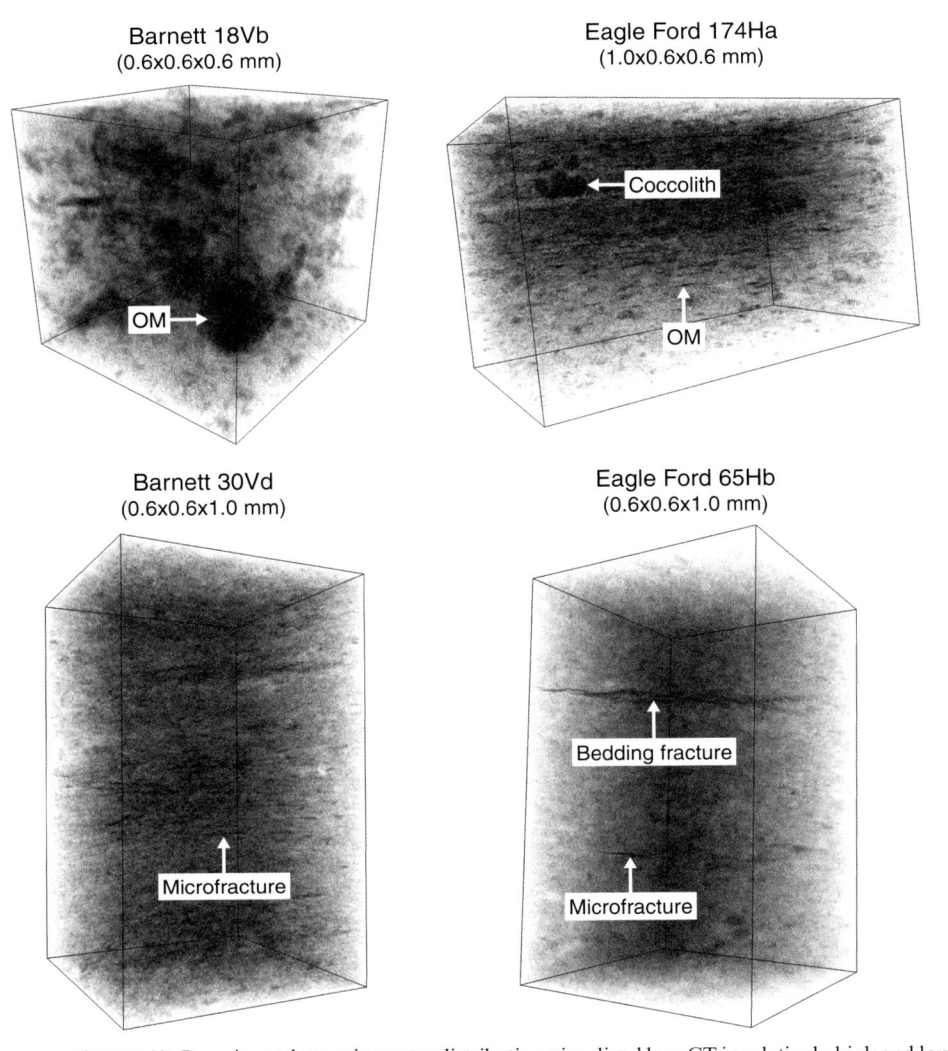

Figure 5.12 Porosity and organic matter distribution visualized by μCT in relatively high and low clay + TOC samples from Barnett and Eagle Ford shales. (a) Barnett 18Vb, (b) Barnett 31 Ha, (c) Eagle Ford 174 Ha, (d) Eagle Ford 210Va. See Table 2.1 for sample compositions. Kohli (Unpublished).

Although imaging methods can provide information on the distribution of porosity within the matrix, it is necessary to use higher resolution methods to quantify the changes in the pore size distribution with composition. Figure 5.14 shows several examples of how pore size distributions determined with N_2 adsorption vary with composition (Al Ismail, 2016). In the Utica samples, all compositions fall in to a narrow range of TOC (1.7–3.3 wt%) and maturity (R_o 0.99–1.18) and show similar pore size distributions with pore sizes ranging from ~2–40 nm (Fig. 5.14a). The major variations between the samples is in the pore volume contained between ~2–4 nm.

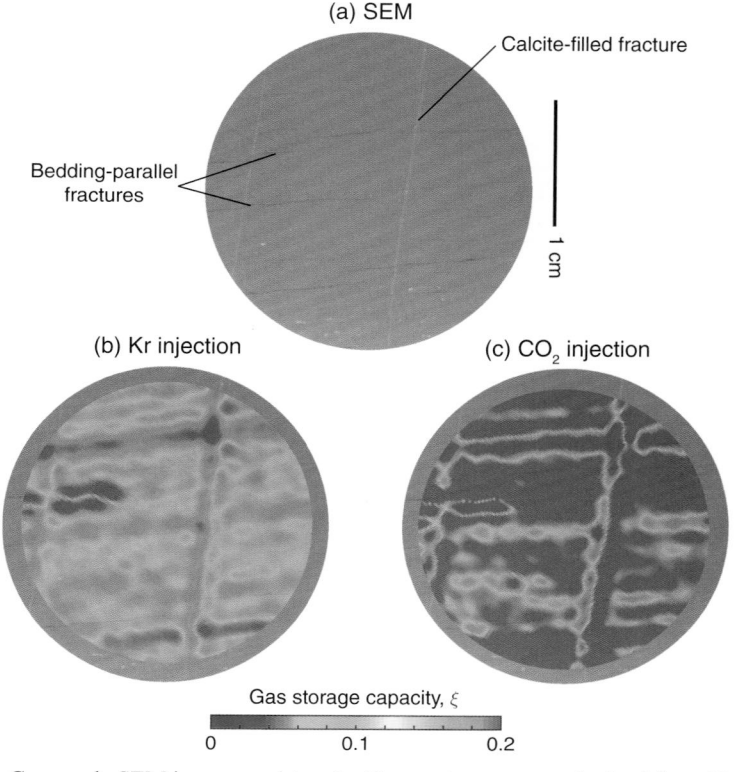

Figure 5.13 Core-scale SEM images registered with gas storage maps obtained from X-ray CT imaging, Barnett 26 Ha. (a) SEM, (b) Krypton gas storage, (c) CO_2 gas storage. From Aljamaan et al. (2017).

Sample U4 (11 wt% clay) shows the most pore volume in this range, nearly double the pore volume in the same range than for clay-rich samples U1 and U2 (49, 41 wt% clay). Sample U3 with intermediate clay content (20 wt%) shows an intermediate value of pore volume, which suggests the porosity contained in pores ~2–4 nm is associated with the replacement of clay by clastic minerals in the matrix, in this case carbonates. In the Permian samples, the majority of the pore volume is contained in pores <10 nm (Fig. 5.14b). Although most of the variations in pore volume occur over a narrow range ~2–8 nm, the variations in the shapes of the pore size distributions indicate porosity does not vary with composition. Sample P4 (53 wt% clay) shows the most pore volume in this range, with an apparent bimodal distribution, while sample P2 (5 wt % clay) shows a similar distribution with only a fraction of the pore volume. Since these samples exhibit similar TOC (0.8–1 wt%) and maturity, it appears the porosity in pores <10 nm is associated with clay minerals, essentially the opposite trend seen in the Utica samples. Sample P1 with intermediate clay content (27 wt%) and much greater TOC (5.6 wt%) shows an intermediate value of pore volume with a narrower, unimodal distribution, which may reflect a combination of organic and clay porosity. All Eagle Ford samples show similar pore size distributions with pore sizes ranging from ~10–300

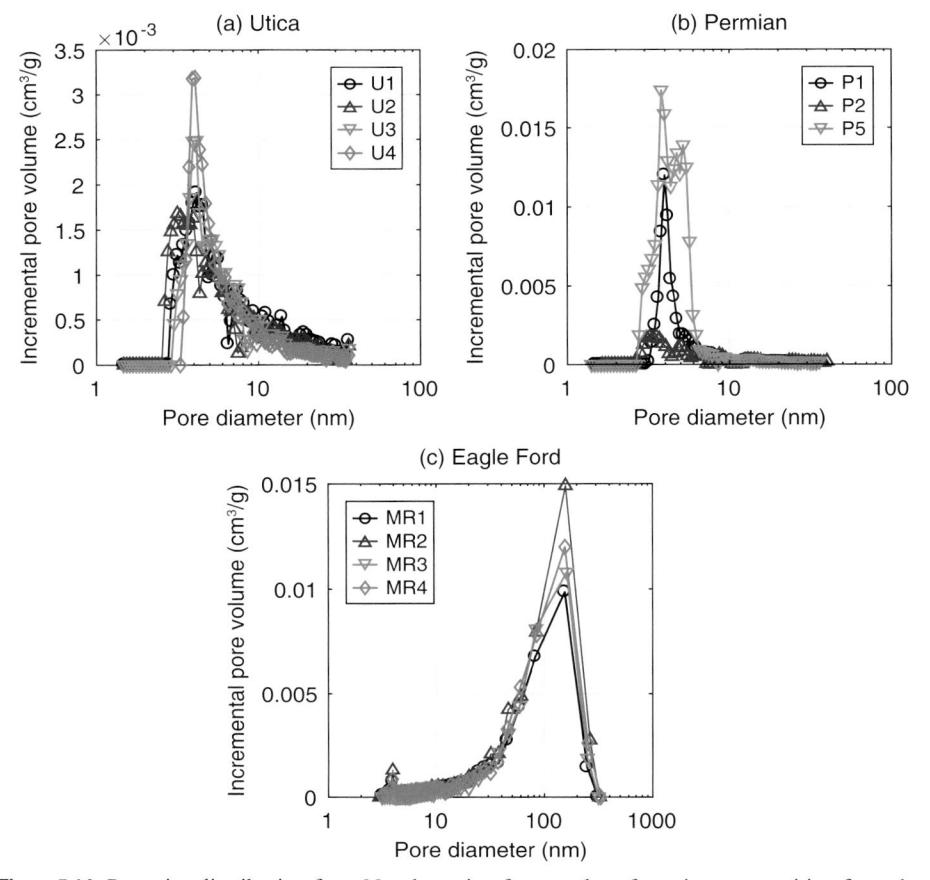

Figure 5.14 Pore size distribution from N_2 adsorption for samples of varying composition from the (a) Utica, (b) Permian and (c) Eagle Ford. Sample compositions are provided in the Table 2.1. From Al Ismail (2016) and Al Alalli (2018).

nm and the major variations in pore volume contained between ~100–200 nm. Although the variations in pore volume appear to reflect a systematic trend, unlike the Utica and Permian samples, there is no obvious correlation with any particular matrix component. This may reflect confounding variations in composition and microstructure that require independent observations of porosity to explain (e.g., imaging, NMR). Although the trends in these datasets are essentially anecdotal (specific to reservoir location), the discussion above provides a conceptual framework for understanding the variations in pore networks with composition between and within unconventional basins.

Upscaling – The variations in nm-scale porosity with composition and the multi-scale, heterogeneous nature of unconventional reservoir rocks make upscaling of pore network properties an immense challenge. Consider a single hydraulic fracture spanning a layered sequence of 100 m. If compositional layering extends to the cm-scale (e.g., Fig. 2.9), there would be 1,000 individual layers with potentially variable pore network properties. Even if core plugs were recovered from the entire interval, it would not be feasible to characterize nm-scale pore structure of each sample using NMR or visualize the pore networks using electron and/or X-ray tomography. The fundamental challenge, particularly for imaging techniques, is that as resolution increases, the field of view decreases. This is evident when considering multi-scale imaging of matrix pore networks. Figure 5.15 illustrates this point by comparing μCT and FIB-SEM images of organic-rich shales with similar TOC and distributions of organic matter. Although the composition and microstructure of these samples is similar, the differences in resolution and field of view illustrate the challenge of upscaling. The composition of the FIB-SEM volume is consistent with the μCT volume in μm-scale, organic-rich regions, but cannot capture the variations on the mm-scale (Fig. 5.15a,b). The same is true when comparing the segmented pore structures (Fig. 5.15c,d). In this case, the loss of resolution with increasing field of view is even more important, since μCT at scale only resolves porosity >1 μm. While it is clear that μCT is not appropriate for quantifying pore size distribution, it is particularly useful for resolving qualitative differences in pore structures between lithofacies (Fig. 5.12),

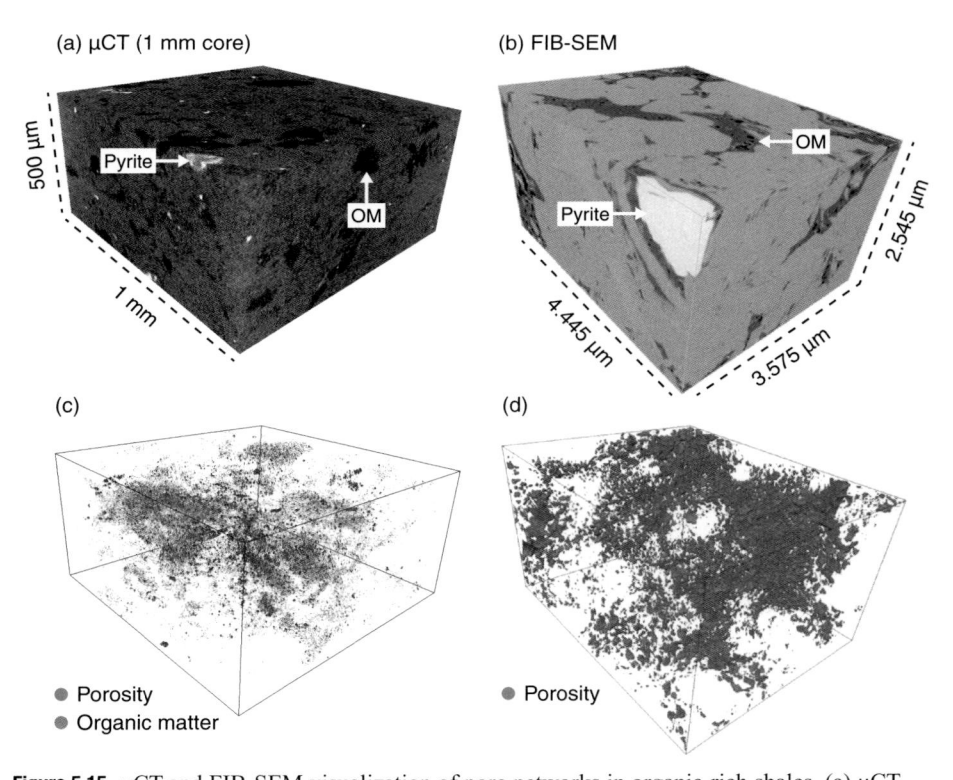

Figure 5.15 μCT and FIB-SEM visualization of pore networks in organic-rich shales. (a) μCT volume of Barnett 18Vb (20.4 vol% TOC). (b) Segmented connected porosity and organic matter. Kohli (Unpublished). (c) FIB-SEM volume of Horn River sample (15.6 vol% TOC). (d) Segmented porosity. Curtis et al. (2012a). *Reprinted by permission of the American Association of Petroleum Geologists whose permission is required for further use.*

and quantifying the distribution and connectivity of potentially porous matrix components, such as organic matter (Fig. 5.15c) and microfractures (Fig. 5.13). Therefore, the use of correlative imaging techniques may be one way to approach the challenges presented by upscaling (Peng et al., 2015; Keller & Holzer, 2018; Ma et al., 2018).

A critical step in understanding how to apply multi-scale imaging is to determine the representative elementary volume (REV), or the minimum volume that captures the characteristic heterogeneity of the matrix. Ma et al. (2016) performed cm- to nm-scale 3D imaging on the Bowland shale to determine the REV of porosity and mineralogy at various scales. The REV is defined as the size of a sub-volume for which the error with the bulk properties is less than 10%. Using high-resolution FIB-SEM (6.7 nm resolution), Ma et al. determine a REV for porosity of ~9 μm within a sample dimension of 14 μm. For lower resolution, mm-scale μCT volumes (0.5–1 μm resolution), the dimension of the REV is ~400 μm. In each case, the encapsulation of the REV with the field of view implies that the image volumes are representative of the microstructure at their specific scales. This is an important check on the use of multi-scale imaging to upscale matrix properties; however, it is also necessary to cross-check quantitative pore information with independent methods that are sensitive at the nm-scale. Ma et al. compare the pore size distribution from relatively high and low resolution FIB-SEM volumes to the results of N_2 adsorption on the same sample material. Although both imaging datasets are limited in their ability to resolve nm-scale porosity, the higher resolution (smaller field of view) dataset captures the specific bimodal distribution evident in N_2 adsorption data. This type of comparison is important for determining the scale at which the REV is representative of pore structure. With this knowledge, it is possible to use FIB-SEM volumes of different microfacies to inform the larger REVs that represent the variations in composition at larger scales.

Although it is important to understand the scales of heterogeneity in the pore network, ultimately, the goal of upscaling is to understand how nm-scale pore networks give rise to mechanical and flow properties on the log- and basin-scale. Therefore, it is necessary to not only evaluate REVs in terms of structure, but also in terms of mechanical and flow properties. For example, Kelly et al. (2016) performed numerical simulations of permeability on the extracted pore structures of FIB-SEM volumes to compare against core-scale laboratory measurements. FIB-SEM volumes up to 5,000 μm^3 are shown to be not representative of porosity and TOC at the core-scale, consistent with the scale of REVs in similar studies (Ma et al., 2016). Although the simulations record permeability anisotropy at the μm-scale, the permeability values span ~3 orders of magnitude for FIB-SEM volumes taken from the same core sample, illustrating controls of variations in the nm-scale pore network on flow properties (e.g., Fig. 5.7). Based on these results, Kelly et al. conclude that although FIB-SEM is useful for understanding the relationships between pore network properties and flow, it is not appropriate for directly upscaling pore network properties for simulating flow (digital rock physics). Therefore, in order to connect the quantitative characteristics of the pore network to core-scale properties and beyond, it necessary to bridge the gap in scale. Recent studies demonstrate the potential of multi-scale 3D imaging for upscaling flow properties by using FIB-SEM to quantify pore network properties and μCT to quantify the distribution of porous components (Peng et al., 2015; Keller & Holzer 2018). Peng et al. apply this approach to an organic-rich Barnett shale sample and use the distribution of organic matter visualized by μCT to estimate permeability at the mm-scale based on the flow properties of nm-scale FIB-SEM volumes. The upscaled permeability is consistent with core-scale measurements on the same sample material, which suggests this approach may be suitable for defining a REV in terms of flow properties. Keller & Holzer (2018) follow a similar approach in upscaling clay matrix hosted porosity in a sample of Opalinus clay and also find agreement between mm-scale upscaled permeability and core-scale measurements. It is worth noting that both these studies use a computational fluid dynamics approach that only accounts for Darcy–Stokes flow, and disregard any diffusive effects. In Chapter 6, we will discuss pore-scale models that incorporate the physics of diffusion and gas adsorption. Regardless, the apparent success of upscaling flow properties in organic and clay pore networks suggests that multi-scale 3D imaging has the potential to relate variation in pore networks and flow (or mechanical properties) across the various lithofacies present in unconventional reservoirs.

In Situ Pore Fluids

In this section, we will briefly review the types of pore fluids in unconventional reservoirs and discuss the role of multi-phase effects on flow. Due to the low porosity, ultra-low permeability character of the matrix, most core-scale experimental studies have been performed with a single fluid or gas phase to understand the physics of flow in nm-scale pore networks (Chapter 6). However, recent studies of wettability and multi-phase flow highlight the potential effects of immiscible fluids (water and hydrocarbons) on *in situ* flow properties, providing insight into the potential for capillary blocking during stimulation and production.

There are two main types of pore fluids in unconventional reservoirs: water (brine) and hydrocarbons (natural gas, oil and condensates). In Chapter 13, we will review brine compositions from major US unconventional basins in the context of recycling and reuse of produced water from hydraulic fracturing. Hydrocarbons in unconventional reservoirs can range from dry gas to wet gas (gas and condensates) to oil, depending on the organic matter composition and maturation, as well as the current pressure–temperature conditions (Fig. 1.7). Because the matrix acts as both the source and the reservoir for hydrocarbons, natural gas content is strongly correlated with TOC (Fig. 5.16). In the absence of organic matter, the gas content represents the amount of free gas present in the non-organic matrix. Due to the tendency for hydrocarbons to adsorb to organic matter and clays (Chapter 6), the gas content at moderate TOC is distributed between an absorbed and a free phase.

Condensates or natural gas liquids are formed when a liquid phase condenses from natural gas. The occurrence of condensates is controlled by the phase relationships for the specific gas composition and the current pressure–temperature conditions of the reservoir (Fig. 5.17). The phase envelope comprises a bubblepoint line and dewpoint line, which connect at the critical pressure–temperature. Crossing the bubblepoint line represents vaporization of gas bubbles from the liquid and crossing the dewpoint line represents condensation of a liquid from the gas phase. The critical point represents the

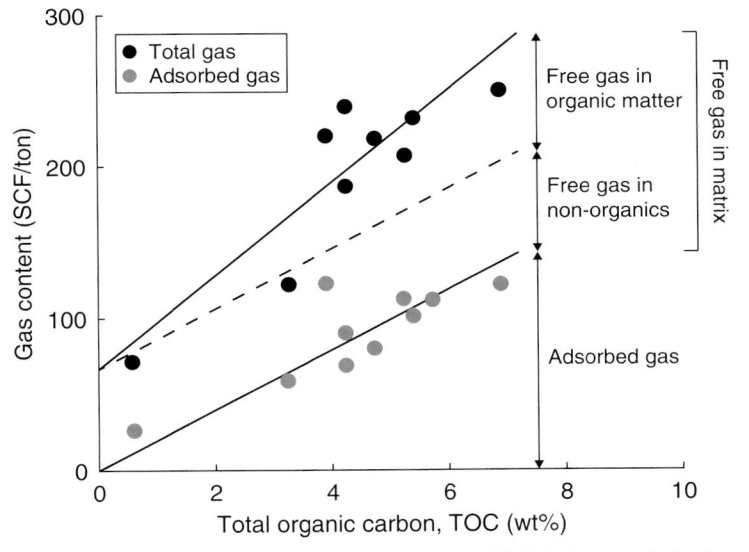

Figure 5.16 Adsorbed and total gas content as a function of wt% TOC, Barnett shale. From Wang & Reed (2009).

pressure–temperature conditions at which the physical properties of the two phases become identical. The formation of condensates in wet gas reservoirs depends on the initial pressure–temperature conditions and how the pressure evolves during operations (injection or depletion). If the initial conditions are beyond the cricondentherm (max dew point temperature), neither injection or depletion (at isothermal conditions) will result in formation of a liquid phase. If the reservoir temperature is between the critical temperature and the cricondentherm, pore pressure depletion may cause the reservoir to enter the two-phase region (red line). As pressure decreases below the dew-point line, condensate saturation will increase to a critical saturation, after which further depletion will cause vaporization of a gas phase. It is also important to note that gas-condensate dynamics may be significantly influenced by confinement with the nm-scale pore networks of unconventional reservoir rocks (Zuo et al., 2018). The increased interactions between pore walls and gas or liquid molecules in small pores can impact the physical properties of gas-condensate systems including the critical pressure–temperature, viscosity and capillary pressure (wettability). In Chapter 6 we will explore the impacts of nm-scale pore networks on the physics of gas flow. Al Ismail (2016) provides a comprehensive review of the physical controls on gas-condensate dynamics in unconventional reservoir rocks.

In reservoirs with lower thermal maturity, the main hydrocarbon phase is oil. Similar to gas-condensate systems, the physical properties of oil depend strongly on organic composition and the current pressure–temperature conditions. For this book, we will not discuss the relationship between the physical properties of oil and flow properties in

Unconventional Reservoir Geomechanics

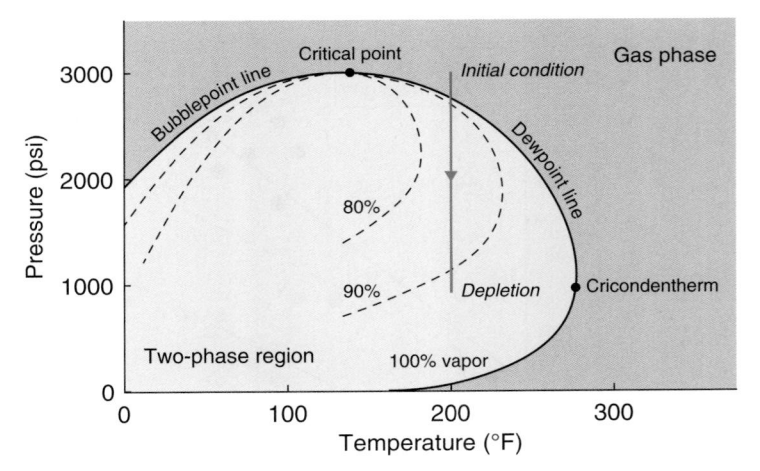

Figure 5.17 Phase diagram for a gas-condensate system (Marcellus shale). The single and two-phase regions are bounded by the bubblepoint and dewpoint lines. Contours of constant phase saturation meet at the critical point. In most gas-condensate reservoirs, the initial condition is above the critical point. Pore pressure depletion causes a liquid phase to condense. Below a certain pressure, the liquid phase begins to evaporate. Adapted from Fan et al. (2005), Al Ismail (2016).

detail, but it is important to understand that variations in viscosity and composition will impact multi-phase behavior.

Wettability, Capillary Pressure and Relative Permeability – It is well known that water and hydrocarbons are immiscible fluids, meaning they do not mix due to the differences in molecular properties. In porous media, immiscibility is represented by the concept of wettability, which is the tendency for one fluid to spread on or adhere to the solid surface (matrix pore walls) in the presence of another immiscible fluid. The interfacial boundary between immiscible fluids is a curved surface. The angle of curvature depends on pore size and the interfacial tension between the fluids, which causes a difference in pressure across the interface, known as the capillary pressure. Wettability of a multiphase fluid-rock system is quantified by the contact angle between the rock and fluid interface. Capillary pressure is estimated by the Young–Laplace equation (Laplace et al., 1829):

$$P_c = \frac{2\sigma\cos\theta}{r} \tag{5.1}$$

in which σ is the interfacial tension, θ is the contact angle and r is the radius of the capillary (pore throat). Contact angles for multiphase systems are often determined by direct observation of droplets on a rock and fluid interface (e.g., Fig. 5.18a). When the water–oil contact angle, θ_{wo}, is less than 90°, the system is considered water-wet and capillary pressures are positive, promoting water imbibition. When $\theta_{wo} > 90°$, the system is considered oil-wet and capillary pressures are negative, opposing water imbibition. In water-wet systems, water preferentially wets pore surfaces and oil occupies relatively

Pore Networks and Pore Fluids 141

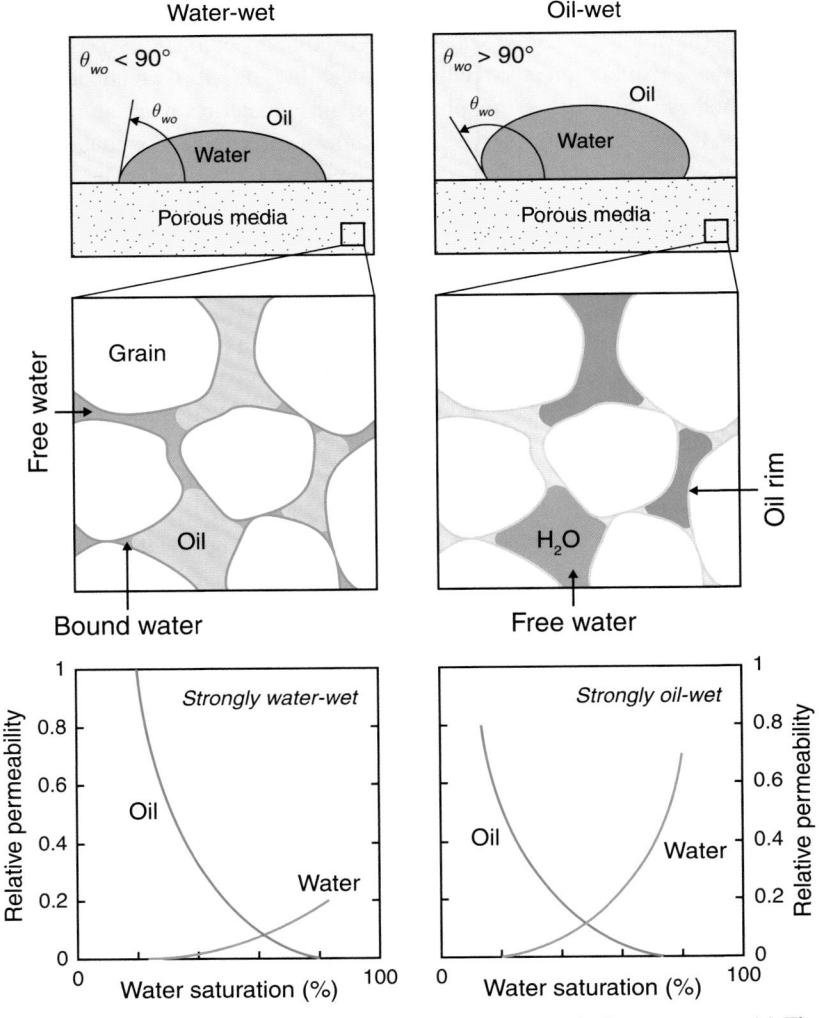

Figure 5.18 Wettability and relative permeability for water-wet and oil-wet systems. (a) The contact angle between the fluid interface and rock surface quantifies the tendency for the fluid to adhere or spread on the solid medium. (b) In water-wet systems, water adheres to pore surfaces and oil occupies pore bodies. In oil-wet systems, oil adheres to pore surfaces and water occupies pore bodies. (c) In conventional reservoir rocks, the non-wetting phase shows greater relative permeability while the wetting phase maintains greater residual saturation. Adapted from Crain (2010).

large pore bodies (due to dependence of capillary pressure on pore size) (Fig. 5.18b). In oil-wet systems, oil preferentially wets pore surfaces and water occupies relatively large pore bodies. In oil–gas or gas–water systems, gas is never the wetting phase. The presence of immiscible fluids reduces the permeability of both the wetting and non-wetting phase relative to a single-phase system. The reduction in permeability as function of fluid saturation is commonly represented by relative permeability, which is normalized by permeability in a single phase system. In conventional reservoir rocks, the non-wetting phase exhibits greater relative permeability, due to its tendency to occupy larger pore bodies, while the wetting phase maintains greater residual saturation (Fig. 5.18c).

Unconventional reservoir rocks show a wide range of wettability from strongly water-wet to mixed-wet to oil-wet. Experiments on bulk samples show spontaneous imbibition of both water and oil (Lan et al., 2015; Singh, 2016). See Siddiqui et al. (2018) for a comprehensive review of wettability data on unconventional reservoir rocks. Many studies of wettability performed at ambient pressure–temperature conditions measure the contact angle of water or oil droplets in the presence of air (θ_{wa} and θ_{oa}). Under *in situ* reservoir conditions, the relevant fluid interface is likely between oil–water, so Siddiqui et al. recalculate values from the literature using a modified form of the Young–Laplace equation to obtain θ_{wo}. Figure 5.19 shows θ_{wo} plotted as a function of both clay and TOC. At low clay and TOC, unconventional reservoir rocks appear strongly water-wet. Water–oil contact angles increase with increasing clay and TOC for samples with moderate clay content (1–50 wt%), crossing over from water- to oil-wet at ~3–4 wt% TOC (Fig. 5.19b). Samples with high clay contents (>50 wt%) exhibit low TOC values (in this dataset) and are water-wet, showing similar contact angles to low clay samples with similar TOC (Fig. 5.18a). The strong dependence of θ_{wo} on TOC reflects the hydrophobic character of organic matter porosity, which favors preferential wetting by hydrocarbons over water. As with many other physical properties, the scatter within this compositional trend can be attributed to a variety of factors related to the matrix microstructure including pore accessibility and connectivity (Lan et al., 2015). In addition, a variety of factors related to the fluid properties can impact wettability including brine composition (ionic strength), pH and pressure-temperature conditions (Siddiqui et al., 2018).

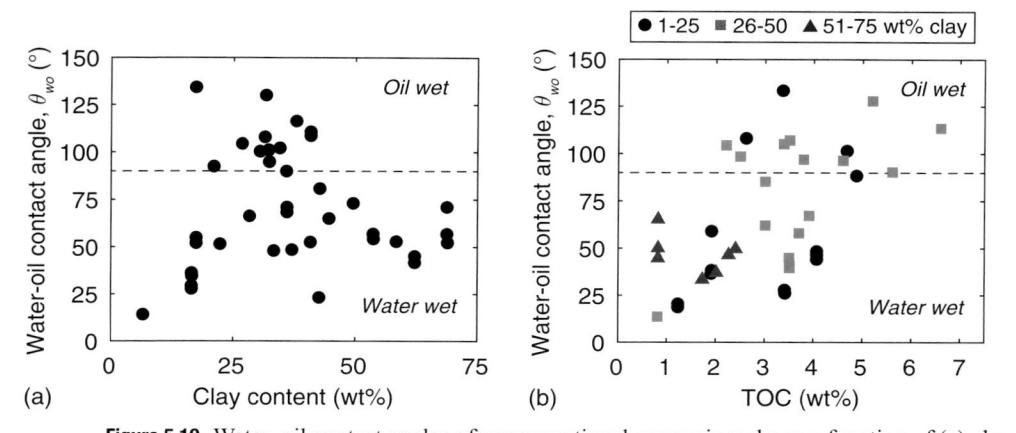

Figure 5.19 Water–oil contact angles of unconventional reservoir rocks as a function of (a) clay content and (b) TOC. From Siddiqui et al. (2018). *Reprinted with permission from Elsevier whose permission is required for further use.*

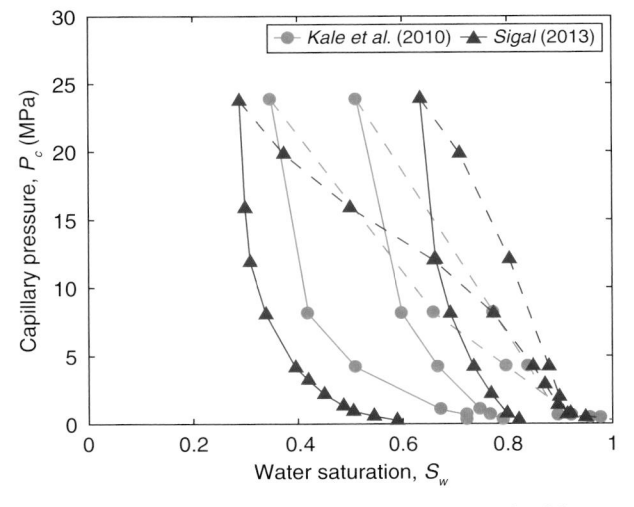

Figure 5.20 Capillary pressure as function of water saturation determined from mercury–air intrusion experiments on several samples of Barnett shale. Solid lines represent drainage, dashed lines represent imbibition. From Edwards & Celia (2018).

In gas–water systems (water-wet), the relationship between water saturation and capillary pressure dictates how water is imbibed and produced. Figure 5.20 shows examples of capillary pressure data modeled from mercury–air intrusion experiments on samples of Barnett shale (Kale et al., 2010; Sigal, 2013). The differences between the imbibition and drainage curves illustrate the hysteresis of capillary pressure with respect to water saturation. It is important to note that these data are truncated at 20 MPa pressure and do not reach residual water saturation. Given that initial water saturations in unconventional reservoirs are estimated as being quite low (~15–40%; EPA, 2016), these data imply very high *in situ* capillary pressures, which is reasonable considering the nm-scale pore networks in the matrix. It also important to note that within samples from the same basin, there is considerable variation in the capillary pressure behavior, likely reflecting variations in composition, pore network characteristics and/or wettability.

The wettability and capillary pressure characteristics of the matrix dictate the relative permeability of multi-phase systems; however, due to the low porosity, ultra-low permeability nature of unconventional reservoir rocks, determining relative permeability experimentally is quite challenging. As we will discuss in Chapter 6, most studies of permeability to date employ only a single phase to study the relationship between flow mechanisms and pore network characteristics. Also, due to the ultra-low permeability of the matrix, steady-state (Darcy) flow techniques that are used to measure relative permeability in conventional reservoir rocks are often not feasible. Although, experimental data is lacking, there have been many recent efforts to model relative permeability in unconventional reservoir rocks using measurements of wettability and pore size distribution (Ojha et al., 2017). Ojha et al. combine pore network characteristics obtained from N_2 adsorption with effective medium and percolation theory to calculate relative permeabilities for samples from the Eagle Ford gas and oil windows and the

Wolfcamp condensate window (Fig. 5.21). In this analysis, the relative permeability curves for gas, oil and condensates are almost identical. The hydrocarbon phase curve for condensates shows slightly higher residual saturation, which may reflect reduced pore accessibility in the Wolfcamp samples. All the hydrocarbon phase curves for unconventionals fall between the fast decline rate of strongly oil-wet conventional rock and slower decline rate of a strongly water-wet rock. In contrast to conventional rocks, at low water saturations ($S_w < 0.8$), the rate of decrease of hydrocarbon permeability (with increasing saturation) is much greater than the rate of increase of water permeability. Also, the hydrocarbon and water relative permeability curves intersect at a higher value of water saturation ($S_w \sim 0.7$) than either of the conventional rocks, which

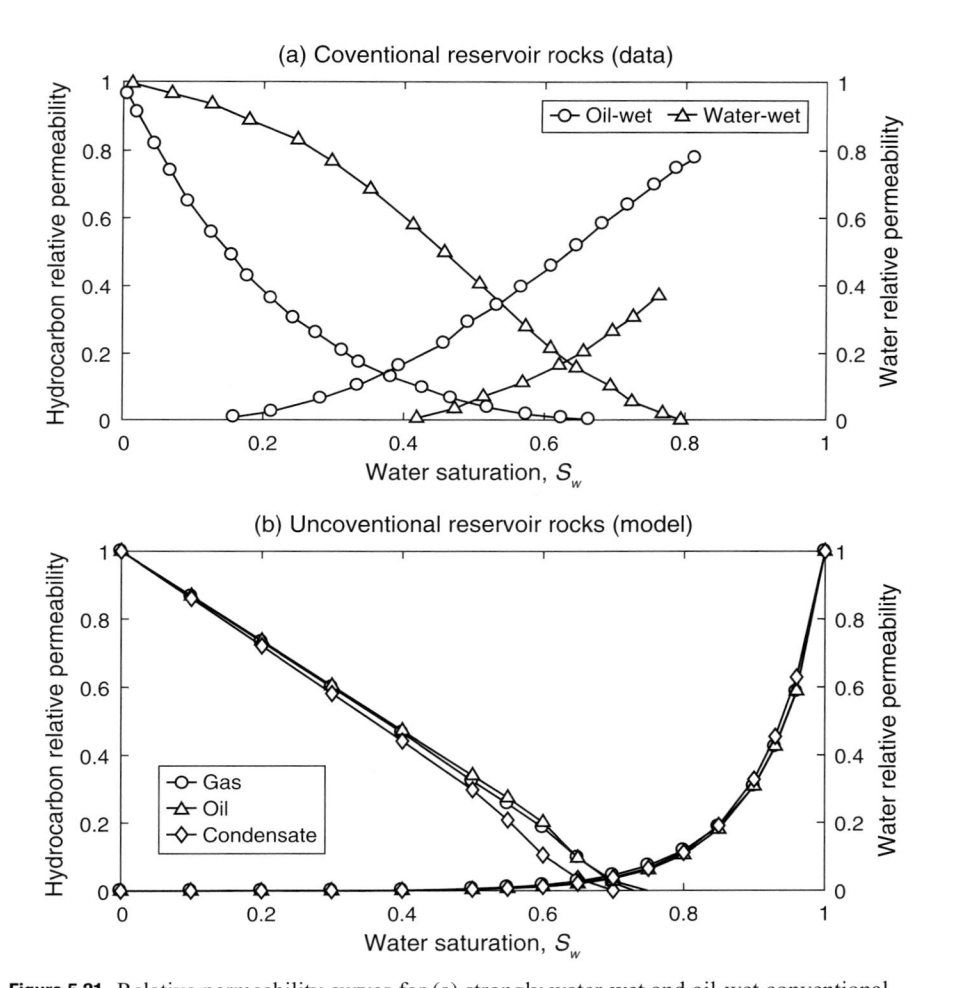

Figure 5.21 Relative permeability curves for (a) strongly water-wet and oil-wet conventional rocks, and (b) samples from the Eagle Ford gas and oil windows and the Wolfcamp condensate window compared. Adapted from Ojha et al. (2017). *Reprinted with permission from Elsevier whose permission is required for further use.*

implies the unconventional samples are strongly water-wet. The residual water saturation is also much greater for unconventional rocks; in this case, no water is produced below $S_w \sim 0.6$. It is also important to note that the intersection of the unconventional relative permeability curves occurs at ~5% of single phase permeability (compared to ~ 15–20% for conventional rocks). Complete lack of intersection between the hydrocarbon and water curves sets up the conditions for a permeability "jail," in which water in small pores (high capillary pressure) blocks the flow of hydrocarbons, but cannot establish connectivity through the pore network (Shanley et al., 2004). Although this phenomenon has been hypothesized for low permeability porous media, there is no direct evidence for complete immobility of both the wetting and non-wetting phase in unconventional reservoir rocks.

Implications of Multi-Phase Effects – The impact of multi-phase effects on the *in situ* flow properties of unconventional reservoir rocks remains a state of the science question. One aspect that merits discussion is how to relate bulk measurements of wettability to the physical mechanisms of flow in the nm-scale pore network. Given the presence of varying proportions of organic (hydrocarbon-wet) and non-organic (water wet) pores in the matrix (Fig. 5.22a), it is important to understand how bulk behavior manifests from pore network properties. As discussed in the section on quantifying porosity, core-scale NMR can be a powerful tool for characterizing pore size distribution and connectivity using different pore fluids. Tinni et al. (2017) performed sequential NMR measurements on cores in "as received," dried, water-saturated and hydrocarbon-saturated states, in order to study water-wet and hydrocarbon-wet pore networks within the same sample. Figure 5.22b shows T_2 distributions for brine and dodecane intake on a sample of Barnett shale. As discussed earlier, T_2 distributions can be interpreted in terms of pore size by assuming pore shape and the surface relaxivity between the solid and fluid interface. In this case, T_2 times <100 ms correspond to pore sizes ≤ 340 nm. Tinni et al. interpret the brine and dodecane distributions as representative of the (connected) water-wet and hydrocarbon-wet pore networks. Applying this method to a range of samples from unconventional basins reveals two classes of pore networks in terms of hydrocarbon flow. In samples with low TOC (<3 wt%), hydrocarbons must flow in series from hydrocarbon-wet pores to water-wet pores, due to the lack of an interconnected organic network. Above >3 wt% TOC, Tinni et al. observe an increase in the amount of dodecane intake as a function of increasing TOC, which suggests this value represents a threshold for the development of an interconnected organic network. Interestingly, Siddiqui et al. (2018) also find a similar threshold for the transition from water-wet to oil-wet behavior in bulk measurements of water–oil contact angle (Fig. 5.19). The emergence of this threshold value in pore-scale and bulk measurements indicates that the development of an interconnected network of porous organic matter is an important control on the wettability of pore network at the core-scale.

Although "upscaling" wettability properties from the pore- to core-scale is important for determining the *in situ* flow properties of the matrix, the ultimate goal is to use this information to understand multi-phase flow at the reservoir-scale. Models of capillary pressure (Fig. 5.20) and relative permeability (Fig. 5.21) can be deployed in reservoir-

146 Unconventional Reservoir Geomechanics

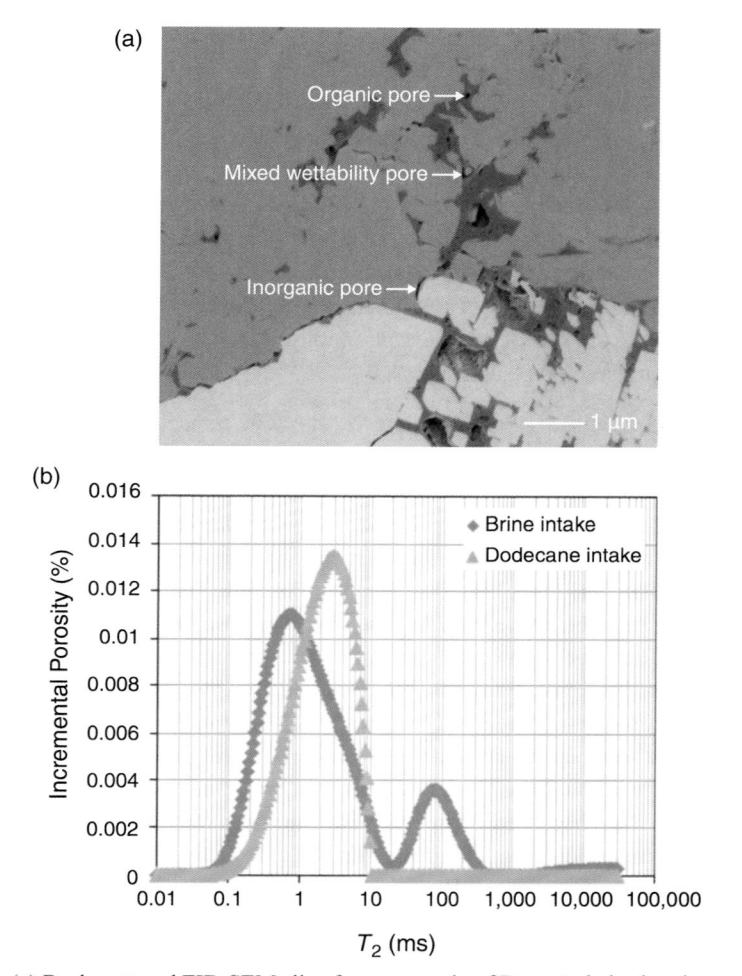

Figure 5.22 (a) Backscattered FIB-SEM slice from a sample of Barnett shale showing a variety of pore types at the μm-scale (Curtis et al., 2012a). *Reprinted by permission of the American Association of Petroleum Geologists whose permission is required for further use.* (b) NMR T_2 distributions for brine and dodecane intake on a sample of Barnett shale (Tinni et al., 2017). Brine intake is calculated by subtracting the "as received" condition from the brine imbibed condition. Dodecane intake is calculated by subtracting the brine imbibed condition from the dodecane imbibed condition. T_2 times <100 ms correspond to pore sizes ≤340 nm. T_2 times <1 ms are attributed to bound fluids. *Reprinted by permission of the Society of Petroleum Engineers.*

scale simulations of flow to estimate the impacts of multi-phase effects on production. Figure 5.23 shows the results of two scenarios of simulated production from a dual porosity matrix-fracture system (Cheng, 2012). In the first scenario (base case), the capillary pressure and relative permeability relationships are derived from empirical correlations (Gdanski et al., 2009) (dashed lines). In the second scenario, Cheng defines an invasion zone where the process of water imbibition alters the capillary pressure and relative permeability relationships (solid lines). In the invasion zone, capillary pressure

Pore Networks and Pore Fluids   147

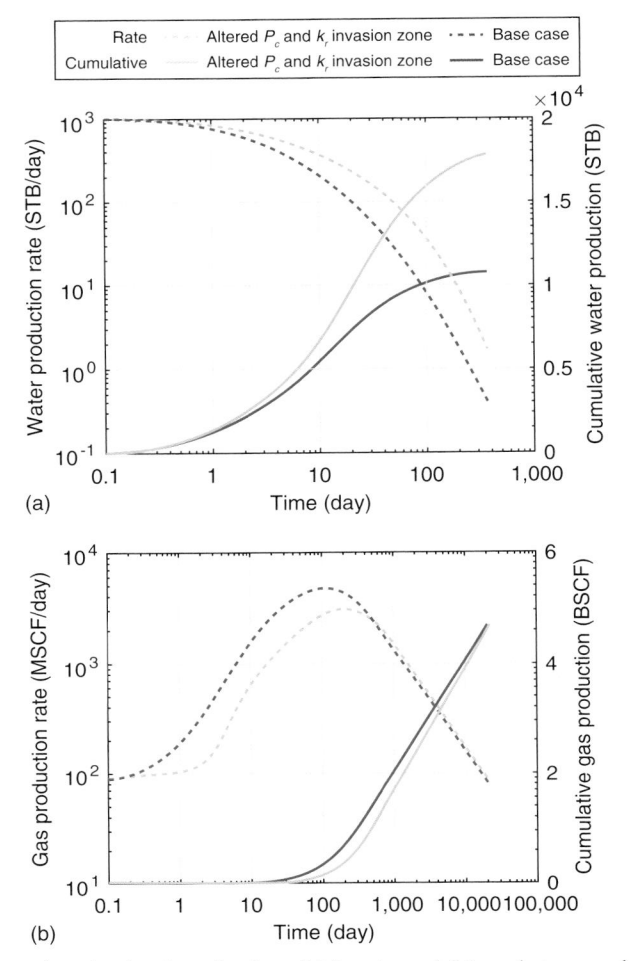

Figure 5.23 Comparing simulated production of (a) water and (b) gas between a base scenario (dashed lines) and a scenario with modified capillary pressure and relative permeability resulting from water imbibition. Adapted from Cheng (2012).

increases with decreasing water saturation at a much greater rate than the base case, which reflects the impact of water imbibition on hydrocarbon transport. Also, relative permeability is increased for water and decreased for hydrocarbons (increased residual water saturation, decreased residual hydrocarbon saturation), again reflecting the effects of capillary blocking. Figure 5.23a compares cumulative water production and production rate for the two scenarios. Compared to the base case, water production rates remain higher for longer and cumulative water production almost doubles. Figure 5.23b compares cumulative gas production and production gas for the two scenarios. Although cumulative production is essentially identical, in the invasion zone scenario, the peak value of the gas production rate is slightly reduced and occurs months later. In both cases, Cheng used relative permeability curves resembling the shape of those from conventional reservoirs (i.e., hydrocarbon and water permeabilities increase/decrease at same rate), so the gas rate peak corresponds with the decline in water rate. Future studies have the potential to directly consider the wettability properties of the matrix by using relative permeability relationships based on direct observation of pore network characteristics (e.g., Fig. 5.21). Regardless, the discrepancies between the two scenarios illustrate the effects of capillary blocking from water imbibition, which is ultimately responsible for the prolonged water production rate and delayed gas rate peak. Using the type of relative permeability curves in Fig. 5.21 could yield additional insights into effects of multi-phase flow on water and gas production. For example, increased residual water saturation may explain a scenario of high initial gas production rates accompanying diminished water production (low water-cut). Also, the rapid decline in hydrocarbon relative permeability compared to the increase in water permeability with increasing saturation may account for a rapid decline in hydrocarbon production during relatively constant water production (Ojha et al., 2017). In Chapter 13, we will revisit the issue of water production in the context of reservoir-scale controls and environmental issues related to disposal and re-injection.

6 Flow and Sorption

Understanding the flow and sorption properties of unconventional reservoir rocks is essential for predicting the movement of water and hydrocarbons during stimulation (injection), production and depletion. Unconventional reservoir rocks are unified (and essentially defined) by their ultra-low matrix permeability, which is approximately a million to a billion times less than that of conventional reservoir rocks. For this reason, hydraulic fracturing and shear stimulation on pre-existing faults (Chapters 8 and 10) are necessary to expose more surface area of the matrix for production. Although the initial flow behavior during injection and production is controlled by the properties of the fracture network, the long-term flow behavior is controlled by the ultra-low permeability matrix.

To build a foundation for understanding laboratory measurements of permeability, we will first consider the role of various flow mechanisms at reservoir conditions in the nm-scale pore networks of unconventional reservoir rocks (Chapter 5). Specifically, we will discuss how pore size and fluid properties affect the relative contributions of viscous (Darcy) and diffusive mechanisms to matrix flow. We will then discuss the pressure-dependence of flow properties, which is critical for understanding the reservoir-scale response to changes in stress from injection and depletion (Chapter 12), as well as the relationship between permeability and the elastic response of the matrix at the pore-scale.

In the next section, we will examine laboratory measurements of matrix flow in unconventional reservoir rocks. We focus on datasets that span a range of pore and confining pressures, which enables independent analyses of pressure-dependent effects and the relative contributions of viscous and diffusive flow mechanisms. We will then review methods from the literature for using estimates of diffusive flux to determine a chacteristic or effective pore size for matrix flow. As with any physical property of the rock matrix, it is important to understand how flow properties vary with composition. We will review data on a wide variety of lithologies (at similar stress conditions) in order to understand compositional and fabric controls on permeability, pressure-dependence and anisotropy.

To predict reservoir-scale flow behavior, it is important to consider not only the response of the pore network to changes in pressure, but also the physical interactions between various pore fluids and the matrix components (sorption). We will review laboratory observations of gas and liquid sorption, as well as the physical models used to describe the pressure-dependence of sorption. As in our discussion of permeability, we will consider compositional variations in order to understand what matrix components are responsible for sorption. In addition, we will examine the coupling of adsorption and deformation in the context of permeability measurements with adsorbing and non-adsorbing gases, which illustrate the relationship between pressure-dependent permeability and adsorption. The chapter will conclude with a discussion of the implications of the integrated effects of pressure-dependent properties (permeability, diffusion and adsorption) for the flow in rock matrix.

Matrix Flow

Flow Mechanisms – In conventional reservoir rocks, flow is governed by Darcy's law, in which the driving force is simply the pressure gradient across the porous medium (viscous flow). This continuum approach assumes that no flow (slip) occurs along pore walls and that interactions between gas molecules are much more frequent than interactions between gas molecules and pore walls (Fig. 6.1).

The relative contributions of molecule–molecule and molecule–wall interactions are encapsulated in the dimensionless Knudsen number, Kn, which is the ratio of the molecular mean free path, λ, and the pore diameter, d_p:

$$Kn = \frac{\lambda}{d_p}.\qquad(6.1)$$

The molecular mean free path represents the average distance between molecular collisions and is a function of molecular diameter, d_m, pressure, temperature and the Boltzmann's constant, K_B:

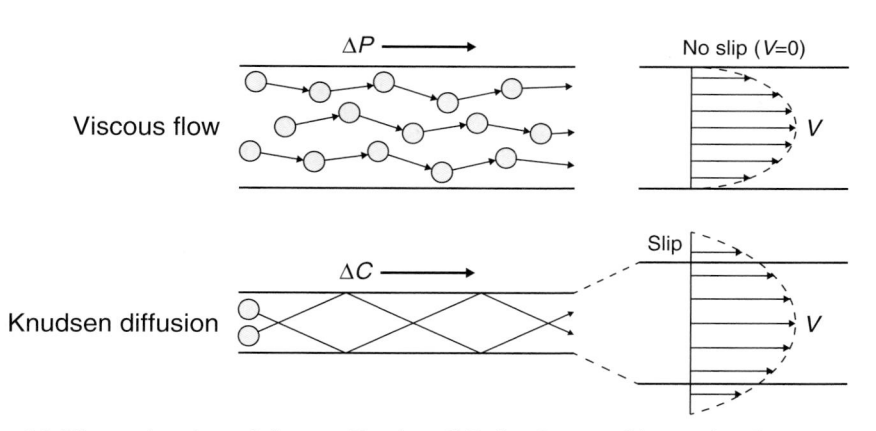

Figure 6.1 Viscous (continuum) flow vs. Knudsen diffusion. Increased interactions between gas molecules and pore walls at small pore sizes promotes diffusive flow mechanisms, resulting in non-zero flow rates along pore walls (slip flow).

Flow and Sorption 151

Table 6.1 Summary of flow regimes as function of Knudsen number (Eqn. 6.1). Adapted from Ziarani & Aguilera (2012) and Heller et al. (2014).

Flow regime	Knudsen number, Kn	Driving force	Model
Continuum (viscous) flow	$Kn < 0.01$	Total pressure gradient	Assumes zero flow velocity along pore walls. No permeability correction required
Slip flow	$0.01 < Kn < 0.1$	Mostly viscous flow with some diffusion	Darcy's law with Klinkenberg correction
Transition flow	$0.1 < Kn < 10$	Mostly diffusion with some viscous flow	Darcy's law with Knudsen correction
Knudsen (molecular) diffusion	$Kn > 10$	Total concentration gradient	Knudsen diffusion equation

$$\lambda = \frac{K_B T}{\sqrt{2\pi}d_m^2 P}. \qquad (6.2)$$

Figure 6.2a shows the variation of mean free path as function of pressure for various gases at 100 °C. Note that mean free path begins to increase significantly at relatively low pressures ~400–500 psi (3 MPa) as gas density decreases. The Knudsen number is used to represent transitions between the various flow regimes described in Table 6.1. At $Kn < 0.01$, molecule–wall interactions are negligible compared to molecule–molecule interactions, so Darcy (continuum) flow is valid. At $Kn > 10$, the increased probability of molecule–wall interactions results in significant gas flow along pore walls (slip), which violates the continuum assumption. Under these conditions, gas flow is represented by molecular (Knudsen) diffusion, in which the driving force is the molecular concentration gradient (Table 6.1). In the diffusion regime, gas composition has no significance because gas molecules move independently, colliding with the pore wall more often than with each other as they move through the porous media (Fig. 6.1). At intermediate values of Kn, both diffusive and viscous mechanisms contribute to flow to some degree, so specific corrections are required to accurately represent flow rates and pressure gradients (Table 6.1).

Figure 6.2b shows the variation of Knudsen number with pore diameter at pressures relevant for unconventional gas reservoirs. Considering that matrix pore sizes range from ~1 to 100 nm (Chapter 5), we expect flow in unconventional reservoirs to occur within the transition, slip and Darcy flow regimes, with the relative contributions depending on the specific pore size distribution. It is important to note that decreasing pressure (depletion) or pore size (compaction or swelling) increases the Knudsen number and the contribution of slip flow.

To account for increased flow rates due to slip flow along pore walls, Klinkenberg (1941) modified Darcy's law in terms of viscous flow through a cylindrical pipe (pore) to develop a linear relationship between the apparent measured permeability, k_a, and the permeability at infinite pressure, k_∞. At infinite pressure, the gas density approximates

152   Unconventional Reservoir Geomechanics

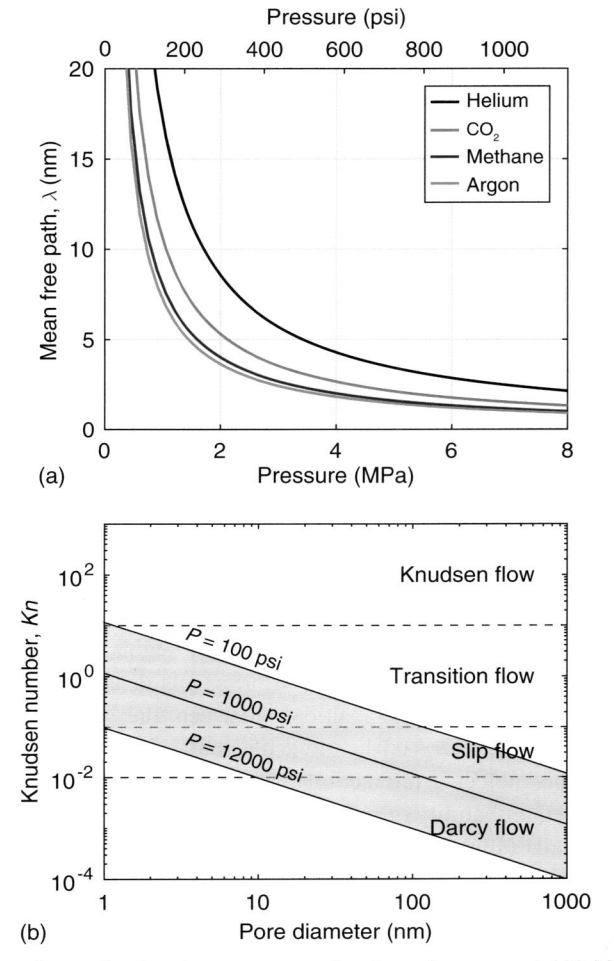

Figure 6.2 (a) Mean free path of various gases as a function of pressure at 100 °C. (b) Knudsen number as a function of pore diameter for methane at 100 °C and pressures typical of unconventional gas reservoirs. Adapted from Heller et al. (2014).

a liquid state and the mean free path is negligible compared to pore size (i.e., no slip flow). Klinkenberg performed permeability measurements at various pore pressures to quantify the contribution of slip flow, b, which represents the increase in permeability as a function of inverse pore pressure:

$$k_a = k_\infty \left(1 + \frac{b}{P} \right). \tag{6.3}$$

Klinkenberg derived this expression for the purposes of estimating liquid permeability from gas permeability measurements, but recent studies of gas flow in unconventional reservoir rocks employ this relationship to quantify the relative contributions of diffusion and Darcy flow as a function of pressure (Alnoaimi et al., 2015; Bhandari et al., 2015; Heller et al., 2014; Mckernan et al., 2017).

Pressure-Dependence – As discussed in Chapter 2, the variations in the physical properties of rocks with confining and pore pressures are described by an effective stress, σ_{eff}. The simple (Terzaghi) effective stress is the difference of confining and pore pressures, but this relationship is often modified by a pore pressure (Biot) coefficient, χ, in order to account for the relative sensitivities of rock properties to changes in confining and pore pressures.

$$\sigma_{eff} = P_c - \chi P_p. \tag{6.4}$$

For elastic properties, the Biot coefficient is α, which describes the relative sensitivity of volumetric strain to changes in confining and pore pressure (Chapter 2). For permeability, the Biot coefficient is represented by the ratio of the change in permeability with pore pressure (at constant confining pressure) to the change in permeability with confining pressure (at constant pore pressure):

$$\chi = \frac{\partial k / \partial P_p}{\partial k / \partial P_c}. \tag{6.5}$$

Most unconventional reservoir rocks exhibit $\chi \leq 1$, which signifies that permeability is more sensitive to changes in confining pressure than changes in pore pressure (Heller et al., 2014; Mckernan et al., 2017). In other words, pore pressure is less effective at expanding pores than confining pressure is at closing them. In contrast, studies on granite (Morrow et al., 1986) and clay-bearing sandstones (Zoback & Byerlee, 1975; Walls & Nur, 1979) demonstrate $\chi \geq 1$, with values reaching high as 7. To explain this behavior, Zoback & Byerlee (1975) proposed a clay-pore model, in which the pore network comprises relatively compressible clays contained within a load-supporting framework of relatively stiff quartz (Fig. 6.3). In this model, the high compressibility of the pore network relative to the rock frame results in pore pressure being more effective at expanding pores than confining pressure is at closing them. Kwon et al. (2001) studied clay-rich (45 wt%) samples from the Wilcox shale and found $\chi \sim 1$. Kwon et al. described this behavior by a clay-matrix, clay-pore model in which the compressibility of pores and the load-supporting framework are equal (Fig. 6.3). Based on observations of $\chi \leq 1$ in unconventional reservoir rocks, Heller et al. (2014) proposed a similar model, in which the pore network is composed of compliant phases (clay and organic matter) with similar elastic properties to the load-bearing frame. Given our discussions of the compositional variations in elastic properties and pore networks between different lithofacies (Chapters 2 and 5), any of these models for effective stress dependence on a particular sample necessitates direct observation of the matrix microstructure.

Unconventional Reservoir Geomechanics

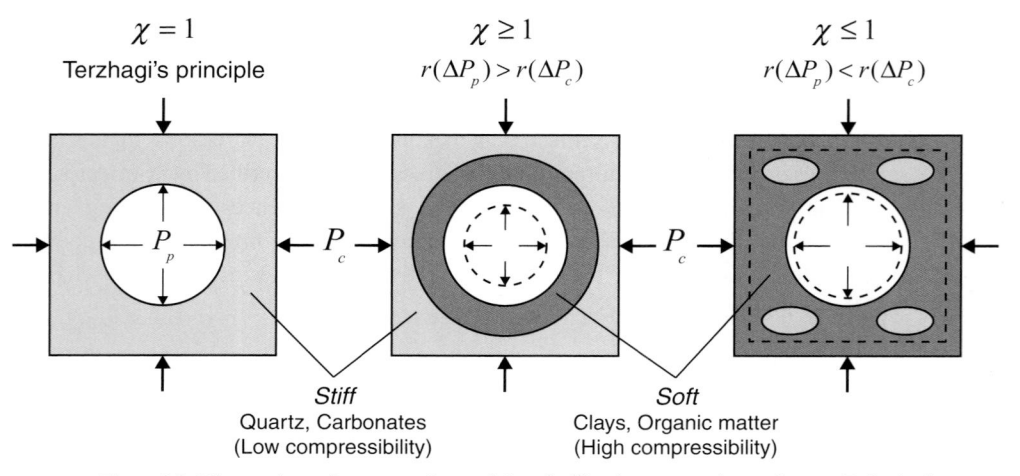

$\chi = 1$
Terzhagi's principle

$\chi \geq 1$
$r(\Delta P_p) > r(\Delta P_c)$

$\chi \leq 1$
$r(\Delta P_p) < r(\Delta P_c)$

Stiff
Quartz, Carbonates
(Low compressibility)

Soft
Clays, Organic matter
(High compressibility)

Figure 6.3 Illustration of pore-scale models of effective stress dependence. Relatively compliant phases (clay and organic matter) are shown in dark gray and stiff clastic minerals (quartz and carbonates) are shown in light gray. The Biot coefficient, χ, defines the relationship between pore pressure and effective stress. Single phase porous media are affected equally by changes in pore and confining pressure, resulting in $\chi = 1$. A stiff rock frame containing a pore network composed of compliant phases may be more affected by changes in pore pressure, resulting in $\chi \geq 1$. A rock frame with similar elastic properties to the pore network may be more affected by changes in confining pressure, resulting in $\chi \leq 1$. Adapted from Gensterblum et al. (2015).

Permeability

Measurement– Permeability quantifies the ease of flow of a fluid or gas through porous media. In Darcy's law (1D), permeability is represented by the relationship between flux, Q, dynamic viscosity, μ, and the differential pressure, $P_1 - P_2$, over a length, L:

$$Q = \frac{kA(P_1 - P_2)}{\mu L} \tag{6.6}$$

where P_1 and P_2 are the upstream and downstream pressures, and A is the cross-sectional area. In SI units, permeability is expressed in m^2, but it is often converted to Darcys ($1\,\text{D} = 9.87 \times 10^{-11}\,\text{m}^2$) to describe reservoir rocks (Fig. 6.4). Conventional reservoir rocks span a wide range of permeability values from $\sim 10^{-10}$ to $10^{-17}\,\text{m}^2$ (100 D–10 μD). All less permeable reservoir rocks are considered unconventionals, which includes tight gas, shale oil and shale gas reservoir rocks.

The permeability of the rock matrix dictates the choice of measurement technique. Commonly applied techniques include steady-state (Darcy) flow, probe permeametery, pore pressure oscillation, crushed measurements (GRI method) and pressure pulse-decay (Fig. 6.4). Steady-state measurements are performed by imposing a flow rate across an intact sample and observing the resultant pressure difference or by imposing a pressure difference and observing the resultant flow rate. This is

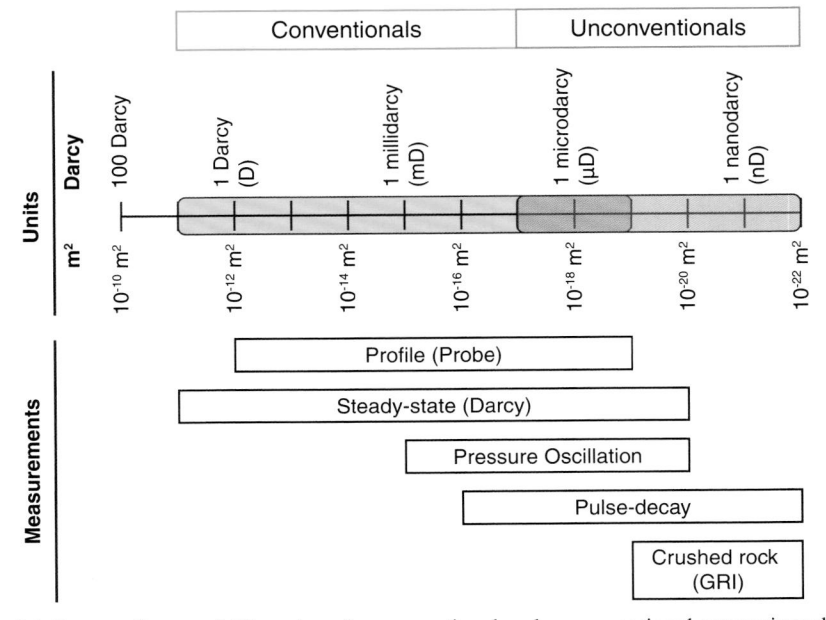

Figure 6.4 Range of permeability values for conventional and unconventional reservoir rocks in the context of measurement techniques. Adapted from Gensterblum et al. (2015).

appropriate for permeabilities as low as ~1 µD, below which the time to reach steady-state is not experimentally practical. Probe permeametery uses the same principle in a slightly different geometry, which slightly reduces "resolution," but enables rapid characterization of core samples. Pore pressure oscillation measures permeability by quantifying the amplitude reduction and phase shift of an oscillatory pressure signal sent across an intact sample (Kranz et al., 1990). It is relatively insensitive to leaks and can provide quasi-continuous measurements of permeability, but is effectively limited to permeability values >0.1 µD due to the impractically long oscillation periods that would be required at lower permeabilities (Mckernan et al., 2017). Crushed permeability measurements are made on rock chips ~0.1–2 mm by observing the fractional uptake of gas expanding from a reference cell to a cell containing the chips. Permeability values are determined based on the time it takes for gas uptake to reach equilibrium (Cui et al., 2009). Although the crushed method is capable of measuring ultra-low permeability values representative of unconventional reservoir rocks, several studies have demonstrated that the results do not correlate with measurements on intact samples, which suggests the act of crushing may destroy important aspects of the matrix pore networks (Sinha et al., 2012; Heller et al., 2014). Therefore, ultra-low permeability reservoir rocks are most often characterized by (transient) pressure pulse-decay, in which a relatively small pressure difference is imposed across an intact sample and the rate of decay of the pressure difference with time is observed (Brace et al., 1968). The logarithm of the pressure difference, ΔP, follows a linear trend in time, in which the decay exponent, α, is proportional to permeability:

$$\Delta P(t) = \Delta P_0 e^{-\alpha t} \tag{6.7}$$

$$\alpha = \frac{kA}{\beta V_{down} L \mu} \tag{6.8}$$

where β is the fluid or gas compressibility and V_{down} is the downstream volume. Figure 6.5 shows an example of a pressure pulse-decay measurement on horizontal and vertical samples of Barnett shale (Bhandari et al., 2015). See Heller et al. (2014) for a detailed discussion of the sources of error and experimental challenges associated with using pressure pulse-decay to measure ultra-low matrix permeability.

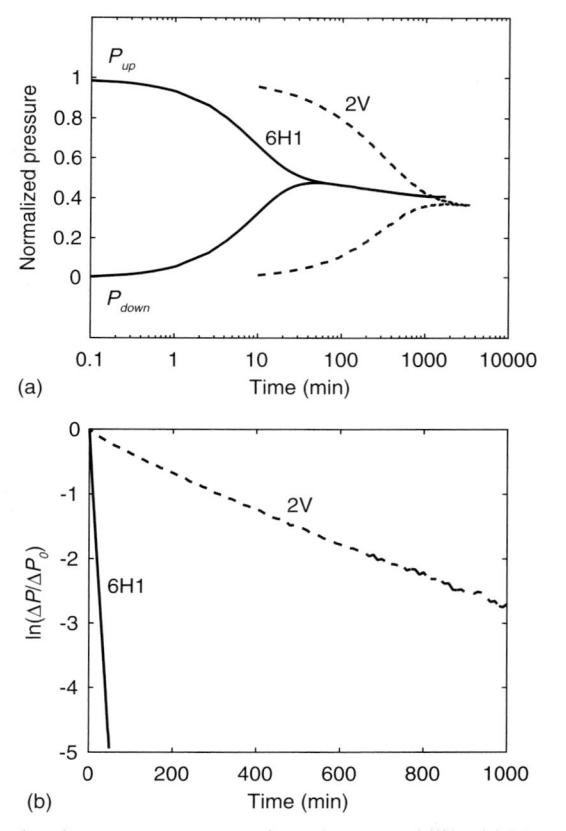

Figure 6.5 Pressure pulse-decay measurement of matrix permeability. (a) Normalized pressure difference as function of time for experiments on vertical (2V) and horizontal (6H1) samples of Barnett shale. (b) The logarithm of the pressure difference as function of time yields a linear relationship, in which the rate of decay is proportional to permeability (Eqns. 6.7, 6.8). From Bhandari et al. (2015).

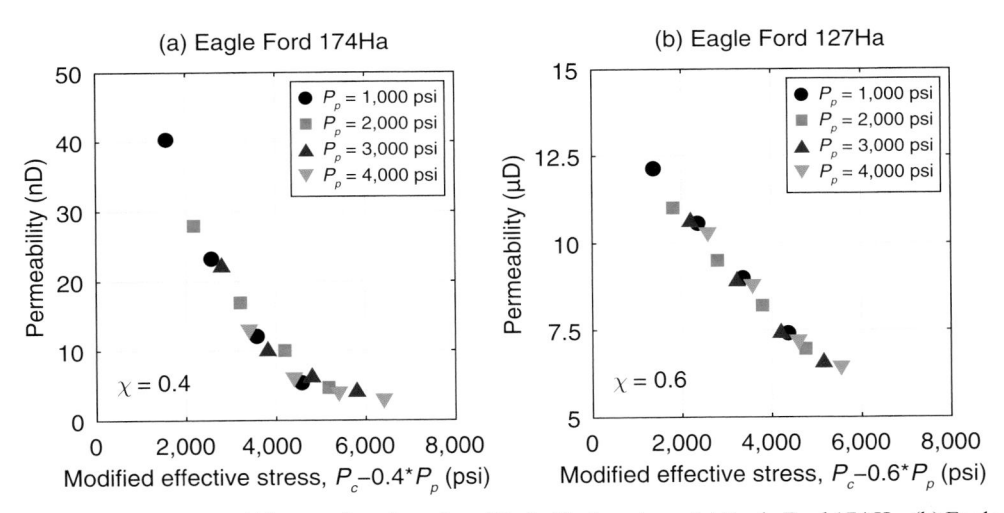

Figure 6.6 Permeability as a function of modified effective stress. (a) Eagle Ford 174 Ha. (b) Eagle Ford 127 Ha. Sample compositions are provided in Table 2.1. From Heller et al. (2014). *Reprinted by permission of the American Association Petroleum Geologists whose permission is required for further use.*

Effective Stress Dependence – To examine how the permeability of unconventional reservoir rocks varies with effective stress, we will focus on the dataset of Heller et al. (2014), which employs experiments at various confining and pore pressures to determine their relative effects on effective stress (Eqn. 6.4). The ratio of the change in permeability with P_p (at constant P_c) to the change in permeability with P_c (at constant P_p) determines the Biot coefficient, χ (Eqn. 6.5). Figure 6.6 shows an example of the relationship between permeability and modified effective stress for clay and TOC-rich and -poor horizontal samples from the Eagle Ford shale.

Both samples show an approximately exponential decrease of permeability with effective stress, which is consistent with experiments on other unconventional reservoir rocks and mudstones (Kwon et al., 2001; Cui et al., 2009; Al Ismail & Zoback, 2016; Mckernan et al., 2017; Al Alalli, 2018). Although the samples both show χ values much less than 1 and differ by only ~20 wt% clay + TOC, clay and TOC-rich sample Eagle Ford 174 Ha is nearly 3 orders of magnitude less permeable than Eagle Ford 127Ha. Later in the chapter, we will review data from the literature in order to understand the compositional and microstructural controls on flow properties.

Quantifying Diffusive Effects – Figure 6.7a shows the same permeability effective stress data in Fig. 6.6 with lower pressure (<1,000 psi) data added. For both samples, the low pressure data deviate from the exponential trend as a function of decreasing pore pressure. This apparent permeability enhancement is attributed to increasing diffusive flux (slip flow) resulting from the dramatic increase in the mean free path between 400 and 500 psi (Fig. 6.2a). To understand how the contribution

158   **Unconventional Reservoir Geomechanics**

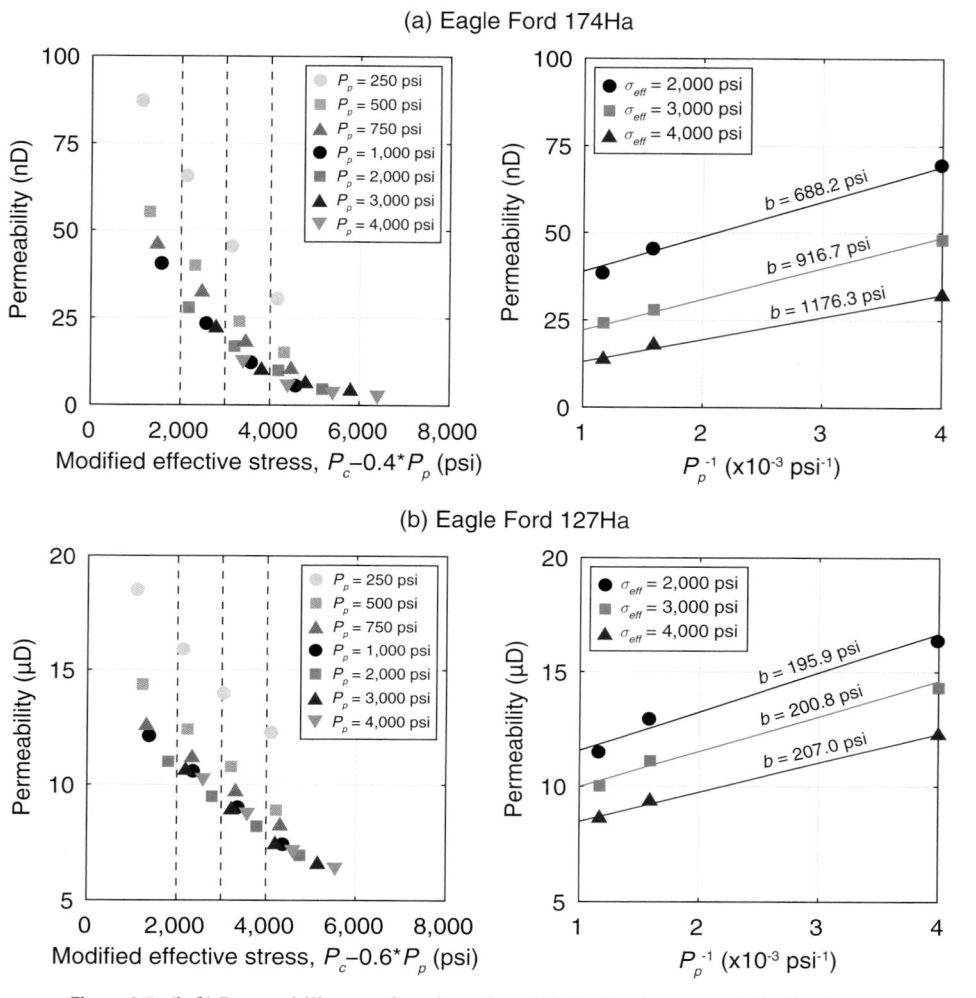

Figure 6.7 (left) Permeability as a function of modified effective stress including low pore pressure data. The intersections of the black dashed lines and the permeability trend for each pore pressure are used to determine the relationship between permeability and pore pressure at constant effective stress. (right) Klinkenberg plot (apparent permeability, k_a vs. P_p^{-1}) for effective stresses 2,000–4,000 psi. Note that the value of b (slope/intercept) increases with increasing effective stress. From Heller et al. (2014). *Reprinted by permission of the American Association of Petroleum Geologists whose permission is required for further use.*

of slip flow varies as a function of pressure, Heller et al. (2014) considered the change in apparent (measured) permeability at constant effective stress for the low pressure data (Fig. 6.7a, black dashed lines). Figure 6.7b shows apparent permeability plotted as function of inverse pore pressure for effective stresses 2,000–4,000 psi. In these Klinkenberg plots, the slope of a line of constant effective stress is proportional to the slip flow coefficient, b, and the intercept represents the liquid (infinite pressure) permeability, k_∞ (Eqn. 6.3). The value of b increases with increasing effective stress, consistent with the idea that the Knudsen number increases with pore compaction. This effect is much more prominent in the relatively low permeability, clay and TOC-rich sample Eagle Ford 174 Ha (b nearly doubles), whereas for Eagle Ford 127 Ha, the value of b only increases by ~10% over the same range of effective stress. This is reasonable considering that lower permeability of Eagle Ford 174 Ha likely results from smaller characteristic flow paths (pores), which, when compacted under stress, would tend to magnify diffusive effects much more than larger flow paths.

To quantify the relative contributions of flow mechanisms with pressure, Heller et al. (2014) determined the ratio of diffusive to Darcy flux at the same values of effective stress, 2,000–4,000 psi (Fig. 6.8). As seen in the Klinkenberg plots (Fig. 6.7), the relative proportion of flow accommodated by diffusion (diffusive flux) increases with decreasing pore pressure (i.e., increasing λ and Kn) and increasing effective stress. For the relatively low permeability, clay and TOC-rich sample Eagle Ford 174 Ha, the ratio of diffusive to Darcy flux exceeds 1 at a pore pressure of ~400–500 psi, whereas for Eagle Ford 127 Ha, this ratio remains <0.6 at lowest pore pressure (250 psi). It is important to note that, at low pore pressures, the increased contribution of diffusion (slip flow) with increasing effective stress opposes reduction of Darcy flux due to pore compaction. Thus, the relative effects of diffusion and effective pressure-dependence must be considered to understand how permeability evolves during depletion.

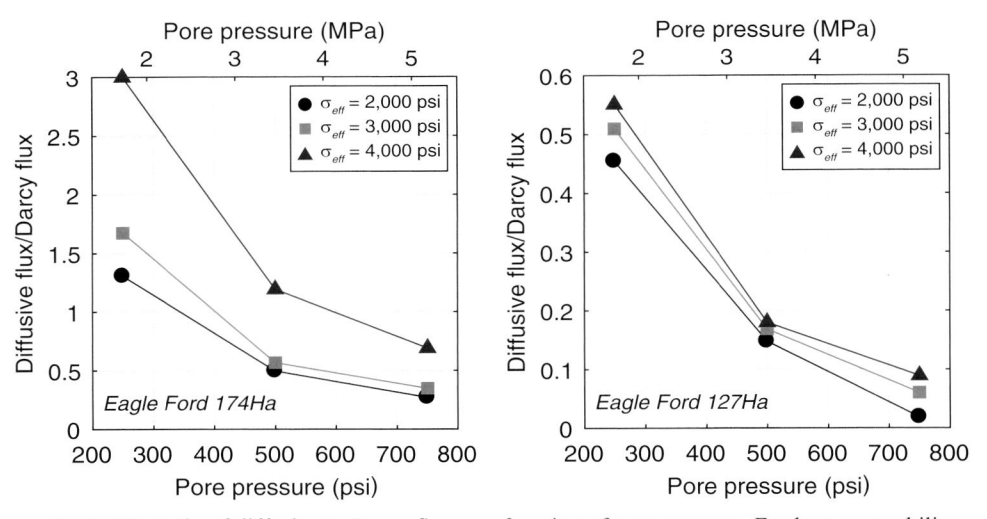

Figure 6.8 Ratio of diffusive to Darcy flux as a function of pore pressure. For low permeability sample Eagle Ford 174, the ratio exceeds 1 at ~400 psi (2.8 MPa), whereas for more permeable sample Eagle Ford 127, this ratio does not exceed 0.6 at the lowest pore pressure. Adapted from Heller et al. (2014). *Reprinted by permission of the American Association of Petroleum Geologists whose permission is required for further use.*

160   **Unconventional Reservoir Geomechanics**

Estimating Effective Pore Size – Many studies of the permeability of unconventional reservoirs utilize the dependence of slip flow on pore diameter to estimate an effective pore size responsible for flow (Heller et al., 2014; Al Ismail & Zoback, 2016; Letham & Bustin, 2016; Mckernan et al., 2017; Al Alalli, 2018). Depending on the specific model used to describe pore shape and/or pore size distribution, this estimate may be termed a characteristic or effective pore size. Randolph et al. (1984) re-cast Poiseuille's law for viscous flow through slit-shaped pores in terms of the Klinkenberg-corrected form of Darcy's law (Eqn. 6.6) to yield an expression relating slit pore width, w, and the slip flow coefficient, b:

$$w = \frac{16c\mu}{b} \left(\frac{2RT}{\pi M} \right)$$

(6.9)

in which c is an empirical constant ~ 1, R is the universal gas constant and M is molar mass. Heller et al. (2014) and Al Ismail & Zoback (2016) use this expression to calculate pore width as function of effective stress (Fig. 6.9a). Effective pore widths range from ~ 20 to 140 nm, which is generally consistent with observations of pore size from imaging (Chapter 5), as well as the range of pore sizes where slip flow is expected at reservoir pressures (Fig. 6.2b). Because effective pore width is inversely proportional to the amount of slip flow, relatively permeable, clay and TOC-poor sample Eagle Ford 127 Ha shows larger effective pore widths and less reduction in pore width with increasing effective stress compared to Eagle 174 Ha. The Marcellus sample in Fig 6.8a is a vertical sample relatively rich in clay (52 wt%) and poor in TOC (1.17 wt%), and exhibits similar permeability and slip flow to Eagle Ford 174 Ha.

A modest correlation is observed between permeability and effective pore size when considering data from literature on a variety of different lithologies using different stress conditions and pore fluids (Fig. 6.9b). In some cases, an average pore size is determined based on the results of mercury injection porosimetry (Yang & Aplin, 2007; Chalmers et al., 2012b; Ghanizadeh et al., 2014a, 2014b;), whereas other studies use the relationship between slip flow and pore size for various pore geometries (Heller et al., 2014; Al Ismail & Zoback, 2016; Letham & Bustin, 2016; Mckernan et al., 2017; Al Allali, 2018). Mckernan et al. (2017) attempted to gain a mechanistic understanding of effective pore size by applying a pore conductivity model based on measurements of changes in acoustic wave velocity and pore volume with pressure. This enables prediction of the evolution of permeability with effective stress in successive loading cycles, and interestingly yields an estimate for effective pore size that is ten times smaller than the value based on the magnitude of slip flow (Eqn. 6.9). Simultaneous recording of elastic properties (ultrasonic velocities) during the flow measurements also enables application of a micro-crack model, which provides estimates of the distribution of crack (pore) aspect ratios and their relative sensitivities to effective stress.

As discussed in Chapter 5, it is important to understand that using flow properties to characterize the matrix pore networks is only an approximate approach meant to provide estimates of pore size. Given the range of pore-scale models used in the literature, it is

Flow and Sorption   161

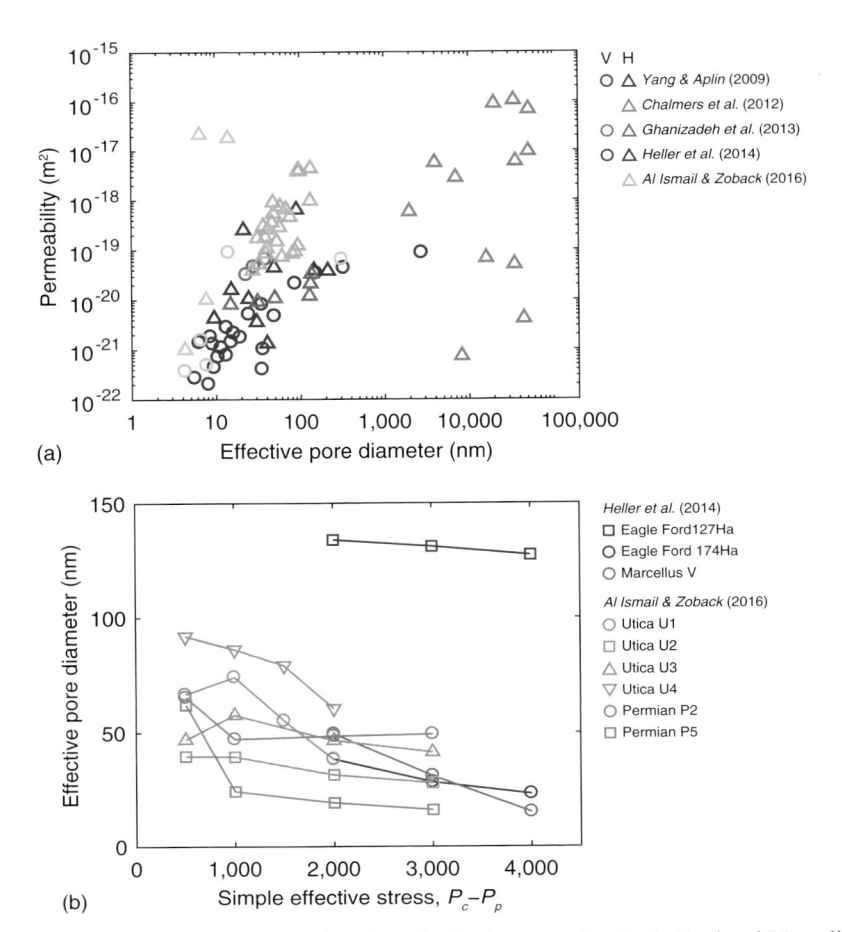

Figure 6.9 (a) Effective pore size as a function of effective stress for Eagle Ford and Marcellus shales. Compiled from Heller et al. (2014), Al Ismail & Zoback (2016). (b) Permeability as a function of effective pore size (diameter). Data are compiled from studies using various models for estimating pore size. See the individual references for details. Open symbols are vertical samples, closed symbols are horizontal samples. Adapted from Gensterblum et al. (2015).

clear that any number of them can be fit to flow data and somehow related to observations of the pore network. Therefore, in order to understand flow properties and pore network characteristics in a given lithofacies, it is important to cross-check pore size estimates with observations of pores from imaging (to quantify pore geometry) and flow/adsorption methods (to distinguish between connected and isolated porosity).

Compositional and Microstructural Controls – As with all other physical properties of unconventional reservoir rocks, it is important to understand how flow properties vary as function of matrix composition and microstructure. Figure 6.10 presents a compilation of

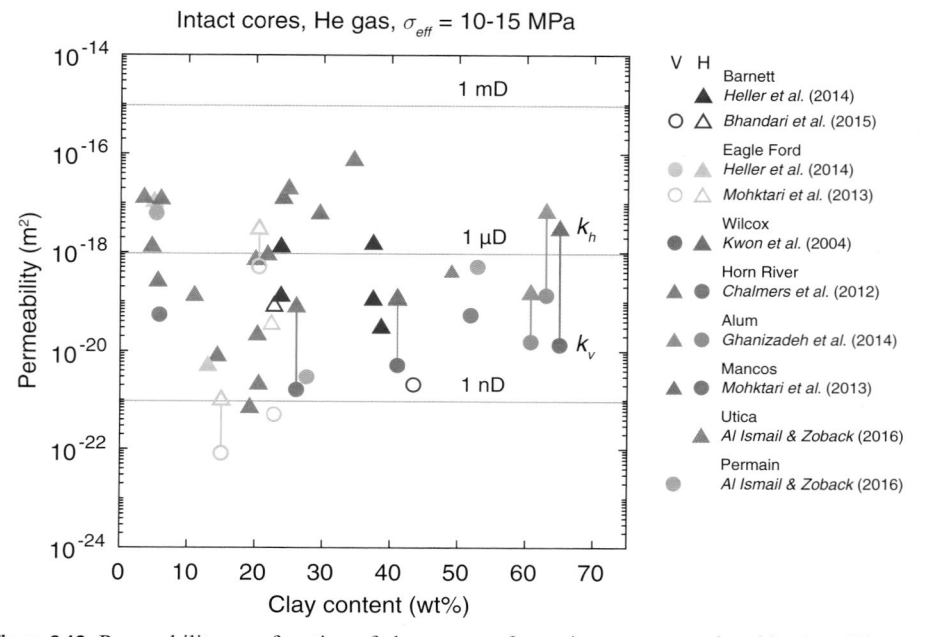

Figure 6.10 Permeability as a function of clay content for various unconventional basins. All measurements are performed on intact cores using He or Ar gas at effective stresses 10–15 MPa (~1,500–2,000 psi). Open symbols are vertical samples, solid symbols are horizontal samples. Lines between symbols represent measurements on the same sample at both orientations.

(pulse-decay) permeability measurements on unconventional reservoir rocks and mudstones as a function of clay content. Permeability values range from ~10^{-16} to 10^{-22} m^2 (10 μD–0.1 nD) and, unlike the mechanical properties of the matrix (Chapters 2–4), show no apparent correlation with clay content. The lack of any correlation between reservoirs implies that permeability is not simply controlled by the composition of the rock matrix, but also depends strongly on the geometric properties of the pore network (pore shape and size, tortuosity).

In some cases, there are in fact correlations between flow properties and composition within individual basins, but interpreting these trends in terms of pore network properties is not necessarily straightforward. For example, Fig. 6.11 shows increasing permeability as a function of increasing organic content for samples from the Utica shale (Al Ismail & Zoback, 2016). We discussed the pore size distributions of these samples in the context of variations in composition in Chapter 5 (Fig. 5.13). Interestingly, these samples show very similar pore size distributions, but pore volume between ~2–4 nm decreases with increasing clay content and does not correlate with TOC. This is somewhat counterintuitive since we might expect that the increase in permeability with TOC reflects increasing porosity within an expanding organic matter network. However, the inverse correlation of porosity with clay content may not necessary reflect pores that are actually flow pathways. In fact, for these samples Al Ismail & Zoback estimate effective pore widths of ~20–100 nm, an

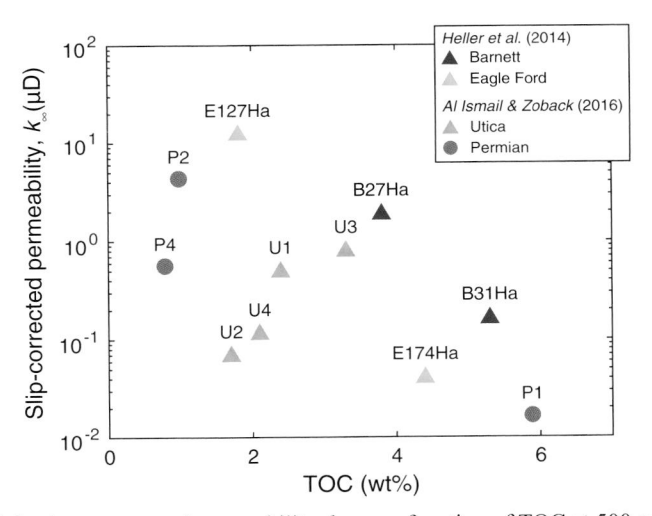

Figure 6.11 Klinkenberg-corrected permeability, k_∞, as a function of TOC at 500 psi (~3.4 MPa) for (a) Utica, (b) Permian. Pore size distributions for these samples are provided in Fig. 5.13. From Al Ismail & Zoback (2016).

order of magnitude larger than the scale of porosity that varies with clay content. In addition, SEM imaging of these samples reveals an abundance of µm-scale bedding-parallel microfractures (e.g., Fig. 5.3), which are likely responsible for the relatively high permeability values (0.1–1 µD). In the context of these supporting analyses, the observed correlation between permeability and TOC may actually reflect increased potential for microfracture development due to hydrocarbon maturation at increased organic content (Vernik, 1994). While it is not necessarily straightforward to establish a mechanistic explanation for such variations in permeability, this discussion of an apparent correlation with composition emphasizes the importance of considering multiple independent lines of evidence to understand how flow properties arise from matrix microstructure.

We might also expect that the effective stress dependence of permeability varies systematically with matrix composition as it is often interpreted to represent the compressibility of the matrix pore network (Kwon et al., 2001). The relationship between permeability and effective stress (Fig. 6.6) is often characterized by exponential decay:

$$k = k_0 e^{-C_m \sigma_{eff}} \tag{6.10}$$

in which C_m is the effective stress sensitivity of permeability (Fig. 6.3). Figure 6.12a shows C_m values as a function of clay + TOC for many of the studies represented in Fig. 6.10. Stress sensitivity generally decreases with increasing clay + TOC, but there is significant scatter and no obvious trends between samples from different reservoirs. The inverse correlation of stress sensitivity is somewhat counterintuitive because increasing the fraction of soft components should increase the compliance of the pore

164 Unconventional Reservoir Geomechanics

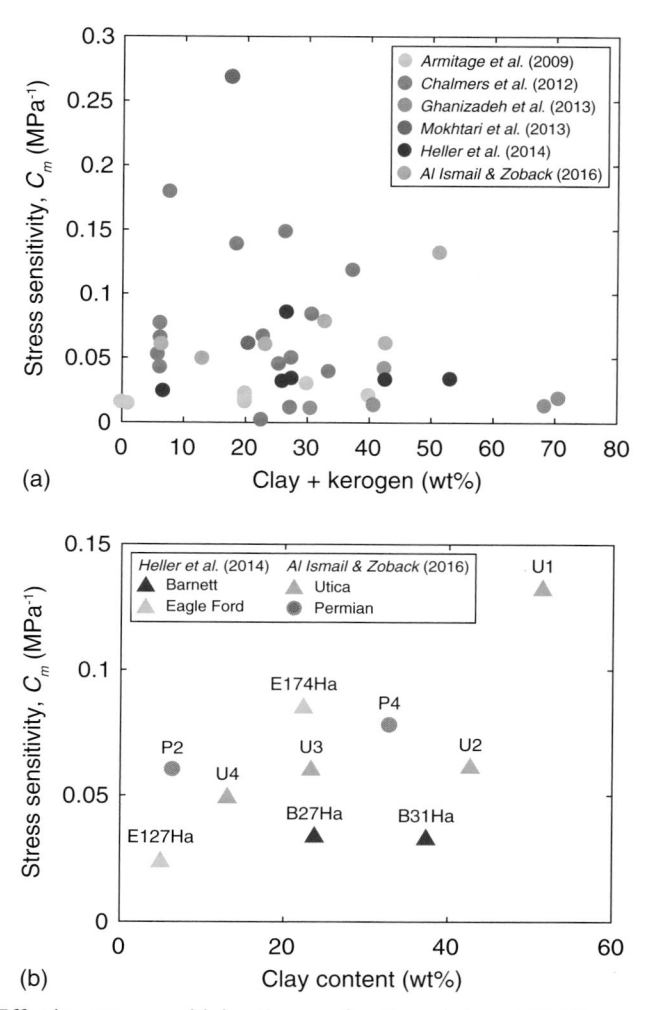

(a)

(b)

Figure 6.12 (a) Effective stress sensitivity, C_m, as a function of clay + TOC for unconventional reservoir rocks. Adapted from Gensterblum et al. (2015). (b) C_m increases as function of clay content for select samples Utica shale. From Al Ismail & Zoback (2016).

network. Again, in some cases, there are correlations observed within individual datasets, but relating these trends to pore network properties simply based on observations of flow and composition is not straightforward. We return again to the Utica samples from Fig. 6.11, which, show a positive correlation between stress sensitivity and clay content (Fig. 6.12b). Recall that these samples show increasing pore volume at the nm-scale as a function of decreasing clay content (increasing carbonate); however, since these pores are likely not responsible for the observed μD-scale permeability, it is unlikely that their closure is responsible for loss of permeability with increasing effective stress. Therefore, the observed correlation between stress sensitivity and clay content could reflect (i) the increased abundance of microfractures at the boundaries of clay-rich and -poor microfacies in clay-rich samples or (ii) differences in the distribution of the organic pore network. Even when equipped with observations of permeability, pore size distribution and composition, it is still difficult to establish a mechanistic model of how permeability varies with effective stress, which limits the interpretation of stress sensitivity as pore compressibility. A much more direct approach is to simply measure deformation during permeability measurements. Mckernan et al. (2017) performed measurements of pore compressibility (volumetric strain) by recording changes in the pore volume, and combined the results with a pore conductivity model to successfully predict permeability evolution over several loading/reloading cycles. This demonstrates how integrating deformation and flow measurements can yield a physical understanding of relationship between permeability and effective stress.

As a corollary to our discussion of how flow properties evolve with stress, it is also worth exploring the relationship between time-dependent deformation and permeability. In Chapter 3, we learned that unconventional reservoirs show appreciable time-dependent deformation (creep) under constant applied stress. Since permeability experiments are performed at constant stress conditions over periods of hours to days, we might expect time-dependent compaction to play a role in permeability evolution over time; however, there are several important issues to consider when trying to relate flow and creep properties. First, most studies of flow properties discussed above use a hydrostatic stress state (isotropic confining pressure). Sone (2012) demonstrated that creep is negligible under hydrostatic stress, so time-dependent compaction may also be niglible in permeability tests. Second, creep experiments are typically vacuumed before being performed on dried and "as received" samples, while samples in permeability tests are saturated with the test fluid or gas. Therefore, creep experiments on saturated samples may also reflect time-dependent poro-viscoelastic effects due to the movement of fluids within the compacting rock matrix. Since the stress conditions are significantly different between creep and flow experiments, any comparisons of ductility and flow properties cannot be used to explain permeability evolution.

Figure 6.13 shows the relationship between flow properties and the power law parameters that are used to characterize viscoelastic behavior (Chapter 3). Power law parameter B is equivalent to the inverse of Young' modulus (Fig. 3.11) and n determines the amount of time-dependent deformation (creep). Although both creep and flow tests were only conducted on a limited number of samples, there are some trends worth discussing. Rocks with low B and high n show higher stress sensitivity (Fig. 6.13a). These samples (i.e., Utica and Eagle Ford shales) also show relatively low permeability values in the 1–100 nD range (Fig. 6.13b). In other words, rocks with low elastic compliance that exhibit significant creep tend to have relatively low permeability and high sensitivity to effective stress. This hints at the possibility that pore compaction in relatively low permeability facies could actually be attributed to time-dependent deformation. For rocks with low B and high n, the elastic response to changing effective stress is relatively small, so the increased stress sensitivity (pore compressibility) may be attributed to creep at the pore-scale. Given the differences in the stress conditions of the experiments, we cannot read too much into these trends, but further investigation of this topic is certainly warranted.

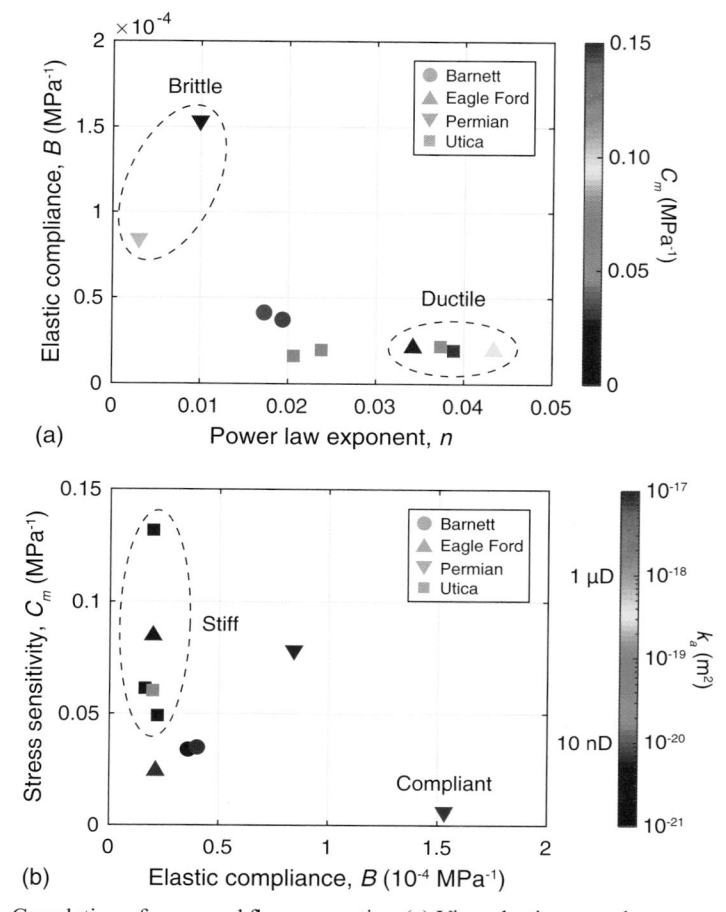

Figure 6.13 Correlation of creep and flow properties. (a) Viscoelastic power law parameters B and n colored by C_m. B is equivalent to the inverse of Young's modulus (elastic stiffness) and n determines the amount of time-dependent deformation (creep). (b) C_m and elastic compliance colored by apparent permeability.

An et al. (2018) integrated the viscoelastic creep model from Sone & Zoback (2014a) with an exponential relationship between permeability and effective stress (i.e., Eqn. 6.10) to predict the evolution of permeability with respect to time and stress. The creep and permeability model is calibrated based on the results of uniaxial creep tests and is successfully applied to describe the time- and stress-dependence of liquid permeability in samples of oil shale. Along with the study of Mckernan et al. (2017), this reinforces the importance (and utility) of considering permeability evolution in the context of deformation behavior. Future studies that simultaneously characterize elastic, viscoelastic and flow properties have the potential to determine the timescales and mechanisms of permeability evolution, which can inform models of reservoir-scale flow behavior during depletion (Chapters 10 and 12).

Flow and Sorption

Anisotropy –

As we discussed in previous chapters, the mechanical and microstructural properties of unconventional reservoir rocks are often anisotropic due to the presence of horizontal fabrics in the rock matrix (Chapter 2) and pore networks (Chapter 5). Given that flow properties manifest from the distribution of flow pathways and the mechanical response of the rock matrix to stress, it is not surprising that many unconventional reservoir rocks show significant permeability anisotropy. Figure 6.14 shows the ratio of horizontal to vertical permeabilities, k_h/k_v, for a range of unconventional reservoir rocks and mudrocks. The majority of the samples are significantly anisotropic ($k_h > k_v$) over a wide range of permeability values (100 µD–0.1 nD). Anisotropy is generally lower for samples with lower horizontal permeabilities. In most studies, increasing effective stress tends to lower permeability more in the vertical than the horizontal direction, thereby increasing anisotropy. It is well known that compaction increases anisotropy through the re-orientation of inherently anisotropic clay and phyllosilicate minerals perpendicular to the maximum compressive stress as well as the aspect ratio of bedding-parallel pores (Faulkner & Rutter, 1998; Kwon et al., 2004; Yang & Aplin, 2007; Daigle & Dugan, 2011). In addition, the development of porosity anisotropy magnifies the impact of compaction on the permeability perpendicular to layering (k_v) due to the increased tortuosity of vertical flow pathways (Arch & Maltman, 1990). However, both experimental and modeling studies indicate that the re-orientation of anisotropic, platy minerals alone is insufficient to produce $k_h/k_v > 10$ (Yang & Aplin, 2007; Daigle &

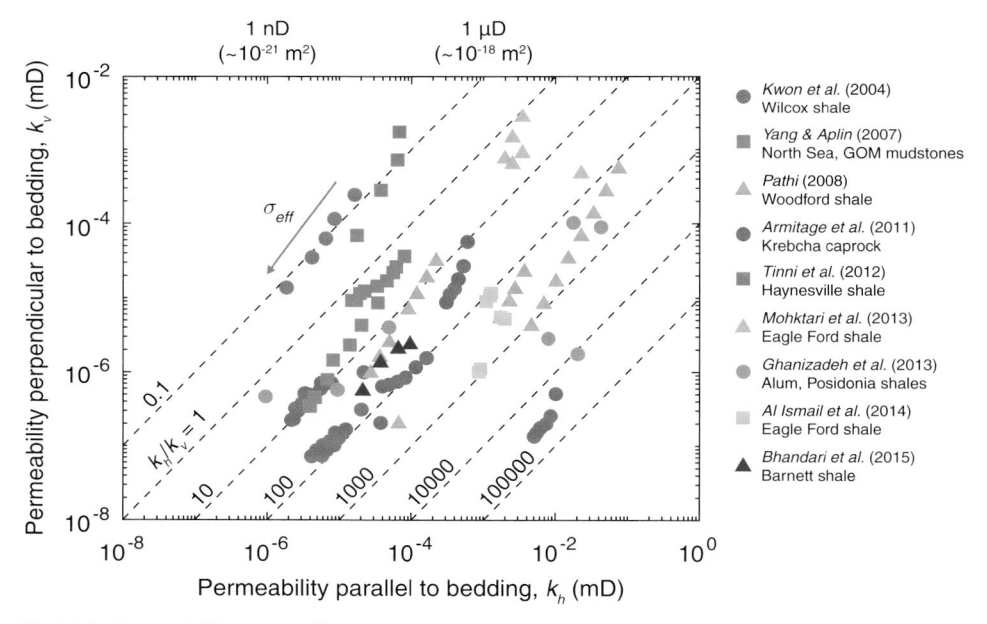

Fig. 6.14 Permeability perpendicular to bedding (k_v) vs. permeability parallel to bedding (k_h). Dashed lines indicate contours of k_h/k_v. Arrows adjacent to symbols indicate increasing effective stress. Adapted from Gensterblum et al. (2015).

Dugan, 2011; Bhandari et al., 2015). Therefore, the relatively large anisotropy ratios observed in unconventional reservoir rocks likely result from other mechanisms such as compositional layering of high and low permeability microfacies (e.g., Fig. 2.9), bioturbation (Aplin & Macquaker, 2011), diagenetic growth of clays and phyllosilicates (Loucks et al., 2012) and/or the formation of bedding-parallel microfractures (e.g., Fig 5.3). In some cases, the contribution of slip flow is also anisotropic ($b_h < b_v$), which may reflect the effects of relatively narrow vertical flow paths and increased tortuosity in the vertical direction (Letham & Bustin, 2016). Given our discussion of the complex relationship between flow properties and matrix composition, determining the specific causes of permeability anisotropy in any one lithofacies requires integrating observations of pore network and mechanical properties.

Sorption

In addition to permeability, the sorption properties of the matrix may also impact reservoir-scale flow behavior, particularly during gas production (depletion). The term sorption encapsulates the physical and chemical processes involved when one material is incorporated into another. Absorption refers to the dissolution of a gas or liquid (absorbate) into a liquid or solid (absorbent). Adsorption is the physical process by which the absorbate adheres to the surface of the absorbent. In unconventional reservoirs, natural gas is stored in the matrix as a free phase and a dense, liquid-like adsorbed phase on the pore surfaces, which results in greater overall storage capacity relative to the pore space being filled by only free gas (Fig. 5.16). Sorption capacity increases as a function of pore pressure, so depletion of the matrix releases adsorbed gas, which may impact the evolution of effective stress in the matrix during production.

Measurement, Mechanisms and Models – The adsorption capacity of a material is quantified by the adsorption isotherm, which is the surface uptake of an absorbate as a function of pressure at isothermal conditions. The magnitude and shape of the adsorption isotherm is determined by the adsorption capacity, as well as the surface properties and pore size distribution of the porous media. Adsorption capacity generally increases with stronger absorbate–absorbent interactions and increasing pore surface area. As pore size decreases, the absorbate particles are more strongly affected by interactions with the pore walls, which increases the affinity to the absorbent surface. Although this represents a distinct mechanism from surface adsorption, it also results in gas being incorporated into a denser phase on the pore surface.

In ultra-low permeability rocks, adsorption isotherms are measured on crushed samples (~50–500 μm particles) because of the impractically long equilibration times required for intact samples. For a detailed discussion of adsorption measurement techniques, see Heller & Zoback (2014). The standard method of measurement is based on Boyle's law for gas expansion, similar to helium pycnometry (Chapter 5). An experiment consists of two pressure cells: a reference cell with a known volume and a sample cell that contains the crushed material. In each pressure step, the reference

Flow and Sorption 169

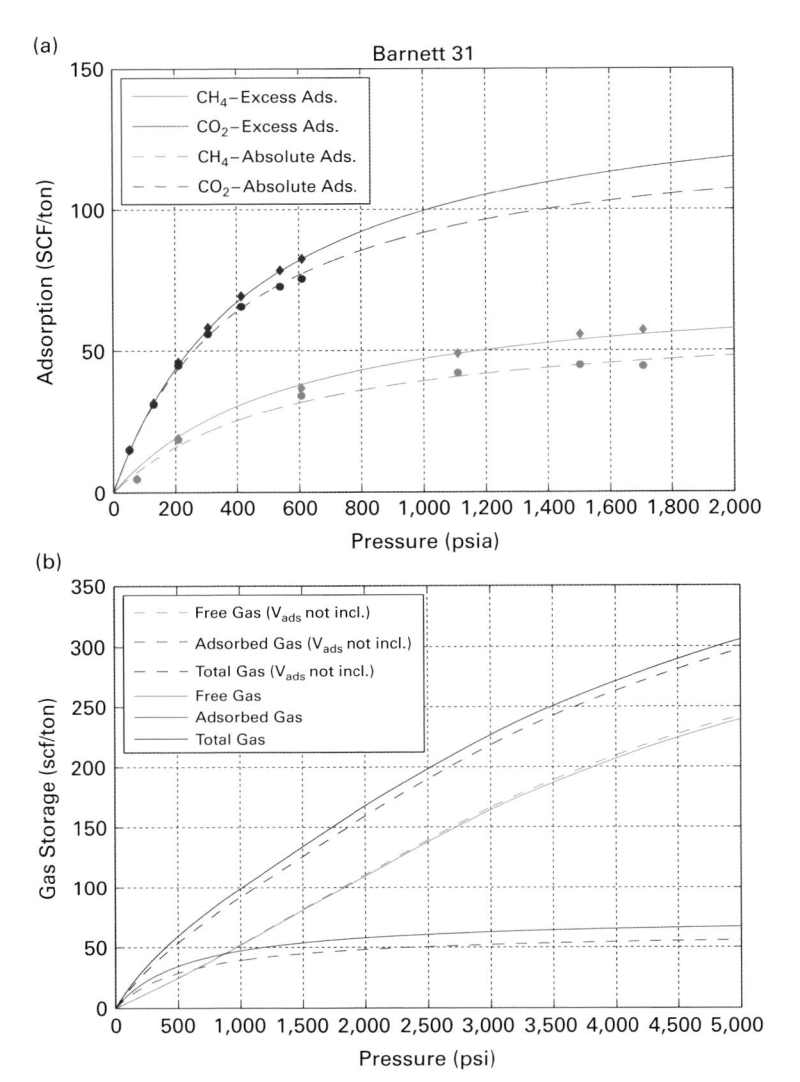

Figure 6.15 (a) Excess and absolute sorption of CH_4, and CO_2 for Barnett 31 Ha at 40 °C fit by Langmuir single layer models. (b) Gas storage of CH_4 and CO_2 as a function of pressure. Solid lines indicate data corrected for the volume of the adsorbed phase. From Heller & Zoback (2014). *Reprinted with permission from Elsevier whose permission is required for further use.*

cell is filled to determine the number of molecules using the ideal gas law, n_{total}, and then is opened to the sample cell, allowing the gas to expand into the void space between and within the sample particles. First, a non-adsorbing gas (e.g., He or Ar) is used to determine the void space in the sample cell. Then, the system is vacuumed and the same procedure is repeated with an adsorbing gas. If adsorption occurs, the pressure drop during gas expansion will be greater because some of the gas molecules will be physically bound to the particle surface, and thus will not contribute to pore pressure. The total amount of adsorption is determined from the difference of the total number of molecules in the reference system and the number of molecules of the adsorbing gas, $n_{total} - n_{free}$. This quantity is known as the excess adsorption as it includes additional gas compared to what would be present if the adsorbed-phase volume were filled with bulk gas. Correcting for the fraction of the pore space filled by the adsorbed phase requires knowledge of the density of the adsorbed phase, and yields what is known as the absolute adsorption. This procedure is performed at various pressures to determine the absorption isotherm.

To extract information about the pore structure and surface properties, the adsorption isotherm is fit to one of many physical models that describes the absorbate-absorbent interaction. In most studies of unconventional reservoir rocks, methane adsorption isotherms are characterized by a Langmuir model, which assumes adsorption is

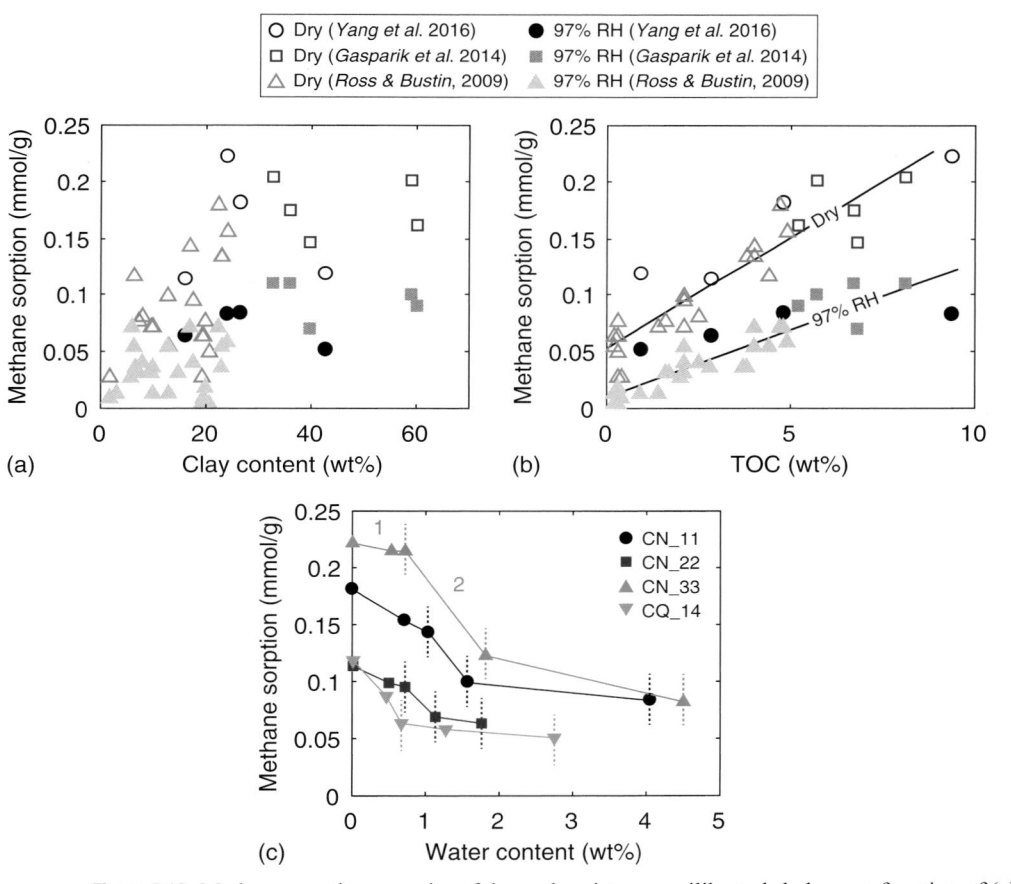

Figure 6.16 Methane sorption capacity of dry and moisture-equilibrated shales as a function of (a) clay content (b) TOC and (c) water content (Sichuan Basin shales). Ross & Bustin (2009) – Western Canadian Sedimentary Basin shales, Gasparik et al. (2014) – Alum and Posidonia shales, Yang et al. (2016) – Sichuan Basin shales. From Yang et al. (2016).

a reversible, isothermal process that manifests as a monolayer of molecules on all available surface sites (Langmuir, 1916). Adsorption increases as a function of pressure until all sites on the pore surfaces are filled. Figure 6.15a shows adsorption isotherms for methane and CO_2 for a sample of Barnett shale. As is observed in similar studies, CO_2 adsorbs much more than methane, nearly double in this case, which reflects its stronger absorbate-absorbent interactions (affinity) with the rock matrix (Nuttall et al., 2005; Kang et al., 2011; Aljamaan (2015). Although beyond the scope of this book, the preferential absorption of CO_2 over methane illustrates the potential for using CO_2 as a fracturing fluid (Middleton et al., 2015) and sequestering CO_2 in depleted unconventional gas reservoirs (Godec et al., 2013; Liu et al., 2013).

Figure 6.15b shows gas storage in Barnett 31 Ha as a function of pressure based on the Langmuir fits to the adsorption data. At low pressures <750 psi (~5 MPa), the amount of adsorbed gas exceeds the amount of free gas, but at the estimated pore pressure of the Barnett shale, ~3,000–4,000 psi (20–27 MPa) (Edwards & Celia, 2018), the adsorbed gas is only ~20–25% of the total.

Compositional and Pore Fluid Controls – To assess the important of adsorption in unconventional reservoirs, it is important to understand the variations in sorption properties as a function of rock matrix composition and the pore fluid environment. Studies of the sorption of capacity of individual matrix components demonstrate that adsorption occurs primarily in clay minerals and organic matter (Ross & Bustin, 2009; Heller & Zoback, 2014). Figure 6.16a,b shows the methane sorption capacity of various shales plotted as function of clay and TOC. Each dataset includes experiments conducted on samples at dry and at 97% relative humidity (RH) conditions. Sorption capacity is much less under 97% RH, which reflects the effects of the competitive sorption of water and capillary blocking (Chapter 5). Devonian-Mississippian shales from Western Canada show increasing sorption capacity with increasing clay content (Ross & Bustin, 2009), but no correlation is observed in any of the other datasets. This behavior is often observed in organic-poor, clay rich shales in which the pore surface area is dominated by clay minerals (Gasparik et al., 2012; Yang et al., 2015). Sorption capacity is strongly correlated with TOC in all datasets for both dry and 97% RH conditions, which indicates the presence of organic matter is the primary control on methane sorption.

Yang et al. (2016) also measured methane sorption capacity at increasing RH from 33% to 97% to evaluate the effect of water content on sorption in marine shales from the Sichuan Basin. Figure 6.16c shows methane sorption declines in a step-wise manner with increasing water content. Yang et al. interpret the reduction in sorption capacity in three stages. The initial, gradual decline is attributed to the preferential adsorption of water on the hydrophilic surfaces of clay minerals. This is supported by the fact that relatively organic-rich, clay poor samples (CQ-14, -22) experience the smallest reduction in sorption capacity during the initial stage. The steeper, secondary decline is attributed to capillary blocking of nm-scale pores by water, which reduces available surface area for adsorption and limits the access of methane within the pore network. Relatively organic-rich samples show a larger effect during this stage, which is consistent with the observation that nm-scale porosity is concentrated within organic matter in these samples. The final, gradual decline is interpreted to represent volumetric displacement of fas by water in macropores that do not contribute significantly to methane sorption capacity. Similar behavior is observed in relatively porous coals (Day et al., 2008; Gensterblum et al., 2013), but has not been widely reported in shales. The variability in the onset of the different stages as function of water content likely reflects the effects of variable matrix composition and pore network characteristics on the processes described above.

Swelling – In addition to measuring sorption capacity of clay minerals and activated carbon, Heller & Zoback (2014) also recorded swelling (volumetric strain) associated with adsorption of methane and CO_2. Figure 6.17 shows volumetric strain as function of the amount of CO_2 and methane adsorption for clay minerals illite and kaolinite, as well as activated carbon, which represents an analog for organic matter. Each dataset shows an approximately linear relationship between swelling and adsorption log–log space for both CO_2 and methane. For a given amount of adsorption, the amount of swelling varies considerably between clays and activated carbon, which Heller & Zoback interpret as the result of the differences in elastic stiffness. For stiffer matrix components, the amount of swelling is relatively less than for more compliant components. It is important to note that the magnitude of adsorption for the isolated matrix components is much greater than if the bulk adsorption of intact shale samples (e.g., Fig. 6.15) was attributed entirely to clay and organic matter. This is likely due to increased accessibility (porosity) and surface area of the pure mineral and carbon samples relative to clays and organic matter contained within chips of the rock matrix.

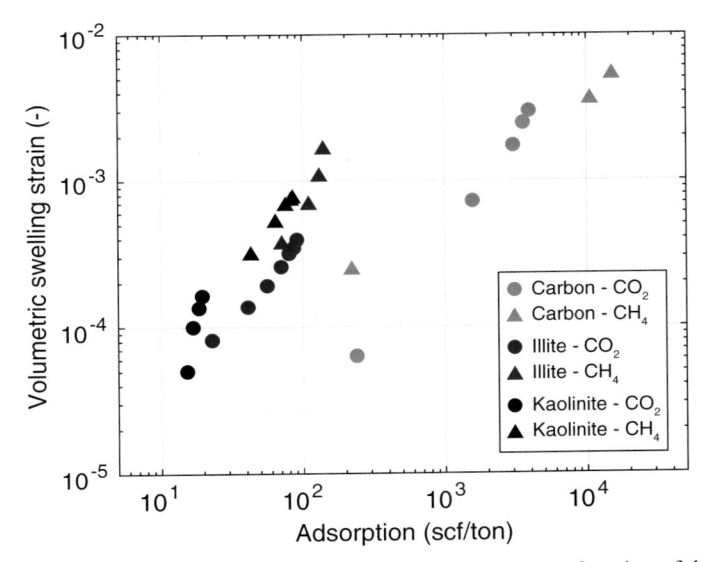

Figure 6.17 Volumetric swelling strain of clays and activated carbon as function of the amount of adsorbed CH_4 and CO_2. For a given amount of adsorption, more compliant phases show greater swelling strain. From Heller & Zoback (2014). *Reprinted with permission from Elsevier whose permission is required for further use.*

It is also well known that clay minerals absorb water and swell in a variety of sedimentary rocks, which can significantly impact permeability, clay mobility (dispersion) and wellbore stability (rock strength). The magnitude of absorption and swelling depends on a variety of factors, including clay mineralogy, fluid composition (ionic strength) and pressure. For a comprehensive review of swelling by water adsorption in unconventional reservoir rocks, see Lyu et al. (2015). As discussed in Chapter 2, swelling clay minerals in unconventional reservoir rocks include smectite and mixed layer illite–smectite (Fig. 2.3). These minerals lack hydrogen bonding between the octahedral layers of the unit cells, allowing for water to enter their structures and cause volumetric expansion (swelling) up to 10–20 times their original volume. Clay swelling also is strongly controlled by the ionic content of adsorbing fluid due to the process of cation exchange between the fluid and clay interlayers. Swelling decreases with increasing ionic strength, but also depends on the specific cation species. Ca- and Mg-exchanged clays show reduced swelling compared to Na- or K-exchanged clays, which reflects differences in ionic size and electrostatic interactions with the octahedral layers. Increasing confining pressure inhibits volumetric expansion by counteracting the swelling pressures generated by water adsorption. Water imbibition and swelling also have significant impacts on the mechanical properties (elasticity, strength and friction) of clay minerals (Moore & Lockner, 2004; Lyu et al., 2018), but for the purposes of this chapter we will focus on the impacts on flow properties.

Effects of Adsorption on Permeability – Swelling of the rock matrix due to imbibition and adsorption is a potentially important control on *in situ* permeability during injection processes (e.g., hydraulic stimulation, CO_2 storage). In Chapter 5, we review the fundamental processes controlling imbibition (wettability and capillarity), as well as the effects of multi-phase saturation on flow properties (relative permeability). Here, we focus on the impacts of gas and fluid adsorption on laboratory measurements of permeability. Figure 6.18a illustrates the effect of adsorption and swelling on gas permeability through a multi-stage experiment that employs both He and CO_2. In the first and second cycles, permeability is measured with He at 2,000 and 500 psi, respectively. Both cycles show permeability hysteresis with effective stress, which reflects inelastic (non-recoverable) pore compaction during initial loading. The lower pressure cycle shows slightly higher initial permeability (possibly due to slip flow), but much greater hysteresis due to the effect of increased effective stress on pore compaction. In the third cycle, the sample is vacuumed and permeability is measured with CO_2 at 377 psi. This pressure is specifically selected in order to obtain a CO_2 mean free path that matches that of He at 2,000 psi, thereby eliminating any confounding effects of slip flow in any comparisons. In the fourth cycle, the sample is vacuumed and the He measurement at 2,000 psi is repeated to assess the permanent effect of CO_2 adsorption on permeability. The reduction in permeability between cycles 1 and 4 represents the effect of irrecoverable pore compaction over the first two cycles of effective stress, while the lack of difference between cycles 2 and 4 shows that the reduction in permeability due to CO_2 adsorption (~80%) is almost entirely recoverable. Al Ismail et al. (2014) also performed this analysis on a complementary horizontal sample and actually observed elevated

174   Unconventional Reservoir Geomechanics

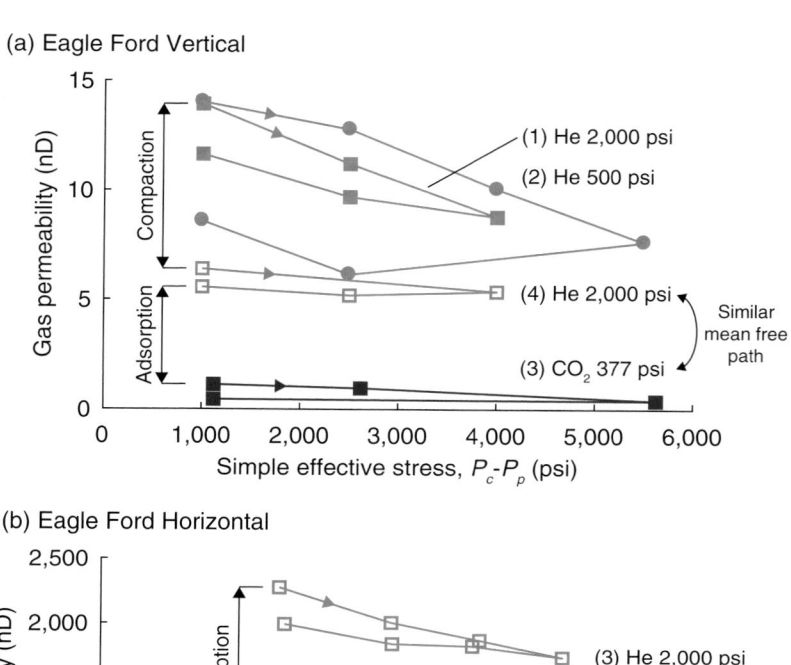

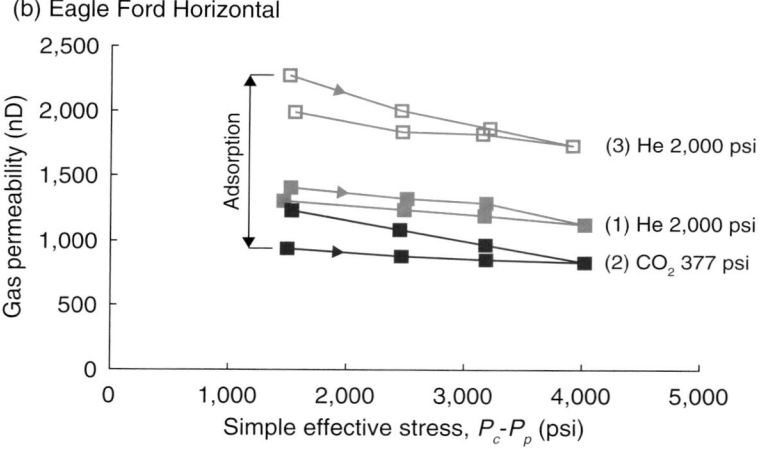

Figure 6.18 Effect of CO_2 adsorption on permeability evolution with effective stress. (a) Eagle Ford AUK6-1V. Cycle 1: He gas at 2,000 psi (13.8 MPa). Cycle 2: He gas at 500 psi (3.4 MPa). Cycle 3: CO_2 at 377 psi (similar mean free path to Cycle 1). Cycle 4: He gas at 2,000 psi. The reduction in permeability between Cycles 1 and 4 represents pore compaction, while the lack of difference between Cycles 2 and 4 implies that permeability reduction from CO_2 adsorption is reversible. (b) Eagle Ford AUK1-1H. See Table 2.1 for sample compositions. Adapted from Al Ismail et al. (2014).

permeability after the measurement with CO_2 (Fig. 6.18b). Note that the permeability of the horizontal sample, Eagle Ford AUK1-1H is nearly 3 orders of magnitude greater than that of the vertical sample. This pre-existing permeability anisotropy is apparently magnified after the CO_2 measurement, which Al Ismail et al. attribute to growth of sub-horizontal microfractures during the first two cycles of effective stress.

    Shen et al. (2017) examined the effects of water imbibition and adsorption on gas permeability by performing *ex situ* imbibition steps in-between permeability measurements. Figure 6.19 shows permeability and water content as a function of time for two horizontal samples from the Longmaxi shale. The samples initially exhibit similar patterns of permeability evolution and water imbibition with respect to time, but ultimately undergo distinct physical changes in the flow pathways with increasing water content. Shale Y1 shows an initial exponential decay of permeability until ~120 min, at which point permeability rapidly increases and decays just as rapidly. Shen et al. interpret the initial decline as the narrowing of flow pathways due to water adsorption and clay swelling, as well as increasing stress sensitivity (pore compliance) with increasing water content. The latter is consistent with observations of reduced bulk compressive strength under water-saturated conditions (Lyu et al., 2018). The sudden, rapid increase in permeability is attributed to microfracture propagation due to the accumulation of swelling pressures with increasing water content. In each sample, Shen et al. document the development of

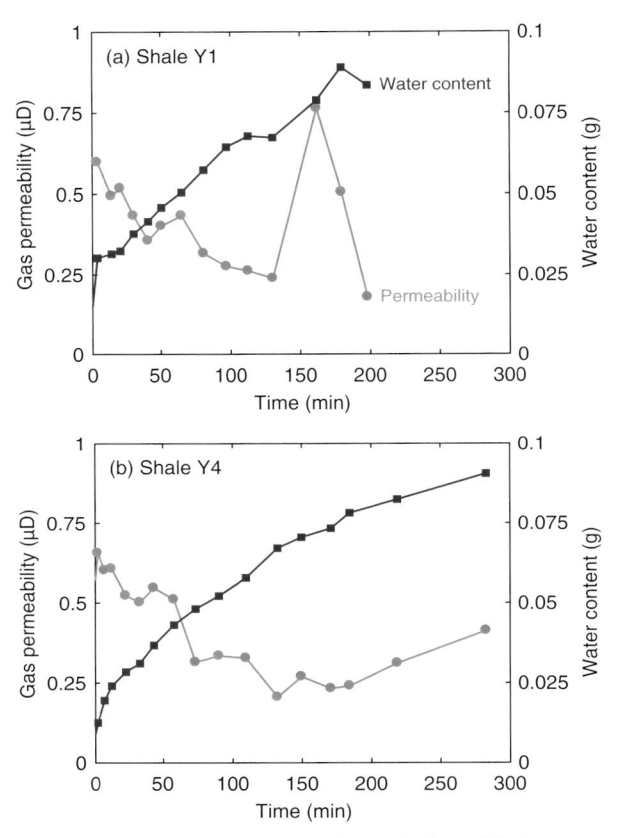

Figure 6.19 Effect of water imbibition on permeability evolution with time, Longmaxi shales. Permeability measurements are performed at 8 MPa confining pressure and 5 MPa pore pressure. Imbibition steps are performed *ex situ* at ambient conditions. (a) Shale Y1 shows an initial exponential decrease in permeability with increasing water imbibition. This is followed by a rapid recovery in permeability which decays just as with increasing imbibition. (b) Shale Y4 shows approximately linear decay during initial imbibition, followed by approximately linear recovery at a slightly lower rate. From Shen et al. (2017).

through-going, bedding-parallel fractures after the complete procedure. Fracture propagation may be magnified due to the unloading/loading cycles inherent to this procedure, in particular if complete depressurization is not achieved between the permeability and imbibition stages. The rapid decay of permeability enhancement likely represents closure of fractures due to the re-application of confining pressure and further capillary blocking due to additional water adsorption. Shale Y4 shows an initial quasi-linear decay in permeability over a similar time period to Y1, which is followed by a slower, linear recovery of permeability with increasing imbibition. In this case, a microfracture network likely formed relatively slowly, with each increment of imbibition enhancing its development. The differences in permeability evolution between these samples illustrate the control of initial microstructure (in this case, microfractures) on the imbibition response of the rock matrix. Shen et al. also perform this procedure using water with an added anionic surfactant design to limit adsorption and swelling – a practice which is often termed "clay stabilization." In this case, permeability decreases only slightly during the initial stage and then remains constant. The overall imbibed mass of the surfactant mixture is only about half of that for distilled water. The lack of a dramatic loss of permeability and any noticeable fracture propagation effects suggest that the surfactant successfully limited clay swelling due to water adsorption. Although not the focus of this book, it is important to understand that during hydraulic stimulation, the fluids penetrating the formation are specifically engineering with clay stabilizers and other components to minimize the loss of permeability (permeability damage). We will discuss fracturing fluids further in Chapter 8 (in terms of the mechanics of hydraulic fracturing), as well as in Chapter 13 (in terms of environmental issues associated with re-use and disposal).

Implications for Hydrocarbon Production – Ultimately, the goal of studying the flow and sorption properties of unconventional reservoirs rocks is to understand the mechanisms and timescales of long-term hydrocarbon production from the ultra-low permeability matrix. Given our discussion of the pressure-dependence of adsorption, it is important to understand to what degree adsorbed gas contributes to production. Figure 6.20 shows the result of a simple model for gas production based on the adsorption isotherm of Barnett 31 Ha (Fig. 6.13). The conditions of this model are the following: 40 °C, 8% matrix porosity and 25% water saturation. At infinitely small pressures, adsorbed gas is ~20% of total production and decays rapidly with increasing pressure. At the estimated *in situ* pore pressure of the Barnett shale (~3,000–4,000 psi), the amount of adsorbed gas produced is negligible compared to free gas. This sample shows relatively large sorption capacity due to its high clay and organic content, so for lithologies on the other end of the composition spectrum, adsorbed gas production would be even less important.

Slip flow is another phenomenon that may be relevant at low pore pressures. It is important to note that the magnitude of slip flow in many of the experimental studies reviewed above is measured using He gas (e.g., Fig. 6.10). At the same pressure–temperature conditions, magnitude of slip flow for methane will be half that of helium due to its smaller mean free path (Fig. 6.2a). While this implies that slip flow is not as important for *in situ* gas production, methane is also capable of a surface diffusion

Flow and Sorption   177

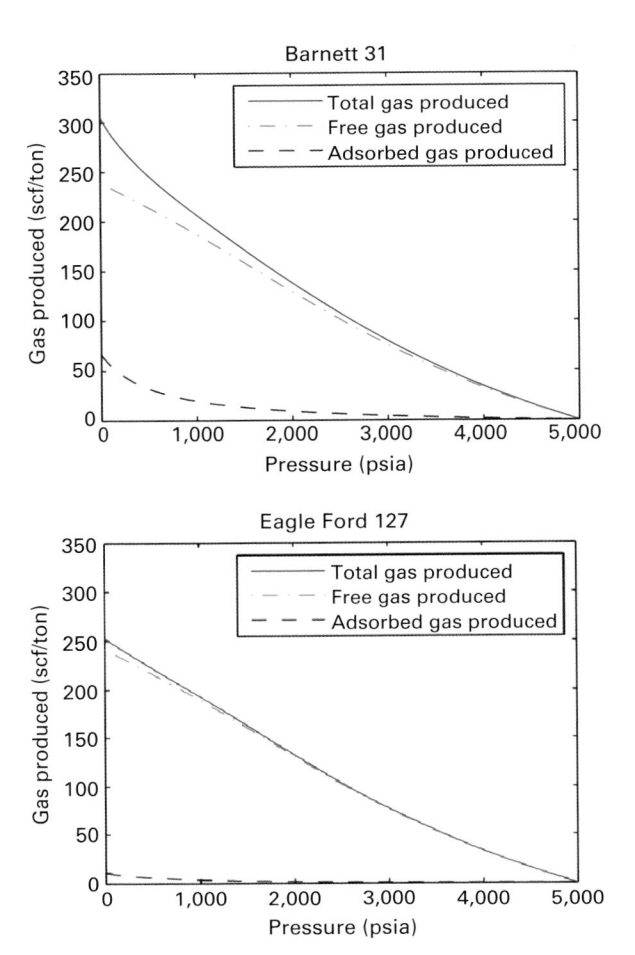

Figure 6.20 Estimated production of free and adsorbed methane as a function of pressure at 40 °C for clay + TOC-rich sample Barnett 31Ha and clay + TOC-poor sample Eagle Ford 127Ha. Adapted from Heller & Zoback (2014). *Reprinted with permission from Elsevier whose permission is required for further use.*

in the presence of an adsorbed phase (Tien, 1994), which may enhance permeability in a similar manner to slip flow.

Experimental studies of the flow and sorption properties of unconventional reservoirs have opened the door for numerical simulations to study the impacts of various physical processes on gas production at various scales. Since this chapter focuses primarily on the ultra-low permeability matrix, we will consider pore-scale simulations that incorporate the physics of diffusion and adsorption. Figure 6.21 shows the results of gas production simulations performed on a μm-scale volume that was digitized from FIB-SEM imaging of a sample of Haynesville-Bossier shale (Guo et al., 2018). This model uniquely couples a continuum (effective medium) approach for the sub-resolution nano-porous constituents (clays, organic matter) with a direct pore-scale approach (Stokes flow) for the larger pores (macropores) that are captured in the imaging dataset. This enables application of the physical models for flow and adsorption discussed earlier in the chapter. There are four models plotted in Fig. 6.21: DS – Darcy–Stokes flow only; DSA – Darcy–Stokes and adsorption; NDSA – Darcy–Stokes, adsorption and Knudsen diffusion; Full model – NDSA and surface diffusion. In the simulations, the rock volume is set at an initial pressure and a lower pressure is set at one face that is open for flow, which drives gas production from the volume. The production rate and time have been non-dimensionalized in the results.

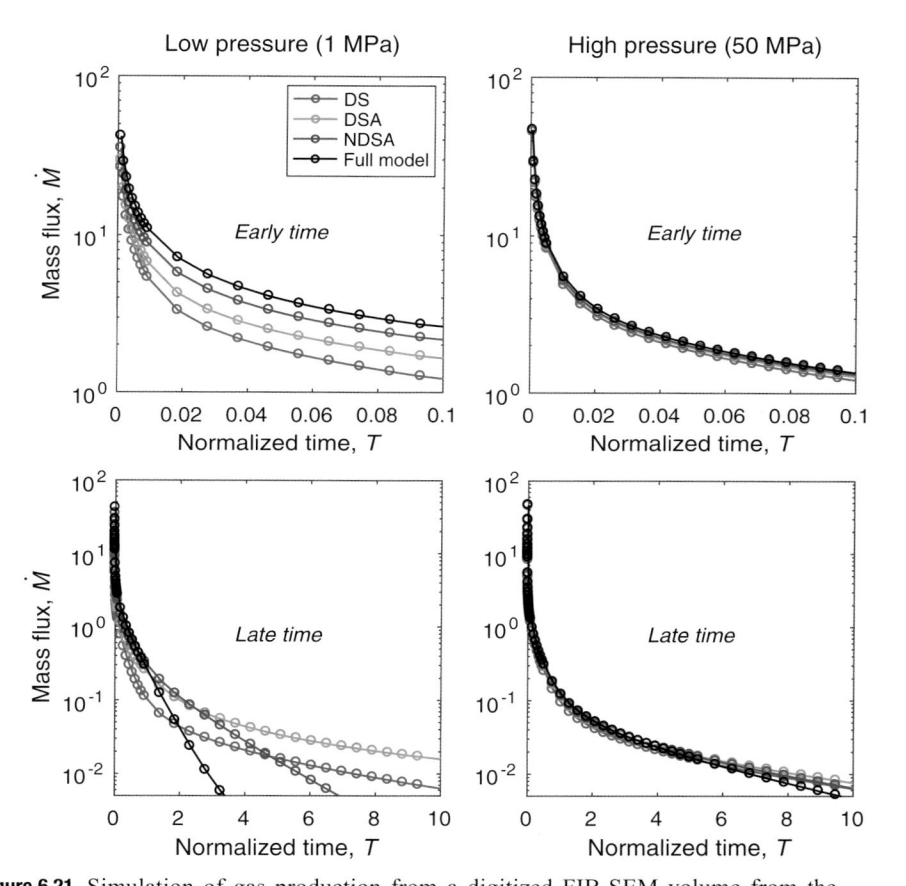

Figure 6.21 Simulation of gas production from a digitized FIB-SEM volume from the Haynesville-Boisser shale as a function of dimensionless time, T, at (a,b) 1 MPa and (c,d) 50 MPa pore pressure. At low pressure and early time, production is entirely due to Stokes flow through the macropore network, resulting in an equal decline rate for all models. At late time, when the macropores are depleted, the NDSA and full model show greater decline rates due to increased diffusive flux from micropores at low pore pressures. At high pressure, desorption and diffusive flux are suppressed, so all models show the same decline rate resulting from Stokes flow. See the text for details of the model abbreviations. From Guo et al. (2018).

At low pressure (1 MPa) and at early time ($0.05 < T < 0.1$), the production rates for all models decline linearly in semi-log plot with similar slope (Fig. 6.21a). This is because the 3D image has a macropore network spanning the entire sample, which allows Stokes flow to dominate the gas production at early time. Note that at a very early time ($T < 0.05$) material heterogeneity (macro-pores intersecting the open boundary) is the dominant control on gas production, which does not represent the behavior of the entire volume. At late time, the production rates differ considerably between the models (Fig. 6.21b). The Darcy–Stokes models show similar decline rates to early time because flow is not pressure-dependent. Models that consider adsorption show higher initial values of mass flux due to their increased gas storage potential. The NDSA and full model show faster decline rates at late time because the gas can escape faster via diffusive mechanisms at low pressure. At high pressure, all models show essentially the same behavior at early and late time because the pressure is too high for adsorbed gas or diffusive mechanisms to contribute to production (Fig. 6.21c,d). Although these models represent a simplified, single phase system, the results confirm that the physical processes interpreted from laboratory measurements of flow and sorption have important implications for gas production from the matrix on the pore-scale. In other words, while a FIB-SEM volume does not necessarily constitute a representative elementary volume (REV) in terms of the matrix pore structure (Chapter 5), it may represent a microcosm of the reservoir in terms of its depletion behavior. Integrating these sort of numerical experiments with approaches for upscaling matrix pore structures has the potential to answer questions related to the scale-dependence of depletion behavior, as well as the effects of mm- to cm-scale compositional layering on reservoir flow properties.

Finally, when considering pressure-dependent mechanisms that may enhance production, it is important to ask the obvious: what in fact is the *in situ* pore pressure in rock matrix during production? It has been widely hypothesized that desorption and diffusion are responsible for maintaining production over the long, flat tails of decline curves (Chapter 1), but there is essentially no direct evidence for how the pore pressure in the matrix evolves during production. Although depletion may be estimated using modeling and geophysical methods (Chapter 12), due to the dual-porosity nature of unconventional reservoir rocks (fractures and matrix), any measurements of *in situ* pore pressure are only likely representative of the interconnected fracture network, not the matrix. The timescale of flow from the matrix to fracture network dictates the area of the matrix that is depleted over time, which, for ultra-low permeability reservoirs, is on the order of tens of meters (Fig. 12.2). This is significant because much of the reservoir does not experience any pressure drawdown from production, so these regions have the potential to replenish the depleted zones around fractures over time. Although pressure in the fracture network declines rapidly during initial production, the slow, constant diffusion of gas from the matrix may explain relatively low, sustained production rates at long production lifetimes (on the order of years) (Chapter 1). If the areas around the fracture network are depleted and replenished at similar rates, the pressure in the matrix may never become low enough for significant contributions from diffusion or desorption. Given that existing studies have extensively characterized the activity of these mechanisms in relatively simple, single-phase systems, future studies have the potential to assess their importance at the reservoir-scale by examining experimental systems that more directly address the *in situ* conditions during production (depletion).

7 Stress, Pore Pressure, Fractures and Faults

There are a number of interrelated topics presented in this chapter that define the geomechanical state of unconventional reservoirs. As alluded to in Chapter 1 (and expanded upon in Chapters 10–12) the process of hydraulic fracturing and stimulating slip on pre-existing fractures and faults is critical to the success of production from unconventional formations with extremely low permeability. This entire process depends on the interplay between the stress field, pre-existing fractures and faults, pore pressure and the perturbation of pore pressure that occurs during hydraulic fracturing. Chapter 8 discusses how this kind of comprehensive geomechanical characterization affects hydraulic fracturing and Chapter 11 discusses how the geomechanical characteristics of underlying and overlying formations affect vertical hydraulic fracture growth.

In the first section of this chapter we define relative stress magnitudes in some detail, and we present detailed stress orientations and relative stress magnitudes in different regions of the US where unconventional reservoirs are currently being produced. By combining both stress orientation and relative stress magnitude data, we present a new generation of stress maps for Texas, Oklahoma and southeastern New Mexico, along with detailed maps and discussion of the state of stress in the Permian and Fort Worth Basins. We also present a detailed map of the part of the Appalachian Basin in the northeastern US where the Marcellus and Utica shale plays are being developed. In the second section we review how stress orientations, and stress magnitudes are measured. The third section reviews pore pressure in a number of active plays in the US and Canada, how it is measured and the linkage between pore pressure and stress magnitude. The fourth section of the chapter addresses fractures and faults in unconventional reservoirs, initially focusing on relatively small fractures within individual lithologic units. As mentioned in Chapter 1, the microseismic events that accompany hydraulic fracturing are associated with slip on small pre-existing fractures (at the ~1 m scale), which are induced to slip by high fluid pressure leaking out of the hydraulic fractures. This process is discussed in detail in Chapter 10 as it is critical to production by creating an interconnected, permeable fracture network (which itself is further discussed in Chapter 12). We then discuss the architecture of larger pad-scale faults (approximately the kilometer scale) and discuss how they influence the process of hydraulic stimulation in a number of important ways. Finally, we discuss how 3D seismic data can be used to map fractures in reservoirs and the frequent confusion about whether it is fractures or stress that control horizontal velocity anisotropy.

182   Unconventional Reservoir Geomechanics

State of Stress in US Unconventional Reservoirs

Relative Stress Magnitudes – In the first part of this section we extend classical Andersonian faulting theory that defines the relationship between relative principal stress magnitudes and their orientation (S_V, S_{Hmax} and S_{hmin}) with the style, and orientation of currently active faults (normal, strike-slip and reverse) as originally defined by Anderson (1951). In the classical formulation, the key relationships between stresses and faults as defined by Anderson involve only pre-existing fractures and faults in a rock mass that might be active in the current stress field. In other words, it is logical that the fractures and faults we observe in a rock mass today, which could have a wide array of orientations, were formed in multiple episodes over very long periods of geologic time (most unconventional formations range in age from early Paleozoic to mid-Mesozoic – roughly 400–100 Ma). Hence, we do not show all the faults that are likely present in Fig. 7.1, nor do we consider how the fractures and faults formed. Rather, we discuss below (and consider throughout this book) which and how pre-existing fractures in the rock mass might be activated to slip in shear as faults in the current stress field, especially as they are stimulated to slip during hydraulic fracturing.

The center column of Fig. 7.1 shows the three familiar Andersonian faulting regimes. In normal faulting stress states, the overburden is the maximum principal stress ($S_V > S_{Hmax} > S_{hmin}$), and active faults are steeply dipping (~60°) and strike subparallel to S_{Hmax}. In strike-slip faulting regimes, the overburden is the intermediate principal stress ($S_{Hmax} > S_V > S_{hmin}$), and active faults are subvertical and strike ~30° in either direction from S_{Hmax}. Finally, under the more compressional stress state of reverse faulting, moderately dipping (~30°) faults striking subparallel to S_{hmin} will be active.

Although this simple formulation is extremely useful for understanding the general styles of faulting in the Earth, it does not explain the full range of possible faulting styles. To extend Anderson faulting theory to all stress conditions, we first introduce two intermediate (or transitional) stress and faulting cases (the right column of Fig. 7.1): The case in which both normal-and strike-slip faults are active at the same time ($S_V \approx S_{Hmax} > S_{hmin}$) and the case in which strike-slip and reverse faults are active at the same time ($S_{Hmax} \approx S_{hmin} > S_V$). Both states of stress are quite common in unconventional reservoirs (and in various parts of the world). There are also two limiting states of stress (the left column of Fig. 7.1), one in which normal faulting can occur on steeply dipping faults (~60°) of any strike ($S_V > S_{Hmax} \approx S_{hmin}$) or where reverse faulting could occur at moderately dipping faults (~30°) of any strike ($S_{Hmax} \approx S_{hmin} > S_V$). Both are relatively rare states of stress, but regrettably the former (very low horizontal stress anisotropy) is sometimes (incorrectly) thought to characterize many unconventional reservoirs.

To represent the relative magnitudes of the three principal stresses (S_V, S_{Hmax} and S_{hmin}) on maps, we use the A_ϕ parameter defined by Simpson (1997), which conveniently describes the ratio between the principal stress magnitudes using a single, readily interpolated value that ranges smoothly from 0 (the most extensional possible stress state) to 3 (the most compressive). The A_ϕ values that correspond to the seven stress states shown in Fig. 7.1 are shown. The parameter is defined mathematically by

$$A_\phi = (n + 0.5) + (-1)^n(\phi - 0.5) \tag{7.1}$$

where

$$\phi = \frac{S_2 - S_3}{S_1 - S_3} \tag{7.2}$$

and S_1, S_2 and S_3 are the magnitudes of the maximum, intermediate and minimum principal stresses, respectively, and n is 0 for normal faulting ($S_V = S_1$, $S_{Hmax} = S_2$

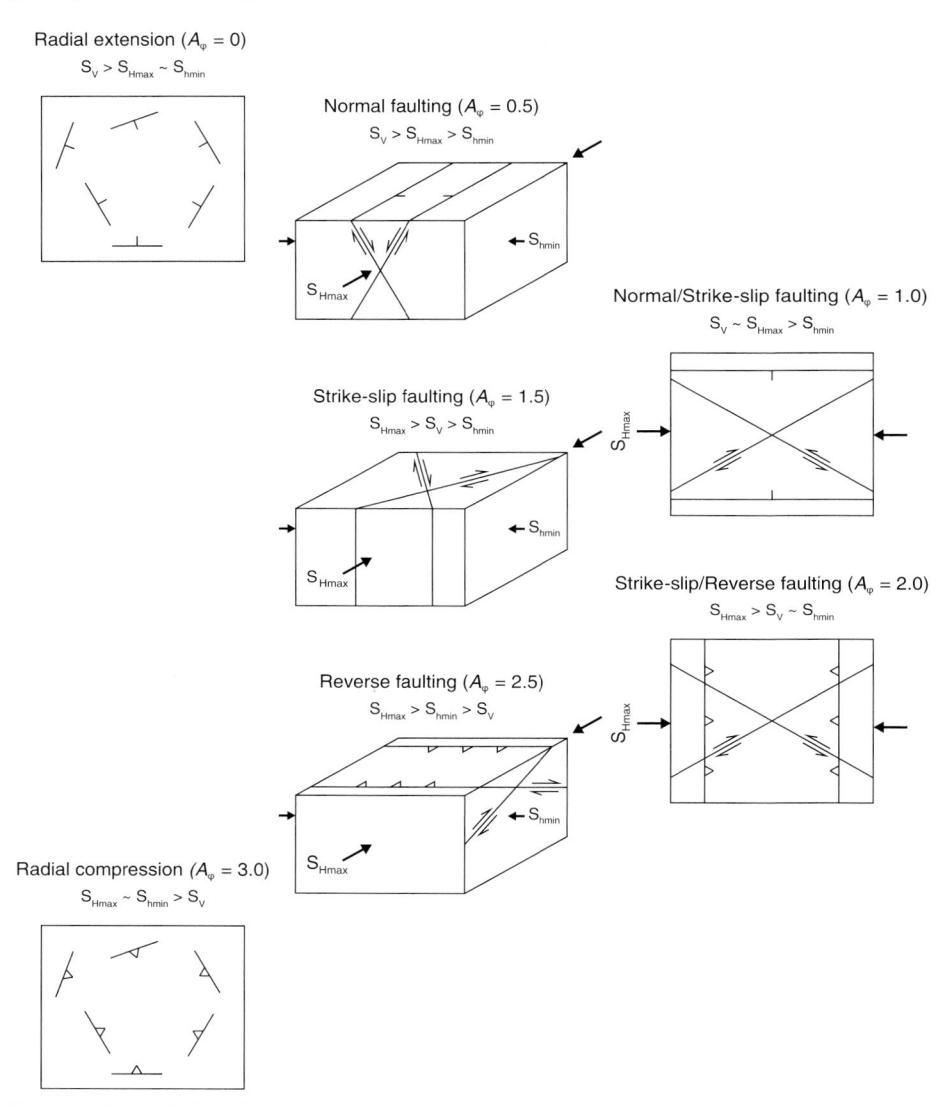

Figure 7.1 Extended Andersonian faulting theory (Anderson, 1951) to incorporate intermediate and limiting stress states (see text). We assign a numerical value to each stress state following. Simpson (1997).

and $S_{hmin} = S_3$), 1 for strike-slip faulting ($S_V = S_2$, $S_{Hmax} = S_1$ and $S_{hmin} = S_3$) and 2 for reverse faulting ($S_V = S_3$, $S_{Hmax} = S_1$ and $S_{hmin} = S_3$). Note that A_ϕ varies smoothly between all of the stress states as the ratio between the principal stresses changes.

A New Generation of Stress Maps and Quality Criteria – The map of the south-central United States shown in Fig. 7.2 (from Lund Snee & Zoback, 2018a) and the subsequent maps in this chapter show both the direction of maximum horizontal stress, S_{Hmax}, as well as relative stress magnitudes, the latter utilizing the A_ϕ parameter defined above. While we only present data for regions in which stress orientation and relative magnitude are reasonably well known for unconventional plays in the United States, our hope

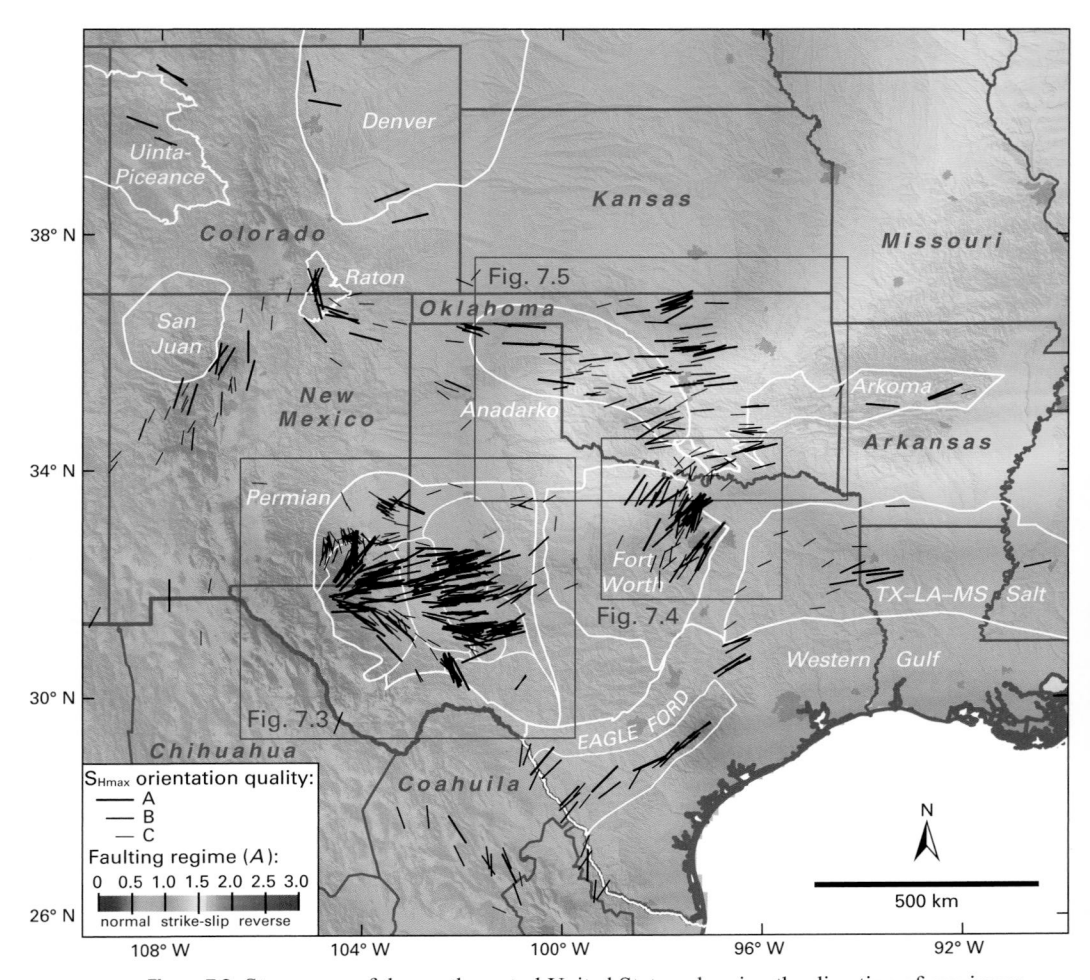

Figure 7.2 Stress map of the south-central United States, showing the direction of maximum horizontal compression, S_{Hmax} (the length of the line indicates data quality) and the magnitude of relative stress magnitude as parameterized by A_ϕ as illustrated in Fig. 7.1. After Lund Snee & Zoback (2018b). Basin outlines are from the US Energy Information Administration.

is that the methodologies defined here will be adopted by other researchers working in other parts of the world.

Methods for determination of stress orientation are well known and are briefly reviewed below. The fact that wellbore stress measurements at depths of 1–5 km are usually consistent with those from recent geologic stress indicators from shallower depths and also with stress values obtained from earthquake focal plane mechanisms from greater depth provided the basis for making comprehensive regional stress maps (M. L. Zoback & Zoback, 1980; M. L. Zoback, 1992a) such as the ones presented above. An important step in using observations of wellbore failure to determine stress orientations is to establish rigorous quality criteria. The quality of each type of stress indicator is shown in Table 7.1 (from Lund Snee & Zoback, 2018a). The quality criteria were originally developed by Zoback & Zoback (1991) but expanded to include additional stress indicators and applied to create the first World Stress Map by M. L. Zoback (1992a) and later applied and adapted to regional stress mapping projects as shown below. These updated criteria are similar to those utilized for the World Stress Map project: http://dc-app3-14.gfz-potsdam.de/pub/introduction/introduction_frame.html. The criteria in Table 7.1 do not include single focal mechanism solutions as reliable indicators because such data do not provide sufficient constraints. When no other data are available in a region, it is helpful to consider the approximate stress orientations indicated by a single, well-constrained focal plane mechanism.

The data shown in Figs. 7.2–7.5 come principally from drilling-induced tensile fractures, wellbore breakouts and crossed-dipole shear velocity logs in vertical wells, supplemented by hydraulic fracture orientations obtained from alignments of microearthquakes during multi-stage hydraulic fracturing in horizontal wells. The background color reflects the relative magnitude of the three principal stresses (warmer colors reflecting increasingly compressive stress state) utilizing the numerical scale for A_ϕ discussed above and earthquake data as discussed below. The map of the south-central US in Fig. 7.2 shows areas of relatively consistent stress orientation and relative magnitude such as the eastern part of the Permian Basin (the Midland Basin), which is characterized by approximately E–W compression (with some minor variations), and a consistent strike-slip to normal/strike-slip stress field. The northern panhandle of Texas and most of Oklahoma are characterized by a similar stress state although a large, pre-historic earthquake that occurred on the Meers fault in southwestern Oklahoma indicates strike-slip and reverse fault movement (Crone & Luza, 1990). In marked contrast with the Midland Basin, extension characterizes both the Rio Grande Rift and Texas Gulf Coast (where the Eagle Ford play is located) with the direction of S_{Hmax} paralleling the active normal faults in each region, as expected (Fig. 7.1). The stress state in central Texas shows a progressive rotation northeastward from the ~E–W compression in the Midland Basin to NE–SW compression in the Fort Worth Basin. Interestingly, the S_{Hmax} direction appears to rotate back to an E–W direction in southeastern Oklahoma. Figures 7.3, 7.4 and 7.5 show detailed maps of some of the areas that are important for unconventional development.

Figure 7.3 is a detailed stress map of the Permian Basin and the sub-basins within it (Lund Snee & Zoback 2018a, 2018b). As noted in Chapter 1, the Permian Basin is likely to be an area of very considerable unconventional development in the years to come.

Table 7.1 Quality control criterion used for the maps presented in this book (after Lund Snee & Zoback, 2018a).

Stress Indicator*		A	B	C
Drilling-induced tensile fractures		Ten or more distinct tensile fractures in a single well with standard deviation (sd) $\leq 12°$ and with highest and lowest observations at least 300 m apart	At least six distinct tensile fractures in a single well with sd $\leq 20°$ and with highest and lowest observations at least 100 m apart	At least four distinct tensile fractures in a single well with sd $\leq 25°$ and with highest and lowest observations at least 30 m apart
Focal mechanism inversions	(Directions)	Formal inversion of ≥ 35 reasonably well-constrained focal mechanisms resulting in stress directions with sd $\leq 12°$	Formal inversion of ≥ 25 reasonably well-constrained focal mechanisms resulting in stress directions with sd $\leq 20°$	Formal inversion of ≥ 20 reasonably well-constrained focal mechanisms resulting in stress directions with sd $\leq 25°$
	(Relative magnitude)	Formal inversion of ≥ 35 reasonably well-constrained focal mechanisms resulting in ϕ with sd ≤ 0.05	Formal inversion of ≥ 25 reasonably well-constrained focal mechanisms resulting in ϕ with sd ≤ 0.1	Formal inversion of ≥ 20 reasonably well-constrained focal mechanisms resulting in ϕ with sd ≤ 0.2
Wellbore breakouts		Ten or more distinct breakout zones in a single well (or breakouts in two or more wells in close proximity) with sd $\leq 12°$ and with highest and lowest observations at least 300 m apart	At least six distinct breakout zones in a single well with sd $\leq 20°$ and with highest and lowest observations at least 100 m apart	At least four distinct breakout zones in a single well with sd $\leq 25°$ and with highest and lowest observations at least 30 m apart
Microseismic alignments along hydraulic fractures		Twelve or more distinct linear zones associated with HF stages, with sd $\leq 12°$	Eight or more distinct linear zones associated with HF stages, with sd $\leq 20°$	Six or more distinct linear zones associated with HF stages, with sd $\leq 25°$
Shear velocity anisotropy from crossed-dipole logs†		Anisotropy $\geq 2\%$ present at a consistent azimuth, with highest and lowest observations at least 300 m apart, and with sd of fast azimuth $\leq 12°$	Anisotropy $\geq 2\%$ present at a consistent azimuth, with highest and lowest observations at least 100 m apart, and with sd of fast azimuth $\leq 20°$	Anisotropy $\geq 2\%$ present at a consistent azimuth, with highest and lowest observations at least 30 m apart, and with sd of fast azimuth $\leq 25°$

* The shallowest measurement must be at least 100 m deep and also sufficiently deep that measurements are not affected by topography.
† In addition to anisotropy $\geq 2\%$, measurements should ideally have an energy difference between fast and slow shear waves $\geq 50\%$ and a minimum energy $\geq 15\%$.

In addition to the maximum horizontal stress orientation and relative stress magnitude data, earthquake epicenters and focal plane mechanisms (discussed in Chapter 9) are also shown, as well as some of the regional-scale geologic structures in the region. Note that the stress state in the Central Basin Platform is quite similar to that described above for the Midland Basin. In marked contrast to the uniform stress field in those areas, the Delaware Basin stress field is locally coherent but rotates dramatically by ~150° clockwise from north to south across the basin. In the western part of Eddy County, New Mexico, S_{Hmax} is ~N–S (consistent with the state of stress in the Rio Grande Rift) but rotates to ~ENE–WSW in southern Lea County, New Mexico. S_{Hmax} continues to rotate clockwise southward in the Delaware Basin to become ~N155°E in western Pecos County, westernmost Val Verde Basin and northern Mexico (Suter, 1991; Lund Snee & Zoback, 2018b). On the Northwest Shelf, A_ϕ varies from ~0.5 (normal faulting) in north Eddy County to ~0.9 (normal and strike-slip faulting) further east. S_{Hmax} rotates significantly across the Northwest Shelf as well, from ~N–S in northwest Eddy County to ~ESE–WNW in northern Lea and Yoakum counties.

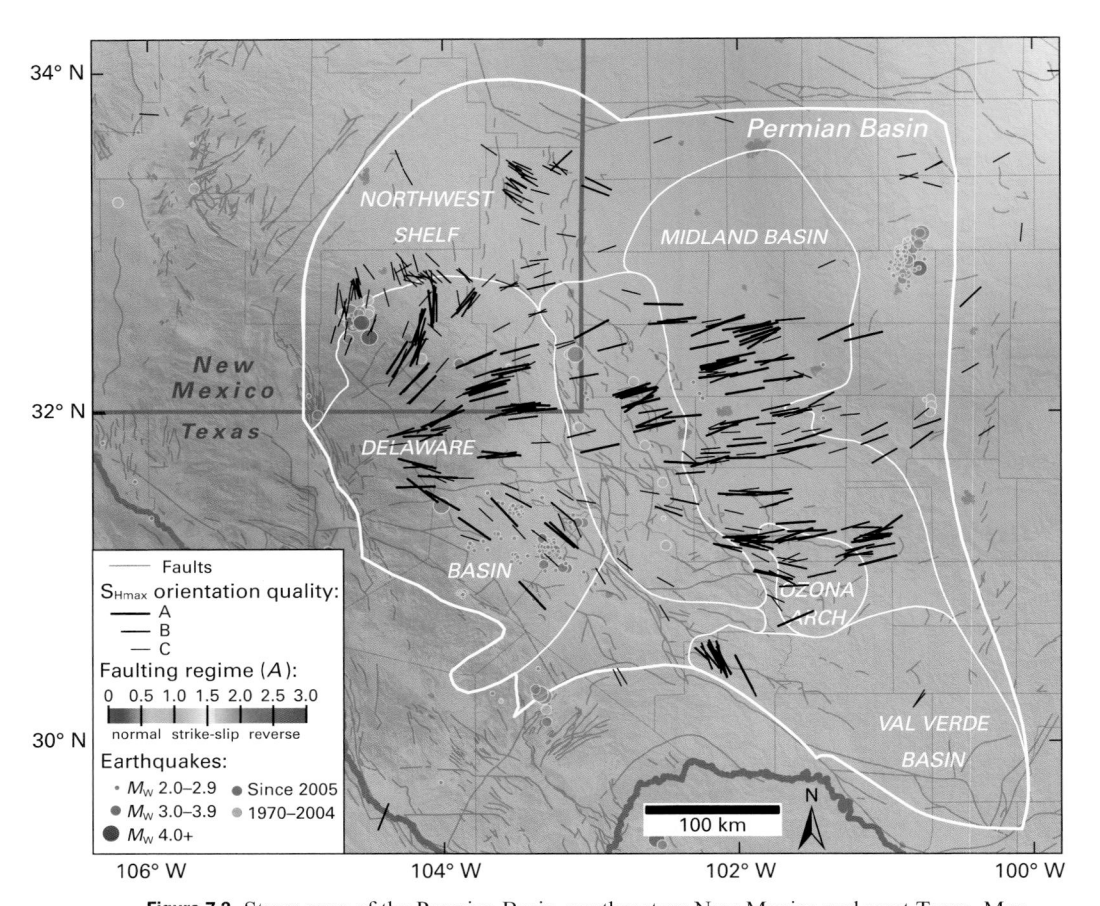

Figure 7.3 Stress map of the Permian Basin, southeastern New Mexico and west Texas. Map location shown in Fig. 7.2. showing the direction of maximum horizontal compression, S_{Hmax} (the length of the line indicates data quality). The background colors are relative stress magnitudes as parameterized by A_ϕ as illustrated in Fig. 7.1. After Lund Snee & Zoback (2018).

While the geologic cause of the progressive rotation of horizontal principal stress in the Delaware Basin is unknown, such a map is quite useful for siting well trajectories and identification of potentially active faults that should be avoided when injecting hydraulic fracturing flow-back or produced water (Chapter 14). It is noteworthy that this stress rotation occurs in an area of relatively low values of A_ϕ (indicative of relatively small differences between the horizontal stresses) and elevated pore pressure. As discussed below, this reduces the differences in the magnitudes of the principal stresses making it possible for relatively minor stress perturbations to cause significant changes in stress orientation (e.g., Moos & Zoback, 1993).

The Fort Worth Basin is an area of relatively consistent stress orientation and relative magnitude (Fig. 7.4, after Hennings et al., 2019). Northeast–southwest oriented S_{Hmax} orientations are seen throughout the basin as well as a strike-slip/normal faulting stress state – A_ϕ is approximately 1.0 essentially everywhere within the basin although stresses are slightly more compressive near Wichita Falls, near the Red River and slightly more

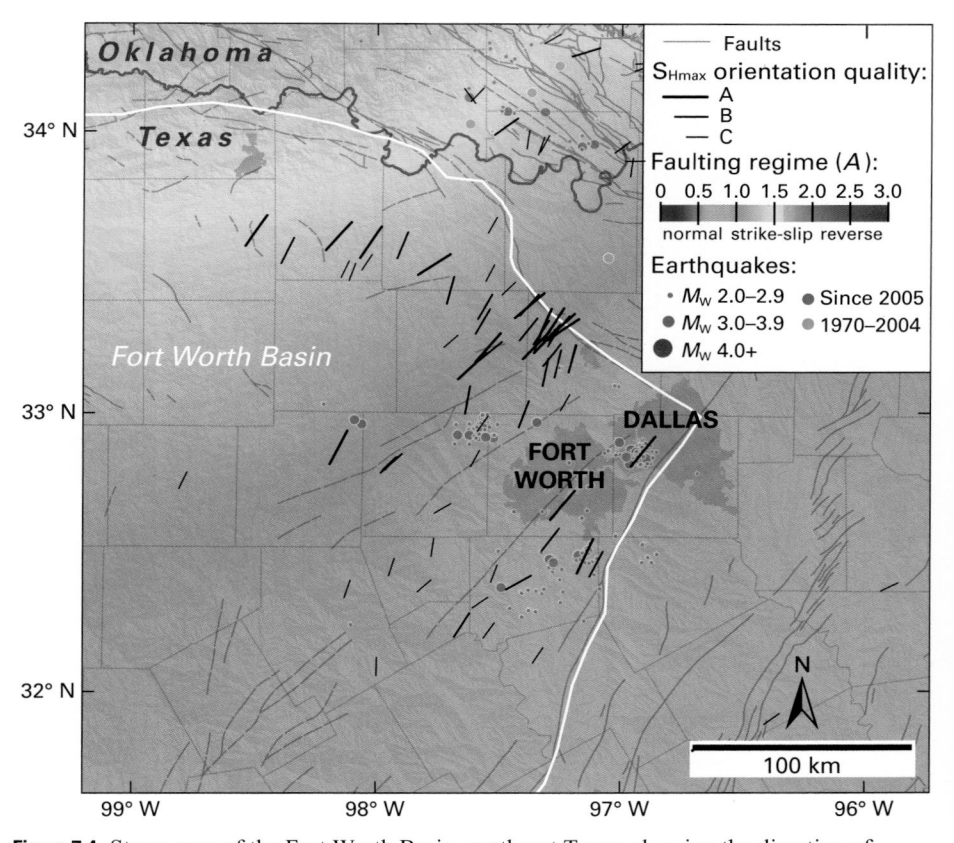

Figure 7.4 Stress map of the Fort Worth Basin, northeast Texas, showing the direction of maximum horizontal compression, S_{Hmax} (the length of the line indicates data quality) and the relative stress magnitudes as parameterized by A_ϕ (illustrated in Fig. 7.1). Data from Hennings et al. (2019).

Stress, Pore Pressure, Fractures and Faults 189

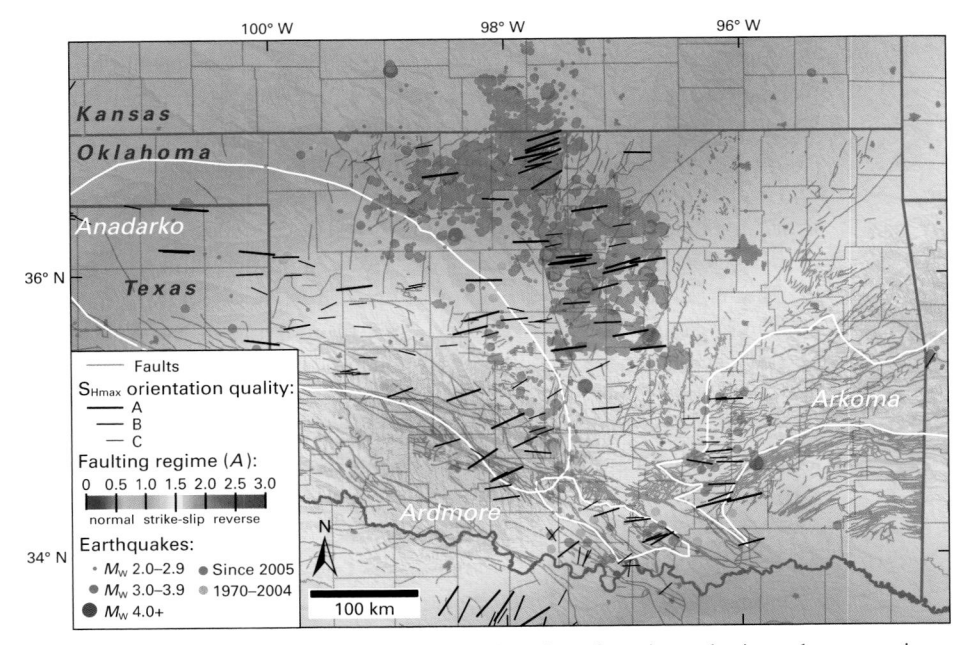

Figure 7.5 Stress map of Oklahoma showing the direction of maximum horizontal compression, S_{Hmax} (the length of the line indicates data quality) and the magnitude of relative stress magnitude as parameterized by A_ϕ as illustrated in Fig. 7.1. The red dots indicate earthquakes that have occurred since 2005, and the yellow dots indicate earthquakes that occurred from 1970 to 2004. Dots are scaled by earthquake magnitude and are from the US Geological Survey. After Lund Snee & Zoback (2018b), modified from Alt & Zoback (2017).

extensional near Dallas and Fort Worth. Several case studies involving production from the Barnett shale in the east central part of the basin and core samples from the Barnett have been discussed earlier in the book. Figure 7.4 also shows earthquakes that have occurred recently in this area, which may have been triggered by wastewater injection that will be discussed in Chapters 13 and 14. As also shown in Fig. 7.4, a series of normal faults trend sub-parallel to the maximum horizontal stress direction and in several areas, earthquake focal plane mechanisms generally indicate slip on planes with a similar orientation.

Figure 7.5 is a stress map of Oklahoma, based principally on Alt & Zoback (2017) and Lund Snee & Zoback (2016, 2018a). Throughout much of Oklahoma, there is a consistent N80°–90°E orientation of S_{Hmax}. This fairly uniform stress field appears to extend west into the Texas panhandle and is seen in southeastern Oklahoma northeast of the Ardmore Basin (Lund Snee & Zoback, 2016). In southern Oklahoma, there is a marked deviation from this uniform stress field, with a counter-clockwise rotation of S_{Hmax} southward, culminating in an approximately N050°E orientation near the Texas border south of the Ardmore Basin. There is a similar counter-clockwise stress rotation near Lawton, where the Meers fault appears to have produced a prehistoric M ~7 earthquake (Crone & Luza, 1990). As indicated, most of north-central Oklahoma is characterized by ~E–W compression and a strike-slip faulting stress field (McNamara et al., 2015). The state of stress in

190   **Unconventional Reservoir Geomechanics**

north-central Oklahoma is discussed in more detail later in this chapter where we compare wellbore stress measurements with those determined from earthquake focal plane mechanisms. We also discuss the stress field in this part of the state in Chapters 13 and 14 where we review the relatively intense seismicity in the region that resulted from disposal of billions of barrels of produced water over just a few years.

It is clear in Fig. 7.5 that throughout much of Oklahoma there is a consistent N80°-90° E orientation of the maximum horizontal principal stress, S_{Hmax}. This uniform stress field appears to extend west into the Texas panhandle and is seen in southeastern Oklahoma northeast of Ardmore (Lund Snee & Zoback, 2018a). In southern Oklahoma, there is a marked deviation from this uniform field, with a counter-clockwise rotation of the S_{Hmax} azimuth, culminating in an approximately N50°E orientation near the Texas border south of Ardmore. There is a similar counter-clockwise stress rotation near Lawton where the Meers fault, as noted above, appears to have produced a prehistoric M ~7 earthquake. A 3 m fault scarp (which was likely formed in a single slip event) indicates an oblique reverse/strike-slip earthquake that occurred 1,100–1,400 years ago (Crone & Luza, 1990). Going south and east from there, the stress field becomes less compressive. Like in the Permian Basin, these regional variations of stress orientation and relative magnitude will influence development of unconventional resources such as the Scoop and Stack.

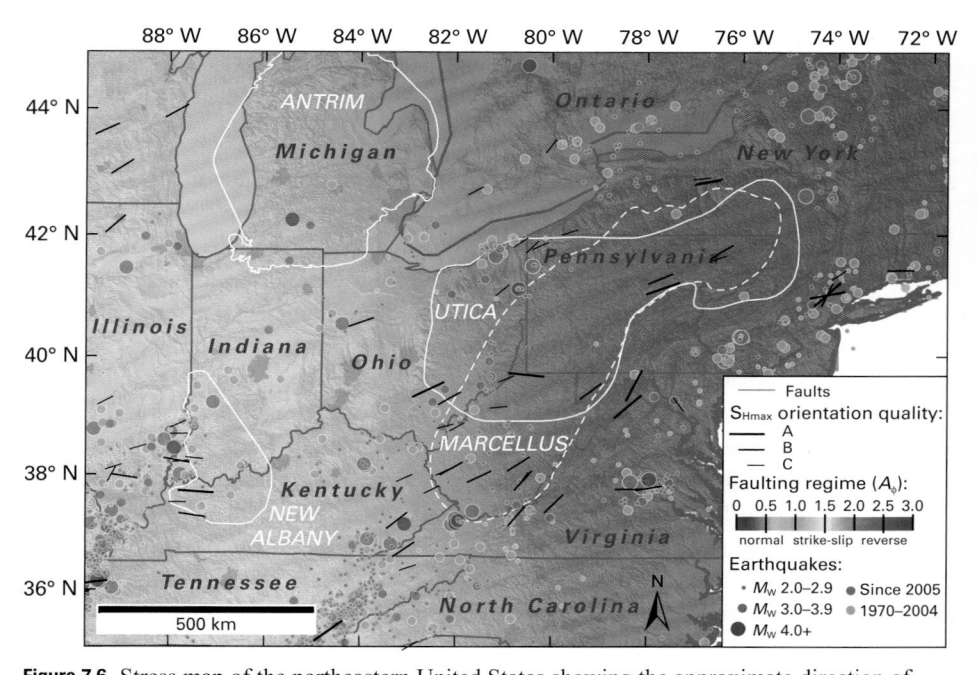

Figure 7.6 Stress map of the northeastern United States showing the approximate direction of maximum horizontal compression from earthquake focal plane mechanisms and the magnitude of relative stress magnitude as parameterized by A_ϕ. Active unconventional plays are shown by the white outlines. Modified after Hurd & Zoback (2012a) and Lund Snee & Zoback (2018a).

Stress, Pore Pressure, Fractures and Faults

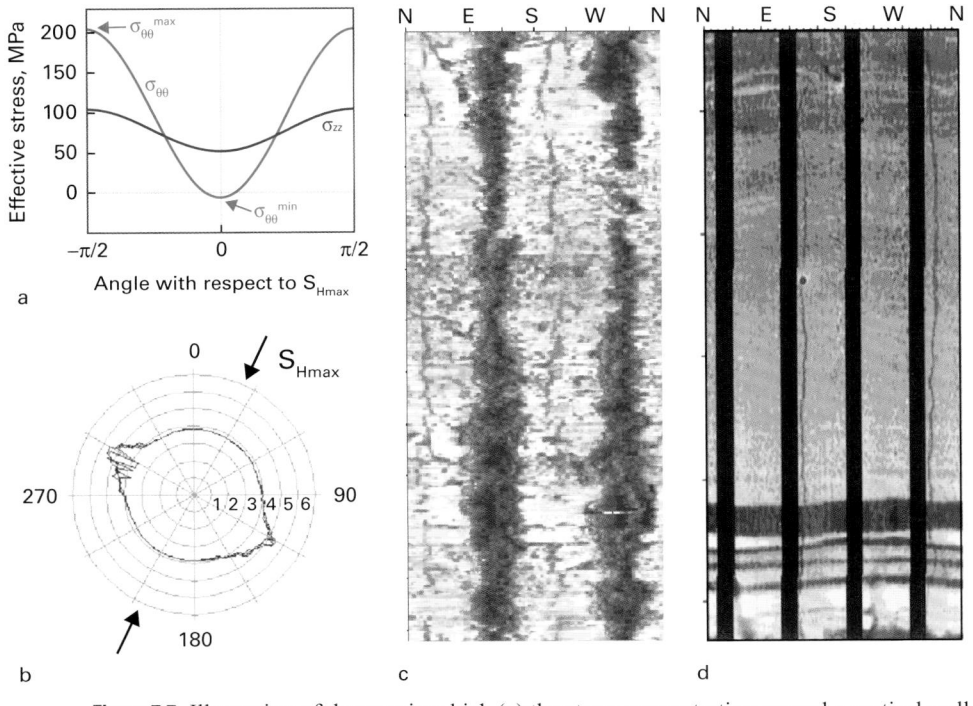

Figure 7.7 Illustration of the way in which (a) the stress concentration around a vertical well can lead to (b, c) stress-induced wellbore breakouts and (c, d) drilling-induced tensile fractures. Figures from Zoback (2007).

Figure 7.6 is a stress map of the northeastern United States (after Hurd & Zoback, 2012a and Lund Snee & Zoback, 2018a) that illustrates the state of stress in the region of the Antrim, New Albany, Utica and Marcellus plays. Note that the state of stress in the vicinities of these plays (mostly strike-slip to strike-slip/reverse) is more compressive than that in the areas discussed above and the stress state becomes generally more compressive northward and eastward from the central US to the north and east. This was pointed out by Zoback & Zoback (1980) on the basis of very preliminary data. While most unconventional development in this region is occurring in areas characterized by strike-slip faulting, areas of West Virginia and central and eastern Pennsylvania are characterized by both strike-slip and reverse faulting.

Measuring Stress Orientation and Magnitude

Observations of Wellbore Failure – The stress orientation data presented in the previous maps were principally obtained from utilizing observations of drilling-induced tensile fractures, stress-induced wellbore breakouts and crossed-dipole sonic logs in vertical wells. Figure 7.7 illustrates the fundamental aspects of utilizing breakouts and

192   **Unconventional Reservoir Geomechanics**

drilling-induced tensile fractures for determination of horizontal principal stress directions in vertical wellbores. Zoback (2007) discusses how and why these wellbore failures occur in vertical and deviated wellbores and the relationships of the wellbore failures to stress orientation and stress magnitude. As discussed below, observations of compressive failure (breakouts) and/or drilling-induced tensile fractures enable one to place constraints on stress magnitudes.

In the context of determining stress orientations, the stress concentration around a vertical well at the surface of the wellbore produces a region of highly compressive hoop stress ($\sigma_{\theta\theta}$, which acts parallel to the wellbore wall) at 90° to S_{Hmax}, the direction of maximum horizontal compression in the far-field (Fig. 7.7a). If the stress concentration exceeds the strength of the rock, a breakout will form that enlarges the well in the direction of S_{hmin}. As long as the zone of failure is not excessive (i.e., the breakouts are not too large), their occurrence will not affect wellbore stability and can go unnoticed during drilling. Breakouts were first studied using 4-arm caliper data, in which case one needs to avoid misinterpreting mechanical erosion of the wellbore wall (called key-seats) from stress-induced breakouts. Figure 7.7b shows a cross-section of a section of a well with breakouts from analysis of an ultrasonic scanning device. As shown, the breakouts occur over a finite span of the wellbore circumference, 180° apart at the azimuth of S_{hmin}. As seen in an unwrapped view of the wellbore wall (Fig. 7.7c) breakouts appear as dark bands on either side of the well as the reflection amplitude of the ultrasonic pulse is severely attenuated by the breakouts.

As indicated in Fig. 7.7a, the hoop stress can go into tension at the wellbore wall in cases when there is a significant difference between the two horizontal stresses. As the tensile strength of rock is extremely low, this can lead to the formation of drilling-induced tensile fractures that form axially on each side of the wellbore wall at the azimuth of S_{Hmax}. Because these fractures occur only at the wellbore wall, they too do not affect drilling and go unnoticed without image logging. Drilling-induced tensile fractures are best viewed using electrical imaging tools, as illustrated by the example shown in Fig. 7.7d. There are no fundamental reasons why breakouts and drilling-induced tensile fractures cannot form simultaneously in a wellbore, which seems to be the case shown in Fig. 7.7c. Note that the axial drilling-induced tensile fractures are 90° from the breakouts, as expected.

Obtaining Stress Orientation and Relative Magnitude from Earthquakes – An important field in earthquake seismology is the study of earthquake focal plane mechanisms from radiated seismic energy. There are many good texts on this topic (e.g., Aki & Richards, 2002; Stein & Wysession, 2003) and our goal in this section is simply to outline some of the basic principles of focal plane mechanisms as they relate to understanding how earthquakes can be used to constrain the stress orientation and faulting styles. Earthquake seismology as applied to the microseismic events associated with hydraulic fracturing in unconventional reservoirs is discussed in more length in Chapter 9.

In simple terms, the radiated seismic energy allows two orthogonal planes to be defined. One is the fault plane and one is an auxiliary plane. Without additional

Stress, Pore Pressure, Fractures and Faults

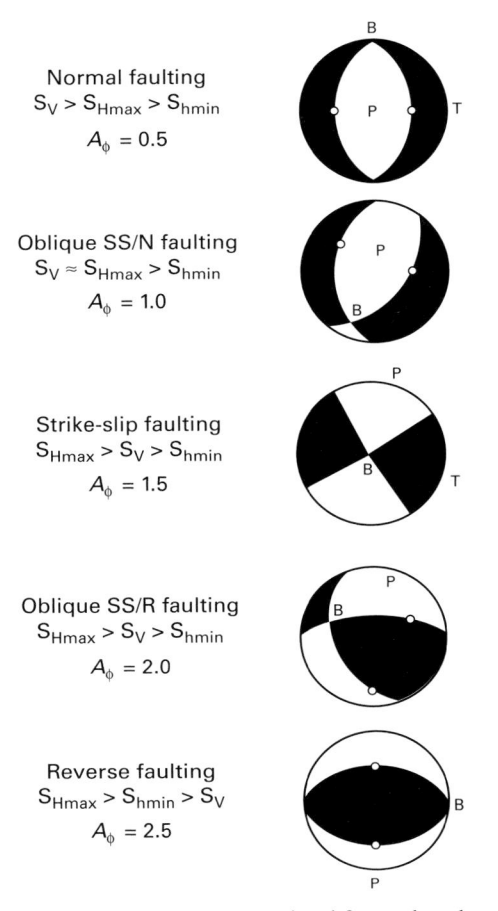

Normal faulting
$S_V > S_{Hmax} > S_{hmin}$
$A_\phi = 0.5$

Oblique SS/N faulting
$S_V \approx S_{Hmax} > S_{hmin}$
$A_\phi = 1.0$

Strike-slip faulting
$S_{Hmax} > S_V > S_{hmin}$
$A_\phi = 1.5$

Oblique SS/R faulting
$S_{Hmax} > S_V > S_{hmin}$
$A_\phi = 2.0$

Reverse faulting
$S_{Hmax} > S_{hmin} > S_V$
$A_\phi = 2.5$

Figure 7.8 Generalized focal plane mechanisms associated for earthquakes associated with the stress states defined in Fig. 7.1 (the rare cases of $A_\phi = 0$ and $A_\phi = 3.0$ are omitted). The mechanisms are oriented to be consistent with a north–south S_{Hmax} orientation. The P, T and B axes are shown as well as slip vectors (small circles) on each plane. Modified from Stein & Wysession (2003).

information such as an alignment of hypocenters, knowledge of the stress field or association of the earthquake with a known fault, it is not possible to define which of the two planes is the fault plane. The compressional and dilatational quadrants are defined by these planes indicating whether the strain produced by the fault slip is compressional or dilatational. Classically, these were defined by the polarities of P-waves traveling from the fault to the seismic station, although more sophisticated techniques are commonly used today, which are referred to in Chapter 9.

It is standard to present focal plane mechanisms in lower hemisphere stereographic projections as shown in Fig. 7.8. These simplified, but illustrative focal plane mechanisms shown in the figure assume that the S_{Hmax} direction is approximately north–south. The dilatational quadrant of the focal sphere is shown in white in Fig. 7.8, the compressional quadrant in black. The characteristic appearance of these *beach-balls* makes it easy to relate particular focal plane mechanisms to the fault orientations and relative stress states shown in Fig. 7.1. For example, the simple case of normal, strike-slip and reverse faulting shown in the center column of Fig. 7.1, correspond to the top, middle and bottom focal plane mechanisms shown in Fig. 7.8. For normal faulting, the causative faults strike north–south and dip steeply to the east and/or west and the dilatational quadrant is in the center of the focal mechanism. If only normal fault mechanisms occur in a region it would suggest an A_ϕ value of ~ 0.5. For strike-slip faults, the causative faults strike northeast–southwest or northwest–southeast and dip vertically. A symmetric pattern of compressional and dilatational quadrants are seen in Fig. 7.8. If only strike-slip fault mechanisms occur in a region it would suggest an A_ϕ value of ~ 1.5. Reverse faulting earthquakes are expected to occur on relatively shallow-dipping east–west striking planes that dip to the north and/or south. The compressional quadrant is in the center of the focal sphere and an A_ϕ value of ~ 2.5 would be indicated if only reverse faulting earthquakes occur in an area. The open dots on the faults shown in Fig. 7.8 indicate the slip vector on each of the associated planes. As it turns out, the slip vector of one plane is the pole of the other.

Actual earthquakes are more complicated in several regards. First, if a group of focal plane mechanisms in an area indicate both strike-slip and normal faulting, they define a transitional stress state shown in Fig. 7.1 with an A_ϕ value of ~ 1.0. Correspondingly, if both strike-slip and reverse faulting mechanisms are observed, this would define the transitional stress state shown in Fig. 7.1 with an A_ϕ value of ~ 2.0. Second, if the stress tensor is tilted from being purely in horizontal and vertical planes and/or a significant fault in an area is not perfectly aligned with the principal stresses in accordance with Coulomb faulting theory, oblique slip, such as a combination of strike-slip and dip slip motion can occur on a fault as illustrated in the second and fourth rows of Fig. 7.8.

By definition, the P-axis (or compressional axis) bisects the dilatational quadrant, the T-axis (or extensional axis) bisects the compressional quadrant and the B-axis (sometimes called the neutral, or intermediate, axis) is orthogonal to P and T. These are indicated for the five schematic focal plane mechanisms shown in Fig. 7.8. However, the exact relationship between well-oriented faults and principal stresses is explained in some detail in Chapter 4. Recalling that the coefficient of friction defines the angular relationships between the principal stress directions and fault orientations, we can see in Fig. 7.8 why focal plane mechanisms approximate, but do not accurately define these relationships. It would seem safer to assume that S_{Hmax} is 60° based on the fault-normal (rather than 45°) if you think you know which nodal plane is the fault plane and if you assume that a coefficient of friction of about 0.6 is applicable to the fault in question. However, if either of these assumptions is incorrect, your assumed stress direction would be wrong.

To better use earthquake focal plane mechanisms for determination of the stress field, it is common to invert a family of focal mechanisms in a given area to find the best-fitting stress field. This technique yields the orientation of the three principal stresses as well as the relative magnitude of the stresses as defined by the parameter ϕ (Eqn. 7.2) or equivalently R, where $R = 1 - \phi$. Over the past 30 years, a number of methods have been proposed for determining tectonic stress from focal mechanisms (see review by Maury et al., 2013). The most commonly used methods were developed over 25 years ago (Michael, 1984; Gephart & Forsyth, 1984; Angelier, 1990) with modifications and extensions proposed by others (Arnold & Townend, 2007; Lund & Townend, 2007), Maury et al., 2013; Vavryčuk, 2014; Martínez-Garzón et al., 2016). These methods assume that tectonic stress is homogeneous in the volume of rock in which the earthquakes occur and that the earthquakes occur on pre-existing faults with varying orientations. The technique is based on the Wallace–Bott hypothesis (Wallace, 1951; Bott, 1959) that slip on a fault will always occur in the direction of maximum resolved shear stress. Hence, the inversion attempts to find a best-fitting stress tensor (and ϕ, or R, value) that best matches those observed in the focal plane mechanisms. Table 7.1 describes the quality ranking procedures for using focal plane mechanism inversions for determination of stress orientation and ϕ.

There are competing priorities when doing stress inversions. For example, one would like to analyze the state of stress in as small a volume of rock as possible to obtain the most detailed picture of the stress field (and possible stress variations). However, it is important to have a sufficient number of well-constrained, and diverse, focal mechanisms as possible to accurately constrain the inversion. It is also important to have reliable criteria for assessing the accuracy of the inversions. These issues are addressed in the papers cited above (and many others). Without going into the details, an intuitive understanding of the usefulness of this technique can be obtained from Figs. 7.9–7.11 that show three applications of focal mechanism inversions relevant to the topics addressed in this book.

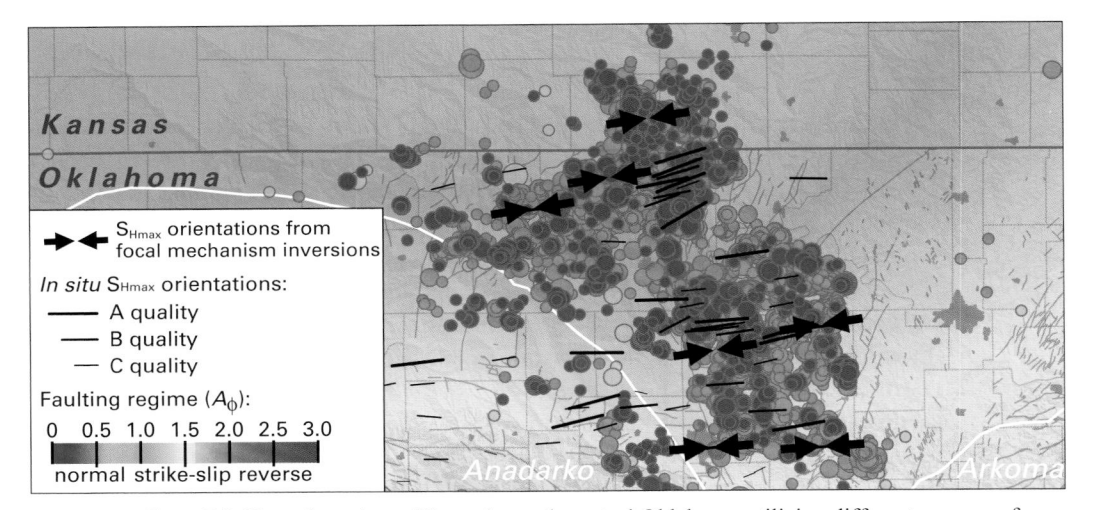

Figure 7.9 The orientation of S_{Hmax} in north-central Oklahoma utilizing different sources of wellbore data (and data quality) shown by straight lines and focal mechanism inversions indicated by inward pointed heavy black arrows. Earthquake epicenters (red dots) of M 2.5 and greater between 2009 and 2015. Data from Alt & Zoback (2017). Faults throughout the state were compiled by the Oklahoma Geological Survey (Darold & Holland, 2015).

196   **Unconventional Reservoir Geomechanics**

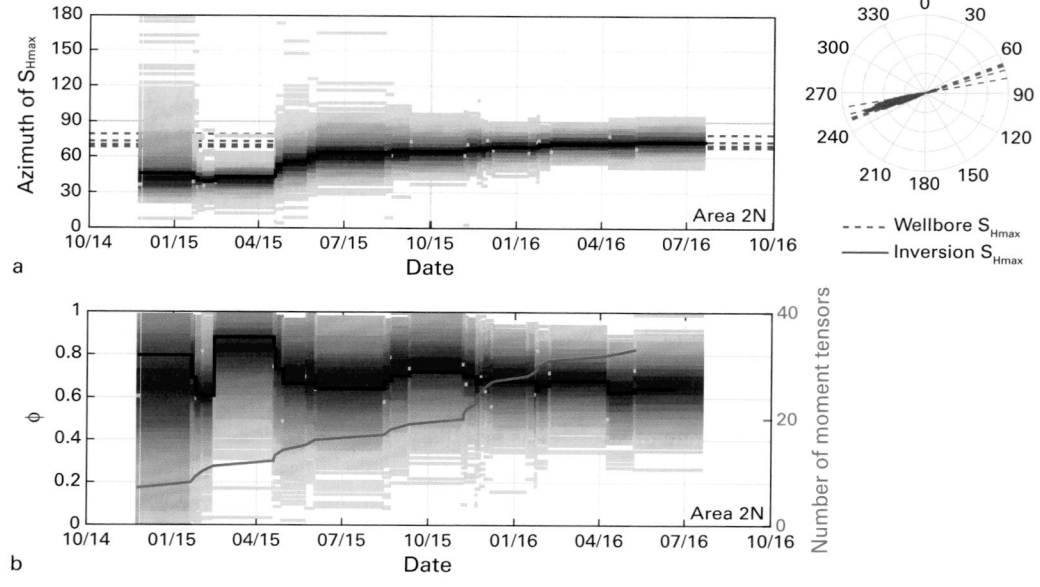

Figure 7.10 (a) Focal mechanism inversion results through time as more earthquakes occurred in area 2 N shown in Fig. 7.9. From the supplemental material of Walsh & Zoback (2016). Black is the deterministic inversion result, and blue-yellow shows the density of bootstrapped statistics results through time. The magenta dashed line shows the S_{Hmax} azimuth from a wellbore measurement in area 2 N. The inversion converges to match the wellbore measurement. The rose diagram is also shown with the final S_{Hmax} azimuths from the final bootstrapped results in blue, and the wellbore measurement as a magenta dashed line. (b) The red line indicates the number of earthquake focal plane mechanisms used in the inversion as a function of time. The black line indicates the ϕ values determined from the inversion with boot strap statistics.

Figure 7.9 (modified from Alt & Zoback, 2017) shows the north-central part of the stress map of Oklahoma previously shown in Fig. 7.5. This map will be utilized again in Chapter 14 when we discuss managing the risk associated with injection-induced seismicity in this region (shown by red dots, scaled by magnitude as discussed in the caption for Fig. 7.5). Note that the direction of S_{Hmax} determined by wellbore indicators (straight lines, which are mostly from drilling-induced tensile fractures observed in electrical image logs) matches the S_{Hmax} orientations determined from focal mechanism inversions (large inward pointed arrows) quite well. This is particularly significant because nearly all the wellbore data come from the upper ~2.5 km in the sedimentary section and the focal plane mechanisms mostly come from 4–6 km depth in crystalline basement.

Figure 7.10 (from the supplemental material of Walsh & Zoback (2016)) shows how the focal mechanism inversions in Fig. 7.9 were obtained. For illustration purposes, the inversion was repeated as a function of time, as more focal mechanisms became available. Note that once 20–25 focal mechanisms (red line) were used in the inversions, the inversions produced stable and well-constrained results for both stress orientation

Stress, Pore Pressure, Fractures and Faults 197

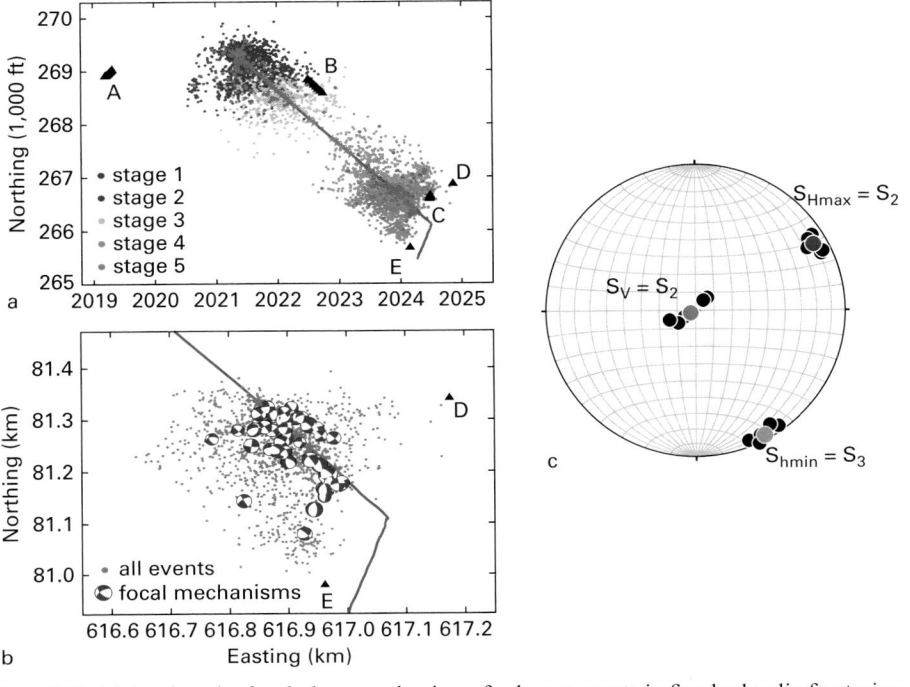

Figure 7.11 (a) Earthquake focal plane mechanisms for larger events in five hydraulic fracturing stages in the Barnett shale. (b) Focal plane mechanisms (and all events) from stage 5. (c) Focal mechanism inversions from all five stages (black dots) gave essentially the same results for the three principal stresses. The colored dots represent averages of all of the stress inversions. Modified from Kuang et al. (2017).

and ϕ. Note how well the stress orientation matches that of the wellbore data (dashed magenta lines). Other investigators have reached similar conclusions about the number of mechanisms required to obtain reliable stress inversions. The uncertainties shown in the inversion results were obtained by boot-strapping the available focal plane mechanisms at each time step. This involves repeatedly recalculating the result by resampling from the available mechanisms with replacement. Sample sizes in each resampling iteration are the same as the available number of focal mechanisms at that time step, meaning that some mechanisms will be sampled multiple times and others won't be sampled at all in each repetition.

Figure 7.11 (modified after Kuang et al., 2017) shows an application of this method for a microseismic dataset in the Barnett shale that will be discussed at greater length in Chapters 9 and 12. Five hydraulic fracturing stages were carried out along the well and each stage was monitored by arrays in two of the five wellbores shown in the figure depending on the proximity of the monitoring well to the stage. Focal plane mechanisms were calculated using a novel technique reviewed in Chapter 9 for the best-recorded events (illustrated in Fig. 7.11b for stage 5, and recorded in monitoring wells D and E). The independent focal plane mechanisms were inverted separately for the five stages. Note in Fig. 7.11c that the five inversions yielded essentially the same stress field. Moreover, as discussed by Kuang et al. (2017) both the stress orientation and ϕ value determined by the focal mechanism inversions match that determined from wellbore data.

As indicated in Table 7.1, data from cross-dipole sonic logs in subvertical wells can be useful for determining stress orientation. In nearly horizontal sedimentary rocks, shear wave anisotropy in a horizontal plane (the polarization of shear waves into fast and slow directions) could occur due to two primary mechanisms: (1) macroscopic fractures that are closed preferentially perpendicular to S_{Hmax} due to appreciable differences between the magnitudes of the two horizontal principal stresses, or (2) structural anisotropy due to the alignment of sub-parallel macroscopic fractures and faults. In the first case, the fast shear direction is parallel to S_{Hmax}; in the latter case, the fast shear direction is parallel to the strike of the structural fabric. Shear velocity anisotropy is commonly modeled with a transversely isotropic (Maxwell et al., 2002) symmetry where the shear waves are polarized parallel and perpendicular to the planes normal to the formation symmetry axis (Thomsen, 1986), as discussed in Chapter 2. Crossed-dipole sonic logs in vertical wells have been used by the industry to determine S_{Hmax} orientations from the fast shear polarization direction when bedding planes are sub-horizontal and aligned fractures are not likely to influence the polarization of the shear waves. Because of the possibility for nonunique results resulting from the two mechanisms that can produce S-wave velocity anisotropy, it is important that operators follow the criteria outlined in Table 7.1, based on those developed by Boness & Zoback (2006).

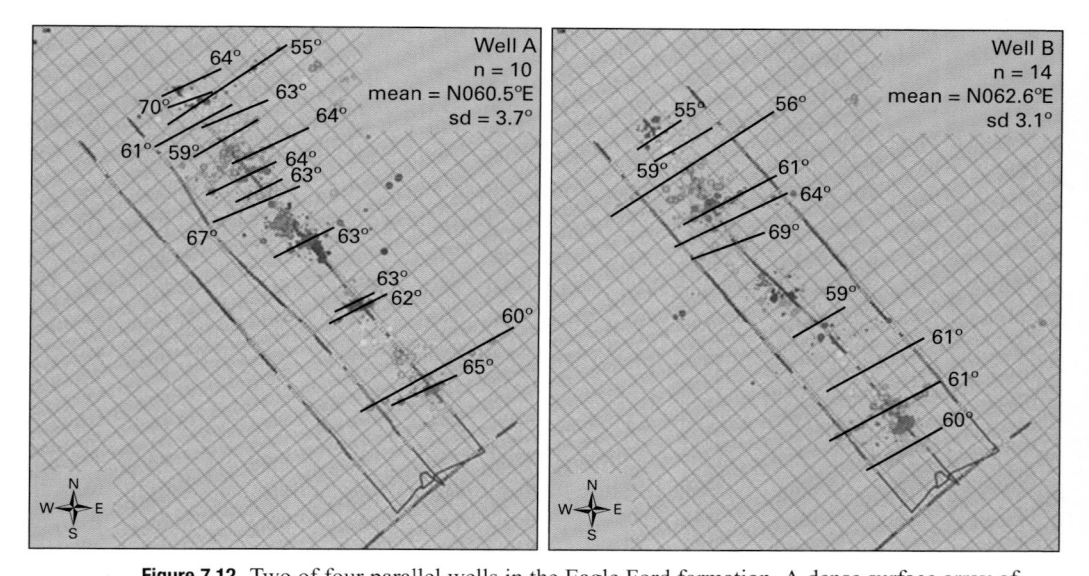

Figure 7.12 Two of four parallel wells in the Eagle Ford formation. A dense surface array of seismometers was used to monitor the microseismic events associated with the hydraulic fracture stages. The hydraulic fractures propagate nearly orthogonal to the well paths (as intended). Forty eight well-defined lineations of microseismic events in the four wells yield a mean orientation of S_{Hmax} of N64.5°E with a standard deviation of 5.7° (data courtesy of XTO and Microseismic Inc.).

The final method for obtaining S_{Hmax} orientations used in the maps above (and referred to in Table 7.1) are measuring the orientations of clear lineations of micro-seismic events associated with multi-stage hydraulic fracturing operations. Figure 7.12 shows the results for two of four parallel wells in the Eagle Ford formation that were drilled in the presumed direction of least principal stress, S_{hmin}. A dense surface array of seismometers was used to monitor the microseismic events associated with the hydraulic fracture stages. As shown, the hydraulic fractures propagate nearly orthogonal to the well paths (as intended). The results from each of the two wells indicate similar stress orientations and uncertainties. Overall, the results from 48 clearly defined lineations of microseismic events in the four wells yield a mean orientation of S_{Hmax} of N64.5°E with a standard deviation 5.7°. The quality ranking system for these types of data are also listed in Table 7.1.

Measurement of the Magnitude of the Least Principal Stress and Pore Pressure –
Ever since Hubbert & Willis (1957) presented compelling physical arguments that hydraulic fractures will always propagate perpendicular to the orientation of the least principal stress, S_3 (see Figs. 8.9 and 8.10) it has been widely recognized that the magnitude of the least principal stress (usually S_{hmin}) can be determined using hydraulic fracturing. In strike-slip and normal faulting environments where $S_3 = S_{hmin}$, hydraulic fracture propagation will be in a vertical plane perpendicular to S_{hmin} (and parallel to S_{Hmax}). In reverse faulting environments where $S_3 = S_V$, hydraulic fracture propagation will be in a horizontal plane.

There are a variety types of hydraulic fracturing tests used for this purpose. In classic hydraulic fracturing stress measurements, a limited interval in an uncased borehole isolated by packers, or a perforated interval in a cased borehole, is pressurized in a carefully controlled way. Low viscosity fluid (usually water) is injected at a low, constant, flow rate injection (~20 liter/min or 0.1 BBL/min, or less) to propagate the fracture away from the wellbore. Because any pressure associated with friction due to viscous pressure losses dissipates, the least principal stress is obtained from the instantaneous shut-in pressure (ISIP) in hydraulic fracturing stress measurements after abruptly stopping flow into the well (Haimson & Fairhurst, 1967). During extended leak-off tests, an analogous procedure is followed by pressurizing a short open hole interval drilled at the end of well after casing is cemented in place (see discussion by Gaarenstroom et al., 1993, and Zoback, 2007).

Diagnostic Fracture Injection Tests (DFITs) are now being performed in low-permeability unconventional formations to estimate the magnitude of the minimum principal stress, formation pore pressure and other parameters. During a DFIT, a small volume of water is injected into the reservoir at relatively low injection rates (usually 1 to 5 BBL/min) in a limited, open hole section at the end of the well as in a minifrac test). A hydraulic fracture is initiated and propagated from the wellbore for a few minutes after which the well is shut in and pressure is monitored for an extended period of time (Cramer & Nguyen, 2013; Araujo et al., 2014). Figure 7.13a, shows a schematic DFIT test. The principal difference between a DFIT and minifrac test is the monitoring of pressure for an extended period of time after shut in.

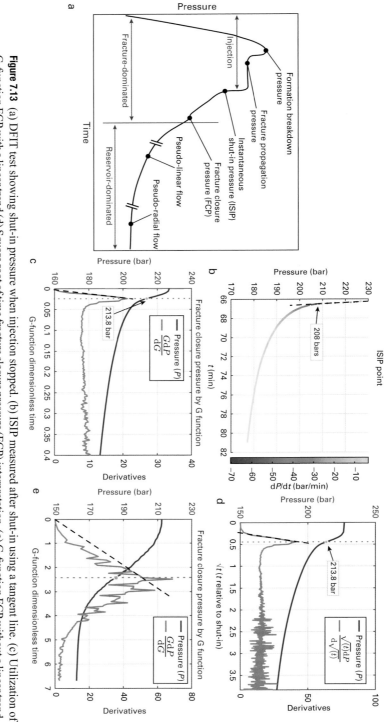

Figure 7.13 (a) DFIT test showing shut-in pressure when injection stopped. (b) ISIP measured after shut-in using a tangent line. (c) Utilization of G-function FCP with a linear trend (d) Square root of time fracture closure pressure (FCP) interpretation. (e) G-function FCP without a linear trend.

Although techniques utilized for pore pressure measurement in conventional reservoirs are well established, because of the extremely low permeability of unconventional reservoirs, direct measurement of pore pressure can be challenging. As implied in Fig. 7.13, long after a hydraulic fracture closes, the long-term equilibration pressure measured during a DFIT can be used to estimate pore pressure because the pressure measured in a shut-in well will gradually decay to a value approaching the formation pore pressure. After sufficient time, utilizing a $t^{-1/2}$ analysis should yield a reasonable estimate of pore pressure (see Nolte, 1979). A growing literature addresses how a variety of well-testing techniques have been applied to the interpretation of DFIT tests for determination of pore pressure, permeability and other leak-off parameters. However, because of the extremely low permeability and heterogenous nature of unconventional reservoirs, there is controversy surrounding interpretation of DFIT tests for the determination of S_{hmin} magnitudes (e.g., Craig et al., 2017; McClure, 2017). Barree et al. (2007) reviewed methodologies for determination of the fracture closure pressure (FCP) using both the $t^{-1/2}$ and Nolte G-function methods (Nolte, 1979) as illustrated in Fig. 7.13c,d. The $t^{-1/2}$ method is shown in Fig. 7.13d where a linear trend aids in the interpretation of fracture closure. In the case of the g-function method, the FCP is chosen when the semi-log derivative of the pressure with respect to the non-dimensional G-function time first deviates from a linear trend (Fig. 7.13c). In the ideal case, the linear trend should start at the origin, but due to the uncertainty of choosing the shut-in point the trend sometimes starts from slightly higher values. Note that the FSP determined from the $t^{1/2}$ and G-function methods are the same. There are some cases in which the G-function method shows a nonlinear trend from the origin which is uninterpretable (Fig. 7.13f).

We argue that as in minifrac tests, when interpreting DFIT data, the ISIP should be used to determine the magnitude of S_{hmin} and shown in Fig. 7.13b. In other words, when a limited amount of water is pumped into a hydraulic fracture, the fracture will immediately attempt to close on the trapped fluid once pumping has stopped (Haimson & Fairhurst, 1967). Both Jung et al. (2016) and McClure (2017) have questioned the determination of the magnitude of the least principal stress using the G-function method. On the basis of numerical modeling of hydraulic fracture propagation, both studies argued that the method illustrated in Fig. 7.13c can sometimes significantly underestimate the magnitude of the least principal stress. An important implication of this is that it would suggest hydraulic fracturing net pressures (the difference between the pumping pressure and least principal stress) are significantly overestimated. Thus, they argue, hydraulic fracturing net pressures are typically in the range of a few hundred psi, not several thousand as is sometimes assumed.

In addition to the arguments of Jung et al. (2016) and McClure (2017), there are two reasons we utilize the ISIP from DFIT and minifrac tests to obtain estimates of the least principal stress. First, it gives physically reasonable numbers. As shown below, in cases in which there appears to be horizontal hydraulic fracture propagation (i.e., $S_{Hmax} > S_{hmin} > S_V$), the ISIP yields values of the least principal stress that are very close to the overburden stress, S_V. If a closure pressure that is considerably below the ISIP were taken as the value of the least principal stress, this S_3 would not be the vertical stress even in cases where independent information indicates a horizontal hydraulic fracture (see, for example, Fig. 11.3 from Alalli & Zoback, 2018). In other cases where S_V is not the minimum principal stress, the value of the least principal stress expected for a normal faulting environment from frictional faulting theory yields a value close to the ISIP (see also Zoback, 2007). Second, the physical situation in a DFIT test is quite similar to that of a minifrac or extended leak-off test. In other words, a relatively small hydraulic fracture is propagated from the well by pumping a relatively small quantity of water. There is no reason why the fracture should not close as soon as the pumping stops. In a recent proprietary study, S_{hmin} was determined from 110 extended leak-off and mini-frac tests in 40 separate wells. The quality of the data for each test was variable and depended to some degree on whether the well operator was able to maintain a constant rate of injection during the test. In over half the tests, the values obtained from the ISIP were within 1 MPa ($\sim$150 psi) of that determined using other techniques. In 90 of the 110 tests, the values obtained for S_{hmin} were within 2 MPa ($\sim$300 psi).

When hydraulic fracturing of conventional reservoirs with viscous gel and prop-pants, it makes sense that the fracture does not immediately close when pumping stops. Hence, one should not consider the ISIP value as the least principal stress. However, in small scale hydraulic fractures such as those produced by DFITs in unconventional reservoirs, the ISIP should yield a reliable value of the least principal stress. In fact, even if gel is used as a frac fluid during stimulation of an unconventional well, a clean water tail is pumped at the end of the hydraulic fracture job such that there is no impediment to fracture closure near the wellbore once pumping ends. Hence, ISIP values obtained under these conditions should provide a reasonable estimate of the least principal stress.

Constraining the Magnitude of the Maximum Horizontal Principal Stress – Perhaps surprisingly, the importance of constraining the magnitude S_{Hmax} to address geomechanical questions related to unconventional development is not nearly as important as it is in the case of conventional oil and gas development as discussed by Zoback (2007). There are several reasons for this. First, the relatively high strength (Chapter 3) and shallow depth of unconventional reservoirs makes wellbore stability only rarely problematic, particularly as horizontal wells are usually drilled in the direction of S_{hmin}, thus minimizing the compressive stress concentration around the wellbore. Instabilities may occur in some formations due to excessive underbalance, or due to slip on pre-existing planes of weakness, such as fractures or bedding. In these cases, wellbore stability analysis for appropriate mud weight selection may still be required (see discussion in Zoback, 2007). In addition, viscoplastic stress relaxation (Chapters 3 and 10), relatively low frictional strength in clay-rich formations (Chapter 4) as well as elevated pore pressure (as described below) reduces overall stress anisotropy in unconventional reservoirs. A third reason why obtaining the magnitude of S_{Hmax} is less important in unconventional reservoirs is that while stress orientation and the magnitude of S_{hmin} are critically important for understanding hydraulic fracture growth, the magnitude S_{Hmax} is essentially irrelevant. As it turns out, knowing the relative magnitude of S_{Hmax} (whether

Stress, Pore Pressure, Fractures and Faults   203

it is below, approximately equal to or greater than S_V) will usually suffice as indicated by the A_ϕ parameter.

The magnitudes of the horizontal principal stresses at a given depth can only range between certain values due to the limited frictional strength of rock (see the lengthy discussion in Zoback, 2007). This leads to three simple equations that define the maximum difference between the maximum and minimum principal stress depending on the pore pressure and coefficient of friction.

$$\text{Normal faulting} \quad \frac{\sigma_1}{\sigma_3} = \frac{S_V - P_p}{S_{hmin} - P_p} \leq [(\mu^2 + 1)^{1/2} + \mu]^2 \tag{7.3}$$

$$\text{Strike-slip faulting} \quad \frac{\sigma_1}{\sigma_3} = \frac{S_{Hmax} - P_p}{S_{hmin} - P_p} \leq [(\mu^2 + 1)^{1/2} + \mu]^2 \tag{7.4}$$

$$\text{Reverse faulting} \quad \frac{\sigma_1}{\sigma_3} = \frac{S_{Hmax} - P_p}{S_V - P_p} = \leq [(\mu^2 + 1)^{1/2} + \mu]^2. \tag{7.5}$$

In the above equations, μ is the coefficient of friction and P_p is the pore pressure. Following Zoback et al. (1987) and Moos & Zoback (1990), these bounds can be graphically portrayed in Fig. 7.14 in a stress polygon that shows the range of possible values of S_{hmin} and S_{Hmax} for a given depth, pore pressure and friction coefficient. By definition, S_{Hmax} on the y-axis must be greater than S_{hmin} on the x-axis, hence only values above the line that defines $S_{hmin} = S_{Hmax}$ are allowed. In addition, S_V is fixed by the weight of the overburden and does not vary. The three triangular areas labeled N, SS and R define areas of normal, strike-slip and reverse faulting, respectively, as defined in Fig. 7.1. The periphery of the figure represents stress states in frictional failure equilibrium as predicted by Eqns. (7.3)–(7.5). In Fig. 7.14, we extend the original stress polygon by adding color that is coded by A_ϕ (as defined in Eqns. 7.1 and 7.2). Stress magnitudes are expressed as gradients in Fig. 7.14 so that it is depth independent. S_V is assumed to increase at 23 MPa/km (1 psi/ft), as appropriate for sedimentary rock, and hydrostatic pore pressure, a friction coefficient of 0.6 and mud weight that is slightly in excess of pore pressure are assumed for illustration purposes. Following Moos & Zoback (1990), Fig. 7.14 shows the stress magnitudes associated with development of wellbore failures in a vertical well. The sub-horizontal lines labelled 20 and 40 MPa indicate the magnitude of S_{Hmax} (as a function of S_{hmin}) that would be required to cause breakouts with a width of $10°$ for the respective rock strengths assuming a depth of 2 km. A line is also shown for the stress states associated with development of drilling-induced tensile fractures (DITF) for which we assume zero tensile strength. As discussed by Zoback (2007), the conditions under which DITF occur during drilling in a vertical well are nearly the same as those when the limiting magnitude of S_{Hmax} is controlled by strike-slip faulting (Eqn. 7.4). This includes, however, strike-slip/normal faulting and strike-slip/reverse faulting. Thus, if one knows the magnitude of the least principal stress and the mud weight, the presence of drilling-induced tensile fractures allows one to estimate the magnitude of S_{Hmax}.

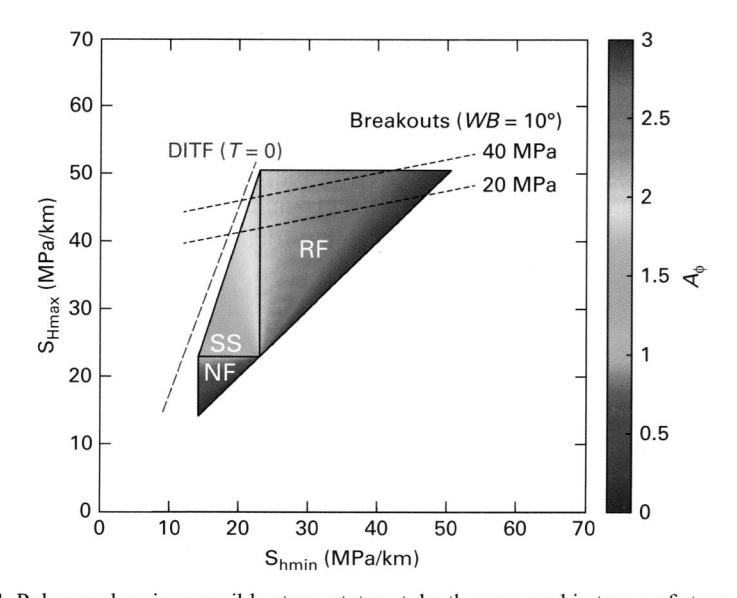

Figure 7.14 Polygon showing possible stress states at depth expressed in terms of stress gradients. The sub-horizontal lines labelled 20 and 40 MPa indicate the magnitude of S_{Hmax} (as a function of S_{hmin}) that would be required to cause breakouts with a width of 10° for the respective rock strengths. The steep diagonal line indicates the stress values associated with the initiation of tensile fractures in the wellbore wall. The colors within the polygon correspond to A_ϕ values as defined in Eqns. (7.1) and (7.2). WB = 10° is the width of the breakouts, $T = 0$ indicates that the tensile strength is assumed to be 0.

Figure 7.14 illustrates that even minor breakouts are expected to form only if the stress state is quite compressive and rock strength is anomalously low. Values for unconfined compressive strength (UCS) of 20 and 40 MPa (with a coefficient of internal friction of 1.0) are illustrated assuming a depth of 2 km. As shown in Chapter 3, the UCS of the formations of general interest are generally 100–200 MPa. Because of the high strength of the rocks and relatively shallow depth of the formations being developed, wellbore stability only rarely affects the development of many unconventional reservoirs.

Pore Pressure in Unconventional Reservoirs

Because of its central role in driving flow, pore pressure at depth in unconventional formations is extremely important for their productivity. While this will be addressed in general in Chapter 12, one example of how elevated pore pressure is correlative with high well productivity is shown in Fig. 7.15 for the Utica play in the Ohio, Pennsylvania and West Virginia. While other factors are clearly important such as total organic content (TOC), thermal maturity, formation thickness, porosity, etc., the correlation between high initial production and high pressure gradients in the figure is striking. High pore pressure gradients in the figure are about 0.9 psi/ft and moderate gradients are around

Stress, Pore Pressure, Fractures and Faults   205

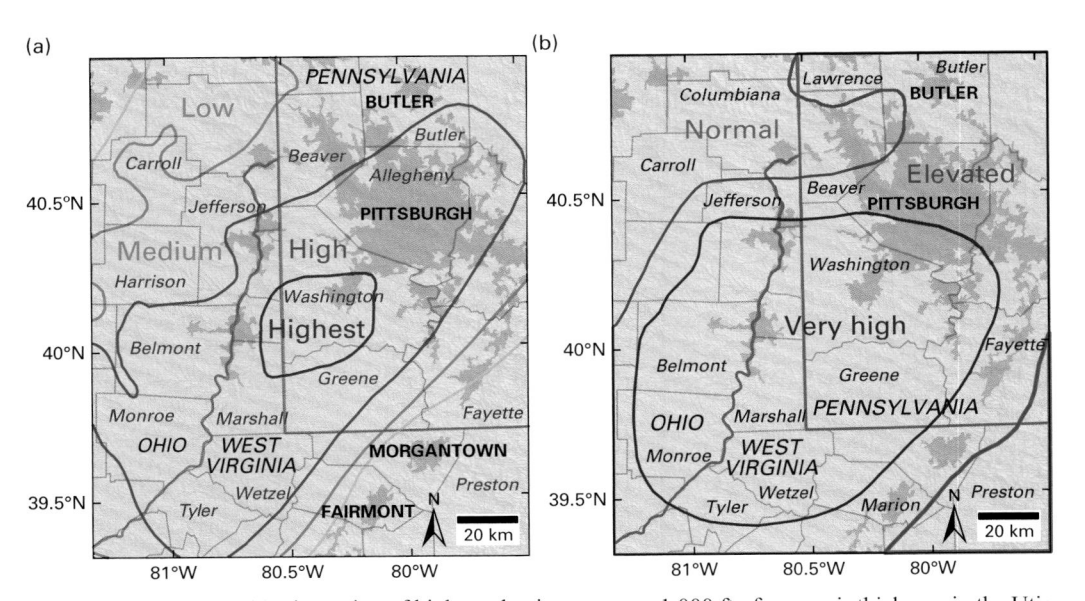

Figure 7.15 (a) The region of high production rates per 1,000 ft of reservoir thickness in the Utica formation of Ohio. (b) Pore pressure gradients in the Utica formation. Patchen & Carter (2015) spatially correlate with the region of elevated pore pressure. High gradients correspond approximately to 0.9 psi/ft, medium gradients are about 0.6 psi/ft, low gradients are close to hydrostatic (0.45 psi/ft).

0.6 psi/ft (Patchen & Carter, 2015). Moreover, because leakage can occur along permeable fault damage zones (as discussed below) there are mechanisms that could cause localized dissipation of pore pressure and hydrocarbon leakage. Thus, pore pressure could be spatially variable at a variety of scales, having a significant impact on production.

As summarized in Table 7.2, available published data shows that, in general, unconventional reservoirs are generally characterized by elevated pore pressure. In some cases, modest overpressure is observed across much of the play (the Barnett, Montney and Marcellus shales) and in other cases considerable overpressure is generally observed (such as in the Haynesville shale in east Texas and Louisiana and the Duvernay, Horn River and Laird Basins of Canada). In other cases, while generally overpressured, there is considerable variability in pressure reported in some formations such as the Eagle Ford and Utica. An exception is the sub-hydrostatic pressure reported in the Marcellus by Wang & Gale (2009). As discussed below, the reported values could reflect true spatial variations in pore pressure, but can also reflect the difficulty of measuring pore pressure in such low permeability formations.

Rittenhouse et al. (2016) compiled thousands of drill stem tests, mud weight data and DFITs to demonstrate that the eastern Delaware Basin in west Texas is overpressured (Fig. 7.16a), and in some areas severely so (Fig. 7.16a). Because the overpressure appears to develop at the depth of organic-rich formations such as the Woodford and Wolfcamp shales (whereas pore pressures in overlying and underlying formations are

Table 7.2 Pore pressure gradients in selected unconventional reservoirs.

Formation	Basin	Age	Pressure gradient (psi/ft)	Data source
Barnett	Fort Worth	Mississippian	0.52	Bowker (2007)
			0.42–0.51	Wang & Gale (2009)
Eagle Ford	East Texas	Cretaceous	0.5–0.9	Bowker (2007), Cander (2012), Gherabati et al. (2016)
Bakken	Williston		0.61–0.7*	Dohmen et al. (2017)
			0.72–0.73**	Dohmen et al. (2017)
Haynesville	East Texas and Northern Louisiana	Jurassic	0.8–0.9	English et al. (2016), Torsch (2012)
Marcellus	Appalachian	Devonian	0.6	English et al. (2016)
			0.29–0.68	Wang & Gale (2009)
Woodford, Wolfcamp, Bone Spring	Delaware (western Permian)	Upper Devonian to Middle Permian	0.8	Luo et al. (1994) Engle et al. (2016) Rittenhouse et al. (2016)
Wolfcamp	Midland (eastern Permian)	Lower Permian	0.5–0.7	Engle et al. (2016)
Utica	Appalachian	Ordovician	0.6–0.9	Patchen & Carter (2015)
Montney	Western Canada Sedimentary Basin	Lower Triassic	0.52+/-0.01	Eaton & Schultz (2018)
Duvernay	Western Canada Sedimentary Basin	Devonian	0.73+/-0.01	Eaton & Schultz (2018)
"Mississippian" Carbonates	Merge Play Anadarko Basin	Devonian Mississippian	0.8	Parshall (2018) after G. Augsberger (pers. comm.)
Silurian gas shale	Sichuan Basin, China	Silurian	0.6–0.7	Li et al. (2016)

*Middle Bakken, ** Three Forks

typically hydrostatic or even underpressured, it does appear that hydrocarbon maturation is likely responsible for the elevated pore pressure.

Another example of overpressure developing in organic-rich formations is seen in Fig. 3.24 from the Multi-well (MWX project) carried out in western, Colorado (after Nelson, 2003) in an alternating sequence of sands and shales. Detailed pore pressure data were obtained from three vertical wells on pore pressure and the least principal stress (as measured by hydraulic fracturing) as a function of depth and lithology. Note that elevated pore pressure (blue triangles) is seen at depths below ~6,000 ft that appears to result from maturation of organic-rich units (including coal) prior to 10 Ma, when regional uplift began.

Current temperature at depth in unconventional reservoirs reflects both the depth of the formation and the heat flow in the area. Figure 7.17a is a heat flow map of the central and eastern US (Blackwell et al., 2011). Current temperatures and past thermal history of

Stress, Pore Pressure, Fractures and Faults

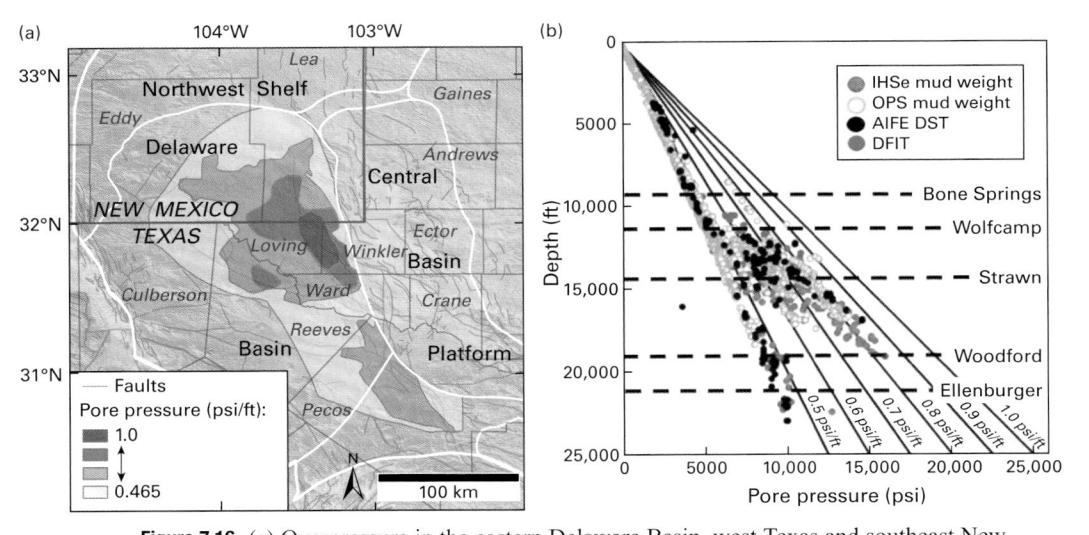

Figure 7.16 (a) Overpressure in the eastern Delaware Basin, west Texas and southeast New Mexico, is observed over a large area. (b) The available information indicates that the overpressure develops below the Wolfcamp formation. After Rittenhouse et al. (2016).

unconventional reservoirs is important in three fundamental ways. When a unconventional formation is at appreciable depth and in an area of relatively high heat flow (such as the Haynesville formation in northwest Louisiana/east Texas) temperatures in the formation can be sufficiently high (Fig. 7.17b) to affect operational issues such as choice of additives to frac fluids and the deployment of geophysical instrumentation. As discussed above, the thermal history of organic-rich formations affects kerogen maturation and pore pressure generation. Basin modeling (as illustrated in Fig. 7.17c, after Amer et al., 2013), illustrates how, depending on heat flow and the history of subsidence and uplift of a given organic-rich formation, maturation can continue for long periods of geologic time. The curve shown in red, corresponds to a heat flow of 50 mW/m^2, a typical value for much of the central and eastern US. Note that almost 40% of organic matter remains to be matured, even 150 MY after deposition. Finally, as noted above, ongoing or geologically recent maturation appears to be responsible for currently elevated pore pressure at depth in unconventional reservoirs which helps drive flow of hydrocarbons from the low permeability matrix to the well.

Stress Magnitudes and Pore Pressure – As pore fluid pressure acts everywhere in porous rocks at depth, it is self-evident that the three principal stresses must always be compressive and the magnitude of the least principal stress must always exceed the pore pressure (in steady state), or the earth would be self hydraulic fracturing. In this regard, Fig. 7.18 (adapted from Zoback, 2007) compares stress magnitudes with depth for normal faulting environments for the cases of hydrostatic pore pressure (Fig. 7.18a)

208   **Unconventional Reservoir Geomechanics**

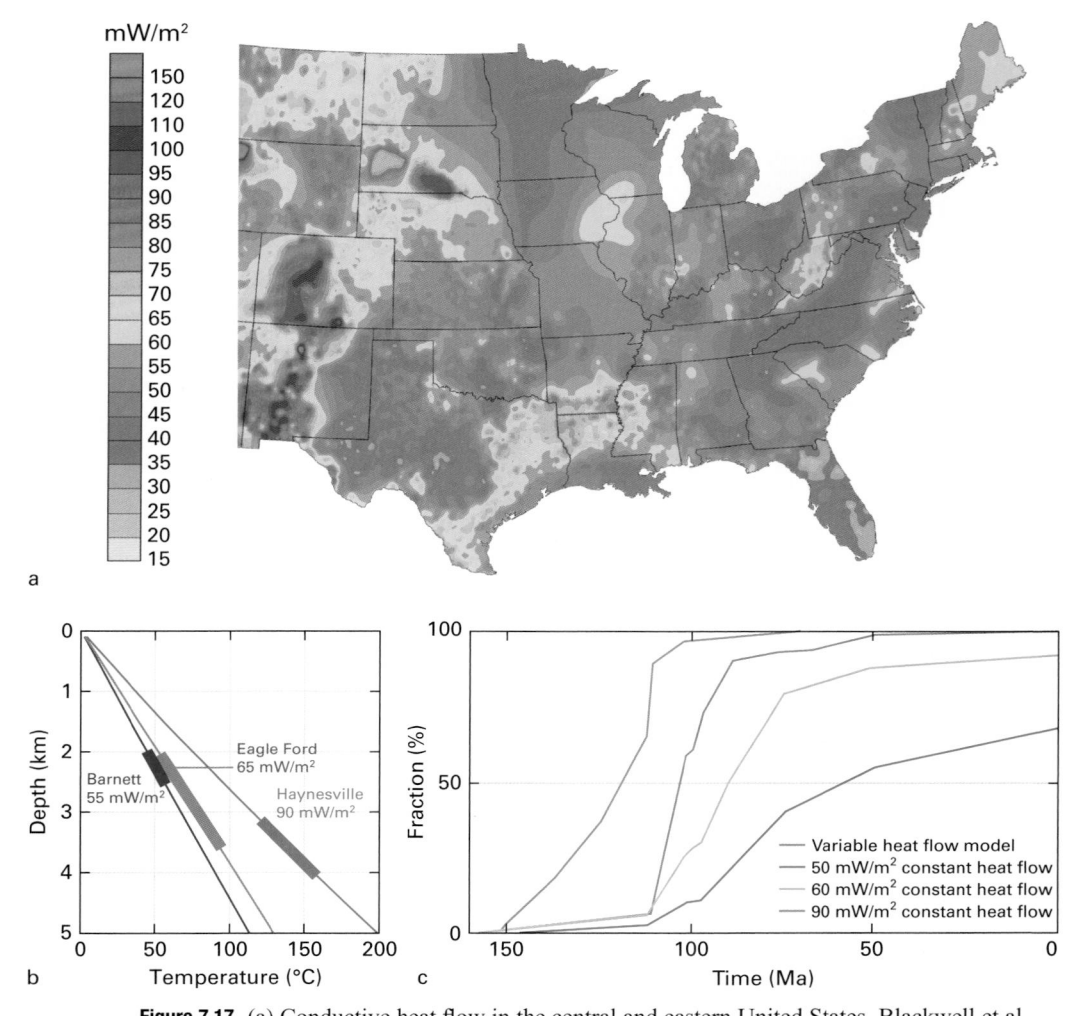

Figure 7.17 (a) Conductive heat flow in the central and eastern United States. Blackwell et al. (2011) (b) Approximate temperatures at depth in the Barnett, Eagle Ford and Haynesville plays based on the heat flows shown and reasonable estimates of thermal conductivity and heat production (c) Models of the fraction of organic matter reaching maturity as a function of geologic time for regions with different heat flows. After Amer et al. (2013).

and increasing overpressure with depth (Fig. 7.18b). The heavy gray dashed line in the figures shows the limiting magnitude of S_{hmin} in accordance with Eqn. (7.3) (that is, at any given depth, overburden stress and pore pressure, there is a lower bound to value of S_{hmin} controlled by the finite frictional strength of the rock). In the case of hydrostatic pore pressure, the lower bound of the S_{hmin} is about $0.6S_V$. Note that in the case of increasing overpressure with depth shown in Fig. 7.18b, the least principal stress not only increases to always be greater than the pore pressure (we discuss pore pressure transients and formation of opening mode fractures later in this chapter), but faulting is

Stress, Pore Pressure, Fractures and Faults

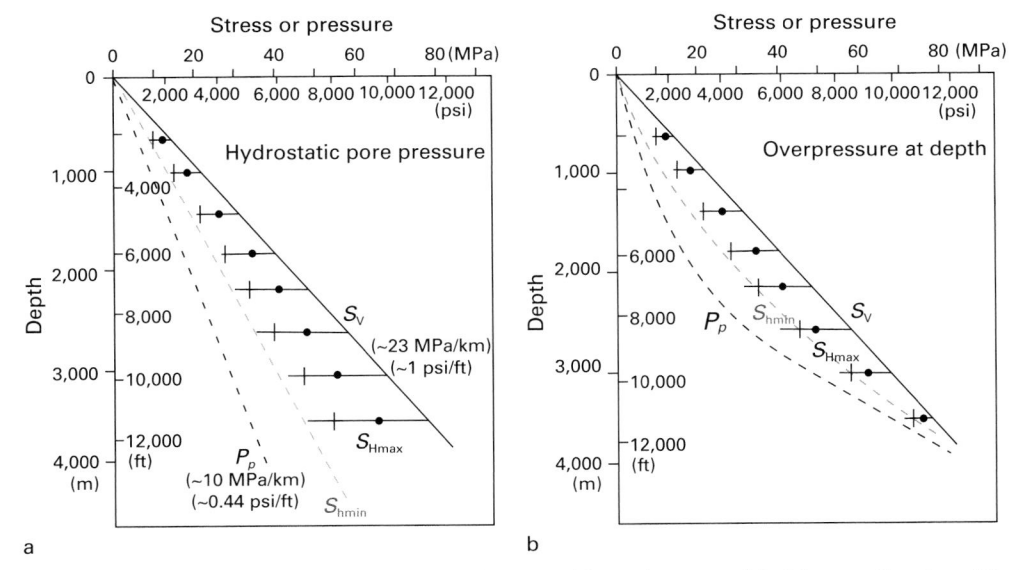

Figure 7.18 Range of possible values of S_{hmin} and S_{Hmax} in a normal faulting as a function of depth under (a) hydrostatic pressure conditions and (b) increasing overpressure with depth. After Zoback (2007).

expected to occur (as indicated by the heavy gray dashed line) when S_{hmin} is only slightly less than the vertical stress. In other words, under conditions of very low stress anisotropy.

As is widely known and discussed in Chapter 4, the frictional strength of faults decreases at elevated pore pressure. While this is generally interpreted in terms of the decrease in shear stress required for faulting with increasing pore pressure due to the decrease in the effective normal stress acting on the fault plane (Hubbert & Rubey, 1959), it is equivalent to think about this in terms of Eqns. (7.3)–(7.5) which show that as pore pressure increases, the differences between the principal stress magnitudes required to cause faulting decreases. The measurements of S_{hmin} in sandstones in the MWX wells present in Fig. 3.19 shows that increases in overpressure below 5,500 ft are associated with an increase in the magnitude of the least principal stress (inverted yellow triangles) that can be predicted reasonably well with Eqn. (7.3) for a coefficient of friction of about 0.6.

In marked contrast with the variation of S_{hmin} with depth in sandstones in the MWX wells, the measured values of the least principal stress in the shales and mudstones in the MWX experiments (dots in Fig. 3.19) are approximately the overburden stress, regardless of the degree to which the pore pressure is in excess of hydrostatic. The cause of such high stresses for the least principal stress appears to be viscoplastic stress relaxation in shales due to high clay content as discussed in Chapter 3. In Chapter 10, the topic of vertical variations of the least principal stress will be considered to address issues surrounding vertical hydraulic fracture growth as controlled by variations of the magnitude of S_{hmin}, the frac gradient from layer to layer. As mentioned above, these can result from variations of pore pressure, viscoplastic stress relaxation or variations of the coefficient of friction as discussed in Chapter 4. As will be discussed, while it is always important to restrict vertical hydraulic fracture growth to pay zones, it is especially important in areas of stacked pay to optimize drilling and hydraulic fracturing strategies in a manner that takes into account the variations of the least principal stress from layer to layer.

In Fig. 7.19 we generalize the decrease in stress anisotropy with increasing pore pressure using the constrain stress diagram shown in Fig. 7.14. From left to right, the diagrams represent allowable stress magnitudes for successively higher pore pressure. The top row of figures show the magnitude of the least principal stress (or frac gradient), while the lower row shows A_ϕ values. There are three important implications of the linkage between pore pressure and stress magnitudes.

First, in cases of very high pore pressure, there are very small differences in the principal stresses, regardless of whether one is in a normal, strike-slip or reverse faulting domain and hydraulic fractures can be propagated at pressures only slightly in excess of the pore pressure. For example, if the pore pressure gradient is about 0.9SV, at a depth of 3 km, the pressure needed to propagate a hydraulic fracture above the pore pressure is

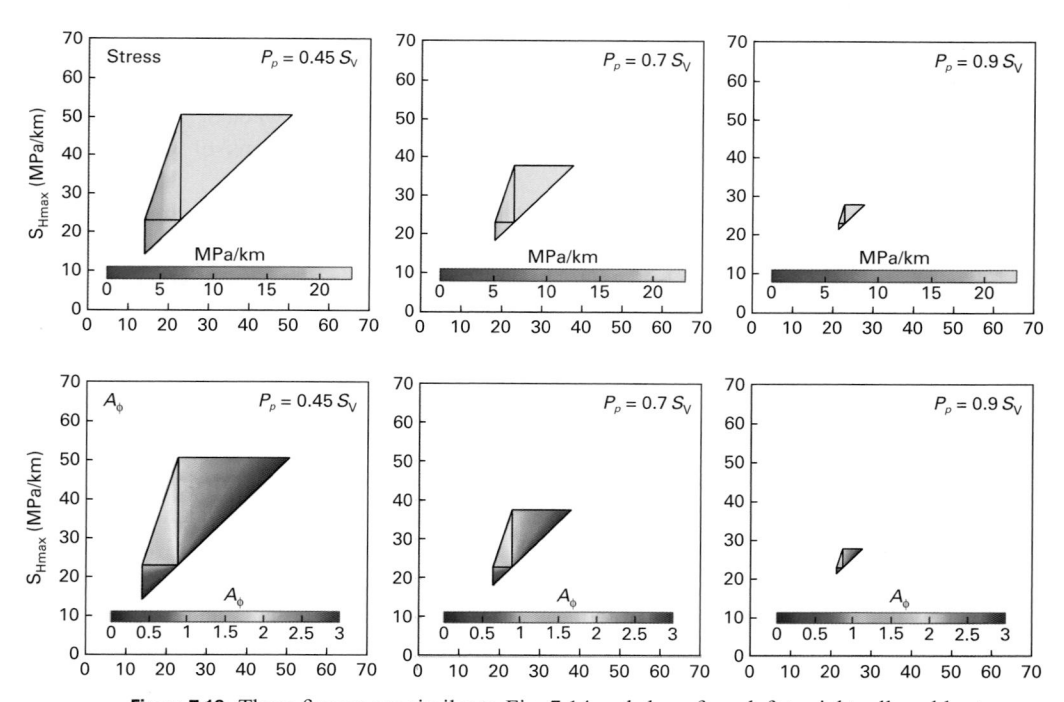

Figure 7.19 These figures are similar to Fig. 7.14 and show from left to right, allowable stress magnitudes for successively higher pore pressure. The polygon is color coded by the magnitude of the least principal stress, or frac gradient, in the top row. The lower row shows A_ϕ values. Note that in cases of very high pore pressure, there are very small differences in the principal stresses, regardless of whether one is in normal, strike-slip or reverse faulting environment.

less than 1,000 psi (~7 MPa). In addition, when pore pressure is elevated, there is relatively little variation of absolute magnitude of S_{Hmax}, which, as mentioned above, is difficult to measure. Knowledge of A_ϕ, further reduces the range of possible values of S_{Hmax}. Hence, as geomechanical issues such as wellbore stability and reservoir compaction are uncommon when exploiting unconventional reservoirs, precise knowledge of the magnitude of S_{Hmax} is not as critical as it often is in conventional reservoirs.

Second, as will be discussed in Chapter 10, when hydraulic fracturing operations are being carried out at elevated pore pressure, even relatively small net pressures can induce shear slip on fractures with a wide range of orientations. In other words, fractures with orientations that are very poorly oriented for slip in the current stress field, can be made to slip. For equivalent stress states and pre-existing fracture orientations, it is easier to induce shear slip and create an interconnected fracture network in over pressured systems than when formation pressure is closer to hydrostatic.

Finally, at highly elevated pore pressure, relatively small perturbations caused by a variety of geologic processes could result in changes in the stress state. When pore pressure is hydrostatic, large stress perturbations are required to cause the stress field to change from place to place. When pore pressure is highly elevated, relatively small perturbations could change the stress state in a significant way, for example from normal faulting to reverse faulting, or the A_ϕ value might change or the direction of horizontal principal stresses. Moos & Zoback (1993) discussed such a case in the Long Valley caldera where a 90° change in the stress direction (and a change from normal to strike-slip faulting) occurred over only a few km.

Fractures and Faults in Unconventional Reservoirs

As first mentioned in Chapter 1, the creation of a network of permeable fractures and faults in contact with the extremely low permeability matrix is a critically important component of unconventional reservoir development. Thus, even the presence of pre-existing old, dead fractures and faults can be important components of successful production, if properly stimulated. In the volume of rock affected by high pressure during hydraulic fracturing, the high pressure can cause slip on these fractures and faults thereby creating a network of interconnected permeable conduits that greatly enhance the permeable surface area in contact with the low permeability matrix. These topics will be discussed further in Chapters 10 and 12. It should be obvious at this point, however, that the occurrence and orientation of pre-existing fractures and faults are quite important.

Throughout this book, when we refer to fractures we are referring to any planar discontinuity in a rock mass that is identifiable at the macroscopic scale (core scale, or larger). We refer to any macroscopic fracture that has moved in shear as a fault, regardless of its size. Microfractures, visible only under high magnification in thin section, CT scan or SEM (as discussed in Chapter 5), are considered part of the rock matrix.

We discuss here four distinct types of fractures and faults. First, near vertical opening mode (or Mode 1) fractures that are orthogonal to bedding. These are frequently found to be filled with calcite in many shales but need not be. As illustrated in Fig. 7.20a (after Gale et al., 2014) for the New Albany shale in eastern Kentucky such fractures may or may not be confined to a single bed and can sometimes cross multiple bed boundaries as shown. Vertical Mode 1 fractures are almost certainly caused by transiently high pore pressures that temporarily exceed the local magnitude of the minimum horizontal stress (Engelder et al., 2009). In other words, they form through a transient process of hydraulic fracturing to relieve pore pressure that builds up faster during maturation than can be dissipated through fluid flow.

As discussed in Chapter 10, the high pore pressure generated during hydraulic fracturing can often make fractures which form as opening mode fractures which slip in shear. Thus, a Mode 1 fracture that formed naturally at some time in the geologic past, could be made to slip in shear by the hydraulic stimulation process. The second type of fracture and fault of interest here are relatively small scale faults, as shown in Fig. 7.20b (after Gale et al., 2014), an outcrop photo of the Vaca Muerta formation in Argentina. Note that there is a few cm of normal slip (hanging-wall down) offset on this fault. There

 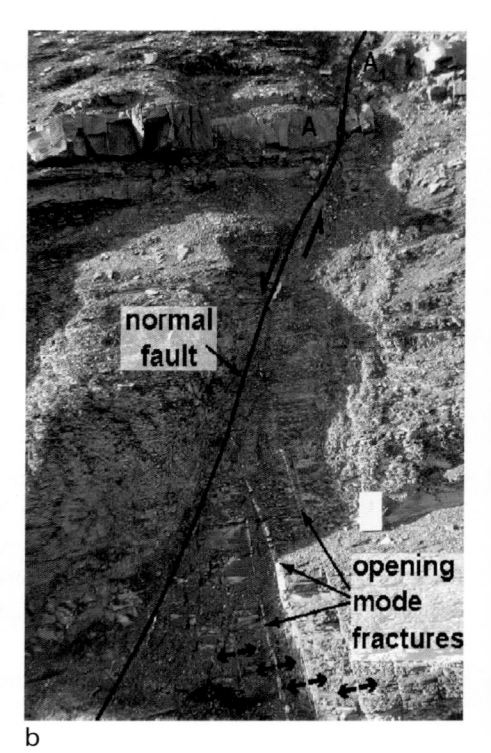

a   b

Figure 7.20 Outcrop photographs of (a) opening mode fractures in the New Albany shale in eastern Kentucky and (b) a normal fault in the Vaca Muerta formation in Argentina with several cm of offset. From Gale et al. (2014). *Reprinted by permission of the American Association of Petroleum Geologists whose permission is required for further use.*

could also be a component of shear slip in the horizontal direction, but it usually cannot be determined from a two-dimensional outcrop. Third, we will also discuss larger scale faults which we will refer to as pad-scale faults to indicate their importance on the scale of several km, the same scale as a well pad. As explained below, such faults are important because of the ways they can affect the stimulation and production process. Finally, because they are planar discontinuities that can potentially move in shear, we will consider sub-horizontal bedding planes as potential faults in Chapter 10.

Observations versus Assumptions – An entrenched, but simplistic notion about pre-existing fractures in unconventional reservoirs is that they are principally near-vertical, Mode 1 joints oriented in orthogonal that are aligned (parallel and perpendicular) to the current orientations of S_{Hmax} and S_{hmin}. Hence, it is assumed that induced hydraulic fractures are surrounded by a regular grid of fractures parallel and perpendicular to them. From a geological perspective it is difficult to understand the basis for this assumption although it makes numerical modeling easier to perform, which may have contributed to the prevalence of this view. Pre-existing fractures on the Appalachian plateau exhibit such a fracture pattern in some places but the actual pattern of fracture orientations in these areas are often more complicated (Engelder et al., 2009). For opening mode fractures to be parallel to modern hydraulic fractures, the stress orientation at the time they formed had to have been similar to that which acts today. Even when Mode 1 fracture sets are sub-parallel to hydraulic fractures forming in the current stress field it would not explain how fractures orthogonal to them would form as they would be opening perpendicular to the maximum horizontal stress. A switch of horizontal principal stress orientations had to have occurred over geologic time. We present below fractures observed in image logs in a number of horizontal wells that generally show a complex pattern, unrelated to current stress orientations.

Some of the history of assuming near vertical, orthogonal fracture sets aligned with the current stress field comes from an early study by Fisher et al. (2002) that made important and seminal contributions in the early days of using microseismic events. Figure 7.21 shows two figures from that paper. One of the important conceptual breakthroughs of this study was that the effects of hydraulic fracturing extend well beyond the hydraulic fracture plane(s), in this case propagating in a NE–SW direction, via some sort of fracture network. Their evidence for this, shown in Fig. 7.21b is that five vertical wells surrounding the well that was hydraulically fractured were severely affected by the hydraulic fracturing. In addition, they correctly surmised that the absence of microseismic events to the southwest probably indicated a relative absence of pre-existing fractures in that area. Unfortunately, their conceptual model of this fracture network (Fig. 7.21a) was two sets of near vertical faults, trending parallel and perpendicular to the current S_{Hmax} directions. One could draw lines representing fractures in almost any direction from the cloud of microseismic events shown in Fig. 7.21b.

With image logs, it is straightforward to show that pre-existing fractures in unconventional reservoirs are not parallel and perpendicular to current hydraulic fractures. In Fig. 7.22, we show stereonets of fracture orientations from three plays in Texas and Oklahoma. The fracture orientation data come from image logs, mostly in horizontal

Unconventional Reservoir Geomechanics

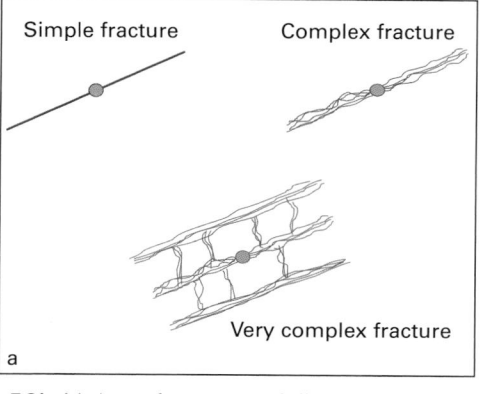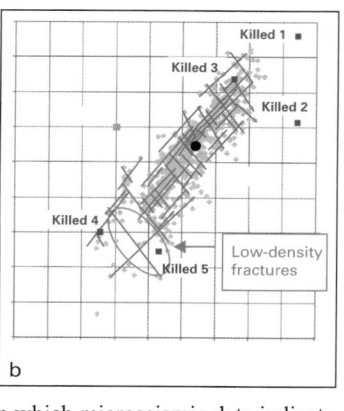

Figure 7.21 (a) An early conceptual diagram of the way in which microseismic data indicate a complex pattern of deformation that occurs during hydraulic fracturing of a vertical well. (b) The distribution of microseismic events led the authors of this study to interpret a network of orthogonal vertical fractures. A vertical well used to deploy the microseismic array is at the yellow square, about 1,000 ft to the west of the well begin stimulated, indicated by the black dot. From Fisher et al. (2002).

wells. In each case the wells were drilled close to the azimuth of S_{hmin}, which would be ideal for detecting fractures parallel to hydraulic fractures. At each site the direction of S_{Hmax} is known from observations of wellbore failures (typically drilling-induced tensile fractures) in nearby vertical wells. Two vertical wells in west Texas show markedly different fracture distributions in the Clearfork and Wichita Albany units. Steeply dipping fractures are seen in each formation but with a wide range of azimuths. As can be seen in the stereonets of fracture observations in three horizontal wells in the Barnett shale of the Fort Worth Basin, each well shows a wide variety of pre-existing fracture orientations. In each case, the orientations of the pre-existing fractures are different in the lower Barnett than the overlaying formations. Although some of the steeply dipping fractures in each well trend NE–SW, the same direction as S_{Hmax}, there is considerable variability in the fracture orientations. Most fractures are steeply dipping, but a wide range of dips is also seen. Although there are fewer fractures in the formations overlying the Barnett, a similarly wide variation of orientations is also seen. Finally, two horizontal wells from the Woodford formation in Oklahoma are also shown. In this case there is a concentration of steeply dipping faults parallel to the current S_{Hmax} direction. The fact that there is a wide range of fracture orientations observed in these wells is conducive for creating an interconnected fracture network during stimulation (see Chapter 10). If there was a limited range of fracture orientations, it would be difficult to create an interconnected fracture network to provide conduits for flow. We show in Chapter 10 that it is essentially impossible for fractures striking normal to S_{Hmax} (i.e., normal to the direction of hydraulic fracture propagation) to be stimulated in shear during hydraulic fracturing.

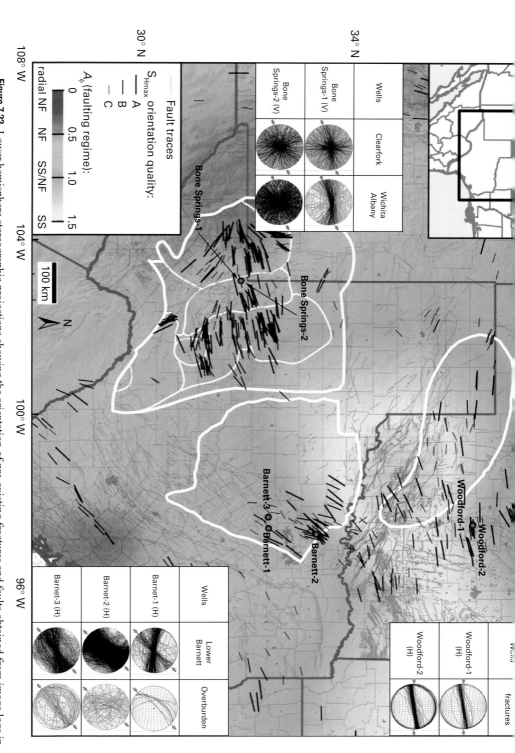

Figure 7.22 Lower hemisphere stereographic projections showing the orientation of pre-existing fractures and faults obtained from image logs in wells drilled approximately orthogonal to the direction of maximum horizontal stress (red arrows in each stereonet) in three areas where unconventional oil and gas are currently being produced. The data in the upper left are from the Permian Basin in west Texas, the upper right are from central Oklahoma and lower right from the Barnett shale in the northeastern Fort Worth Basin of Texas.

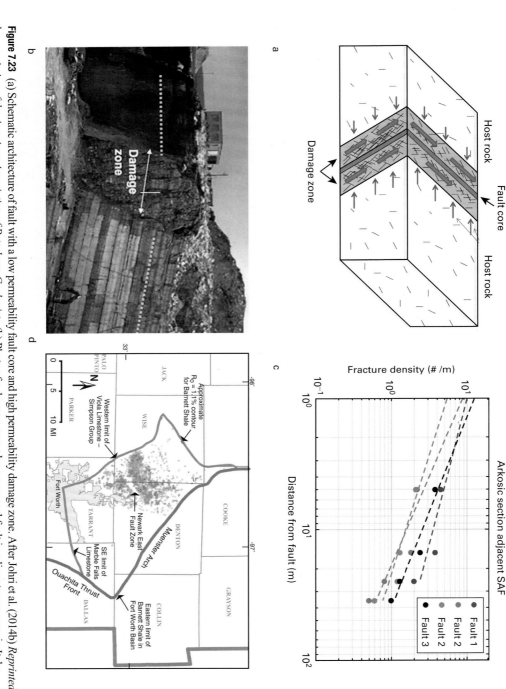

Figure 7.23 (a) Schematic architecture of fault with a low permeability fault core and high permeability damage zone. After Johri et al. (2014b) Reprinted by permission of the American Association of Petroleum Geologists. (b) Photo of a damage zone along a normal fault in a limestone quarry in Italy. Courtesy of Peter Hennings, Univ. of Texas. (c) The decay of fracture density with distance from faults in arkosic sandstone near the San Andreas fault. The red dashed line indicates the *background* fracture density away from the faults. After Johri et al. (2014b) Reprinted by permission of the American Association of Petroleum Geologists. (d) There was a pronounced absence of wells producing from the Barnett shale in proximity to the Newark East fault zone. From Pollastro (2007) Reprinted by permission of the American Association of Petroleum Geologists whose permission is required for

Kaluder et al. (2014) compared stress orientations and fracture orientations in the Bazenhov shale of the western Siberian Em-Egovskoe tight-oil field. In six wells, the maximum horizontal stress orientation was found to be consistently oriented NNW–SSE but the fracture strikes differ widely from well to well and only in one case is there a predominant fracture strike within 20° of the current S_{Hmax} direction. A similarly poor correlation between natural fracture strike and current stress orientation were reported by Laubach et al. (2004) in the East Basin and Val Verde Basins of Texas and the Green River and Powder River Basins of Wyoming. In two areas, the Val Verde and Green River Basins, there is no correlation between the fracture strike and maximum stress orientation. In the other two areas, a number of fractures strike subparallel to S_{Hmax}, but overall, the fracture strikes have considerable variability.

Pad Scale Faults and Damage Zones – The third type of fracture and fault of interest here are relatively large, faults which we will allude to as pad-scale faults, because they are large enough to affect stimulation of multiple wells on a drill pad. It is important to recognize that these faults are not simple planar discontinuities. Figure 7.23a (from Johri et al., 2014b), shows a generalized morphology of fault zones based on field mapping (Chester & Logan, 1986; Chester et al., 1993; Faulkner et al., 2003; Mitchell & Faulkner, 2009; Savage & Brodsky, 2011). Two distinct structural elements are shown – the narrow fault core where slip is localized and a much wider damage zone containing numerous macroscopic fractures and faults, extending several tens of meters from the fault core. Because of extensive comminution and the generation of fault gouge, and the mechanical and chemical alteration of the host rock to clay, the fault core may be relatively impermeable with respect to the host rock. In fact, in conventional oil and gas reservoirs, it is important to try predict whether faults visible in seismic data are, or are not, "sealed" to cross-fault flow which can compartmentalize a reservoir. In contrast, the broad damage zone consists of a concentration of small-scale faults that are likely conduits for flow parallel to the main fault trend (Zhang & Sanderson, 1995; Caine et al., 1996; Paul et al., 2009; Hennings et al., 2012). The study by Hennings et al. (2012) in the Suban gas field in South Sumatra is especially noteworthy because tremendous gas flow was channeled along damage zones of large-scale faults located in extremely low permeability granite and dolomite at the edge of a major horst block.

Figure 7.23b is a photograph of a ~10 m wide damage zone seen in a limestone quarry in Cretaceous age carbonates in eastern Italy (courtesy Peter Hennings). It is obvious that the pervasive fracturing in the damage zone would augment fluid flow parallel to the fault plane. Figure 7.23c (Johri et al., 2014b) shows the density of macroscopic fractures seen in a wellbore image log in the SAFOD research well. The data shown come from an arkosic sandstone with eight identifiable faults just to the west of the San Andreas fault in central California (data from only four of faults are shown). Note that the fracture density (F) in the damage zone drops off rapidly with distance from the fault (r), reaching the background level of fracture density about 50–80 m from the fault. Johri et al. (2014b) found the fracture density to decrease with distance from the fault to follow an exponential decay proposed by Savage and Brodsky (2011) of the form

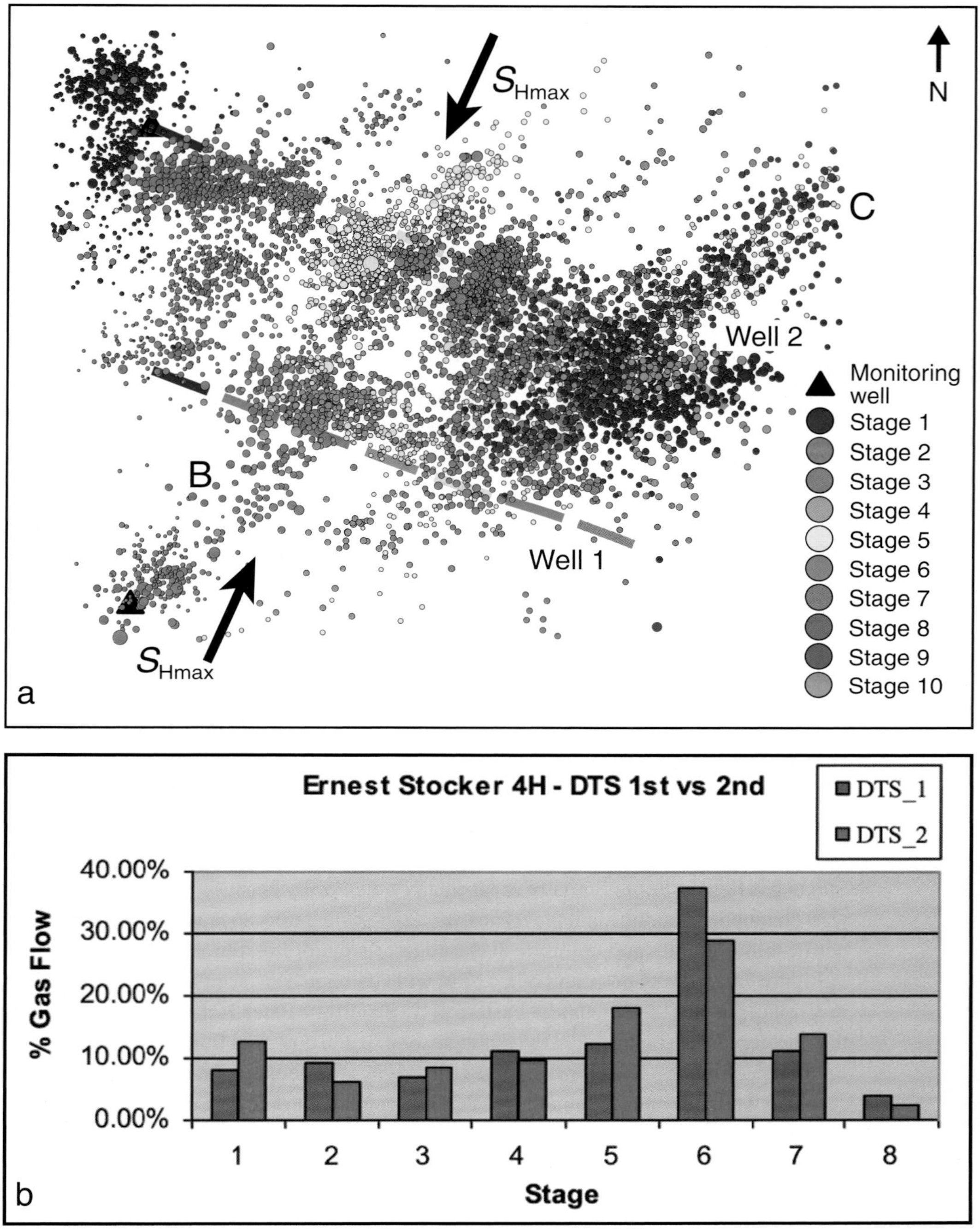

218   Unconventional Reservoir Geomechanics

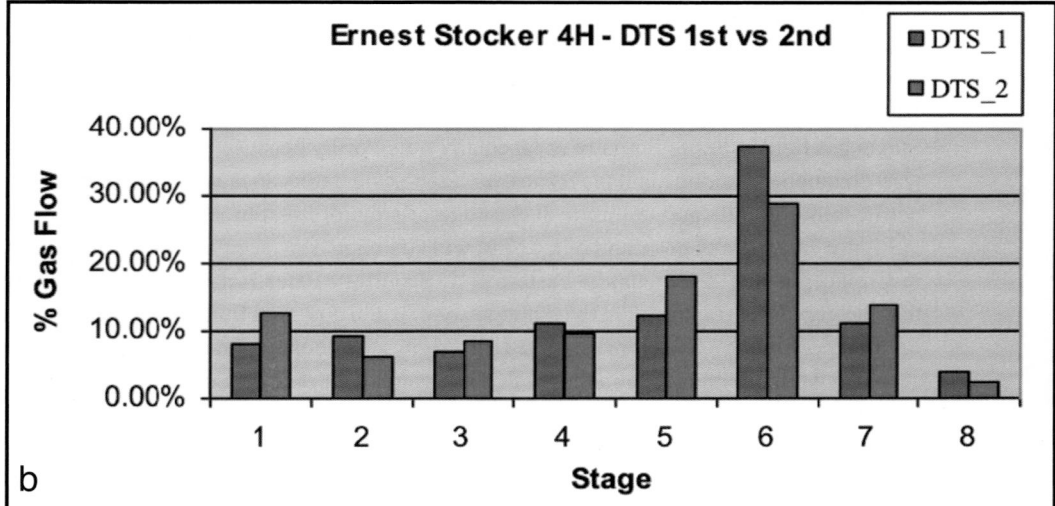

Figure 7.24 (a) Microseismic epicenters recorded during hydraulic fracturing stimulation of two wells in the Barnett shale. The data were recorded by two microseismic arrays in vertical wells indicated by the black triangles. One was located near the toe of the well (the northwest end), the other in the southwest corner of the figure. (b) Relative flow rates associated with various stages from distributed temperature sensor (DTS) fiber optics sensor deployed in well 1. The surveys were carried out several months after production began. After Roy et al. (2014).

$$F(r) = F_0 \, r^{-n} \tag{7.6}$$

where F_0 is the fracture density at a unit distance from the fault. Interestingly, the value of n found for the damage zones for the faults in the arkosic sandstone adjacent to the San Andreas fault ranges from about 0.40 to 0.75, generally consistent with the value of 0.8 reported for faults with total slip less than ~150 m by Savage and Brodsky (2011) who found n to decrease for faults with larger displacements. The corresponding values of F_0 and n for sixteen faults in four wells in the Suban field by Johri et al. look quite similar to Fig. 7.23c with values of n between 0.6 and 1.1.

Figure 7.23d (from Pollastro, 2007) shows a significant gap in the distribution of producing wells in the Barnett in the region immediately surrounding the Newark East Fault Zone. It is possible that due to the high permeability of the faults associated within the damage zone surrounding this major fault, much of the natural gas leaked out of the Barnett prior to development. This could be happening at smaller scale in unconventional plays and could be important to consider during development.

Damage Zones and Multi-Stage Hydraulic Fracturing – Another important aspect of permeable fault damage zones is how they can affect hydraulic fracturing operations, as illustrated in Fig. 7.24 (modified after Farghal & Zoback, 2015). In this case, microseismic events were observed to propagate rapidly along pre-existing faults zones as shown. Of the unusual lineations of microseismic events seen in Fig. 7.24a, perhaps the most dramatic are the events along fault C which propagated far to the northeast at an azimuth about 25° from that expected for a hydraulic fracture. This orientation is what would be expected for a strike-slip fault, which is not unexpected as this is a strike-slip/ normal faulting area ($S_V \sim S_{Hmax} > S_{hmin}$). Events along the sub-parallel fault B are also clearly visible which propagate to the southwest from stage 3 in well 1. In fact, microseismic events occured along fault C northeast of well 2 when well 1 was being stimulated during stages 5 and 6 and pressure was observed to increase in well 2. More microseismic events along this fault occurred when well 2 was being stimulated during stages 8–10. All of these observations clearly indicate hydraulic connectivity along the fault damage zone from well 1 to well 2 and hundreds of meters beyond.

Another way of describing what is seen in Fig. 7.24 is that the pre-existing faults appear to be hijacking a number of the hydraulic fracturing stages, by which we mean that fluid pressure is being diverted along pre-existing fault damage zones. An example of this was presented by Maxwell et al. (2008) where the propagation of microseismic events clearly defines a propagating hydraulic fracture that runs into a pre-existing fault. If this happens relatively close to a well, one would not achieve the intended hydraulic fracture stimulation. The microseismic events seen to the southwest of well 1 along fault B were associated with pressurization of an existing vertical monitoring well near the southwest end of the microseismic zone.

A fiber optics distributed temperature sensor (DTS) in the wells was surveyed in the months following the onset of production (Fig. 7.24b). The first two surveys in well 1, indicated that maximum flow into the well was near stages 5 and 6 where fault C intersected the well. In well 2, the first two DTS surveys indicated high flow rates in

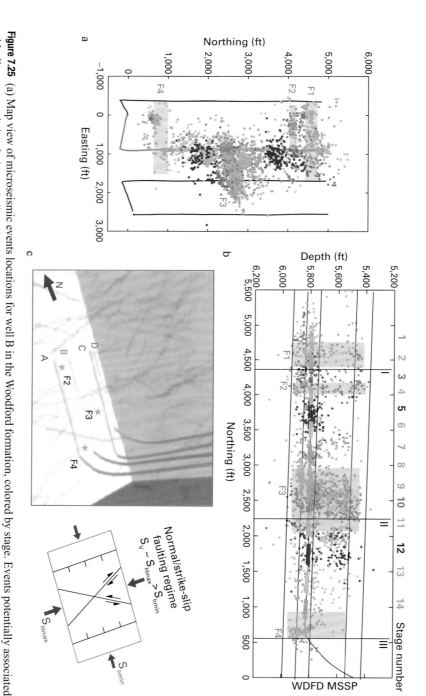

Figure 7.25 (a) Map view of microseismic events locations for well B in the Woodford formation, colored by stage. Events potentially associated with slip on pre-existing faults are indicated by the red shaded areas and discussed in the text. Red stars indicate the locations of vertical observation wells where seismic arrays were deployed. (b) Cross-section view of microseismicity associated with well B as shown in map (c) Three-dimensional rendering of the enhanced discontinuities using ant tracking (based on the variance attribute) of 3D seismic data. The locations where the observation wells intersect the shown horizon are marked with stars. As shown by the cartoon on the right, the trend of the faults are consistent with the observation wells in a normal/strike-slip faulting stress regime representative of this part of Oklahoma. Modified from Ma & Zoback (2017b).

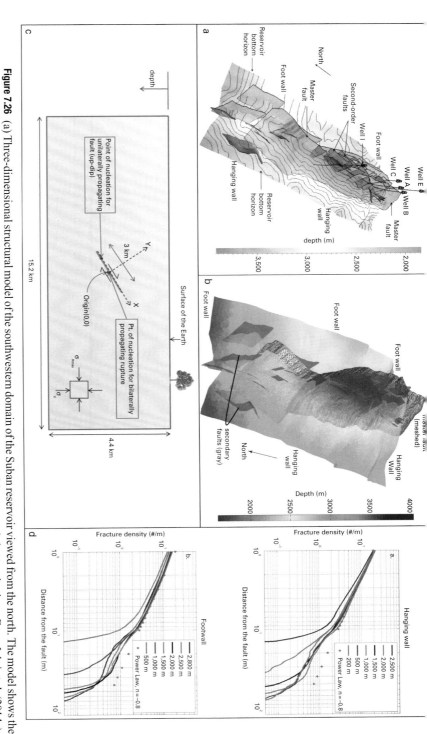

Figure 7.26 (a) Three-dimensional structural model of the Suban reservoir viewed from the north. The model shows the master fault, reservoir-scale second-order faults, well trajectories and the depth structure of the reservoir-bottom horizon. From Johri et al. (2014a). Reprinted by permission of the American Association of Petroleum Geologists whose permission is required for further use. Contour interval is 100 m (328 ft). (b) Three-dimensional structural model of the southwestern domain, viewed from the north. The model shows the master fault (first-order fault) and second-order faults. The colored surface represents the depths of the upper horizon of the reservoir. (c) Two-dimensional idealization to model slip on a buried thrust fault dipping at 30°. The green line represents the fault surface. X axis is aligned along the fault in the up-dip direction. (d) Fracture decay profiles in the (a) hanging wall and (b) footwall of the damage zone modeled. Various lines represent fracture density decay profiles across transects at the mentioned distances from the base of the fault. Dots represent a power law decay (decay rate of 0.8) reported for damage zones identified using image logs reported in Johri et al. (2014a). Figures b–d from Johri et al. (2014b).

the vicinity of stages 6 and 7. The third survey indicated maximum flow with stages 8, 9 and 10, where fault C cuts through the well. It seems clear that gas flow along the damage zones was beneficial to production in both wells. What is not clear is whether this would have been the case had these stages not been stimulated or stimulated at lower flow rates and pressures. These data are discussed in more detail in Chapters 10 and 12.

Another example of fluid flow along the damage zone of pre-existing faults is shown in Fig. 7.25 where four parallel wells were drilled in Oklahoma, two in the Woodford shale formation and two in the overlying Mississippi limestone (after Ma & Zoback, 2017b). This is a strike-slip/normal faulting area ($S_V \sim S_{Hmax} > S_{hmin}$) in which the minimum horizontal stress is oriented approximately north–south. As shown in Fig. 7.25a, b there were several lineations of microseismic events that are associated

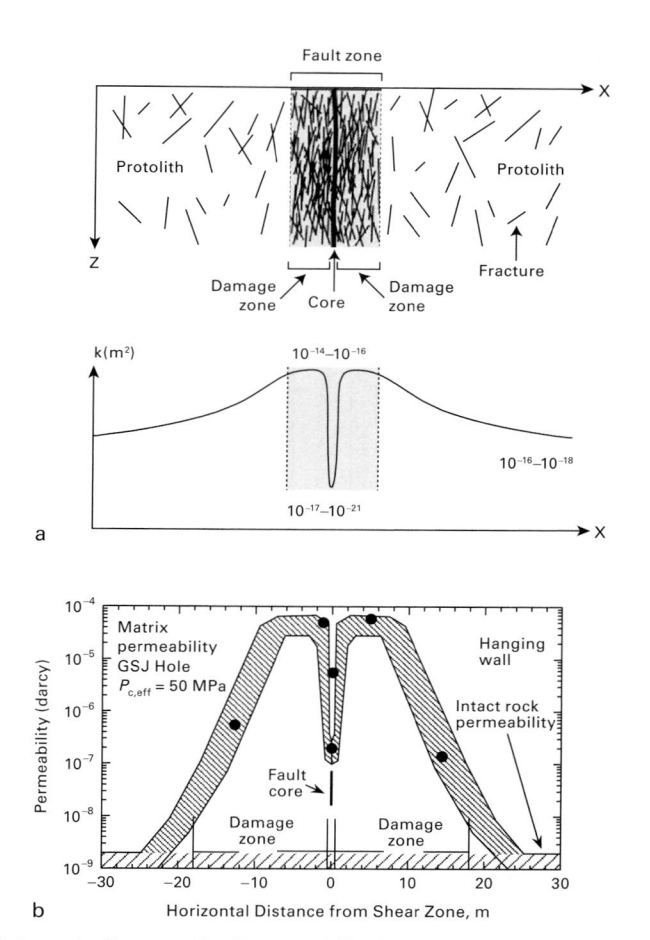

Figure 7.27 (a) Schematic diagram of bulk permeability in the core and damage zone of a fault, and its decay with distance to the background permeability of host rock. Modified after Cappa & Rutqvist (2011) (b) Variations of the matrix permeability of cores taken at different distances to the Nojiima fault in Japan. After Lockner et al. (2009).

Stress, Pore Pressure, Fractures and Faults   223

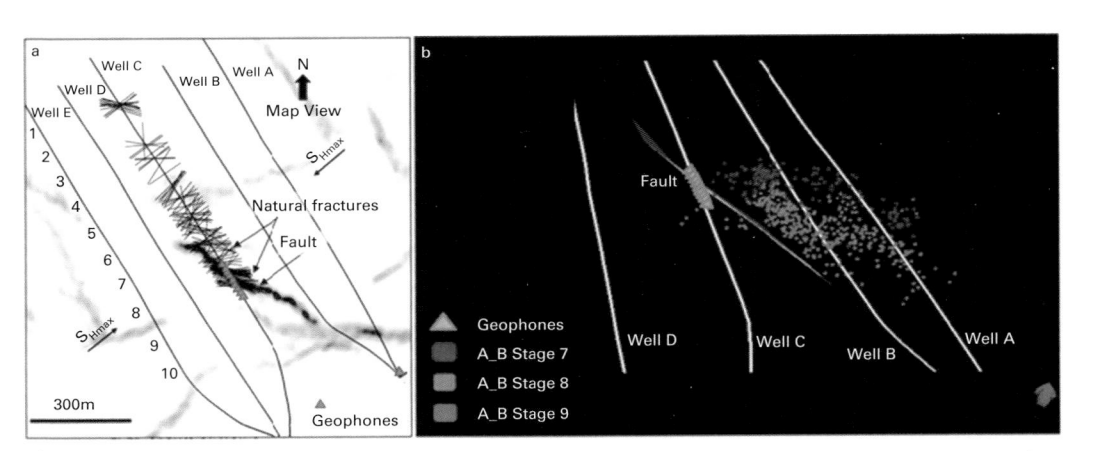

Figure 7.28 (a) Ant-tracking time slice showing a fault intersecting Well C in a dataset from the Barnett shale where a dense cluster of pre-existing fractures were indicated by an image log. (b) A perspective view of microseismicity associated with Wells A and B abruptly end where they encounter the fault cutting through Well C. Modified after Farghal & Zoback (2015).

with fault zones. In fact, events were occurring near well A associated with fault F1 when all four wells were being stimulated. As indicated in Fig. 7.25c, ant tracking revealed a pattern of strike-slip and normal faults that are in accordance with that shown in Fig. 7.1 for a strike-slip/normal faulting area.

While there have been a number of studies of damage zones in the geological literature as noted above, there have been few studies of how they formed. One idea is that they form during earthquakes due to the concentration of stress that propagates along a fault as it slips. Dynamic rupture modeling by Paul et al. (2009) utilized analytic formulas to estimate the nature of damage zones associated with faults in the Bayu Udan field of Indonesia. The basic idea is that the stress concentration associated with a rupture pulse traveling close to the shear velocity down a fault plane is large enough to break the rock. Following up this idea, Johri et al. (2014a) utilized numerical methods to study dynamic rupture propagation for faults analogous to those in the Suban field mentioned above while incorporating complexities such as roughness of the fault plane, direction of rupture propagation, etc. The faults in question are referred to as secondary faults in the perspective map shown in Fig. 7.26a (after Johri et al., 2014a) which were modeled utilizing the numerical grid shown in Fig. 7.26b (Johri et al., 2014b). A simplified view of single fault in the model is shown in Fig. 7.26c. What turns out to be extremely interesting is that damage zones resulting from this geometrically simplified dynamic rupture propagation model produces realistic values for the widths of the damage zone (tens of meters), the decay of damaged rock with distance (the parameter n) and the degree of damage close to the fault (the parameter F_0). Figure 7.26d shows the model predictions for both the hanging wall and foot wall of one of the modeled faults at various positions along its length. Depending on the profile position, the damage zone width varies between ~10 and ~50 m (but is not symmetric in the hanging wall and footwall). Regardless, the F_0 and n values are quite similar for all profile positions.

224   **Unconventional Reservoir Geomechanics**

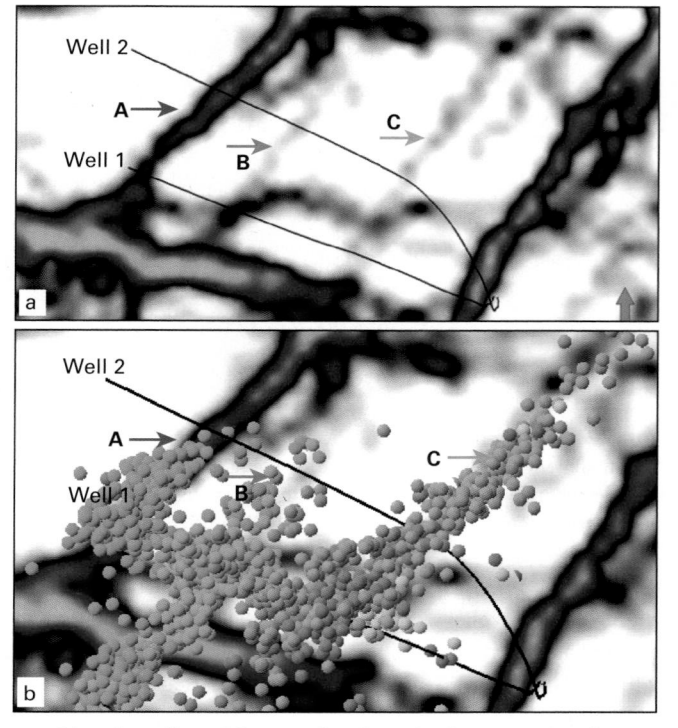

Figure 7.29 Ant-tracking time slice of the post-frac 3D seismic survey showing apparent faults A, B and C and the association of microseismic events with these faults during hydraulic fracturing in Well 1. Modified after Farghal & Zoback (2015).

One interesting observation about damage zones is that not only does it contain a high concentration of macroscopic fractures and faults, but the matrix permeability of the host rock is also higher than it is outside the damage zone. This is illustrated schematically in Fig. 7.27a (after Cappa & Rutqvist, 2011). Figure 7.27b shows a series of matrix permeability measurements adjacent to the Nojiima fault in Japan that slipped in the 1995 Kobe earthquake. Using core samples from a scientific research well drilled through the fault, Lockner et al. (2009) found more than an order-of-magnitude 4 increase in permeability immediately adjacent within 10 m of the fault core and more than a two order-of-magnitude increase compared to the protolith at greater distance. There were insufficient samples to document the distance at which the matrix permeability reaches the value of that of the host rock protolith outside the damage zone. Note that the comminution and alteration of the gouge within the fault core itself results in a markedly lower permeability than in the damage zone.

Utilizing 3D Seismic Data to Map Fault zones and Fractures

In the sections above, the presence of faults and fractures are shown to have significant influence on production from unconventional reservoirs. To simplify complex issues, one could argue that one should target distributed fractures and avoid larger scale faults. Targeting pre-existing fractures and small faults would promote stimulation of a shear fracture network that would be beneficial to production (Chapters 10 and 12). Avoiding larger scale faults would prevent hydraulic fracture stages from being diverted along fault damage zones or otherwise affect stimulation (or trying to produce hydrocarbons from an area where they have leaked out over time). In some cases, pressurizing larger scale faults could possibly connect hydraulic fractures to potentially problematic faults and induce potentially damaging earthquakes (Chapter 14).

Ant Tracking and Variance – A number of techniques have been developed in the seismic reflection industry to identify structural discontinuities such as fractures and faults. Figures 7.28 and 7.29 expand on the case studies in the Barnett shale where ant tracking of the variance attribute of the seismic volume was used to identify potentially active faults discussed by Farghal & Zoback (2015). The ant tracking algorithm of a 3D seismic reflection survey as described by Randen et al. (2001). Figure 7.28a shows a time slice near the temporal depth of the five wells shown in the Barnett shale. The 3D reflection seismic data provided for the study was a post-stack migrated volume. To locate faults in this area, they calculated the variance attribute of the seismic data followed by two passes of ant tracking for discontinuity/edge enhancement. Note that an apparent fault cuts right through Well C, exactly where there is a concentration of fractures seen in an image log in this well (see also Das & Zoback, 2013b). Both the fracture concentration seen in the image log and the fault where it cuts through the well have a similar orientation. Figure 7.28b shows a perspective view of microseismic events resulting from stages 7–9 in Wells A and B, which were hydraulically fractured simultaneously. Note the position of the seismometer array in Well C is close to the location of the microseismic events from Wells A and B. It is quite interesting that the events associated with these three stages seem to be terminated by the fault (as the array is quite close, this is not a detection artifact). We interpret this to be a result of the permeable damage zone associated with this fault channelizing fluids along this fault which prevents hydraulic fracture propagation across it.

Figure 7.29 takes a closer look at the case study shown in Fig. 7.24, also from the Barnett shale. The upper part of the figure shows the apparent fault zones identified by ant tracking following hydraulic fracturing. After data processing and migration, the variance attribute of the post-frac survey was computed before two passes of ant tracking were performed. Of particular note are the three northeast trending fault zones labeled A, B and C shown in the upper part of the figure; these are clearly associated with concentrations of microseismic events in the lower part of the figure. The lower part of the figure shows the concentration of the microseismic events generated during pressurization of Well 1. Note that the microseismic events extend beyond Well 2, apparently due to the high permeability of the damage zones associated with these

226   **Unconventional Reservoir Geomechanics**

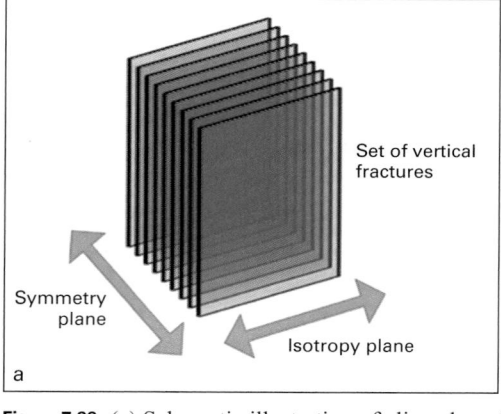
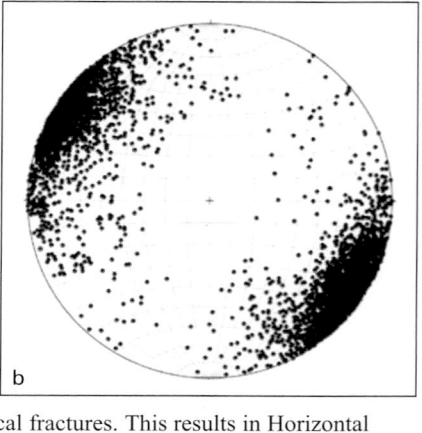

Figure 7.30 (a) Schematic illustration of aligned vertical fractures. This results in Horizontal Transverse Isotropy. (b) The orientation of pre-existing fractures observed in an image log in well 1 in the dataset shown in Figs. 7.24 and 7.28.

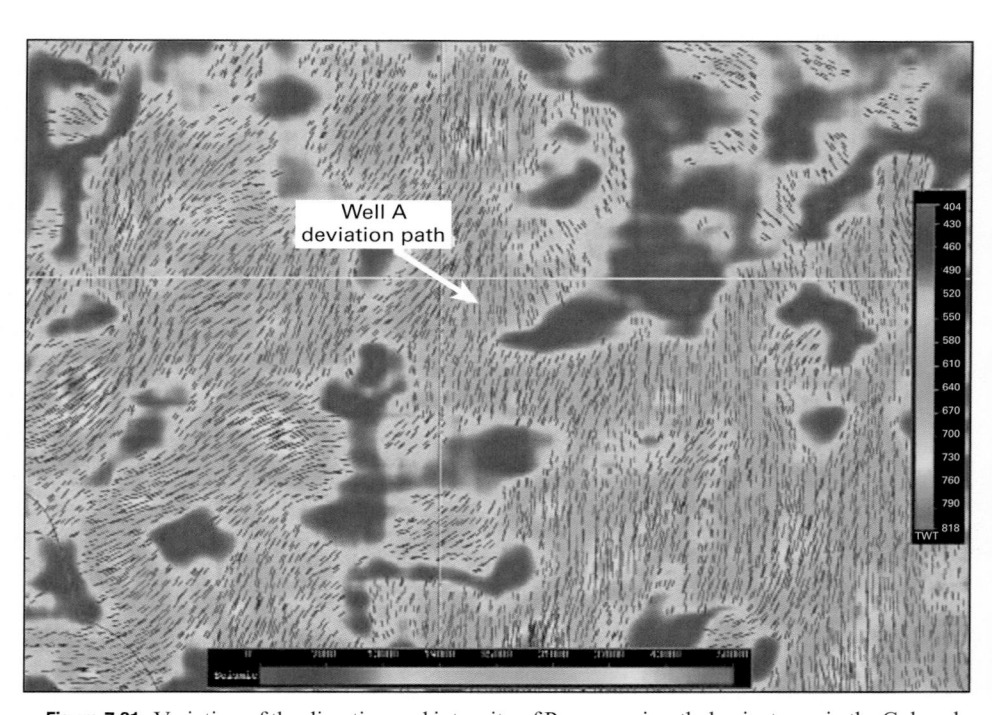

Figure 7.31 Variation of the direction and intensity of P-wave azimuthal anisotropy in the Colorado gas shale. After Goodway et al. (2006)

faults. Apparently, the large apparent northeast trending fault that cuts through the near vertical sections of the wells is not associated with microseismicity as it was not pressurized.

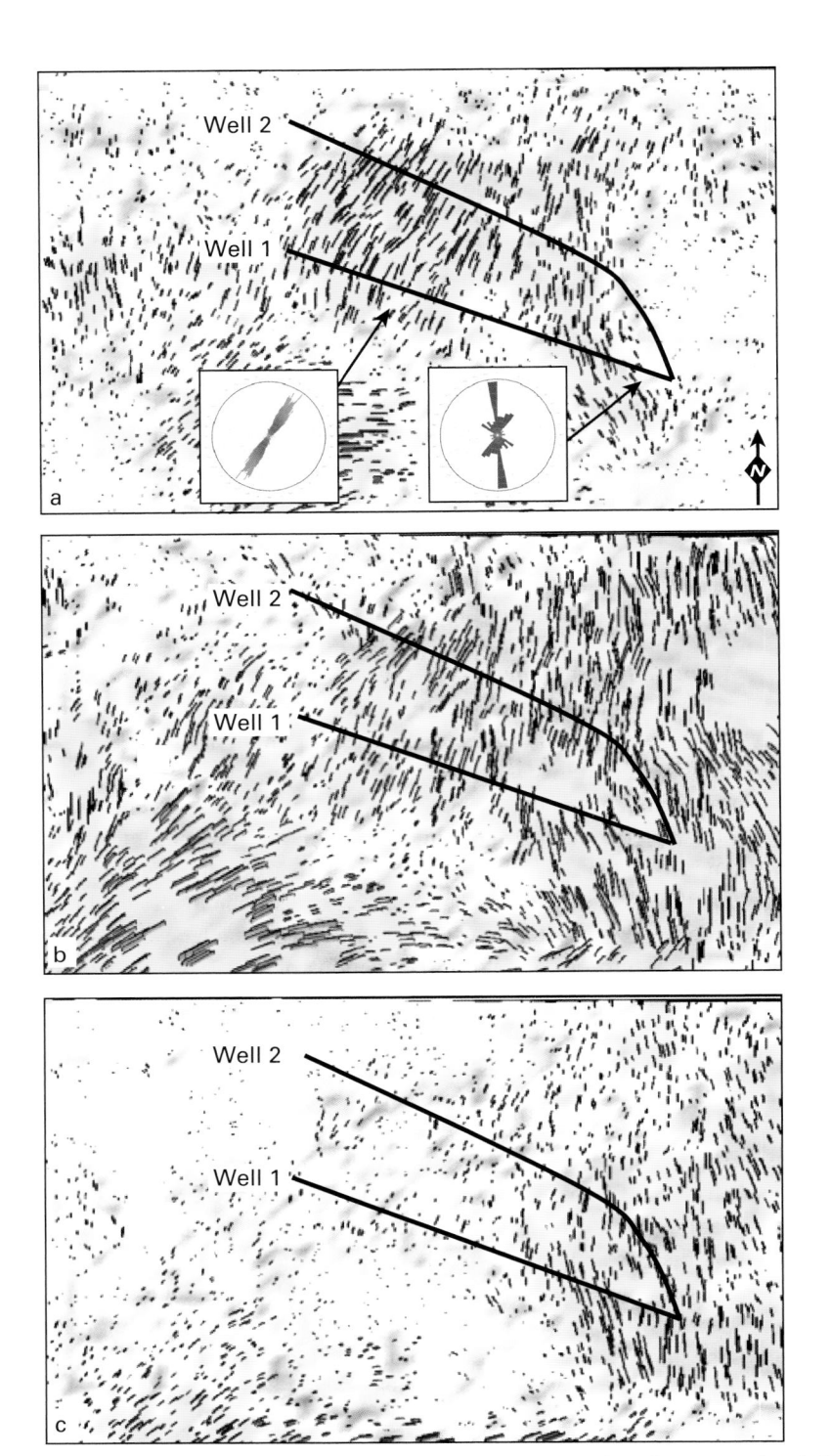

Figure 7.32 Color scale indicates variation of the direction and intensity of P-wave azimuthal anisotropy in the Barnett shale. After Farghal (2018). (a) Baseline survey carried out prior to drilling and stimulation (b) Post-frac survey carried out soon after the two wells were hydraulically fractured. (c) Post-production survey carried out three months after production.

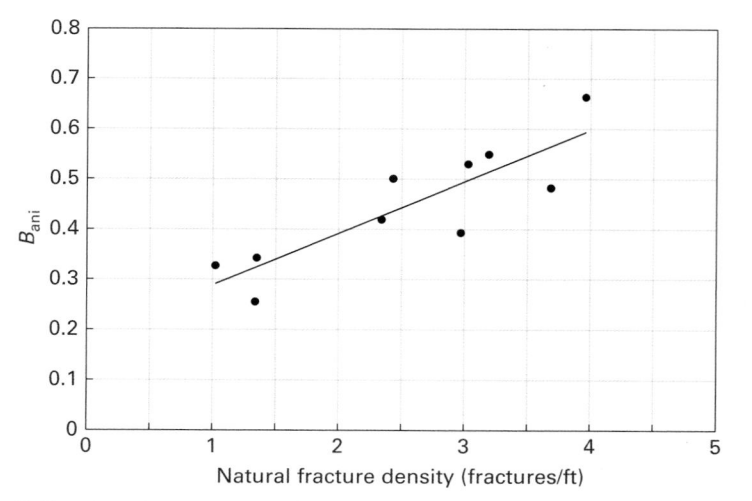

Figure 7.33 Calibration of the observed shear velocity anisotropy to the observed fracture density in Well 1. After Farghal (2018).

Azimuthal Amplitude versus Offset – The importance of stimulating slip on preexisting fractures for successful exploitation of unconventional reservoirs was alluded to in Chapter 1 and is discussed at greater length in Chapters 10 and 12. It would be of obvious importance if seismic reflection data could be used to identify areas of greater fracturing, in a manner analogous to identifying larger scale faults as discussed above. In Chapter 2, the sub-horizontal bedding and layering of unconventional reservoirs lends itself to be characterized in terms of Vertical Transverse Isotropy (VTI), meaning that the stiffness in a vertical direction is less than that in the horizontal direction, and parameterized in terms of the Thomsen parameters, ε, γ and δ (Thomsen, 1986). We now turn our attention to variations of stiffness with azimuth. It has long been recognized that there are two principal mechanisms that could cause horizontal velocity anisotropy in the upper crust. In rock in which relatively compliant, steeply dipping pre-existing fractures that are highly aligned (as illustrated in Fig. 7.30a) the medium will be compliant in the direction normal to the fractures and relatively stiff in the direction parallel to the fractures. Such formations are said to be characterized by Horizontal Tranverse Isotropy (HTI). In rock with more randomly oriented fractures, the orientation of stress would tend to close the pre-existing fractures normal to S_{Hmax}, making this direction preferentially stiff. There have been numerous published studies that discuss azimuthal variations of seismic velocity in terms of either aligned fractures or anisotropic stress (see, for example, Lynn, 2004).

Goodway et al. (2006) presented a successful case study in the Colorado gas shale shown in Fig. 7.31 that utilized horizontal variations of P-wave velocity (AVAZ) to map variations of fracture intensity and orientation. Starting with the theory of anisotropy and Thomsen parameters as defined for VTI, Goodway et al. redefined the Thomsen parameters for HTI. As shown by the colors in the figure, there is almost a factor of two variation of the degree to which P-wave velocity changes with azimuth (suggesting

Stress, Pore Pressure, Fractures and Faults

comparable variations of fracture intensity) as well as variation of the direction of the fractures. Well A in the figure was drilled perpendicular to the orientation of fractures predicted from AVAZ and coherence, and encountered many fractures as predicted. These fractures were confirmed by a post-drill image log.

Using a different, but analogous, method, Farghal (2018) used azimuthal variations of shear wave velocity as determined from Azimuthal AVO (amplitude versus offset) data with a time lapse 3D dataset obtained as part of the Barnett dataset shown in Figs. 7.24 and 7.28 and described by Roy et al. (2014). Rüger & Gray (2014) compare azimuthal P-wave anisotropy (Fig. 7.31) and Pb-wave anisotropy (Fig. 7.32).

The metric used for P-wave velocity anisotropy, B_{ani}, is defined for a dry gas reservoir (assuming penny-shaped cracks) by Bakulin et al. (2000) as

$$B_{dry}^{ani} = \frac{4(-8g^2 + 12g - 3)}{3(3 - 2g)(1 - g)} e \tag{7.7}$$

where g is the square of the ratio S-wave velocity to P-wave velocity and e is given by

$$e = N_f \left(\frac{a^3}{8} \right) \tag{7.8}$$

where N_f is the density of fractures and a is the diameter of penny-shaped crack (as defined below).

Figure 7.32a shows the variation of the intensity and orientation of shear velocity anisotropy in the baseline survey carried out prior to stimulation of the two wells shown in the figure. Note that as in Fig. 7.31, the variation in the azimuthal dependence of shear velocity anisotropy is about a factor of two (suggesting comparable variations in the concentration of natural fractures) as well as the orientation of the fractures. Rose diagrams showing the orientation of fractures observed in an image log in well 1 are also shown in Fig. 7.32a. Near the toe and along the central part of the lateral, the observed fractures in the image log trend about N30°E, the same direction implied by shear velocity anisotropy. The same is true near the heel of the lateral, where both the image log and shear velocity anisotropy indicate fractures striking NNW.

In addition to the good comparison in inferred fracture orientations from shear velocity anisotropy and the image log in well 1, there are two other observations important to note in Fig. 7.32. First, each of the three images show the same orientation of shear velocity anisotropy and inferred fracture strike. Thus, B_{ani}, as calculated using the methodology described by Farghal seems to be a stable attribute indicative of *in situ* fracture orientation. Second, the region of high B_{ani} grows considerably following hydraulic fracturing and shrinks considerably after production. This implies, as one would expect, shear velocity anisotropy increases when pore pressure is high and decreases when pore pressure is low.

One surprising observation comparing the baseline and post-frac data (Figs. 7.32a and 7.32b) is the large spatial extent of the area affected by hydraulic fracturing. Some of the areas affected by stimulation are somewhat predictable in terms of the northeast trending fault damage zones shown in Figs. 7.24 and 7.28. Comparing the baseline and post-frac data with the post production data (Fig. 7.32c), it is clear that the area in which shear velocity anisotropy decreases is principally around the producing wells. It is interesting to speculate whether mapping B_{ani} could be used to track depletion surrounding producing wells.

Using the image log for calibration, Farghal compared the degree of anisotropy observed in the baseline survey to the observed fracture density in Fig. 7.33. As the figure indicates a factor of two variation of B_{ani} corresponds to a factor of 4 variation in fracture density. This implies that azimuthal AVO data may be a sensitive indicator of variations of fracture density in formations in which the pre-existing fractures are well aligned.

Part II

Stimulating Production from Unconventional Reservoirs

8 Horizontal Drilling and Multi-Stage Hydraulic Fracturing

In this chapter we review several key aspects of horizontal drilling and multi-stage hydraulic fracturing. While this is not an engineering text, it is necessary to briefly cover several operational procedures associated with horizontal drilling and hydraulic fracturing to provide readers with a basic understanding of what is being typically done in the field, and why. More detailed information about the topics in this chapter is available from Economides & Nolte (2000), Ahmed & Meehan (2016), Smith & Montgomery (2015) and other sources. Detournay (2016) offers a comprehensive review of the mechanics of hydraulic fracturing from the perspective of theoretical fracture mechanics.

Another objective of this chapter is to build a foundation for linking operational practices to the geological and geomechanical setting. In Chapter 11 we will consider selection of optimal landing zones for horizontal wells and hydraulic fracturing procedures to optimize vertical propagation of hydraulic fractures and proppant placement. We illustrate in that chapter how practical questions such as the separation between perforations, perforation diameter, pumping fluids, proppants and rates, etc. are linked (through the variation of the least principal stress with depth) to optimizing hydraulic fracture growth and the distribution of proppant within the hydraulic fractures.

Figure 8.1 (from LaFollette et al., 2012) shows the evolution of operational procedures over a 20-year period that led to the success of economically viable gas production from the Barnett shale in terms of peak monthly gas production during the first year of operations. The marked increase in production results from three key operational procedures: horizontal drilling, multi-stage hydraulic fracturing and utilization of slick-water as a fracturing fluid. These three technologies provide the basis for unconventional development everywhere in the world.

Horizontal Drilling

As shown in Fig. 8.2 for a well in the Woodford formation, construction of a typical horizontal well has a vertical section, a build section and a near-horizontal lateral. While well construction differs from place to place based on both geologic factors and regulations, there are several common characteristics important to note. First, surface casing typically extends beyond the depth of potable aquifers and is cemented to the surface to prevent drilling fluids from contaminating aquifers. An intermediate string of casing extends to a depth just above the kick-off point, where the well begins to build angle.

234   Unconventional Reservoir Geomechanics

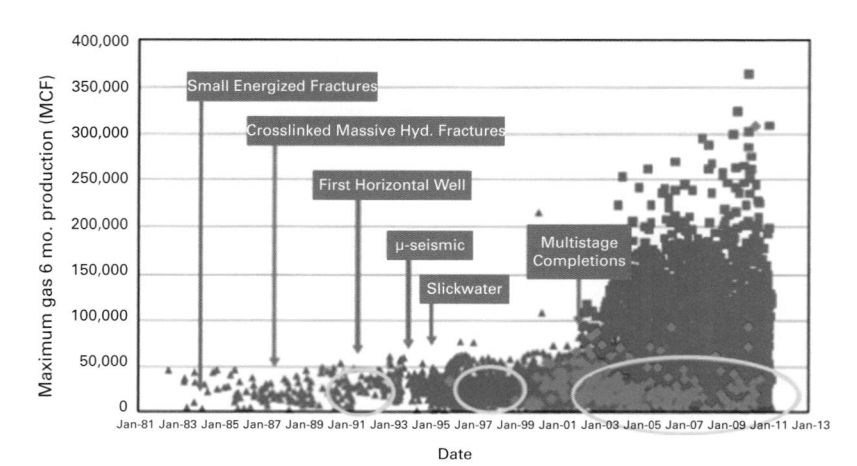

Figure 8.1 Barnett shale development over time showing gas production in the peak monthly gas production for over 15,000 wells. Vertical wells are in blue, deviated wells are in green and horizontal wells are in red. From Kennedy et al. (2016). Data originally from LaFollette et al. (2012).

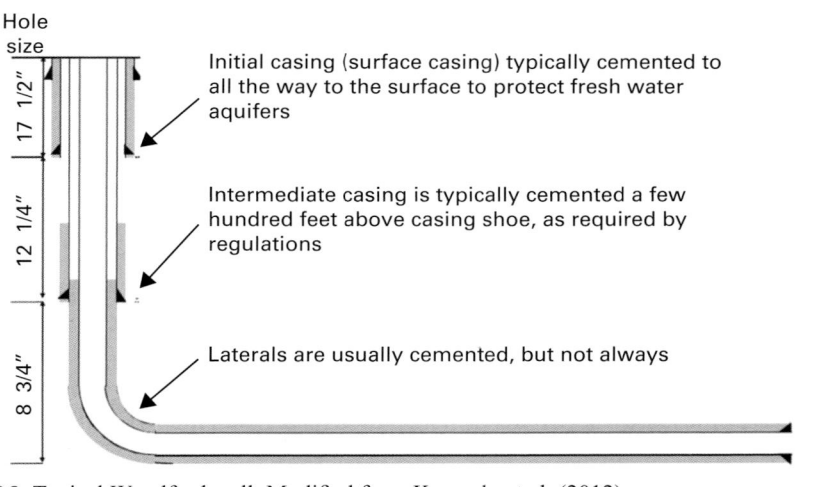

Figure 8.2 Typical Woodford well. Modified from Kennedy et al. (2012).

This casing is cemented up 100 m or more to create a hydraulic barrier to flow up behind the casing in the event that hydrocarbons leak past the final string of casing deployed all the way to the toe of the well. As shown, this string of casing is cemented up into an intermediate string of casing. In the case shown, there are three barriers to hydrocarbons escaping up the well from the producing along the lateral – the production casing, the cement behind this casing and the cement behind the intermediate casing string. The manner in which a well such as this is hydraulically fractured is shown below.

    While there are a number of techniques for drilling deviated and horizontal wells, the most common type of drilling system used involves a *bent sub* above a downhole mud motor and bottom hole assembly (which includes the drill bit, stabilizers, etc.). The bent sub makes it possible to turn the well a given amount over a certain distance (expressed in terms of °/100 ft). When the drill string is not rotating, adjusting the *tool face* (which indicates the direction the bottom hole assembly is pointed) makes it possible to alter the well path in a desired direction at the rate of curvature defined by the bent sub. Thus, when drilling in the *slide* mode of operation, only the assembly below the downhole motor is turning while the remainder of the drill string simply slides down the hole. This allows the well to progressively build angle at a desired azimuth. When drilling in the *rotating* mode of operation, both the drill string and downhole motor are turning such that drilling simply extends the well in a constant direction. Drilling in the rotating mode results in a more rapid rate of penetration than sliding. An alternative drilling technology is to use interactive rotary-steerable systems that allow a well to be continuously steered by a system that can automatically adjust the amount and direction of deflection of the bottom hole assembly. While these systems work quite well, they are not being used frequently because of the additional cost and lower rate of penetration.

Drilling mud is used for hole cleaning, pressure balancing, lubrication (to avoid torque and drag), clay stability, wellbore stability and powering the downhole drilling motor and electrical power generator. While water-based drilling muds are usually used to drill the vertical sections, most shale wells in the US convert to some form of oil-based mud for drilling the curve and lateral to reduce friction and drag. While it is beyond the scope of this book to discuss drilling mud in detail, drilling mud selection is important for optimizing drilling conditions.

Drilling Direction – As discussed initially in Chapter 1 and previously illustrated in Figs. 1.15 and 7.12, it is typical to drill in the direction of the minimum horizontal stress, S_{hmin}, such that hydraulic fractures propagate in a plane orthogonal to the well path. While this has an appealing geometrical simplicity for laying out patterns of wells (such as illustrated in Fig. 1.14), it appears also to be critical for optimizing production.

Stephenson et al. (2018) describe an interesting recent experiment in the Duvernay shale in Alberta. As shown in Fig. 8.3, they drilled two wells, one *on-azimuth* meaning that it was drilled parallel to the S_{hmin} direction, the other was drilled *off-azimuth*, about 45° from the S_{hmin} direction. They also experimented with different hydraulic fracturing fluids. As can be seen in the figure, microseismic data shows that hydraulic fracturing of the on-azimuth well produced far fewer shear events, but had double the productivity. There is also a difference in treatment fluid, with the on-azimuth well having 1/3 more gel and 1/3 less slickwater, although total injected volumes are equivalent. There appear to be two important take-aways from this very interesting experiment. First, use of low viscosity slickwater clearly does a better job of stimulating a shear fracture network surrounding the hydraulic fractures. Shear stimulation during hydraulic fracturing is discussed in detail in Chapter 10. The importance of fracturing fluid viscosity is discussed further below. Second, near-wellbore tortuosity of the hydraulic fracture planes in the off-azimuth well may have been sufficiently severe that it strongly affected distribution of proppant in the hydraulic fractures and thus restricted production.

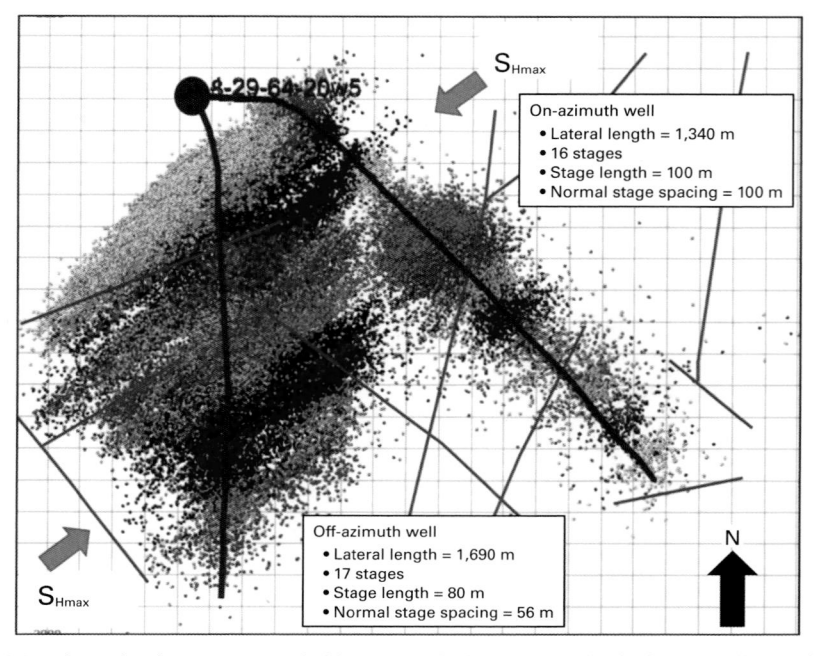

Figure 8.3 Microseismic events recorded by arrays deployed in the heel of each well associated with two wells drilled in the Duvernay formation. Events are colored by stage. Red lines represent lineaments (possible faults) interpreted from seismic data. From Stephenson et al. (2018)

Staying in Zone – Staying in the appropriate lithofacies during horizontal drilling is a challenging endeavor. There are fundamentally two ways of doing this. One can use a limited number of geophysical measurements (such as natural gamma or resistivity) which is generally referred to as measurement while drilling (MWD). Logging while drilling (LWD) is an extension of this technology utilizing a more comprehensive suite of geophysical measurements. The other method used for staying in zone while drilling is to carefully analyze the composition of drilling mud. Generally known as *mud logging*, tracing mineralogic and elemental composition of cuttings and drilling mud can be used to determine when the drill path is drifting out of the optimal lithofacies.

Using real-time geophysical measurements to guide horizontal drilling faces two challenges – placing sensor(s) close enough to the bit that decisions can be made before the well path drifts out of the desired lithofacies, and having sufficient instrumentation (and telemetry bandwidth) to adequately characterize the formation being drilled through. Figure 1.16b showed a profile of a horizontal well drilled in the Woodford formation in Oklahoma which illustrates the first problem. As shown in the upper part of the figure, a gamma log was used to assure that the drill path was not only staying within the Woodford, but also staying within the WDFD-2 lithofacies where it was optimal to complete the well (as discussed in Chapter 11). As this sensor was mounted about 30 m behind the bit, the well path could drift out of this lithofacies before being detected. Thus, even though the well path was corrected each time the well drifted up into the

Table 8.1 Types of LWD measurements used in unconventional wells

Geophysical measurement	Purpose
Resistivity	Fluid saturation, TOC
Compensated density or neutron	Total porosity, lithology, TOC
Cross-dipole acoustic	Mechanical properties
Gamma or spectral gamma	Mineralogy, TOC
NMR	Porosity, TOC, fluid volume and type
Elemental spectroscopy	Lithology, mineralogy, TOC
Resistivity imaging	Geologic structures, bed boundary and fracture identification
Acoustic imaging	Geologic structures, bed boundaries, fracture identification, hole conditions

WDFD-1 lithofacies, considerable sections of the well were in the less-desirable lithofacies which was characterized by unusually high frac gradients (Ma & Zoback, 2017b).

There are two contexts in which LWD technologies are utilized during horizontal drilling. The first is to utilize real-time data provided by sensors in the bottom hole assembly to keep drilling in the correct lithofacies. This data is telemetered up the hole, typically using a mud-pulse system to provide drillers with real-time information about the properties of the formation being drilled through. Second, LWD can be used to provide a comprehensive suite of geophysical measurements for reservoir characterization as would be acquired in a conventional oil or gas well. In this case the data is typically saved in a memory unit in the bottom hole assembly and downloaded when the equipment is brought to the surface. Using LWD tools that scan the formations surrounding the horizontal well can be helpful in detecting formation boundaries as they are approached.

Reservoir characterization using geophysical logging tools in unconventional wells offers nearly as many options as those associated with conventional wells. Table 8.1 (after Bratovich & Walles, 2016) describes a number of the commercially available LWD measurements that are currently available. Of course, while it is often most efficient (and less risky) to obtain as much information as possible via LWD, it is also possible to convey conventional geophysical logging tools in a horizontal well through use of special conveyance techniques such as wellbore tractors. There are a number of commercial service companies offering these types of measurements.

The lower part of Fig. 1.16b showed the variation of clay and kerogen content along the well path in the Woodford obtained from analysis of data from an elemental spectroscopy tool. Note that lithofacies WDFD-1 is characterized by distinctly higher clay plus kerogen and that the total gamma log does a reasonable job as a proxy for clay content as clay content is typically much higher than kerogen.

For completeness, it is important to note that a subset of the technologies used for detailed characterization of unconventional formations (and formation fluids) as outlined in Chapters 2, 5 and 6 can be applied in real-time during drilling through sophisticated mud logging. One challenge is that the common use of PDC drill bits in laterals tends to produce rock powder, rather than cuttings, which limits applicability of

Multi-Stage Hydraulic Fracturing

In this section we briefly review a number of general attributes of multi-stage hydraulic fracturing. There are several excellent sources of information about hydraulic fracturing available. The book by Economides & Nolte (2000) consists of 20 chapters written by experts on different aspects of this technology. The book by Smith & Montgomery (2015) is also a helpful reference.

We start with a brief overview of the common methods used to create multiple hydraulic fractures along a well path as well as issues concerning *zonal isolation*, restricting pressurization to only the intended positions along a well path. We then consider issues of hydraulic fracture propagation – both laterally away from the well and vertically. As the most common hydraulic fracturing technique currently being used, the *plug-and-perf* method, is based on simultaneously propagating multiple hydraulic fractures in close proximity to each other, we discuss the phenomenon known as *stress shadows*, which refers to the effect of one propagating hydraulic fracture on another as they simultaneously propagate. After considering this topic from a theoretical perspective, we briefly review the results of two very interesting experiments where cores obtained from multiple wells reveal a great deal of information about the hydraulic fracturing process. The topic of vertical hydraulic fracture growth is discussed briefly in this chapter, but will be the subject of a greatly expanded discussion in Chapter 11. The topics of fracture initiation near the wellbore and proppant distribution in fractures will also be discussed in that chapter in terms of layer-to-layer variations of stress magnitude with depth resulting from various processes.

A typical hydraulic fracturing operation involves several steps, the first being to isolate the section of the well to be pressurized and hydraulically fractured. The most common ways to do this are described in the next section. Fluid is pumped in several stages, initially a low viscosity *pad* is pumped at relatively low rate to initiate the fracture propagation process. As pumping rates are increased, a slurry is then pumped consisting of the hydraulic fracturing fluid (discussed below) mixed with proppant (which is also discussed later in this chapter). Following the slurry, a relatively clean *tail* fluid is pumped to clear the wellbore and near wellbore area. After all the hydraulic fractures in a well have been generated, the well is allowed to flow back in an attempt to clear up flow paths and allow the hydrocarbons to flow to the well without difficulty.

As illustrated in Table 8.2 (from Edwards & Celia, 2018 and the sources referenced therein), hydraulic fracturing requires a great deal of water, which can be an issue in arid regions. As discussed in Chapter 13, when potable water is scarce it is possible to use brackish or even saline fluids for hydraulic fracturing. Also, the flow back water that comes to the surface after hydraulic fracturing needs to be disposed of properly (also discussed in Chapter 13 as well as Chapter 14) or recycled. It should be noted that the values presented in Table 8.2 represent information in the public domain and practices

Table 8.2 Hydraulic fracturing in different formations.

Parameter	Value	Formation
Total injected fracturing fluid volume	20,000 m^3 (16,000–26,000 m^3)	Marcellus
	19,000 m^3 (11,000–23,000 m^3)	Barnett
	77,000 m^3 (mean), 66,000 m^3 (median)	Horn River
	(35 wells) (2013–2014)	Haynesville
	64,000 m^3 (2010–2012)	Eagle Ford
	19,000 m^3 (6,000–25,000 m^3)	
	23,000 m^3	
Injected fluid volume normalized by horizontal well length	14 m^3/m (235 wells)	Marcellus
	25 m^3/m (2004)	Barnett
	19 m^3/m (2006)	Horn River
	15 m^3/m (2008–2012)	
	27 m^3/m (35 wells) (2012–2014)	
Injected volume flowback recovery	1–50%	Marcellus
	65% (1 year)	Barnett
	90% (2 years)	Horn River
	100% (3 years)	Haynesville
	13% (8 wells)	
	5%	
Surface injection pressure	45–62 MPa	Marcellus
	54 MPa (max. 22 wells)	Horn River
	49 MPa (avg. 22 wells)	
Bottom-hole injection pressure	55–83 MPa (30–55 MPa surface injection pressure)	Woodford
		Unspecified
	48–85 MPa	
Number of stages	12 (7–24) (184 wells)	Marcellus
	18	Horn River
Fluid injection duration per stage	2–3 h	Marcellus
	3–4 h	Horn River
	2.5-3h	Woodford
Average injection flow rate (for the duration of each stage)	12 m^3/min	Marcellus
	8–16 m^3/min	Barnett
	16 m^3/min (35 wells)	Horn River
	15 m^3/min	Woodford
Injected proppant mass (per well)	2,100 tonnes (400–3,600 tonnes) (187 wells)	Marcellus
	3,000 tonnes (48 wells)	Horn River
	4,000 tonnes	
Fracture height inferred from microseismic measurements	~160 m (median), ~500 m (max.)	Marcellus
	~160 m (median)	Barnett
	250 m (12 wells)	Horn River
	~130 m (median)	Woodford
	~100 n (median)	Eagle Ford
Fracture horizontal length inferred from microseismic measurements	~300–400 m	Marcellus
	~600–900 m (12 wells)	Horn River

change with time as operational strategies evolve. For example, it is currently common to stimulate far more hydraulic fracture stages per well than the table indicates.

Well Completion Methods –The two most common types of well completion methods for multi-stage hydraulic fracturing are the *plug-and-perf* and *sliding sleeve* methods which are illustrated schematically in Fig. 8.4. Plug-and-perf is done in a cased and cemented horizontal well. Starting near the toe of the well and working toward the heel, a small frac plug, normally made out of an easily drilled composite, is electrically actuated and set at the depth where the pressurization zone is to end. The frac plug initially allows fluid to pass through it so that tools can be pumped down the well. The same wireline tool used to set the frac plug contains a perforation gun which is used to create clusters of perforations at several places along the well (usually tens of meters apart) before hydraulic fracturing. Sometimes multiple runs with perforating guns are needed. A ball is dropped to land in a seat in the frac plug, to block fluid pressure from passing through it and to allow for pressurization of the interval with open perforations. The simultaneous pressurization of the clusters is referred to as a stage. As discussed

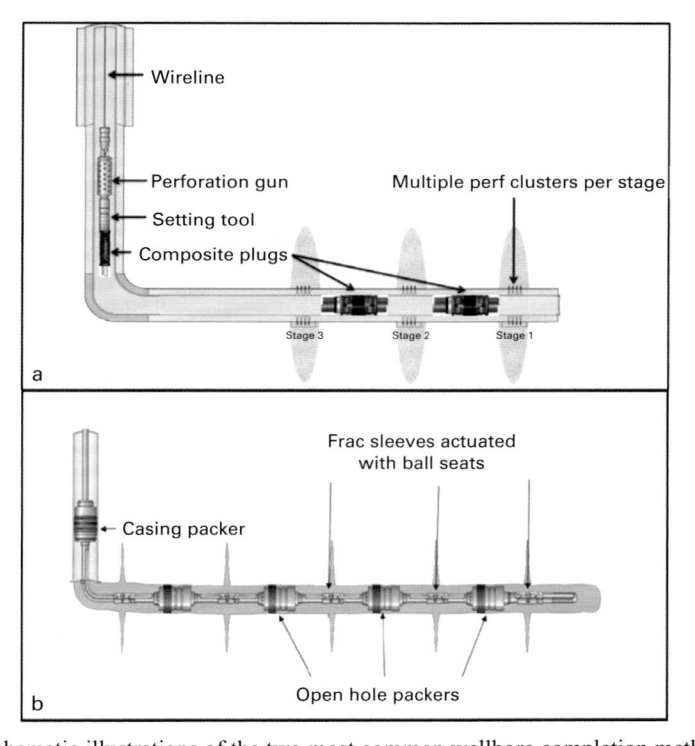

Figure 8.4 Schematic illustrations of the two most common wellbore completion methods. (a) The plug-and-perf method which utilizes separately deployed frac plugs to isolate sections of a cased and cemented well. After setting the plug, clusters of perforations are made at several places (usually tens of meters apart) before hydraulic fracturing. (b) The sliding sleeve method is usually used in open holes. A single piece of tubing with multiple packers is deployed. A given interval is pressurized by dropping a ball into a valve which slides open when pressurized. After Burton (2016).

below, the intent of having multiple perforation clusters is to propagate hydraulic fractures from each of them. Thus, if there are 20 stages in a well, each with five perforation clusters, the intent is to generate 100 hydraulic fractures.

Note that during plug-and-perf hydraulic fracturing operations nearly the entire well, from the well head to the frac plug, is pressurized. Ideally, only the perforated section above the plug provides access of the fluid pressure to the formation. Fracture initiation from the perforations near the wellbore is discussed in Chapter 11. After all the stages have been completed in a given well, the plugs are drilled out (often by a relatively inexpensive coil-tubing unit) and the well is allowed to flow-back to recover the hydraulic fracturing fluid and make it easier for hydrocarbons to flow into the well.

The sliding sleeve method shown in Fig. 8.4b usually involves running a single tubing string into an uncased lateral with packers to seal the well into multiple intervals. Swell packers are usually used that expand with time after contact with the fluid in the well. After numerous intervals have been isolated they are sequentially hydraulically fractured starting at the toe and working back toward the heel. Valves located between the packers start in a closed position which separates the fluid in the tubing from the open hole. To isolate an interval to be stimulated, a ball is dropped to land in a seat and open the valve by sliding a sleeve when pressurized. The balls are of increasing size as you go from the toe to the heel of the well and matched with the size of the landing seat in each interval. In some cases, balls are used that are made of a material that disintegrates with time.

The sliding sleeve method is more efficient as it does not require the lateral to be cased and cemented, nor does it require multiple wireline runs with the setting/perforating tool. However, as it is likely that only a single hydraulic fracture is generated from each stage, it may not be as effective as the plug-and-perf method for stimulation as the low matrix permeability prevents pore fluid diffusion over significant distances in unconventional reservoirs (see discussion in Chapter 12).

There are a number of variations of these two techniques. For example, sliding sleeves can be used in a cased and cemented well that has been perforated. What is sometimes called *controlled entry point* hydraulic fracturing is similar to plug-and-perf but involves pressurization of only one or two perforation clusters to limit interactions among hydraulic fractures propagating simultaneously. In a detailed case study in the Utica formation, Cipolla et al. (2018) argue that controlled entry point completions were no better for production than plug-and-perf completions and ultimately, it was the number of perforations that mattered most. This will be discussed in Chapter 11.

Zonal Isolation – The issue of restricting pressurization to the interval desired (zonal isolation) affects both plug-and-perf and sliding sleeve operations. Figure 8.5a illustrates a well fractured using plug-and-perf in the Barnett shale (well 1 shown in Fig. 7.27). While the main feature illustrated by the distribution of microseismic events in this dataset is the degree to which they are dominated by propagation along pre-existing fault damage zones (as discussed in Chapter 7), poor zonal isolation is especially clear for stage 2 (green dots). Note that many stage 2 events occur in the vicinity of stage 1, the result of leakage past the frac plug separating the two stages or the cement behind the casing. Figure 8.5b shows a sliding sleeve example from Shaffner et al. (2011). Note that

242   **Unconventional Reservoir Geomechanics**

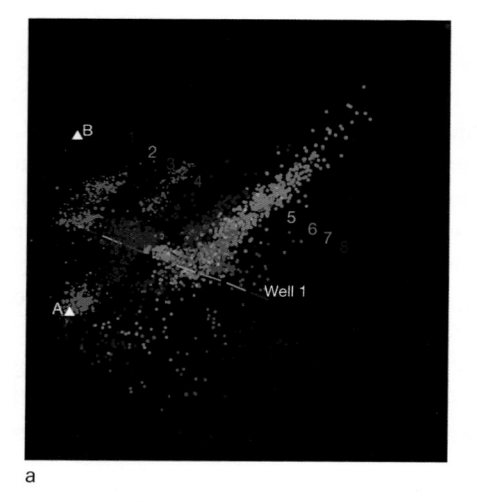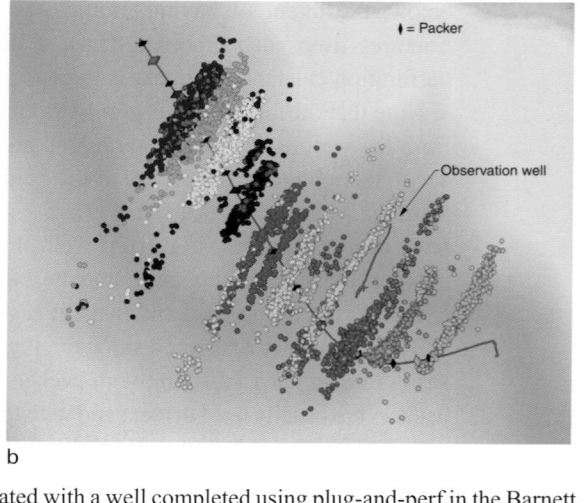

a   b

Figure 8.5 (a) Microseismicity associated with a well completed using plug-and-perf in the Barnett shale (previously shown as well 1 in Fig. 7.27) which illustrates zonal isolation problems. (b) A well in the Duvernay shale completed with the sliding sleeve method that indicates zonal isolation problems. After Shaffner et al. (2011).

the stage 1 events (red), occur in the same position as the stage 2 events, apparently indicating leakage past the packer separating the two stages. Similar observations indicate that leakage occurred in several other stages.

Raterman et al. (2017) and Ugueto et al. (2018) report recent studies investigating zonal isolation with fiber optic and other technologies in multi-stage hydraulic fracturing experiments. The fibers are used for both Distributed Temperature Sensing (DTS) and Distributed Acoustic Sensing (DAS). In the drill-through experiment in the Eagle Ford described in more detail below, Raterman et al. report different amounts of leakage past frac plugs as much as 10% of the injected volumes in six of the seven stages for which the fiber data was available. Ugueto et al. present a case study in which DAS and DTS fiber technology (and other techniques) were used to evaluate zonal isolation of controlled entry point completions in cemented casing. Some degree of communication was noticed in about half of the 69 hydraulic fracture stages carried out. As these were single entry point completions, Ugueto et al. concluded that leakage behind casing due to poor casing cement was the principal cause of the problems encountered. There seems that, regardless of the completion method used, achieving zonal isolation is an outstanding issue limiting the overall effectiveness of multi-stage hydraulic fracturing,

Models of Horizontal Hydraulic Fracture Propagation – For many years, the principles of hydraulic fracture propagation were considered in the context of 2D analytical models known as the PKN (Perkins, Kern and Nordgren) and KGD (Khristianovich, Geertsma and de Klerk), which are illustrated in Fig. 8.6. As shown, the PKN model considers a hydraulic fracture of fixed height, representative of a laterally propagating

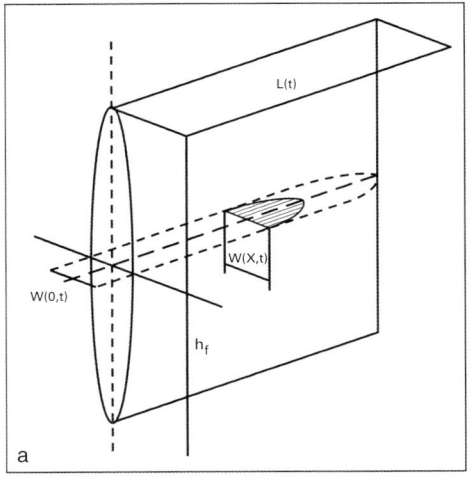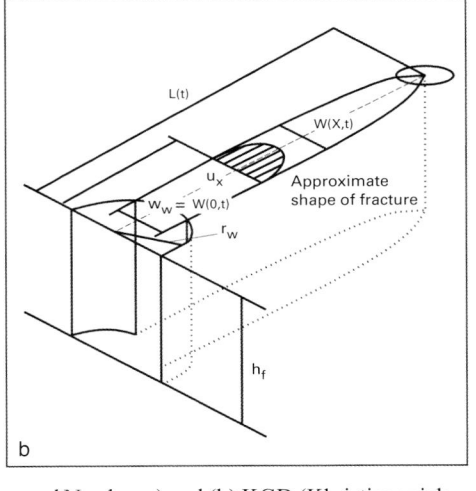

Figure 8.6 (a) Illustration of the PKN (Perkins, Kern and Nordgren) and (b) KGD (Khristianovich, Geertsma and de Klerk) two-dimensional models of hydraulic fracturing. From Mack & Warpinski (1989).

hydrofrac that might be confined between two stress barriers. Mack and Warpinski (1989) present a comprehensive review of these (and extended models). There is a thorough review of these models in Smith & Montgomery (2015) and many other texts.

A considerable benefit of models such as these is that it is relatively straightforward to use analytical formulas to investigate some of the first-order linkages between some of the operational and geological factors that affect hydraulic fracture propagation and address important questions such as fracture width which controls proppant transport. Although some of the simplifications in the original PKN and KGD models were addressed in subsequent studies, a number of assumptions limit their usefulness. For example, they consider only 1D flow of fluid along the hydraulic fracture and they disregard the layered nature of the rock through which they propagate as they assume it is a homogeneous, isotropic linear elastic solid. This will be examined in greater detail in Chapter 11. As a two-dimensional model, the PKN model assumes the fracture to be completely confined to a single layer. The two-dimensional KGD model assumes slip at the top and bottom of the pressurized layer to allow the fracture to have full width through the entire interval. The models assume that either the height is large (KGD) or small (PKN) relative to length. Also, the KGD model includes the assumption that tip processes dominate fracture propagation, whereas the PKN model neglects fracture mechanics altogether. Detournay (2004) improves upon this by considering the transition between early propagation time when fracture toughness (rock strength) controls propagation to later time when viscous processes within the fracture are more important.

Over the past 25 years, there has been a progressive evolution away from relatively simple analytical models such as the PKN and KGD model and toward 2D and 3D numerical codes. Warpinski et al. (1994) reviewed some of the early numerical hydraulic fracturing models in the context of data obtained from a detailed hydraulic fracturing experiment in the Waskom field in Texas. Perhaps not surprisingly, there was often a factor of 2 to 3 difference in predicted frac heights and lengths among models. Needless to say, as computational power has increased dramatically over time the models have become increasingly sophisticated. As reviewed by McClure et al. (2016), three-dimensional numerical models have been developed to overcome the inability of analytical models to consider both geologic and operational complexities. Importantly, such models can consider detailed pumping histories, non-Newtonian fluid rheologies, fracture initiation from perforations near the wellbore, viscous pressure drops in the well, across perforations and in the fracture, proppant distribution within the fracture, etc. Several authors have reviewed the numerical methods being used. See for example, Adachi et al. (2007), Lecampion et al. (2017). At the time of this writing, there are almost a dozen models in commercial use. It is beyond the scope of this book to review these models or enumerate their respective advantages and disadvantages.

Drill-Through, Core-Through Experiments – An interesting set of observations related to hydraulic propagation was described by Raterman et al. (2017) based on the results of drilling, coring and logging in five monitoring and data wells drilled to study hydraulic fractures produced during multi-stage hydraulic fracturing in four producing wells spaced approximately 1,000 ft apart. The plug-and-perf method was used with cluster spacings between 43 and 47 ft. One of the producing wells, P3, was instrumented with optical fibers for temperature and acoustic sensing. This is the well referenced above in the discussion of zonal isolation.

Figure 8.7 shows a perspective view of the P3 well and the four horizontal data wells that are within a few hundred feet of it. A radioactive tracer used in P3 indicates that proppant was distributed in all clusters (shown in yellow in the well profile). Raterman et al. discuss at length how core and logs from pre-stimulation S2 core were used to identify drilling-induced fractures (obtained with significantly over balanced mud system) as well as infrequent fractures.

One of the most surprising findings of this experiment is the great number of hydraulic fractures that were encountered as indicated by the white disks in the cross-sections. Using image logs calibrated by cores (as shown in Fig. 8.8), between 397 and 966 hydraulic fractures were observed in the four data wells shown in Fig. 8.7 over lengths ranging between 1,378 and 1,748 ft. If these hydraulic fractures were all generated by the nearby P3 well, there are about 10–25 times more hydraulic fractures observed than the number of clusters over the same interval. (Even if some of the hydraulic fractures come from stimulation of well P2 offset more than 1,000 ft from S3 STO3, the 966 fractures observed over a 1,735 ft length represents over 10 times the number of clusters in wells P2 and P3.) Once must conclude therefore that multiple hydraulic fractures are propagating from each perforated interval. The manner in which this occurs is unknown, but the hydraulic fractures were distributed unevenly in the observation wells with clusters of fractures seen in a number of 1 ft intervals (Fig. 8.8). Because of the stress shadow effect (discussed below) it is difficult to see how very much fluid pressure could be

Horizontal Drilling and Multi-Stage Hydraulic Fracturing 245

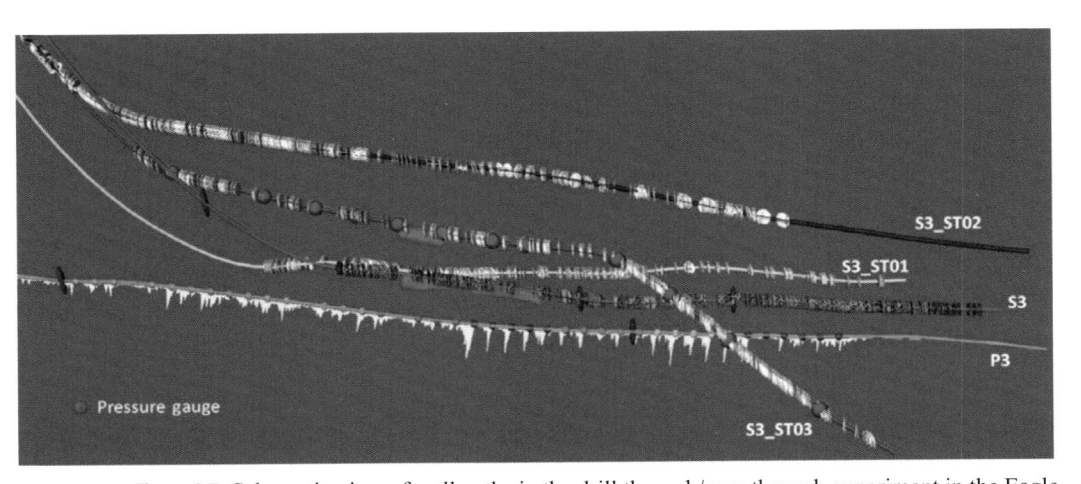

Figure 8.7 Schematic view of well paths in the drill through/core through experiment in the Eagle Ford formation reported by Raterman et al. (2017). Hydraulic fractures as identified in either core or calibrated logs are shown as white discs. Cored intervals are shown in pink. The yellow filled log on P3 is Scandium RA tracer log. Blue discs show locations of iridium RA tracer from offset producer P2.

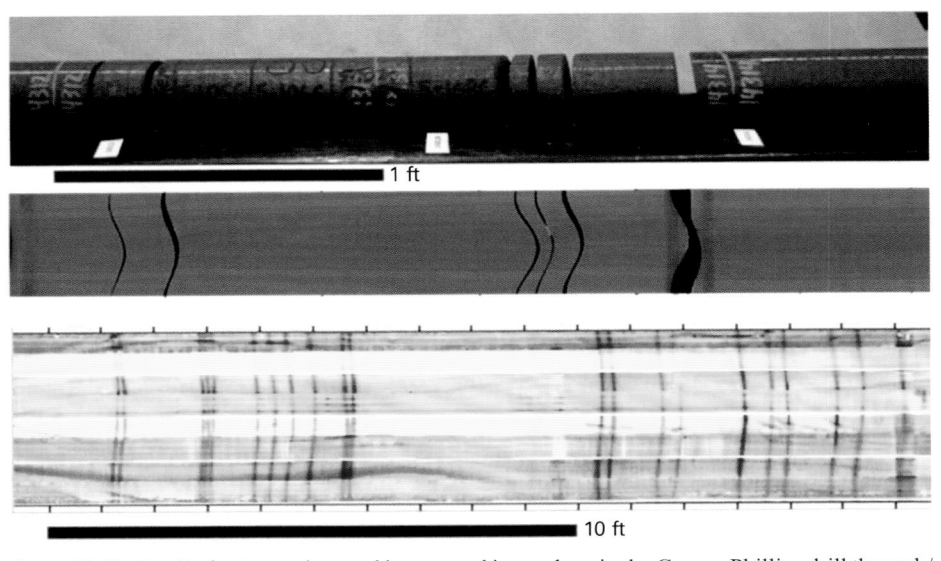

Figure 8.8 Hydraulic fractures observed in core and image logs in the ConocoPhillips drill through/core through experiment. After Raterman et al. (2017).

acting in the fractures if they were propagating simultaneously. An alternative hypothesis is that they formed sequentially during the hydraulic fracturing process.

It is equally interesting that there was so little evidence of proppant in cuttings taken at 20 ft intervals along the four data wells. While proppant was relatively abundant in well S3 (the well closest to P3) – it appeared in 76% of the cuttings (and on 25% of the fractures observed in core) – it was rarely seen in the other wells. Only 5% of the cuttings in well S3 ST02 (which is 100 ft above the P3) showed evidence of cuttings and just three fractures in core from S3 STO3 (5%) showed proppant.

What is striking about the thousands of hydraulic fractures documented in the four observation wells is how aligned they are with the stress field. When discussing Figs. 1.14b, 7.12 and 8.5b we referred to the manner in which the linear cloud of microseismic events indicated the overall direction of hydraulic fracture propagation. The stereonets and rose diagrams shown in Fig. 8.9 from the Eagle Ford experiment shows how remarkably aligned the actual hydraulic fractures are, pointing in a direction normal to the minimum horizontal principal stress in this area (Fig. 7.2).

Very similar findings come from the HFTS-1 experiment in the Wolfcamp formation of the Midland Basin (see overview by Ciezobka et al., 2018). Figure 8.10a (from Shrivastava et al., 2018) shows the cores obtained from two cored intervals in a slant well (SCW) that passes two wells stimulated in the Upper Wolfcamp (3 stages, cores 1, 2, 3 and 4) and Middle Wolfcamp (2 stages, cores 5, 6). Hundreds of hydraulic fractures were observed in the cores, with a strike parallel to the direction of maximum horizontal stress in the region (Fig. 8.10b, after Gale et al., 2018).

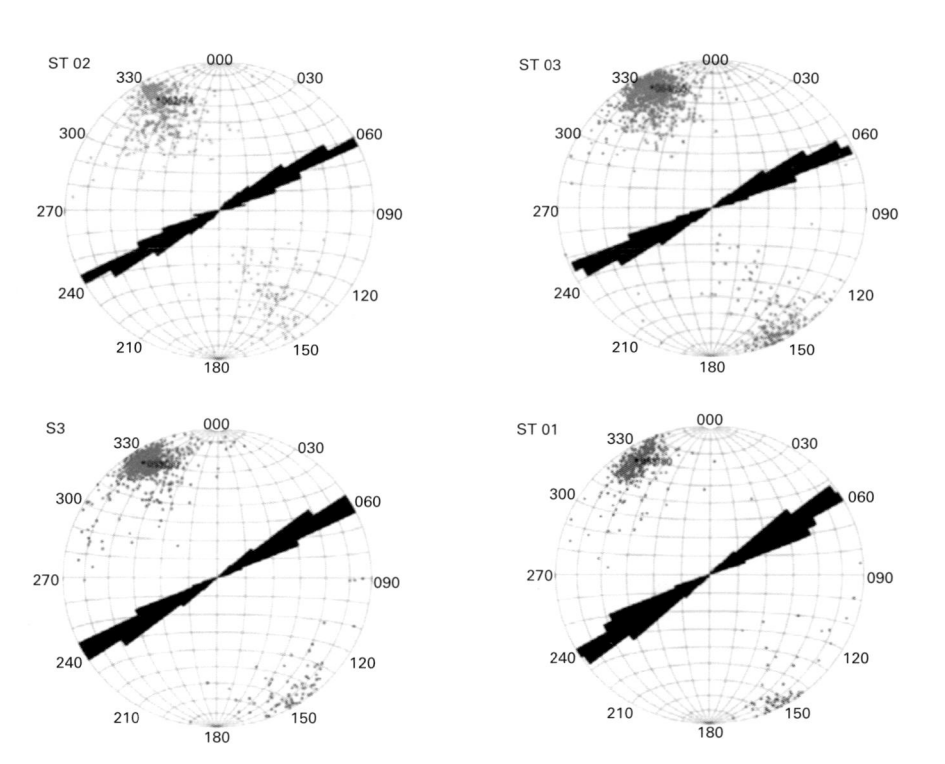

Figure 8.9 The orientation of hydraulic fractures observed in core and image logs in the ConocoPhillips drill through/core through experiment. After Raterman et al. (2017). Note that the hydraulic fracture orientations are perfectly aligned with the stress field in this region as shown in Fig. 7.2.

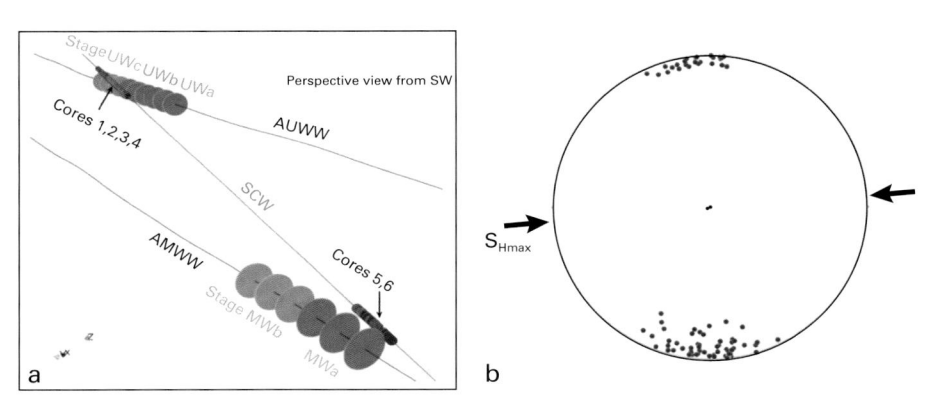

Figure 8.10 (a) Perspective view of core positions in slant well SCW and stimulation zones in the wells in the upper Wolfcamp (AUWW) and lower Wolfcamp (AMWW). After Shrivastava et al. (2018). (b) The orientation of hydraulic fractures observed in core are perfectly aligned with the stress field in this region as shown in Fig. 7.3. From Gale et al. (2018).

Hence, these two core-through experiments reveal an aspect of hydraulic fracture propagation not previously known (nor speculated about) in the literature. Either the hydraulic fractures repeatedly bifurcate and multiply as they propagate from a perforation cluster or multiple hydraulic fractures are created over time. In either case it is clear that at distances of hundreds of feet (or less), there are many more hydraulic fractures than perforations. As there are no significant variations in the direction of propagation, the direction of propagation is principally controlled by the stress field.

Stress Shadows during Simultaneous Hydraulic Fracture Growth – One of the governing factors that affects multiple hydraulic fractures propagating in close proximity to one another is often referred to as the stress shadow. Fundamentally, a stress shadow means that when a hydraulic fracture opens, it increases the stress normal to the fracture face (Warpinski & Branagan, 1989; Fisher et al., 2004). Thus, while hydraulic fractures are propagating simultaneously from during plug-and-perf completions, the opening of a fracture propagating from one perforation cluster increases the stress normal to that of a hydraulic fracture propagating from clusters nearby.

The magnitude of the stress shadow affect was discussed by Warpinski et al. (2013) in the context of a 3D elliptical crack representing a hydraulic fracture that was 400 ft in length and 100 ft high. As shown in Fig. 8.11, when the pressure in the hydraulic fracture is 1,000 psi above the least principal stress (that is, the net pressure is 1,000 psi), the stress normal to a hydraulic fracture propagating from a cluster 50 ft away at the fracture mid-point, is about 700 psi higher than it would have been if the first hydraulic fracture had not been present. Nonetheless, taken at face value, stress shadow effects might seem to suggest that it would be unlikely for plug-and-perf to work unless cluster spacings were quite far apart. Rainbolt et al. (2018) show evidence of increasing treating pressures with closely spaced hydraulic fracture stages. If single entry point hydraulic fracturing is being done, the propping open of one hydraulic fracture would affect the

248 Unconventional Reservoir Geomechanics

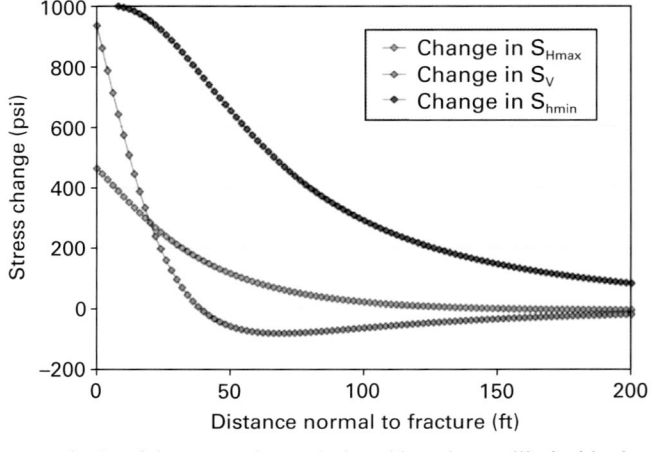

Figure 8.11 The magnitude of the stress change induced by a large elliptical hydraulic fracture as a function of distance normal to the fracture assuming the pressure in the fracture is 1,000 psi above the least principal stress. From Warpinski et al. (2013).

stress acting normal to the next one. Hence, depending on the distribution of proppant within the fracture (which would have to be modeled), the spacing between hydraulic fractures would need to be carefully evaluated.

One limitation of the calculations shown in Fig. 8.11 is that during plug-and-perf hydraulic fracturing, the hydraulic fractures that are propagating more or less simultaneously from each cluster. There is no single large hydraulic fracture (as assumed in the model) casting its stress shadow on hydraulic fractures trying to propagate nearby. Another is that it is more likely that net pressures are more on the order of a hundred psi than a 1,000 psi (as discussed in Chapter 7). Raterman et al. (2107) documented proppant going into multiple perforation clusters only about 45 ft apart. Other studies report similar findings. Thus, while there is definitely competition between hydraulic fractures propagating simultaneously, the effect is much less severe than that shown in Fig. 8.11. As illustrated in Fig. 8.12, Agarwal et al. (2012) present a study of four fractures propagating simultaneously from clusters 30 m apart. The stress shadow effect results in shorter hydraulic fractures in the center of the stage than at the outside as those fractures are affected by the stress shadows associated with hydraulic fractures on both sides. Wu & Olson (2013) address this phenomenon for different stress states, fracturing fluids, etc.

Another important finding from the modeling shown in Fig. 8.12 is relevant to the question of how microseismic events are triggered during hydraulic fracturing. The area shown in blue in Fig. 8.12a indicates where mean stress is becoming more compressive and the tendency for inducing shear on pre-existing fractures is less likely. In other words, if the microseismic events that occur during hydraulic fracturing resulted only from the stress concentrations from propagating hydraulic fractures, they would only occur out ahead of the propagating fractures as shown in red. As shown in Fig. 8.12b, leak off is likely to induce shear stimulation throughout the regions surrounding the

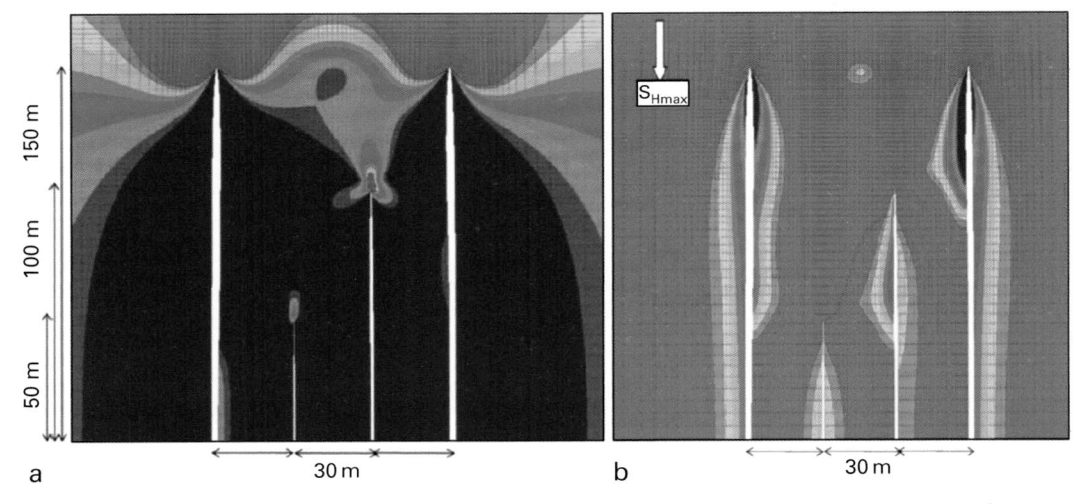

Figure 8.12 (a) Numerical model of four hydraulic fractures propagating simultaneously from four perforation clusters 30 m apart with no leak-off of pressure from the fractures. (b) The same model with leak-off. The color indicates the tendency of the induced change in stress and pressure to stimulate shear on optimally oriented pre-existing fractures. From Agarwal et al. (2012).

hydraulic fractures. The analysis in Fig. 8.12 considers the potential for inducing shear slip on pre-existing faults with an optimal orientation for slipping in the current stress field.

The topic of triggering shear slip on pre-existing faults due to pressure leaking out of hydraulic fractures is considered at some length in Chapter 10. It should be pointed out that such pressurization would also cause stress magnitudes to increase poroelastically in the rock volumes adjacent to the hydraulic fractures. This led Vermylen & Zoback (2011) to suggest that poroelastic effects were responsible for the increase in ISIPs for successive fracturing stages in five wells in the Barnett shale (Fig. 8.13). Wells A and B were hydraulically fractured simultaneously such that pressure build-up between stages (and associated stress increases) occurred quickly. Ten stages in each of the two wells were carried out in about 90 hours. Wells D and E were fractured in an alternating sequence (termed *zipper fracturing*). The stages in the two wells took almost twice as long to carry out. Well C was hydraulically fractured in sequence from toe to heel in about 120 hours. Although the fracturing stages were otherwise the same (flow rates, fluid and proppant volumes, etc.), it is clear that the faster hydraulic fracturing was carried out along a well, the larger the stress change from stage to stage. This implies that when pressure between the stages had time to dissipate, poroelastic stress changes were smaller.

A final word about stress shadow effects is that some workers have argued that there are such significant interactions among fractures propagating simultaneously during plug-and-perf operations that their trajectories were severely distorted. However, as shown in Fig. 8.14 (after Wu & Olson, 2013), bending of hydraulic fracture trajectories during plug-and-perf operations is minor, even when there are small differences between

250   **Unconventional Reservoir Geomechanics**

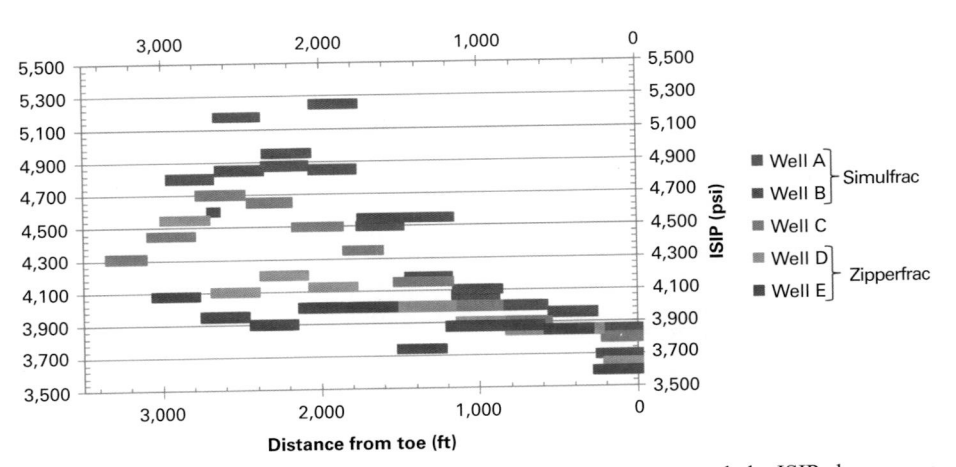

Figure 8.13 ISIP pressure in successive stages in five wells in the Barnett shale. ISIP changes are largest in wells A and B which fractured most rapidly. Wells D and E were fractured most slowly. As discussed in the text, the increase in shut-in pressure appears to be due to poroelastic stress changes. From Vermylen & Zoback (2011).

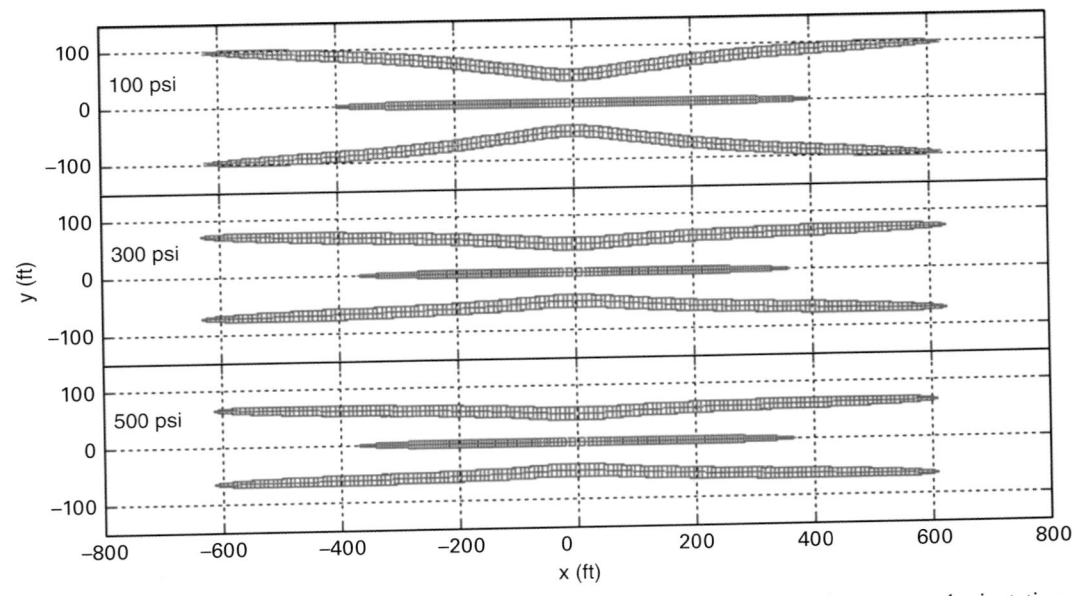

Figure 8.14 Investigation of the dependence of stress shadow effects on the extent and orientation of three simultaneously propagating hydraulic fractures from perforations 50 ft apart. With only modest differences between the two horizontal principal stresses, the angular deflection of the fracture propagation directions is essentially insignificant. From Wu & Olson (2013).

the two horizontal stresses. The modeling in Fig. 8.14 shows that when the difference between horizontal stress is 100 psi (top panel), there could be a few degrees of orientation change due to interactions among three hydraulic fractures propagating from perforations 50 feet apart. When the stress difference is 300–500 psi, the deflection is insignificant (middle and lower panel). Like the case shown in Fig. 8.12, the length and width of the center hydraulic fracture is impeded by the stress shadows associated with the outer ones. Observationally, the results presented in Figs. 8.9 and 8.10 from the drill/through core/through experiments described above confirm the consistency of the fracture propagation directions.

Vertical Hydraulic Fracture Growth – The only observational data generally available that provides information about vertical hydraulic fracture growth comes from microseismic data. Putting aside uncertainties in microseismic event locations (see Chapter 9), the occurrence of a microseismic event is indirect evidence of hydraulic growth. As mentioned previously and discussed at length in Chapter 10, microseismic events are caused by pore pressure changes during hydraulic fracturing that induce slip on pre-existing fractures and faults (Chapter 10). Thus, while there must be a hydraulic connection between a hydraulic fracture and slipping fracture or fault, it would likely be through an interconnected fracture network which could extend beyond the hydraulic fractures. In fact, we presented in the previous chapter several examples of fluid pressure during hydraulic fracturing getting channelized along pre-existing fault damage zones. Still another issue is the difference between where fluid pressure is acting in a hydraulic fracture (which could induce a microseismic event nearby) and the parts of the hydraulic fracture in which there is proppant acting to hold open the hydraulic fracture as depletion is occurring and the effective normal stress on the hydraulic fracture increases.

Figures 8.15 (from Xu et al., 2017) and 8.16 (from Fisher & Warpinski, 2012) are examples in which the maximum hydraulic fracture height might be inferred from the vertical distribution of microseismic events. In the first case, four horizontal wells were drilled in the upper (well B), middle (wells C and D) and lower (well A) Wolfcamp formations in the Permian Basin. Figure 8.15a a shows compositional data as a function of depth as well as frac gradients measured from DFIT tests in three of the wells. Figure 8.15b shows cross-section views of the distribution of microseismic events associated with all of the hydrofracs in each well. As the wells were drilled in a north–south direction (parallel to the local orientation of S_{hmin}), the cross-sections are looking down the well to the north, from the toe to the heel. The solid vertical line in the figure shows the position of the seismic monitoring array in a vertical well. Because the sensors were both above and below most of the microseismicity, the depths of the microseismic events are well constrained.

Note in Fig. 8.15 that the great majority of events do seem to track the vertical propagation of the hydraulic fracture in a manner consistent with the three measurements of the frac gradient. For example, the microseismic events associated with well A (near the top of the lower Wolfcamp) propagate upward through the middle Wolfcamp and stop abruptly in the middle of the upper Wolfcamp where the frac gradient is higher. It is not known why there was no downward propagation as there is no information available about the frac gradient below well A. Similarly, very few microseismic events from wells C and D in the middle Wolfcamp are seen in the upper Wolfcamp, suggesting

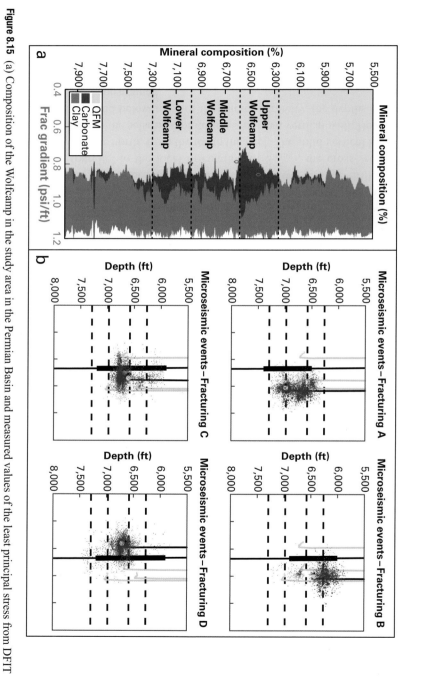

Figure 8.15 (a) Composition of the Wolfcamp in the study area in the Permian Basin and measured values of the least principal stress from DFIT measurements expressed as the frac gradient (red circles). Note the distinctly higher frac gradient in the Upper Wolfcamp) (b) Gun-barrel view of microseismic events associated with stimulation of four wells located at different positions in the Wolfcamp sequence (indicated by stars). Wells A and B are directly above each other. Wells C and D are offset laterally. The vertical line in the middle indicates a vertical well and the microseismic monitoring array. From Xu et al. (2017).

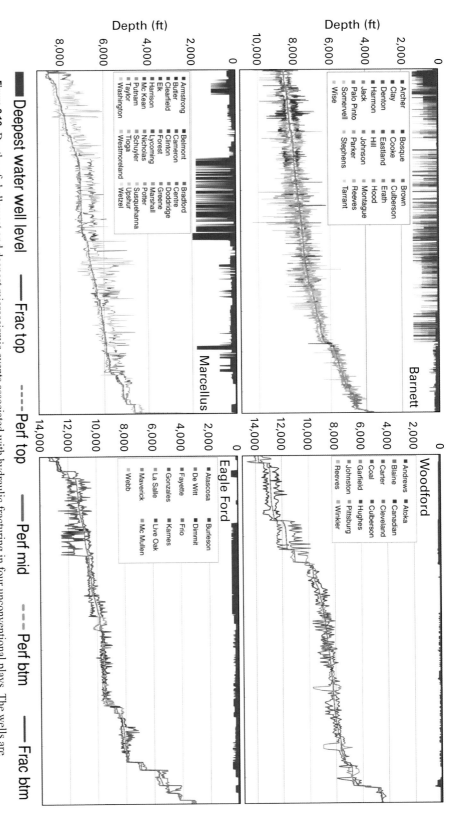

Figure 8.16 Depths of shallowest and deepest microseismic events associated with hydraulic fracturing in four unconventional plays. The counties in which the measurements are made is shown in each inset and are used to color the lines representing microseismic event depths. The blue bars indicate the depth of water wells in each area. From Fisher & Warpinski (2012).

it is acting as a hydraulic fracture barrier. While one would expect downward propagation when well B was stimulated in the upper Wolfcamp, well A (located directly below well B) had been stimulated first. Thus, it could be that few events were seen in the middle Wolfcamp when well B was stimulated because they had already been stimulated during stimulation of well A. Thus, despite the cautionary notes above, it seems as if microseismic events are occurring in the region surrounding the hydraulic fractures and provide a realistic sense of vertical frac height. We examine below the issue of proppant distribution within hydraulic fractures as the hydraulic fracture height and the propped fracture height are likely two different things.

Figure 8.16 (Fisher & Warpinski, 2012) is a compilation of now iconic figures indicating the range of microseismic events associated with hydraulic fracturing in four different plays – the Barnett (upper left), Woodford (upper right), Marcellus (lower left) and Eagle Ford (lower right). The wells are ordered from right to left with the red line indicating the depth of the perforations being stimulated. The blue bars at the top indicate the depth of water wells in the area where the hydraulic fracturing was being carried out. The two rapidly fluctuating lines represent either the shallowest or deepest microseismic events associated with hydraulic fracturing. Perhaps there are two generalizations that can derived from viewing the data at this relatively coarse scale. First, there are some formations, such as the Marcellus, for out-of-zone hydraulic fracture growth. In this case, there appears to be consistent evidence for hydraulic fracture growth, hundreds of feet above the Marcellus and almost no growth downward. Second, in each of the four areas, there are *spikes* of microseismicity that seem to indicate localized upward or downward growth. One possibility is that these represent events propagating upward and/or downward along specific faults intersected during hydraulic fracturing in a given area. These spikes are most pronounced in the case of the Barnett, but spikes are seen in the other areas as well. In these cases the microseismicity is not indicating that the hydraulic fractures are propagating 1,000–2,000 ft above, or below, the stimulation zone. Rather, they indicate that pressure is being transmitted vertically through a fracture network.

One of the most important implications of Fig. 8.16 is that hydraulic fracturing is not directly affecting near surface aquifers. This will be discussed at greater length in the environmental impacts of unconventional oil and gas development in Chapter 12.

Fracturing Fluids and Proppants

It is beyond the scope of this book to thoroughly discuss hydraulic fracturing fluids and proppants. The reader is referred to the appropriate chapters in Economides & Nolte (2000); Ahmed & Meehan (2016), Smith & Montgomery (2015) and other sources. There are, however, a few issues to discuss that are relevant to topics we address later in this chapter and in subsequent chapters. Basically, an ideal frac fluid is environmentally benign, easy to use, inexpensive and provides the appropriate additives to achieve the desired viscosity, lubricity, biocides, etc.

Some particular attributes of fracturing fluids are worth noting, as summarized by Martin et al. (2016). These include:

- *Gelling Agents or Viscosifiers* – used to raise viscosity for proppant transport or, in low concentrations to reduce friction
- *Buffers* – pH control
- *Crosslinkers* – to dramatically increase fluid viscosity when hydrated
- *Biocides* – to prevent growth of bacterial colonies
- *Surfactants* – to reduce surface tension between fracturing fluids and formation fluids
- *Friction Reducers* – to reduce pumping pressures
- *Gel Stabilizers* – to increase stability of crosslinkers with temperature
- *Breakers* – to break crosslinks and reduce viscosity for clean up at the end of a hydraulic fracture stage
- *Clay Stabilizers* – to prevent interactions with clays in the formation, swelling, etc.

Needless to say, the exact combination of additives and how they are varied at different times in a hydraulic fracturing stage is a complicated process that is not addressed here. Of general note is that there has been a steady progression to lower viscosity and higher pumping rates to meet the combined challenges of the extremely low matrix permeability of unconventionals and the need to stimulate a complex fracture network (Chong et al., 2010). According to Martin et al. (2016) the most common hydraulic fracturing fluids being used in unconventional development is slickwater using a relatively small amount of linear gelling agents such as guar (a common thickener used in food products) and several of the additives listed above such as a biocide. Hybrid fluid systems are also used, in which case slickwater might be used at the beginning and end of a hydraulic fracturing stage but a linear, or crosslinked gel, is used to facilitate proppant transport during the main pumping stage. There are many such scenarios that have been used.

Proppants are also a complex topic, with the type of material, its shape and size distribution being the main variables. While sand is most commonly used as a proppant, there are a number of different types of sand, a wide variety of grain sizes (and shapes). Bauxite and ceramic alternatives are also used as proppant depending on the size distribution and crushing strength desired. Smaller diameter and lighter proppants are expected to travel further when hydraulic fracturing with low viscosity fracturing fluids; the selection of the ideal proppant for a given case also depends on the length of the lateral and is complicated by the desire to have the largest area of the hydrofrac propped and the highest conductivity of the propped area (Saldungaray & Palisch, 2012).

Slickwater versus Gel – The importance of shear stimulation of pre-existing fractures and faults surrounding hydraulic fractures has been mentioned several times and is discussed in some detail in Chapter 10. Key to this process is the pressure that leaks-off from hydraulic fractures as they propagate. In fact, one could argue that the use of slickwater as a hydraulic fracturing fluid was a critical technology contributing to the success of unconventional development. Figure 8.17 (from Cipolla et al., 2008) shows that microseismicity associated with slickwater persists much longer and extends much

256   **Unconventional Reservoir Geomechanics**

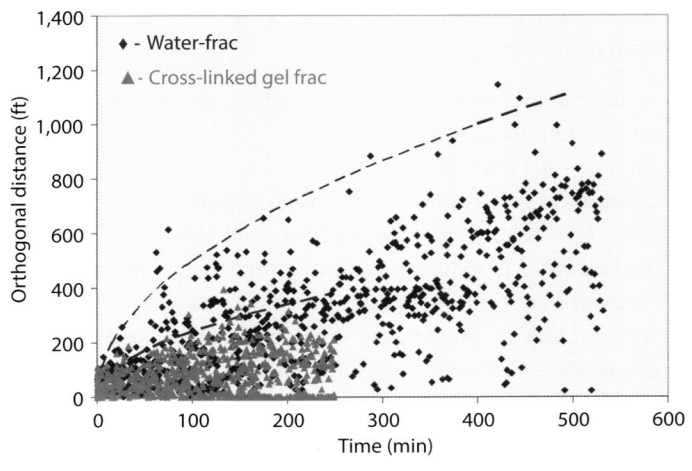

Figure 8.17 The number of microseismic events accompanying fracturing with water or cross-linked gel as a function of time and orthogonal distance from perforations. After Cipolla et al. (2008).

further than higher viscosity gel. They note that the seismicity spreads out as $t^{-1/2}$, as expected for linear diffusion. The widespread seismicity associated with slickwater fracturing in the off-azimuth well (compared with the nearby on-azimuth well) in Fig. 8.4 is also clear evidence of the ability of pressure surrounding hydraulic fractures to increase over a large area when using a low viscosity fracturing fluids. This process is described in more detail in Chapter 10. In this context, one can envisage the potential advantages of using even lower viscosity fluids such as supercritical CO_2 or CO_2 foams as a hydraulic fracturing fluid.

Of course, the penalty one pays when using slickwater is the reduced ability to transport proppant along the hydraulic fracture. In a laboratory study of propped and unpropped hydraulic fractures in the Niobrara formation (Suarez-Rivera et al., 2013) showed the considerable difference in conductivities at elevated effective normal stress (Fig. 8.18a). Note that as effective normal stress increases (as would be expected to accompany depletion) the conductivity of the unpropped hydraulic fracture is about three orders of magnitude lower than the propped hydraulic fractures. Aybar et al. (2015) developed a numerical model of this process (calibrated by the laboratory data of Suarez-Rivera et al.) shown in (Fig. 8.18b) that generalizes the laboratory results. For stress and pore pressure values expected in unconventional reservoirs, there is no question that the propped portion of induced hydraulic fractures is much more permeable and is expected to remain permeable with depletion.

The ability of different types of hydraulic fracturing fluids is expressed in Table 8.3 (after Brannon & Bell, 2011) in terms of the velocity required to transport proppant along a fracture as a function of the type of hydraulic fracturing fluid. Fluid velocity within a fracture is assumed to decline exponentially with increasing distance from the wellbore. For purposes of comparison the velocity at the wellbore is assumed to be

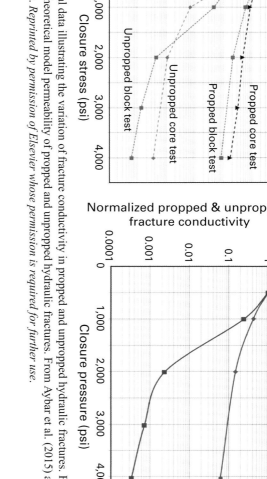

Figure 8.18 (a) Experimental data illustrating the variation of fracture conductivity in propped and unpropped hydraulic fractures. From Suarez-Rivera et al. (2013) (b) A theoretical model permeability of propped and unpropped hydraulic fractures. From Aybar et al. (2015) and calibrated using the data shown in (a). *Reprinted by permission of Elsevier whose permission is required for further use.*

Table 8.3 Minimum horizontal fluid viscosity required for suspension of 40/70 sand proppant for given hydraulic fracturing fluids with the specified viscosities in ft/min. From Brannon & Bell (2011).

	Wellbore	100 ft	200 ft	300 ft	400 ft	500 ft	1000 ft
Calculated velocity (ft/sec)	10	1	0.1	0.01	0.001	0.0001	0.000000001
Slickwater 1 gpt	0.016	0.032	0.032	0.032	0.032	0.032	0.032
Linear 10 pptg* guar	0.0032	0.004	0.005	0.0064	0.0064	0.0064	0.0048
Conv. borate Xlinked 20 pptg guar	0.000064	0.000064	0.00008	0.00008	0.00008	0.00008	0.00008
Conv. delayed borate Xlinked 20 pptg guar	0.00064	0.000064	0.00008	0.00008	0.00008	0.00008	0.00008
New borate Xlinked 10 pptg guar	0.0008	0.0016	0.0032	0.00107	0.016	0.024	0.032
Alternate new borate Xlinked 10 pptg guar	0.0016	0.0032	0.0064	0.016	0.016	0.032	0.032

* pounds per thousand gallons

10 ft/s and declines by an order of magnitude per each 100 ft laterally from the wellbore. As shown in Table 8.3, if the proppant slurry flow velocity at the perforations is 10 ft/s, it will decline to 1 ft/s at 100 ft from the wellbore, and so on. The red text in Table 8.3 indicates fluids that are no longer capable of suspension transport. In other words, the minimum velocity determined as needed for suspension transport of the proppant slurry is greater than the velocity calculated to exist at that distance from the wellbore. In fact, if one were to incorporate leak-off into these calculations the required velocities would have been even higher as also would have been the case if multiple hydraulic fractures were propagating simultaneously as suggested by the drill-through experiments discussed above.

Figure 8.19 illustrates the proppant transport issue. The calculations were carried out using the ResFrac software described by McClure & Kang (2017). It is a three-dimensional wellbore, hydraulic fracture and reservoir simulator that we will use to illustrate several principles of hydraulic fracture growth and proppant transport, etc. In addition to the complexities cited above that complicate the calculations shown in Table 8.3, there are a number of competing processes that affect proppant transport (e.g., Wu & Sharma, 2016; Lee et al., 2017). Our intent in showing the simulations below is to isolate a number of issues that are germane to the subject matter in this book. In Chapter 11, for example, we will consider the way in which operational procedures affect proppant transport in the context of vertical hydraulic fracture growth – the latter depending on variations of the magnitude of the least principal stress within, above and below the interval being hydraulically fractured.

The model shown in Fig. 8.19 is a single, strongly confined hydraulic fracture propagating away from a horizontal well that is drilled in the center of a 200 ft thick

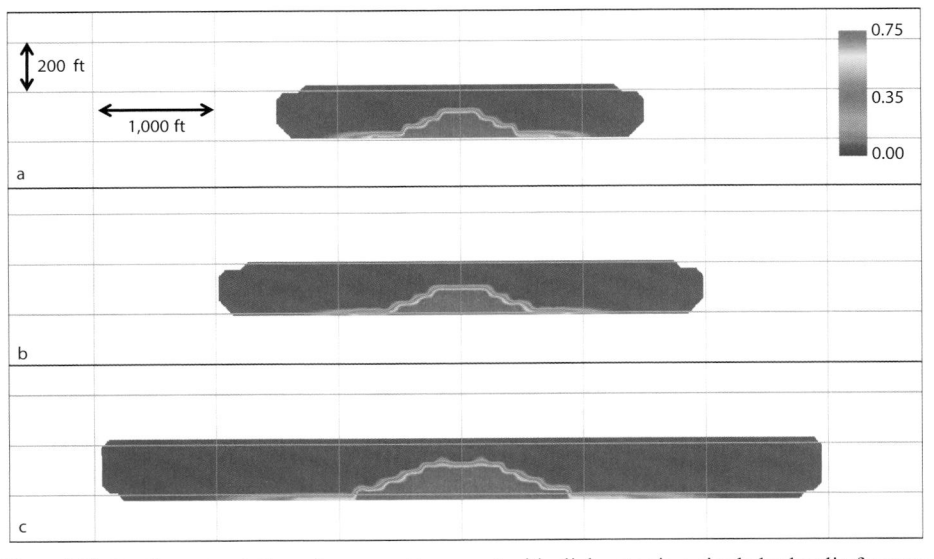

Figure 8.19 ResFrac simulation of proppant transport with slickwater in a single hydraulic fracture propagated from a horizontal well with different diameter perforations and flow rates. The color scale indicates the volume fraction of proppant (see text). The hydraulic fracture is strongly confined by stress barriers above and below the interval being stimulated.

formation. The least principal stress in the formation is 8000 psi, the formations above and below have a 500 psi higher value of the least principal stress. Pumping was carried out for 2.5 hours at various rates, as described below. The color scale indicates the proppant volume fraction defined as the volume of proppant at any point divided by local volume of the fracture (area × aperture). The proppant mix assumed equal amounts of 40, 55 and 70 mesh size particles. In each case, proppant was added to the slurry at an initial rate of 1 ppg, which then was raised in steps to 3 ppg. As this simple model might be representative of a controlled entry point completion, we also varied the diameter of the perforations, assuming 12 perforations uniformly distributed around the wellbore. The pressure drop across the perforations is given by Eqn. (8.1)

$$\Delta P_{pf} = \frac{0.24 Q^2 \rho}{C_{pf}^2 N_{pf}^2 D_{pf}^4} \tag{8.1}$$

(after Veatch, 1983), where Q is the flow rate in bbls/min, ρ is density in pounds per gallon, C_{pf} is a coefficient of discharge for the perforations (set to 0.5 for these calculations), N_{pf} is the number of perforations and D_{pf} is the perforation diameter in inches.

The first case considered (Fig. 8.19a) assumed perforations 0.32 inches in diameter. The maximum injection rate was 40 bbl/min due to extremely high wellbore pressures that result from perforation friction. Note the importance of perforation diameter in

Eqn. (8.1). Although the model predicts appreciable lateral hydraulic fracture growth (half lengths of about 1,500 ft) the propped half length of the hydraulic fracture is mostly less than 500 ft, with proppant only below the well due to settling. The second case considered (Fig. 8.19b) increased the perforation diameter to 0.64 inches. In this case, a rate of 50 bbl/min was easily achieved with very little excess pressure in the wellbore due to the larger perforations. In this case, the total fracture size and the total propped area are only slightly more than in the first case. In the third case considered (Fig. 8.19c), the proppant diameter was increased to 1 inch and the flow rate was increased to 100 bbl/min without excess pressurization of the wellbore. While the lateral length of the fracture increased considerably, there was only an incremental increase in the size of the propped area despite having injected twice as much proppant as in the case shown in Fig. 8.19b.

Figure 8.20 is a comparison of proppant distributions using slickwater and gel. Most of the parameters are the same as those use in Fig. 8.19. A perforation diameter of 0.64 inch was used in the model and the injection rate was 10 bbl/min for 20 min, 60 bbl/min for 30 min and 80 bbl/min for 90 min, for a total injection period of 2.33 hours. The proppant was added according to the following schedule – 1 ppg for 15 min, 1.5 ppg for 30 min, 2.25 ppg for 30 min and 3 ppg for 30 min. The base case considered (Fig. 8.20c) is a slickwater case (0.3 cp viscosity in the well and hydraulic fracture) after 5 hours, approximately 2.7 hours after the stimulation ended. The proppant has largely settled, almost achieving the same distribution seen in Fig. 8.20d, which is 1,000 hours after the stimulation ended. Hence, proppant settling is quite fast when hydraulic fracturing with slickwater. Stimulation with cross-linked viscous gel (100 cp viscosity) is shown in Fig. 8.20a at the end of 5 hours. Notice that the proppant settling is much slower. The cross-linked gel attains high viscosity only in the formation, while flowing with 0.3 cp in the wellbore. The crosslinking is controlled by the assumed formation temperature of 200 °F. Figure 8.20b shows the viscous gel

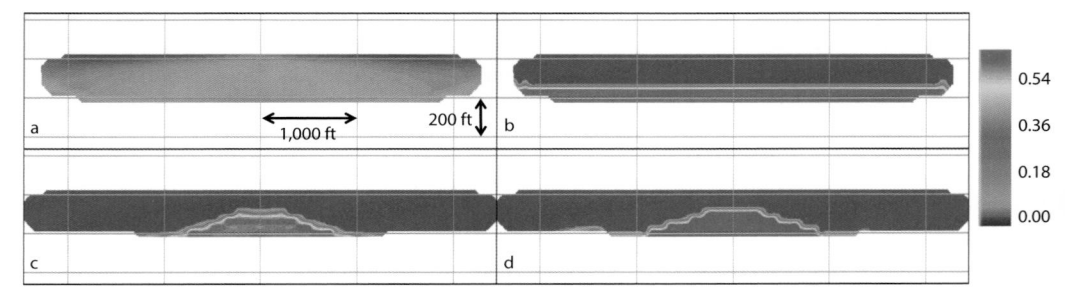

Figure 8.20 ResFrac simulation of proppant transport comparing gel and slickwater in a single hydraulic fracture propagated from a horizontal well. The color scale indicates the volume fraction of proppant (see text). As in Fig. 8.19, the hydraulic fracture is strongly confined by stress barriers above and below the interval being stimulated. (a) Distribution of proppant from a gel fracture after 5 hours. (b) The distribution of proppant from a gel fracture after 1,000 hours. (c) Distribution of proppant from a slickwater fracture after 5 hours. (d) Distribution of proppant from a slickwater fracture after 1,000 hours.

Horizontal Drilling and Multi-Stage Hydraulic Fracturing

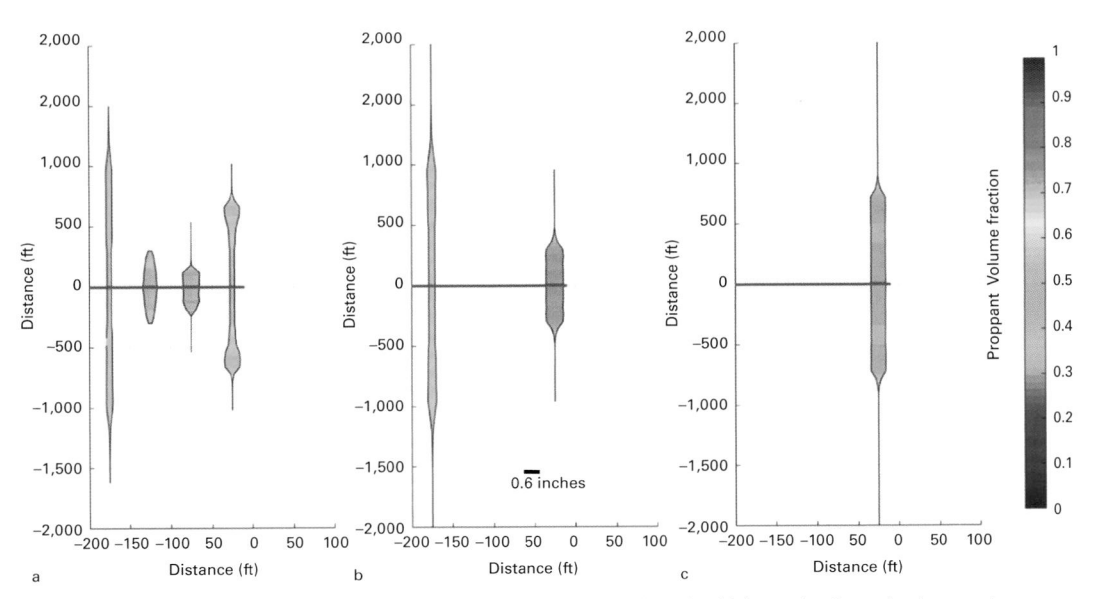

Figure 8.21 The variation of hydraulic fracture length and width results from the interaction between the stress shadow associated with each and wellbore pressures related to perforation diameter and flow rate (see text). (a) Plan view of a ResFrac simulation of four hydraulic fractures growing from four clusters with 0.32 inch perforations. (b) Only two hydraulic fractures grow from four clusters with 0.64 inch perforations. (c) Same as (b) but with a single perforation cluster.

stimulation at the end of 1,000 hours. The proppant has spread out much further from the well, but it is still highly concentrated in the lower part of the hydraulic fracture such that there is little proppant in the hydraulic fracture closest to the wellbore.

While the case shown in Fig. 8.19 would seem to argue that using small diameter perforation clusters is problematic because it limits injection rates, there are cases in which high wellbore pressure can be beneficial. Figure 8.21a,b,c illustrate fractures propagating from four perforations in a stage 200 ft long. The simulation parameters are similar to those used in Figs. 8.18 and 8.19 (depth, stress state, vertical confinement, number of perforations, proppant sizes, etc.). One exception is that the flow rate was progressively increased from 10 to 60 to 80 bbl/min. Figure 8.21a had 0.32 inch perforations whereas Fig. 8.21b had 0.64 inch perforations. Note that by achieving higher wellbore pressure with the smaller perforations there was sufficient pressure in the wellbore to overcome the stress shadows associated with the adjacent hydraulic fractures making it possible to propagate fractures from each of the perforation clusters. When the perforations were larger and wellbore pressures lower, there are only two hydraulic fractures growing from the outermost perforations. The unusual shape of some of the fractures (where they are wider near the tip than near the wellbore), is the result of the stress shadows of nearby hydraulic fractures. Figure 8.21c represents the same case as Fig. 8.21b but for a single perforation cluster. Note that while the length of the single hydraulic fracture in Fig. 8.21c is much greater than three of the four fractures in Fig. 8.21a (and one of the two fractures in Fig. 8.21b) and the width of the fracture is greater than in the other cases, overall, the area of propped fracture is considerably less than in the other cases. As area of contact between highly permeable fractures and low permeability matrix is critical for production (which is discussed at length in Chapter 12), it would seem that the situation shown in Fig. 8.21a would be best for production.

9 Reservoir Seismology

As previously discussed, microseismic events are generated when pore pressure reaches a pre-existing fracture plane and induces slip. This process is described in more detail in Chapter 10 as well as how microseismic data can be used to better understand the stimulation process.

To use the microseismic data properly, it is important to understand what can be determined and the limitations of such information. To this end, the topics considered in this chapter briefly consider how microseismic monitoring is carried out, how we know the events reflect shear slip on pre-existing faults, how accurately we know the locations of the seismic events, what can be determined about the seismic sources in terms of the size of the faults that slip (and the distribution of fault sizes) and the geometry of slip as defined by focal plane mechanisms (first introduced in Chapter 7). Thus, the topics considered in this chapter are intended to inform the reader about significant issues related to monitoring microseismicity triggered during hydraulic fracturing, not to comprehensively discuss any of the topics considered.

Fortunately, a number of recent books have been published that go into these issues in great detail. The reader is referred to the Society of Exploration Geophysicists (SEG) short course on *Microseismic Imaging and Hydraulic Fracturing* by Shawn Maxwell (Maxwell, 2014), the SEG geophysical reference on *Microseismic Monitoring* by Vladimir Grechka and Werner Heigl (Grechka & Heigl, 2017) and the book on *Passive Seismic Monitoring and Induced Seismicity* by David Eaton (Eaton, 2018). As many topics of importance in microseismic monitoring are related to fundamental principles of earthquake seismology, the books by Aki & Richards (2002) and Stein & Wysession (2003) are recommended. In that regard, while earthquake locations, magnitudes, focal plane mechanisms and source properties of sizeable earthquakes are of obvious importance in the context of injection-induced earthquakes and are discussed in Chapters 14 and 15, the focus in this chapter is on the microseismic events that accompany multi-stage hydraulic fracturing.

Microseismic Monitoring during Reservoir Stimulation

The recording of seismograms (recorded movements of the earth due to passing waves) can be accomplished in a number of ways, each with advantages and disadvantages. It is helpful to remember that a seismogram represents the seismic source, passage of the seismic waves through the earth and the response of the seismic instrument. Following Stein & Wysession (2003), a seismogram can be expressed as the convolution (indicated by the symbol $*$) of the source time function $x(t)$ with the earth structure, $e(t)$ and $q(t)$, and the instrument response, $i(t)$.

$$u(t) = x(t) * e(t) * q(t) * i(t) \qquad (9.1)$$

where convolution of two arbitrary time-series is written as

$$s(t) = w(t) * r(t) = \int_{-\infty}^{\infty} w(t - \tau)r(t)d\tau$$

which is quite convenient as convolution in the time domain is equivalent to multiplication in the frequency domain such that Eqn. (9.1) becomes

$$U(\omega) = X(\omega)E(\omega)Q(\omega)I(\omega) \qquad (9.2)$$

where each term represents the Fourier transform of the respective time series. One reason this perspective is important to keep in mind is that the challenge of microseismic monitoring is to utilize seismograms to derive information about high frequency, small amplitude seismic sources (as will be shown below), typically at depths of 2–3 km, in the presence of large-scale sources of surface noise (pumps and engines). In addition, to locate events one must know the velocity structure of the earth (represented by $e(t)$) which is intrinsically anisotropic, as described in Chapters 1 and 2, and acknowledging that geometric spreading of the seismic waves, scattering and attenuation (represented by $q(t)$) will further reduce the size of the initially small seismic signals.

Borehole Arrays – Because of the extremely small size of the microearthquakes occurring during hydraulic fracturing, the most common technique used for recording microseismic events is through deployment of three-component geophone arrays in nearby pre-existing wells, typically within 600 m of the stimulation well. Geophones are mechanically or hydraulically clamped to the wellbore casing for good coupling and typically record particle motion over a limited range of frequencies. It is also possible to deploy geophone arrays in horizontal wells. The advantages and disadvantages of downhole monitoring are somewhat obvious. Seismic signals are easier to detect because the geophones are closer to the seismic sources and located in a relatively noise free environment. Because the seismometers are close to the microseismic events, uncertainties in velocity structure have less effect on determining event locations, especially their depth as long as the length of the array is comparable to the distance to the microseismic events. Of course, the disadvantages include the fact that pre-existing wells need to be available in the area of interest, the spatial coverage of the arrays is limited, which affects both the ability to detect and locate the events and determination of focal plane mechanisms, as discussed below. Another disadvantage is that wave propagation is complicated by horizontal layering (which results in complex waveforms) and the intrinsically anisotropic nature of the sedimentary layers between

Reservoir Seismology   265

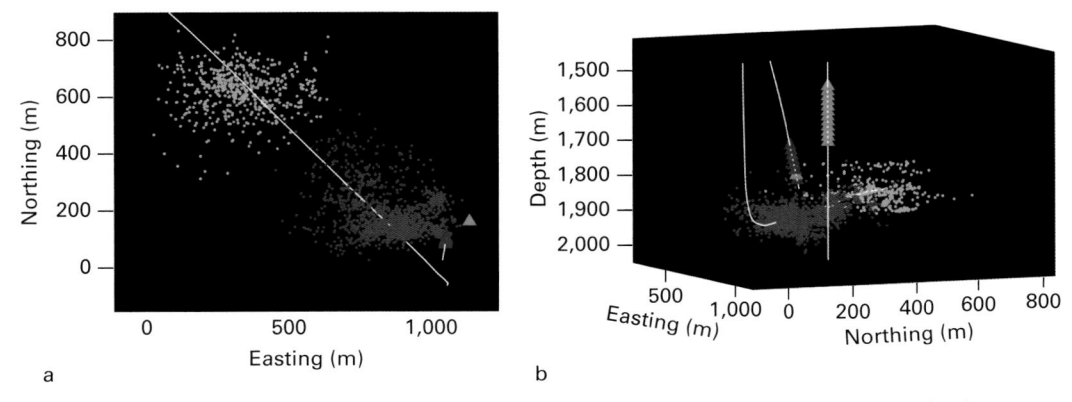

a   b

Figure 9.1 Two seismometer arrays were deployed in near vertical wells to microseismic events associated with stages 3 (in blue) and stage 4 (in red) in a horizontal well in the Barnett shale. After Hakso & Zoback (2017).

the microseismic sources and receiver. The importance of anisotropy is discussed at length by Grechka & Heigl (2017).

Figure 9.1 shows an example of microseismic events monitoring associated with two stages of a multi-stage hydraulic fracturing in the Barnett shale (after Hakso & Zoback, 2017). Two near-vertical wells were used to deploy 20-level, 3-component geophones with a spacing of 15 meters. As shown in the map view on the left, despite the proximity of these wells, distances to microseismic events are sometimes more than 500 meters, which makes detection of small events difficult, as discussed below. Also, illustrated in the perspective view on the right, ray paths are quite complicated in that nearby events from stage 4 (in red) are associated with nearly vertically traveling rays whereas distant events associated with stage 3 (in blue) principally involve horizontal ray propagation. Thus, markedly different ray paths (and velocities along the ray paths) affect the location accuracy of these seismic events in two adjacent stages, as discussed below.

Surface Arrays – Deployment of geophones at the surface have the potential for addressing some of the problems associated with downhole recording. Obviously, no pre-existing wells are needed and there is good coverage of the focal sphere to facilitate focal plane mechanisms as discussed below. Moreover, ray propagation is principally vertical and quite similar to ray paths routinely obtained from 3D seismic reflection or VSP data. Figure 9.2 (courtesy Microseismic, Inc.) shows deployment of surface geophones along radial lines. It is not uncommon for more than 1000 stations to be deployed with 6–12 geophones per site. When geologic and noise conditions are favorable, surface monitoring can produce accurate locations, especially horizontally (Chambers et al., 2010) and focal plane mechanisms for relatively large events. Unfortunately, there are areas where near surface conditions are such that surface monitoring works poorly. The hydraulic fracture orientations shown in Fig. 7.12 were obtained with the type of array shown in Fig. 9.2. Because the seismometers are

Figure 9.2 Deployment of surface seismic array in the Eagle Ford. It consists of approximately 1200 channels, each with six, vertical component seismometers recorded in series in a ten-arm star configuration. The array extends approximately 6.7 km in the north–south direction and 5.7 km in the east–west direction, centered on the well pad to monitor stimulation of the two wells shown in blue (courtesy Microseismic Inc.).

several km from the microseismic events, there is often difficulty identifying coherent P- and S-wave arrivals visually from the stacked seismic records, especially for smaller events. Also, surface noise can sometimes be severe (and not easily canceled) and near surface formations can sometimes seriously attenuate seismic signals and require significant static corrections. Hence, only the relatively large events will be detected, and specialized signal processing techniques need to be utilized and migration methods (as described below) are commonly used to locate the events. Special array designs are deployed to help with the removal of surface noise, such as using clusters of geophones.

All of this said, Fig. 9.3 compares the locations of microseismic events recorded during hydraulic fracturing in the Haynesville formation, obtained with a single downhole array of ten three-component geophones (inverted triangles) located 1,000–2,000 ft east of the heel of the wells and that recorded by a shallow borehole arrays (discussed below) as shown in Fig. 9.4. As seen in Fig. 9.3b, the deepest seismometer was about 500 ft above the laterals due to temperature limitations of

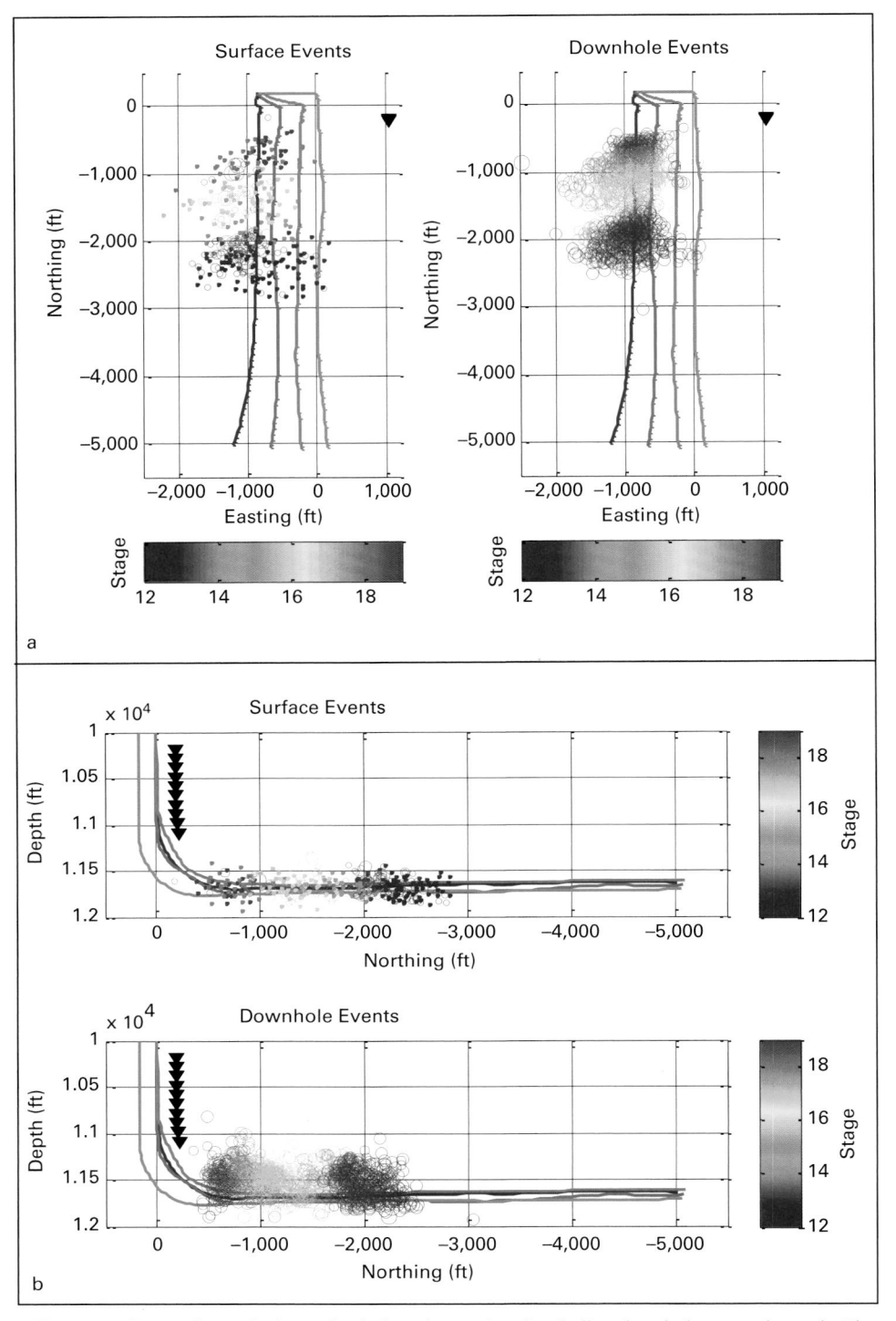

Figure 9.3 Comparison of microseismic locations using the shallow borehole array shown in Fig. 9.3 with events recorded by a 10-level array of three-component seismometers in a vertical well (inverted triangles). (a) Map view, only stages 12–19 of well H2 are shown. (b) Cross-section (looking east). Note that the events located with the downhole array are systematically shallower than those located with the surface array (see text).

268   Unconventional Reservoir Geomechanics

Figure 9.4 An array of approximately 100 shallow boreholes designed to monitor hydraulic fracturing operations in the Haynesville formation. Each borehole is approximately 100 m deep. The four horizontal wells shown are approximately 5,000 ft in length (courtesy Microseismic, Inc.).

the geophones. Only stages 12–19 of well 2 are shown. There are three noteworthy features in this comparison: First, not surprisingly, approximately ten times as many events are seen with the downhole array than the surface array (e.g., Maxwell et al., 2012). Second, in this case, both the spatial locations and depths of the larger events seem to be better constrained by the surface array. The velocity model used for locating the events recorded at the surface was based on a check-shot survey and adjusted to locate the recorded perforation shots at the known locations. Because only P-waves were used (as there were single, vertical component geophones in the shallow boreholes), migration techniques were used to locate the events (see below). Comparison of the locations of the larger events show an average of 150 ft difference in spatial location but as much as 800 ft for all events. Also, the events located with the shallow borehole array are systematically deeper than those located with the downhole array. A comparison of event locations demonstrates that the events located on the borehole array seem to be systematically shifted in the direction of the downhole array. As alluded to above, the ray paths from the events to the downhole arrays involves both horizontal and vertical wave propagation, possible refraction at formation boundaries and complications due to velocity anisotropy. It certainly appears that because of the simpler ray paths and spatial extent of the surface array, the microseismic locations in this case are more accurate than those with the downhole array. Finally, when viewed in 3D, the locations of the events recorded at the surface seem to define a number of planes striking in the S_{Hmax} direction but concentrating along planes dipping around 60°. As this is a normal faulting environment, the events could be representing the cloud of events expected to surround hydraulic fractures or along normal faults, as discussed in Chapter 7.

Shallow Borehole Arrays – A third deployment strategy is something of a hybrid between downhole and surface arrays which uses numerous, relatively shallow boreholes to deploy geophones at depths on the order of 100 m. Significant reduction of surface noise can be achieved in holes inexpensively drilled by truck-mounted drill rigs. Moreover, the geophones can be installed permanently in the holes to monitor activities through time without redeploying the network. Typically, the geophones are deployed semi-permanently and recording equipment is re-deployed each time a well is being stimulated. In addition, shallow borehole arrays may be less expensive than deployment of over 1,000 stations at the surface, there is the potential for improvement of signal to noise over geophones at the surface and shallow borehole arrays offer much better spatial sampling of the wavefield than downhole arrays that can improve earthquake locations and focal plane mechanisms if the events are large enough to be recorded at the surface. An array deployed in the Haynesville formation is shown in Fig. 9.4 In this case, approximately 100 holes were used to deploy three, single-component geophones at different depths, the deepest being almost 100 m deep.

Optical Fibers and Distributed Acoustic Sensing – An interesting new developing technology is to use optical fibers that are cemented behind casing to act as a distributed acoustic sensor (DAS) or deployed temporarily on coiled tubing. Fundamentally, the flaws built into the fiber cause backscattered energy from a laser interrogator to calculate strain as a function of time (which is converted to ground motion) along a length of the fiber referred to as the gauge length, which determines the frequency resolution. Although ground motion is measured only in the direction parallel to the fiber, the advantages of such fibers is that they can be deployed temporarily or permanently and have the potential to measure ground motion all along the length of the fiber thereby yielding detailed information about the seismic wavefield. As pointed out by Miller et al. (2012) and Hull et al. (2017b), it is possible to record seismic wavefields that are quite similar to those recorded with conventional geophones and in many cases over thousands of feet of the wellbore. The deployment of optical fibers also allows for distributed temperature sensing (DTS) that is useful to investigate zonal isolation (as discussed in Chapter 8) and stage-by-stage production (as discussed in Chapter 12). Strain measurements along the wellbore are also possible when the fiber is cemented in behind casing. Jin & Roy (2017) and Hull et al. (2017b) discuss the multiple uses of deployed fibers.

Seismic Wave Radiation

When microseismic events were discovered in association with hydraulic fracturing observations, an obvious question is whether they were showing the opening of a hydraulic fracture as it propagated, or, like earthquakes, represented shear on distributed small faults. As alluded to previously, they are the latter – extremely small earthquakes that are triggered by the increase in pore pressure during hydraulic fracturing operations. The reason this is known comes from the character of seismic waves propagating from the source. To begin this discussion, we need to first consider the radiation of seismic energy away from the source, which begins with first considering forces acting in a body and the seismic energy that emanates from them.

As illustrated in Fig. 9.5 (modified from Stein & Wysession, 2003) one could consider a single body force (perhaps representative of a landslide), a single dipole force couple (which could represent an opening fracture, although it is likely more complicated than this, as discussed below) or a double couple, representing slip on a fault. Note that the double couple can be represented equivalently by two dipole couples of equal magnitude and opposite sign, P and T, representing a compressional and extensional dipole respectively.

In order to evaluate seismic radiation without any pre-conceived ideas about the origin of the seismic waves, we next need to consider the moment tensor of the event. As the next step in this discussion, let us first consider the components of the seismic moment tensor, a description of the force couples that might be responsible for a seismic event. In general,

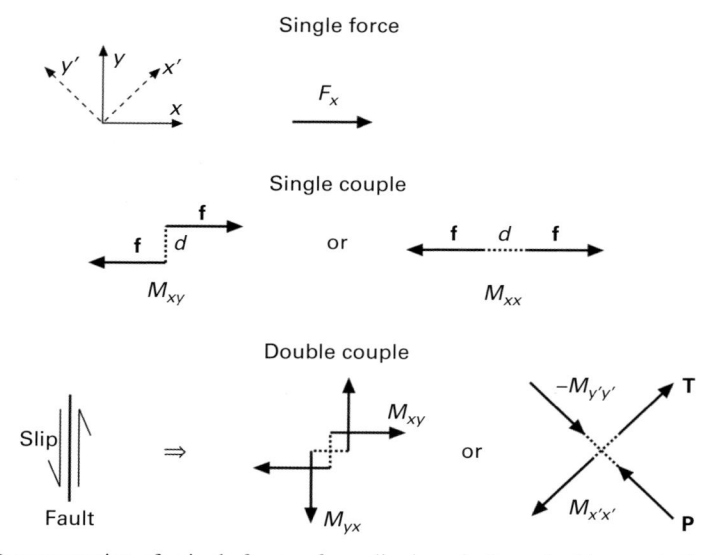

Figure 9.5 Representation of a single force, a force dipole and a force double-couple that represents slip on a fault. Note that fault shear can be represented by either a shear force double couple or an orthogonal set of dipoles, one compressional and one extensional. The corresponding components of the moment tensor (discussed below) are indicated. From Stein and Wysession (2003).

this second rank moment tensor can be written as a 9-component array of force dipoles, representing six independent force couples as $M_{ij} = M_{ji}$ to satisfy the equations of equilibrium (i.e., zero net torque). In an x,y,z coordinate system, the moment tensor is expressed as:

$$\begin{pmatrix} M_{xx} & M_{xy} & M_{xz} \\ M_{yx} & M_{yy} & M_{yz} \\ M_{zx} & M_{zy} & M_{zz} \end{pmatrix}. \tag{9.3}$$

These force dipoles can be visualized as shown in Fig. 9.6.

Before going any further, we need to define the scalar moment tensor because it is the most accurate representation of the size of an earthquake. The scalar moment, M_0 is given by

$$M_0 = G\overline{D}S \tag{9.4}$$

where G is the shear modulus in the source area, $\overline{D}$ is the average displacement and S is the area over which slip occurs. The relationship between the scalar moment and earthquake magnitude is discussed below.

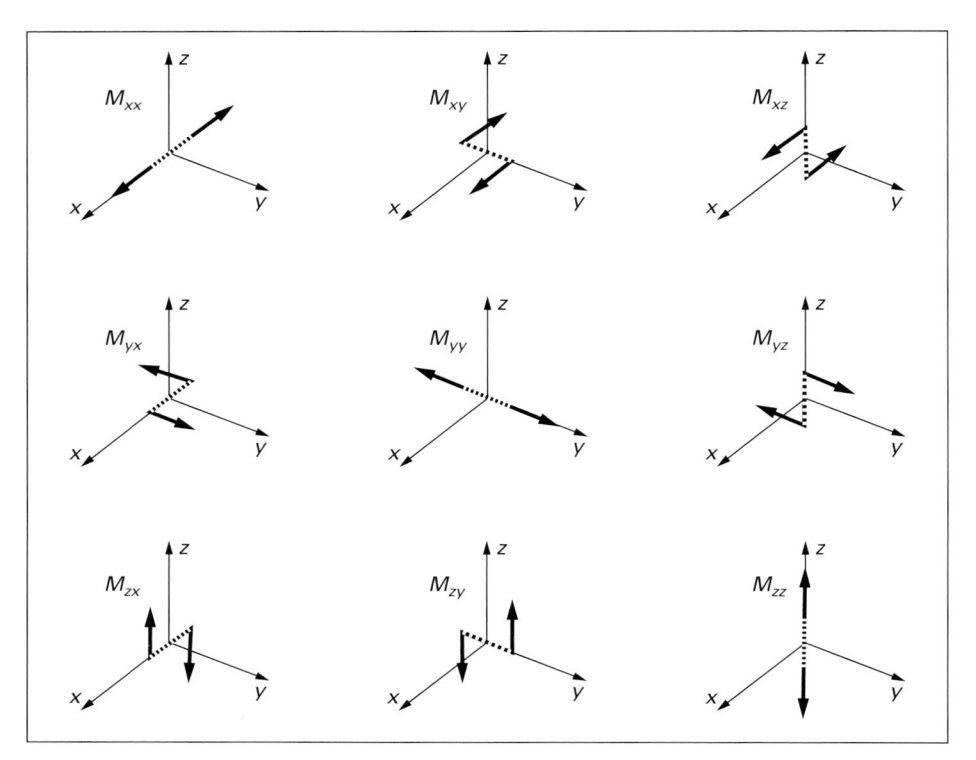

Figure 9.6 Illustration of the dipole forces that correspond to the nine components of the moment tensor. From Stein and Wysession (2003).

Unconventional Reservoir Geomechanics

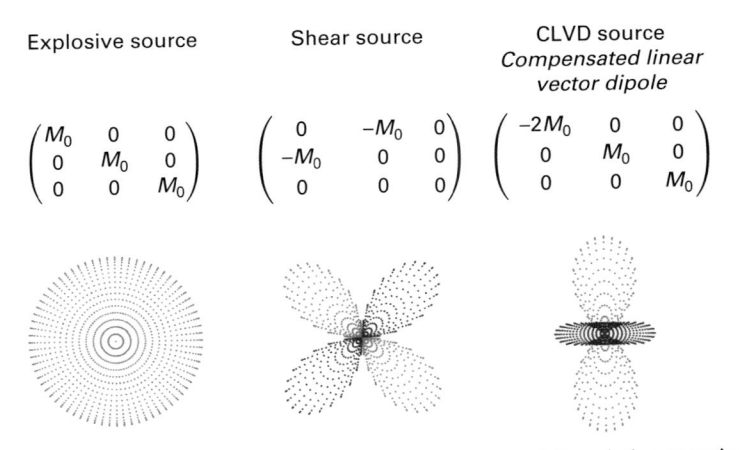

Figure 9.7 Illustration of the components of the moment tensor and the relative magnitude of compressional wave radiation in different directions (compression in red, extension in blue) associated with an explosive source, a right lateral strike-slip fault striking north–south (or a left-lateral strike-slip fault striking east–west) and one type of CLVD. Figure courtesy Leo Eisner, Seismik.

With these fundamental concepts established, Fig. 9.7 (courtesy L. Eisner) shows the P-wave radiation pattern associated with an explosive source (uniform radiation in all directions due to three extensional dipoles, M_{xx}, M_{yy} and M_{zz}), a double couple shear source (M_{xy} and M_{yx}) and compensated linear vector dipole (CLVD), which is analogous to a cylinder being squeezed radially and expanded axially an equivalent amount (or vice versa). Any seismic signal can be decomposed into components of these three general moment tensors although the decomposition is non-unique.

It is intuitive that an explosion generates a compressional P-wave in all directions from the source. An implosive source (perhaps corresponding to collapse of a cavity) represents the opposite – an extensional P-wave propagating in all directions. Note that for a shear source (the figure shown corresponds to a right-lateral strike-slip earthquake on a vertical plane striking N45°W), extensional P-waves (indicated in blue), are maximum 45° from the shear plane and compressional and extensional P-waves have symmetric radiation patterns 90° apart.

Before proceeding further, it is important to recognize that there is a distinctive pattern for shear wave propagation that is illustrated in Fig. 9.8 for horizontal shear slip on a vertical plane striking north-south that goes through the origin. Note that SH waves (waves in which particle motion is parallel to Earth's surface) show a similar pattern as P-waves but rotated 45°. This means that in the direction 45° from the strike of the fault plane, the ratio of P to SH waves is infinitely large (as SH waves have zero amplitude) but this same ratio is zero (P waves have zero amplitude) 45° away. This strong variation of amplitude ratios will be exploited in observational data below as it is diagnostic of shear sources but neither isotropic nor CLVD sources. The radiation of SV waves (with particle motion orthogonal to SH- and P-waves) of a vertical shear fault is much more complicated.

Reservoir Seismology 273

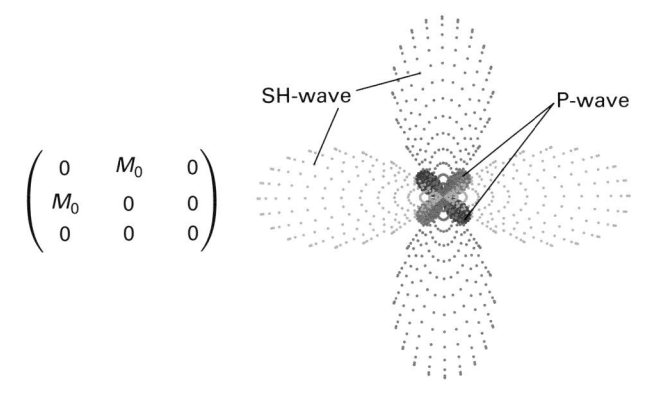

Figure 9.8 The relative magnitudes of the compressional and shear wave radiation in different directions (compression in red, extension in blue) and the corresponding moment tensor for a vertical, north-trending strike-slip fault. Figure courtesy Leo Eisner, Seismik.

Focal Plane Mechanisms from P- and S-Wave Polarities – The typical way to represent the observed P- or S-wave polarity recorded on seismometers is to use lower hemisphere stereonets (readers unfamiliar with such representations for earthquake data are referred to Stein & Wysession, 2003). Thus, Fig. 9.9 expands upon the discussion above by showing stereographic projections representing wave polarities (beachballs, the white area indicates the extensional quadrants) that are characteristic of the three types of moment tensors introduced above. Note that CLVDs appear as either eyeballs or base-balls (in the parlance of Stein & Wysession) depending on whether the extending long axis of a cylinder is vertical (third column from the right) or horizontal and trending north–south (fourth column).

The stereonets shown in Fig. 9.9 lead naturally to the illustration of focal plane mechanisms for several idealized faults illustrated in Fig. 9.10 (modified from Stein & Wysession, 2003) along with the P-wave radiation pattern – in plan view in Fig. 9.10a for a vertical strike-slip fault and in cross-section on the right for idealized dip-slip faults. What is clear is how construction of a focal plane mechanism from, typically, the polarity of P-waves and, more rarely, that of S-waves, can be used to infer relative stress magnitudes from the style of faulting. Obviously, the spatial coverage of the focal sphere is quite important if accurate focal plane mechanisms are to be determined from polarities.

Because of the symmetry of the radiation patterns there is an inherent ambiguity about which of the two focal planes correspond to the actual fault plane and which is an orthogonal auxiliary plane. When microseismic events line up along a plane clearly defining a fault (as in Fig. 4.4) it is straightforward to know which plane is the fault. In other cases, it is more ambiguous. However, when there is independent information about the state of stress in a region, it is reasonable to assume that the fault is the plane with the higher ratio of shear to normal stress.

An example of the challenges of determining accurate earthquake focal plane mechanisms of microseismic events using P-wave polarities recorded on downhole

274 Unconventional Reservoir Geomechanics

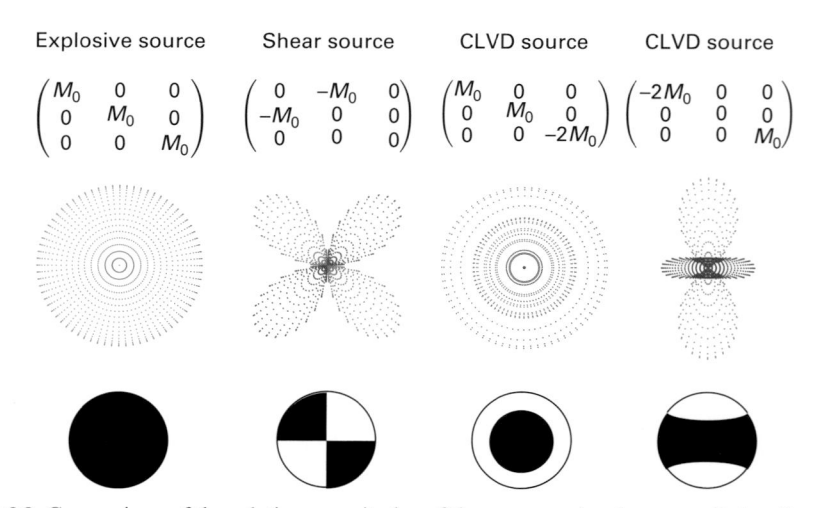

Figure 9.9 Comparison of the relative magnitudes of the compressional wave radiation for an explosive source, a shear source (as in Fig. 9.7) and both types of CLVD sources (compression in red, extension in blue). The middle row is a plan view looking down. The corresponding display of wave polarities are shown as focal plane mechanisms in lower hemisphere stereographic projections (compression is dark, extension is white). Figure courtesy Leo Eisner, Seismik.

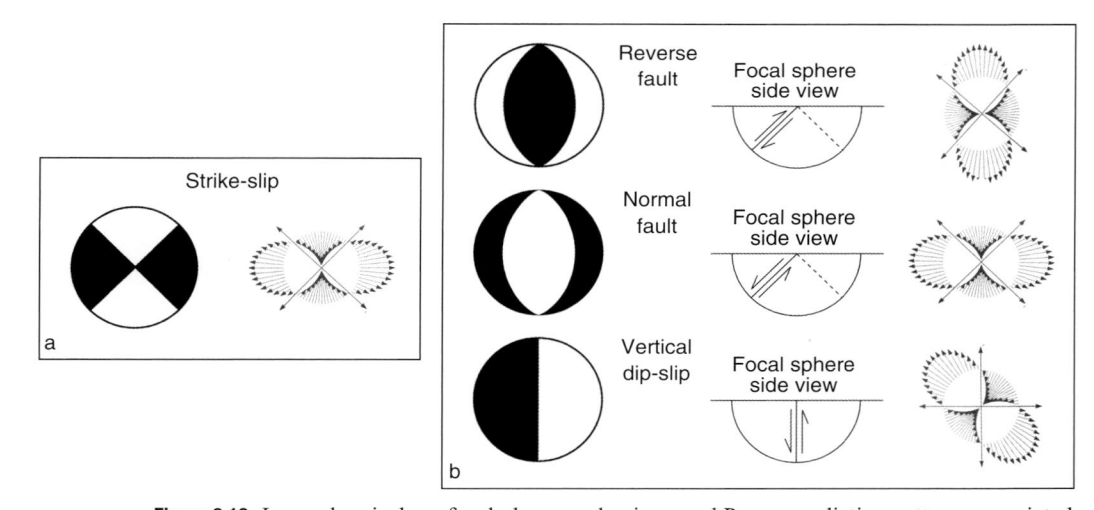

Figure 9.10 Lower hemisphere focal plane mechanisms and P-wave radiation patterns associated with idealized faults. (a) A vertical strike-slip fault that strikes either N45° E or N45° W. (b) North–south striking thrust, normal or vertical faults. Note that the latter case could be related to slip on a horizontal fault. Modified from Stein & Wysession (2003).

seismic arrays is presented in Fig. 9.11 from a study in the Bakken formation by Yang & Zoback (2014). In this case the microseismic events were recorded on four downhole arrays in vertical wells, each consisting of 40, 3-component geophones. By standards at the time, this represents a remarkable level of microseismic monitoring. This said, in

Reservoir Seismology 275

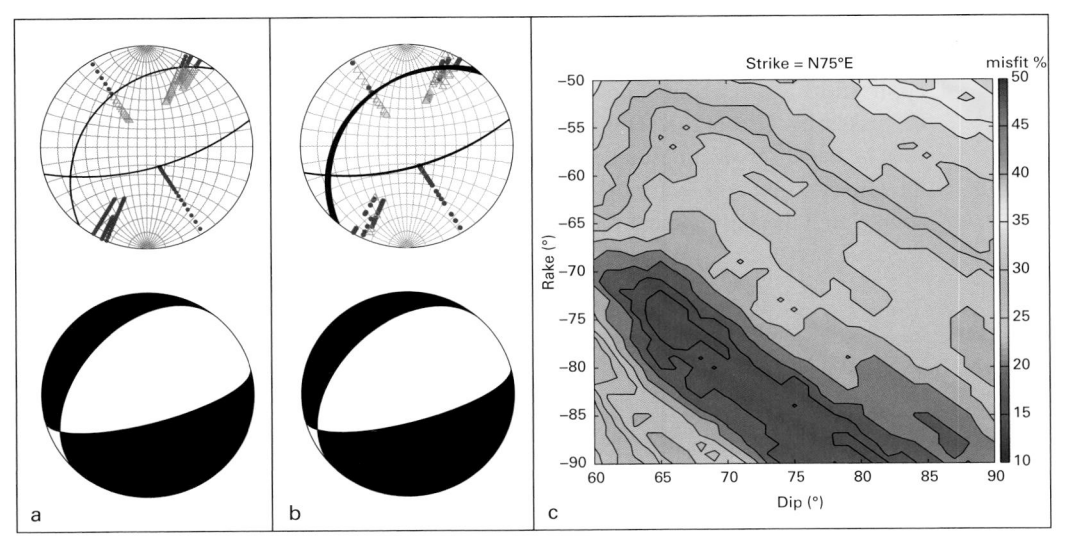

Figure 9.11 Theoretical (a) and observed (b) P-wave polarities for a microearthquake in the Bakken formation as recorded in four vertical monitoring wells, each with a 40-level, three-component seismic array. (c) Contour map of misfit between predicted and observed P-wave first motions for various fault dip and rake assuming the fault strike is known. From Yang & Zoback (2014). *Reprinted with permission from the American Association of Petroleum Geologists whose permission is required for further use.*

order to accurately constrain the focal plane mechanism of one of the larger events, Yang & Zoback (2014) created a synthetic focal plane mechanism based on knowledge of the fault orientation and relative stress magnitudes illustrating the expected P-wave polarities on the four arrays (Fig. 9.11a), which they compare with the observed polarities. Figure 9.11c shows the variation of dip and rake (rake is the slip direction in the plane of the fault) assuming the fault strike is well constrained by a lineation of seismic events along a well-defined fault crossing the wellbores. Thus, even in this case of having polarities available from four vertical seismic arrays, there is still a considerable degree of uncertainty in the focal plane mechanism.

Figure 7.8 illustrates how the faulting styles discussed in Fig. 7.1 are represented by different types of focal plane mechanisms. Because the P-axis of the focal mechanism bisects the extensional P-wave quadrant of a focal plane mechanism, the T-axis bisects the compressional P-wave quadrant and the B-axis is in the plane of the fault, orthogonal to the P- and T-axes. The P- and T-axes are sometimes used as proxies for principal stress directions. As pointed out in Chapter 7 these axes are similar to, but not the same as principal stress directions but once one has a group of well-constrained mechanisms it is often possible to invert them to obtain reliable stress orientations.

Focal Plane Mechanisms Using Full Waveforms – Because of the limited coverage of the focal sphere during microseismic monitoring, techniques such as full waveform inversion must be used to determine earthquake focal plane mechanisms. Fundamentally, full waveform inversion means modeling the entire waveform in order to obtain information about an earthquake focal plane mechanism and location. Waveform inversion is well developed in the field of earthquake seismology and its application for determination of focal plane mechanisms of microseismic events is discussed by Grechka & Heigl (2017). Following the nomenclature of Staněk et al. (2017), the observed amplitudes of direct P- and S-wave displacement d can be related to the Green's function matrix, $\mathbf{G}$, which is the seismic waveform resulting from a point source, and the full moment tensor, $\mathbf{M}$, are given by

$$\mathbf{d} = \mathbf{G} * \mathbf{M} \tag{9.5}$$

where an example of matrices $\mathbf{d}$ and $\mathbf{G}$ for one three-component seismometer for both P- and S-waves is given by

$$\mathbf{d} = \begin{bmatrix} A_{PN} \\ A_{PE} \\ A_{PZ} \\ A_{SN} \\ A_{SE} \\ A_{SZ} \end{bmatrix} \text{ and } \mathbf{G} = \begin{bmatrix} G_{PN}(1)G_{PN}(2)...G_{PN}(6) \\ G_{PE}(1)G_{PE}(2)...G_{PE}(6) \\ G_{PZ}(1)G_{PZ}(2)...G_{PZ}(6) \\ G_{SN}(1)G_{SN}(2)...G_{SN}(6) \\ G_{SE}(1)G_{SE}(2)...G_{SE}(6) \\ G_{SZ}(1)G_{SZ}(2)...G_{SZ}(6) \end{bmatrix}$$

where A_{PN} is the amplitude of the P-wave displacement measured on the north component and $G_{PN}(1)$ is the P-wave Green's function derivative for a far-field ray approximation of the north component amplitude due to the first component of the moment tensor. (1)–(6) denote the six components of the moment tensor, $\mathbf{M}$. The moment tensor inversion is given by

$$\mathbf{M} = (\mathbf{G}^T\mathbf{G})^{-1}\mathbf{G}^T. \tag{9.6}$$

The technique used to determine $\mathbf{M}$ is to find the best-fitting moment tensor that fits the observed seismograms via a least squares inversion (e.g., Eaton & Forouhideh, 2011; Song & Toksöz, 2011; Eyre & van der Baan, 2017). Earthquake data contains low frequency seismic energy which is a significant advantage over microseismic data. Vavryčuk (2007) and Vavrycuk et al. (2008) discuss some of the challenges of utilizing inversion for determining focal plane mechanisms of microseismic events from borehole monitoring. These challenges arise from a variety of sources; perhaps most important is the limited portion of the radiation sampled by downhole seismic arrays which is partly overcome by using full waveform inversion assuming good knowledge of the velocity structure and the ability to model at the frequencies of interest. As pointed out by Stein & Wysession (2003), the ability to invert a family of waveforms for the moment tensor depends on knowledge of the Green's function that relates the seismogram to the moment tensor.

To illustrate the use of waveform inversion for focal mechanism determination applied to the dataset shown in Fig. 9.1, we will utilize the results of a study carried out by Kuang et al. (2017) that was novel in two ways. Kuang et al. extended the method of Li et al. (2011) for matching synthetic P- and S-waveforms as recorded on 3C geophones to include SV/P and SH/P ratios. The method of Li et al. only considered the SV/P amplitude ratio of a single component of the seismogram. Thus, the method utilizes two seismically determined parameters as observed on each of the three components of the seismograms. After pre-computing numerous possible locations and focal plane mechanisms, they utilize an advanced search algorithm to identify the best-fitting solutions. They empirically found that data from at least two downhole arrays are needed for the technique to work well and a reasonably well-constrained velocity model is needed for the technique to work although uncertainties in the velocity and attenuation model (and event locations) are taken into account.

The method of Kuang et al. minimized the following objective function

$$obj = a_1 * f_1(pol(obs), pol(syn)) + a_2 * f_2(obs \otimes syn) + a_3 * f_3 \left(\left(rat \left(\frac{SV_{obs}}{P_{obs}} \right), \right. \right.$$
$$rat \left(\frac{SV_{syn}}{P_{syn}} \right) \right) + a_4 * f_4 \left(\left(rat \left(\frac{SH_{obs}}{P_{obs}} \right), rat \left(\frac{SH_{syn}}{P_{syn}} \right) \right) \right) \qquad (9.7)$$

where f_1, f_2, f_3 and f_4 are different general functions for each term; a_1, a_2, a_3 and a_4 are the weighting factors; obs is the observed data; syn is the synthetic data; pol is the polarity of the first motion; rat is the amplitude ratio between the SV, SH and P-waves.

Using data from the Barnett study shown in Fig. 9.1, Kuang et al. evaluated 23,328 synthetic focal plane mechanisms and compared the goodness of fit to the observed waveforms. The 1,000 best-fitting mechanisms are shown in Fig. 9.12a. Note that there is considerable variability among these focal mechanisms. By the point where the 10 best solutions have been determined (Fig. 9.12b) the goodness of fit improves very little and there is considerable consistency among the solutions. In other words, any of the first 10 best search results could be regarded as well-constrained solutions with uncertainties in strike, dip and rake within 10°, fundamentally limited by both the recording geometry and discretization and efficiency of the computations.

Figure 9.13 shows the excellent comparison between the observed (blue) and modeled (red) P- and S-waves. The P- and S-waves are windowed in the inversion using specified time windows based on the total duration of the P- and S-wave trains. For the P-waves, the transverse components are theoretically zero after the 3C waveforms were rotated into radial, transverse and vertical components. For the purposes of illustration, the waveforms have been aligned by removing time shifts. Location and velocity errors mainly affect the arrival times. Uncertainty in attenuation affects the amplitude of waveforms, as we can see from the S-wave comparison. However, the amplitude effect is relatively small so that a 20% attenuation error does not affect the SV/P and SH/P ratios significantly. Consequently, the inverted focal mechanism matches the reference focal mechanism well, even when attenuation is uncertain at frequencies of about 100 Hz.

Kuang et al. (2017) determined focal plane mechanisms for 123 of the larger microseismic events in this dataset that were used for the stress inversions shown in Fig. 7.11. By knowing the stress state, the nodal plane that corresponds to the fault was determined as the nodal plane with the highest ratio of shear to normal stress. In other words,

278   Unconventional Reservoir Geomechanics

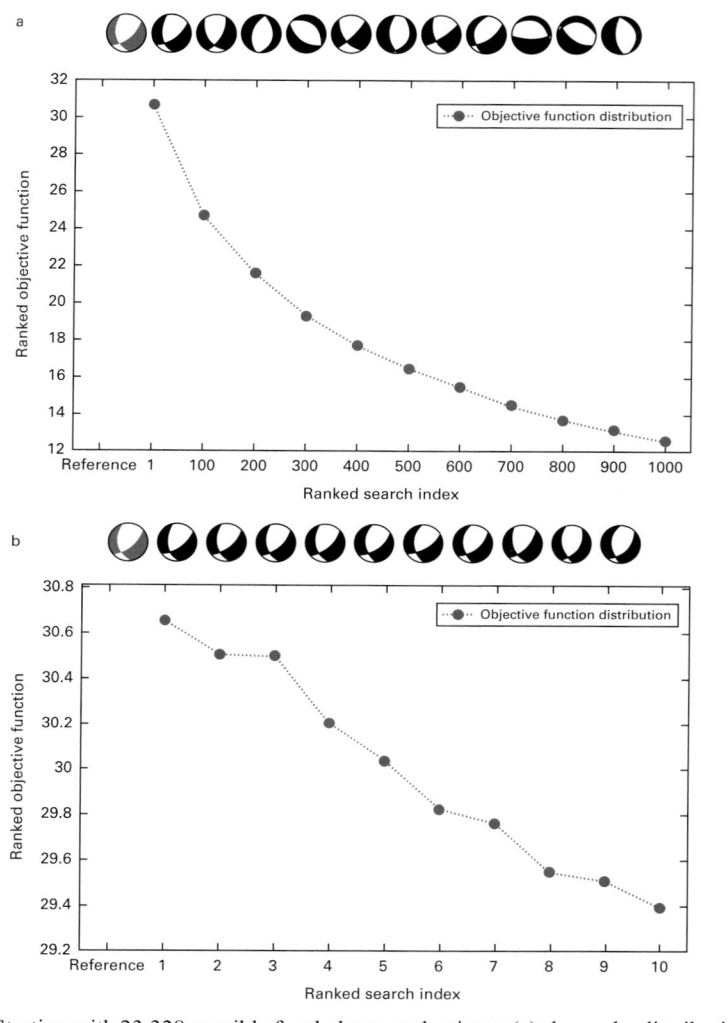

Figure 9.12 Starting with 23,328 possible focal plane mechanisms, (a) shows the distribution of the best 1,000 best solutions after ranking (along with their corresponding focal mechanisms for every 100th result). (b) The distribution of the objective function for the first 10 best solutions after ranking, along with their corresponding focal mechanisms. The reference beach ball (red) is the one that is used to generate the reference data in the synthetic test. From Kuang et al. (2017).

regardless of its orientation, this plane would be expected to slip at a lower pore pressure perturbation than the auxiliary plane (as discussed in Chapter 10). Utilizing this information is important when creating network models from microseismic data (Chapter 12).

Do Microearthquakes Represent Shear or Tensile Fractures? – To consider whether microseismic events recorded during hydraulic fracturing might represent opening of the hydraulic fracture, the radiation pattern for an east–west trending vertical tensile fracture

Reservoir Seismology   279

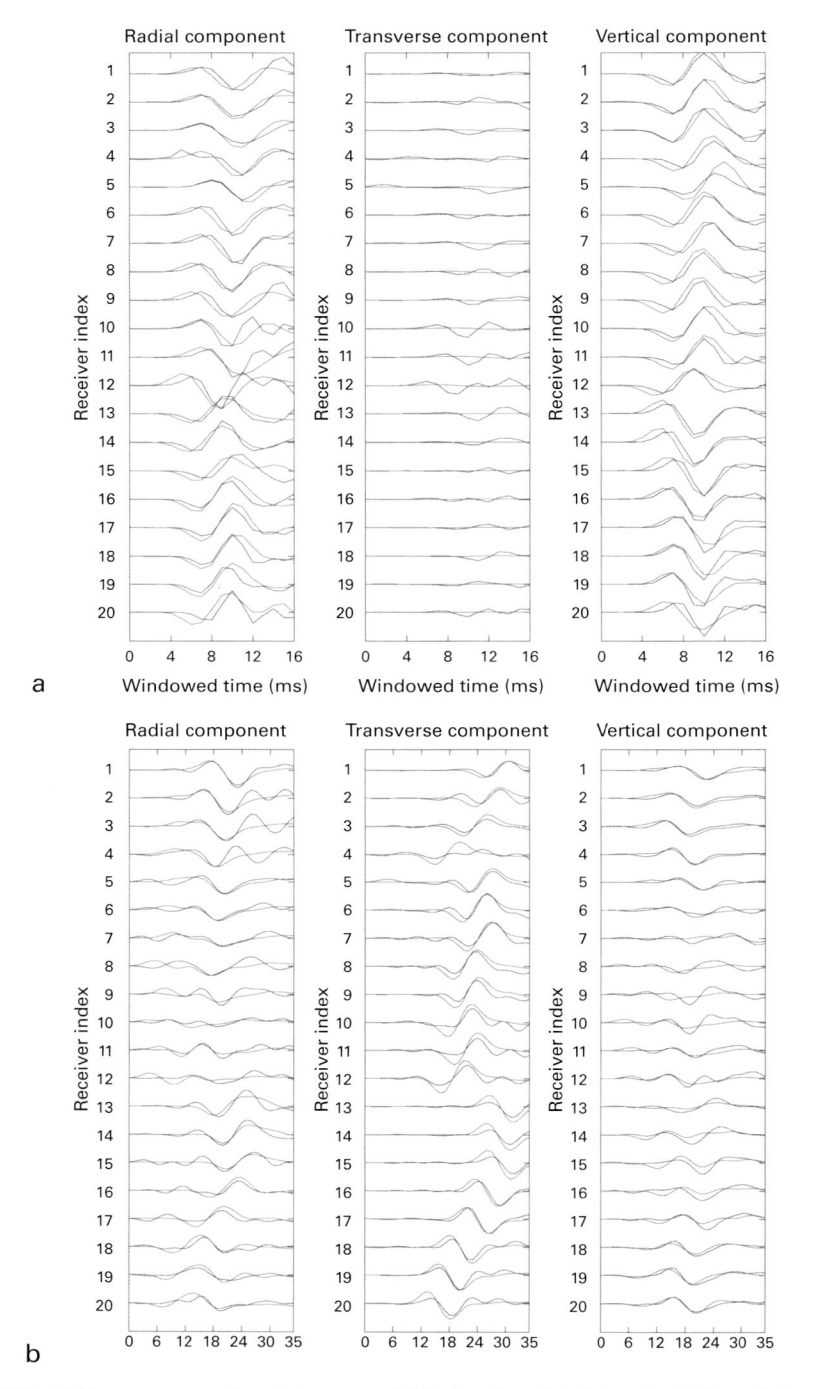

Figure 9.13 (a) P-wave comparison between real (blue) and modeled data (red). From left to the right are the radial components, transverse components and vertical components. (b) S-wave comparison between real (blue) and modeled data (red). From left to right are the radial, transverse and vertical components. From Kuang et al. (2017).

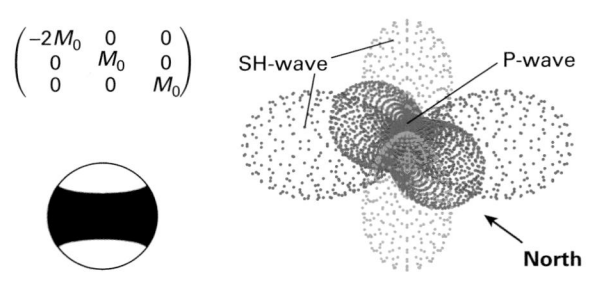

Figure 9.14 Comparison of the P-wave and SH-wave radiation patterns from an east–west trending vertical tensile fracture modeled as CLVD. The smaller, dilational lobes extending east–west are not shown for clarity. Figure courtesy Leo Eisner, Seismik.

is shown in Fig. 9.14 as a CLVD. Note that the shear waves have maximum amplitudes that are about twice that of the P waves. As mentioned above, all mechanisms can be decomposed into three components: an isotropic (or volumetric), shear and CLVD sources. In this case, the volumetric component is 55%, the CLVD component is 45% and the shear component is zero. An alternative to Fig.. 9.14 would be the representation of opening hydraulic fracture as a tensile dipole point source, which would produce a focal plane mechanism indicating compression everywhere. In practice, neither the CLVD nor point source radiation patterns are seen.

Stepping back, there are three general observations that argue for microseismic events recorded during hydraulic fracturing principally representing shear slip and not tensile opening. The first set of observations is that the time and location of the microseismic events do not correlate with the tip of a propagating hydraulic fractures (Shaffner et al., 2011; Warpinski et al., 2013; Rutledge et al., 2013). Warpinski et al. made the additional argument that as fluid in a propagating hydraulic fracture is not likely to reach the fracture tip, there is likely shear deformation occurring in the vicinity of the fracture tip not opening (as illustrated in Fig. 8.11a). Figure 9.15 is an example of time–distance microseismic event locations that demonstrate that the events do not monotonically move away from a wellbore as the hydraulic fractures propagate. This figure presents a time–distance plot for events associated with an early hydraulic fracturing experiment in the Cotton Valley formation (from Rutledge et al., 2004). Over time, the great majority of events occur in the volume of rock adjacent to the hydraulic fracture not at its propagating tip. For example, if the rapid propagation of microseismic events from the perforations during the first hour of pumping (at a rate of ~200 m/hour) indicates a few events from the near-tip region, events persist for several hours at distances much closer to the perforations. As illustrated in Fig. 8.11b and discussed at length in Chapter 10, this suggests the events are occurring in the rock volume surrounding the hydraulic fractures. A hydraulic fracture propagation velocity on the order of 200 m/hour is consistent with theoretical predictions of Detournay (2016) and fracture propagation simulations shown in Chapter 8.

The second line of evidence indicating that microseismic events are shear events is the radiation pattern of seismic energy as shown by variation of P- and S-wave amplitudes

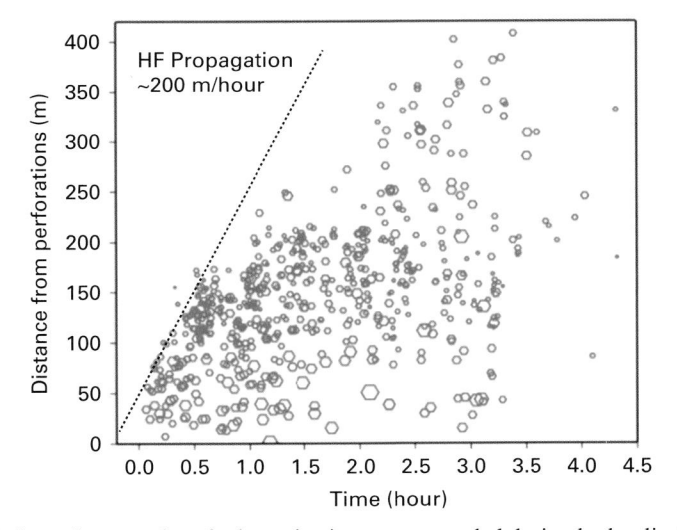

Figure 9.15 A time–distance plot of microseismic events recorded during hydraulic fracturing in the Cotton Valley formation. Symbol size is proportional to relative magnitude. Modified from Rutledge et al. (2004).

and ratios indicative of shear faulting, not tensile fracture opening. Not long after microseismic monitoring came into use for monitoring hydraulic fracturing Rutledge et al. (2004) presented a critically important analysis of waveforms of microseismic data from the Cotton Valley formation (Fig. 9.16). It showed that despite the tight, linear trend of microseismic hypocenters at the azimuth expected for hydraulic fractures, the focal plane mechanisms of the microseismic events based on P-wave polarities and P- to SH-wave ratios (Fig. 9.8) were consistent with left-lateral strike-slip earthquakes. They also showed the ratio of the magnitude of SH to P waves as a function of azimuth from the perforations from which the hydraulic fracture was propagating (Fig. 9.16c). They interpreted the shear events as resulting from small faults that offset the hydraulic fractures as they propagate (Fig. 9.16b). The radiation pattern shown in Fig. 9.14 for an opening mode fractures is not seen.

Third, modeling of the complete seismic waveforms (as illustrated above in Fig. 9.13) is consistent with double couple focal plane mechanisms. Moreover, inverting a family of focal mechanisms in a given region for the best-fitting stress tensor (as discussed in Chapter 7) reliably yields principal stress orientations and relative magnitudes (as illustrated in Fig. 7.11) consistent with independent data.

Since publication of the Rutledge et al. (2004) paper, there is a broad consensus among researchers that the microseismic events associated with hydraulic fracturing are principally due to shear on pre-existing fractures and faults (see also Warpinski et al., 2013; Maxwell, 2014). The possibility that there may be bedding plane slip associated with hydraulic fracturing is discussed in Chapter 10. While an opening (non-double couple) component of deformation associated with earthquakes induced during hydraulic fracturing has been proposed by various workers (e.g., Sileny et al., 2009), more

282   **Unconventional Reservoir Geomechanics**

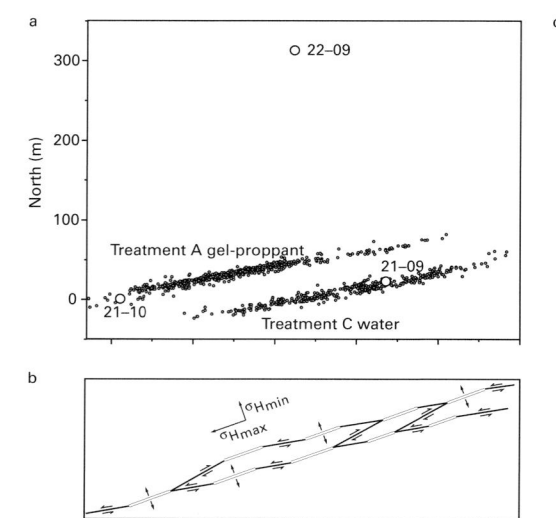

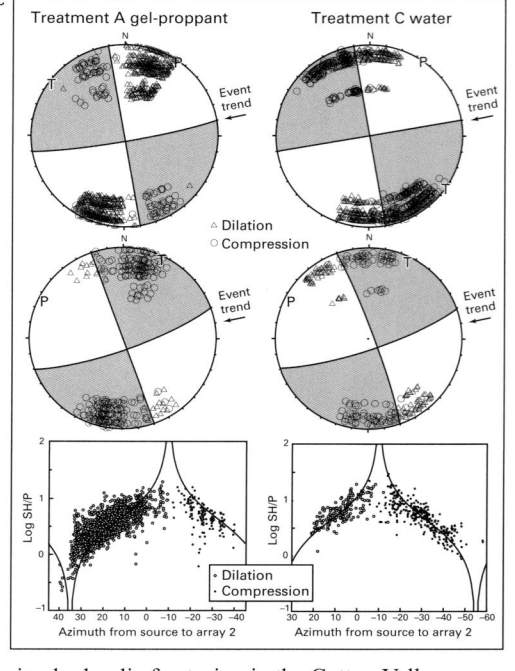

Figure 9.16 Microseismic events recorded during hydraulic fracturing in the Cotton Valley formation. (a) Microseismic epicenters appear to define a NNE-trending hydraulic fracture. (b) Interpretation of anastomosing shear and opening mode fractures. (c) Focal plane mechanisms and SH/P wave amplitudes as a function of angle indicate that the microseismic events correspond to shear slip on faults. From Rutledge et al. (2004).

recent work by that group contradicts their earlier conclusion. For example, Vavrycuk et al. (2008) point out that the non-double component of seismograms associated with small earthquakes generated during pressurization of the ultra-deep KTB borehole in Germany likely results from anisotropy of the rock rather than tensile opening. They base this conclusion on the fact that the negative CLVD component is inconsistent with tensile fracturing. Moreover, these events were triggered by fluid pressures far below the frac gradient (Zoback & Harjes, 1997) such that tensile opening would not be expected. It is important to note the similarities to microseismic events associated with hydraulic fracturing in unconventional reservoirs. The medium in which the events are occurring is quite anisotropic and the events are being triggered at pressures well below the frac gradient (Chapter 10).

Earthquake Source Parameters and Scaling Relationships

The relationships among different seismically determined parameters related to earthquake size have been well established for a number of years. Stein & Wysession (2003) and Eaton (2018) review these relations at some length.

Reservoir Seismology

Energy Release and Magnitude – The magnitude of an earthquake represents the amount of energy released by slip on a fault. As work is equal to force times distance, the strain energy released in an earthquake, W, represents the average stress resisting slip, $\bar{\tau}$, times the average displacement times the fault area as defined above:

$$W = \bar{\tau}\bar{D}S. \tag{9.8}$$

Earthquakes potentially large enough to cause damage or injury are of most concern and will be discussed in Chapters 14 and 15 which address induced seismicity associated with unconventional oil and gas production. In terms of the microearthquakes that occur during hydraulic stimulation, a typical magnitude is about −2. A magnitude +2 earthquake (the smallest earthquake that might be felt at the surface) releases about 63 million joules (equivalent to an explosion of about 15 kg of TNT), whereas a magnitude −2 earthquake releases about 63 joules, equivalent to the energy associated with a gallon of milk hitting the floor after falling off a kitchen counter. Each magnitude unit represents a difference in energy release of a factor of 33. Hence, two magnitude units represent a factor of about 1,000 in strain energy release. Therefore, four magnitude units (the difference between a magnitude +2 and magnitude −2 earthquake) is a factor of one million in energy release.

The magnitude scale developed by Charles Richter was based on the amplitude of ground motion as measured on a certain type of seismometer at a certain distance from the earthquake. Utilization of similar methods are usually referred to as local magnitudes, M_L, because they are calibrated for local geologic conditions that affect the attenuation of seismic waves with distance.

It's widely recognized that the best method for calculating magnitude as a measure of strain energy release is based on the scalar seismic moment (note the similarity of Eqns. 9.4 and 9.8). The scalar moment is often determined from the low frequency level of the displacement spectra of a shear wave in a seismogram, Ω_0, as illustrated in Fig. 9.17 (from Warpinski et al., 2013), utilizing the following relation:

$$M_0 = \frac{4\pi\rho c^3 R\Omega_0}{f_c} \tag{9.9}$$

where R is the distance from the microearthquake to the seismogram and f_c is the corner frequency as illustrated in Fig. 9.17. Eisner et al. (2013) discuss source spectra for microseismic events and methods to determine corner frequencies and attenuation.

Once the moment is determined from the spectra, the moment magnitude is given by:

$$M_W = \frac{2}{3}(log_{10}M_0 - 9) \tag{9.10}$$

where M_0 has the units N m. As the magnitudes that are reported are often local magnitudes, M_L, it is useful to express seismic moment in terms of M_L as

$$log_{10}M_0 = 0.77M_L - 10.27. \tag{9.11}$$

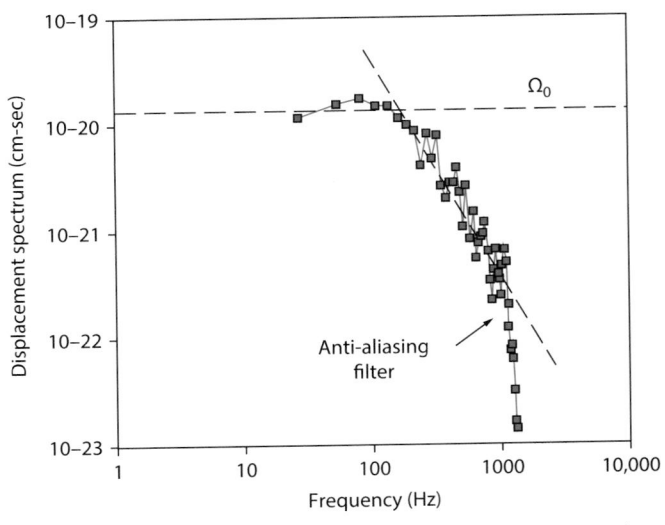

Figure 9.17 An example of the displacement spectra of a seismogram from a microearthquake in the Barnett shale. From Warpinski et al. (2013). The scalar moment, M_0, is the low frequency level of displacement spectrum. The corner frequency, f_c, is related to the source dimension of the earthquake, as discussed below.

For microseismic data recorded during hydraulic fracturing, it is common for magnitudes reported by different investigators to differ whether they are reporting local magnitude or moment magnitude. Shemeta & Anderson (2010) discuss the variance of moment magnitudes determined by different service providers, most likely due to the attenuation relations that are used.

Stress Drop – For a circular fault of radius, r, the relationship between stress drop, the average change of shear stress on the fault after slip, moment, fault radius, average slip and shear modulus, G, is shown by the relations in Eqn. (9.12).

$$\Delta\tau = \frac{7}{16}\frac{M_0}{r^3} = \frac{7}{16}\frac{G\pi r^2 \overline{D}}{r^3} = \frac{7}{16}\frac{G\pi \overline{D}}{r}.$$

(9.12)

The relationship between the corner frequency as defined in Fig. 9.17 the source dimension of the circular fault patch that slips and shear velocity, r and β is given by:

$$f_c = \frac{2.34\beta}{2\pi r}$$

(9.13)

such that the relationship between the stress drop, moment and source dimension is given by

$$\Delta\tau = 8.47 M_0 \left(\frac{f_c}{\beta}\right)^3.$$

(9.14)

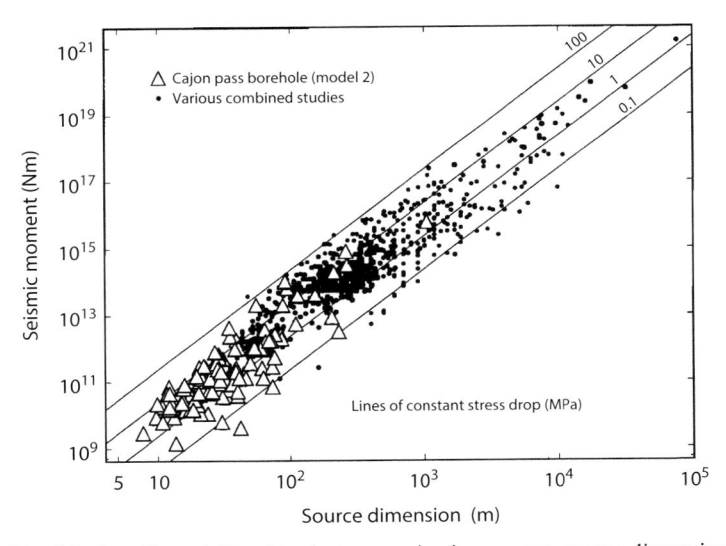

Figure 9.18 Empirical scaling relationships between seismic moment, source dimension and stress drop (in MPa) as determined from downhole seismic recordings in accordance with Eqns. (9.9)–(9.12). From Abercrombie (1995).

Because of the dependence on f_c^3 in this relationship (determined from attenuation-corrected spectra, as illustrated in Fig. 9.17), there is a fair degree of uncertainty in seismically determined stress drops (which generally vary between 0.1 and 10 MPa). Figure 9.18 (from Abercrombie, 1995) shows the relationships illustrated in Eqn. (9.12). In general, while stress drops are poorly constrained, it is remarkable that earthquake scaling is invariant over a wide range of sizes. Figure 9.18 shows local magnitudes that range from −1 to 5.5. Note that this figure was compiled from data recorded at depth, which removes sources of noise and near surface attenuation.

Figure 9.19 (from Cocco et al., 2016) shows an even more remarkable range of moment magnitudes that range from some of the largest earthquakes that have ever occurred on Earth ($M_W \sim 9$) to *lab quakes* ($M_W -7$). Hence, over 16 orders of magnitude, a factor of about 2×10^{24} (33^{16}) in energy release, stress drops do not change systematically with earthquake size. Thus, empirical earthquake scaling relationships defined for relatively large, naturally occurring earthquakes are equally relevant to much smaller microseismic events. This suggests that we can use the earthquake scaling relations defined above whether we are addressing questions of the very small microearthquakes associated with hydraulic fracturing stimulation as in Chapters 10–12 or induced seismicity in Chapters 13 and 14.

Relating Magnitude to Fault Size and Slip – Based on widely used earthquake scaling relationships, Fig. 9.20 (after Walters et al., 2015) summarizes relationships among earthquake size (expressed as either magnitude or moment), the size of the fault that slips and the amount of slip based on stress drop. Despite the scatter in stress drop observations, a magnitude −2 microseismic event is expected to be associated with a source dimension of about 1 m and slip of less than 0.1 mm. Hence, slip on fractures and faults

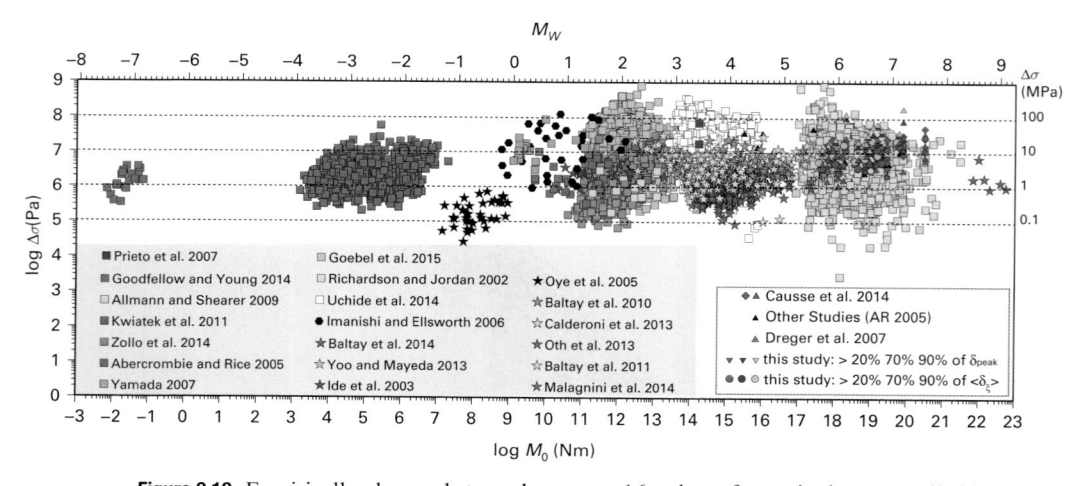

Figure 9.19 Empirically observed stress drops over 16 orders of magnitude as compiled by Cocco et al. (2016). The different colors and symbols correspond to the different tectonic settings and methodologies as indicated in the legend.

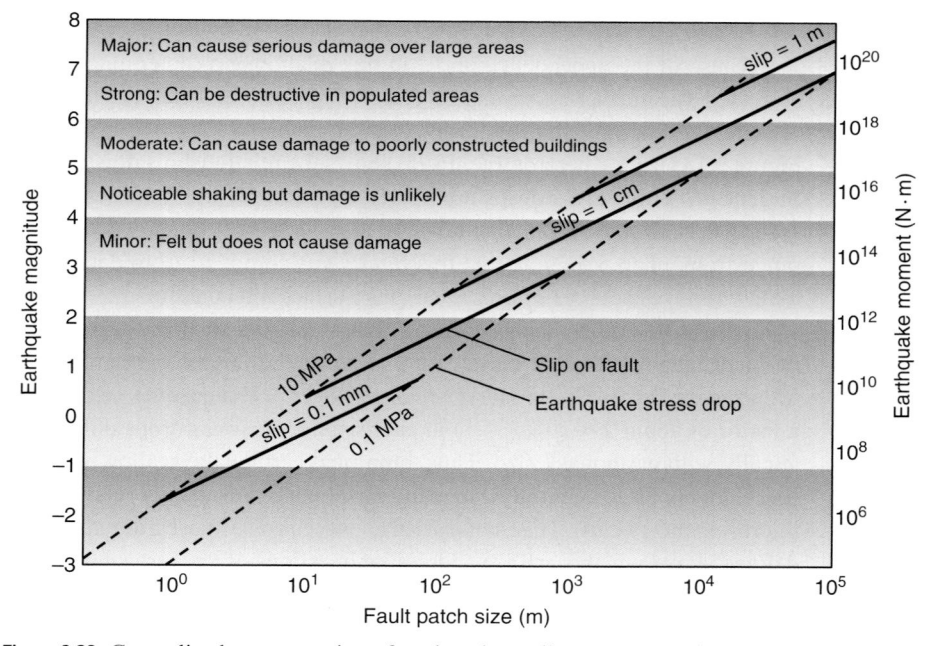

Figure 9.20 Generalized representation of earthquake scaling parameters in Eqn. (9.9). After Walters et al. (2015).

on the scale of the thickness of individual beds (as illustrated in Chapter 7) are sufficient to account for many of the observed microseismic events that occur during hydraulic fracturing stimulation. At the other end of the magnitude scale of interest in this book in the context of induced seismicity, potentially damaging M 5 to 6 earthquakes require fault slip (on a portion of larger faults) of the order of 10 km in extent, meaning that the slip is occurring on faults that extend down into basement. As will be discussed at length in Chapters 14 and 15 the hazard associated with induced seismicity often depends on whether injection might stimulate slip on faults extending into crystalline basement.

Earthquake Statistics

The Gutenberg–Richter Relation – One of the best-known relations in seismology is the Gutenberg–Richter (G–R) frequency–magnitude relation that illustrates that the logarithm of the number of earthquakes (N) of magnitude M and larger in a given area over a given period of time follows the empirical rule:

$$\log N = a - bM \tag{9.15}$$

Although originally developed for earthquakes occurring in California, the relationship has been found to be generally valid regardless of the region, faulting style or level of seismicity. Moreover, the parameter "b" has been universally found (for natural earthquakes) to be close to 1. In other words, there are 10 times as many M 5 earthquakes as M 6, 10 times as many M 4 as M 5, and so on. Hence, there are 100 times as many M 4 earthquakes as M 6 earthquakes. In terms of energy, however, a M 4 earthquake releases one thousand times less energy than a M 6. Thus, 100 M 4 earthquakes release only 10% of the energy associated with a M 6 earthquake.

There are two general points about the G–R relation that should be noted. In areas with few earthquakes it is obviously necessary to make sure the earthquake catalog is *complete* down to a known magnitude level, a topic revisited below in the context of microseismicity that accompanies hydraulic fracturing. The parameter "a" in the G–R relation literally corresponds to the number of M 0 earthquakes. More generally, it corresponds to the level of seismic activity in a given area. For example, while the rates of M 4, 5 and 6 earthquakes in seismically active southern California and relatively stable central and eastern United States is markedly different as represented by the "a" value, the relative numbers of M 4, 5 and 6 events are the same (as indicated above) as the "b" value for both areas are the same ($b \approx 1$).

Catalog Completion and Detection Capability – An incremental G–R plot for the microearthquakes associated with the dataset illustrated in Fig. 9.1 is shown in Fig. 9.21 (from Hakso & Zoback, 2019). As shown, there is a linear relationship between earthquake number and magnitude above the magnitude after catalog completeness, in this case ~M −1.8 as shown in blue. If one were to extrapolate down to M −2.25, about 2000 more seismic events are indicated (see Chapter 10). Unlike natural earthquakes, this dataset indicates a very high b value of 2.9, meaning there are statistically many more small earthquakes compared to larger ones. High b values (often ≥ 2) are a common observation in microseismic data (see Hurd & Zoback, 2012b; Maxwell, 2014; Eaton, 2018). One possible explanation for this is that shear stimulation in

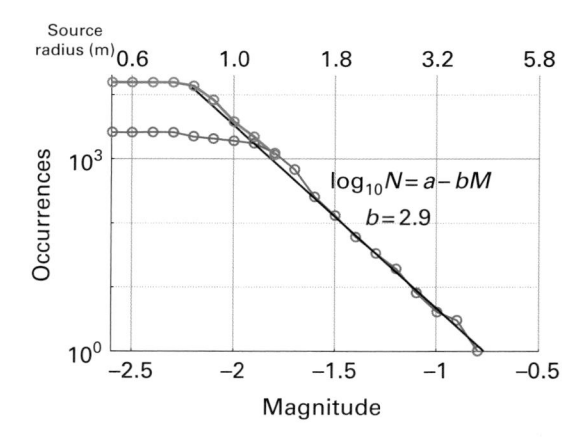

Figure 9.21 Incremental Gutenberg–Richter plot associated with the microseismic dataset illustrated in Fig. 9.1. After Hakso & Zoback (2019). Note the very high *b* value. The catalog of microearthquakes is complete down to about M −1.8 (blue). Extrapolating down to M −2.25 would involve inclusion of about 2,000 more events.

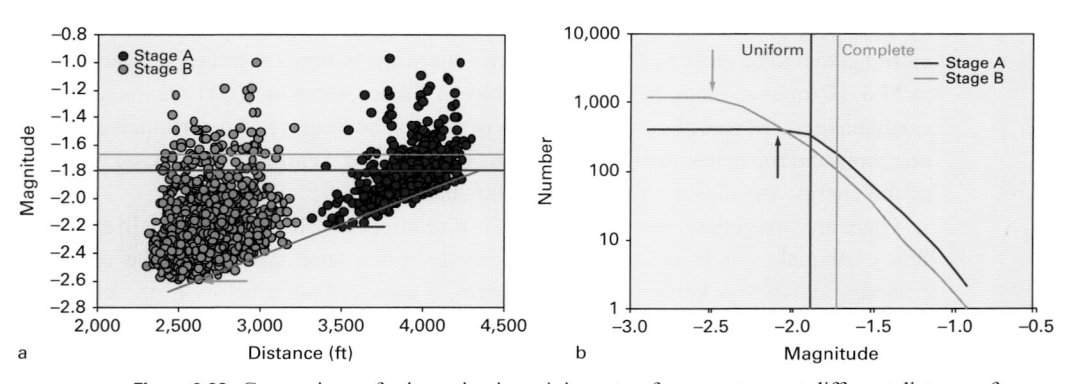

Figure 9.22 Comparison of microseismic activity rates, for two stages at different distances from a monitoring array (a) and the corresponding cumulative frequency–magnitude plot (b). Stage B has more overall events because it is closer to the array and thus has a lower detection limit. The event count is compared at levels for either uniform detection (blue lower line, consistent with detecting some events at all distances) or complete (green top line: all events detected). After Maxwell (2012).

individual hydraulic fracturing stages affects a relatively small volume of rock over a relatively short period of time and thus does not sample the overall fractal distribution of fracture sizes in the crust as whole. In other words, shear stimulation is limited to relatively small fractures, as illustrated in Fig. 7.20. In fact, Hallo et al. (2014) argue that $b \geq 1.5$ is not physically realistic such that the observations of high *b* values associated with microseismic events are an artifact of the small range of magnitudes considered.

The most common method used for detecting catalog completion associated with microseismic monitoring is to plot magnitude as a function of distance as shown in Fig. 9.22a (after Maxwell, 2012). Such plots give one a qualitative sense of the ability to detect events of a certain magnitude at a certain distance from the monitoring array for various stages. For example, M −2 events can be detected at a distance of 4,000 ft, while at a distance of 2,500 ft, events as small as M −2.5 could be detected. However, such plots do not reflect true catalog completeness, and so do not allow the seismicity recorded with various stages to be quantitatively compared. This was pointed out by Vermylen & Zoback (2011) who showed that the variable distances of successive stages from the monitoring array in the Barnett shale affected the total number of events detected. This is shown in Fig. 9.22b for the data shown in Fig. 9.22a. Because Stage A is further from the monitoring array, there were fewer total events detected. However, more microseismic events occurred at all magnitudes above the completeness level in Stage A. The importance of this will be clear in Chapter 10 where we discuss creation of fracture networks utilizing microseismic data. It should be noted the *b* value associated with these data is slightly greater than 2.

Locating Microearthquakes

The topic of earthquake locations has been discussed extensively by a number of authors. In this section, we briefly discuss the techniques commonly utilized to locate microearthquakes associated with hydraulic fracturing stimulation. We start by discussing the two techniques that are used commonly with downhole and surface seismic recordings, respectively. Both techniques are used to determine the location of seismic events in space and time. Finally, we discuss relative location methods utilizing the double-difference method of Waldhauser & Ellsworth (2000) which can be used to achieve highly accurate microseismic events with respect to each other.

Travel Time Methods – The traditional way in which earthquakes are located first utilize the differential travel time of P-waves and S-waves to determine the distance of the earthquake from a given seismometer. When the earthquake is recorded at multiple sites, the location is obtained by fitting the data from all of the seismometers using *Geiger's* method. As defined in Chapter 2, P-waves and S-waves travel with different velocities. Hence, in an isotropic, homogeneous medium, the arrival times of P-waves and S-waves at a seismometer are given by

$$t_p = t_0 + \frac{x}{V_p}$$
$$t_s = t_0 + \frac{x}{V_s}$$

(9.16)

where x is the distance to the seismometer and t_0 is the origin time of the earthquake. As t_p and t_s are determined from the seismogram but the origin time is unknown, these equations lead to the distance being given by

$$x = (t_s - t_p)\frac{V_p V_s}{(V_p - V_s)}.$$

(9.17)

Figure 9.23 illustrates (for an isotropic and homogeneous medium) how successively more seismometers allow for the determination of a precise location of a microseismic event using Eqn. (9.17). A single seismometer is capable of determining the distance to an earthquake. Utilization of the fact that motion of a P-wave is polarized in the direction of the ray path (Fig. 9.24), one could use the polarization direction to determine the direction of ray propagation. Thus, in theory, one could use a single seismometer to determine the location of an event assuming t_s and t_p have been determined accurately and the velocity model is known. Eisner et al. (2009) present a method for determining S-wave velocity from borehole arrays.

In practice, earthquakes are located using travel times recorded at multiple stations. Obviously, when only a single, or a few, seismic stations are used, uncertainties in

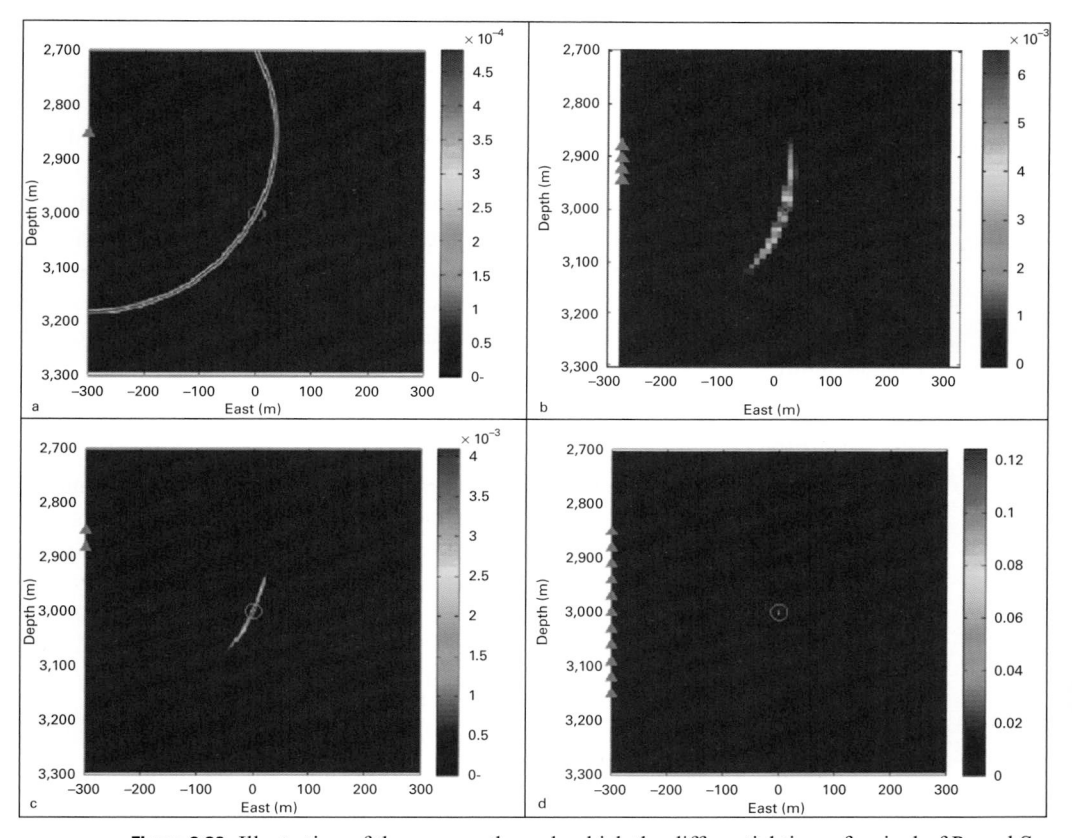

Figure 9.23 Illustration of the manner through which the differential time of arrival of P- and S-waves allows the distance between a seismometer to be determined by an array of seismometers. A geometry corresponding to downhole monitoring in an isotropic and homogeneous medium is used for illustration. In the upper left, the event is recorded at a single downhole seismometer. The location of the event could occur anywhere along the hyperbola shown. In the other images, 2, 4 and 11 seismometers are used, which help determine a precise location. Figure courtesy Leo Eisner, Seismik.

Reservoir Seismology   291

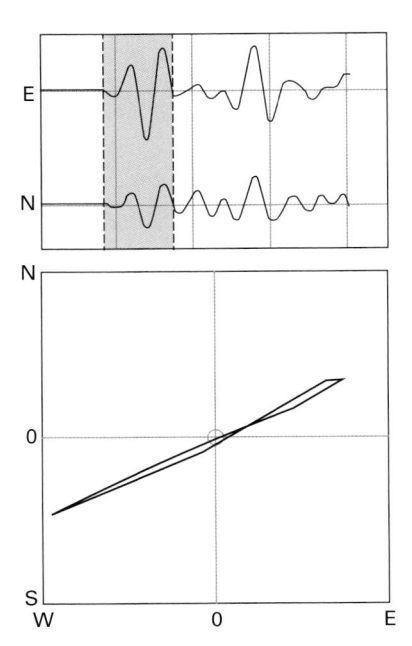

Figure 9.24 As P-waves are polarized in the direction of ray propagation, determination of the polarization direction as shown allows one to determine the location the direction of ray propagation. Figure courtesy of J. Shemeta, MEQ Geo.

picking P- and S-wave arrival times, polarization directions, the velocity model, etc. turn into significant uncertainties in location accuracy. As mentioned above, Geiger's method is the most common method used to locate earthquakes using a network of seismic stations. It involves solving for the earthquake location (x,y,z) and origin time by minimizing the residual between observed and theoretical arrival times at all of the stations where an event is detected and travel times can be determined via the formula

$$RMS_{residual} = \sqrt{\frac{\sum_{i=1}^{n} (T_{Pobs}^i - T_{Pcalc}^i)^2 + (T_{Sobs}^i - T_{Scalc}^i)^2}{n_{picks}}} \qquad (9.18)$$

where n is the number of stations and the terms correspond to observed and calculated P- and S-wave arrival times. Equation (9.18) is solved iteratively until this residual reaches a suitably small value.

While Geiger's method has been used successfully for over 100 years, there are a number of challenges – accurately picking P- and S-wave arrival times can be imprecise, especially when signal-to-noise ratios are low which is often the case with microseismic monitoring. Techniques using cross-correlation are helpful to improve travel time picking accuracy in reducing this uncertainty (Rutledge & Phillips, 2003). Utilizing correct velocity models is quite important as is network design. A general rule of thumb

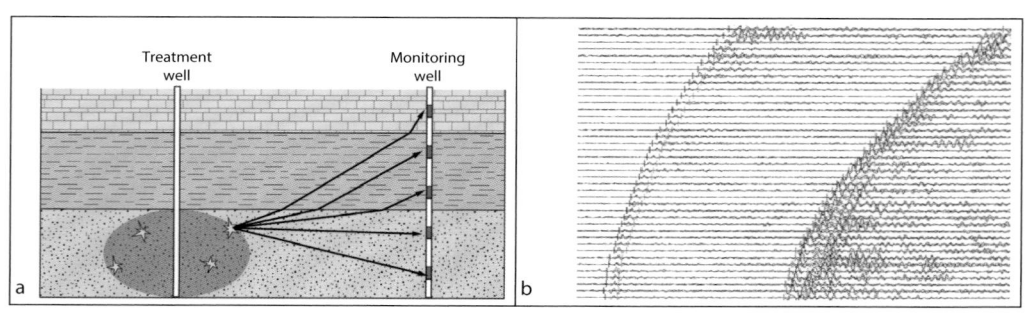

Figure 9.25 Schematic illustration of monitoring microseismic events with a single downhole monitoring array. (a) Note the complicated ray paths that include direct arrivals, refracted arrivals and different amounts of horizontal and vertical wavy propagation. (b) Moveout along a vertical geophone array illustrating P-wave and S-wave arrivals with the three seismograms superimposed on one another. Figure courtesy of J. Shemeta, MEQGeo.

in earthquake seismology is that events that occur outside the spatial extent of a network will be prone to relatively large errors in location, yet this is almost always the case with downhole monitoring of microseismic data.

Figure 9.25a schematically illustrates the basic geometry of downhole monitoring and Fig. 9.25b shows data from a 40-level, three-component geophone array with the three seismograms superimposed on one another. Note the increasing time of arrival of the P-waves (and the increasing $(t_s - t_p)$) as the distance between the microseismic event and receivers increases. Despite the apparent simplicity of this case, note the complicated ray paths caused only by the layering of formations with different physical properties and the position of the monitoring array with respect to the microseismic events. These include direct arrivals, refracted arrivals and rays that not only sample different formations, but formations with different amounts of anisotropy with different amounts of horizontal and vertical wave propagation. On top of this, uncertainties in the velocity model affects location accuracy as well as uncertainties in picking precise arrival times. In general, the arrival of time of a P- or S-wave is given by the equation

$$t_{arrival} = t_{origin} + \int_{path} \frac{ds}{V} \tag{9.19}$$

indicating that the travel time depends on the cumulative path of the seismic wave and the velocity structure along the path. This emphasizes the influence of complex ray paths and velocity structures on achieving accurate event locations. In practice, it is common to use grid search techniques to avoid local residual minima.

Migration Methods – A number of technologies in reflection seismology have been used to correct for formation dip and migrate reflected seismic energy to their true spatial position. The application of such techniques to locate microseismic events to their true spatial position and origin time is illustrated in Fig. 9.26 (from Bardainne, 2011). The three components of the seismogram are rotated into a coordinate system parallel to the

Reservoir Seismology

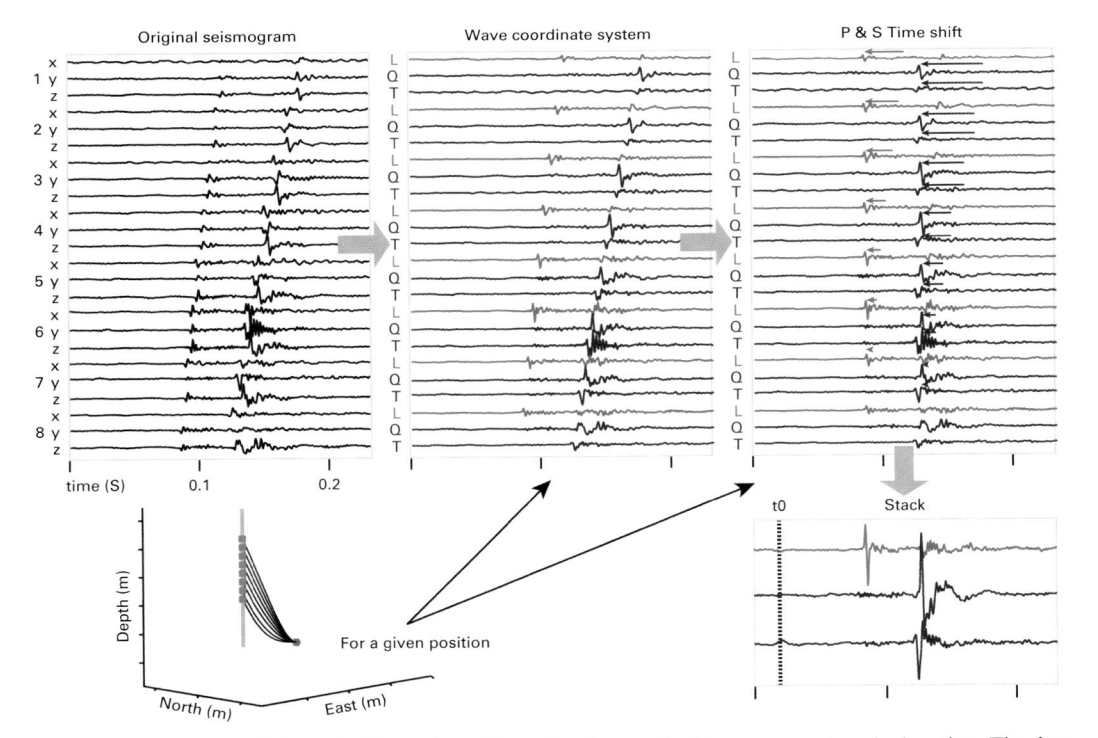

Figure 9.26 Schematic illustration of the migration method for microearthquake location. The data processing procedure is discussed in the text. A 4D grid search (x, y, z and origin time) is used to find the optimal location. From Bardainne (2011). Figure courtesy of Magnitude, a BHGE company).

ray path as defined by P-wave polarization (L), and transverse components in a vertical (Q) and horizontal (T) orientation. An automated 4D grid search was used to examine different P- and S-wave time shifts corresponding to different earthquake locations (x, y, z) and origin times to determine the values that produce maximum stacked energy. In that particular study they also discussed a method to augment automated phase picking algorithms to produce event locations with greater precision.

As illustrated in Fig. 9.26 for downhole seismic data, migration techniques are also widely used for locating microseismic events recorded with surface seismic data where it is commonly difficult to identify clear P- and S-wave arrivals. A comparison of locations determined for one of the stages of the microseismic dataset shown in Fig. 9.1 showed location differences that typically ranged from 100 to 300 ft. Migration techniques also depend on the velocity model, so it needs to be remembered that the accuracy of absolute event locations depends on the accuracy of the velocity model used and significantly impacted if the correct phase is not identified.

Eisner et al. (2010) presented a detailed comparison of microseismic events located with borehole arrays and surface arrays. For the case they investigated, they found the final location uncertainty was related to estimation of the back azimuth for a vertical

294   **Unconventional Reservoir Geomechanics**

array of receivers in a single monitoring well. They also found the uncertainty to increase with distance between the events and receivers. The locations of the events determined with the downhole array showed a systematic shift due to uncertainties in the velocity model. For surface monitoring, which utilized migration methods utilizing only P-wave arrival times, lateral positions were estimated robustly and were less sensitive to the velocity model than the borehole recordings and generally had less scatter in both vertical and horizontal directions.

Using Multiplets to Determine Relative Event Locations – As a way of independently determining the relative microseismic location uncertainty (Hakso & Zoback, 2017) utilized multiplets, which are microseismic events that occur in essentially the same place with the same source characteristics. Figure 9.27a shows five located multiplet

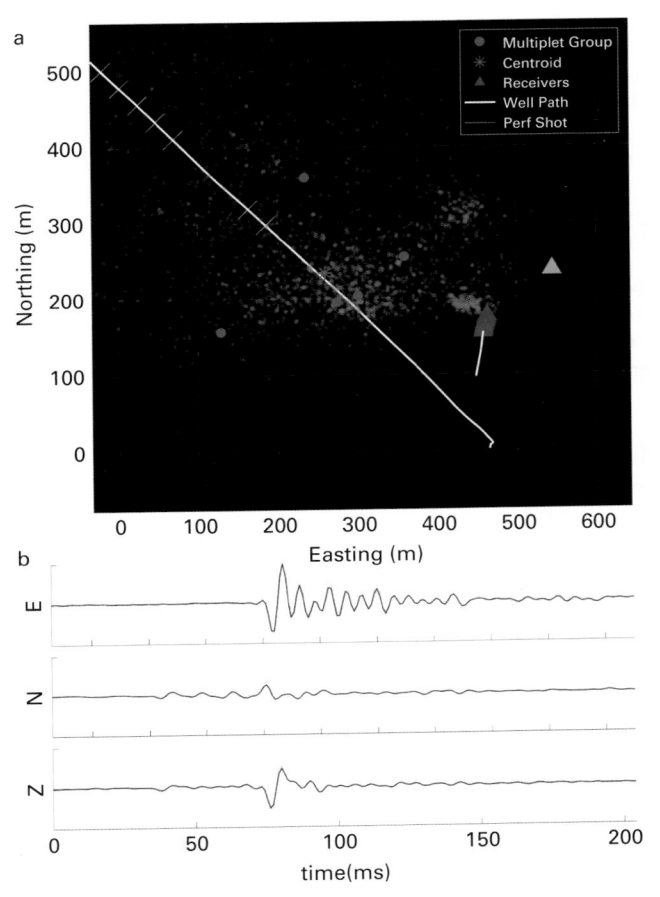

Figure 9.27 (a) Illustration of five located multiplet events that occurred during the stimulation of stage 5 are plotted, superposed on all contractor locations of microseismic events in the stage (light gray). (b) Their superimposed waveforms, displaying similarity on all components from shear-wave onset into the coda. From Hakso & Zoback (2017).

events (shown in green) that occurred during the stimulation of stage 5 of the dataset illustrated in Fig. 9.1. The superimposed waveforms show the remarkable similarity of the three components of the seismograms – from arrival of the shear-waves through the coda, as well as the seismic arrivals that follow that result from arrival of scattered energy.

Hakso & Zoback (2017) identified hundreds of multiplet groups in their study. Locations of different groups of multiplets that occurred during stage 4 are shown in Fig. 9.28. The scatter in contractor-provided event locations from the multiplet centroid is about 60 m for the most well-located multiplet events with source receiver distances of 150–200 m (twice the stated absolute location uncertainty of about 30 m reported by the microseismic contractor). The scatter in event locations increases to 120 m when the source receiver distance is more than 250 m.

To determine how close multiplets needed to be in space to produce such similar waveforms, Hakso & Zoback (2017) used synthetic seismograms to calculate the

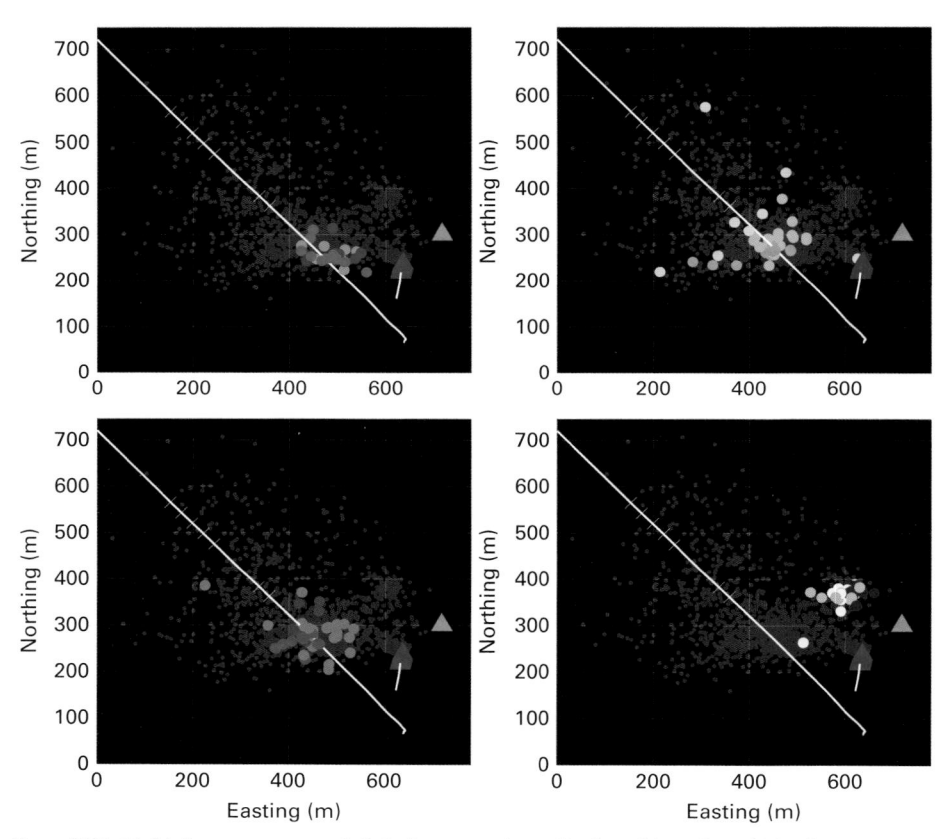

Figure 9.28 Multiplet groups recorded during stage 4 are displayed in order of 30 minute increments from the beginning of pumping (clockwise from upper left). Each color represents a multiplet group. The groups overlay all contractor locations of microseismic events in the stage (light gray). The intragroup scatter in reported locations ranges from ~75 to 200 m. From Hakso & Zoback (2017).

296   Unconventional Reservoir Geomechanics

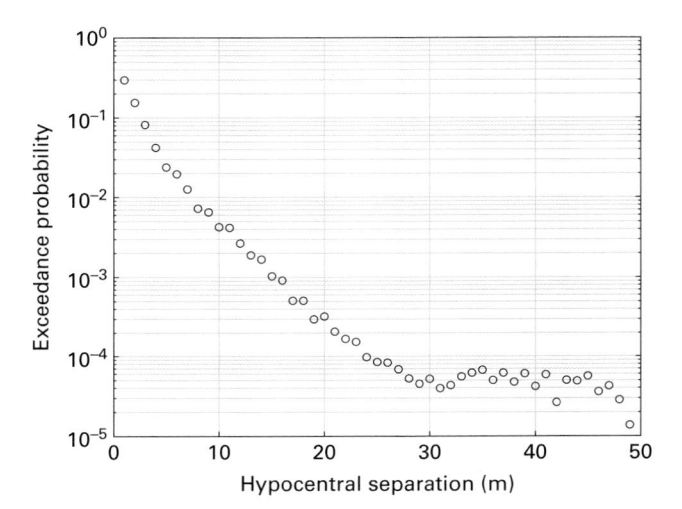

Figure 9.29 For a similarity threshold of 0.90 there is a rapidly decreasing exceedance probability with hypocentral separation. Hence, the probability is less than 0.1% that the events are more than 15 m apart. From Hakso & Zoback (2017).

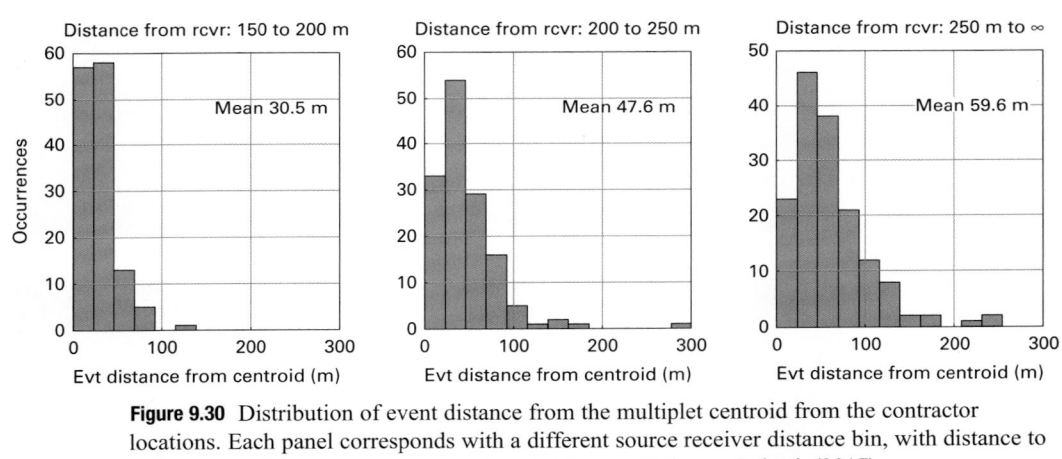

Figure 9.30 Distribution of event distance from the multiplet centroid from the contractor locations. Each panel corresponds with a different source receiver distance bin, with distance to reference receiver increasing from left to right. From Hakso & Zoback (2017).

probability for several similarity thresholds across a range of hypocentral separation values. Similarity is given by the signal-to-noise weighted cross-correlation coefficient across all components. Figure 9.29 shows the relationship between exceedance probability and hypocentral separation for a similarity threshold of 0.9. This suggests that the probability of categorizing two events with highly similar seismograms separated by a significant distance is very low. In other words, they demonstrated that the events must be within 15 m of each other to achieve a degree of waveform similarity of 0.9, or higher.

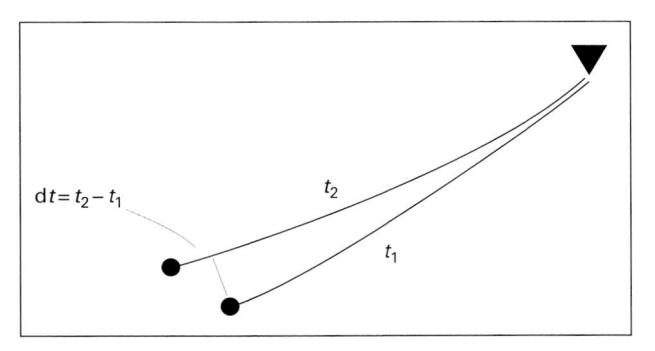

Figure 9.31 Illustration that the double-difference algorithm of Waldhauser & Ellsworth (2000) is based on analysis of the difference in travel times between two earthquakes in close proximity to each other (dots). The great majority of the travel path from the earthquakes to the seismic station (inverted triangle) is the same, canceling out the effects of uncertainties in velocity along the travel path.

Knowing that microseismic events within a multiplet group must be located within 15 m of each other, the scatter of the reported locations is a measure of relative location uncertainty. As shown in Fig. 9.30, multiplet groups farther from the receiver arrays show more scatter, indicating greater location uncertainty. The scatter for events relatively close to the receiver arrays have a mean scatter of about 30 m, the uncertainty reported by the microseismic contractor determined by perforation shot relocations. However, the uncertainty in event locations more than 250 m from the receivers is twice that. Looking across all multiplet groups, the median distance between each event and the centroid of its multiplet cluster is approximately 45 m.

Double-Difference Relative Locations – The double-difference (DD) earthquake location technique of Waldhauser & Ellsworth (2000) is designed to obtain accurate relative locations between events in general proximity to each other. This technique is the basis for the methodologies used to determine the extremely accurate relative locations shown in Fig. 4.4. As illustrated in Fig. 9.31, the basis of the technique is to focus on the difference in travel time between events in proximity to each other. Because the great majority of the ray paths from the earthquakes to the seismic station are essentially the same, this circumvents the problem in absolute earthquake location caused by incomplete knowledge of seismic velocity along the ray paths between earthquakes and the seismometer.

Figure 9.32 (from Waldhauser & Ellsworth, 2000) illustrates the technique using earthquakes located along the Calaveras fault by the Northern California Seismic Network. They illustrated the technique using catalog picks of P- and S-wave arrival times as well as more accurately determined travel times obtained through cross-correlation. The results are shown in map view (Fig. 9.32a) and cross-section (Fig. 9.32b) looking to the northeast at the plane of the vertical Calaveras fault. As shown, location uncertainties on the order of 100–200 m could be reduced considerably.

Hurd (2012) attempted to use the double-difference technique to improve the accuracy of microearthquake locations associated with a dataset acquired in the Horn River

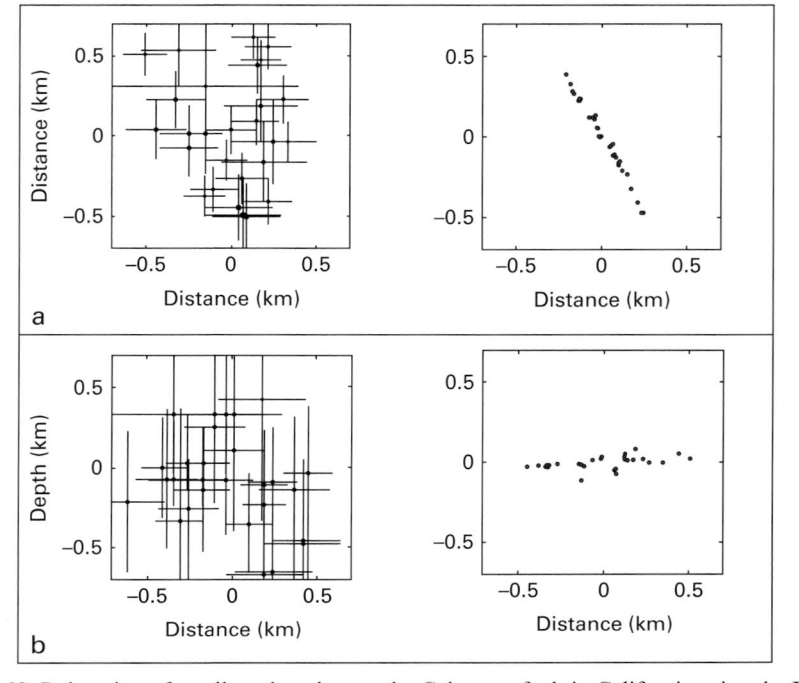

Figure 9.32 Relocation of small earthquakes on the Calaveras fault in California using the DD technique and P- and S-wave arrival times determined with cross-correlation. (a) Earthquakes located by the NCSN in map view and relative locations using DD on the right. (b) Cross-sectional view looking northeast, at the plane of the Calaveras fault. From Waldhauser & Ellsworth (2000).

area of British Columbia. As illustrated in Fig. 9.33a, these data were recorded on two recording arrays – one in a horizontal well (array 1) and one in a vertical well (array 2). As can be seen in the figure, the original microearthquake locations were quite scattered – both spatially and with depth. The image on the left side of Fig. 9.33b shows the result of using the double-difference algorithm. As shown, the events line up in the direction expected for a hydraulic fracture propagating normal to the direction of the least principal stress as expected.

Unfortunately, the northeast trending lineation shown is an artifact associated with the recording geometry. The middle image in Fig. 9.33b shows a synthetic test with the same recording geometry. The actual locations of the synthetic events are at the sites of the small green circles whereas the algorithm spread the events out in a northeast trending lineation as shown. After carefully looking at this problem, the artifact turned out to be a consequence of using only two recording arrays. When a third seismic array was added to the synthetic test mathematically, the events were properly located. Clearly, there is great potential in using techniques such as the double-difference algorithm but like all techniques, they should be used with caution.

Reservoir Seismology

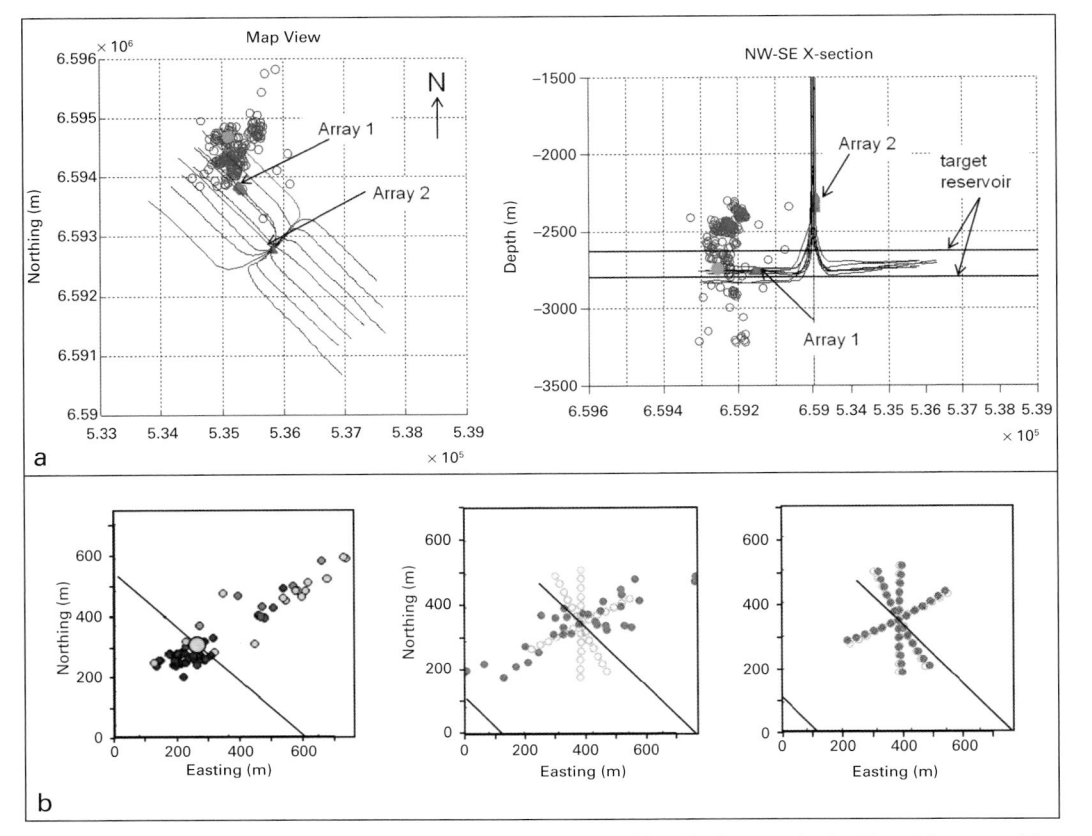

Figure 9.33 (a) A microseismic dataset associated with a single stage in the Horn River area. Two arrays were used to record the dataset. Note the scatter in locations both in plan view (left) and cross-section (right). (b) Application of the DD algorithm initially seemed to produce a realistic pattern of events propagating in the expected direction of hydraulic fractures (left). A synthetic microseismic location effort (center) showed that actual locations (green circles) were not reproduced and the northeast trending locations DD (red dots) were an artifact of the location methodology. After mathematically including a third monitoring array (right), the locations were accurately reproduced by the DD algorithm. From Hurd & Zoback (2012b).

Concluding Statements – Microseismic monitoring established the importance of shear slip on pre-existing fractures and faults surrounding hydraulic fractures as a key element of the stimulation of production. The importance of stimulated shear fracture network on production is discussed at some length in Chapters 10 and 12. One of the peculiarities of assessing the value of microseismic data seems to sometimes boil down to the question *Does the distribution of microseismic events accurately reflect the stimulated rock volume?* There is no simple answer to this question. As discussed in Chapter 10, an accurately located microseismic event is evidence of elevated fluid pressure reaching the fracture or fault that was induced to slip. So, the event *is* evidence of stimulation at a distance from hydraulic fractures. However, as we discuss in Chapter 12, the very low matrix permeability of unconventional reservoirs results in production limited to distances of only a few meters from a stimulated fracture. Thus, drawing an envelope around scattered microseismic events does imply production from a significant rock volume. There is no question that microseismic monitoring is an imperfect tool, but there is also no question that it is extremely valuable in:

- Identifying cases in which depletion from older wells affects stimulation of infill wells (as discussed in Chapter 12).
- Confirming when the azimuth of hydraulic fracture propagation is perpendicular to the least principal stress or affected by the presence of pre-existing faults, as illustrated in Chapter 7.
- Selecting appropriate stage spacing and well spacing.
- Documenting the vertical extent of hydraulic fracture propagation to determine whether hydraulic fracture propagation is extending out-of-zone. Factors affecting vertical hydraulic fracture growth are addressed in Chapter 11.
- Improving understanding of the mechanisms responsible for well-to-well communication associated with the *frac hit* and *parent/child* interactions discussed in Chapter 12.
- Illustrating problems with zonal isolation during successive hydraulic fracturing stages as illustrated in Fig. 8.6a,b.
- Establishing the degree to which permeable surface area has been created by stimulation that enables statistical reservoir models to be developed to help assess relative production (also discussed in Chapter 12).

10 Induced Shear Slip during Hydraulic Fracturing

In this chapter we first discuss a variety of aspects of triggered slip on pre-existing fractures and faults during multi-stage hydraulic fracturing. The importance of inducing shear slip on pre-existing faults during multi-stage hydraulic fracturing was introduced in Chapters 1, 7 and 8. After briefly motivating this topic, we discuss how, in the context of Coulomb faulting theory introduced in Chapter 4, the high pore pressure perturbation associated with multi-stage hydraulic fracturing is capable of triggering slip on pre-existing fractures and faults in the formations surrounding the hydraulic fractures. These fractures and faults are generally not related to the current stress state as they likely formed as the result of multiple episodes of tectonic deformation, potentially over tens (sometimes hundreds) of million years. The fact that many of these *old, dead* fractures and faults (often mineralized with calcite) often have a wide variety of orientations is critical for the pressures associated with multi-stage hydraulic fracturing to create an interconnected permeable fracture network, which, in turn, is critical to facilitate production (Chapter 12).

To investigate the relationship between shear deformation and production, we address in this chapter the relationship between the stress state, the orientation of pre-existing fractures and faults and the pore pressure perturbation induced by multi-stage hydraulic fracturing. We also discuss the likelihood of induced slip on sub-horizontal bedding planes and whether it is possible for there to be both opening mode (Mode I) and shear (Mode II and III) deformation on near-vertical fractures oriented sub-parallel to the direction of hydraulic fracture propagation (normal to S_{hmin}) or nearly perpendicular to the hydraulic fractures (normal to S_{Hmax}).

We next discuss the permeability enhancement associated with inducing slip on pre-existing fractures and faults. We discuss why slip enhances permeability, and how permeability is likely to decrease with depletion based on a limited number of laboratory experiments. In Chapter 12 we consider the permeability of these shear planes in the context of an integrated view of reservoir production.

Shear Stimulation and Production

Based on empirical observations (and intuition), it has been argued that the degree to which shear stimulation of pre-existing faults occurs during hydraulic fracturing affects the success of production. One metric for this that has been widely discussed is to equate the volume in which shear stimulation has occurred to the volume of rock encompassed by the microseismic cloud, often referred to as the stimulated rock volume, or SRV. Figure 10.1a shows data from the Barnett shale in terms of cumulative 3-year production as a function of the size of the SRV (from Mayerhofer et al., 2010) and Fig. 10.1b shows average daily production rate as a function of the size of the SRV (from Fisher et al., 2004). While such correlations make sense, because of the uncertainties in relative locations as discussed in Chapter 9, incorporating all microseismic events into a theoretical SRV can imply a much larger volume of the reservoir is being stimulated than is actually the case (Maxwell, 2009). We discuss this issue in some detail in Chapter 12. A number of other complicating factors can affect production than just the size of the envelope encompassing the microseismic events (Cipolla & Wallace, 2014). A variety of other metrics have been used to characterize the stimulated volume such as cumulative moment density and "*b*" values from Gutenberg–Richter plots (discussed in Chapter 9). There have also been attempts to capture the spatial distribution of event magnitudes (Haege et al., 2012) as possibly better metrics for the amount of rock being stimulated. As mentioned above, we will focus on the total area associated with the stimulated shear network in Chapter 12.

This being said, there is definitely a general correlation between stage by stage production and the number of microearthquakes in the Barnett case study presented in Fig. 7.24. In Fig. 10.2, we show stage by stage production data along with the number of microearthquakes per stage from Roy et al. (2014). While the correlation is not exact, the stages of relatively high and low production correlate with the positions along the laterals with relatively more or less microseismicity. As discussed in Chapter 7, the relatively large number of microseismic events seen in the

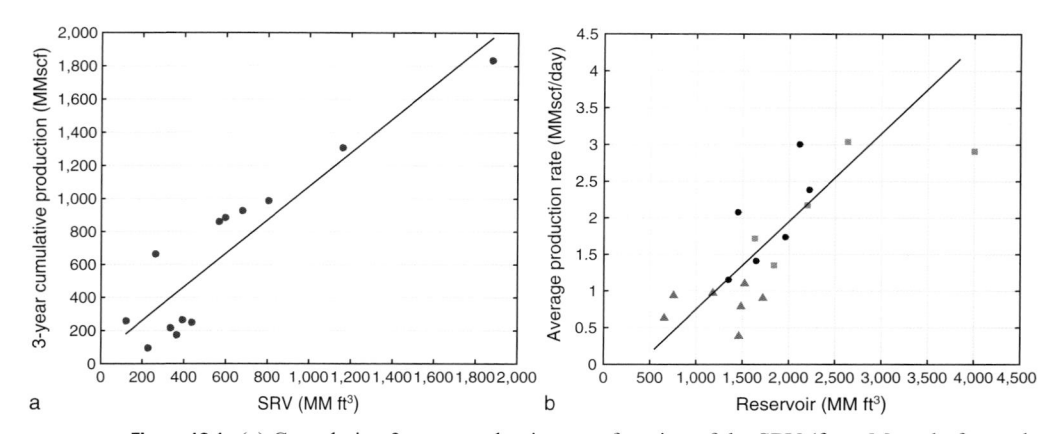

Figure 10.1 (a) Cumulative 3-year production as a function of the SRV (from Mayerhofer et al., 2010). (b) Average production rate as a function of the apparent SRV in the Barnett shale. Symbols indicate various completions, including un-cemented/sliding sleeve completions (circles) and cemented/plug-and-perf completions (squares) horizontal wells as well as vertical wells (triangles). From Fisher et al. (2004).

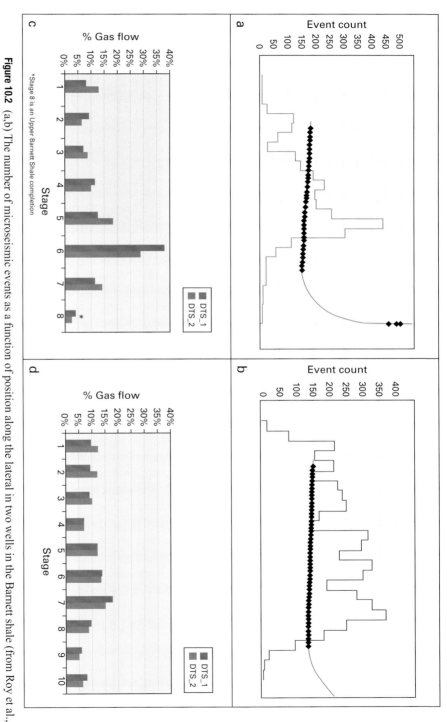

Figure 10.2 (a,b) The number of microseismic events as a function of position along the lateral in two wells in the Barnett shale (from Roy et al., 2014) and (c,d) stage by stage production from DTS data as presented in Fig. 7.24. After Roy et al. (2014).

304 Unconventional Reservoir Geomechanics

vicinity of stages 5 and 6 in well 1 and stages 8 and 9 in well 2 are associated with slip on small fractures in fault damage zones seen in seismic reflection data. This dataset will be revisited in Chapter 12.

Coulomb Faulting and Slip on Poorly Oriented Fracture and Fault Planes

As reviewed in Chapter 4, Coulomb faulting theory has wide applicability for understanding fault slip and controlling stress magnitudes in brittle rock, as discussed in Chapter 7. In this section, we first utilize a case study from the Barnett shale in which a fracture distribution from an image log in a horizontal well and well-recorded microseismic events for which focal plane mechanisms are available to quantitatively investigate how slip on pre-existing fractures and faults is triggered. The dataset in question is shown in both plan and perspective views in Fig. 10.3. The same dataset was considered previously in relation to determination of (1) focal plane mechanisms (Figs. 9.12 and 9.13), (2) stress from inversion of focal plane mechanisms (Fig. 7.11) and (3) the relative uncertainty of microseismic locations (Fig. 9.30). An unusual aspect of this dataset is that there were only five hydraulic fracturing stages that were not spaced equally along the length of the wellbore.

Figure 10.4 illustrates the manner in which elevated pore pressure during hydraulic fracturing triggers slip using both Mohr circles and stereonets. Note that in the Mohr circles shown, the abscissa indicates total normal stress acting on each plane, not the effective normal stress. Thus, the Mohr circles were constructed using the principal stresses S_{Hmax}, S_{hmin} and S_V. Presentation in this manner makes the intercept of the friction faulting line initially correspond to the unperturbed pore pressure. To consider the effect of increasing pore pressure due to leak off, we simply need to shift the friction line to the right. For the purposes of illustration, we assume near hydrostatic pore

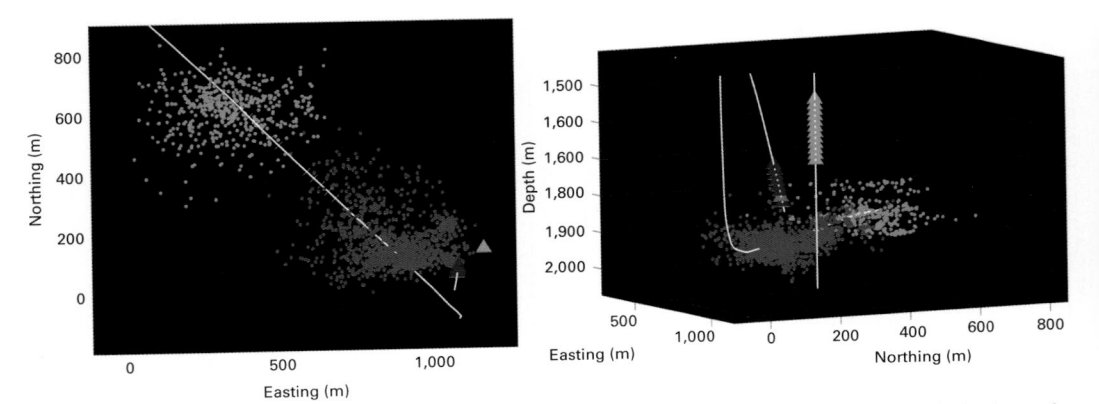

Figure 10.3 Microseismic events associated with five hydraulic fracturing stages of a horizontal well in the Barnett shale. Stages are colored by number. Each stage was recorded by two microseismic arrays. This dataset was discussed previously in Chapters 7 and 9.

Induced Shear Slip during Hydraulic Fracturing

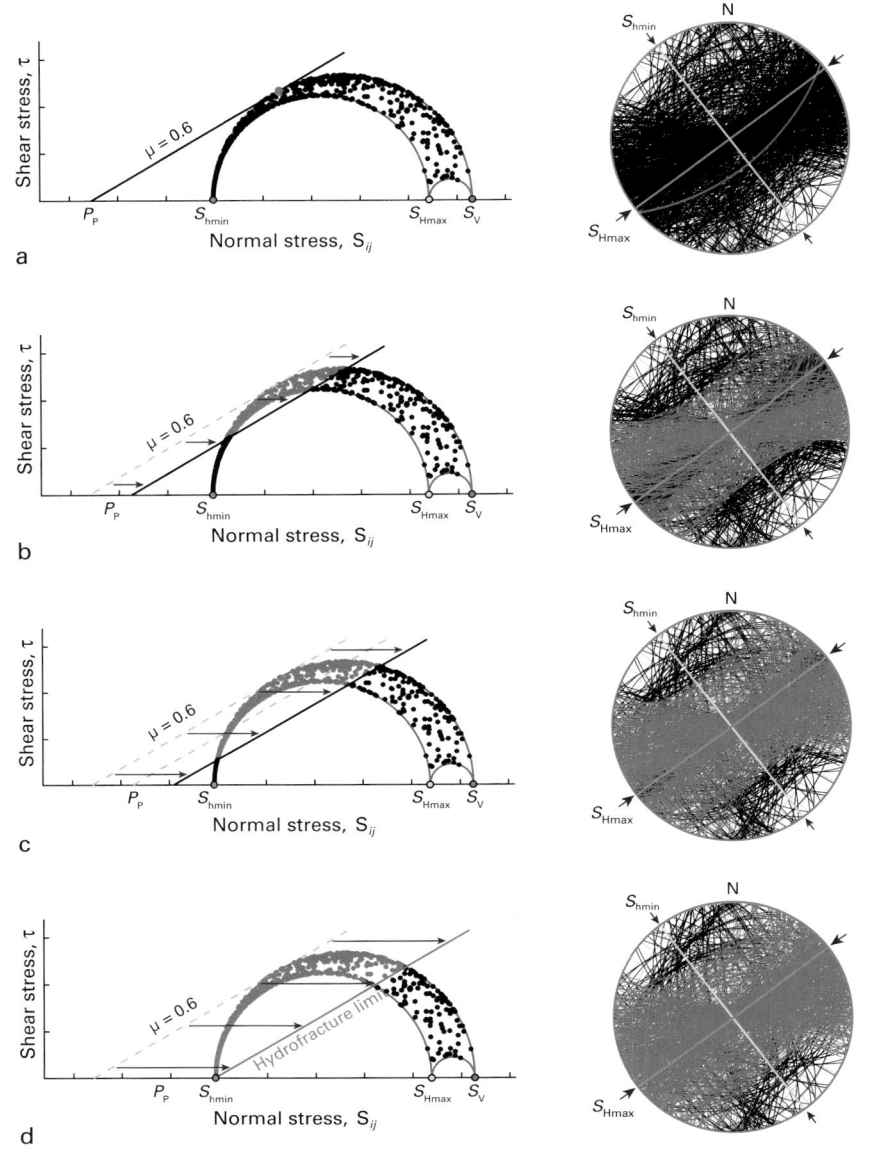

Figure 10.4 Illustration of the manner in which elevated pore pressure during hydraulic fracturing triggers slip on pre-existing fractures and faults. The fracture planes and stress state are from an image log in a horizontal well in the Barnett shale previously discussed in Chapter 9 characterized by a normal/strike-slip stress state ($S_{hmin} \ll S_{Hmax} \sim S_V$). The blue, yellow and green planes are hypothetical planes normal to S_{hmin} (like a hydraulic fracture), S_{Hmax} (normal to hydraulic fractures) and S_V (parallel to horizontal bedding planes) respectively. (a) Reference case assuming near hydrostatic pore pressure. (b) Slightly elevated pore pressure indicates a group of NE-trending planes is activated – those relatively well oriented for slip in a SS/NF stress state. (c) At a relatively high pore pressure perturbation, shear is expected on more planes. (d) At a pressure corresponding to the least principal stress, shear is expected on many planes of highly varied orientation. From Zoback & Lund Snee (2018).

pressure and a coefficient of friction of 0.6 and we ignore cohesion, which is likely quite small. Although friction can vary depending on clay content, an average value of about 0.6 is quite reasonable (Chapter 4).

The fracture planes shown in the stereonets and Mohr circles in Fig. 10.4 were obtained from the image log in the horizontal well. A normal/strike-slip faulting stress state is appropriate for this location (see Fig. 7.11 and associated discussion). This stress state is illustrated on the right side of Fig. 7.1. In blue and yellow we also show hypothetical planes normal to S_{hmin} (like a hydraulic fracture) and normal to S_{Hmax} (normal to hydraulic fractures). We consider slip on these hypothetical planes because, as discussed in Chapter 7, it is commonly (but incorrectly) assumed that pre-existing fractures in unconventional reservoirs are generally aligned parallel and perpendicular to the current principal stresses. We also show in green a hypothetical plane normal to S_V (simulating horizontal bedding planes) as it is often debated whether bedding plane slip is significant during hydraulic fracturing.

Figure 10.4a represents the reference state prior to stimulation. As the frictional strength of pre-existing, well-oriented normal faults is expected to control the stress state (in accordance with Eqn. (7.3)), there is only a single fault that is perfectly oriented for slip in the ambient field. In other words, the great majority of planes are currently inactive fractures and faults, often mineralized with calcite in the Barnett (Gale et al., 2007). Figure 10.4b demonstrates that as pore pressure is elevated by a relatively small amount due to fluid leak-off during hydraulic fracturing, a number of steeply dipping normal faults sub-parallel to S_{Hmax} are activated (the dots in the Mohr diagram and corresponding planes in the stereonet turn from black to red), as are near-vertical strike-slip faults trending approximately ±30° from the S_{Hmax} direction. When pore pressure is further elevated (Fig. 10.1c) shear is expected to occur on far more planes because as pressure increases, more and more poorly oriented planes begin to slip as the CFF → 0 (Chapter 4). Note that once pressure reaches the minimum horizontal stress (Fig. 10.1d), the majority of planes are expected to slip, but a significant number will not. As explained below, these are the planes that are roughly perpendicular to the vertical stress or S_{Hmax} (striking NW–SE with steep dip). For the purposes of illustration, we neglected *net pressure* in Fig. 10.4, the amount that the pressure during hydraulic fracturing exceeds the least principal stress. This is typically on the order of several MPa (a few hundred psi) and could be important close to the hydraulic fractures in the vicinity of the wellbore.

In summary, Fig. 10.4 is a simple illustration of the critical importance of the process of shear stimulation during slickwater hydraulic fracturing. Many *old dead* fractures and faults can be stimulated in shear because of the very large pressure perturbation during hydraulic fracturing. By moving in shear, they become permeable (as discussed below) and their highly varied orientations result in an interconnected permeable fracture network. This helps us understand why hydraulic fracturing with low viscosity slick-water is so important. In order for slip to occur on mis-oriented faults, pressure has to reach them. Hydraulic fracturing with a viscous gel with little leak-off would not allow this to happen. It also helps us understand the importance of the distribution of pre-existing faults. When they are not present, they are not available to be stimulated, as first pointed out by Fisher et al. (2004). Moreover, it is the distribution of pre-existing fractures and faults (and the ability for the high pressure acting in the hydraulic fractures to reach them) that defines the nature of the microseismic cloud. In fact, if pre-existing fractures have a limited range of strikes (for example, if they were sub-parallel to the direction of maximum horizontal stress and sub-parallel to hydraulic fractures), fluid pressure would have a difficult time propagating away from the hydraulic fracture and would be expected to remain very close to the hydraulic fracture plane (as in the case reported by Rutledge et al., 2004). When multiple fractures and faults are present with a wide range of orientations, it is possible to create a diffuse cloud of microseismic events and an interconnected fracture network.

We also discuss below how the network of natural fractures with highly varied orientations evolves through time during the five respective fracturing stages in this dataset. While evolution of this network is usually considered in terms of its spatial evolution with time, the analysis is hampered by the uncertainty in the exact location of the microseismic events as was discussed in Chapter 9. Rather, we consider the evolution of the total amount of area of permeable planes (fractures and faults that move in shear and propped hydraulic fracture planes) that is in contact with the very low permeability reservoir. As mentioned above, this topic is revisited in the context of reservoir production in Chapter 12.

While Fig. 10.4 utilizes real fracture and stress data from a representative case study in the Barnett shale, the prediction of which planes are likely to slip in response to increases in pore pressure resulting from leak off from hydraulic fractures is obviously conceptual. Fortunately, it turns out that we can independently confirm this prediction using the focal plane mechanisms of the microseismic data (discussed previously and shown in Fig. 7.11). In Fig. 10.5a we show again the prediction of which planes will slip at a pressure equivalent to the frac gradient based on image log data and previously presented in Fig. 10.4d. Figure 10.5b is based on the fracture and fault population that slipped as obtained from analysis of the focal plane mechanisms of the recorded microearthquakes. Knowledge of the stress state obtained from the stress inversion and independent analysis of the wellbore data were used to select which nodal plane of the focal plane mechanism corresponded to the plane that slipped (see Kuang et al., 2017). As the focal mechanism analysis was limited to only 123 of the largest magnitude events, the total number of planes shown in Fig. 10.2b was scaled to magnitude -2.5 to show a comparable number of events as shown in Fig. 10.4a. As discussed in Chapter 9, a magnitude -2.5 event has a source dimension smaller than ~1 m, a reasonable lower bound for consideration. Note in Fig. 10.5 the overall consistency between the two analyses. The planes shown in Fig. 10.5b did slip and produced microseismic events. Thus, the excellent comparison between these figures provides support for the fact that much of the pressure acting in the hydraulic fractures during stimulation is reaching each of the fault planes to initiate slip, thereby indicating creation of an interconnected, permeable fracture network in contact with the relatively impermeable shale matrix.

Importance of Stress State – It is obvious that the nature of shear stimulation as illustrated in the figures above results from the interactions among the orientation of

308   **Unconventional Reservoir Geomechanics**

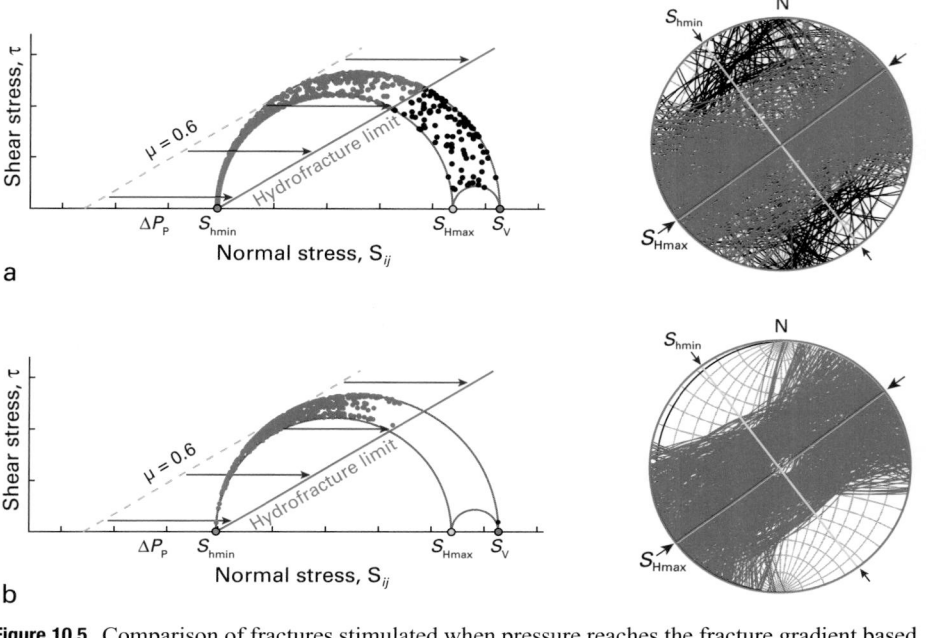

Figure 10.5 Comparison of fractures stimulated when pressure reaches the fracture gradient based on (a) fractures observed in image logs (figure same as 10.4d) and (b) planes obtained from analysis of focal plane mechanisms in the same well. Modified from Zoback and Lund Snee (2018).

fractures and fault planes that are present, the stress state and the ability for the high fluid acting in the hydraulic fractures to reach the fractures and faults. To illustrate the importance of the stress state, Fig. 10.6 revisits the heuristic arguments illustrated in Fig. 10.4 keeping everything constant except the stress state. Instead of a SS/NF stress state ($S_{hmin} \ll S_{Hmax} \approx S_V$) typical of the Fort Worth Basin, we use a SS/RF stress state representative of parts of the Appalachian and Alberta Basins ($S_{hmin} \approx S_V \ll S_{Hmax}$). Coincidentally, the orientation of the maximum horizontal stress in both of these areas is approximately NE–SW. While there are many similarities between Figs. 10.4 and 10.6, note that with moderate increases in pore pressure in Fig. 10.6b, only strike-slip faults (very steeply dipping faults striking $\pm 30°$ from the S_{Hmax} direction) are activated. At higher pressures (Figs. 10.6c,d), slip is expected on nearly all the available planes, some at very high angle to the S_{Hmax} direction and some sub-horizontal.

Slip (and Opening) of Horizontal and Vertical Planes – There have been many studies in which slip and/or opening is hypothesized to have occurred on either sub-horizontal bedding planes and/or near vertical fractures striking sub-parallel to the direction of maximum horizontal stress (see Hull et al., 2017a; Kahn et al., 2017; Stanek & Eisner, 2017) and the references therein). This is illustrated schematically in Fig. 10.7. At the left, a hypothetical focal plane mechanism is shown that implies that dip slip (west side up) occurred on a near vertical plane striking approximately N–S, or slip occurred on an

Induced Shear Slip during Hydraulic Fracturing   309

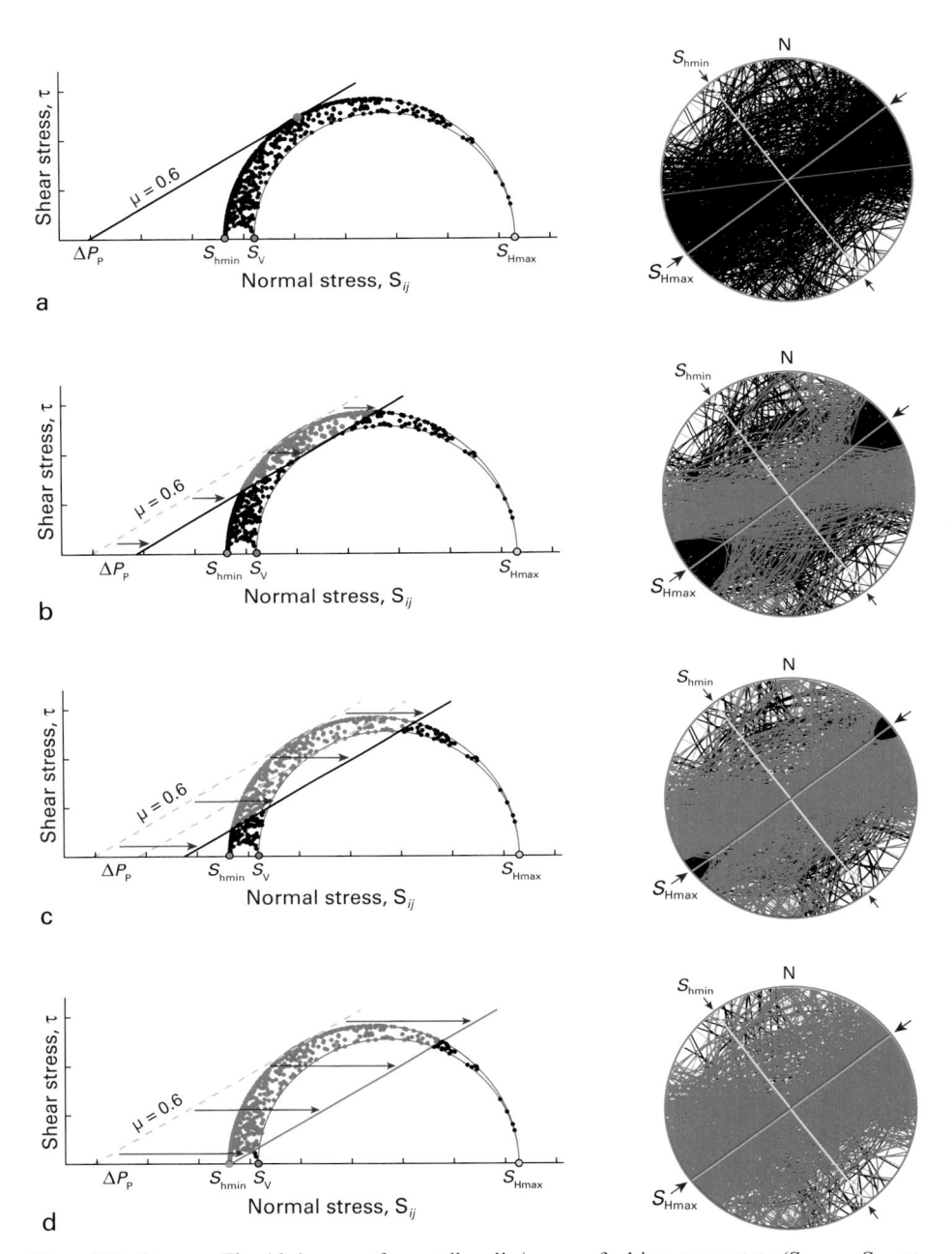

Figure 10.6 Same as Fig. 10.4 except for a strike-slip/reverse faulting stress state ($S_{hmin} \sim S_V \ll S_{Hmax}$) as observed, for example, in parts of the Appalachian and Alberta Basins. From Zoback & Lund Snee (2018).

Dip-slip on vertical planes
or
horizontal slip on bedding planes?

Opening mode on vertical planes
or
bedding planes?

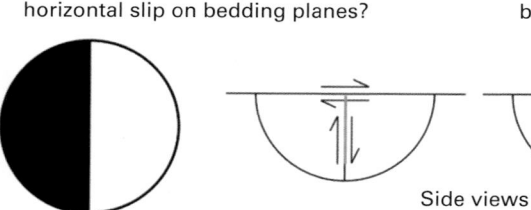

Side views

Focal plane mechanism

Figure 10.7 Schematic diagram illustrating the tendency for slip and/or opening on sub-horizontal bedding planes or near vertical fractures striking sub-parallel to the direction of maximum horizontal compression.

orthogonal horizontal (presumably bedding) plane (rock above the bedding plane moving to the east). This is illustrated in the cross-section in the center. The possibility of opening mode deformation is shown schematically on the right. Of course, it might be theoretically possible for both slip and opening to occur on the same plane.

Stanek & Eisner (2013), Rutledge et al. (2014) and Rutledge et al. (2016) all make a case for bedding plane slip, but despite the many heated conversations surrounding these possibilities (the issue of open mode deformation, or non-double couple focal plane mechanisms was discussed in Chapter 9), one can extend the analysis illustrated in fig. 10.4 to address this issue. In terms of the Coulomb Failure Function (Eqn. 4.2), for slip to occur, $CFF \equiv \tau - \mu\sigma_n - S_0 \rightarrow 0$. Another way of saying this in terms of the Mohr diagrams shown above, is that the point representing a potential slip plane must touch the frictional failure line for sliding to occur. Similarly, for opening to occur, the effective normal stress on the fault must be less than zero, $S_n - P_p < 0$. Another way of saying this is that a point representing a potential opening plane must be to the left of the intersection of the frictional failure line with the abscissa in Figs. 10.4–10.6.

Figure 10.8 considers whether these conditions might be met in the normal/strike-slip stress state considered in Fig. 10.4d ($S_{hmin} \ll S_{Hmax} \approx S_V$) and strike-slip/reverse stress field considered in Fig. 10.6d ($S_{hmin} \approx S_V \ll S_{Hmax}$). For a normal/strike-slip stress state, Fig. 10.8a demonstrates that it is essentially impossible for planes sub-parallel to bedding (nearly normal to the vertical stress) or those almost normal to S_{Hmax} (the green and yellow circles, respectively) to either slip or open, even when pore pressure reaches the magnitude of the least principal stress. In a strict sense this is also true for planes sub-parallel to S_{hmin} (blue circle). However, recognizing that during hydraulic stimulation, the net pressure can be a few MPa (a few hundred psi) above the least principal stress, the frictional failure line would shift slightly to the right making it possible for both the slip criterion and opening mode criterion to be satisfied on planes sub-parallel to the hydraulic fractures. Hence, in a normal/strike-slip stress environment, it is possible that both shear and opening of fractures and faults sub-parallel to hydraulic fractures may occur when the pressure exceeds S_{hmin}, but it is essentially impossible for bedding planes to open or shear as a consequence of elevated pore pressure alone.

Induced Shear Slip during Hydraulic Fracturing

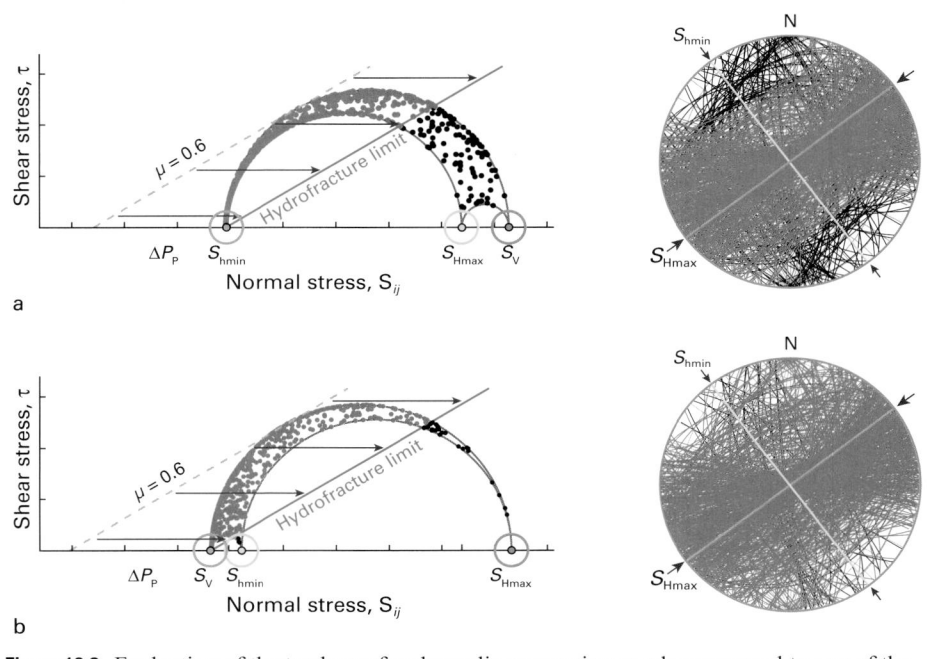

Figure 10.8 Evaluation of the tendency for shear slip or opening on planes normal to one of the principal stresses. Sub-horizontal bedding planes are illustrated by the green dot on left and the plane shown in green on the stereonet. Planes normal to S_{hmin} (sub-parallel to hydraulic fractures) are shown in blue and those normal to S_{Hmax} are shown in yellow. (a) A normal/strike-slip stress state ($S_{hmin} \ll S_{Hmax} \sim S_V$). (b) A strike-slip/reverse stress state ($S_{hmin} \sim S_V \ll S_{Hmax}$). Modified from Zoback & Lund Snee (2018).

The situation is completely different in a strike-slip/reverse faults stress state shown in Fig. 10.8b. When pressures reach values slightly in excess of S_{hmin}, it would be possible for opening and shear to occur on planes sub-parallel to hydraulic fractures *and* it might also occur on sub-horizontal planes, depending on the difference in the magnitude of S_{hmin} and S_V and the magnitude of the net pressure. Just like the normal/strike-slip stress state, it is essentially impossible for slip, or opening, to occur on planes approximately normal to S_{Hmax}.

In order to consider these principles in practice, it is important to remember that there are two conditions under which the differences among the principal stresses decrease markedly (i.e., the Mohr circle would be much smaller): formations with highly elevated pore pressure (as illustrated in Fig. 7.18) and those in which considerable viscoplastic stress relaxation occurs as discussed in Chapter 3 (and as will be discussed in the context of hydraulic fracturing in Chapter 11). In the case of relatively isotropic stress states resulting from highly elevated pore pressure, it is very easy to induce slip on pre-existing planes. In the case of relatively isotropic stress states resulting from highly viscoplastic formations, it is very difficult to induce slip or initiate hydraulic fractures, unless the formation is also overpressured.

312   Unconventional Reservoir Geomechanics

Shear Stimulation in a Regionally Varying Stress Field – As discussed in Chapter 8, it
is generally the case that wells are drilled in the direction of the minimum horizontal
stress for successful exploitation of unconventional reservoirs. As mentioned above,
the manner in which pre-existing fractures and faults are re-activated in shear depends
on the stress state. The Permian Basin offers an interesting example of how these
issues play out as both the stress field and fracture systems vary considerably across
the basin. Figure 10.9b shows a generalized map of the direction of maximum
horizontal stress in the region. We showed this in more detail in Fig. 7.3. The stress
orientation in the Central Basin Platform and Midland Basin is consistently east–west
and the stress state is generally a strike-slip/normal faulting. In marked contrast, the
S_{Hmax} direction varies from north-south in the north western part of the Delaware
Basin, rotates clockwise to east–west in the central part of the basin and continues to
rotate clockwise to northwest–southeast in the southern part of the basin. This means,

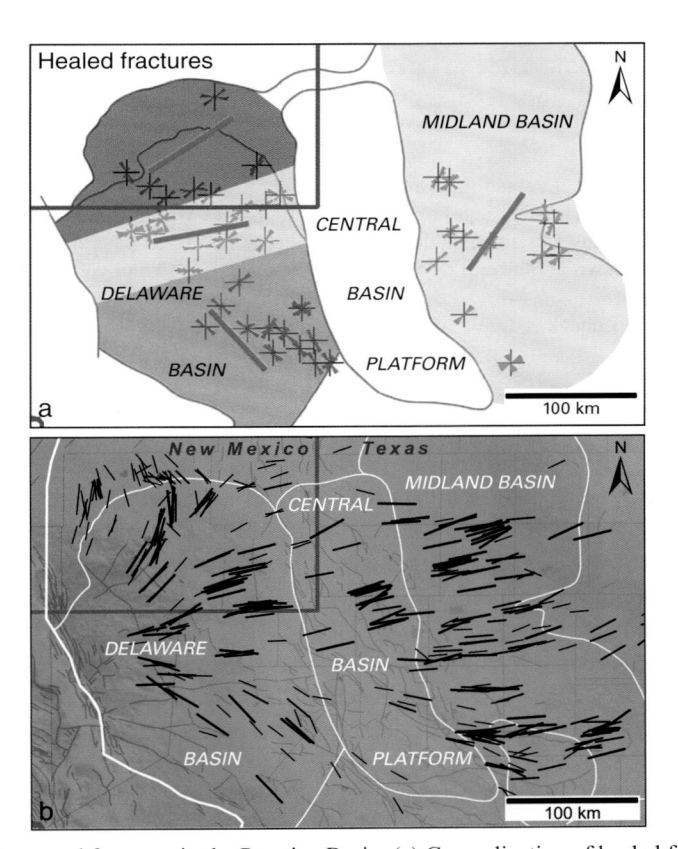

Figure 10.9 Stress and fractures in the Permian Basin. (a) Generalization of healed fracture
orientations in the Delaware and Midland Basins. The heavy red lines indicate the average strike of
healed fractures in each of the four regions shown. From Forand et al. (2017). (b) A detailed stress
map of the Permian Basin (excerpted from Fig. 7.3) that shows both stress directions and relative
stress magnitudes (see text).

of course, that horizontal wells in the Delaware Basin should ideally be drilled at azimuths that rotate from north to south as does the direction of maximum horizontal stress. In other words, wells should be drilled generally east–west in the northern part of the basin, approximately north–south in the central part of the basin and the trajectory of the wells needs to progressively rotate clockwise as one considers more southerly parts of the basin. Lease-holding constraints may make these optimal well orientations impractical in some areas, however.

The variation of relative stress magnitudes shown in Fig. 10.9b is quite significant with respect to triggering slip on pre-existing fractures and faults. Going from west to east, we see that the stress state changes from normal faulting in the Delaware Basin to normal/strike-slip in the Central Basin Platform, to strike-slip in the Midland Basin. Interestingly, Forand et al. (2017) show a generalization of healed fracture orientations in Fig. 10.9a. They note that healed fractures are generally aligned with the S_{Hmax} direction in the Delaware Basin, but consistently about 30° from the direction of S_{Hmax} in the Midland Basin. As it turns out, these directions are exactly the orientations of fractures that are most well oriented for slip during stimulation, regardless of the fact they appear to be healed in the image logs. As illustrated in Fig. 7.1, in the Delaware Basin a normal faulting area, the fractures most easily stimulated are those striking sub-parallel to the direction of maximum horizontal stress (Fig. 10.9a). Thus, the healed fracture orientations in the northern, central and southern parts of the Delaware shown in Fig. 10.9a are at the ideal orientation to be stimulated during hydraulic fracturing. Similarly, the orientation of healed fractures in the in the strike-slip faulting environment of the Midland Basin, are easily stimulated as they strike about 30° from the S_{Hmax} direction, which is characterized by a strike-slip stress state.

Microseismic Indications of Apparent Local Stress Variations – In addition to the misunderstanding about the orientation of pre-existing faults in unconventional reservoirs discussed in Chapter 7, there are two unfortunate misunderstandings about micro-seismic clouds and their relation to variations of stress magnitudes at the scale of an individual well or well pad. The first is illustrated in Fig. 10.10 (modified from Daniels et al., 2007) based on micro-seismic events associated with four hydraulic fracture stages in the Barnett shale. On the basis of unreliable estimates of the magnitude of S_{Hmax} from shear velocity logs in the horizontal well, Daniels et al. mistakenly argued that the linear microseismic cloud associated with stages 1 and 2 indicates a high degree of stress anisotropy (a large difference between S_{Hmax} and S_{hmin}) whereas the more distributed microseismic events seen in stages 3 and 4 indicates a more isotropic stress state (a smaller difference between S_V, S_{Hmax} and S_{hmin}). This is simply not true. As mentioned above, fault slip is difficult to initiate when stress magnitudes are relatively isotropic and widely distributed events could simply indicate the variations of the number and orientation of pre-existing faults encountered along that part of the well and have nothing to do with stress magnitudes. The purpose of Figs. 10.4–10.6 is to illustrate that the fractures and faults that are activated in shear during stimulation result from the interaction between stress, fracture orientations and pore pressure perturbations. The data from the well shown in Fig. 10.10 indicate that stress anisotropy is actually higher

314   **Unconventional Reservoir Geomechanics**

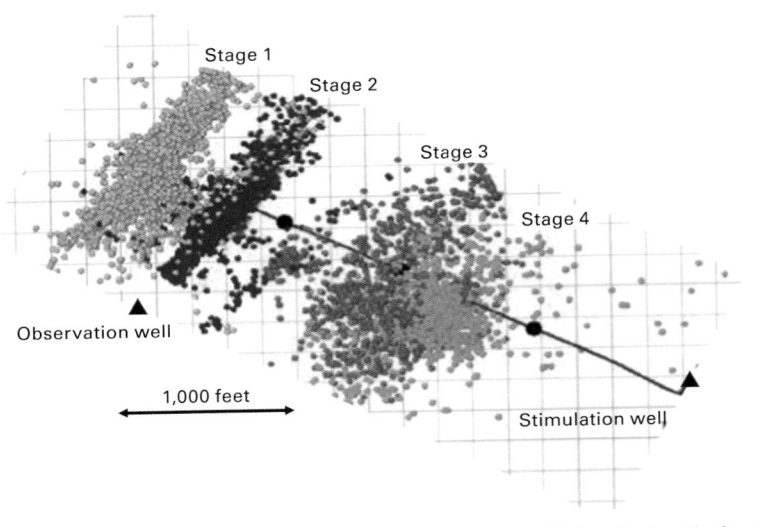

Figure 10.10 The distribution of microseismic events associated with four hydraulic fracturing stages in the Barnett shale (modified from Daniels et al., 2007). As discussed in the text, the distribution of microseismic events cannot be used to infer horizontal stress anisotropy.

for stages 3 and 4 than for stages 1 and 2. The magnitude of the reported least principal stress from the ISIPs associated with the four hydraulic fracture stages correspond to frac gradients of 0.7, 0.64, 0.65 and 0.62 psi/ft, respectively. As this is a normal/strike-slip faulting environment ($S_{Hmax} \approx S_V$), the most horizontal stress anisotropy is associated with stage 4 (frac gradient of 0.62 psi/ft) and the least is for stage 1 (frac gradient of 0.7 psi/ft). One cannot simply infer stress magnitude changes from the gross pattern of microseismicity.

This criticism aside, a significant contribution of the Daniels et al. paper is the importance of paying careful attention to well trajectories when evaluating microseismic clouds. As stress magnitudes are expected to be more isotropic in formations that are viscoplastic (as introduced in Chapter 3) or more overpressured (as discussed in Chapter 7), hydraulic fracture stages along a well trajectory that intersects different lithofacies would be carried out under different stress conditions, which is discussed at some length in Chapter 11. Hence, what might appear to be lateral variations of formation properties may simply be variations with depth as a well encounters different lithofacies.

A second common pitfall about stress states inferred from microseismicity data comes from the interpretation of focal plane mechanisms in transitional stress environments illustrated in Fig. 7.1. This could be either extensional (normal faulting/strike-slip where $S_V \approx S_{Hmax} > S_{hmin}$) transitional stress states as widely observed in Texas, northern Oklahoma and other areas or compressional (strike-slip/reverse faulting where $S_{Hmax} > S_V \approx S_{hmin}$) transitional stress states as observed in parts of the Appalachian and Alberta Basins. If we consider the first case for illustration, the occurrence of normal faulting focal mechanisms in one area along a stimulated well and strike-slip faulting focal mechanisms in another does not indicate a change of stress from one place to another.

Induced Shear Slip during Hydraulic Fracturing   315

Rather, it likely indicates a change in the orientation of the pre-existing fractures and faults being stimulated. In fact, the example shown in Fig. 7.11 illustrates that within a consistent strike-slip/normal faulting stress state and uniform stress orientations seen along the length of a well, focal plane mechanisms indicate both strike-slip and normal faulting on planes of different orientation, as illustrated schematically in Fig. 7.1.

Shear Slip and Permeability

Implicit in the discussion above is the fact that shearing on pre-existing faults increases their permeability. Figure 10.11 (after Barton et al., 2009) illustrates this point schematically. A natural fracture prior to stimulation (Fig. 10.10) is potentially healed or mineralized, and has relatively low permeability. Moreover, the fracture permeability would be expected to decrease rapidly with depletion due to the decrease in effective aperture, a, resulting from the increase in effective stress acting perpendicular to the fracture plane. Cho et al. (2013) review many of the relations that have been proposed to estimate the change in fracture permeability with effective normal stress. This change in permeability is easily understood in terms of flow through parallel plates. For plates separated with aperture, a, the flow rate, Q, of a fluid with viscosity, η, in response to a gradient in pressure, ∇P, is given by:

$$Q = \frac{a^3}{12\eta}\nabla P. \tag{10.1}$$

In contrast, after shearing, not only would the initial permeability of the fracture be expected to increase, the mis-alignment of the fracture walls would likely decrease the sensitivity of the fracture permeability to depletion (Fig. 10.11). In other words, natural fractures moving in shear would act as being self-propped to a degree.

Figure 10.12 (after Rutter & Mecklenburgh, 2018) presents a quantitative perspective of the phenomenon illustrated in Fig. 10.11 based on laboratory permeability experiments. Depending on the orientation of the fracture or fault moving in shear, the

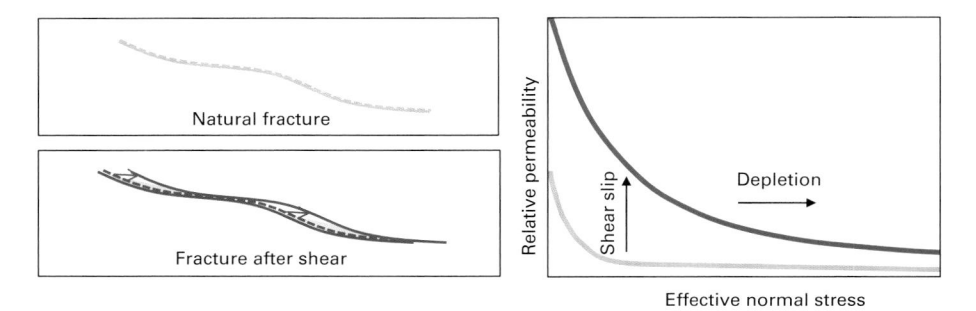

Figure 10.11 Generalization of the way in which slip on a pre-existing natural fracture is likely to increase its initial permeability and decrease the sensitivity of the permeability to depletion (after Barton et al., 2009).

316   **Unconventional Reservoir Geomechanics**

enhancement of permeability varies – both as a function of stress state (it is greatest at low mean effective stress) and the orientation of the plane to the stress field (which affects the normal stress acting on the fault). Note that the ratio of the effective maximum and minimum stress is constant for the three stress states considered but the variation of permeability increases with mean normal stress.

Most published research investigating how shear slip affects the permeability of fractures focuses on relatively strong, brittle rocks (granite, sandstone). In these rocks it common to see permeability consistently increase with several mm of shear (Esaki et al., 1999; Rong et al., 2016). In contrast, experiments on clay and phyllosilicate-rich rocks show loss of permeability with slip, particularly when the effective normal stress is quite high (Gutierrez & Nyga, 2000; Crawford et al., 2008; Fang et al., 2017; Rutter & Mecklenburgh, 2018). Here, we review recent experimental observations to understand how induced shear slip may affect permeability under different conditions.

Ye et al. (2017) report an experiment where they induced shear on a small natural fault within a core sample from the Eagle Ford shale. Similar to experiments on intact core samples (Chapter 6), permeability of the fault decreases exponentially with increasing

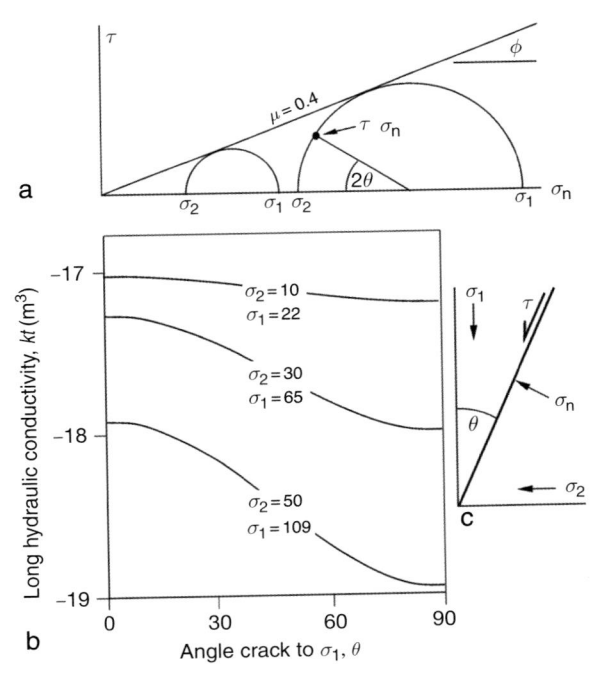

Figure 10.12 Generalization of how a non-hydrostatic stress state produces variations in transmissivity with variation of fault orientation change. (a) Two stress states arbitrarily constrained to lie below a friction limit of $\mu = 0.4$ to prevent sliding. (b) Angular relations between principal stresses and fault orientation θ. (c) Resultant variation of transmissivity with θ for three different stress states (values of principal stresses in MPa). The permeability range over 90° variation in crack orientation is greater for higher values of σ_2. From Rutter & Mecklenburgh (2018).

effective stress (Fig. 10.13a). In response to ~0.1 mm of shear and ~0.05 mm dilation induced by increased fluid pressure, the fault permeability increased by about a factor of 6 (Fig. 10.13b,c). Permeability hysteresis is observed upon unloading, which indicates that not all the permeability enhancement from shear slip is permanent. Ye et al. also found a factor of 15 increase in permeability associated with shear on an induced tensile fracture in a similar sample. From the earthquake scaling relations presented in Chapter 9, the expected slip in a magnitude −2 earthquake is ~0.1 mm, so the displacements considered in this experiment are quite reasonable to compare to microseismic events and generally consistent with the schematic diagram on the right side of Fig. 10.11. It is also important to note that the slip rates in these experiments are on the order of 0.1 μm/s, much slower than the expected dynamic slip rates (~1 m/s) during microearthquakes. Scuderi et al. (2017b) also observed significant permeability enhancement due to induced cycles of aseismic slip in a clay-rich shale fault gouge (Chapter 4). These observations suggest that slow slip on faults also has the potential to contribute to permeability enhancement during hydraulic stimulation. Interestingly, Scuderi et al. (2017a) also observe permeability enhancement in a calcareous fault gouge that undergoes dynamic (unstable) failure in response to the same injection procedure as the clay-

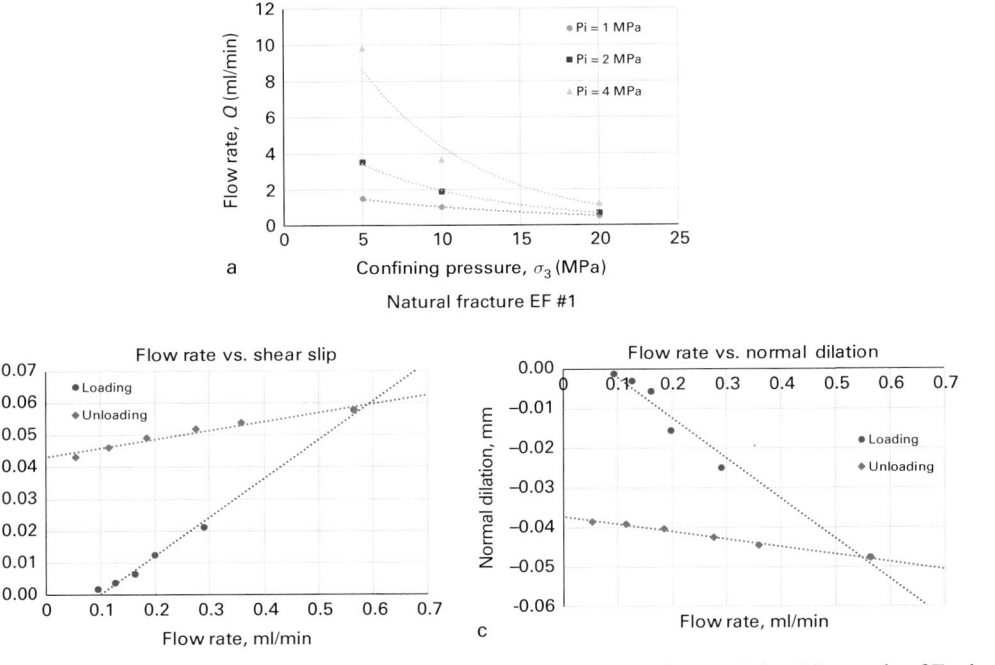

Figure 10.13 Induced shear on natural fracture in clay and organic-poor (~9 wt%) sample of Eagle Ford shale. (a) Permeability (flow rate) decreases exponentially with increasing effective stress. (b,c) Permeability increases linearly with shear slip and dilation of the fault. Permeability hysteresis in shear slip indicates that the enhancement due to offset is permanent. From Ye et al. (2017).

rich shale, which emphasizes the importance of rock composition in determining the frictional response to stimulation. This is discussed at length in Chapter 4.

Crandall et al. (2017) report inferred permeability changes of several orders of magnitude with several mm of shearing of three Marcellus shale cores containing natural fractures. They infer the permeability changes from fracture aperture measurements made by CT scanning. While such large displacements are not likely representative of the amount of slip accompanying microseismic events, they document significant permeability enhancement in two samples with <1 mm or less displacement.

Kassis & Sondergeld (2010) report permeability measurement on axial fractures induced in cylindrical core samples of the Barnett shale. They considered four samples characterized by different amounts of roughness, and measured permeability before and after offsetting the two faces of the fracture with metal foil shims 0.05 mm thick (Fig. 10.14). The normal stress on the fractures was 5.5 MPa (800 psi). The permeability of the samples before and after 0.05 mm and 0.25 mm of offset is shown in Fig. 10.14a. Note that an offset of only 0.05 mm results in a 400-fold increase in permeability. An offset of 0.25 mm results in permeability enhancement by as much as a factor of 8,000. Again, the scale of these offsets are reasonable when considering the amount of displacement expected from microseismic events during hydraulic stimulation (although manually offsetting the faces of the fracture faces is not necessarily representative of frictional shearing).

Another question of obvious importance is how much permeability is retained as depletion occurs. Kassis & Sondergeld (2010) found a relatively modest decrease in permeability with increasing fracture normal stress, P, that is generally consistent with a theory developed by Walsh (1981), in which the change in permeability $\dfrac{k}{k_0}$ is proportional to $\left(\dfrac{P^3}{P_0}\right)$. Cho et al. (2013) report a factor of 2–3 reduction of permeability on roughened, saw-cut faults in Bakken shales samples over an increase of effective pressure of ~20 MPa (~3,000 psi). Britt et al. (2016) measured fracture permeability changes with effective normal stress on samples from a number of unconventional reservoirs. They found a one to two order of magnitude reduction in permeability with several thousand psi of increase in effective stress – much faster than that observed by Cho et al. (2013) or Kassis & Sondergeld (2010). Unfortunately, Britt et al. provide no information about either the fractures studied nor the design of their experiments. Zhang et al. (2015) studied the permeability of natural fractures in samples of the Barnett shale obtained from outcrops. As shown in Fig. 10.15, they too found over an order of magnitude of permeability loss with an increase in effective normal stress, but again there was no shear or offset of the samples faces prior to testing. Ye et al. (2017) note that after shearing, the permeability enhancement is retained, which suggests the fault is permanently propped open due to shear slip (Fig. 10.13b,c). Overall, although the available laboratory data is somewhat sparse, it does suggest that sheared fractures may not lose permeability as rapidly with increases in

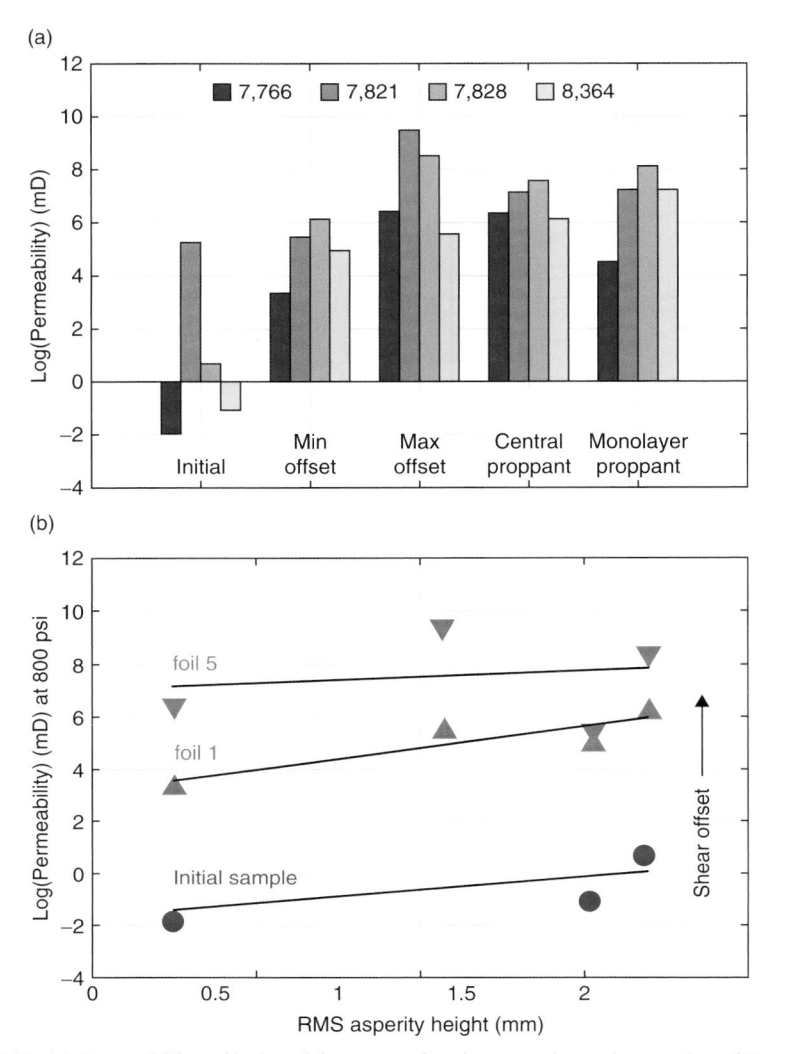

Figure 10.14 (a) Permeability of induced fractures of various roughness in samples of Barnett shale. The increase in permeability with 0.05 mm of offset is slight greater than simply loading the fractures with either a centralized or mono layer or proppant. (b) Permeability enhancement with offset is greatest in the roughest fractures. From Kassis & Sondergeld (2010).

normal stress as un-sheared fractures, which is illustrated schematically on the right side of Fig. 10.15.

Figure 10.16 (Wu et al., 2017) shows a series of experiments investigating the permeability of natural and saw-cut faults as a function of both confining pressure (dotted line) and shear (horizontal red lines) at constant pore pressure. Both clay-rich and calcareous (low clay) samples were tested. In each case, a 10–15 MPa increase of confining pressure (effective stress) resulted in a significant decrease in permeability of

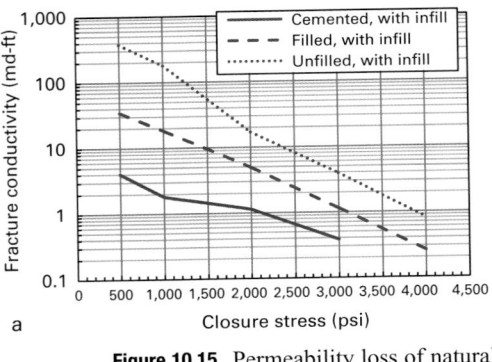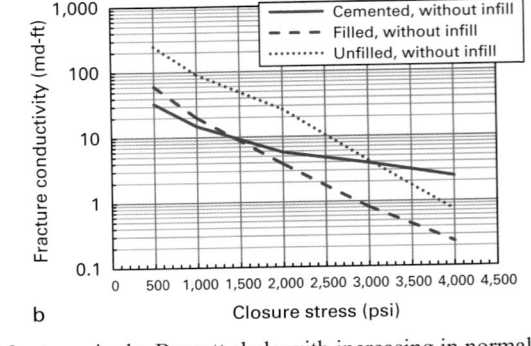

Figure 10.15 Permeability loss of natural fractures in the Barnett shale with increasing in normal stress for fractures with (a) and without infill (b). Adapted from Zhang et al. (2015).

almost an order-of-magnitude in most cases and slightly more for the case of the saw-cut in a clay-rich rock (Fig. 10.16a). This is similar to, but somewhat less than, what is observed by Zhang et al. (2015) (Fig. 10.14). Also note that the permeability loss with increasing effective confining stress is almost completely reversible in essentially every sample, indicating that the permeability loss appears to be associated with elastic deformation reducing the fracture aperture (a in Eqn. (10.1)).

What is perhaps most surprising in the results of Wu et al. (2017) is that shear slip has such a small effect on fault permeability and can either enhance or reduce permeability in different samples at the same stress conditions. On saw-cut faults (Fig. 10.16a) permeability increases with slip in both clay-rich and calcareous shales, whereas on natural faults permeability decreases with slip. This may reflect differences in the generation of wear products during slip. For rougher, natural faults, the formation of fault gouge (crushed material) may clog the fault aperture, counteracting the effects of shear slip and dilation. Gouge formation also depends on effective normal stress, which controls the sliding contact area and thereby controls the stress on the sliding contacts (Chapter 4). Figure 10.16b shows an experiment on a saw-cut, clay-rich fault intended to assess the impact of effective normal stress on permeability evolution due to shear. At the low effective normal stress of the other tests (~2.5 MPa), permeability increases slightly. At greater normal stresses (>5 MPa), permeability decreases over the same increment of slip, which may reflect enhanced production of gouge due to contact stresses exceeding contact strength. In similar experiments at much higher effective normal stress (93.5 MPa) on samples of Bowland shale, Rutter & Mecklenburgh (2018) observe an exponential loss of hydraulic transmissivity (permeability × layer thickness) of nearly two orders of magnitude over the same increment of slip. Similar behavior is observed at much lower normal stresses for relatively strong, brittle lithologies (sandstone and granite), which suggests the relationship between normal stress and contact strength strongly determines how permeability evolves during slip. It also important to note that in all the experiments of Wu et al., permeability increases during the first slip increment of ~0.1 mm. One inference that might be drawn from these results is that it is

Induced Shear Slip during Hydraulic Fracturing 321

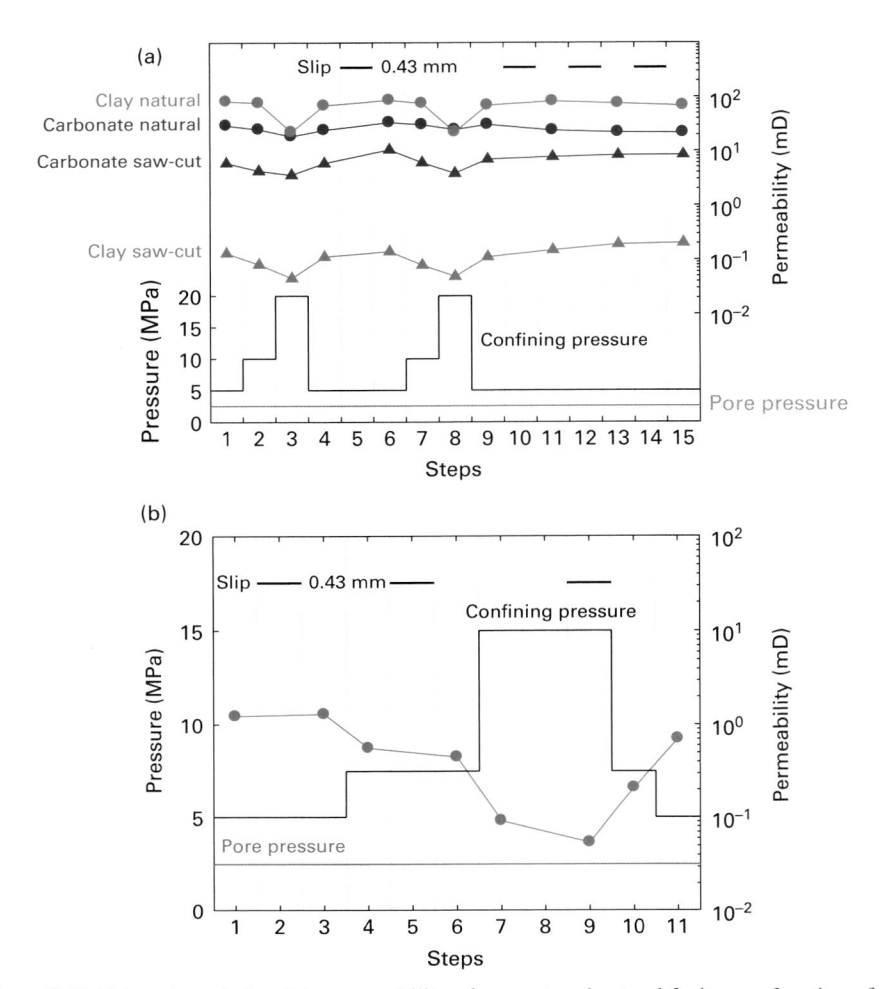

Figure 10.16 Laboratory study of the permeability of saw-cut and natural faults as a function of confining pressure and shear slip, Eagle Ford shale. (a) Saw-cut fault in a clay-rich sample (19–43–31 wt% Qtz–Carb–Clay + TOC). Saw-cut fault in a calcareous sample (9–86–4 wt%). Natural fault in a clay-rich sample. Natural fault in a calcite-rich sample. (b) Saw-cut fault in a clay-rich sample. Permeability is measured before and after slip at increasing effective stress. Adapted from Wu et al. (2017).

the initial slip when reactivation first occurs that has the greatest effect on permeability (as implied schematically in Fig. 10.11 and observed experimentally by Ye et al., 2017). Interestingly, at relative high normal stress, Rutter and Mecklenburg observe the inverse, i.e., the majority of the reduction in permeability occurs at <0.1 mm slip displacement. Another inference one might draw is that while permeability enhancement due to shear and dilation may be permanent (Fig. 10.13; Fig. 10.16), permeability reduction associated with depletion could be largely reversible.

11 Geomechanics and Stimulation Optimization

In this chapter we illustrate several ways in which geomechanical issues affect the success of particular well completion strategies. As it is impossible to cover the wide range of possibilities for each of the active unconventional plays, we attempt to illustrate several fundamental principles affecting completions that are related to the other topics we consider in this book. We present several instructive case studies and review modeling results that illuminate several ways in which seemingly subtle geomechanical principles have a significant effect on hydraulic fracture growth and distribution of proppant in hydraulic fractures. While the periods of trial-and-error testing used to inform optimal well completion parameters are critically important, it is equally important that the testing be done in a context in which the maximum amount of information as possible is obtained.

It is self-evident that a number of reservoir properties control the success of development. As shown in Table 1.2, the best wells in a play are far better than average wells. This could result from many factors – the thickness of the reservoir, pore pressure (as shown for the Utica play in Fig. 7.15), the amount and type of hydrocarbons in place and maturation (as shown for the Barnett in Fig. 1.8). It is obvious that variations of reservoir quality from one area to another will inevitably lead to different completion strategies; however, what is less obvious is how the magnitude of the least principal stress in the formations above and below the pay zone affect vertical (versus lateral) hydraulic fracture growth. This effect, coupled with stress shadow effects (which depend not only on cluster spacing but other parameters such as perforation diameter and flow rate) have a critically important role in determining where proppant is placed. Regardless, the issues of vertical fracture growth and proppant placement become extremely important when exploiting stacked pay and strategies that need to be developed and deciding how to drill the fewest number of wells to access the greatest number of productive intervals.

In the sections below, we start by addressing the issue of identification of optimal landing zones. While the search for *brittleness* (as discussed in Chapter 3) has frequently dictated this topic, we offer some possibly useful new insights into why stress magnitudes change from one formation (or lithofacies) to another and how this might affect choice of an optimal landing zone. In the following section we consider several studies addressing the topic of choosing optimal drilling and completion strategies, first, from the perspective of reservoir simulation and pilot tests. After discussing issues related to vertical hydraulic fracture growth, we revisit the topic of finding optimal drilling and completion strategies from the perspective of the coupling between geomechanics and completion methodologies.

The final two topics considered in the chapter are issues related to the lithologic factors that affect vertical fracture growth such as the variation of the magnitude of the least principal stress with depth due to viscoplastic stress relaxation (introduced in Chapter 3) and whether or not it might be practical to try to use geophysical techniques to target areas which might be more productive (introduced at the end of Chapters 2 and 7).

Landing Zones

In some ways, the issue of choosing optimal landing zones for horizontal wells drilled in unconventional reservoirs would seem straightforward. One would logically want to be stimulating the most hydrocarbon rich interval in a formation that was relatively brittle. Brittle formations may be easier to drill and hydraulically fracture than ductile clay-rich formations and may also contain more pre-existing fractures. An example of this is shown in Fig. 11.1 (from Patel et al., 2013), in which the stages carried out in a clay-rich zone (near stages 9 and 10) in well A near the lower Eagle Ford were characterized by fewer microseismic events and less gas production.

Expanding on the straightforward message in Fig. 11.1 in a study in the Utica formation, Gourjon & Bertoncello (2018) pointed out that *"completions indicate that the productivity decrease originates from limited fractures propagations when the completion is initiated in the clay-rich facies. Stages completed in facies A show a high near well-bore pressure loss and a low net pressure, which is consistent with the notion of shale choke, where fracture propagation is limited to the near well-bore. On the contrary, stages placed in the brittle facies C show high net pressures and low near well bore pressure losses, consistent with well-developed fracture geometry in the far-field."*

Figure 11.2 (from Ma & Zoback, 2017b) revisits one of the cases introduced in Chapter 1 (Fig. 1.16) to illustrate the importance of paying close attention to well paths during drilling and completions. The figure shows two wells drilled in the Woodford formation in Oklahoma. The upper part of each figure shows the clay + TOC content as measured with an elemental spectroscopy logging tool (the fluctuating gray line with the black line representing the median), the measured ISIP for stages (red dots and sticks) and the amount of proppant injected during each stage (blue histogram). The center panel represents the location of fractures seen in an image log. The lower panel shows the well path, position of the stages, the median clay + TOC content and three distinct lithofacies of the Woodford (WDFD-1, WDFD-2 and WDFD-3) that were designated from compositional data obtained from a nearby vertical well. The apparently severe undulations of the well is an artifact of the extreme vertical exaggeration of the vertical scale. The toe of the 4000 ft lateral well is only 40 ft above the heel. In both wells, note that whenever the well trace deviates out of the WDFD-2 lithofacies, there is a marked increase in the magnitude of the least principal stress, or frac gradient. In the case of both wells, no proppant was injected during these stages because of the high pumping pressure and concern about screening out (proppant filling the wellbore). Note also the excellent correlation between the magnitude of the least principal stress/frac gradient with clay + TOC content. As discussed in Chapter 3 and later in this chapter, these high frac gradients are likely

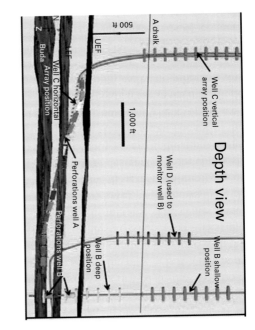
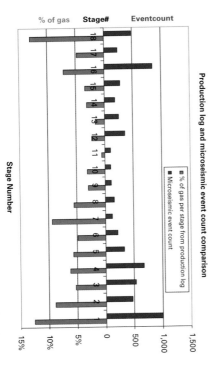
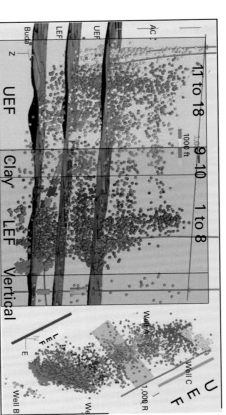

Figure 11.1 Cross-section view of the well path of horizontal well A in the Eagle Ford, the well path and map view of microseismic events (each stage is indicated by a different color) and the number of microseismic events and relative gas production from each stage. From Patel et al. (2013).

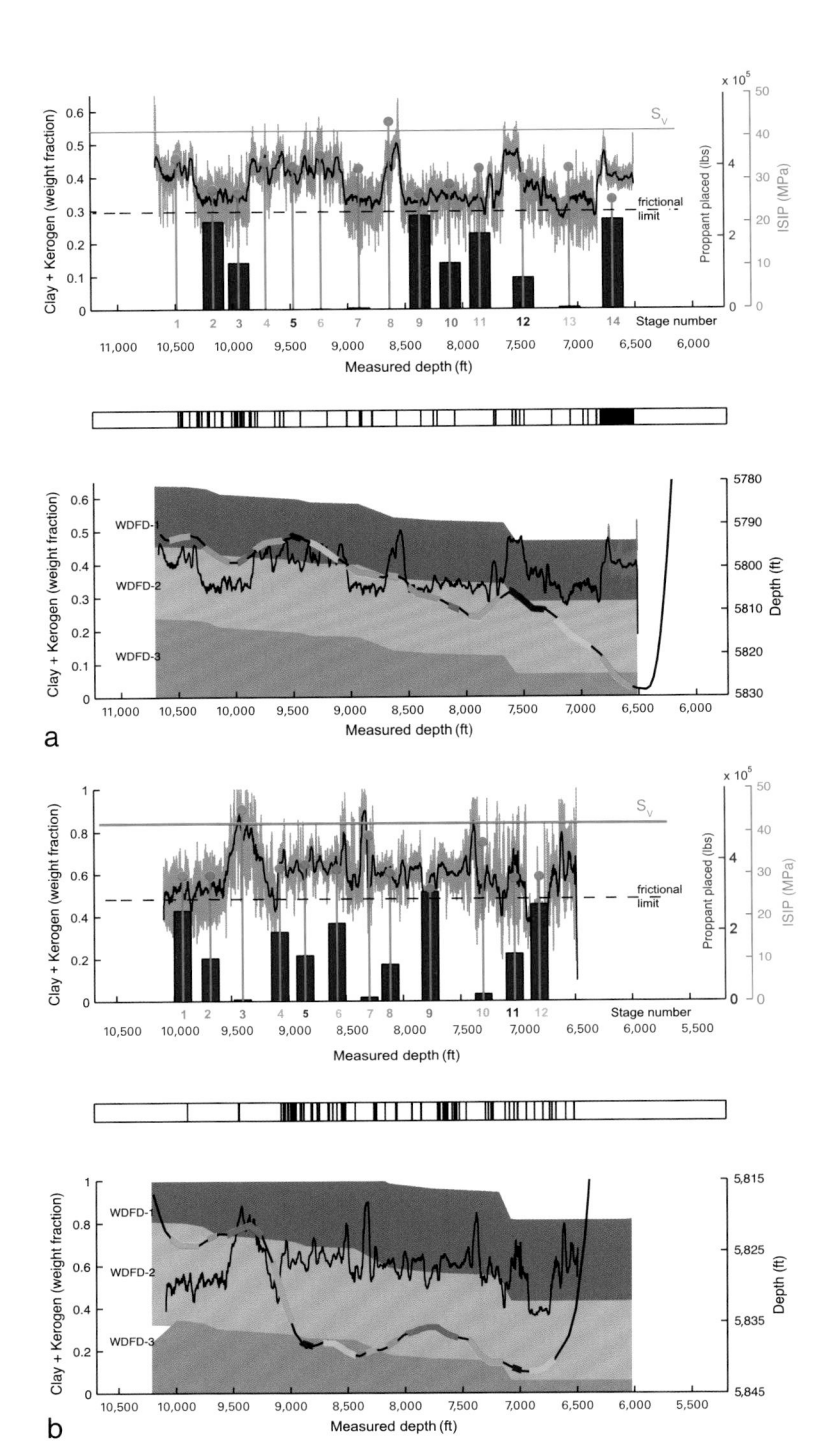

Figure 11.2 Correlated well information, treatment data and lithology of two horizontal wells in the Woodford formation, Oklahoma. The upper panel in each shows the variation of clay plus kerogen content (black curve represents the smoothed log readings in gray). The ISIP values (shown by red dots) and the amount of proppant placed during each HF stage (blue bars). The ISIP and placed proppant amount are positioned at the middle of the respective stages. The horizontal red line indicates the magnitude of overburden stress. The dashed black line indicates the magnitude of S_{hmin} at which normal faulting would occur. The middle panel shows the locations of steeply dipping natural fractures along the well from an image log. The lower panel shows the well trajectory (note the exaggerated vertical scale) and the content of clay plus kerogen along the well and separation of the Woodford into the WDFD-1, -2 and -3 lithofacies. From Ma & Zoback (2017b).

the result of viscoplastic stress relaxation. In fact, as first seen in the clay-rich rocks encountered in the Multiwell Experiment project (Fig. 3.19), complete stress relaxation would result in frac gradient essentially equivalent to the overburden stress. This is seen in stage 8 for the well shown in Fig. 11.2a for the stage where the clay plus kerogen content was highest. Other stages where the clay plus kerogen is high in this well also have high values of the frac gradient. In the well shown in Fig. 11.2b, there also is an excellent correlation between the clay and kerogen content and the frac gradient; the stages with the highest frac gradient are those with the highest clay plus kerogen content.

Figure 11.3 revisits the other case introduced in Chapter 1 from the Marcellus play in West Virginia (from Alalli & Zoback, 2018). The top panel of Fig. 11.3a shows the microseismic event distribution for a well in which stages 11–17 show the events to be limited within the Lower Marcellus shale (LMRC) and with a vertical extent of less than 100 ft, while the events from stages 18–21 show microseismic events extending upward over 500 ft and past the Upper Marcellus shale (UMRC). The middle panel shows the well trajectory. It is clear that perforation locations for stages 1–17 were placed in the Lower Marcellus shale, while perforations along stages 18–21 were in the Cherry Valley limestone. The lower panel of Fig. 11.3 shows the measured ISIPs. Note that for stages 1–16, the ISIPs were approximately equal to (or slightly greater than) S_V, while stages 17–20 show the measured ISIPs to be less than S_V, thus corresponding to S_{hmin}. Because hydraulic fractures propagate in a plane perpendicular to the least principal stress, the values of the least principal stress for stages 1–16 as well as the distribution of micro-seismic events for stages 11–16 indicate horizontal fracture propagation. Analogous data for another well is shown in Fig. 11.3b and yields similar conclusions. Stages 10–16 show the microseismic events are confined to the Lower Marcellus shale and show limited vertical extent, while stages 17–21 show microseismic events extending upward over 500 ft, past the Upper Marcellus shale. Looking closely at the well path, it can be seen that stages are split between being placed in the lower part of the Lower Marcellus shale and the Onondaga limestone, with stages 1–4 and 16–21 in the Onondaga lime-stone and stages 5–15 in the Lower Marcellus shale. When comparing the ISIPs across the horizontal well, stages 4–16 show measured ISIPs to be approximately equal to or slightly greater than S_V, while stages 1–3 and 17–20 show measured ISIPs less than S_V. We thus interpret the least principal stress to be S_V, resulting in horizontal hydraulic fractures for stages 5–15 in the Lower Marcellus shale.

Both of the case studies shown in Figs. 11.2 and 11.3 highlight the importance of changes in stress magnitude with depth. Below, we will show that vertical variations in stress magnitude can have a profound effect on vertical hydraulic fracture propagation which, in turn, affects horizontal propagation and proppant placement. The fact that ISIPs for the stages associated with horizontal hydraulic fractures approximate the overburden stress suggests that it is a reliable indicator of the least principal stress, and represents perhaps 2–3 MPa (a few hundred psi) higher than the pressure at which the hydraulic fracture actually closes (McClure et al., 2016). Another take-away from these two case studies is the importance of distinguishing vertical variations of the state of stress and formation properties from variations along the lateral.

Geomechanics and Stimulation Optimization

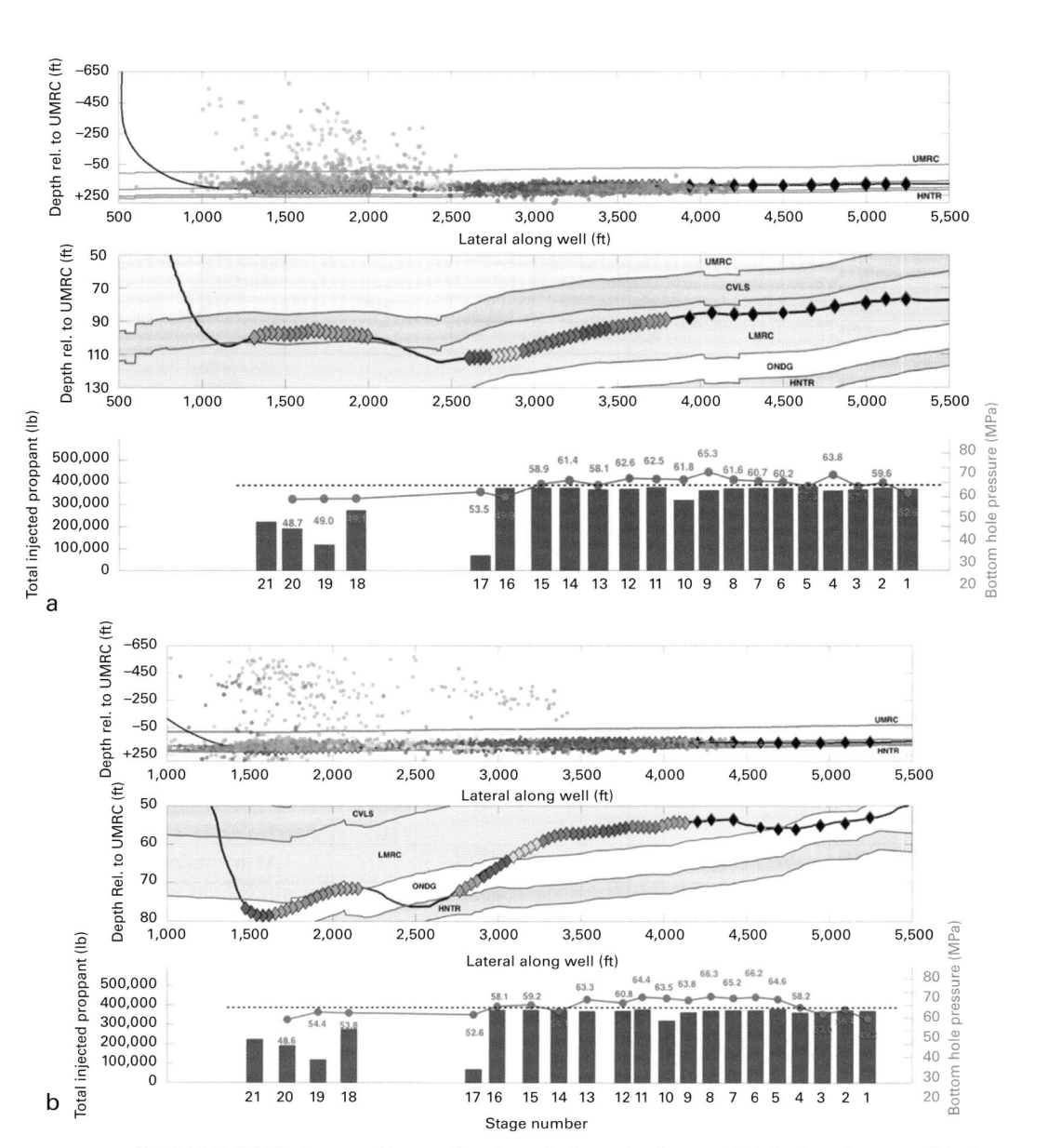

Figure 11.3 (a) The top panel is a profile view of microseismic event distribution along a well in the Marcellus formation for each fracturing stage; only stages 11–21 were monitored for microseismic activity. The middle panel is a vertically exaggerated section showing the well trajectory. The bottom panel plots total injected proppant volume (bar plot) and measured ISIP (red dots) relative to S_V (black dashed line) across all stages. (b) Similar plot to (a) for another well in the area. From Alalli & Zoback (2018).

328   **Unconventional Reservoir Geomechanics**

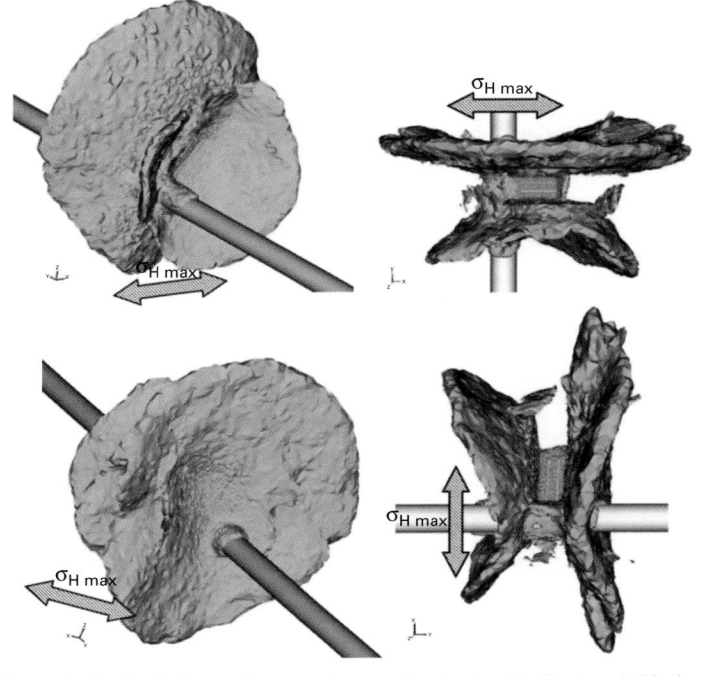

Figure 11.4 Numerical calculations of extremely complex hydraulic fracture initiation from perforations in a horizontal well drilled in shale with a nearly isotropic stresses. The iso-surface represents regions of Mode I failure. The gray arrows indicate the direction of maximum horizontal stress. From Ferguson et al. (2018).

Hydraulic Fracturing in a Near Isotropic Stress Field – As mentioned above, if the variation of the least principal stress is related to viscoplastic stress relaxation, the limiting case is when all three principal stresses become essentially equal to the overburden stress. Ferguson et al. (2018) modeled near wellbore hydraulic fracture initiation in which the relative magnitudes of S_V, S_{Hmax} and S_{hmin} were 1.0, 0.94 and 0.93, respectively. Figure 11.4 shows several views of the calculated opening mode strain surrounding perforations in a cased and cemented horizontal well after approximately 3 minutes of pumping. Note the extraordinarily complex pattern of fractures. Such a complex pattern of fractures could explain the concept of near wellbore *shale choke* referred to by Gourjon & Bertoncello (2018) and why hydraulic fracturing crews were nervous about screening out in the case study illustrated in Fig. 11.2 when the treating pressure suggested a near isotropic stress state.

Optimizing Completions I: Field Tests and Reservoir Simulation

As mentioned above, there is typically a great deal of trial-and-error testing of completion practices as companies attempt to develop different plays, or try to adapt completion practices within a play as conditions change from one area to another. Liang et al. (2017) present a very instructive overview of the approaches used by different investigators in different plays to determine optimal well spacing. Table 11.1 was adapted from Liang et al. (2017). Note the wide dispersion of *optimal* well spacing. For example, in the Niobrara, Li et al. (2017) suggest an optimal well spacing of 2,000 ft based on simulation, but Rucker et al. (2016) concluded that less than 200 ft is optimal based on field tests. In contrast, the simulation studies of Lalehrokh and Bouma (2014) and Siddiqui and Kumar (2016) yield similar results for retrograde gas in the Eagle Ford. The particular focus of the Liang et al. (2017) study was well spacing in the Midland and Delaware Basins. They compared (via simulation, field pilots and economic analysis) 660 ft and 880 ft well spacings (8 wells/ section and 6 wells/section, respectively) and proppant densities ranging from 300 to 1,000 lb/ft of well. For both the Midland and Delaware Basins, the larger well spacing and

Table 11.1 Estimates of optimal well spacing (modified from Liang et al., 2017).

Approach	Authors	Formation	Fluid type	Well spacing conclusions
Field pilot test and direct measurement of microseismic, pressure and tracer, and DNA sequencing to identify time-dependent drainage volume and inter-well communication	Friedrich & Milliken (2013)	Wolfcamp in Midland	Oil	400 ft
	Rucker et al. (2016)	Niobrara and Codell	Oil	Niobrara: < 200 ft Codell: < 700 ft
	Pettegrew & Qiu (2016)	Wolfcamp in Delaware	Oil	1,320 ft with 1,628 lb/ft
Operator data analytics	Sahai et al. (2012)	Marcellus	Gas	Variable for Marcellus
Numerical and analytical simulation. Fracture geometry or Stimulated Reservoir Volume (SRV) is directly assumed, i.e. planar and symmetric fracture with half length, conductivity and height are known or in a range	Sahai et al. (2012)	Haynesville	Gas	1,056 ft for Haynesville
	Lalehrokh & Bouma (2014)	Eagle Ford	Black oil; retrograde gas	330–400 ft for blackoil 440–450 ft for retrograde
	Yu & Sepehrnoori (2014)	Bakken	Black Oil	880 ft
	Siddiqui & Kumar (2016)	Eagle Ford	Retrograde gas	400 ft for single landing
Numerical and analytical simulation. Fracture geometry or Stimulated Reservoir Volume (SRV) is derived from Rate Transient Analysis (RTA). Microseismic data and production log	Belyadi et al. (2016)	Utica	Gas	1,200–1,300 ft
	Li et al. (2017)	Niobrara	Gas	2,000 ft
Numerical and analytical simulation.	Ramanathan et al. (2015)	Duvernay	Retrograde gas	200 m is not optimum

more dense proppant concentration was found to be preferable on the basis of the simulation, pilot testing and economic analyses.

In a study of well spacing in the Eagle Ford based on simulation and economics, Lalehrokh & Bouma (2014) concluded that reservoir permeability and fracture density were the two most important parameters controlling optimal well spacing. Their conclusions about optimal spacing are included in Table 11.1. This said, it is obviously difficult to single out any specific completion metric. In the case discussed above, both well spacing and proppant density were considered. But perforation cluster spacing, total treatment volumes and other parameters are also important, as is the time interval over which a given parameter is optimal.

In a comprehensive set of 3D reservoir simulations of several unconventional plays in the US, Sen et al. (2018) investigated optimal well spacing of wells, the spacing of perforation clusters and the size of completions over different periods of production. In some ways, their reservoir model is quite simple – they modeled the stimulated reservoir volume as a zone of enhanced matrix permeability that is symmetrically distributed around hydraulic fractures propagating away from a perforation cluster. The enhanced matrix permeability represents an effective permeability that simulates the bulk properties of the stimulated reservoir. The enhanced permeability in this stimulated reservoir volume is used to represent many small faults that slipped in shear due to leak off from the hydraulic fractures. Despite the simple geometry of their models, their approach is unique in several ways. First, they allow the hydraulic fractures to grow during stimulation – both in length and height. Second, they consider the area and magnitude of enhanced permeability to decrease with time – simulating the loss of matrix, shear fracture and hydraulic fracture permeability with depletion (Chapters 6, 8, 10). Hence, they refer to a dynamic SRV, or DSRV. Third, rather than simply attempt to history match well production, they sought models that would match stimulation treatment pressures, flowback rates, pressure build up tests and production, as well as microseismic event locations where available.

While the models are inherently non-unique, there are several interesting findings worthy of note. Figure 11.5 illustrates how the geologic setting of four different unconventional plays result in different DSRVs that respond differently to the completion methodology shown. Each panel represents, in plan view, a snapshot of maximum created SRV at the end of pumping for a single stage. As the hydraulic fractures are assumed to propagate symmetrically from the wellbore (located along the top edge of every panel) only one side of the well is shown. As illustrated in the figure, there is a significant variation of the number and spacing of perforation clusters per stage in the four plays. The colored area around each hydraulic fracture represents the region of enhanced permeability due to leak off (red indicates the greatest enhancement of effective permeability). As production is limited to the regions of enhanced permeability during the first several years (first and fourth cases), the perforation clusters are so far apart that there is likely no production from areas mid-way between the clusters. Once this was recognized, cluster spacing was decreased in these plays. It is also noteworthy how variable the areas of permeability enhancement are. In the first and fourth cases, these zones are immediately adjacent to the hydraulic fractures. In the second and third

Geomechanics and Stimulation Optimization   331

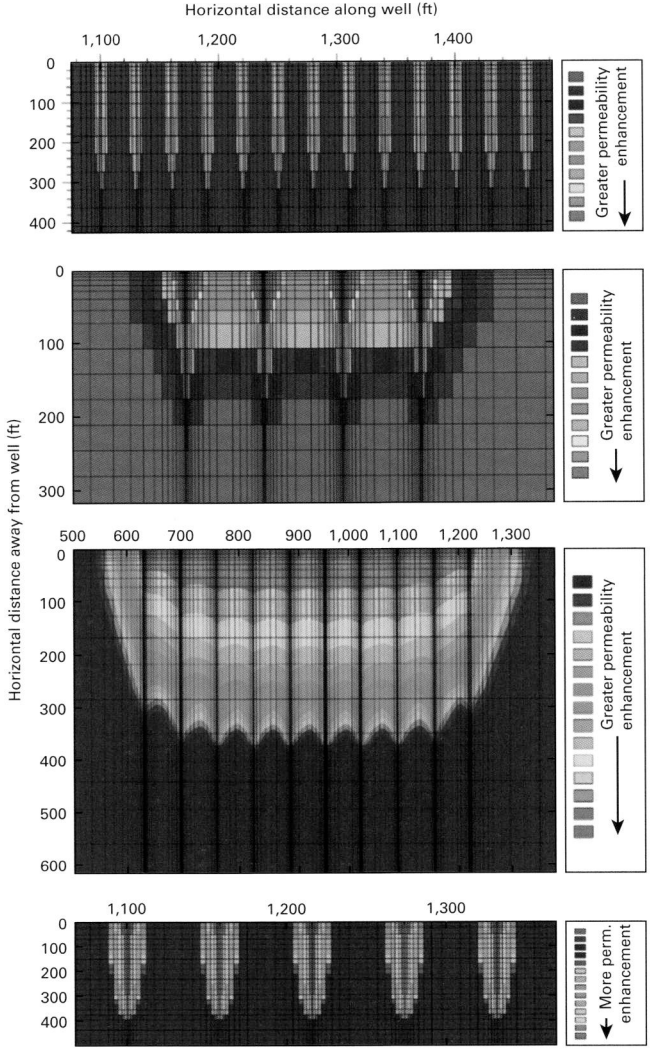

Figure 11.5 Plan views of models of permeability stimulation in four US onshore unconventional plays. Each panel represents a single stage. Note the strong variability in stimulated reservoir volumes surrounding a single hydraulic fracture propagating from a perforation cluster.
The position of the well is along the top of edge of each figure. Because of symmetry, only half the hydraulic fractures are shown. The degree of permeability enhancement is indicated by the colors. Each panel represents a single stage, with variable numbers of clusters. From Sen et al. (2018).

cases, there is more stimulation between the perforation clusters but the areas of greatest permeability enhancement (and likely production) are very close to the wells, which has obvious implications for well spacing. It is interesting to speculate whether the plays in which there are wider SRVs surrounding the hydraulic fractures are characterized by higher leakoff – perhaps due to the presence of more pre-existing fractures.

Unconventional Reservoir Geomechanics

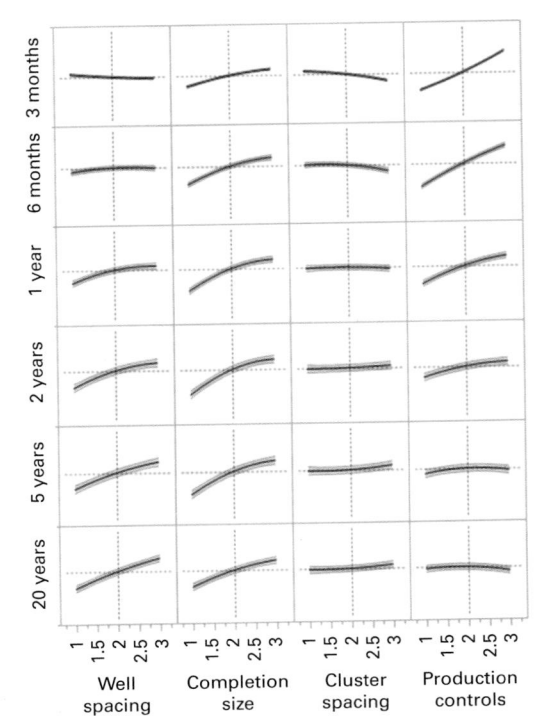

Figure 11.6 Generated prediction profiles for multiple wells in a single play. For each completion parameter, there are 3 month, 6 month, 1 year, 2 year, 5 year and 20 year production forecasts. For each curve, the vertical axis represents normalized cumulative production for the appropriate forecast window. The horizontal axis represents a range for a certain parameter, usually increasing from left to right. From Sen et al. (2018).

Applying the single-stage, single-well methodology shown in Fig. 11.5 to a number of wells in a single play, Sen et al. (2018) investigated parameters such as well spacing, completion size, cluster spacing and well control on cumulative production after time periods ranging from 3 months to 20 years (Fig. 11.6). Again, several interesting trends are observed. For example, the influences of both well spacing and completion size become increasingly important with time. Somewhat surprisingly, in this play, cluster spacing has a relatively second-order effect on production. In contrast, production parameters such as choke sizes and drawdown become less important as a function of time.

One clear benefit of the kind of analysis shown in Fig. 11.6 is that the development of this play with increased well spacing likely resulted in significant cost savings (offset slightly by increased completion costs) and more cumulative production.

Vertical Hydraulic Fracture Growth

Like the landing zone issue, there are aspects of vertical hydraulic fracture growth that seem somewhat self-evident. For example, Fig. 11.7 shows an example from Maxwell (2014) based on a case study reported by Inamdar et al. (2010) in the Eagle Ford. When the stages were carried out at a flow rate of 120 barrels/min (stages 1, 2 and 9), microseismic data indicates that the bottom hole pressures were sufficiently high that the hydraulic fractures propagated upwards, into and even above, the overlaying Austin chalk. At lower flow rates (stages 3–8), the microseismic events indicate that hydraulic fracturing was contained to the desired interval. There was also a slight tendency for downward propagation from the Buda Lime into the formation below (likely the Del Rio shale) when the stages were carried out at the higher flow rates. This can be easily explained if the net pressures at higher flow rates exceeded the stress barrier at the Upper Eagle Ford/Austin chalk contact, but did not exceed (or, did not exceed by much) the stress barrier at the base of the Buda Lime.

The utilization of microseismic data to assess vertical hydraulic fracture growth is not without potential pitfalls. In this book, we have presented several examples in which the vertical extent of microseismicity is affected by faults (e.g., the case study shown in Fig. 7.25). Many of the spikes representing anomalously shallow, or deep, microseismic events in the compilations of Fisher & Warpinski (2012) are likely due to pressure transmission along faults (Fig. 8.16). In other case studies, such as that shown in Fig. 8.15 (and that discussed above), the microseismic events seem to indicate that vertical hydraulic fracture growth is controlled by variations of stress from layer to layer with respect to the net pressure.

In the Earth, the issue of upward and/or downward hydraulic fracture growth is controlled by the stress variation from layer to layer as well as variations in fracture toughness or K_{IC}, the critical stress intensity factor for Mode I fracture propagation. During hydraulic fracturing, the net pressure controls whether a fracture will propagate

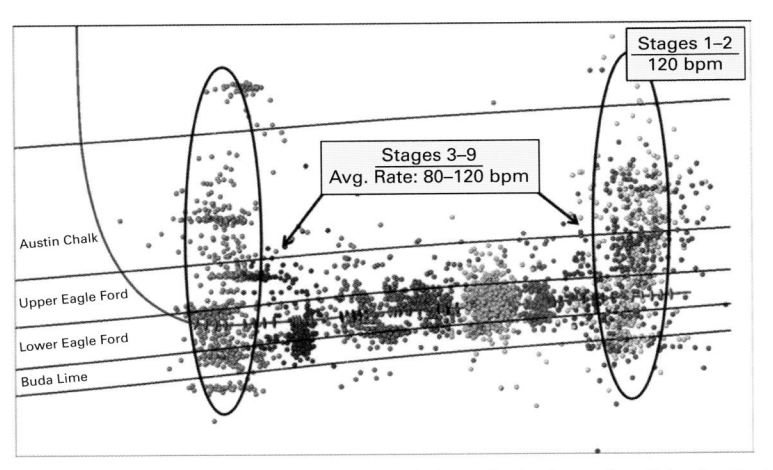

Figure 11.7 Cross-section view of multi-stage stimulation of a horizontal well in the Eagle Ford shale formation. Different colors represent different stages. The first two stages were carried out at 120 bpm which resulted in microseismicity well above the Austin chalk. In stage 3, the injection rate was decreased to 80 bpm and slowly increased for the remaining stages, resulting in microseismicity contained to the target lower Eagle Ford. Stage 9 was again at 120 bpm and again showed significant upward height growth. From Maxwell (2014).

through a layer with a change of stress or toughness. While variations in elastic properties do not affect fracture propagation, they do impact proppant transport, particularly Young's modulus, which controls how much a hydraulic fracture opens for a given net pressure. For the sake of simplicity, we are going to ignore decoupling at bed interfaces. While this assumption may seem restrictive, the successful exploitation of unconventional plays characterized by relatively thin bedding (such as the Eagle Ford, for example), seems to indicate that hydraulic fractures propagate across thinly bedded interfaces.

Liu & Valko (2015) considered hydraulic fracture growth as a function of net pressures in layered sequences with varying stress magnitude and fracture toughness using a PKN-type model (Fig. 8.6a). Their 5-layer model was based on variations of stress reported by Warpinski et al. (1994) in which layer-to-layer stress varied from about 1,500 to 2,400 psi, but Liu & Valko present calculations for a number of combinations of stress magnitude and fracture toughness. For each given model, they evaluated height growth as a function of the net pressure. While they showed many calculations, a general observation in their calculations is that for the layer-to-layer stress variations, net pressures less than about 1,000 psi resulted in limited hydraulic fracture height growth. When net pressures exceeded 1,000 psi, they often found significant upward or downward hydraulic fracture growth. We will revisit this issue below through modeling of an actual case study.

Incorporating variations of K_{IC} in hydraulic fracture models is difficult for multiple reasons – there is a limited number of measured values for representative rocks and there are no generally accepted correlations between K_{IC} and parameters from geophysical logs, for example. In the calculations of Liu & Valko, they show that stress variations are more important than variations in K_{IC} when they used what is referred to as *relatively low* values of K_{IC} of ~2,000 psi ft$^{1/2}$. Data from Zhang (2002) suggests that even lower values of K_{IC} are representative of sedimentary rocks such as shales, limestone and sandstone.

In a study utilizing a 3D fully coupled numerical code, Huang et al. (2018) investigated vertical hydraulic fracture growth that took into account not only the variation of stress with depth, but the presence of pre-existing faults which may be activated in shear during hydraulic stimulation. While it is difficult to generalize their results, the modeling makes very interesting predictions related to the way in which shear slip on pre-existing faults can either encourage, or discourage, the vertical growth of the hydraulic fractures. In the cases which shear slip encourages hydraulic fracture growth, their modeling illustrates how the slip helps overcome stress barriers. Hence, while we distinguish between observations of vertical migration of microseismic events to represent either vertical hydraulic fracture growth or pressure transmission along a fault, this study illustrates the possibility that both phenomena could be happening simultaneously.

Optimizing Completions II: Reservoir Simulation and 3D Geomechanics

In this section, we revisit the topic of optimization of operational parameters, now incorporating two important geomechanical effects – the variation of the least principal stress, or frac gradient, from layer to layer (as discussed above) and the stress shadow effect that arises simultaneously with propagating hydraulic fractures during plug-and-perf operations (Chapter 8). To carry out the calculations, we will use ResFrac, a software package that was used in the modeling presented in Figs. 8.19–8.21. As input to the models, we use a 5-layer, idealized geomechanical model that corresponds to an actual case study. The key geomechanical parameters are given in Table 11.2. The layer thicknesses and measured values of S_{hmin} from DFIT tests are realistic. The porosity, permeability and Young's modulus values are reasonable, but uniform, values. As the organic content is concentrated in formation D, the operator lands the laterals in that formation to achieve maximum production.

In interpreting the following simulations it is important to recognize that there are many geologic complexities that could affect hydraulic fracture propagation and proppant transport. There are also assumptions made in the calculations to make the incorporation of complex physical processes and interactions possible. Moreover, we now know there are processes such as propagation of multiple hydraulic fractures in close proximity (as discovered in the core through/drill through experiments discussed in Chapter 8) that are not yet sufficiently well-understood to be incorporated into models such as ResFrac. That said, we present the following simulations with the expectation that many of the large-scale patterns that are observed are representative of what is happening during plug-and-perf operations, and therefore are applicable for the area where the geomechanical model was developed.

The model results presented in Figs. 11.8–11.12 represent a single plug-and-perf stage and have a number of common features: (1) Injection occurs in a 200 ft stage with perforation clusters 50 ft apart and 25 ft from the plugs at each end of the stage. (2) There are 12 perforations per cluster over a 2 ft interval, spaced uniformly 60° apart. (3) Slickwater with a specified concentration of proppant is injected at a specified constant flow rate for 2 hours. As shown in the top left of Figs. 11.8–11.10, we assume a slowly increasing value of the least principal stress (frac gradient) in each layer. In Figs. 11.11 and 11.12 we investigate the impact of this assumption.

Figures 11.8, 11.9 and 11.11 are presented as follows: Each figure is a cross-sectional view of a hydraulic fracture propagating from one of the perforation clusters with a looking-down-the-well perspective. The hydraulic fracture cross-section on the left

Table 11.2 Values used in 3D fracture growth calculations.

Formation	Top depth (ft)	Thickness (ft)	Measured S_{hmin} (psi)	Permeability (nD)	Porosity (%)	Young's modulus (10^6 psi)
A	9,500	60	10,440	50	5	4
B	10,010	240	10,000	50	5	4
C	10,250	247	10,590	50	5	4
D	10,497	85	11,080	50	5	4
E	10,582	418	11,430	50	5	4

corresponds to the perforation cluster closest to the toe of the well. Each figure is a snapshot taken two weeks after the end of stimulation (to give the proppant plenty of time to settle), but before production began. The color indicates the hydraulic fracture conductivity in units of mD-ft, which is controlled by the amount of proppant in that part of the hydraulic fracture. Note that it is a log scale such that the portions of the hydraulic fractures that are well-propped (as shown in red) have conductivities that are two orders of magnitude higher than portions of the hydraulic fractures that are not propped (shown in green). The unpropped hydraulic conductivity values in these models are similar to values reported by Zhang et al. (2014) and Wang & Sharma (2018) for similar effective normal stresses.

Figure 11.8 shows three sets of calculations that demonstrate the sensitivity to various parameters. The figures in the top row of Fig. 11.8a show the four fractures expected for a flow rate of 50 bbl/min, perforations that are 0.32 inches in diameter with a proppant load of 1 ppg, which is equivalent to 6,000 barrels of fluid and 252,000 lb of proppant per stage. The well is located 15 ft from the bottom of formation D. We refer to this as the base case. Note that this case is shown in the top rows of Figs. 11.8a,b,c, 11.9a,b and 11.11 for easy reference.

The most dramatic aspect of the patterns of hydraulic fracture growth seen in Fig. 11.8a is that it is principally upward because, as shown in Table 11.2, the frac gradient in formation C is less than it is in formation D and less in formation B than it is in formation C. There is essentially no downward propagation of the hydraulic fracture into formation E, nor propagation upward into formation A because the frac gradients are higher in E than D and higher in A than B. Because of the upward hydraulic fracture growth, with the exception of the hydraulic fracture closest to the toe, there is relatively little proppant distributed laterally in formation D. Importantly, each of these observations has been confirmed by microseismic data. There are extremely few microseismic events below formation D, many microseismic events at the depths of formations B and C and none in formation A. Moreover, due to strong upward, not outward propagation of the hydraulic fractures, the microseismic events extend a limited horizontal distance away from the well.

Note also that the well-propped portions of the hydraulic fractures are mostly in formation C where there is relatively little organic matter. In addition, there are *pinch points*, areas of low hydraulic fracturing conductivity between well-propped portions of the hydraulic fractures and the well. The lower row of figures in Fig. 11.8a assumes all parameters are the same except that the well is positioned in the middle of formation D rather than near the bottom. In this case, the middle two fractures show that more proppant is distributed in formation D.

Figure 11.8b compares the base case (top row) with a perforation diameter of 0.64 inches in the lower row. Note that there is still severe upward growth of the hydraulic fractures and little proppant in formation D. Note also that no fracture formed in the second perforation from the heel end of the stage. Two factors contributed to this. In the base case, the maximum bottom hole pressure during stimulation was 12,100 psi. With the larger perforation size in the lower row of Fig. 11.8b, the maximum bottom hole pressure was only 11,700 psi. At this lower pressure the stress shadow appears to have prevented propagation.

Geomechanics and Stimulation Optimization

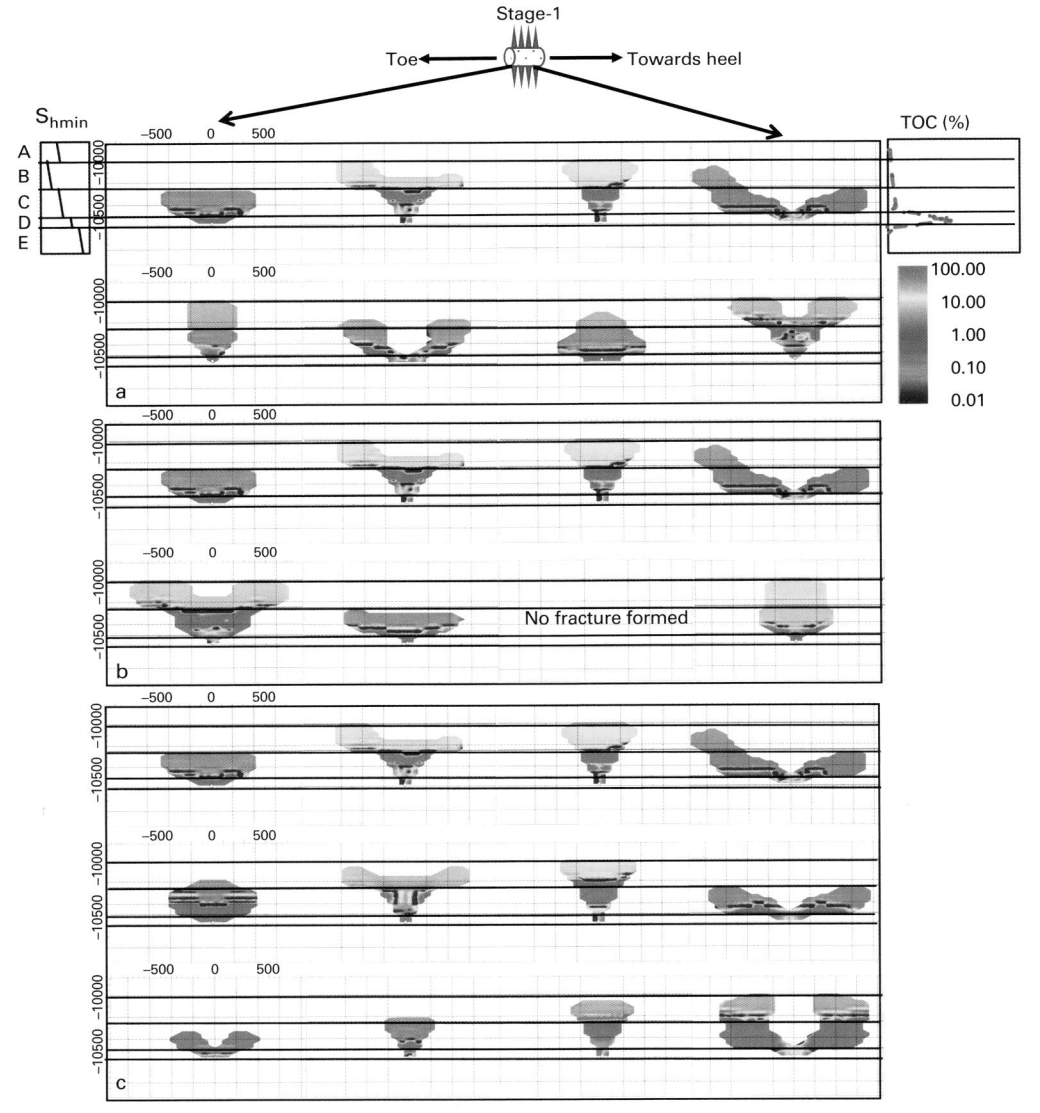

Figure 11.8 Cross-sections of hydraulic fracture conductivities (not the log scale) for three cases discussed in the text as computed with ResFrac. The top row of (a), (b) and (c) are the base case corresponding to a flow rate of 50 bbl/min, perforation diameter of 0.32 inches and a proppant load of 1 ppg with the well 15 ft above the base of formation D. (a) Comparison of well placement – the top row is the base case, the lower row places the well in the center of formation D. (b) Comparison of perforation diameters. the top row is the base case, the lower row uses perforations of 0.64 inches. (c) Comparison of proppant loads. The top row is the base case, the middle row is 3 ppg and the bottom row is 5 ppg.

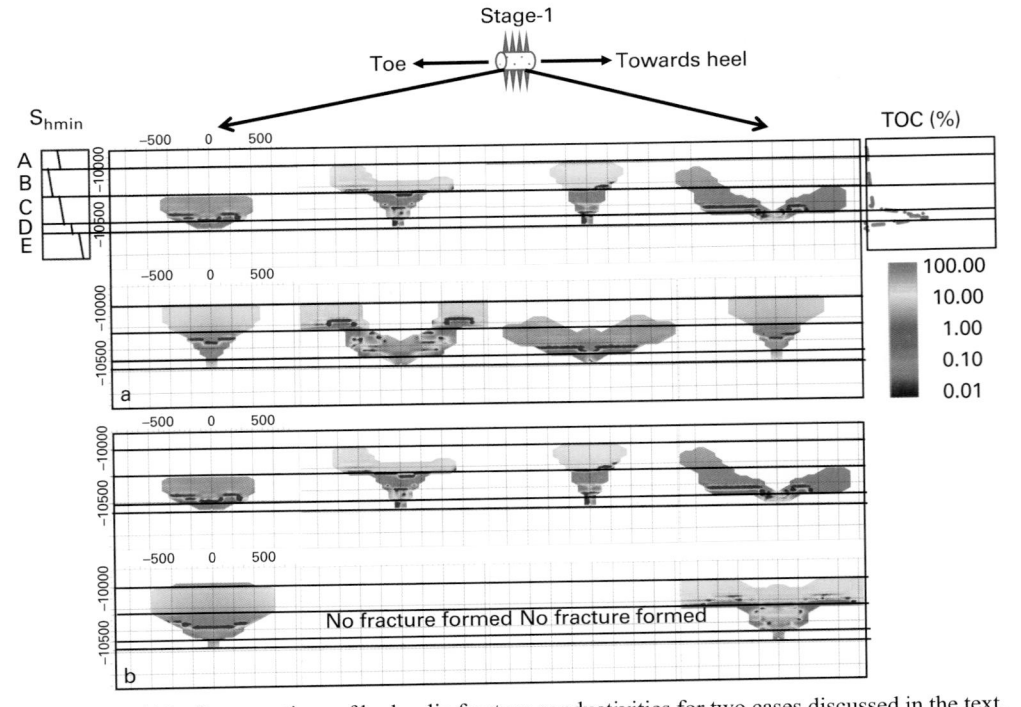

Figure 11.9 Cross-sections of hydraulic fracture conductivities for two cases discussed in the text. (a) Comparison of flow rates. The top row is the base case, the lower row is 100 bbl/min. (b) Comparison of perforation diameters. the top row is the base case, the lower row uses perforations of 0.64 inches and 100 bbl/min.

As shown in Eqn. (8.1), perforation friction is inversely proportional to the fourth power of perforation diameter, D. Hence, it has a significant effect on net pressures.

Figure 11.8c compares hydraulic propagation with different densities of proppant. The top row is the base case of 1 ppg, the middle row utilizes 3 ppg and the bottom row utilizes 5 ppg. Although there is considerably more proppant injected in the latter two cases, there is not a significant increase in the amount of propped area in formation D as one might hope. Visually, it appears that 3 ppg (the middle row of Fig. 11.8c) yields the best results.

Figure 11.9 makes two additional comparisons. Figure 11.9a compares the base case with an increase of the injection rate to 100 bbl/min. Note that the higher pumping rate seems to exacerbate the upward growth of the hydraulic fractures. Thus, even though there is twice as much fluid and proppant pumped, there is no significant improvement of the degree to which formation D is hydraulically fractured and propped. Figure 11.9b compares the base case (top row) with both an increased flow rate and an increased perforation size of 0.64 inches. Note that this combination of parameters yields even poorer results. There is more upward growth, no hydraulic fractures forming at the inner two perforations and no improvement in the propping of the hydraulic fracture in formation D.

Figure 11.10 shows relative production over a 5 year period – all produced volumes are normalized to the base case. Figure 11.10a compares production from the base case (with the well near the bottom of formation D), with placing the well near the middle or placing the well near the top (which is not shown in Fig. 11.8a). It is clear that placing the well in the middle of formation D yielded the best results. Figure 11.10b shows that both the 5 ppg case and the 3 ppg case result in increased production compared to the base case with the 5 ppg yielding the best results. Figure 11.10c compares production from the base case with increasing the diameter of the perforations, increasing the flow rate and increasing both the diameter of the perforations and the flow rate. Note that there is only an increase in production when the flow rate is doubled (as long as the perforation diameter remains 0.32 inches), but the other two cases result in less production.

Little Things Matter – As stress magnitudes generally tend to increase with depth in the upper crust, a slight increase in the least principal stress used in the calculations within each of the five formations above is not unreasonable. That said, a logical question is whether assuming an increasing frac gradient within each of the formations contributed to the strong upward growth of the hydraulic fractures. To test this, Fig. 11.11 compares the base case with constant values of the least principal stress in each formation that correspond to the five measured S_{hmin} values. There are clear differences between the base case (top row) and the stress profile with constant frac gradients in each formation (bottom row). However, both cases show considerable upward growth of the hydraulic fractures, little lateral propagation or proppant placement in formation D.

Figure 11.12 shows cumulative production from the two cases shown in Fig. 11.11. There is a negligible difference in the production between the two cases. It is probably reasonable to conclude that while the manner in which we represent the variation in stress with depth makes a difference, the overall fracture propagation and production trends are not an artifact of how the stress variation was modeled. As indicated in Table 11.2, the direct measurements of the least principal stress in each layer shows there is a progressive decrease in stress magnitudes moving upward from formation D. The frac gradient in formation C is less than it is in D, and it is less in B than it is in C. As mentioned above, microseismic observations confirm that formations A and E are good hydraulic fracture barriers. However, the relative magnitude of stresses in formations B, C and D suggests that careful modeling and careful operational control will be required to optimize production from formation D.

Viscoplastic Stress Relaxation and Varying Stress Magnitudes with Depth

Three mechanisms have been discussed in previous chapters of this book that could lead to variations of the magnitude of the least principal stress from one formation (or lithofacies) to another. As illustrated through case studies and modeling, these stress variations can have a first-order effect on stimulation. Stress magnitudes can vary because the frictional strength of well-oriented faults vary with clay content (Chapter 4), because pore pressure is changing (Chapter 7) or because of viscoplastic stress relaxation (Chapter 3).

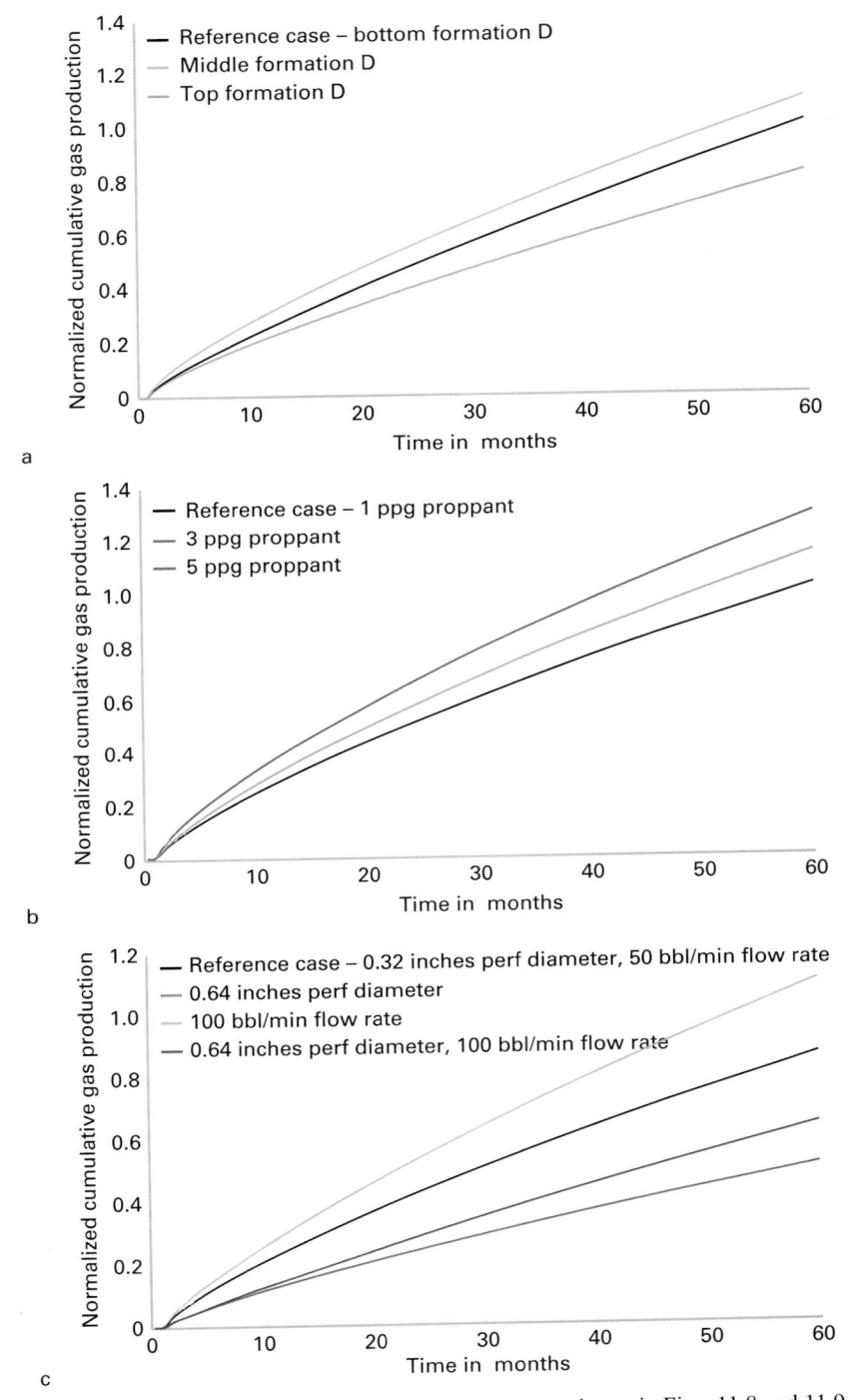

Figure 11.10 Cumulative production over 5 years for the cases shown in Figs. 11.8 and 11.9. (a) Landing the well in the middle of formation D yielded the most production. (b) A proppant concentration of 5 ppg yielded the most production. (c) The pumping rate (100 bbl/min) and perforation diameter (0.32 inches) yields the same production as doubling the flow rate. Increasing the perforation diameter results in less production, regardless of the pumping rate.

Geomechanics and Stimulation Optimization   341

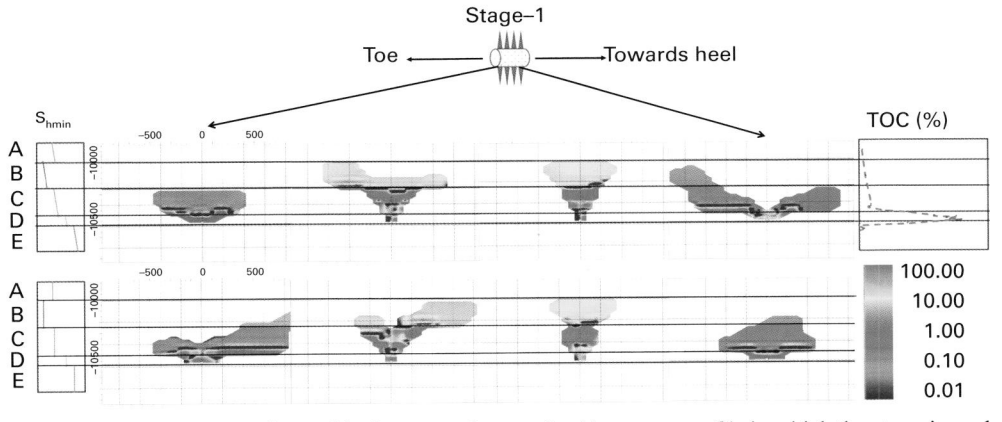

Figure 11.11 Comparison of the base case (top row) with a stress profile in which the stress in each layer is kept constant. Computed with ResFrac.

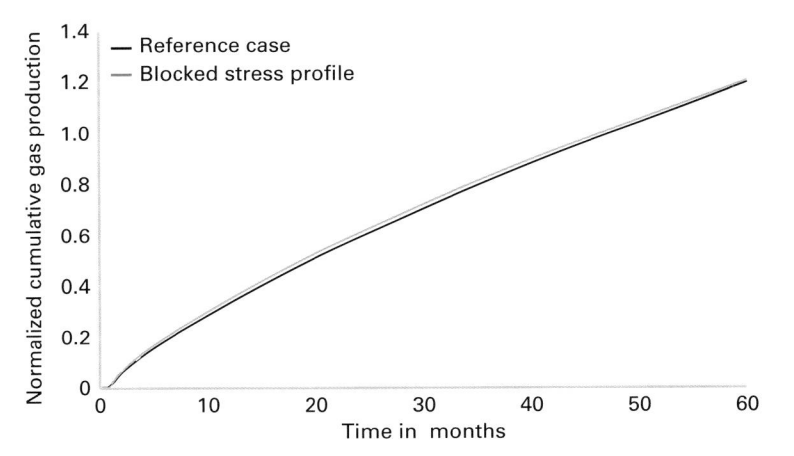

Figure 11.12 Cumulative production over 5 years for the cases shown in Fig. 11.11. From the perspective of cumulative production, the detailed manner in which the stress varies from layer to layer does not matter.

As discussed by Zoback (2007) models that assume stress is varying because of changes in elastic properties are based on faulty assumptions, burdened with numerous unconstrained parameters (such as anisotropic elastic parameters and arbitrary boundary conditions such as an instantaneously prescribed tectonic stress or strain).

Ma & Zoback (2018) attempted to explain the variations of stress with lithology seen in Fig. 11.2 with the viscoplastic stress relaxation model presented in Chapter 3. Neither variations of the coefficient of friction nor variations of pore pressure could explain the strong variation of stress between the three lithofacies of the Woodford. For example, when the frac gradient is approximately equal to the overburden in the high clay plus kerogen sections of Fig. 11.2 (which was previously shown for shale units in the MWX

342   **Unconventional Reservoir Geomechanics**

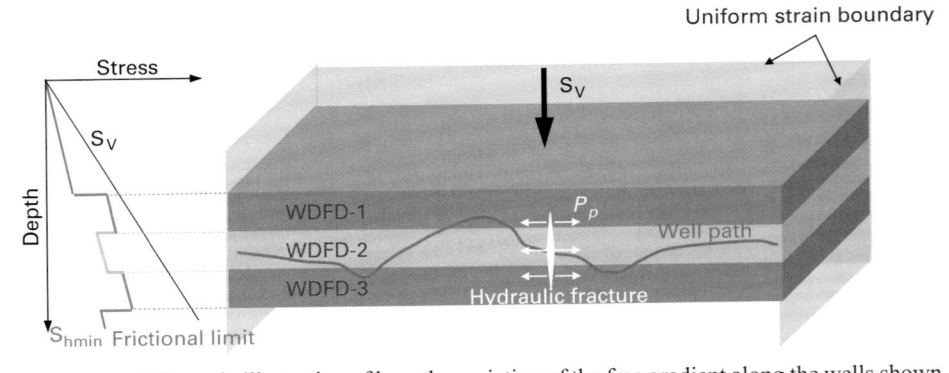

Figure 11.13 Schematic illustration of how the variation of the frac gradient along the wells shown in Fig. 11.2 results from variations of the degree of stress relaxation in the three lithofacies encountered along the well path. From Ma & Zoback (2018).

project (Fig. 3.19), or the Lower Marcellus in Fig. 11.3) there would need to be either lithostatic pore pressure or rocks with a coefficient of friction equal to zero to explain the observations. Figure 11.13 provides a conceptual explanation. As the well porpoised from lithofacies to another, there is more, or less, viscoplastic stress relaxation, resulting in a variation of least principal stress with depth. As more stress relaxation occurs (increased ductility), the state of stress in the rock becomes more isotropic. In normal, normal/strike-slip and strike-slip stress regimes, this means that the least principal stress, or frac gradient, becomes closer to the overburden stress, as previously illustrated in Fig. 3.17.

Unfortunately, for this area of the Woodford, no cores were available for laboratory testing. This required Ma and Zoback to establish an empirical relationship to estimate an upper and lower bounds for values of the viscoelastic power law parameter n (Eqn. 3.14), as well as to understand the relationship between n and the time it takes for viscoplastic stress relaxation to occur. Figure 11.14 addresses both of these issues. Figure 11.14a points out that in order for near complete stress relaxation to occur, n must be relatively high and stress relaxation must occur over tens of millions of years. As shown in Fig. 3.17, for highest n measured in the laboratory (0.08), less than 10% of the initial stress anisotropy would remain. Of course, rocks with higher values n may exist, but have not been tested.

Figure 11.14b shows an upper and lower bound of an empirical relationship between Young's modulus and n. While there is obviously considerable scatter, the resultant upper and lower bounds on predicted stress magnitudes is not particularly severe (Fig. 11.15). This suggests that the estimated degree of viscoplastic stress relaxation in the three Woodford lithofacies does a good job of explaining the observed variations of the least principal stress along the length of the horizontal wells.

Geomechanics and Stimulation Optimization   343

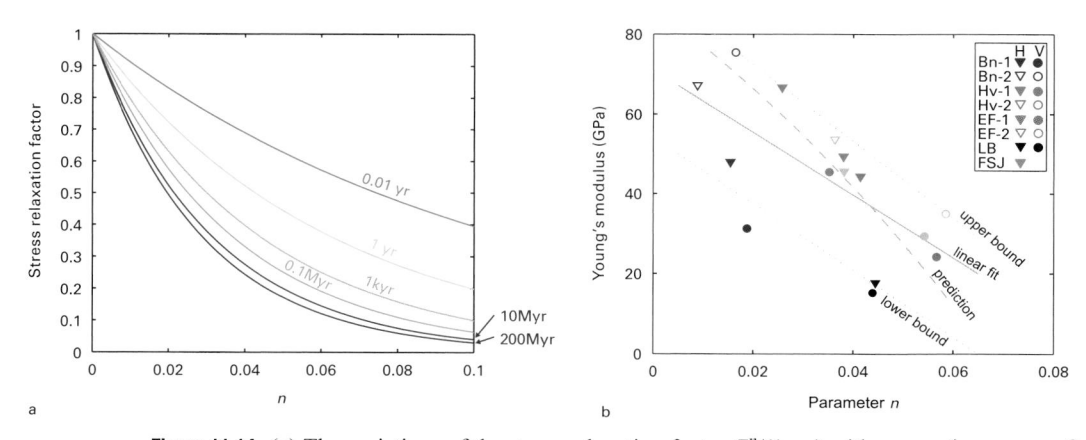

Figure 11.14 (a) The variations of the stress relaxation factor $t^{-n}/(1-n)$ with n over time spans of several days to years and to millions of years. (b) The correlation between Young's modulus and the relaxation parameter n for several shale reservoir samples. The upper and lower bounds are used in Fig. 11.15 to estimate upper and lower bounds on the magnitude of the least principal stress. From Ma & Zoback (2018).

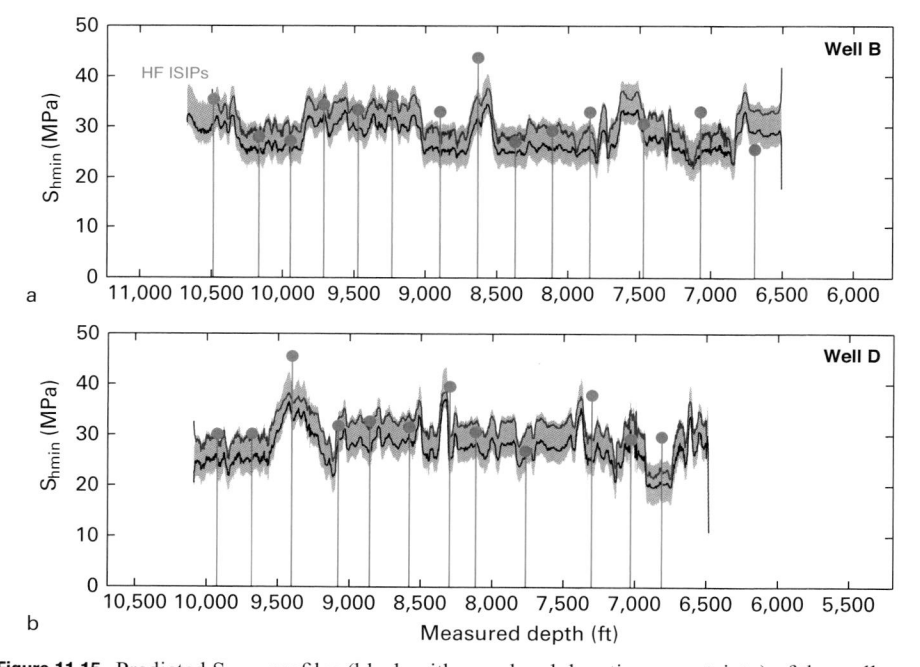

Figure 11.15 Predicted S_{hmin} profiles (black with gray band denoting uncertainty) of the wells shown in Fig. 11.2 based on the empirical relationships as compared with ISIPs (red dots) and predictions based on viscoplastic stress (blue). From Ma & Zoback (2018).

Targeting Geomechanical Sweet Spots: Fractures, Faults and Pore Pressure

To summarize the principal relationships between geomechanics and optimization of stimulation, there are at least five distinct areas that are important to highlight.

First, vertical variations of the magnitude of the least principal stress, whether from formation to formation, or from lithofacies to lithofacies within a given formation, have a direct impact on hydraulic fracture growth and thus the effectiveness of stimulation. While there are multiple mechanisms that could cause variations of stress magnitudes, lithology-dependent stress relaxation may be one of the most important. While prediction of stress magnitudes from log-derived rock properties may be possible someday, it is definitely possible to measure stress variations using mini-fracs or DFIT's as illustrated in Table 11.2. Stress magnitudes above and below the zone in which the lateral is located is just as important as the stress magnitude within it. Moreover, determination of optimal hydraulic fracturing procedures must utilize models that allow the impact of these stress variations on hydraulic fracture growth to be taken into account.

Second, well trajectories and landing zones must also consider the issue of hydraulic fracture growth and proppant placement. While finding optimally *brittle* formations has obvious merit, landing zones need to be chosen in the context of hydraulic fracture growth and the distribution of proppant.

Third, the presence of pre-existing fractures can be helpful in the stimulation process. While this is discussed in Chapter 10 from the perspective of how shear slip is triggered on these fractures during stimulation, Chapter 12 will address the importance of this process on production. At the end of Chapter 7, we also discussed the possible use of seismic reflection data for targeting areas with more pre-existing fractures. Such methods warrant further application in the characterization of unconventional plays.

Fourth, the presence of larger-scale faults (those that can be seen in seismic reflection data) is generally problematic. At the very least, they can *hijack* hydraulic fracture stages as demonstrated in a number of case studies presented in Chapter 7. More problematic are cases when pressurization of faults could exacerbate well-to-well communication (Chapter 12) or induce well-damage and/or seismicity (Chapter 13). Routine utilization of seismic reflection data to avoid pre-existing faults would seem prudent in some areas.

Finally, pore pressure is very important. As pointed out in Chapter 7, all active unconventional plays are overpressured, and to one degree or another, pore pressure is linked to stress magnitudes. It is logical that overpressure is important for driving flow from the pore space in formations with such extremely low matrix permeability (Chapter 6). As shown in Fig. 7.15 there is an excellent correlation between the area of very high pore pressure and production in the Utica. The considerable overpressure in parts of the Delaware Basin (Fig. 7.16) is also undoubtedly an important contributor to production there. That said, spatial heterogeneities of pore pressure, whether due to the source (i.e., related to the distribution of organic matter) or leakage (perhaps along large-scale faults), could certainly be one factor affecting the success of individual wells and whole pads. In Chapter 12, we discuss heterogenous depletion surrounding unconventional wells that can influence the *frac hit* or *parent/child* phenomenon associated with infill drilling.

12 Production and Depletion

As introduced in Chapter 1, production rates from unconventional wells decline very rapidly during the first 2–3 years of production. In this chapter we first demonstrate that production rates (and cumulative production) are dominated by linear flow from the almost impermeable matrix into much more permeable fracture planes. The permeable fracture planes consist of the hydraulic fractures themselves and the pre-existing fractures and faults that have slipped in shear during stimulation. As we show, the rapid decrease in production rates is a natural consequence of depletion in these extremely low permeability formations. We argue that the cumulative area of permeable fracture planes created during stimulation is a key factor influencing ultimate resource recovery. In a case study from the Barnett shale, we assess the total fracture area created during stimulation on a stage-by-stage basis and compare it with relative stage-by-stage production data from distributed temperature sensing.

The next section of this chapter uses a coupled poroelastic model with discrete fractures to illustrate how fracture networks evolve during stimulation. We discuss how the stimulated shear fractures that produce microseismic events may locally enhance the matrix permeability of the rock surrounding the slipping fractures. In the next section of this chapter we return to the question of seismic vs. aseismic slip introduced in Chapter 4. In other words, once the Coulomb failure condition is met and frictional sliding occurs on a pre-existing fault occurs, will there be seismic radiation? This question is essential to understanding whether there is an appreciable amount of slow shear slip on pre-existing faults that augments production but is not reflected by the microseismic cloud that surrounds hydraulic fractures. This was discussed in terms of shale composition in the context of rate and state friction in Chapter 4. In this chapter we discuss this question in the context of slip on highly mis-oriented fracture and fault planes that would not be expected to slip unless stimulated by very high fluid pressure. This idea also will be addressed in the context of induced seismicity in Chapter 14.

In the remainder of this chapter we address a unique aspect of depletion from unconventional formations. Because hydrocarbons are being produced from extremely low permeability formations via a network of interconnected, highly permeable fractures, the patterns of depletion are highly heterogeneous – which affects both production and production-induced stress changes. This allows us to better visualize how heterogeneous depletion is likely to be and provide insights into why the stimulated rock volume (SRV) determined from microseismicity overpredicts the likely production.

The discussion of depletion leads directly to understanding issues associated with well-to-well communication during hydraulic fracturing and the frac hit phenomenon

Unconventional Reservoir Geomechanics

mentioned previously. While there are different reasons why well-to-well communication may occur, one important mechanism is the poroelastic stress changes accompanying depletion. As infill drilling is occurring in many areas, this topic is of growing significance. The topic of well-to-well communication during infill drilling is sometimes referred to as *parent/child* well interactions. After reviewing a couple of illustrative case histories, we consider a theoretical model of poroelastic stress changes that help us understand when such stress changes could have a significant impact on infill wells. We conclude this chapter by discussing pressure-dependent reservoir properties such as matrix, hydraulic fracture and shear fracture permeability.

Production Decline Curves and One-Dimensional Flow

Figure 12.1 shows the average production data for thousands of unconventional oil and gas wells in four areas that was previously shown in Fig. 1.12. In Fig. 12.1 we have added the right column, which shows monthly production rates in log–log space to demonstrate that production rates decline with a rate similar to $t^{-1/2}$ (as indicated by the dashed lines). To first order, the production rate data indicate linear flow from extremely low matrix permeability to much more permeable planes. In linear flow, production rate is proportional to time$^{-1/2}$ as given by

$$q = \frac{1}{2}\frac{\alpha}{\sqrt{t}} \tag{12.1}$$

where, for dry gas reservoirs,

$$q = A\left(\frac{P_r^2 - P_{bhf}^2}{P_s}\right)\sqrt{\frac{c_g \phi_m k_m}{\pi \eta}} \tag{12.2}$$

where A is the total surface area of all fractures (the hydraulic fractures and shear fractures) in contact with the matrix, P_r is the reservoir pressure, P_{bhf} is the bottom hole flowing pressure, P_s is atmospheric pressure, c_g is gas compressibility, ϕ_m is matrix porosity, η is the gas viscosity and k_m is matrix permeability. For oil, c_g is replaced by a constant with units pressure^{-1} and the driving pressure term is slightly different (Katz, 1959). Cumulative production is obtaining by integrating Eqn. (12.1) and is thus given by

$$Q = \alpha\sqrt{t} \tag{12.3}$$

which is shown in the center column of Fig. 12.1. In log–log space, this implies that $\log(q)$ should decline with a slope of 1/2 as indicated by

$$\log(q) = \log\left(\frac{1}{2}\alpha\right) - \frac{1}{2}\log(t). \tag{12.4}$$

Production and Depletion 347

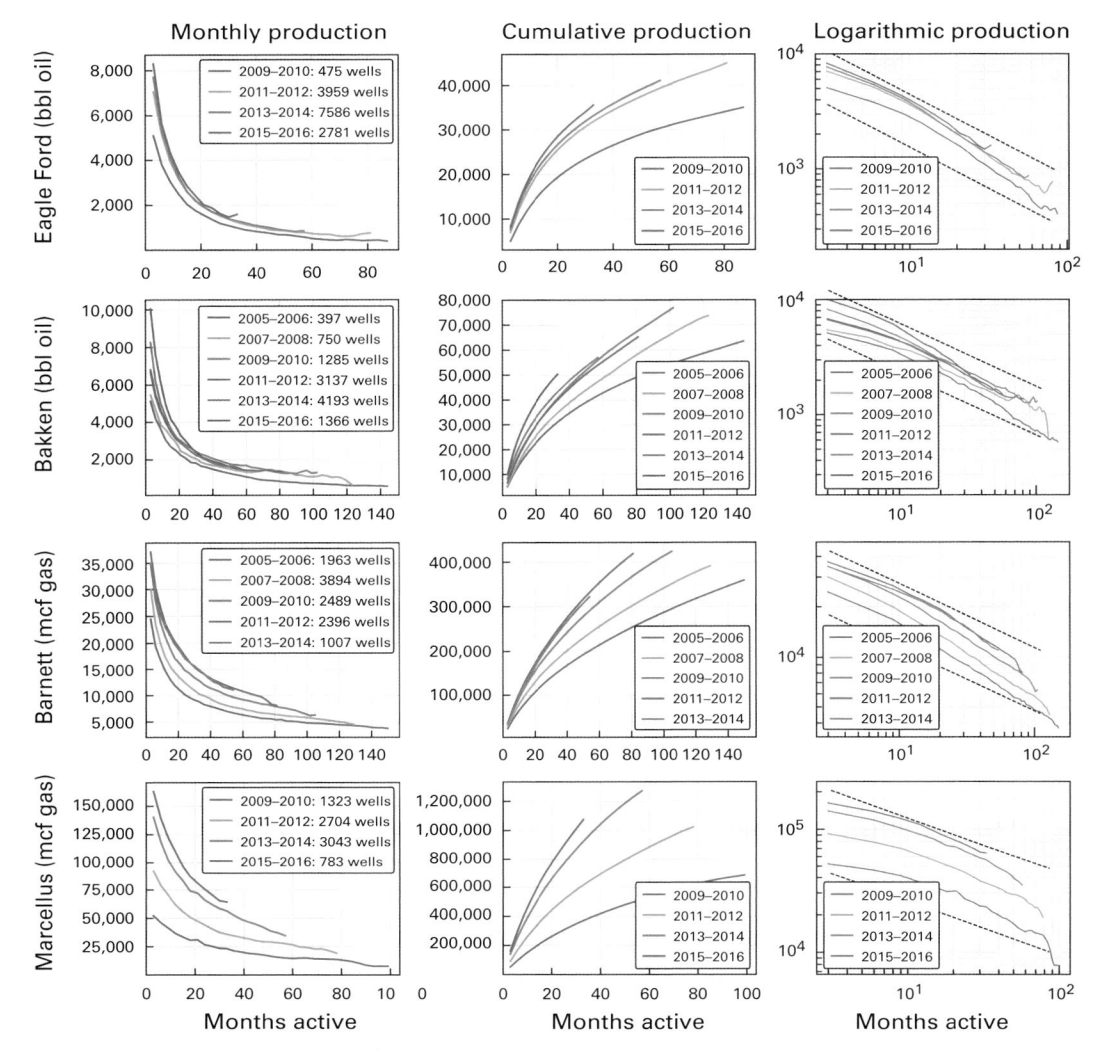

Figure 12.1 Production from four unconventional oil and gas plays is grouped and averaged in two-year increments (after Hakso & Zoback, 2019). The left column shows monthly production rates, the center column shows cumulative production and the right column shows monthly production rates in log–log space to demonstrate that production rated decline with a rate similar to time$^{-1/2}$, as indicated by the dashed lines.

During the first three years of production, the average slope for all two-year increments for all reservoirs shown in Fig. 12.1 is -0.503, with an R^2 value of 0.98, indicating linear flow during the first several years of production.

Linear Flow into Permeable Fractures – With data from fewer wells, Patzek et al. (2014) also noted average production rates decreasing as time$^{-1/2}$ in the Barnett shale. As indicated in Eqn. (12.2), if one considers stimulation of an average well for each of these reservoirs, cumulative production is going to depend on the total surface area in

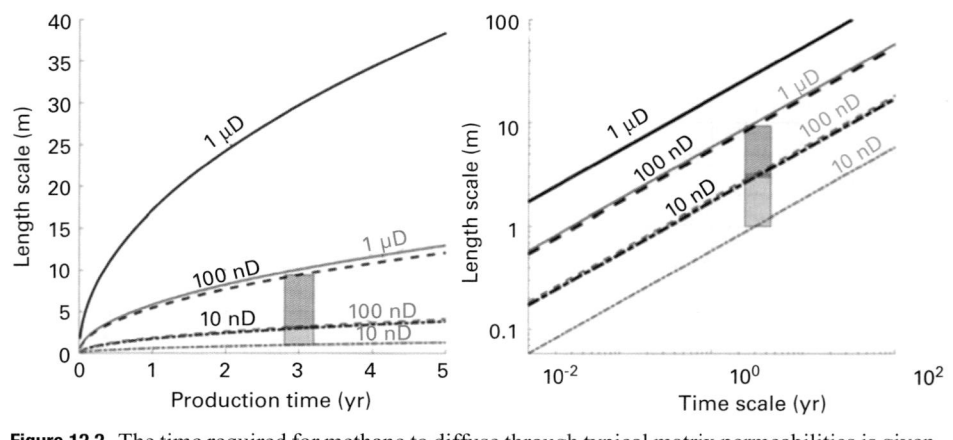

Figure 12.2 The time required for methane to diffuse through typical matrix permeabilities is given on a linear scale on the left and log–log scale on the right. The gray region indicates that approximately three years are required for gas to diffuse approximately 3–10 m through 10–100 nanodarcy matrix to a high permeability pathway. The corresponding time/distance relationship for oil is shown in red, with an approximately factor of 10 higher viscosity resulting in a corresponding decrease in diffusion distances. The gray and pink boxes reflect a representative range of matrix permeabilities for unconventional reservoirs. From Hakso & Zoback (2019).

contact with the reservoir. Linear flow implies flow from the matrix to a highly permeable plane in the direction normal to that plane, whether it is the hydraulic fracture or a stimulated shear fracture. The high permeability of the planes is such that pressure in the plane can generally be considered constant (Walton & McLennan, 2013). In the context of linear flow, particle transport can be modeled by a 1D diffusive process, for which the characteristic diffusion time, τ, is given by Eq. (12.5),

$$\tau = \frac{l^2}{\kappa} = \frac{(\varphi B_f + B_r)\eta l^2}{k} \qquad (12.5)$$

where l is the characteristic diffusion distance, $\kappa \approx k/\eta(\phi B_f + B_r)$ is the hydraulic diffusivity, B_f and B_r are fluid and rock compressibilities, ϕ is rock porosity, and η is fluid viscosity. Using mechanical properties of Barnett cores and the geomechanical setting, the characteristic diffusion times can be constructed for various permeabilities using $B_f = 3 \times 10^{-8} Pa^{-1}$, $B_r = 6 \times 10^{-11} Pa^{-1}$, $\phi = 0.1$ and $\eta = 3.5 \times 10^{-5} Pa\ s$.

The resulting relationships for natural gas and oil in linear and log space are shown black and red, respectively, in Fig. 12.2 (after Hakso & Zoback, 2019). The 3-year characteristic diffusion distance for gas in a 100 nanodarcy reservoir is approximately 10 m. Fundamentally, this means that in three years, gas can flow only about 10 m from a matrix pore to a permeable fracture. Hence, if a sufficiently dense network of fractures is not formed by stimulation, there will not be sufficient production in three years to justify the cost of development. Because of its higher viscosity, the situation is much worse for oil. For a matrix permeability of 100 nanodarcies, the diffusion distance is only about 2.5 m.

Using Microseismicity to Estimate Total Fracture Area

From Eqn. (12.2) and the small diffusion distances in the low matrix permeability of unconventional reservoirs, introducing as much surface area as possible through stimulation appears to be a critical step in increasing production. In the next section, we analyze the microseismic dataset shown in Fig. 10.3. Even though there were two monitoring arrays deployed in five different wells to decrease the distances to each of the five stages, microseismic event intensity varied significantly by stage, and event locations scattered across multiple stages (Hakso & Zoback, 2019). This could have been the result of (1) inaccurate event locations (as discussed in Chapter 9), (2) microseismic events propagating along pre-existing fault systems (as discussed in Chapter 7), or poor zonal isolation (as discussed in Chapter 8).

The area, S, associated with each event is calculated through the seismic moment, M_o, and stress drop, $\Delta\tau$. For a circular fault, area is given as (Stein & Wysession, 2003):

$$S = \pi \left(\frac{7M_o}{16\Delta\tau}\right)^{2/3} \tag{12.6}$$

Note that, perhaps counterintuitively, stress drop and fault patch size are inversely related. Calculating stress drops for microseismic data is difficult due to the limited bandwidth of the recorded seismograms. Because of the extremely small size of microseismic events, data recorded at frequencies of several thousand Hz would be needed to determine the corner frequency used to calculate stress drop (Fig. 9.17). From other studies, we know that stress drop distributions are scale invariant and generally follow a log-normal distribution (Imanishi & Ellsworth, 2006; Allmann & Shearer, 2009). A number of studies (e.g., Goertz-Allmann et al., 2011; Clerc et al., 2016; Cocco et al., 2016; Huang et al., 2016) have examined stress drops over an extremely wide range of earthquake sizes. Each study finds values in the range from 10^{-1} to 10^2 MPa (Fig. 9.18), with single digit median stress drops and no discernible trend with earthquake size as illustrated in Fig. 9.19. Hakso & Zoback (2019) calculated the areas of the microseismic events shown in Fig. 10.3 using a distribution centered on a stress drop of 0.5 MPa. Note that the radius is proportional to the cube root of the stress drop, mitigating the impact of any difference between assumed and actual stress drop distributions.

The cover image of this book is a visualization of the fracture networks implied by a portion of the microseismicity shown in Fig. 10.3. It shows a $100 \times 100 \times 100$ m visualization in the area with the most dense concentration of seismicity in the vicinity of stage 4. Fracture locations and orientations were assigned from microseismic event locations reported by the service company and focal mechanisms reported by Kuang et al. (2017). Independently knowing the orientation and relative magnitude of the principal stresses made it possible to select which nodal plane in the focal mechanism was the likely fault. Note that there is a wide distribution of shear fractures and fracture orientations in the area between the hydraulic fractures, implying a well-connected

350   **Unconventional Reservoir Geomechanics**

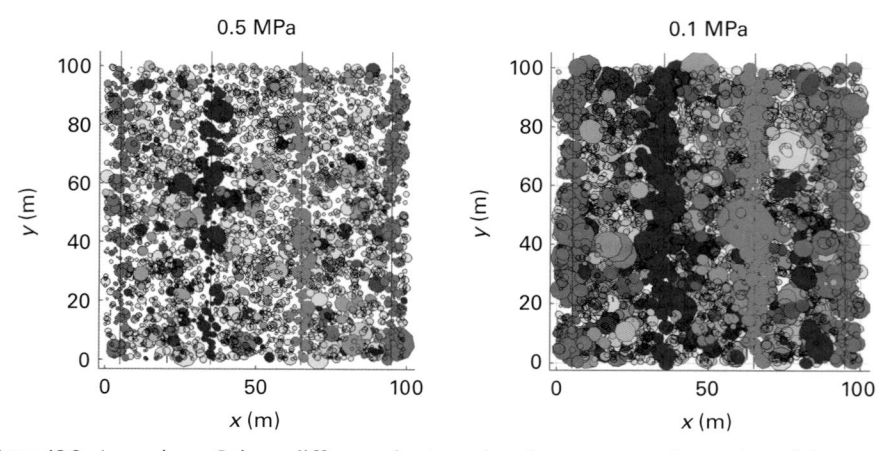

Figure 12.3 Assuming a 5 times difference in stress drop in a representative region of dense seismicity. The larger stress drop results in a fracture network with 40% smaller fracture radii, but otherwise identical characteristics. From Hakso & Zoback (2019).

network of shear fractures in contact with the low permeability matrix. Figure 10.3 shows plan views of fracture networks generated from stress drop distributions with median stress drops of 0.5 and 0.1 MPa to illustrate the impact of a 5-fold change in the median stress drop.

Unfortunately, given the large relative location uncertainties in the microseismic dataset, it is not feasible to consider these images as a precise discrete fracture network model (Hakso & Zoback, 2017). Because of this, and because the total production is controlled by the total area of permeable fractures in contact with the reservoir, the fracture networks modeled in this study should be viewed as statistically representative rather than a direct representation of the fracture network. In the discussion below, we will focus on the relative area created from stage to stage.

Correlating Area with Production – The eight histograms in Fig. 12.4 display the observed number of microearthquakes in the Barnett dataset for well 1 shown in Fig. 7.24 as well as the cumulative area created, calculated in the manner described above. The left side of the figure shows the relative stage-by-stage production data calculated from DTS data obtained at two different times during production (Roy et al., 2014). The total size of the stimulated fracture network associated with each stage varies dramatically, ranging from less than 100 m² in stages 1 and 8 to nearly 30,000 m² in stage 5. The total surface area created in all the stages was at least 76,000 m². As aseismic slip may be occurring during stimulation (as discussed below), the area we calculate from the microseismic events represents a lower bound of the created surface area.

It is clear that the stages with most production (4, 5 and 6) are also the stages in which the most surface area was created. While very little surface area was created by shearing in several stages, there is still a small amount of production, perhaps associated with the area of the hydraulic fractures themselves (or aseismic slip). The duration of injection required to achieve a significant amount of permeable area

Production and Depletion   351

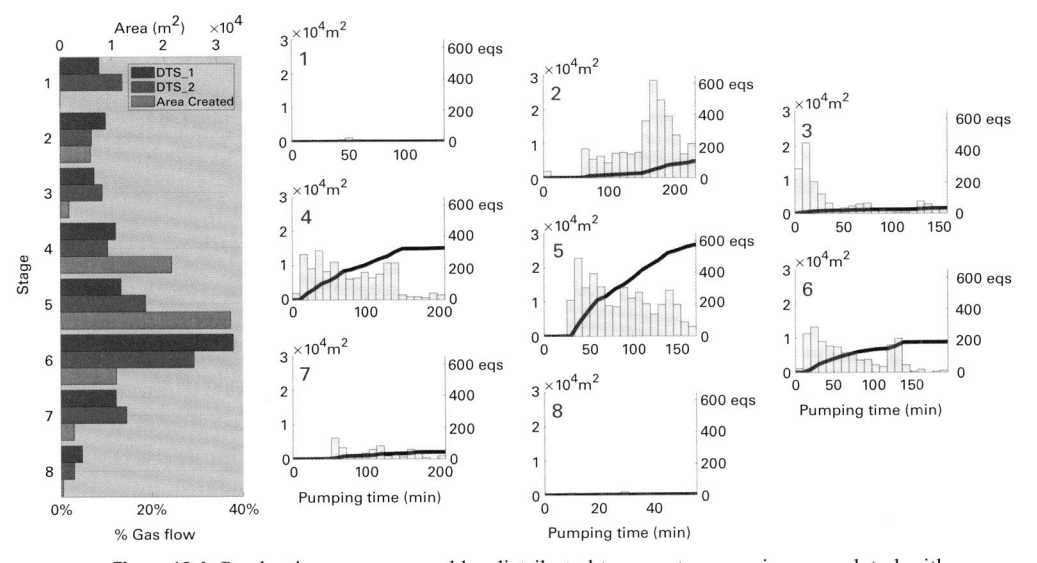

Figure 12.4 Production, as measured by distributed temperature sensing, correlated with area created per stage (left). For each stage, the temporal distribution of microseismic events during injection is shown by the gray histogram. Cumulative area is indicated on the same axes by the black line. From Hakso & Zoback (2019).

varies from stage to stage. In stage 2, the limited amount of surface area created occurred after 140 minutes of pumping, whereas almost none occurred after that amount of time in stage 4. It should be noted that area is assigned to a specific stage if the microseismic event occurred during that stage, independent of the location of the event. It was pretty clear after a relatively short pumping time that stages 1, 7 and 8 were unlikely to produce much area.

The correlation between production and cumulative microseismic source area is significantly stronger than the correlation between production and the number of detected events. Note that stage 2 has more microearthquakes than stages 4 or 6, but much less area was created and there was much less production. As Fig. 12.4 does not take into account the location of the events, the average magnitude of events during stage 2 must be smaller than that of stages 4 and 6. It is also important to note the area created during stimulation of a given stage may not occur in the same place as the stage and thus not contribute to the stage's production. As noted in Chapter 8, zonal isolation seems to affect stage 2 events, and Farghal & Zoback (2015) noted that some of the events associated with stage 2 appear to have been channelized along a NE–SW trending pre-existing fault damage zone (note the events near monitoring well A). Thus, while the number of events created during stage 2 might have led one to expect higher production than what was observed, both the small size of the events and their scattered locations resulted in relatively little production.

As argued by Hakso & Zoback, this analysis provides insight into the importance of low viscosity frac fluids and leak off during the stimulation process as the propped area of the hydraulic fractures is likely comparable to the lower bound estimate of the surface area of the stimulated fracture network. However, the diffusion distance between hydraulic fractures 50 meters apart would make the diffusion times for significant flow far too long to be economically practical. Said yet another way, while large hydraulic fractures could create sufficient area to produce hydrocarbons, it would only be economically practical if the hydraulic fractures were spaced about 10 meters apart as the diffusion distances are so short. This will be illustrated below. This example re-emphasizes the point made previously; on time scales relevant to production (several years), stimulated faults and fractures in low permeability reservoirs can generally be considered to have such high permeability relative to the matrix, that production rates are limited by the rate of diffusion from the matrix to the high permeability network (Walton & McLennan, 2013). Thus, it is not only important to create surface area during stimulation, it is important to create this area distributed throughout the target reservoir volume in order to have permeable fracture planes relatively close to the majority of hydrocarbon-bearing pores in the matrix. Ideally, the stimulated fracture network is relatively homogeneously distributed in the target volume, resulting in relatively short diffusion distances to the nearest fracture.

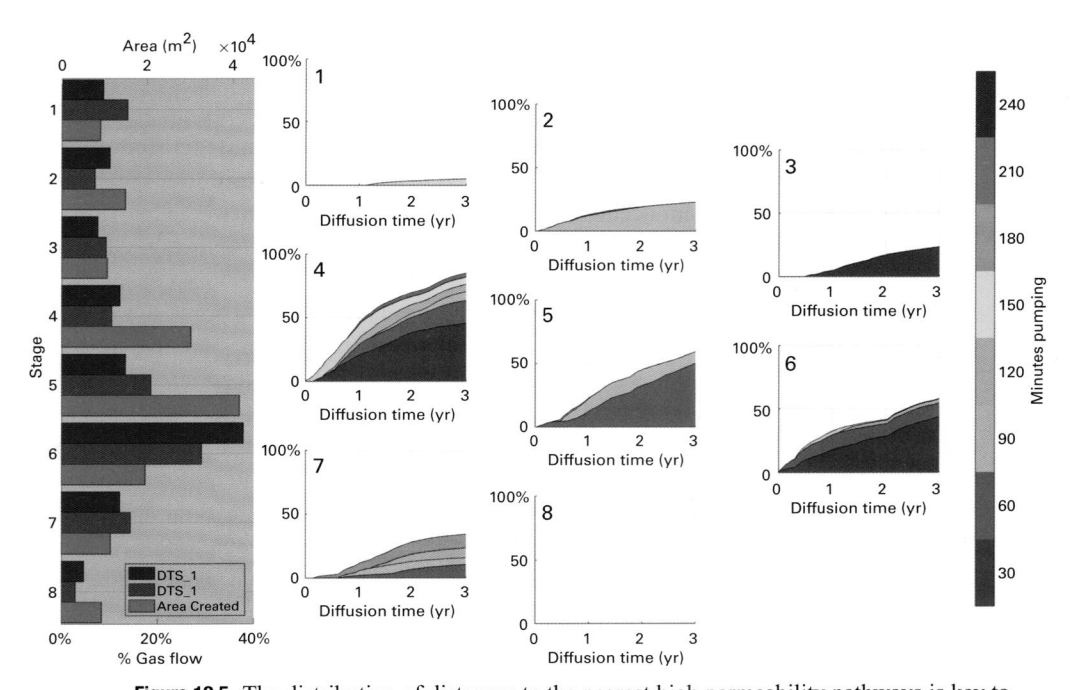

Figure 12.5 The distribution of distances to the nearest high permeability pathways is key to understanding production. By modeling the fracture network and calculating the distribution of distances to permeable zones across the reservoir volume, an estimate of the drained portion of the target volume is calculated as a function of time assuming an average matrix permeability of 100 nD. From Hakso & Zoback (2019).

How Much of the Reservoir is Being Produced? – Figure 12.5 (also from Hakso & Zoback, 2019) attempts to combine the concepts of area and diffusion time by characterizing the efficacy of each stage in terms of distribution diffusion times to the high permeability fracture network in a volume with an average matrix permeability of 100 nD. The percentage of the theoretical stimulated reservoir volume of each stage (on the *y*-axis) within a mean diffusion time (on the *x*-axis) is displayed as a function of time during stimulation (color contours). The estimated reservoir volume is fixed for each stage as given by the well spacing, stage spacing and thickness of the target formation, with the center of the perforation cluster at the center of the volume. Following generation of the fracture network from microseismicity, a distribution of distances to the nearest stimulated fracture was calculated. Converting from distance to mean diffusion time then provides insight into production over time. Note that the majority of the target volume from which diffusion occurs within 3 years is accessed in the first 90 minutes of pumping. Stage 7 is the only exception, with target volume penetration roughly doubling between 90 and 180 minutes of pumping. In other words, this analysis suggests that far more fluid than necessary was pumped in 7 of the 8 stages.

Evolution of a Shear Fracture Network

In this section we present two complementary approaches to clarify the mechanisms associated with creation of stimulated shear fracture network during hydraulic fracturing. First, we review the work of Jin & Zoback (2018) who present a fully coupled analysis of triggered microseismicity in an arbitrary fractured and fluid-saturated poroelastic solid that extends the coupled poroelastic model developed by Jin & Zoback (2017). The advantage of their approach is that it is possible to include a discrete fracture network (consisting of large-scale deterministic fractures and small-scale stochastic fractures) that follow reasonable size distributions and orientations. As an example, we illustrate below a numerical experiment to generate a synthetic catalog of induced seismic events, that includes the seismicity distribution in relation to the fluid pressure, poroelastic stress and fracture distribution, the spatial-temporal characteristics of the seismicity and seismic source parameters and permeability changes.

Without presenting the full details here, we outline several key governing equations. As elsewhere in this book, compressive stress is positive. The fully coupled mass conservation law and quasi-static force-balance law for the fluid and solid phases are:

$$\left(\Lambda_0 \phi_{m0}(C_m + C_p) + (1 - \Lambda_0)\phi_{f0}(C_f + C_p)\right)\dot{p} - \alpha\nabla \cdot \dot{u} + \nabla \cdot v = s \qquad (12.7)$$

$$\nabla \cdot \sigma'_p + \alpha\nabla p = 0 \qquad (12.8)$$

The two fluid flow equations for the matrix and fractures are given by Darcy's law and a nonlinear cubic law, respectively, as:

$$v = -\eta^{-1} k \cdot \nabla p \tag{12.9}$$

$$v = -\eta^{-1} \frac{1}{12} \left(b_0 (1 + C_f p) \right)^2 \nabla_\tau p \tag{12.10}$$

The solid constitutive law for the entire fractured rock is a generalized Hooke's law:

$$\sigma'_p = \mathbb{C} : \nabla^s u \tag{12.11}$$

In Eqns. (12.7)–(12.11), subscripts m and f indicate quantities associated with the hosting rock (porous matrix) and discrete fractures, subscript 0 denotes the initial value of a quantity, subscript and τ indicates the fracture tangential direction ϕ is the intrinsic porosity, Λ is a fracture-dependent parameter enabling the definition of a so-called partial porosity, C is the compressibility, p is the fluid overpressure, v is the fluid velocity vector, s is the external fluid source normalized by the initial fluid density, η is the fluid viscosity, k is the permeability tensor, b is the fracture hydraulic aperture, σ'_p is the solid effective stress (i.e., the poroelastic stress) tensor, u is the solid displacement vector, α is the Biot–Willis coefficient and $\mathbb{C}$ is the elastic stiffness tensor under a plane strain assumption. Note that ∇, ∇^s and ∇_τ are operators for computing the gradient, the symmetric gradient and the fracture-tangential gradient, respectively, and $\nabla \cdot$ is the divergence operator.

The left column of Fig. 12.6 (from Jin & Zoback, 2018) illustrates the pressure evolution during hypothetical hydraulic fracturing operations after 10, 20, 40 and 80 minutes. This figure can be considered as looking down the wellbore at the plane of a hydraulic fracture 100 m on each side. As this is a two-dimensional model, the fractures shown extend in and out of the plane of the page parallel to the wellbore with various dips (randomly distributed). Therefore, some are well oriented for slip, but many are not. The fractures range in scale from 0.1 m to 10 m following a power law distribution. Odling et al. (1999) have reported that the fracture lengths appear to follow a power law distribution ($N(I) = L_{min} I^{-n}$) with a decay rate n of ~2.1 over 4–5 orders of magnitude of fracture lengths. In this study, we assume that fracture lengths follow such a power law distribution over the range of sizes noted above.

Obviously, this simplified penny-shaped fracture geometry is not representative of an actual hydraulic fracture during multi-stage hydraulic fracturing in a horizontal well in an unconventional reservoir. In fact, as this model was not developed for the purpose of modeling stimulation in unconventional reservoirs, the matrix permeability is much higher than would be representative. Nonetheless, several relevant features can be seen. First, as expected, pressure spreads out in the plane of the hydraulic fracture in a quasi-radial way, with pressure concentrated along the discrete fractures with the elevated pressure penetrating the discrete fractures. Microseismic events occur along these planes in accordance with Coulomb failure. As shown in the center column of Fig. 12.6, as time progresses a cloud of microseismicity develops along the plane of the hydraulic fracture, as a direct result of the increase in pore pressure, but also the result of poroelastic stress changes that were not considered in Figs. 10.1 through Fig. 10.5. As a result of the microseismicity along the discrete fractures, the permeability of these

Production and Depletion   355

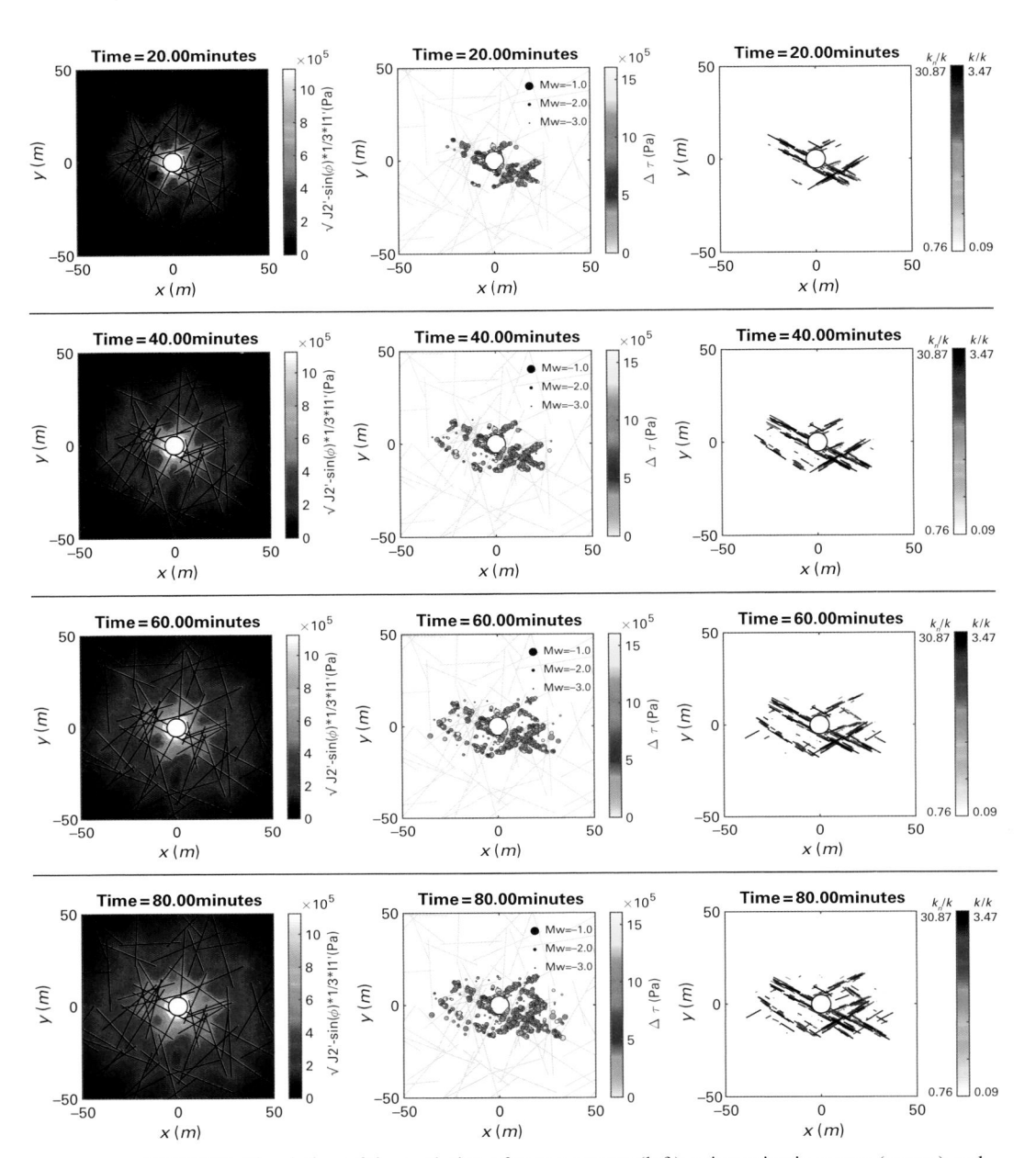

Figure 12.6 Simulation of the evolution of pore pressure (left), microseismic events (center) and permeability enhancement (right) during a modeled hydraulic fracturing operation. From Jin & Zoback (2018). Each row represents snapshots after 10, 20, 40 and 80 minutes (see text).

features increases markedly. For the sake of illustration, the methodology of Ishibashi et al. (2016) was used to estimate permeability changes parallel and normal to the faults as shown in the right column, although it should be noted that the Ishibashi formulation was originally developed for crystalline rock.

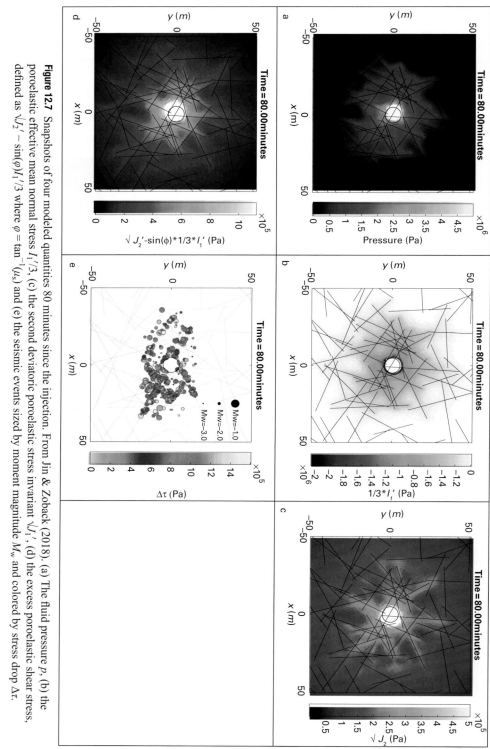

Figure 12.7 Snapshots of four modeled quantities 80 minutes since the injection. From Jin & Zoback (2018) (a) The fluid pressure p, (b) the poroelastic effective mean normal stress $I_1'/3$, (c) the second deviatoric poroelastic stress invariant $\sqrt{J_2'}$, (d) the excess poroelastic shear stress, defined as $\sqrt{J_2'} - \sin(\varphi)I_1'/3$ where $\varphi = \tan^{-1}(\mu_s)$ and (e) the seismic events sized by moment magnitude M_w and colored by stress drop $\Delta\tau$.

Production and Depletion 357

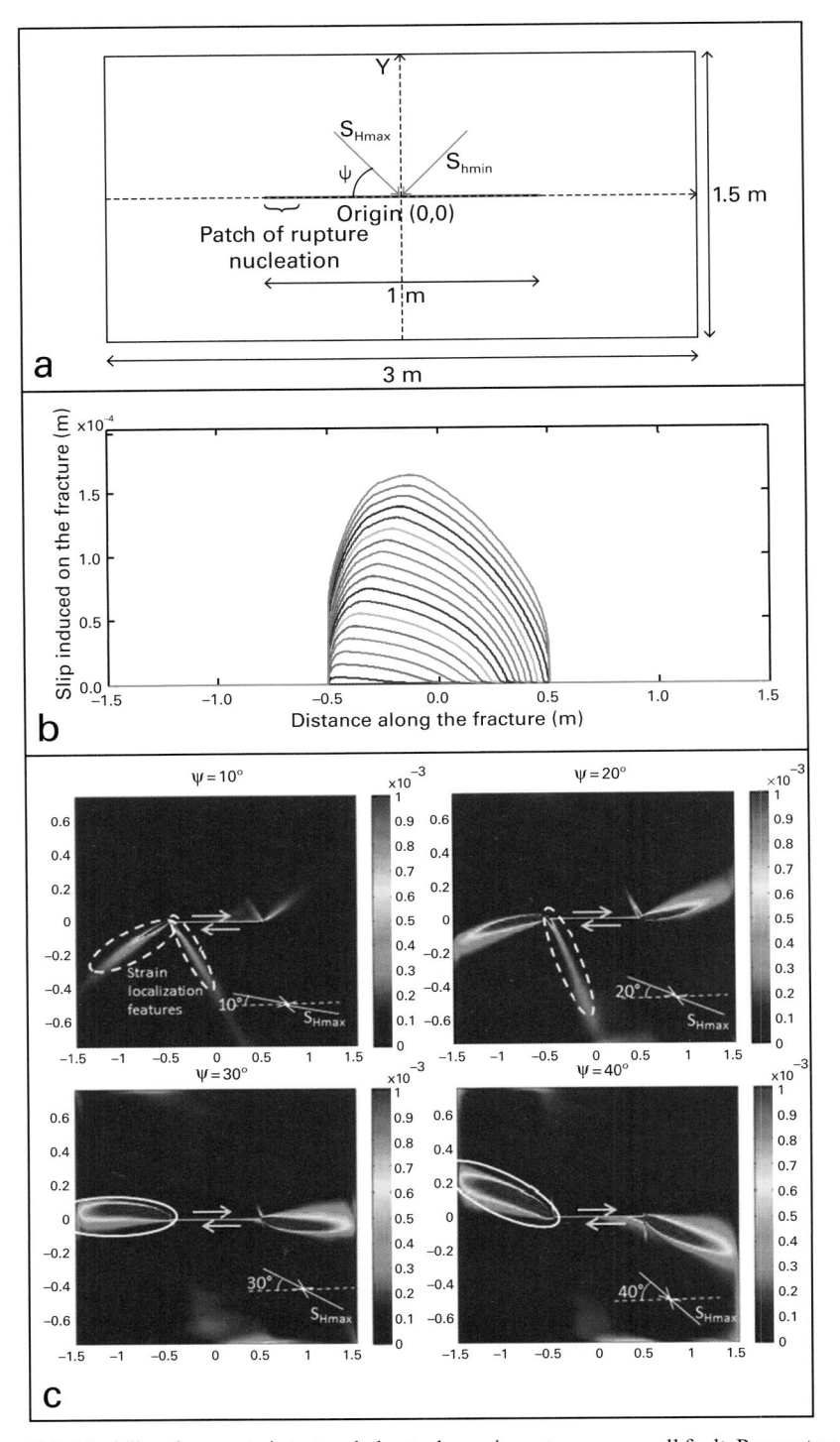

Figure 12.8 Modeling damage to intact rock due to dynamic rupture on a small fault. Parameters used correspond to a typical magnitude −2 microseismic event. (a) Model set up. (b) Contours of slip evolution with contours every 0.4 microsecond. (c) Dilatant strains (damage) for faults at different orientation to the direction of maximum horizontal compression. From Johri et al. (2014a).

Note that the change in pore pressure (Fig. 12.7a) is much larger than the change in the effective Coulomb Failure Function, shown in Fig. 12.7d. While the *CFF* as normally expressed is the difference between shear stress and the product of effective normal stress and friction (Chapter 4), the expression $\sqrt{J_2'} - \sin(\varphi)I_1'/3$ where $\varphi = \tan^{-1}(\mu_s)$ is essentially the same when expressed in terms of the stress invariants, I_1 and J_2. The reason that the poroelastic effect tends to reduce the *CFF* is because when pumping is going on the induced poroelastic stress at distance from the wellbore is generally compressive (Fig. 12.7b), thus inhibiting failure. If pumping ends abruptly and the poroelastic stress decreases faster than the fluid pressure, shear slip would be promoted, as pointed out by Chang & Segall (2016).

Matrix Damage and Permeability Enhancement

In addition to the direct effect of shearing on the permeability of the slipping fault, Johri et al. (2014a) point out that dynamic fracture propagation can damage and increase the permeability in the matrix rock surrounding the slipping fault. The dynamic rupture propagation methodology of Dunham et al. (2011) was used for the calculations.

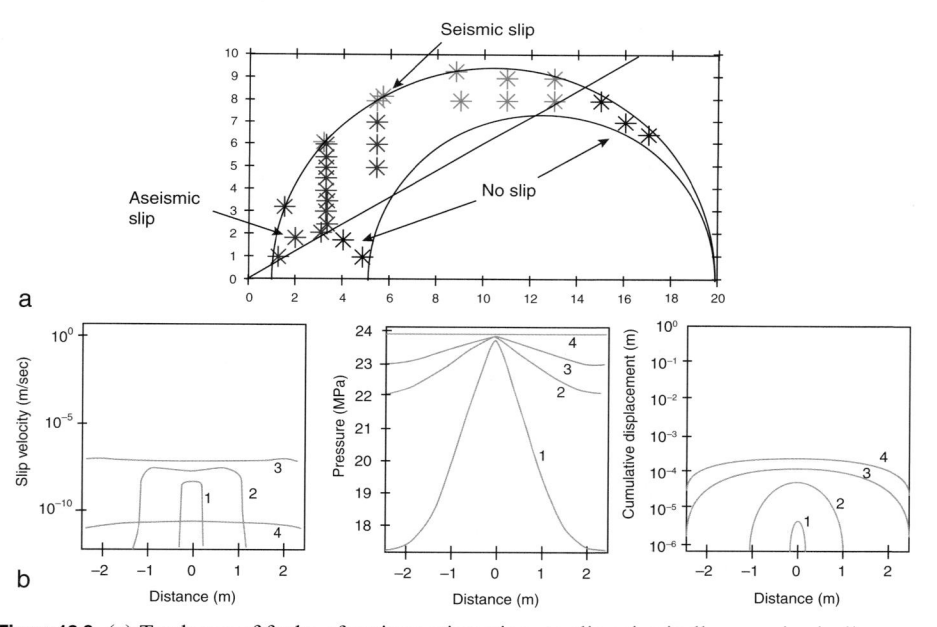

Figure 12.9 (a) Tendency of faults of various orientations to slip seismically or aseismically. In general, well-oriented (critically stressed) faults are expected to slip seismically whereas mis-oriented faults slip slowly. (b) Evolution of slip on a mis-oriented fault. The left panel shows slip velocity, the center panel shows pore pressure and the right panel shows displacement. 1, 2, 3 and 4 indicate successive times as pore pressure spreads out along the fault and displacement increases. While the exact values in these figures are highly dependent on parameters used in the model, they illustrate how slip evolution is dependent on the rate at which pore pressure increases along the fault. From Zoback et al. (2012).

The case considered was for a 1 meter long fracture slipping in shear by about 0.1 mm, corresponding to a typical magnitude −2 microseismic event. The model set up is shown in Fig. 12.8a, the evolution of slip is shown in Fig. 12.8b with contours of slip every 0.4 microsecond. The damage surrounding slipping faults of different orientation to the current direction of maximum horizontal stress is shown in Fig. 12.8c. Damage is expressed in terms of dilatant extensional strains. Note that in every case, multiple zones of damage result from rupture propagation on the scale of the slipping fault patch. Note that this damage results from dynamic rupture propagation and is not the result of static stress changes surrounding a fault with uniform slip (e.g., King et al., 1994).

Seismic and Aseismic Fault Slip

On the basis of the transition from unstable to stable sliding with increasing clay content seen in laboratory experiments of Kohli & Zoback (2013) on unconventional reservoir rocks, Zoback et al. (2012) hypothesized that there could be a considerable amount of aseismic slip during hydraulic fracturing operations. After all, the velocity-strengthening (stable or aseismic slip) *high clay* samples and velocity-weakening (conditionally unstable or seismic slip) *low clay* samples from the Barnett, Haynesville and Eagle Ford came from cores in the same wells, sometimes only 10–20 meters apart (Fig. 4.7). This suggests that seismic and aseismic slip could occur concomitantly within the same hydraulic stimulation stage.

Another mechanism leading to aseismic slip is illustrated in modeling shown in Fig. 12.9 (Zoback et al., 2012), which utilizes the method described by McClure & Horne (2011). Slip was considered on faults of varied orientation in a hypothetical stress field in response to highly elevated pore pressure (Fig. 12.9a). Some faults are relatively well oriented for slip, some are very poorly oriented and only slipped because of the high pore pressure perturbation and some do not slip even when the pore pressure is close to the least principal stress (as discussed above). Because fault slip (diagram on the right in Fig. 12.9b) can only propagate along the fault as fast as high pore pressure diffuses along it (Fig. 12.9b, center), slip would intrinsically be aseismic. In order for seismic waves to be emanated from a propagating shear slip on a fault, the rupture velocity has to be close to the shear velocity (several km/s). As shown by the color of the symbols representing faults of different orientation in the Mohr diagram Fig. 12.9a (note that this Mohr diagram utilizes effective normal stress, $S_n - P_p$), the tendency for seismic slip (red symbols) is limited to faults that are relatively well oriented and faults with relatively high shear stress. While the high pore pressure causes the faults shown in purple to slip, they are expected to slip slowly.

These calculations would seem to suggest that a large fraction of poorly oriented faults would be expected to slip slowly when triggered by high pore pressure during hydraulic fracturing. Thus, whether due to high clay content or the poor orientation of many fractures relative to the current stress field, the degree to which aseismic slip occurs during hydraulic stimulation indicates that estimates of the amount of surface area created during stimulation from microseismic data as discussed above are lower bound estimates. Aseismic fault slip may also account for wellbore damage, as discussed in Chapter 13.

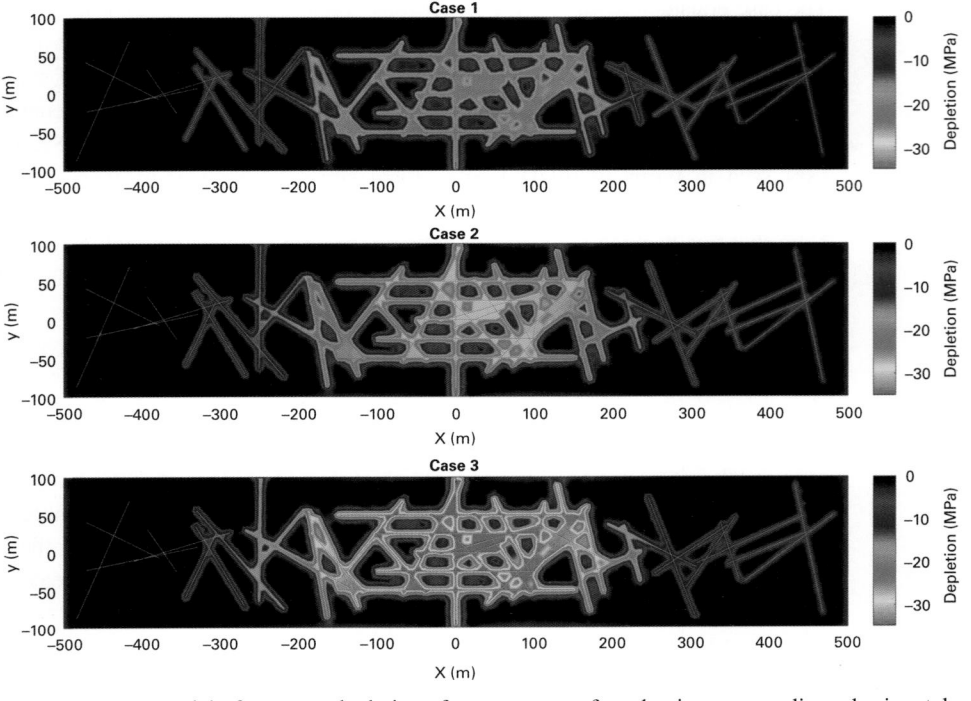

Figure 12.10 Model of pressure depletion after two years of production surrounding a horizontal well with hydraulic fractures that have propagated from five perforations and pre-existing fractures. As described in the text, the three cases represent different initial pore pressures. From Jin & Zoback (2019).

It is difficult to document whether aseismic slip is occurring during multi-stage hydraulic fracturing in unconventional reservoirs. Das & Zoback (2013a) and Das & Zoback (2013b) reported unusual long period, long duration (LPLD) seismic events in two sets of wells in the Barnett shale during multi-stage hydraulic fracturing. They interpret these events as resulting from slow slip on pre-existing faults. Kumar & Hammack (2016) report similar events during hydraulic fracturing in the Marcellus formation. However, Zecevic et al. (2016) point out that in one of the cases reported by Das & Zoback, the LPLD events correlate with very small earthquakes occurring nearby that were not included in the seismic catalogs available to Das & Zoback at the time of their study. In light of this, the occurrence of appreciable aseismic slip occurring during hydraulic fracturing still seems likely, based both on the frictional properties of the rocks and the analysis shown in Fig. 12.9, but has yet to be documented.

Depletion of Ultra-Low Permeability Formations with High Permeability Fractures

Using the coupled poroelastic 2D code described by Jin & Zoback above, Jin & Zoback (2019) examined the effect of depletion and poroelastic stress changes that could affect infill drilling. Figure 12.10 presents three models of depletion for a single stage with 5 perforations and pre-existing fractures. The model assumes no flow and a traction free outer boundary. The hydraulic fractures are assumed to have very high permeability equivalent to an effective hydraulic aperture of about 1 mm (1,000 D). The stimulated natural fractures have a high permeability equivalent to an effective hydraulic aperture of about 0.3 mm (100 D) and the matrix permeability is assumed to be 100 nD (10^{-19} m^2). As discussed above, the rate of flow from the matrix into the hydraulically permeable planes control flow rates and depletion. While the model is 2D, it is hopefully representative of what is occurring if one considers the image the view of a horizontal plane that passes through the propped areas of multiple hydraulic fractures propagating from a well.

The reason there are three models in Fig. 12.10 will be obvious when we consider how poroelastic stress changes caused by depletion affect the initial stress state later in this chapter. All three models consider a normal/strike-slip faulting stress state in which $S_V \geq S_{Hmax} > S_{hmin}$ and a P_p of 1,000 psi (6.9 MPa) in the well to drive production. The first case considers moderate overpressure (P_p = 4,000 psi (27.6 MPa); S_{hmin} = 5,100 psi (35.2 MPa); S_{Hmax} = 6,500 psi (44.8 MPa); S_V = 7,000 psi (48.3 MPa)). The second case considers moderate overpressure with resultantly less initial stress anisotropy as discussed in Chapter 7 (P_p = 5,000 psi (34.5 MPa); S_{hmin} = 5,650 psi (39.0 MPa); S_{Hmax} = 6500 psi (44.8 MPa); S_V = 7,000 psi (48.3 MPa)) and consequently more depletion. The third case considers even higher pore pressure, less initial stress anisotropy (P_p = 6,000 psi (41.4 MPa); S_{hmin} = 6,325 psi (43.6 MPa); S_{Hmax} = 6,500 psi (44.8 MPa); S_V = 7,000 psi (48.3 MPa)) and consequently even greater depletion. It should be noted that in this particular analysis we do not incorporate pressure-dependent matrix or fracture permeability as discussed in Chapters 5 and 10, respectively. Also, relatively few stimulated shear fractures are shown for computational practicality. While many more small fractures are likely present (see the discussion at the end of Chapter 7), the purpose of the modeling shown in Fig. 12.10 is to illustrate several principles about depletion, especially in the areas distant from the near well region.

Needless to say, the combination of high permeability fractures and extremely low permeability matrix results in very heterogenous depletion after two years of production. As can be seen in Fig. 12.10, the depletion is concentrated in the narrow zones adjacent to the permeable hydraulic fractures and stimulated shear fractures. This is consistent with the small diffusion distances indicated in Fig. 12.2. To the degree this model is representative of unconventional reservoirs in nature, it seems obvious why recovery factors are so low. To effectively drain unconventional reservoirs, closely spaced permeable fractures would be needed.

Implications for the SRV – The somewhat schematic view of production from an unconventional reservoir shown in Fig. 12.10 has interesting implications for the concept that the cloud of microseismic events can be used to define a stimulated rock volume (SRV) from which production is occurring. For the sake of discussion, we set aside the issue that the uncertainty in the locations of microseismic events could falsely inflate the size of the SRV. Thus, we assume that the microseismic events recorded during stimulation represent evidence that elevated fluid pressure reached that fracture.

However, Fig. 12.10 emphasizes the fact that because of the extremely low matrix permeability of unconventional reservoirs, production is coming only from distances within meters of the permeable hydraulic fractures and the fractures stimulated in shear. Thus, the places where the concept of producing from a volume defined by microseismic events seems to make sense are only those areas where the permeable fractures are spaced very closely. Unless this close fracture spacing can be justified (e.g., from image log in a horizontal wells), drawing an envelope around a scattered cloud of microseismic events would significantly overestimate the volume of rock being drained.

Support for this assertion may be found in Fig. 10.1 (showing cumulative 3-year production from the Barnett shale as a function of the SRV, from Mayerhofer et al., 2010) and Fig. 12.1 that shows that after three years, approximately 70% of total production from the Barnett has been achieved. Applying typical volume expansion factors for methane, one would expect the methane to expand by a factor of 238 going from reservoir to standard conditions for typical downhole Barnett temperatures and pressures (Aguilera, 2016). Assuming a porosity of ~5% for high TOC Barnett (Chapter 2), the relationship between the volume of gas that can be produced and the size of the SRV shown in Fig. 10.1 implies that the SRV overestimates that amount of producible gas by about a factor of eight.

The Frac Hit Phenomenon and Well-to-Well Communication

Hydraulic communication between and among wells during multi-stage hydraulic fracturing is a topic of growing significance, whether in the context of infill drilling (*parent/child* interactions), or when production from an existing well is impacted when a nearby well is stimulated (the *frac hit* phenomenon). King et al. (2017), Rainbolt et al. (2018) and Miller et al. (2016) offer interesting reviews of this phenomenon.

Well-to-well communication occurs in two general ways. The high pressure associated with stimulation in a new well could be transmitted to a pre-existing well, regardless of the amount of production from the older well. Alternatively, depletion around the parent well and associated poroelastic stress changes effectively *attract* the hydraulic fractures generated by the new well because it is easier for hydraulic fractures to propagate into the depleted zone with lower stress. In this case production from the new well would be less than it would be normally because the hydraulic fractures from the new well (the *child* well) would be propagating into the depleted part of the reservoir associated with the older well (the *parent* well).

An example of the high pressure associated with stimulation of a child well affecting a parent well is reported in detail by Rainbolt et al. (2018) for a case study in the Wolfcamp formation of the Permian Basin (Fig. 12.11). They noted that when stimulation of the child well affected the parent well it generally had a negative effect on production from the parent well. As shown in Fig. 12.11b, the pressure in the parent well increases over 3,500 psi (potentially getting close to the frac gradient) as the child well was hydraulically fractured. Following stimulation of the child well, production of the parent well was severely affected (Fig. 12.11c) but recovered over several months.

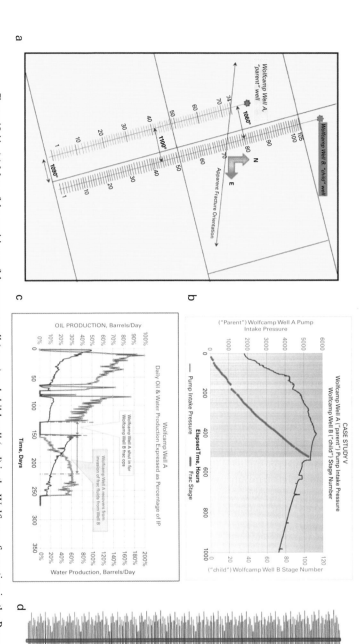

Figure 12.11 (a) Map of the positions of the parent well (cyan) and child well (red) in the Wolfcamp formation in the Permian Basin. (b) Increase in pressure in parent well (blue) as the number of stages in the child well accumulates with time. (c) Oil and water production in the parent well before and after the stimulation of the child well. (d) Schematic representation of dense hydraulic fracture interaction between two wells . After Rainbolt et al. (2018).

364　　Unconventional Reservoir Geomechanics

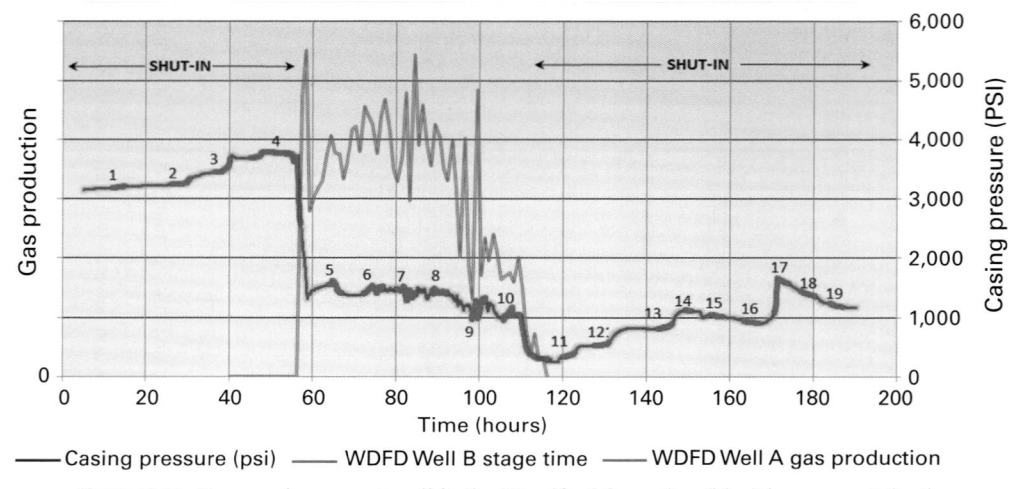

Figure 12.12 Pressure in a parent well in the Woodford formation (blue) increases at the times hydraulic fracturing is being carried out in a child well (red). After stage 4, the parent well was put on production. The production rate (green) dropped markedly after stage 9 in the child well. After shutting in the parent well following stage 10, pressure again increased in the parent well as the child well was being hydraulically fractured. From King et al. (2017).

King et al. (2017) present a case from the Woodford formation in which stimulation of a child well approximately 1,500 ft away negatively affects production from the parent well (Fig. 12.12). Note that the pressure in the parent well (shown in blue) increases about 700 psi during stages 1–4 in the child well (stimulation times associated with different stages are shown in red). When the parent well is reopened, production starts to drop dramatically after stage 8. After the parent well is shut in a second time, pressure builds about 1,000 psi during stages 11–19. King et al. and Rainbolt et al. demonstrated that there was a delay of approximately one to two hours between pressurization of a stage in the child well and rapid increases in the rate of pressure change in the parent well.

Examples of well-to-well communication where depletion appears to play an important role come from the Permian Basin where Ajisafe et al. (2017) studied production from parent and child wells drilled in the Avalon shale in the Delaware Basin. Study of 200 wells showed that child wells were about 30% less productive than parent wells, apparently due to the depletion of the parent well affecting the stimulation of the child well. Ajani & Kelkar (2012) analyzed daily gas and water production data from 179 horizontal gas wells in the Woodford formation of the Arkoma Basin to investigate the impact of interference between infill wells on older wells. As might be expected, the results are very dependent on well spacing. When wells are less than 1,000 ft apart, there is a 50% probability that gas production will be negatively affected. They hypothesize that this results from the poroelastic stress change surrounding the older, depleted wells *attracting* the hydraulic fractures from newer wells into the depleted zone. They noted that as the offset between the infill wells and older wells gets larger, the probability that the depleted zone will affect the infill wells drops markedly.

Table 12.1 Probability that the effects from child wells fractured at least 18 months after parent well's date of first production (from Miller et al., 2016). *Reprinted by permission of the Society of Petroleum Engineers.*

	Bakken	Eagle Ford	Haynesville	Woodford	Niobrara
Positive – long term	17%	9%	20%	2%	0%
Positive – short term	33%	14%	38%	2%	6%
Positive – total	50%	24%	58%	4%	6%
No change	35%	36%	24%	32%	38%
Negative total	15%	41%	19%	64%	56%
Negative – short term	7%	13%	5%	20%	19%
Negative – long term	6%	17%	5%	41%	31%
Cases considered	649	1210	366	259	32

Miller et al. (2016) studied more than 3,000 fracture interferences in five major basins to determine the degree to which well-to-well communication effected infill wells, either negatively or positively. As summarized in Table 12.1, the Haynesville and Bakken plays have more positive fracture hits than the other basins, while the majority of the fracture hits in the Woodford and Niobrara are negative and 41% of interactions in the Eagle Ford had negative consequences. A majority of the negative hits in these two basins are long-term negative changes in production. Overall, only 24–38% of wells were unaffected, thus understanding the what, why and how of well-to-well communication is critically important during unconventional development.

Mechanisms of Well-to-Well Communication – As mentioned above there are two general mechanisms that contribute to well-to-well communication. Perhaps easiest to understand is what happens when well spacing is too close and the hydraulic fractures associated with adjacent wells overlap and interact. Sardinha et al. (2014) present a study of a ten-well pad in the Horn River Basin exploiting pay at three depths prior to depletion. In the pad studied, all wells had multiple hits although interference tests indicate that the connections between wells diminish over time and not all frac hits had a negative effect on production. Their analysis of the frequency of frac hits showed that the dominant factor was distance between the stage being stimulated and the nearby well. When separated by distances less than 200 m, in eight of the ten wells, frac hits occurred more than 77% of the time. At separations ranging between 200 m and 600 m, frac hits were still common, but less probable. At distances greater than 600 m, frac hits rarely occurred. This high degree of interaction between closely spaced hydraulic fractures was confirmed by high-resolution microseismic observations. This phenomenon is shown schematically for another case study in Fig. 12.11d by Rainbolt et al. (2018) where the combination of closely spaced stages and closely spaced wells may promote well-to-well communication (although they suggest that depletion around the parent well may have also played a role in promoting communication).

Another mechanism that could cause well-to-well communication is the existence of through-going pad-scale faults. As discussed in Chapter 7, pad-scale faults are likely to have highly permeable damage zones that can channelize flow and pressure build-up during hydraulic stimulation. An example of this was shown in Fig. 7.24 where during stimulation of well 1 in the vicinity of fault C (stages 5 and 6), pressure in well 2 suddenly increased, accompanied by microseismicity that not only connected the two wells along fault C, but continued farther to the northeast (Fig. 7.24c). Moreover, when well 1 was being stimulated near fault B (stages 3 and 4), microseismicity propagated to the southwest and pressured up a pre-existing vertical well that was being used for one of the microseismic monitoring arrays. This is a case where depletion in the vicinity of an older vertical well may have played a role in facilitating flow along the damage zone.

Figure 12.13 (from Ma & Zoback, 2017b) presents another example of well-to-well communication resulting from pad-scale faults in the Woodford formation introduced previously (Fig. 7.25). Three vertical wells were used for microseismic monitoring such that the detection capability and location accuracy of the microseismic events were relatively good. When Well A was being stimulated (Fig. 12.13a) there were a number of east–west clouds of microearthquakes as would expected to surround hydraulic fractures (the direction of S_{Hmax} is approximately east–west). However, a pronounced east–west trends is concentrated between wells A and B near both the toe and heel of the wells (Fig. 12.13b) when well C was being stimulated. When well D was being stimulated, there were many events along an east west trend near the toe of well A, about 3000 ft away. While one cannot rule out pressure communication along hydraulic fractures, the overall pattern of seismicity associated with stimulation of these four wells suggests the importance of longer-ranging pressure transmission along east–west trending normal faults.

Poroelastic Stress Changes and Infill Drilling – The final important mechanism to discuss that could cause well-to-well communication is the poroelastic stress changes caused by depletion in the vicinity of pre-existing, parent wells. As mentioned above, as pore pressure decreases with production in a formation (e.g., Engelder & Fischer, 1994; Segall et al., 1994; Chan & Zoback, 2002; Chang & Segall, 2016) so do the magnitudes of the horizontal stresses. The vertical stress remains unchanged when the area of depletion is large compared to the thickness of the producing interval. In the case of infill drilling, poroelastic stress decreases make it easier for hydraulic fractures to propagate from a child well into a depleted region around a parent well than into the undepleted rock surrounding the child well where stress magnitudes are higher.

An excellent example of poroelastic effects causing well-to-well communication is shown for the Bakken formation in Fig. 12.14 (from Dohmen et al., 2017). The direction of maximum horizontal stress in this area is N55°E, so that is the direction hydraulic fractures are expected to propagate. At the time when infill wells H2, H3 and H4 were drilled and stimulated, an older well, H1, had been in production. Pore pressure had decreased over 5,700 psi around well H1 from an initial value of 7,540 psi to a value of about 1,860 psi. Because of the poroelastic stress change surrounding well H1, hydraulic fracture propagation (as indicated by microseismicity) from wells H2 and H4 were extremely asymmetric. The hydraulic fractures from the H2 well propagated almost unilaterally to the southwest toward H1 (left side of Fig. 12.14) while those from the H4 well (right side of Fig. 12.14) propagated unilaterally to the northeast toward well H1.

Production and Depletion 367

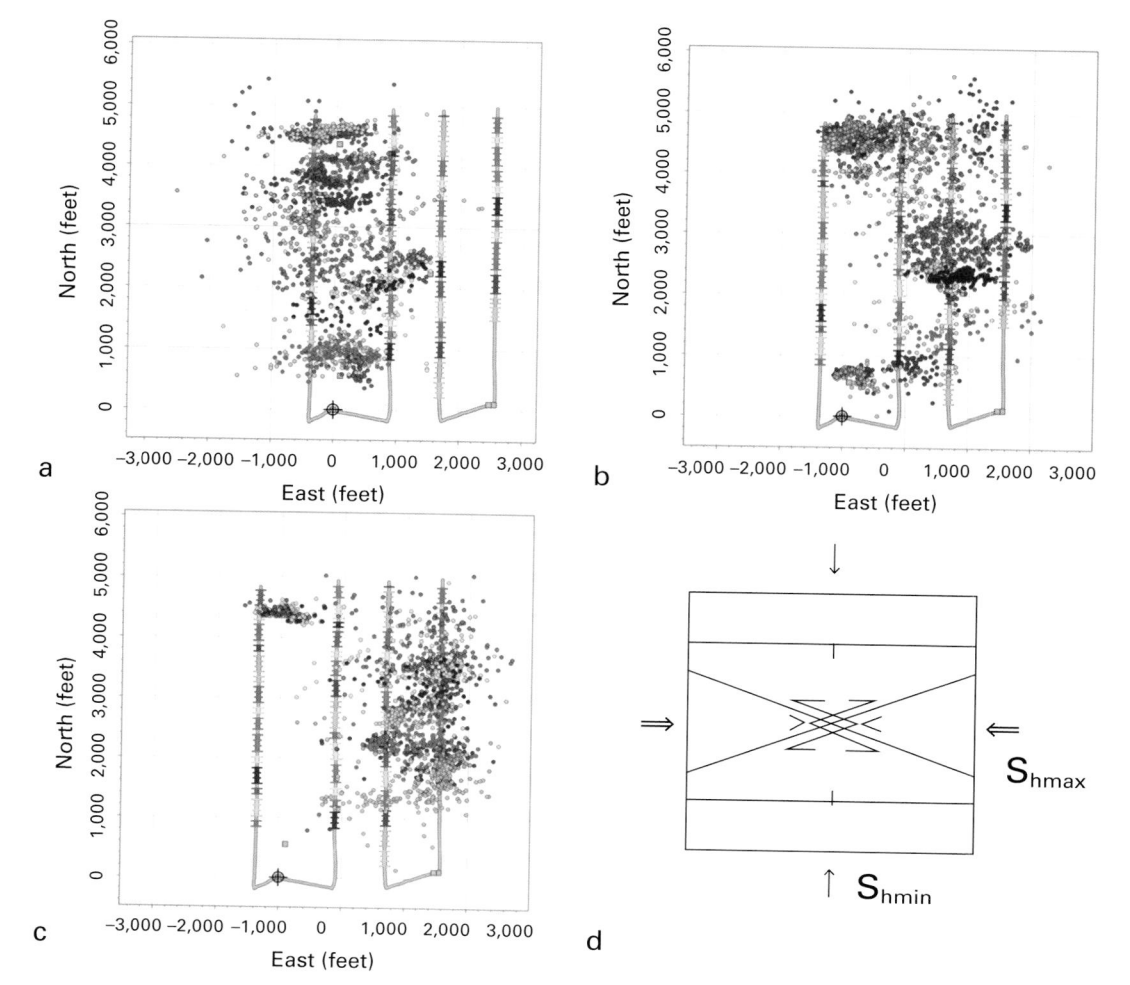

Figure 12.13 Multi-stage hydraulic fracturing in the Woodford formation in central Oklahoma suggests considerable well-to-well pressure communication caused by through-going faults. (a) When well A was being stimulated, microseismic events occurs 2,000 ft to the east. (b) When well C was being stimulated, microseismicity is seen 2,000 ft to the west near well A. (c) When well D was being stimulated microseismicity was observed near well A, almost 3,000 ft to the west. (d) Expected orientation of potentially active, pad-scale, faults in a normal, strike-slip faulting environments when S_{Hmax} is oriented approximately east–west. Modified from Ma & Zoback (2017b).

Note also at the bottom of Fig. 12.14, the downhole pressure recording in the H1 well showing the frac hits from wells H2 and H4.

It is important to note that there appears to be no variation of hydraulic fracture orientation (that is, no change in the stress directions) in the vicinity of well H1. As the hydraulic fractures propagated toward well H1 from both sides the hydraulic fracture orientations remains essentially constant. Hence, it appears that while stress magnitudes clearly changed with depletion around well H1, there was little to no change in stress direction. On the basis of a focal plane mechanism study, Yang & Zoback (2014) argued

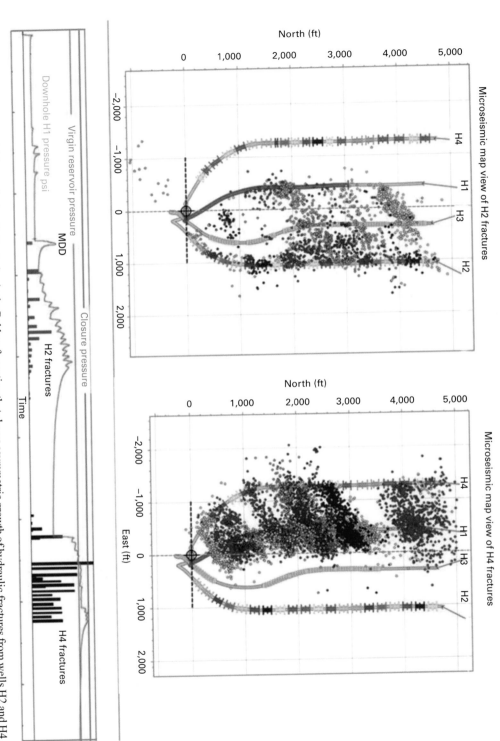

Figure 12.14 Multi-stage hydraulic fracturing in the Bakken formation that shows asymmetric growth of hydraulic fractures from wells H2 and H4 toward the depleted zone surrounding well H1. From Dohmen et al. (2017).

Production and Depletion   369

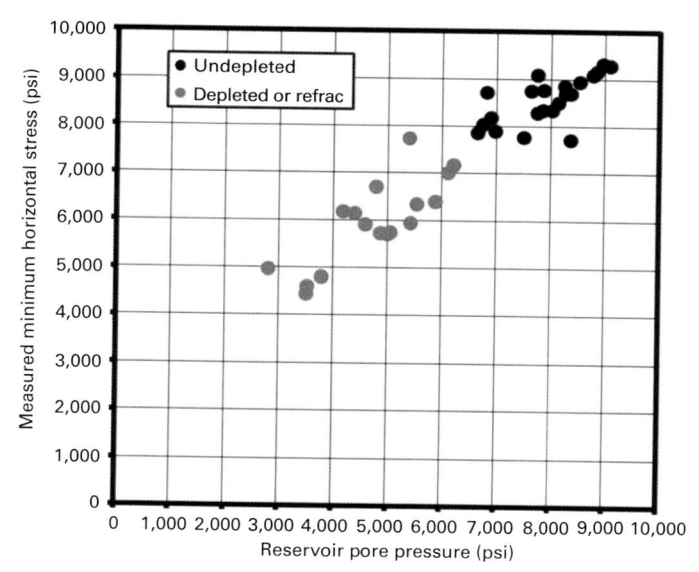

Figure 12.15 DFIT pore pressure and least principal stress measurements before and after depletion in the Bakken and Three Forks formation in the Bakken field. From Dohmen et al. (2017).

that the Bakken is characterized by a normal/strike-slip faulting stress state with the maximum horizontal stress slightly less than the vertical stress. In other words, appreciable horizontal stress anisotropy would be expected.

Dohmen et al. (2017) presented DFIT data on pore pressure and the magnitude of the least principal stress for the Bakken and Three Forks formations from throughout the Bakken field (Fig. 12.15). In some areas the field is overpressured, with gradients as high as 0.7 psi/ft in the Bakken formation and 0.73 psi/ft in the Three Forks. The stress path (the change in the magnitude of the least principal stress for a given change in pore pressure) implied by the data is 0.73, which, following Chan & Zoback (2002), would tend to promote normal faulting. As argued by Dohmen et al. (2017), this means that patterns of microseismicity could be an indicator of depletion. Chan and Zoback (2002) cautioned that stress paths should be determined from stress and pore pressure measurements made at the same location through time. Nevertheless, the Dohmen et al. data clearly show that both pore pressure and the minimum horizontal stress are lower in depleted wells, by thousands of psi.

Miller et al. (2016) reviewed a strategy for attempting to diminish the role of depletion by re-pressuring the parent well. As they point out, the refracturing of a parent well may be a benefit to production, and may minimize the poroelastic stress changes that could potentially attract hydraulic fractures from infill wells drilled at a later time. Lindsay et al. (2016) report incremental production gains for 12 wells that were refractured in the Eagle Ford for parent well protection and project significant increases in oil production over time due to refracturing (see also, Marongio-Porcu et al., 2015; Morales et al., 2015). However, in the case shown in Fig. 12.14, pressurization of well H1 to 5,230 psi

Unconventional Reservoir Geomechanics

over a 6-hour period did not prevent the hydraulic fractures from wells H2 and H4 from propagating into the depleted zone surrounding well H1.

Modeling Poroelastic Stress Changes

Modeling depletion induced stress changes is challenging. Because of the low matrix permeability contrasted with the extremely high permeability of the shear fracture network, depletion is likely to be extremely non-uniform as illustrated in Fig. 12.10. Utilizing appropriate values for key parameters such as the Biot coefficient is also difficult (see discussion in Chapter 2). Moreover, computational techniques that correctly incorporate the poroelastic coupling are not widely available (see discussion in Jin & Zoback, 2017). One approach is to assume relatively uniform depletion surrounding hydraulic fractures, which leads to symmetric depletion shadows that affect stress magnitudes but not orientation. The reason for this is that as most wells are drilled in the direction of the least horizontal principal stress, S_{hmin}, and hydraulic fractures propagate in the direction of S_{Hmax}, symmetric depletion around the hydraulic fractures would create a perturbation that affects the magnitudes of S_{hmin} and S_{Hmax} by different amounts, but would not rotate the principal stress directions. There is a discussion on the related topic of stress rotations resulting from depletion on one side of sealing fault in Chapter 12 of Zoback (2007). There is no stress rotation when the stress perturbation is aligned with the horizontal principal stresses. In a study by Gupta et al. (2012) which assumed uniform depletion, stress reorientation near the depleted well was predicted because it was not drilled in the direction of the minimum horizontal stress. In that study the stress contrast between the two horizontal principal stresses considered was very small – ranging between 1 and 5%.

In marked contrast to the assumption of uniform depletion around hydraulic fractures, Ajisafe et al. (2017), Marongio-Porcu et al. (2015) and Morales et al. (2016) start with models with complex fault patterns and predict extraordinarily complex stress patterns in the depleted areas. The methods used in these calculations are not described in detail (e.g., how pore pressure and stress are coupled, the boundary conditions for the numerical models, etc.). Nor are critical parameters in these models specified such as poroelastic parameters (e.g., the Biot coefficient), the assumed permeabilities of the matrix, stimulated shear faults and hydraulic fractures or the assumed magnitudes of principal stresses and pore pressure prior to depletion. We show below the importance of many of these parameters. Obviously, if the two horizontal stresses are assumed to be essentially equal (which, as discussed in Chapter 7, is almost never the case), even minor stress perturbations have the potential to cause significant stress rotations.

In a follow-up to the study shown in Fig. 10.24, Cipolla et al. (2018a) modeled poroelastic stress changes associated with depletion around well H1. As shown in Fig. 12.16a, the pressure change associated with production from well H1 is as much as 6,000 psi. The zone of depletion is more-or-less uniformly distributed around the well and symmetrically distributed around NE trending hydraulic fractures in the model. Figure 12.16b shows calculated stress changes. As the change in least principal stress near H1 is predicted to be almost 4,000 psi below the unperturbed stress state near well

Production and Depletion   371

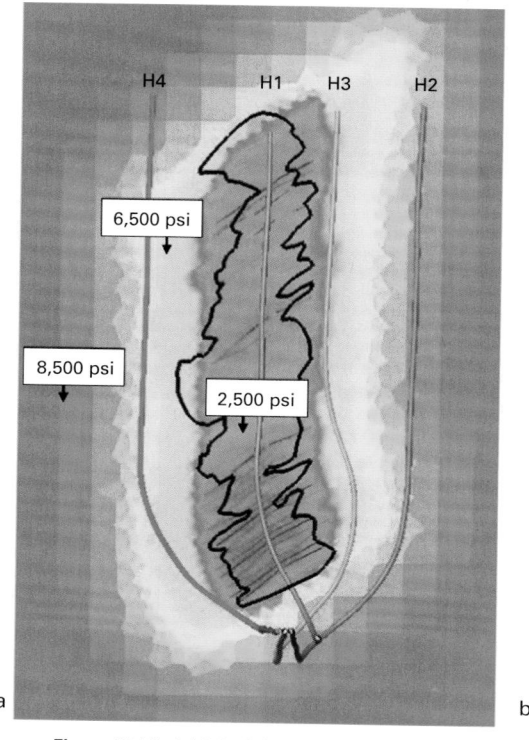 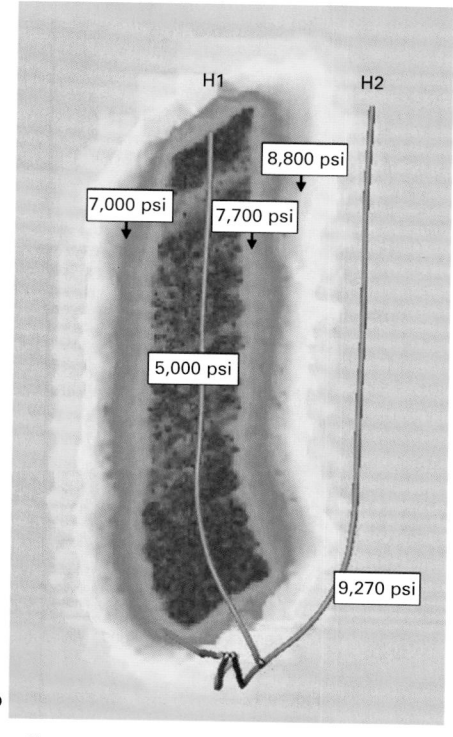

Figure 12.16 (a) Model of pressure depletion surrounding well H1. (b) Modeled stress magnitudes surrounding well H1 due to poroelasticity that affect the propagation of hydraulic fractures from wells H2 and H4. After Cipolla et al. (2018a).

H2, it is not surprising that the zone of depletion around H1 effectively attracted the hydraulic fractures propagating from wells H2 and H4.

Figure 12.17 (after Jin & Zoback, 2019) shows how stress orientations vary with depletion as shown in Fig. 12.10. In all cases, a constant Biot coefficient of 0.8 was used and a pore pressure in the well of 1,000 psi was assumed to drive flow for a two year period. In the model, the initial S_{hmin} and S_{Hmax} directions are referred to as S_{yy} and S_{xx}, respectively. For case 1, where the initial horizontal stress difference is 1,400 psi (9.7 MPa) and the depletion is 3,000 psi (20.7 MPa), there is almost no change in stress orientation. Localized stress orientation changes are observed near the intersections of the hydraulic fractures and pre-existing fractures. In case 2, where the initial horizontal stress difference is 950 psi and the depletion is 4,000 psi (27.6 MPa), the stress orientation changes in the vicinity of intersections between hydraulic fractures and pre-existing fractures are exacerbated and there are modest stress orientation changes near the ends of the hydraulic fractures. In case 3 where the initial stress difference is only 175 psi (1.2 MPa) and the depletion is 5,000 psi (34.5 MPa), significant changes in stress orientation occur both in the area surrounding the hydraulic fractures and the areas away from the wells at both ends of the model. In fact, the marked change in stress orientation at the left and right sides of the model indicate a reversal of the S_{hmin} and S_{Hmax} directions.

Unconventional Reservoir Geomechanics

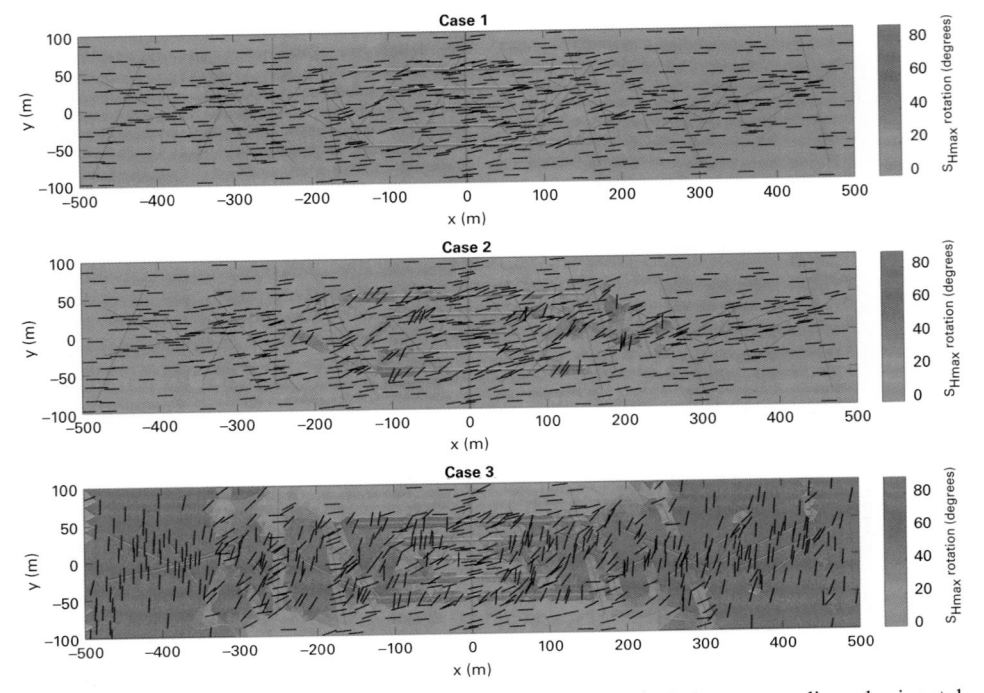

Figure 12.17 Model of stress orientation changes with pressure depletion surrounding a horizontal well with hydraulic fractures that have propagated from five perforations as well as pre-existing fractures. As described in the text, the three cases consider, respectively, models with increasing initial pore pressure and decreasing initial stress anisotropy. From Jin & Zoback (2019).

Figure 12.18 provides insight into why the heterogeneous depletion shown in Fig. 12.10 results in the dramatic changes in stress orientation seen in case 3 in Fig. 12.17. Figure 12.18a shows the changes in the S_{yy} stress (S_{hmin} in the initial model) that accompanies depletion and Fig. 12.18b shows the changes in the S_{xx} stress (S_{Hmax} in the initial model). The changes in stress magnitudes are largely concentrated in the areas of most depletion, along the permeable fractures, but broadly decrease in the model domain. Obviously, this poroelastic stress decrease could lead to asymmetric hydraulic fracture growth as seen in Fig. 12.14. There are some areas of local stress increase due to the complex geometry. The reason the orientations of S_{hmin} and S_{Hmax} reverse at the left and right side of the model as shown in Fig. 12.17 is that the change in the S_{xx} stress is larger than the S_{yy} stress. This is a direct result of the assumed distribution of stimulated shear fractures and heterogeneous depletion as well as the initially small differences in the horizontal principal stresses and large amount of depletion. It is also important to emphasize that as a 2D model, depletion-induced stress changes are greater than they would normally be in actual reservoirs.

In summary, asymmetric hydraulic fracture growth and the frac hit phenomenon can be a significant issue affecting well spacing and infill drilling. In some cases, when modest depletion is occurring uniformly around a well in an area with significant differences between the two horizontal principal stresses, it is relatively straightforward

Production and Depletion 373

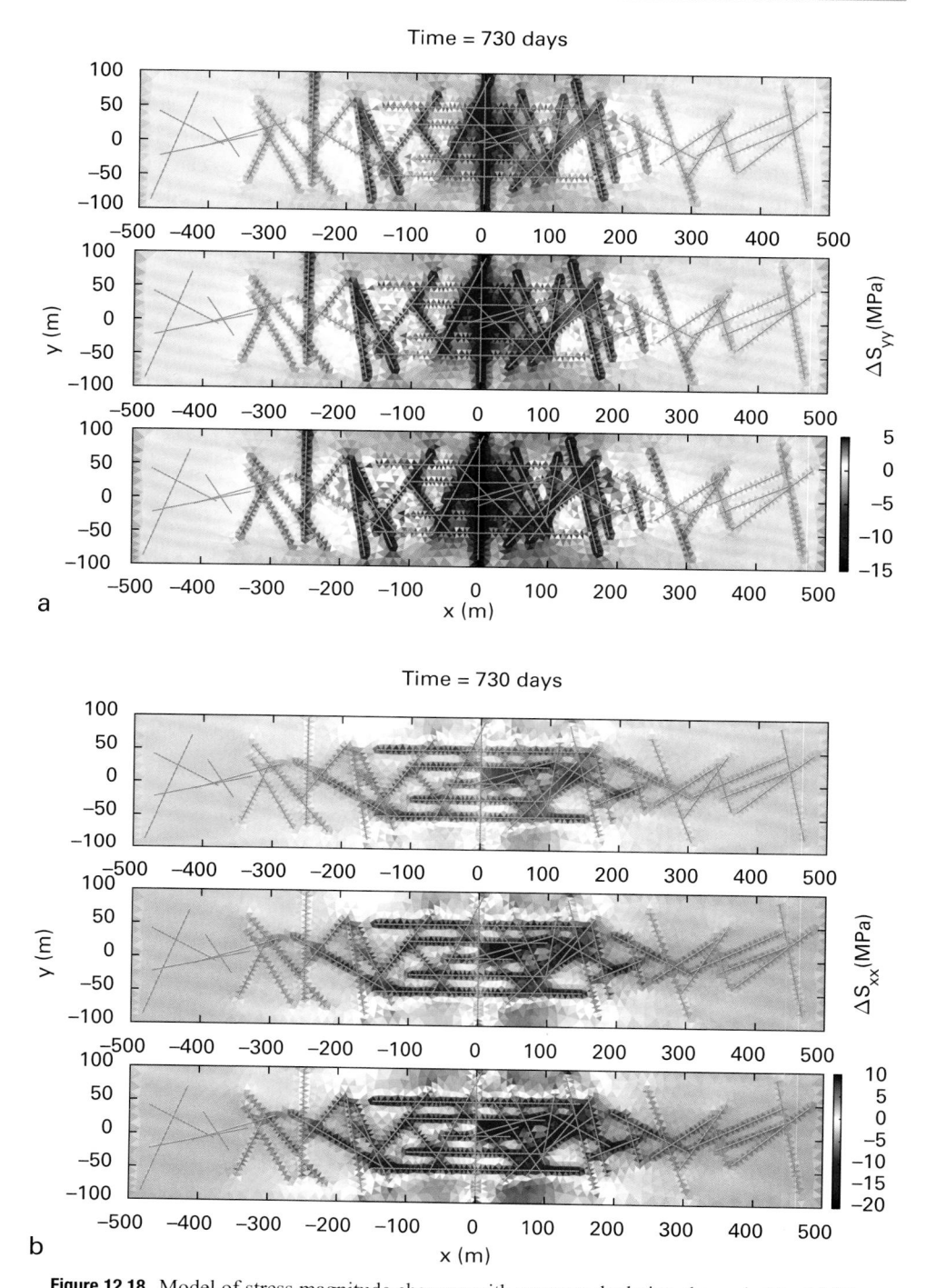

Figure 12.18 Model of stress magnitude changes with pressure depletion shown in Fig. 10.33. The three cases shown are the same as used in Figs. 12.10 and 12.17. (a) The change in S_{yy}, equivalent to S_{hmin} in the initial stress state. (b) The change in S_{xx}, equivalent to S_{Hmax} in the initial stress state. Note that for case 3, the decrease in S_{xx} is larger than the decrease in S_{yy}, resulting in a switch of S_{Hmax} and S_{hmin} in some areas as shown in Fig. 10.32. From Jin & Zoback (2019).

to predict what is likely to happen. That said, Figs. 12.17 and 12.18 illustrate that significant stress re-orientation associated with depletion is most likely to occur in strongly overpressured formations where the initial differences between the principal stress are initially quite small and depletion-induced stresses can be quite large. Because the details of how these stress orientation changes occur in space and time are a direct result of the complexities built into the model, it is difficult to generalize these results and unclear (at this point in time) if processes such as significant stress rotations have significant impact on subsequent stimulation of adjacent wells.

Part III

Environmental Impacts and Induced Seismicity

13 Environmental Impacts and Induced Seismicity

In this chapter we first provide a brief overview of some of the environmental impacts associated with large-scale development of unconventional oil and gas reservoirs. While it is beyond the scope of this book to address all of the potential environmental issues that could arise, there are several which are related to the topics considered elsewhere in the book. We then focus on the topic of induced seismicity associated with unconventional reservoir development – a significant, if somewhat unexpected, environmental impact.

As shown in Fig. 13.1 (from Zoback & Arent, 2014), there are a number of potential risk factors associated with unconventional development that were grouped into four main categories. While many of these issues are also common to conventional oil and gas development, one aspect of unconventional hydrocarbon development that is somewhat unique is the very large number of wells being drilled, sometimes in populated areas. The Barnett shale development (approximately 15,000 wells) took place in the Dallas/Fort Worth metroplex, the Utica and Marcellus developments are occurring in Ohio, Pennsylvania and West Virginia, frequently near populated areas and the Niobrara development in the DJ Basin of Colorado is located near suburbs of Denver. While oil and gas development in populated areas may seem unique, it is often forgotten that there are over 3,000 oil wells currently operating in Los Angeles county, home to over 10 million people. The Texas Academy of Medicine, Engineering and Science (2017) recently published a comprehensive report evaluating impacts of unconventional oil and gas development in Texas on air, water, transportation, seismicity as well as economic and social impacts.

As shown in Fig. 13.2 (updated from Rubinstein & Mahani, 2015), an upsurge of seismic activity began around 2009 throughout the central and eastern US and parts of Canada (not shown). The number of magnitude 3 and larger earthquakes increased from an average of 34 per year to almost 1,000 in 2015. There is now little question that the significant increase in magnitude 3 and larger earthquakes is the result of fluid injection associated with unconventional oil and gas development. We discuss below the primary mechanisms by which this occurs – hydraulic fracturing, flowback water disposal and produced water disposal. In Chapter 14 we discuss ways to reduce the hazard associated with induced seismicity, including the steps taken which resulted in the significant drop in the number of earthquakes after 2015.

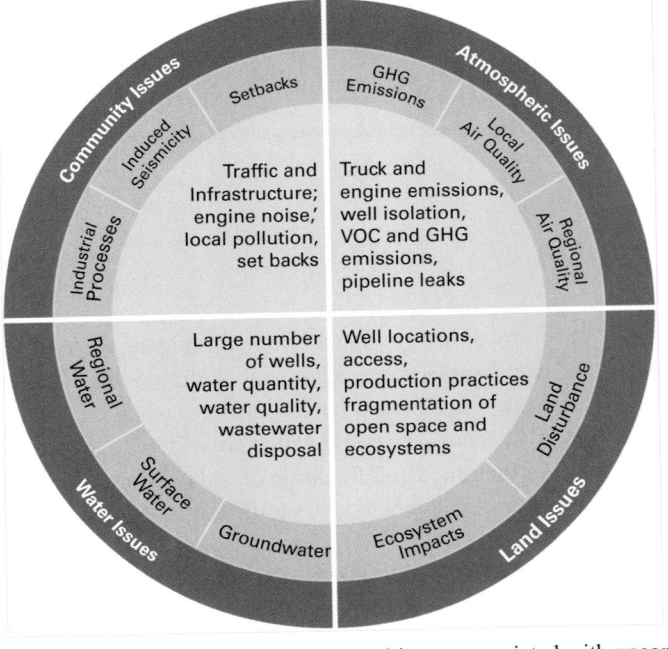

Figure 13.1 Graphical illustration of the environmental issues associated with unconventional oil and gas development. After Zoback and Arent (2014).

Overview of Environmental Issues

The environmental issues discussed below are those that are most germane to the subjects considered in the other chapters in this book. These involve water issues that range from the use of water resources for hydraulic fracturing to the chemicals used in hydraulic fracturing fluids to the disposal of flowback water after hydraulic fracturing as well as disposal of produced formation pore water once wells go into production. Obviously, the topic of water disposal is closely related to the issue of induced seismicity. Although potential contamination of fresh water aquifers is principally related to issues of well construction (introduced in Chapter 8) it is often confused with the potential for hydraulic fractures to propagate to the near surface and contaminate near surface water supplies. Well construction is also related to the issue of methane leakage. As methane is a potent greenhouse gas, sources of methane emissions associated with natural gas wells, pipeline and processing facilities, and municipal distribution systems need to be identified and remedied. There is no question that switching from coal to natural gas for electrical power generation would be a positive development both to lower emissions of CO_2 and alleviate severe air pollution problems affecting hundreds of millions of people around the world (and its impact on public health). It has been argued, however, that methane leakage into the atmosphere could obviate such benefits. Resultantly, there is an extensive literature, and frequently a fair degree of debate, about the environmental impacts (versus benefits) of unconventional reservoir

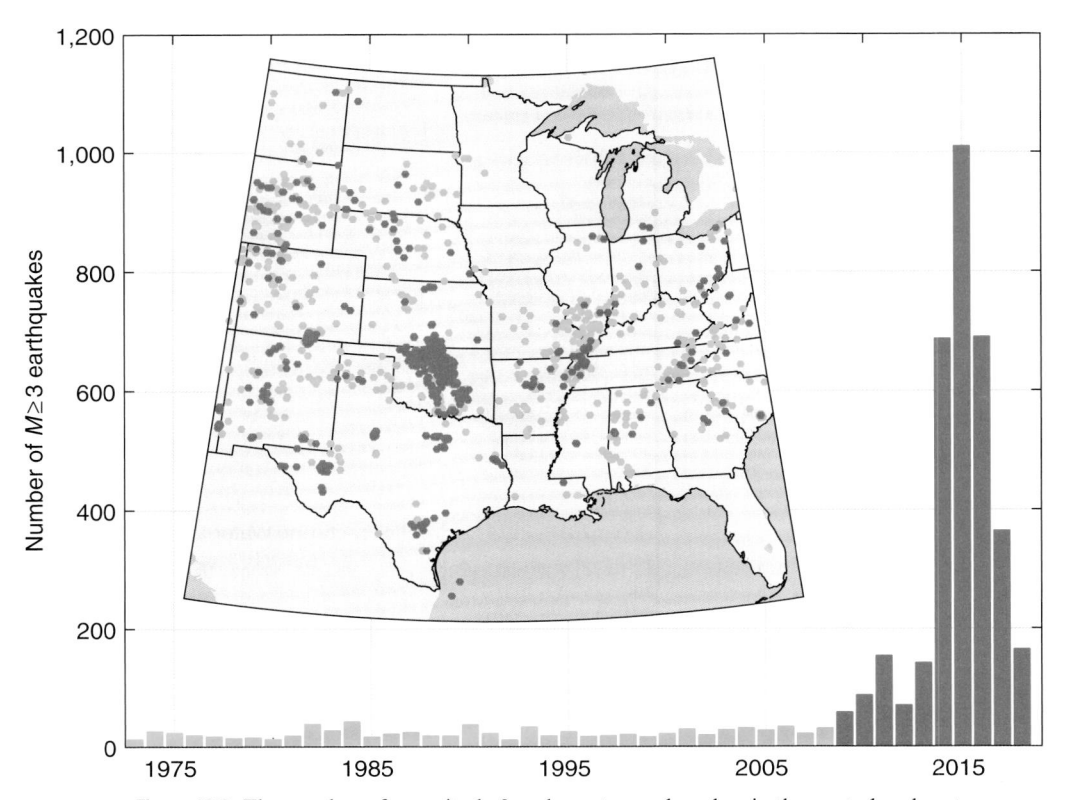

Figure 13.2 The number of magnitude 3 and greater earthquakes in the central and eastern US began to increase markedly around 2009, peaked in 2015, but continues at a level much higher than that during the previous 30 years. Updated from Rubinstein & Mahani (2015).

development. Krupnick et al. (2013) presented an analysis of the environmental risks related to shale gas development based on a survey of experts from government agencies, industry, academia and nongovernmental organizations to identify and, to some degree, prioritize risks from various sources. A report focusing on water resources and water quality was published by US Environmental Protection Agency in 2016 (EPA, 2016). Charlez & Baylocq (2015) present an interesting overview of some of the impressions of the environmental impacts of shale gas production that greatly affected public opinion, whether justified or not. Zhang & Yang (2015) evaluated the environmental impacts of shale gas development in the US from the perspective of what can be learned from US experience as it might apply to shale gas development in China.

Water Use – Water use associated with multi-stage hydraulic fracturing is appreciable. In some areas in the US, such as where the Utica and Marcellus plays are being developed, water is relatively abundant. In these areas, access to fresh water for drilling and hydraulic fracturing is not a major issue although water disposal might be. In the arid west, the situation is quite different. For example, in the Eagle Ford play of south Texas and the Bakken play of North Dakota, Scanlon et al. (2014) determined that

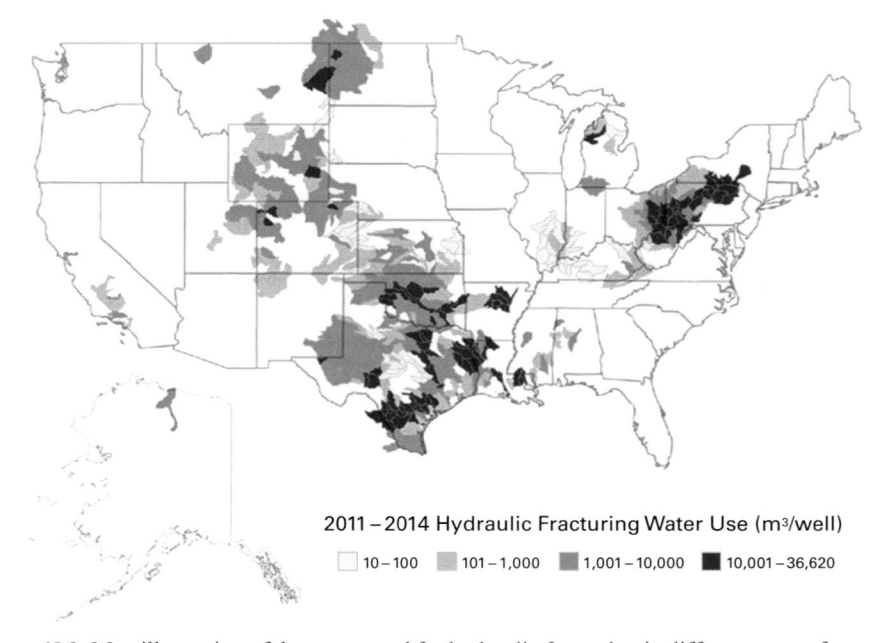

2011 – 2014 Hydraulic Fracturing Water Use (m³/well)

☐ 10 – 100 ▨ 101 – 1,000 ▦ 1,001 – 10,000 ■ 10,001 – 36,620

Figure 13.3 Map illustration of the water used for hydraulic fracturing in different parts of country. (Data from US Geological Survey, map from Columbus Business First.)

unconventional development wells utilized, on average, 4.82 and 2.01 million gallons of water per well, respectively. As these volumes apply to approximately 13,000 and 15,000 wells in the Eagle Ford and Bakken (see Fig. 1.3), this makes total water use in just these two areas approximately 62 and 30 billion gallons, respectively, over periods between 5 and 9 years. Figure 13.3 illustrates the variability of water use for hydraulic fracturing in different parts of the US utilizing data available in 2015 ($1m^3 =$ 6.3 barrels = 264 gallons). While this situation changes with time, the high water use in arid regions is an issue for concern when one considers the large numbers of wells that will be drilled in many of these areas in the future.

To the degree the use of fresh water creates shortages of water for irrigation, raising livestock or domestic use, two actions seem warranted. First, it is important to enact processes that consider management of water resources on a regional scale. As noted above, one of the principal differences between unconventional reservoir development and conventional reservoir development is the large number of wells and the large regions affected. Clearly, management of water resources at the watershed scale seems highly warranted. An example of this for the Susquehana River Basin in Pennsylvania is presented by Rahm & Riha (2012). The second step that seems warranted is using saline or brackish water instead of fresh water for hydraulic fracturing. Figure 13.4 (after Zhu & Tomson, 2013) illustrates the abundance of saline aquifers in areas of unconventional development.

Water Reuse and Recycling – When wells are flowed back after hydraulic fracturing, anywhere between 0 and 50% of the injected hydraulic fracturing fluids are recovered.

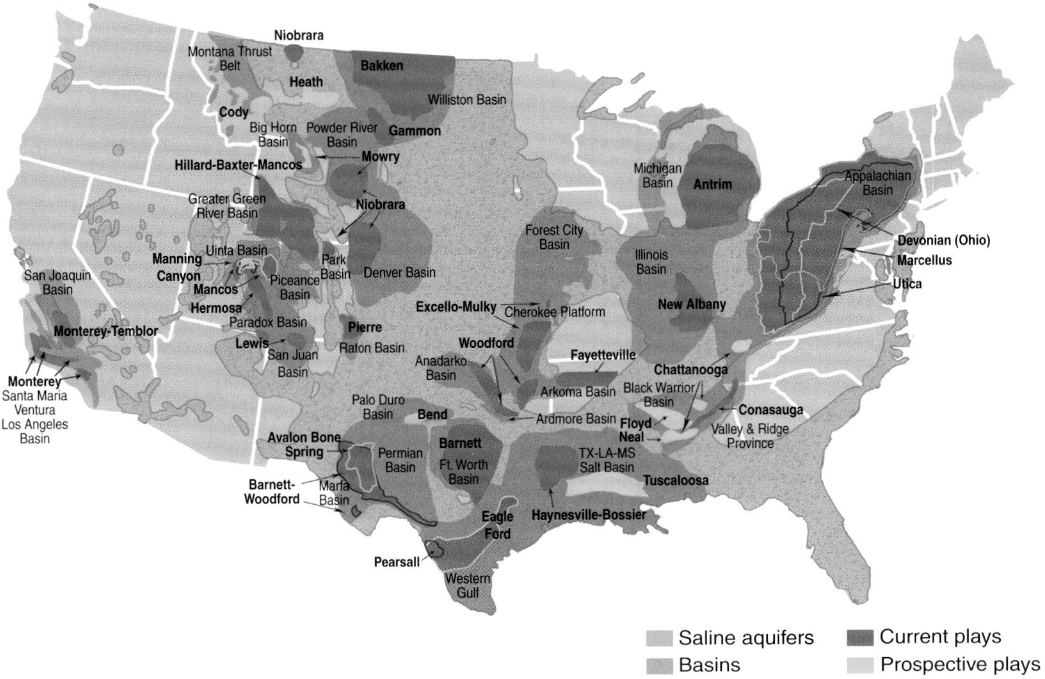

Environmental Impacts and Induced Seismicity   381

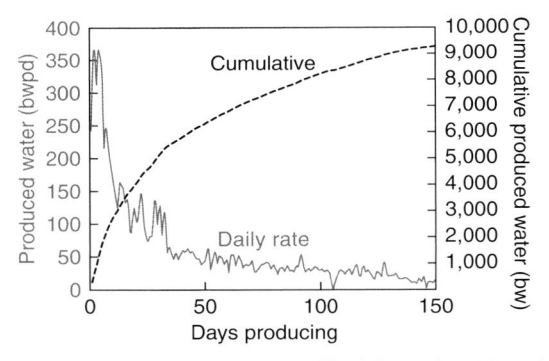

Figure 13.4 Map showing the presence of saline aquifers which could supply water for drilling and hydraulic fracturing associated with unconventional reservoir development. After Zhu & Tomson (2013).

Figure 13.5 Flowback from a typical well in the Woodford formation. Note that about 30% of the injected volume during hydraulic fracturing comes back in the first 15 days.

The US National Academies (National Academies Press, 2017) published a comprehensive review of information on flowback and produced water quality, approaches for water recycling and reuse as well as the challenges of reuse. Figure 13.5 shows the average flowback from thirteen wells in the Utica shale in Ohio. Almost 10,000 barrels of hydraulic fracturing fluid (roughly 10% of the injected volume) flows back in the first 150 days of production. Although hydraulic fracturing fluid continues to flow back after that, it is at a very low, and diminishing, rate. The recovered fracturing fluid is more saline than it was when injected and can contain contaminants such as iron, arsenic, selenium and naturally occurring radioactive material (NORM). As production continues, the water coming from the wells consists of less hydraulic fracturing fluid and more pore water that is co-produced with the oil or gas. For example, in the Barnett shale, in the first year an average well produces approximately equal amounts of hydraulic fracturing fluid and formation water, but after four years, twice as much of the water produced is formation water (Nicot et al., 2014).

Figure 13.6 (from Scanlon et al., 2017) illustrates alternative scenarios for dealing with the water that flows back after hydraulic fracturing and produced water. The figure illustrates the three main oil producing regions in the Permian Basin – from west to east the Delaware Basin, the Central Basin Platform and the Midland Basin. What is shown in the figure, however, is also generally representative of many other regions, as discussed later in this chapter. As shown in Fig. 13.6, in the Delaware Basin the water that is co-produced with oil from the Bone Spring and Wolfcamp formations is disposed of through salt water disposal (SWD) into the relatively shallow Delaware Mountain

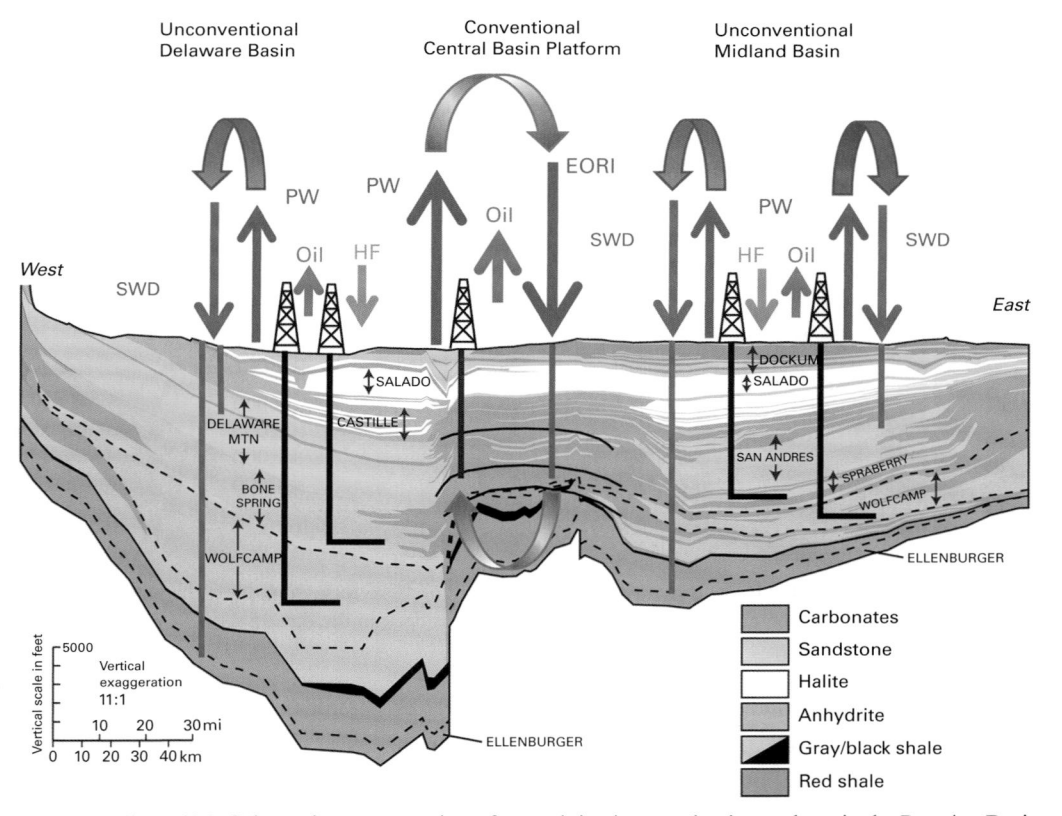

Figure 13.6 Schematic representation of water injection, production and use in the Permian Basin. SWD – salt water disposal, PW – produced water, HF – formations being hydraulically fractured, EORI – enhanced oil recovery injection. After Scanlon et al. (2017).

Table 13.1 Components of average hydraulic fracturing fluid.

Component	Typical abundance (%)	Comments
Water	99.2	Usually fresh water, brackish and saline water could be used
Acid	0.07	Hydrochloric acid may be used to clean up the cement around perforations and reduce fracture initiation pressure. The acid is used up within inches of the hydraulic fracture entry point and is not returned to the surface
Gelling agent	0.5	Thickeners such as guar gum and cellulose polymers may be used in hybrid fracs as referred to in Chapter 8. Guar gum is a common food additive
Corrosion inhibitor	0.05	Several organic compounds that may be toxic are used but only used if acid is used. The inhibitor is adsorbed on steel and then in the formation. About 5 to 10% total (about a gallon in a million gallons of water) returns to surface in the backflow
Friction reducer	0.05	Polyacrylamide is widely used to reduce friction pressure of water during high rate pumping. Also used as an adsorbent in baby diapers and as a flocculent in drinking water preparation
Clay control	0.034	Stabilizes clays in the formation through a sodium–potassium ion exchange. Results in sodium chloride, common salt, which is returned in flowback water
Crosslinker	0.032	Maintains viscosity as temperature increases. Combines with breaker to return salts in flowback water
Scale inhibitor	0.023	Used to prevent mineral scale precipitates and possible blockage of tubing and equipment. Common scale inhibitors are non-toxic and used in very low concentrations
Breaker	0.02	Allows a delayed breakdown of gelling agents and crosslinkers, when used
Iron control	0.004	Prevents precipitation of metal oxides. Reacts with minerals in the formation to form simple salts which are recovered with flowback water.
Biocide	0.005% to 0.05%	Glutaraldehyde is commonly used to control bacterial growth that could be harmful to additives in the fluid or cause generation of H_2S. Glutaraldehyde is a medical disinfectant. UV light, ozone and low concentration chlorine dioxide are being used more
Surfactant	0.05 to 0.2 %	Surfactants modify surface or interfacial tension in order to break or prevent emulsions

group or the deeper Ellenburger carbonate, which sits directly on crystalline basement. The deep injection into the Ellenburger is also occurring in the Fort Worth Basin where the Barnett has been developed. Similar basal carbonates are used for flowback and produced water disposal in parts of Oklahoma and Arkansas where earthquake triggering has been occurring, as discussed below. Note that there is appreciable anhydrite at shallow depth in this area which provides a good top seal for SWD in the Delaware Mountain group.

In the Central Basin Platform area, there has mostly been conventional oil development to date. Thus, produced water is being returned to the producing formations as part of enhanced oil recovery (EOR) operations. As will be discussed in Chapter 14 in the context of addressing induced seismicity caused by disposal of produced water, returning produced water to the formations that produced it can be done in a sustainable manner (i.e., without triggering seismicity). In the Midland Basin, SWD (both hydraulic fracturing flowback water and produced water) is being disposed of in the relatively shallow San Andres group and deeper Ellenburger.

One issue of concern to the public are the chemicals used in hydraulic fracturing and whether there is the potential for these chemicals to contaminate fresh water supplies. The most comprehensive source of data about hydraulic fracturing fluids is available from the FracFocus website (www.fracfocus.org). FracFocus is a US national hydraulic fracturing chemical registry. FracFocus is managed by the Ground Water Protection Council and Interstate Oil and Gas Compact Commission, two organizations whose missions revolve around conservation and environmental protection. The FracFocus website provides a comprehensive overview of the chemicals used in hydraulic fracturing as well as data from almost 130,000 registered wells. While the complete list of chemicals used is somewhat daunting, Table 13.1 illustrates the composition of average hydraulic fracturing fluids in the US as well as their purpose (from FracFocus and King, 2012). Two important issues to keep in mind are that there has been a continuous effort to replace potentially harmful chemical additives with safer additives (Environmental Protection Agency, 2015) and, perhaps more importantly, because of the large separation between the depth of hydraulic fracturing operations and fresh water aquifers in the unconventional plays of the US (Fisher & Warpinski, 2012; Fig. 8.16) there have been no documented cases in which hydraulic fractures have propagated to shallow enough depth to cause contamination of near surface aquifers. As discussed below, when contamination has occurred, it has been the result of poor well construction or spills at the surface (see Jackson et al., 2014; EPA, 2016).

Produced Water – There are major differences in the quantity and composition of produced water in the United States as well as how it is dealt with. As shown in Fig. 13.7 (from Environmental Protection Agency, 2015), the amount of water used for hydraulic fracturing in an average well in the Marcellus shale in the Susquehana River Basin of Pennsylvania (Fig. 13.7a) is similar to that used by an average well in the Barnett shale (Fig. 13.7b). However, there is far more produced water from the Barnett, which is principally disposed of via disposal wells drilled into the Ellenburger formation as noted above. In contrast, there is much less water produced from the Marcellus, but the great majority of it is recycled and re-used. In fact, because produced water disposal is so rarely done in Pennsylvania, injection-related induced seismicity has been quite rare there.

Shaffer et al. (2013) developed a map (shown in Fig. 13.8) that shows the ranges of produced water total dissolved solids (TDS) for shale plays in the US principally using data from the US Geological Survey. The histograms show the produced water TDS concentrations for the Bakken and Gammon shales in the Williston Basin, the Mowry

Environmental Impacts and Induced Seismicity

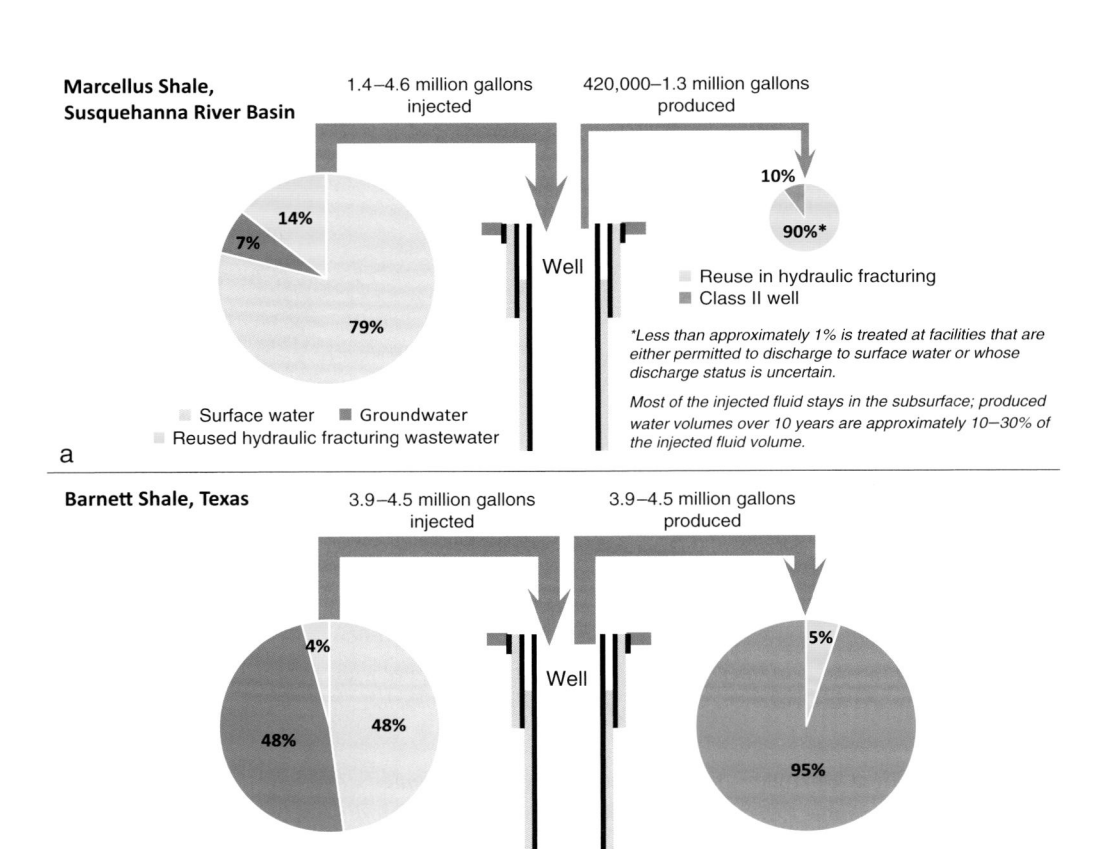

Figure 13.7 Water budgets illustrative of hydraulic fracturing water management practices in (a) the Marcellus shale in the Susquehanna River Basin between approximately 2008 and 2013 and (b) the Barnett shale in Texas between approximately 2011 and 2013. Class II wells are used to inject wastewater associated with oil and gas production underground and are regulated under the Underground Injection Control Program of the Safe Drinking Water Act. After EPA (2016).

shale in the Powder River Basin, the Avalon-Bone Spring and Barnett-Woodford shale plays in the Permian Basin and the Devonian, Marcellus and Utica shale plays in the Appalachian Basin. Note that with few exceptions (such as the Powder River Basin), pore water is considerably more saline than sea water (approximately 20,000 mg/L). Such saline water could be used as a hydraulic fracturing fluid, but it is so saline that treatment is not economically viable so it must be disposed of through injection.

Well Construction – It is widely acknowledged that when oil and gas drilling and completion practices result in contamination of aquifers, the reason contamination occurred is related to poor well construction practices, deterioration of cement and casing over time or damage to the casing and cement during multi-stage hydraulic fracturing. In 2011, a subcommittee of the Department of Energy charged by then

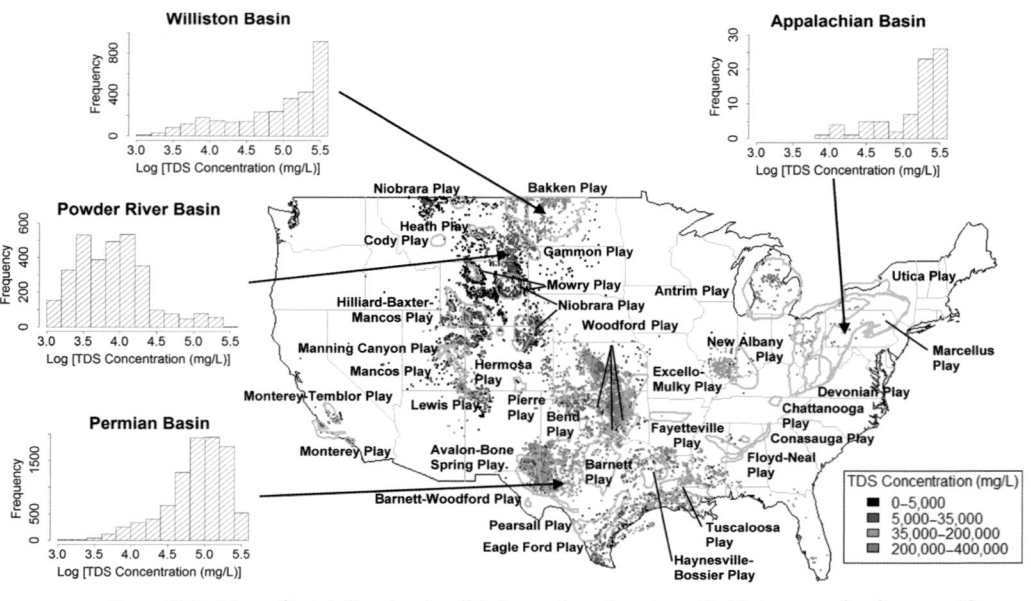

Figure 13.8 Map of total dissolved solids in produced water with histograms for five specific basins. After Shaffer et al. (2013). *Reprinted with permission from American Chemical Society.*

President Obama to provide "*consensus recommended advice to the agencies on practices for shale extraction to ensure the protection of public health and the environment.*" After many meetings with non-governmental and environmental groups, representatives from the oil and gas industry and State and Federal regulatory agencies, the committee concluded that to protect water quality it was necessary to "*Adopt best practices in well development and construction, especially casing, cementing, and pressure management. Pressure testing of cemented casing and state-of-the-art cement bond logs should be used to confirm formation isolation*" (Deutch et al., 2011). Needless to say, the same principles apply to prevent methane leakage from wells.

Table 13.2 indicates the percentage of experts from four different cohorts who considered the types of activities that they were most concerned about related to accidents associated with unconventional reservoir development (from Krupnick et al., 2013). It is clear that potential cement and casing failure are consensus areas of considerable concern. King and King (2013) provide a helpful overview of risks arising from well construction and Carey (2013) reviews mechanisms by which wellbore integrity (both casing and cement) can be compromised..

To put the issues surrounding casing and cement in context, Fig. 13.9 (from King, 2012) illustrates generic well construction practices. One key aspect of well construction is that the surface casing extends below the depth of surface aquifers. Regulations vary but the surface casing might extend from a couple of hundred to several thousand feet. The surface casing should be completely cemented to the surface. The integrity of this casing and cement are critical elements of well construction. Another key aspect of well construction is that the cemented portion of the production casing extend above the

Table 13.2 Percentage of experts concerned with different sources of potential accidents. Bold underlined font – most often selected, bold italic font – second most selected.

Accidents	NGO	Industry	Academia	Gov't	All experts
Cement failure	**80.0**	**58.7**	57.1	**66.7**	**63.3**
Casing failure	68.6	*46.7*	**61.9**	*57.1*	*56.7*
Impoundment failure	*71.4*	33.3	**61.9**	45.2	50.2
Surface blowout	54.3	34.7	49.2	40.5	43.3
Storage tank spills	42.9	30.7	46.0	28.6	36.7
Truck accidents	37.1	40.0	34.9	28.6	35.8
Pipeline ruptures	42.9	30.7	38.1	33.3	35.3
Surface valve failure	40.0	21.3	27.0	26.2	27.0
Underground well comm.	37.1	14.7	28.6	23.8	24.2
Other spills	22.9	20.0	20.6	23.8	21.4
Underground blowout	31.4	14.7	20.6	23.8	20.9
Hose bursts	22.9	17.3	14.3	16.7	17.2
Other fires or explosions	8.6	13.3	7.9	14.3	11.2
Other not listed here	8.6	5.3	11.1	2.4	7.0
All 14 accidents	40.6	27.2	34.2	30.8	32.2
Average # of accidents selected as high priority	5.69	3.81	4.79	4.31	4.50

shallowest gas producing zone (again, regulations vary). As the uncemented section of the production casing can often extend for thousands of feet, care should be taken to ensure that all gas producing horizons are cased and cemented. We discuss below the issue of casing integrity in the context of methane emissions.

Methane Leakage – As a potent greenhouse gas, methane in the atmosphere is of great concern because it is thought to be responsible for about 25% of global warming (see Global Carbon Project, www.globalcarbonproject.org/index.htm). A recent study by Alvarez et al. (2018) of methane emissions associated with natural gas facilities (wells, processing plants, pipelines, etc.) in areas that produce about 30% of US production, has estimated appreciably more methane in the atmosphere than previously estimated. They estimate about 2.3% of gross US gas production is lost along the gas supply chain. Moreover, the impact of this leakage on global warming is comparable to that of the CO_2 produced during natural gas combustion.

Of most relevance here is any association of methane leakage with wells, whether it was caused by poor well construction, or well construction that was adversely affected by multi-stage hydraulic fracturing operations. From studies of soil gases in 31 wells in three different basins in Utah, Lyman et al. (2017) estimated that an extremely small portion (approximately 1 part per million) of the hydrocarbons emitted from oil and gas operations in Utah comes from the wells. A recent study reported by Barkley et al. (2017) utilized ten airborne surveys (and modeling) in north-eastern Pennsylvania to find that leakage from upstream operations is approximately 0.4% of production. If these studies are representative of operations more generally, wells and upstream operations

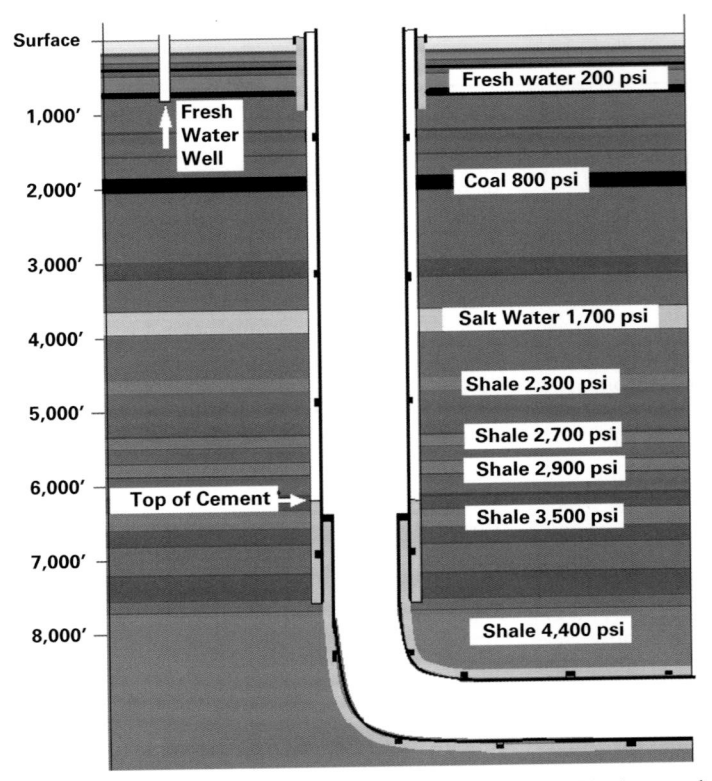

Figure 13.9 Schematic diagram of an unconventional well. As explained in the text the depth and integrity of cemented surface casing and the need to cement well above gas producing zones are critical elements of well construction. From King (2012).

account for only a small fraction of the methane being emitted by the oil and gas industry.

Of course, results could be quite different elsewhere, but it's also important to recognize that a little less than half the methane in the atmosphere comes from the oil and gas industry, with the majority coming from biogenic sources (agriculture, wetlands, waste/landfills, permafrost, livestock, etc.) and other human-related activities (Saunois et al., 2016). It is also sometimes difficult to quantify the origin of methane in the atmosphere. *Top-down* studies utilizing sensors on aircraft or satellites and *bottom-up* studies based on ground-based measurements of specific facilities have obvious advantages and disadvantages. It is sometimes difficult to identify the specific source of detected methane in top-down studies and bottom-up studies could miss important sources of leaks. In the oil and gas industry, the latter problem is of particular concern as it has been shown that a very small fraction of the components leak (about 1%). Of those that leak, a small fraction (~5%) leak a significant amount and are responsible for over half of all methane leakage (Brandt et al., 2016). In other words, most of the methane in the atmosphere related to oil and gas activities come from super-emitters, approximately 1/2,000 of all components, that are usually associated with natural gas processing facilities or pipelines.

Induced Seismicity

From an earthquake perspective, 2011 was a remarkable year. Although the devastation accompanying the magnitude-9.0 Tohoku earthquake that occurred off the coast of Japan captured worldwide attention, the relatively stable interior of the US was struck by a surprising number of small-to-moderate earthquakes. Some of these were natural events, the types of earthquakes that occur from time to time in all intraplate regions. For example, the magnitude-5.8 earthquake that occurred in the central Virginia zone was felt throughout the northeast, damaged the Washington Monument and caused the temporary shutdown of a nuclear power plant. However, a number of small-to-moderate earthquakes that occurred in the US interior in 2011 appeared to be associated with the disposal of wastewater, at least in part related to horizontal drilling, multi-stage hydraulic fracturing and shale gas production. A number of earthquakes apparently caused by injection of flowback water after hydraulic fracturing in the Fayetteville shale occurred near Guy, Arkansas. The largest earthquake was a M 4.7. In the Trinidad/Raton area near the border of Colorado and New Mexico, injection of wastewater associated with coalbed methane production seems to have caused a M 5.3 event. Earthquakes were triggered by flowback water injection on Christmas Eve and New Year's Eve near Youngstown, Ohio, the largest of which was M 4.0. Most significantly, three magnitude 5+ earthquakes occurred near Prague, Oklahoma in November that, we now know were triggered by produced water disposal in deep injection wells.

There is a long history of earthquakes that appear to have been induced by activities associated with fluid injection, reservoir impoundment and mining activities. The occurrence of earthquakes in response to fluid injection was well documented in the pioneering work of the US Geological Survey at the Rocky Mountain Arsenal near Denver, Colorado (Healy et al., 1968) and in the Rangely oil field, also in Colorado (Raleigh et al., 1976). Nicholson & Wesson (1992) reviewed known or suspected cases of induced seismicity associated with deep well injection around the world. In the past several years, there have been a number of review papers on this topic. The US National Academies (National Research Council, 2013) carried out a comprehensive review of induced seismicity associated with a number of energy-related activities. This report reviewed numerous case histories, discussed mechanisms by which earthquakes could be induced by different kinds of anthropogenic activities and suggested ways in which to evaluate (and reduce) the risk of inducing seismicity. In the same year, Ellsworth (2013) reviewed a number of sites of recent injection-induced earthquakes including the central and eastern US, Basel, Switzerland and the Paradox Valley, Colorado. The Environmental Protection Agency (EPA, 2014) updated the information in the NRC 2013 study with a particular emphasis on injection-induced seismicity associated with unconventional oil and gas development. The Groundwater Protection Council working with the Interstate Oil and Gas Compact Commission (GWPC & IOGCC, 2017) provided a useful update to the previous reviews,

390   **Unconventional Reservoir Geomechanics**

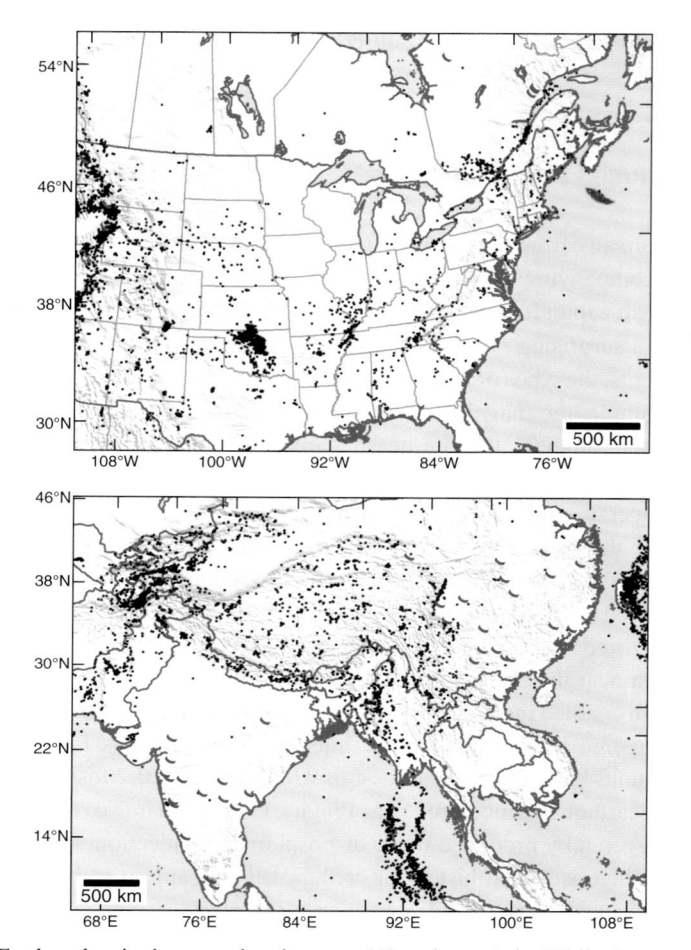

Figure 13.10 Earthquakes in the central and eastern US and east Asia. While the rate of intraplate seismicity varies greatly from one region to another, earthquakes are seen essentially everywhere as are sites where dam construction and reservoir impoundment has induced seismicity (red curved symbols). Data from US Geological Survey.

especially with respect to earthquake induced during hydraulic fracturing. Foulger et al. (2018) present a global review of induced seismicity (and a global data base of events) that includes many examples of reservoir-induced seismicity.

The Critically Stressed Crust – While we will focus in this book on earthquakes induced by activities associated with unconventional reservoir development, it is helpful to consider a concept that is fundamental in continental dynamics – that the crystalline basement rocks that comprise the upper ~20 km of the earth are in a state frictional failure equilibrium. Another way of saying this is that the stress levels in crystalline basement rocks are in equilibrium with the frictional strength of optimally oriented faults. This concept is supported by the observation that intraplate earthquakes are seen to occur essentially everywhere on Earth as shown in Fig. 13.10. Intraplate earthquakes occur at markedly different rates in different areas due to variations of heat flow and crustal composition (Zoback et al., 2002). It is also clear that when an intraplate earthquake occurs, the sense of slip in the earthquake is consistent with relatively large-scale tectonic stresses as defined by independent data (Zoback & Zoback, 1989; M. L. Zoback, 1992b; Hurd & Zoback, 2012a). In other words, the forces that drive the plates are transmitted through the plates and cause small amounts of internal deformation. Sites of reservoir-induced seismicity (i.e., water impoundment after dam construction), shown by the red symbols in Fig. 13.10, also occur nearly everywhere in intraplate regions, including areas with very infrequent natural earthquakes such as eastern China and the Indian and Canadian shields, characterized by thick, cold lithosphere. As the stress and pore pressure perturbation caused by reservoir impoundment is quite small, in order for reservoir-induced seismicity to occur, the initial state of stress must be close to failure. Additional evidence that supports a crystalline crust in frictional failure equilibrium are direct stress measurements in deep boreholes drilled into crystalline rock in intraplate areas (as reviewed by Townend & Zoback, 2000).

The same concept of frictional failure equilibrium applies to sedimentary rocks that are brittle and not prone to viscoplastic stress relaxation as discussed in Chapters 3 and 11. A number of case studies in which *in situ* stress measurements in sedimentary rock that illustrate the frictional equilibrium concept were presented in Zoback (2007) and was shown for the Haynesville shale in Fig. 3.18.

Conceptualizing Earthquake Triggering – In the context of Coulomb friction theory, Eqn. (4.2) pointed out that when pore pressure is raised, it reduces the effective normal stress acting on a fault thereby increasing the likelihood of fault slip. Figure 13.11 uses the analogy between earthquakes and a block on a table pulled by a spring to illustrate earthquake triggering. Force in the spring (analogous to shear stress on a fault) increases through time. On a plate boundary like the San Andreas fault, this occurs relatively rapidly due to relative motion between the Pacific and North American plates. In an intraplate area this occurs quite slowly as plate driving forces cause internal deformation of plates at an extremely low rate. When the frictional resistance to sliding is overcome (as defined by the Coulomb criterion), the block (or fault) slides, and force in the spring (or stress in the earth) drops. Earthquake stress drops were discussed in Chapter 9. This process repeats itself over and over through geologic time. As shown in the lower part of the diagram, when fluid pressure is increased, it lowers the frictional resistance to sliding allowing slip to be triggered on faults that are already near the failure stress but might not be expected to slip for hundreds or thousands of years. A key concept is that this process is one in which earthquakes are triggered by fluid pressure increases; the stress on the fault is already present as a result of natural geologic processes. In Chapter 10 we discussed slip on small faults during hydraulic fracturing that produce a cloud of microseismic events in the surrounding rock masses. In the remainder of this book, we will focus on triggering of larger earthquakes, those that are potentially harmful to people or facilities.

Injection-Induced Seismicity and Unconventional Reservoir Development – As mentioned above, there are three basic mechanisms by which earthquakes are

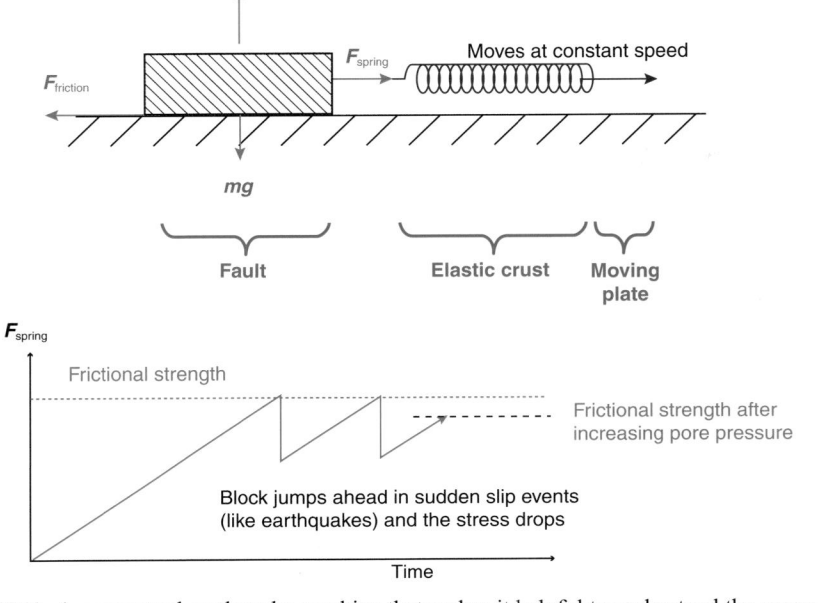

Figure 13.11 A conceptual earthquake machine that makes it helpful to understand the concept of earthquake triggering. (Modified from a schematic developed by Paul Segall.)

triggered by fluid injection during unconventional oil and gas development – hydraulic fracturing, disposal of flowback water and disposal of produced water as conceptually illustrated in Fig. 13.12. In addition to the microseismic events that accompany multi-stage hydraulic fracturing, it is possible for the pressures associated with multi-stage hydraulic fracturing to stimulate slip on pre-existing fault. Examples of this were presented by Maxwell et al. (2008) and Yoon et al. (2017) and are shown in Fig. 13.13. In Fig. 13.13a, microseismic events track the propagation of a hydraulic fracture to the northwest, the direction expected from knowledge of the local stress field. When the hydraulic fracture intersects a pre-existing fault trending north-northeast, microseismic events indicate that this fault was induced to slip by the high pressure acting in the hydraulic fracture. Figure 13.13b shows east–west trending faults near the Guy-Greenbrier fault (discussed below). Slip on these east–west faults was activated during hydraulic fracturing of the wells shown. Microseismic events trending NE–SW, parallel to the current S_{Hmax} direction were also observed but are not shown in the figure. Yoon et al. (2017) presented a composite focal plane mechanism for a number of the events along the east–west faults that are consistent with left-lateral strike-slip motion, as expected in the local stress state (Hurd & Zoback, 2012a).

Figure 13.12 also illustrates the possible occurrence of earthquakes associated with injection of flowback water after hydraulic fracturing or produced water injection. Disposal of these saline waters takes place in what are termed Class II injection wells, regulated in the US by the Environmental Protection Agency. Like in the case of wells

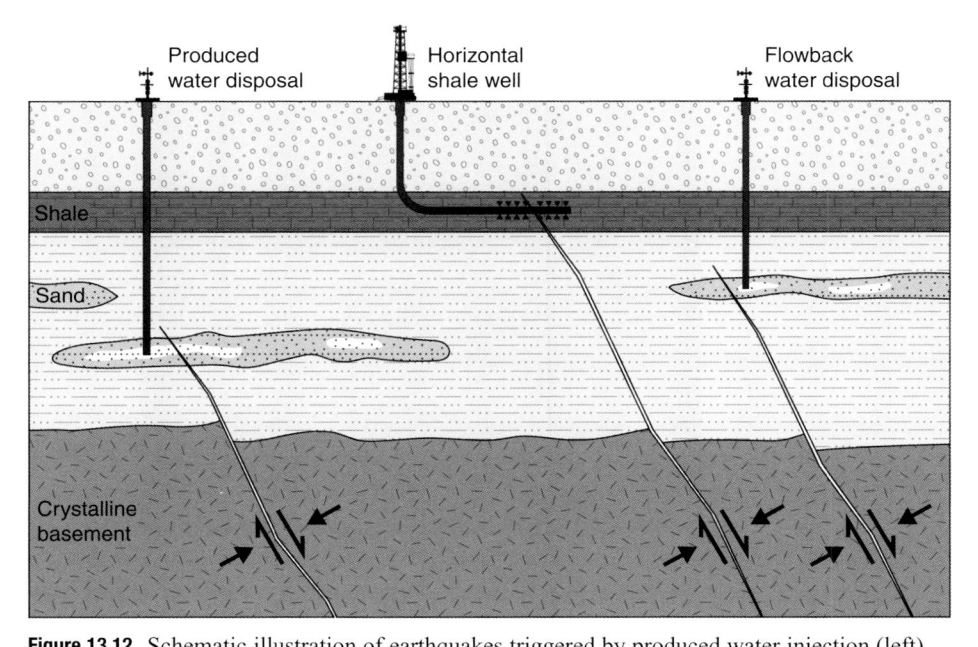

Figure 13.12 Schematic illustration of earthquakes triggered by produced water injection (left), wells being hydraulically fractured (center) and injection of water that flows back after hydraulic fracturing (right) (modified from Southwestern Energy).

used for hydraulic fracturing, well construction requirements for disposal wells are designed to prevent contamination of fresh water aquifers. In most cases, the pressure change caused by the water injection is quite small compared to the high pressures associated with hydraulic fracturing. As mentioned above, there is both hydraulic fracturing flowback water and formation pore water being produced, but as the well depletes, it becomes mostly formation water. Thus, the only significant difference between disposing of flowback water and produced water is that the latter is expected to persist for many years. In general, the faults which are stimulated to slip must be already close to failure. In the sections below, examples of earthquakes triggered by all three of the processes are presented.

The Role of Basement Faults – One of the most important geologic controls on the chance of triggering potentially damaging earthquakes during unconventional development is proximity of the injection zone to crystalline basement. Essentially all of the earthquakes shown in Fig. 13.10 (for which depth could be estimated) occurred in crystalline basement. Moreover, as shown in Fig. 9.20, earthquakes of magnitude ≥ 4 require slip on a fault to occur on a scale of several km. Intuitively, this means that relatively large earthquakes only occur on relatively large faults. In practice, this essentially means that the faults capable of producing magnitude ≥ 4 events are likely to extend into basement. In fact, experience has shown just that.

Figure 13.14a shows the seismicity associated with the 2011 sequence of earthquakes near Prague, Oklahoma that included three magnitude ≥ 5 events (from Keranen et al., 2013). As discussed below, massive amounts of produced water

394 Unconventional Reservoir Geomechanics

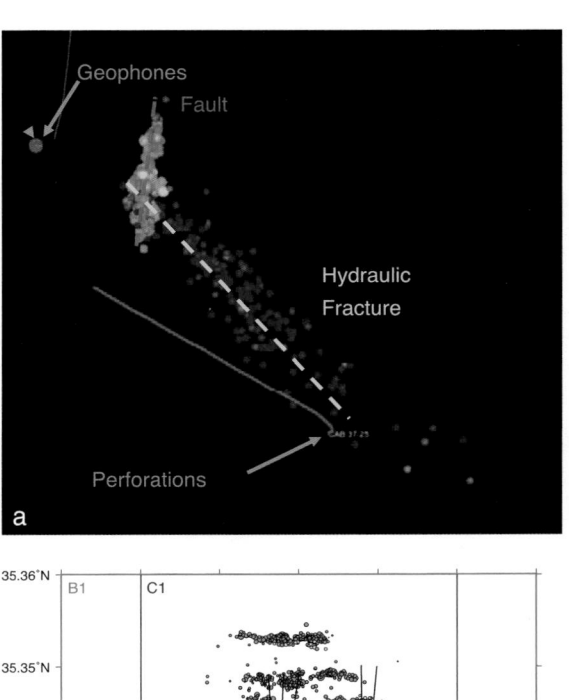

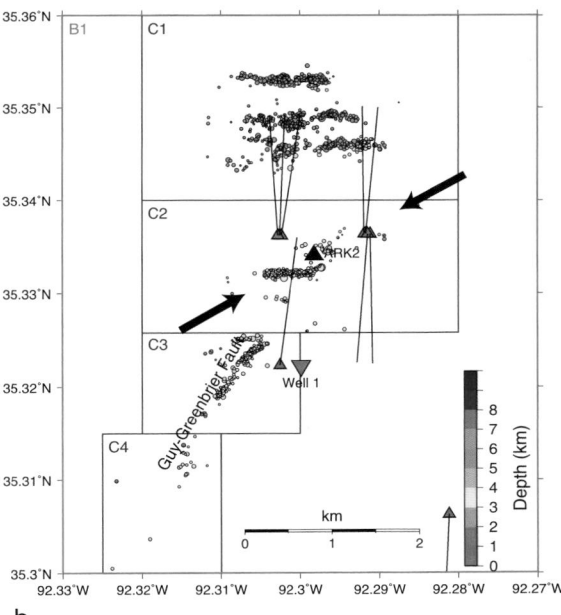

Figure 13.13 Map views of two case studies in which microearthquakes indicate that hydraulic fractures (extending in the expected direction) trigger seismicity on nearby faults. (a) A hydraulic fracture extending to the northwest from a deviated well intersects a pre-existing fault trending NNE. From Maxwell et al. (2008). (b) Small microearthquakes on east–west trending faults triggered by hydraulic fracturing near the Guy-Greenbrier fault in Arkansas. Inward pointed arrows indicate direction of maximum horizontal principal stress. From Yoon et al. (2017).

was injected in the Arbuckle formation (shown in yellow), but the great majority of earthquakes occurred in crystalline basement. Figure 13.14b (from Schoenball & Ellsworth, 2017) shows earthquake depths throughout north central Oklahoma and

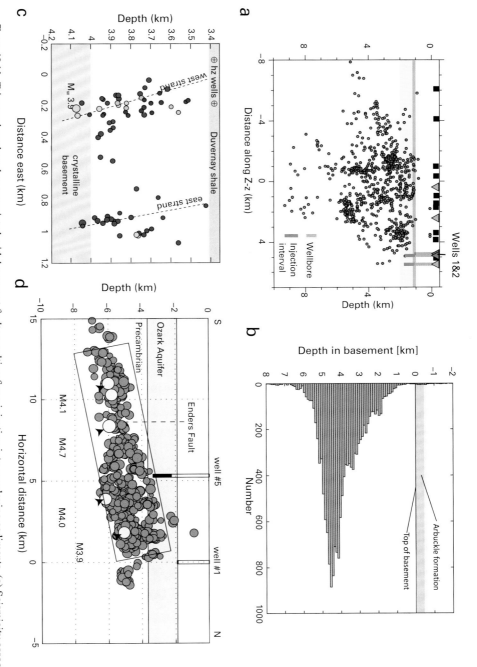

Figure 13.14 Triggered earthquakes associated with basement faults resulting from injection into overlaying sediments. (a) Seismicity associated with produced water injection into the Arbuckle formation in the 2011 Prague earthquake sequence in Oklahoma. From Keranen et al. (2013). (b) Earthquake depths throughout north central Oklahoma and southern Kansas resulting from widespread produced water disposal into the Arbuckle. From Schoenball & Ellsworth (2017). (c) Seismicity associated with a M 3.9 event that occurred several weeks after hydraulic fracturing operations in the Duvernay shale near Fox Creek, Alberta. From Bao & Eaton (2016). (d) Earthquakes on the Guy-Greenbrier fault triggered by flowback water disposal into the Ozark aquifer near Guy, Arkansas. From Horton (2012).

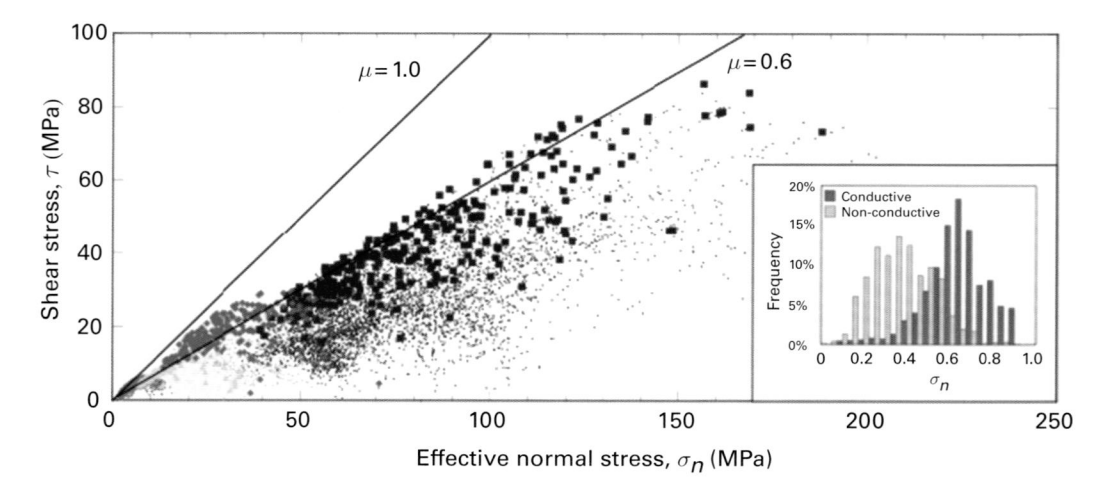

Figure 13.15 The ratio of shear to effective normal stress on fractures and faults encountered in a Nevada Test site (green), Long Valley (yellow), Cajon Pass (red) and KTB (blue) boreholes. The orientation of the fractures and faults were determined from image logs. Hydraulically conductive fractures and faults (bold symbols) were determined from precise temperature logs. The state of stress was determined independently. After Zoback & Townend (2001). *Reprinted by permission of Elsevier whose permission is required for further use.*

southern Kansas are occurring well into the basement. The intense seismicity induced by produced water injection in north central Oklahoma and southern Kansas are discussed at some length later in this chapter as well as in Chapter 14. Figure 13.14c (from Bao & Eaton, 2016) shows a sequence of earthquakes triggered by hydraulic fracturing of the Duvernay shale in the Fox Creek area of Alberta, Canada. Note that seismicity appears to have been triggered by hydraulic fracturing on two fault strands that extend into basement. The largest event was M 3.9, which occurred in the basement along the west fault strand several weeks after injection ended. Figure 13.14d (from Horton, 2012) shows seismicity associated with disposal of flowback water from development of the Fayetteville shale near Guy, Arkansas. Similar to what occurred near Prague, Oklahoma, injection was into the Ozark aquifer, a thick carbonate sitting on top of basement. While a few earthquakes were located in the sedimentary section, the great majority of the events, as well as the larger events (the largest being M4.7) occurred in the basement.

Another reason seismicity occurs on basement faults due to injection into over-laying sediments is that the critically stressed faults in the basement (those well-oriented for failure in the current stress field) are relatively permeable compared to faults that are not critically stressed. As shown in Fig. 13.15 (after Zoback & Townend, 2001) faults in crystalline rock that are hydrologically active are mechanically active. Image logs in four deep scientific boreholes were used to determine the orientation of thousands of fractures and faults encountered at depth. The magnitude and orientation of *in situ* stress was determined in each borehole. Precise temperature logs were used to identify which of the fractures and faults were hydraulically

Environmental Impacts and Induced Seismicity 397

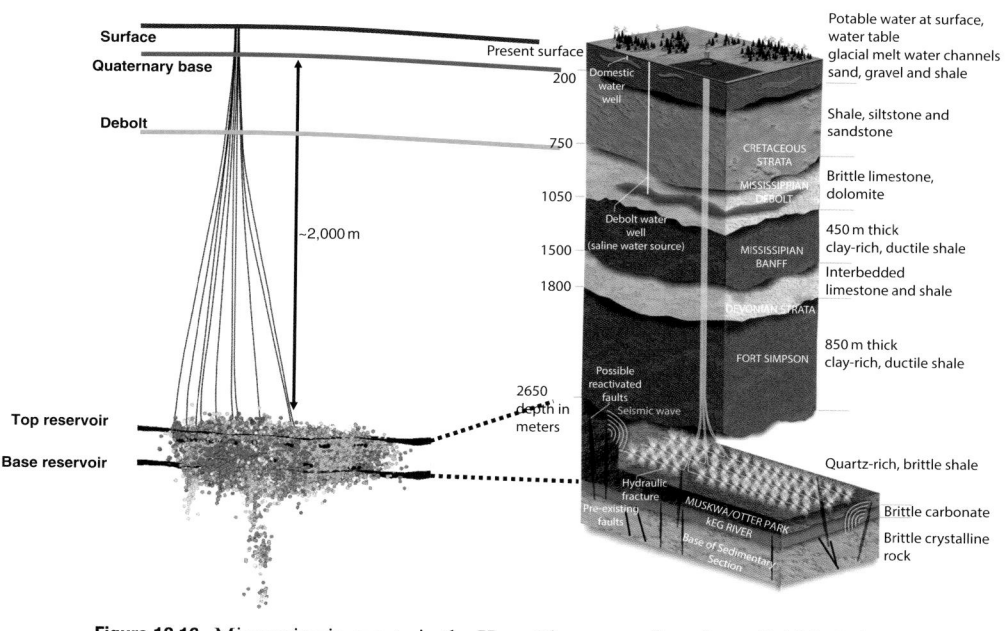

Figure 13.16 Microseismic events in the Horn River area of northeast British Columbia are typically confined to the producing interval. Occasionally, a fault connecting the producing interval is intersected and larger earthquakes (up to M 4) have been triggered. From British Columbia Oil and Gas Commission.

conductive (the original data were described by Barton et al., 1995; Townend & Zoback, 2000; Ito & Zoback, 2000). As shown in the figure, the permeable/hydraulically conductive fractures and faults (filled symbols) are preferentially those with a high ratio of shear to effective normal stress close. The histogram shows that the average ratio for the conductive faults is 0.6, a reasonable value of the coefficient of friction of these rocks (see Chapter 4). The non-conductive fractures and fault have a lower ratio of shear to normal stress. They are not expected to be active in the current stress field.

As shown in Fig. 13.16, a process similar to that illustrated in Fig. 13.14c seems to be occurring in the Horn River Basin of British Colombia. As shown in the figure, extensive hydraulic fracturing in the Muskwa, Otter Park and Evie formations is associated with microseismic events largely limited to those formations. Occasionally, however, hydraulic fracturing stages intersect a potentially active fault connecting to the basement, consequently producing larger earthquakes. Earthquakes of magnitude 2–3 were detected in Horn River as early as 2009. As soon as local seismic networks were installed in the area, these events could be associated in space and time with hydraulic fracturing activities. The largest earthquake to have occurred there is M 3.6.

Because of the importance basement faults, it is perhaps not surprising that Skoumal et al. (2018a) found that in the Appalachian Basin, induced seismicity is more likely when hydraulic fracturing was occurring close to basement.

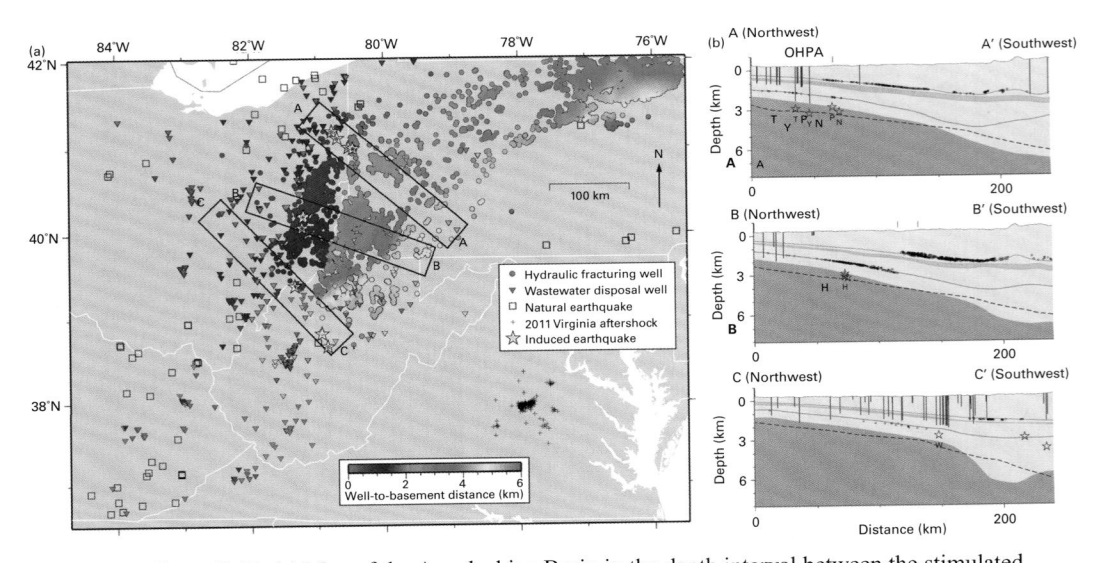

Figure 13.17 (a) Map of the Appalachian Basin in the depth interval between the stimulated formation and the basement for hydraulic fracture wells (colored circles) and wastewater disposal wells (colored triangles). The base of the Ordovician Point Pleasant Formation–Utica shale and Devonian Marcellus shale are used to estimate stimulation depths. Stars are earthquake swarms correlated with hydraulic fracturing or wastewater disposal and squares are earthquakes that appear to be naturally occurring in Ohio. Skoumal et al. (2018a). Crosses show the natural 2011 central Virginia aftershock sequence. Boxes show the location of the cross-sections in (b). (b) Simplified geological cross-sections showing the Marcellus shale (green), Salina Group evaporite (orange), Point Pleasant Formation–Utica shale (red), Precambrian basement (dark gray shading), and strata above the basement (light gray shading). Depth values are relative to sea level. Small circles are depths and locations of hydraulic fracture stimulations, and blue lines are wastewater disposal wells. Stars are induced earthquake sequences: T – Trumbull County, Ohio (OH); Y – Youngstown, OH; P – Poland Township, OH; N – North Beaver Township, Pennsylvania (PA); H – Harrison County, OH; B – Braxton County, West Virginia (WV); G – Gilmer County, WV; W – Washington County, OH. The dashed line is basement depth.

Figure 13.17 shows their study area in Ohio, southwestern Pennsylvania and northeastern Ohio. Comparing cross-sections A and B in Fig. 13.17b (the locations of the cross-sections are shown in Fig. 13.17a) shows that earthquakes in the basement occur more frequently where hydraulic fracturing was occurring in the deeper Point Pleasant-Utica (shown in red) than in the shallower Marcellus formation (shown in green). Basement is only about 1 km deeper than the Pt. Pleasant. Several earthquakes in West Virginia seen in cross-section C occur in the deeper sedimentary section which appear to be associated with hydraulic fracturing and water injection. Skoumal et al. also point out that there appears to be very little basement seismicity associated with the Bakken formation because an evaporite deposit that lies between the basement rocks and stimulated zone may prevent pressure communication from the stimulation zone to basement faults.

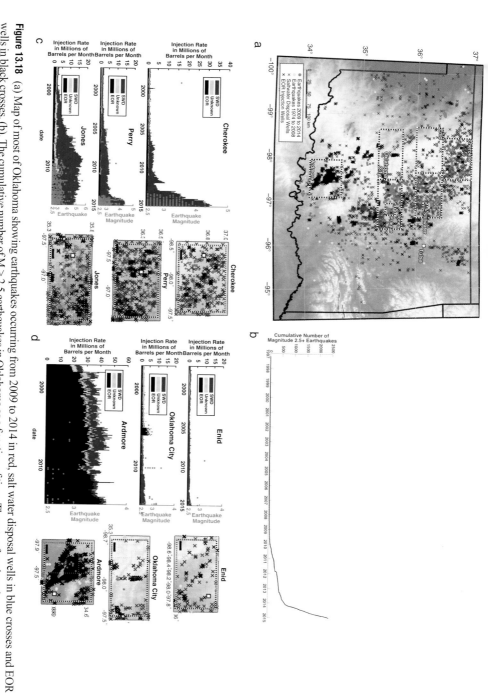

Figure 13.18 (a) Map of most of Oklahoma showing earthquakes occurring from 2009 to 2014 in red, salt water disposal wells in blue crosses and EOR wells in black crosses. (b) The cumulative number of M ≥ 2.5 earthquakes in Oklahoma as a function of time. The rate of earthquake occurrence began to increase in 2009 and increased dramatically around 2014. (c) Monthly injection rates from EOR, SWD and unknown wells within the Cherokee, Perry and Jones seismically active areas, as well as the times and magnitudes of earthquakes in each area. Detailed maps of each study area are also shown. The symbols for earthquakes, SWD and EOR wells is the same as in (a). (d) Monthly injection rates from EOR, SWD and unknown wells within the Enid, Oklahoma city and Ardmore areas with much less seismicity. All of the figures in Fig. 13.18 are from Walsh and Zoback (2015) although the data in Figs. 13.18c,d have been updated with an additional year of data.

400   Unconventional Reservoir Geomechanics

All of this said, it needs to be recognized that induced earthquakes are not restricted to basement rocks. In Chapter 4 we presented experimental data on shales, typically those with relatively low clay content, are velocity weakening. In other words they would be expected to slip in an earthquake if slip was triggered by increased fluid pressure. In fact, in a recent experiment where there was excellent seismic coverage, Eaton et al. (2018) documented a sequence of induced earthquakes along a well-developed fault within and above the Duvernay shale. The largest earthquake was M_w 3.2.

Earthquakes Associated with Produced Water Injection – Most of the earthquakes shown in Fig. 13.1 occurred in north-central Oklahoma and southern Kansas. As shown by the red dots in Fig. 13.18a (all of the figures in Fig. 13.18 are from Walsh and Zoback, 2015 although the data in Figs. 13.18c,d were updated with an additional year of data), thousands of earthquakes $M \geq 2.5$ occurred in an area where massive amounts of highly saline produced water, principally from the Mississippian limestone and Hunton plays, was being disposed of by injection into the deeper Arbuckle formation. The salt water disposal wells that injected more than 30,000 barrels ($\sim$4,800 m^3) in any month are shown by blue crosses. The black crosses are injection wells in conventional reservoirs where produced water is reinjected into the producing formation as part of standard water flooding EOR operations. Figure 13.18b shows the dramatic increase in seismicity in Oklahoma (the number of earthquakes greater than $M \geq 2.5$ are shown) that began with significant increases in water disposal. More important than the magnitude 2.5s are the widely felt $M \geq 4$ earthquakes. Prior to 2009, there was one $M \geq 4$ earthquake in this region per decade. At the time of peak activity (the end of 2015, beginning of 2016 – for reasons which will be explained later), there was an average rate of one widely felt $M \geq 4$ earthquake per week and one potentially damaging $M \geq 5$ earthquake every 2–3 months.

The dramatic increase in seismicity in Oklahoma was initially quite perplexing. There had never been a case where induced earthquakes occurred throughout such a large region and the b-slope of the earthquakes was approximately 1.0, as commonly observed for natural earthquake sequences (Holland, 2013). To establish the correlation between the increase in seismicity and salt water disposal into the Arbuckle, Walsh and Zoback defined six regions in north-central Oklahoma, each 5,000 km^2 in size. The regions are defined by the dotted rectangles in Fig. 13.18a. Three were located where seismicity had been occurring, three were located where it had not been. As shown in Fig. 13.18c, the seismicity in the three seismically active areas increased at the same time that SWD into the Arbuckle formation increased (shown by the blue histogram in the time series). In the Cherokee and Perry areas, the seismicity increased abruptly as the SWD in the area increased. In both areas, the maps show that both the earthquakes and disposal wells were distributed throughout the areas. In the Jones area, the SWD and earthquakes gradually increased with time. In all three of these areas there was comparatively negligible injection related to EOR (shown in black) or hydraulic fracturing flowback. SWD in the most seismically active areas was approaching 20 million barrels per month during this time period, compared with much lower rates prior to the increase in seismicity.

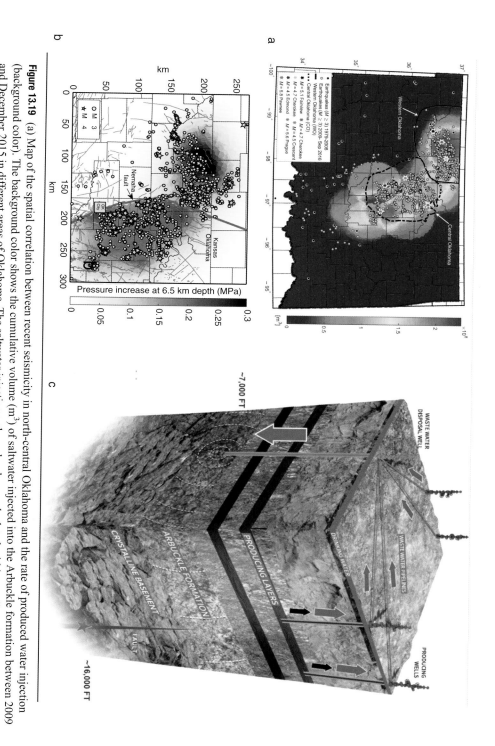

Figure 13.19 (a) Map of the spatial correlation between recent seismicity in north-central Oklahoma and the rate of produced water injection (background color). The background color shows the cumulative volume (m³) of saltwater injected into the Arbuckle formation between 2009 and December 2015 in different areas of Oklahoma. The saltwater injection volume has been calculated within a radius of 0.5° around a given location on the map and is plotted at the center of the areas (~8,000 km²). The dashed lines show areas of interest defined by Oklahoma State regulators. From Langenbruch and Zoback (2016). *Reprinted with permission from American Association for the Advancement of Science.* (b) A hydrologic model of pore pressure changes at 6 km depth. From Langenbruch et al. (2018). *Reprinted with permission from Springer.* (c) A schematic model of produced saline water from the Mississippi limestone being injected into the Arbuckle formation and earthquakes being triggered in the crystalline basement rocks below.

Unconventional Reservoir Geomechanics

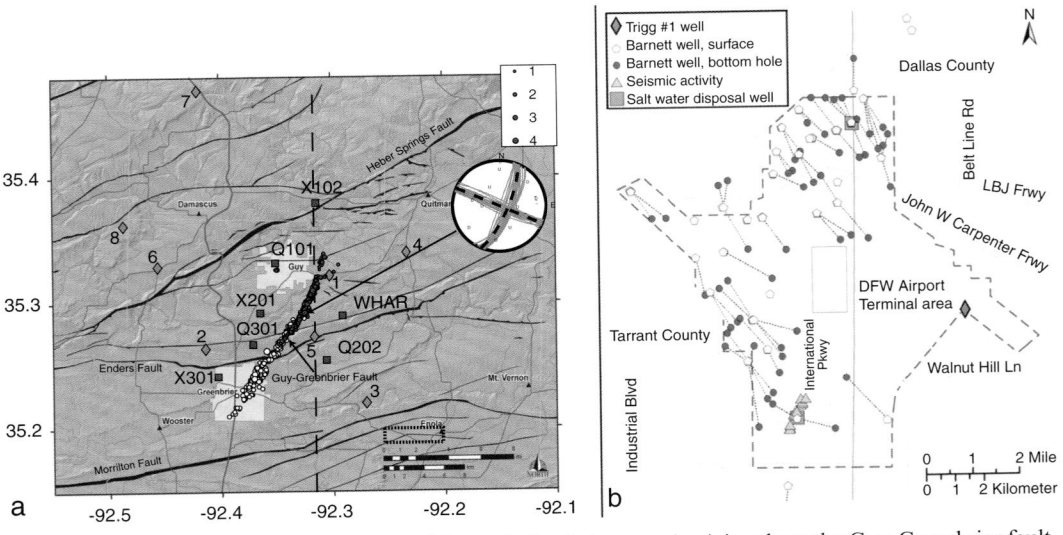

Figure 13.20 (a) Map of the spatial correlation between seismicity along the Guy-Greenbrier fault and wells 1 and 5, that were injecting hydraulic fracturing flow back water. From Horton (2012). (b) In the Dallas-Fort Worth airport area, seismicity along a NE trending fault occurred within 200 m of a well injecting hydraulic fracturing flowback water. From Frohlich et al. (2011).

In contrast to the areas where there has been a great deal of seismicity, Fig. 13.18d shows the three areas with few recent earthquakes. In the Enid and Oklahoma City areas there is very little SWD and very little induced seismicity although both areas are adjacent to more seismically active areas (the vertical axis of all volume histograms are at the same scale in Fig. 13.18c). The geologic reasons for this are explored in Chapter 14. In the Ardmore area the vertical axis is changed to accommodate the very large amount of water injection associated with EOR. There are relatively few earthquakes in this area because pressure decreases with time in producing horizons. The few earthquakes that are occurring in this area are likely associated with the SWD occurring there.

Figure 13.19 summarizes a number of key issues associated with the earthquakes induced by the massive SWD in north-central Oklahoma. As shown in Fig. 13.19a (from Langenbruch & Zoback, 2016), the area of seismicity generally correlates with the area of greatest SWD, as indicated by the background colors. As the Arbuckle group is relatively thick (~300 m) and highly permeable, pressure spreads out rapidly from the injection wells such that the resultant pore pressure increase at depth as shown in Fig. 13.19b (Langenbruch et al., 2018) is quite small, less than 0.3 MPa. Figure 13.19c shows a schematic diagram of the processes associated with the SWD into the Arbuckle that is responsible for the seismicity. Water production from a number of producing wells in the Mississippi limestone is collected and piped to relatively large diameter SWD wells drilled down into the Arbuckle group, which sits directly on top of basement. Because the Arbuckle is underpressured in this area (Nelson et al., 2015), the water would essentially flow in under its own weight although some wells were injecting at such high rates (up to ~80,000 barrels per day) that pumping was needed to overcome wellbore friction. Because of the high permeability of the Arbuckle, pressure spreads out over large areas. When critically stressed and permeable faults in basement are encountered, pressure was transmitted to depth causing earthquakes to be triggered by relatively small pore pressure changes. As shown in Fig. 4.4, the seismicity defines planes that are critically stressed (well oriented) for faulting in the current stress field. However, the majority of faults in the 1.2 billion year old crystalline basement rocks are neither critically stressed nor permeable. Zhang et al. (2013) presented a numerical model of fluid pressure spreading out in a basal aquifer that induces slip on a permeable basement fault several km away from the injection site (see Fig. 4D in that paper).

Flowback Water Injection – As pointed out by Zoback (2012) many of the earthquakes that occurred in previous years which seemed to be associated with unconventional oil and gas development were caused by injection of hydraulic fracturing flow back water. This includes the M 4.7 earthquake near Guy, Arkansas shown in Fig. 13.14d, the events near Youngstown, Ohio that included a M 4.0 event and several years of seismicity in the area of the Dallas-Ft. Worth (DFW) airport studied by Frohlich et al. (2011), the largest of which was M 3.3. Figure 13.20 shows maps of the Guy-Greenbrier earthquake sequence in Arkansas (from Horton, 2012) and some of the earthquakes on the Dallas-Fort Worth airport (from Frohlich et al., 2011) with respect to nearby wells used to inject flowback water after hydraulic fracturing. In the case of the Guy-Greenbrier sequence, there are two disposal wells (labeled 1 and 5) that were injecting within a few km of the fault. In the case of the DFW airport, the earthquakes occurred on a fault within 200 m of the SWD well. In both cases, the NE-trending faults (a strike-slip fault in the case of Guy-Greenbrier and a normal fault in the case of DFW) occurred on faults that could have been identified as potentially active faults from knowledge of the local stress state. The same is true in the case of the earthquakes that occurred near Azle, Texas, that appear to have been triggered by flowback water injection (Hornbach et al., 2015). This will be discussed at greater length in Chapter 14.

Earthquakes Associated with Hydraulic Fracturing – Several cases above were presented in which hydraulic fracturing appears to have triggered earthquakes on pre-existing faults. There are several other cases that should be mentioned. Friberg et al. (2014) and Skoumal et al. (2015) discuss earthquakes related to hydraulic fracturing in Ohio. Regulators in Ohio have implemented a real-time seismic monitoring requirement and traffic-light system (discussed in Chapter 14) to manage the risk associated with hydraulic fracturing-induced seismicity. Earthquakes associated with hydraulic fracturing of the Bowland shales in Lancashire, England (Clarke et al., 2014) have essentially shut down development activities in the United Kingdom, despite the fact that the largest event was only M 2.3. Holland (2013) reported a case of earthquakes triggered by hydraulic fracturing operations in Oklahoma. As the majority of the induced seismicity in Oklahoma has been associated with produced water disposal as discussed above, seismicity associated with hydraulic fracturing was thought to be a second-order issue.

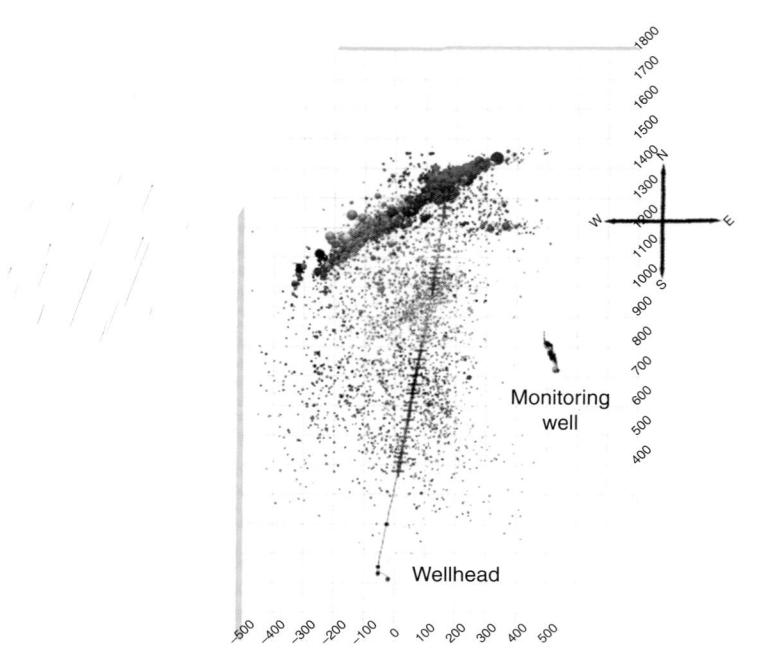

Figure 13.21 Microseismicity associated with a horizontal well in the Sichuan Basin of China. The stages near the toe appear to have activated slip on small faults in the damage zone of a larger fault that cuts across the well. The casing was sheared by slip on this fault. Modified from Chen et al. (2018).

However, Skoumal et al. (2018b) report as many as 500 earthquakes in Oklahoma (M 2.0–3.5) that seem to be best explained by hydraulic fracturing. Within the SCOOP and STACK plays in the Arkoma Basin, more than 90% of the observed seismicity appears to be associated with hydraulic fracturing. Overall, they find that while only about 2% of the hydraulic fracturing stages are associated with induced earthquakes, in some areas as much as 50% of the hydraulic fracturing stages are associated with induced events. Thus, the seismogenic response to hydraulic fracturing in the SCOOP and STACK is quite spatially variable. One geologic explanation of this variability is presented in Chapter 14.

It is instructive to compare these results with a statistical analysis of induced seismicity in western Canadian Sedimentary Basin from Atkinson et al. (2016). In summary, they found that about 1% of disposal wells were associated with M ≥ 3 earthquakes and 0.3% of wells being hydraulically fractured were associated with M ≥ 3 earthquakes. As the limitation of a higher detection threshold for events in Canada (M ≥ 3) does not allow for direct comparison with the numbers quoted above for Oklahoma (M ≥ 2), it is clear that earthquakes associated with hydraulic fracturing increased markedly in 2008, the same time the number of wells being hydraulically fractured increased rapidly. Thus, earthquakes triggered by hydraulic fracturing is a general concern that must be considered in many regions around the world.

The EPA induced seismicity report (EPA, 2014) provides a summary of the USGS report on four hundred years of historical seismicity in the US (Stover & Coffman, 1989) noting that there are very rare occurrences of earthquakes affecting well integrity. The most notable and relevant were from the 1983 M6.2 Coalinga, California earthquake with a hypocenter directly below the oil fields of Coalinga. Approximately 14 of 1,725 wells showed evidence of casing damage. The British Columbia Oil and Gas Commission (BCOGC) investigated the series of earthquakes between 2009 and 2011 in the Horn River Basin attributed to hydraulic fracturing operations. The BCOGC report highlights that "*No casing deformation was reported in the vertical portion of wellbores and no reservoir containment issues were identified. Minor casing deformation within the horizontal well portion of target shale formations occurred in 2 instances. The cause of the casing deformation could not be conclusively linked to the seismicity.*" The BCOGC report also alludes to casing deformation associated with the seismicity in the Bowland Basin in the United Kingdom that was attributed to hydraulic fracturing operations. Casing deformation in the horizontal section of the well was noted, but no loss of well integrity.

Induced Aseismic Fault Slip and Well Shearing – The final example of how faults are affected by hydraulic fracturing comes from a horizontal well in the Longmaxi shale in the Sichuan Basin, China. This is the most commercially successful shale gas play in China but 32 of 101 wells that have been drilled in this block experience casing deformation. One case is shown in Fig. 13.21 (after Chen et al., 2018). Multi-stage horizontal fracturing near the toe of the well generated numerous microseismic events at an orientation consistent with active strike-slip faulting in the local stress field. Casing deformation was observed close to this fault. A number of significant earthquakes of magnitude ~1 occurred during many of the 10 stages. However, casing shear appears to be too large to be explained by earthquakes of magnitude ~1. It appears that the majority of deformation on this fault (as well as other faults affecting other wells in the area) is aseismic. The Longmaxi shale contains 30–50% clay. As discussed in Chapter 4, it is likely that with this much clay, faults in the shale would be expected to be velocity strengthening and would therefore shear aseismically. Thus, while faulting induced by hydraulic fracturing in this well did not produce felt earthquakes, it did cause damage to wells. As this fault was identifiable using ant tracking of 3D seismic data (as in the cases illustrated in Figs. 7.25, 7.28 and 7.29), it would have been beneficial not to have pressurized this fault during the multi-stage hydraulic fracturing.

14 Managing the Risk of Injection-Induced Seismicity

In the previous chapter we presented examples of earthquakes triggered during hydraulic fracturing, injection of flowback water after hydraulic fracturing and injection of produced water. In this chapter we address steps that can be taken to minimize the occurrence of such events. Of course, one of the most obvious ways to avoid injection-induced earthquakes is to minimize injection volumes. It's not a coincidence that areas in Pennsylvania where nearly all the hydraulic fracturing flowback water is recycled have very few injection-induced earthquakes. In the sections that follow, we first discuss the issue of avoiding injection into potentially active faults. In Fig. 13.20 we presented two examples where injection wells were sited very close to potentially problematic faults. The same was true in the case of the earthquakes that occurred near Azle, Texas (Hornbach et al., 2015). As explained in the first section that follows, if there is knowledge of a fault in the subsurface, it is possible to assess, in the context of Coulomb faulting theory and knowledge of the local stress state, whether the fault might potentially be activated in response to an increase in pore pressure resulting from fluid injection. In the second section of this chapter, we extend this discussion to consider the same topic utilizing a probabilistic methodology that allows one to incorporate uncertainty of key parameters into the assessment.

In practice, the conundrum is that it is often not possible to know whether a critically stressed fault is actually present, due to lack of available seismic data or the inability of seismic data to image faults, especially in the basement. Risk management and mitigation approaches are then crucial for addressing the range of uncertainties that may be present. Thus, the next sections discuss three specific management tools that can be used to minimize the risk of injection-induced seismicity – first using traffic light systems (a real-time management tool), second, utilizing the seismogenic index model to evaluate scenarios for management of produced water injection and third development of site characterization risk frameworks as a tool for identifying, pro-actively, the degree to which induced seismicity should be a concern.

Avoiding Injection Near Potentially Active Faults

In Chapter 4 we pointed out that a large and growing body of evidence supports the applicability of Coulomb faulting theory to faults in the earth. Thus, if we know the orientation of a fault at depth, the state of stress in the area (as illustrated by the maps presented in Chapter 7) and the *in situ* pore pressure, we can assess whether the fault could potentially be activated by an increase in fluid pressure. With specific application to areas where earthquakes induced by fluid injection may have occurred in Texas, Lund Snee & Zoback (2016) analyzed four sites in Texas where earthquakes with magnitudes in the range of 3.8–4.8 have occurred in the vicinity of oil and gas activities. These include events near the towns of Azle, Karnes City, Snyder and Timpson. The study areas are shown by the red boxes on the stress map of Texas in Fig. 14.1. Detailed maps of each area in Fig. 14.1 include earthquake locations, indicators of the orientation of S_{Hmax}, known faults (from publicly available sources) and focal plane mechanisms of varying reliability, due, in most cases to sparse seismic monitoring. Note that there are no known faults in the vicinity of the earthquakes near Snyder or Timpson.

Lund Snee & Zoback found that in three of the four earthquake sequences considered, at least one of the fault planes (and sense of slip) defined in the focal mechanisms is consistent with the local stress field and would have required relatively small changes of pore pressure to be triggered. This was quite clear for the Azle and Karnes City events where both relatively consistent stress measurements and the orientation of faults in the area made the analysis straightforward. However, the 1978–2016 earthquakes near Snyder, Texas occurred in an area where the stress orientation was poorly constrained and where there were no mapped faults in the vicinity of the earthquakes. Although the epicenters of the Snyder events trend north-northeast, the focal plane mechanisms had nodal planes striking mostly north and east for the strike-slip events and northeast for the oblique normal faulting events. These fault orientations are consistent with nearby *in situ* measurements showing NE–SW S_{Hmax} orientations and a strike-slip/normal faulting stress state. Thus, if steeply dipping faults had been mapped with orientations similar to the ones that apparently slipped in the earthquakes, they would have been identified as potentially active faults. Unfortunately, very little seismic reflection data that could be used to identify faults not on the state fault maps is publicly available.

While three of the four areas studied by Lund Snee & Zoback (2016) were easily understood in terms of Coulomb faulting theory, seismicity that occurred in 2012 near Timpson, Texas, is more difficult to understand. There are two strike-slip focal plane mechanisms with northwest and east-northeast striking nodal planes. While there were no mapped faults in the immediate area of the seismicity, the east-northeast striking nodal plane orientation is quite similar to the orientation of faults mapped just to the north of the seismicity. However, aftershock hypocenters of one of the larger events define a fault that strikes to the northwest, suggesting slip along a previously unmapped northwest-striking fault that dips to the southwest (Frohlich et al., 2014). As available stress data show that S_{Hmax} orientations range from N068°E to N080°E, Lund Snee & Zoback demonstrated that slip on an east northeast striking nodal plane (which is sub-parallel to the mapped faults just to the north) would be expected in response to a modest increase in fluid pressure (see Mohr diagram in Fig. 14.1). However, slip on the northwest-striking plane defined by the hypocenters would not be expected. This unusual scenario perhaps implies a complex pattern of faults in the subsurface. While it is straightforward to reconcile the stress state, focal plane mechanisms and

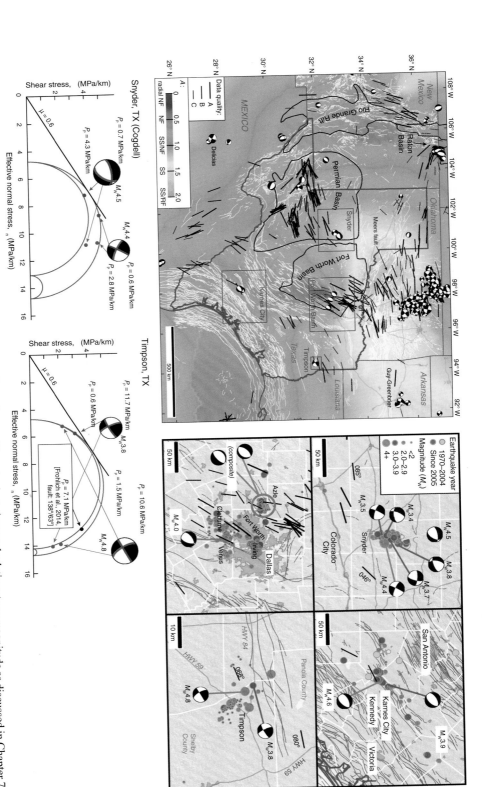

Figure 14.1 Stress map of Texas showing maximum horizontal stress (S_{Hmax}) orientations and relative stress magnitude as discussed in Chapter 7. Basin boundaries are from the US Energy Information Administration. The red boxes define four study areas shown in more detailed maps. As explained in the text, the Mohr circles represent a frictional failure analysis in two of the study areas. From Lund Snee & Zoback (2016).

mapped faults nearby, it is not easy to explain the northwest alignment of hypo-centers. Using a geomechanical model that accounted for poroelastic stress changes, Fan et al. (2016) found that slip on the NW striking mapped fault could have been triggered by an unusually large pore pressure perturbation 12.9 MPa resulting from nearby injection of saline wastewater. Regardless, this is clearly a case in which better knowledge of the geometry of faults at depth and accurate earthquake locations and focal plane mechanisms would clarify the relationships among the local stress state, fault distribution and fluid pressure changes at depth.

Probabilistic Assessment of Fault Slip Potential – These case studies clearly illustrated how Coulomb faulting theory could be used to assess whether an identified fault is potentially problematic if pore pressure was increased due to injection. However, there is uncertainty in each of the critical parameters needed to make this assessment – the strike and dip of the fault, the orientation and relative magnitudes of the three principal stresses as well as the pore pressure and likely pore pressure perturbation, etc. In an attempt to address these uncertainties, Walsh & Zoback (2016) extended the applicability of Coulomb faulting theory by putting forth a probabilistic methodology to formally account for the uncertainty in each of the parameters that go into the assessment. As a result, they are able to express the potential for fault slip probabilistically. Walsh and Zoback utilized quantitative risk assessment (QRA), a Monte-Carlo technique to calculate the conditional probability of slip on mapped faults in response to injection-related increases in pore pressure. Leakage of CO_2 from carbon sequestration reservoirs through wells, faults and fractured caprock was investigated using QRA by Chae and Lee (2015). Chiaramonte et al. (2008) used QRA to evaluate if increases in pore pressure from a CO2 injection pilot project might induce fault slip. The Chiaramonte et al. analysis assumed either purely strike-slip or normal faulting and only considered uncertainty in fault orientation. Walsh & Zoback generalized this type of analysis to include potential slip in any direction on mapped faults from uncertain stresses. They developed and initially applied this technique in north-central Oklahoma (USA) where widespread injection of produced saltwater has triggered thousands of small to medium-sized earthquakes as discussed in Chapter 13. The conditional probability incorporates the uncertainty in each Mohr–Coulomb parameter (stress tensor, pore pressure, coefficient of friction and fault orientation) through QRA. The result is a cumulative distribution function of the pore pressure required to cause slip on each fault segment. The result can be used to assess the probability of induced slip on a known fault from a given injection related pore pressure increase.

After dividing north-central Oklahoma into six study areas (Fig. 14.2, from Alt & Zoback, 2017) Walsh & Zoback applied QRA to the mapped faults based on the uncertainties of each parameter in each area. An example of the uncertainty distributions for area 6 is shown in Fig. 14.3. They evaluated 10,000 random combinations of parameters for each mapped fault segment to evaluate the conditional probability of slip as a function of pore pressure perturbation given the model assumptions described above in the context of Coulomb faulting theory.

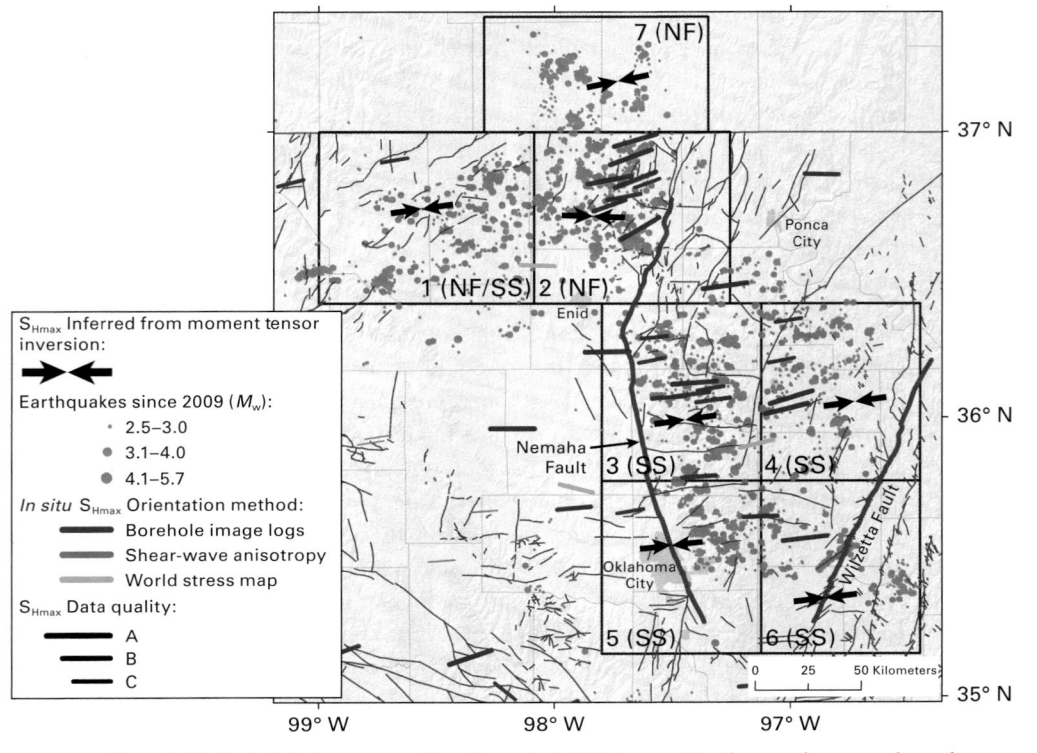

Figure 14.2 Detailed stress map of north-central Oklahoma. The lines and arrows show the orientation of S_{Hmax}, utilizing different sources of data (and data quality) as described in the legend. Earthquake epicenters for earthquakes (red dots) of M ≥ 2.5 between 2009 and 2015 (US Geological Survey) and faults compiled by the Oklahoma Geological Survey (Darold & Holland, 2015). Six study areas in Oklahoma and one in Kansas were defined on the basis of the stress orientation and style of faulting in each. Note that most of the earthquakes are not associated with the mapped faults. From Alt & Zoback (2017).

Figure 14.3 shows the distributions of friction, pore pressure and stress used in the QRA of study area 6. Similar figures for the other study areas were presented by Walsh & Zoback, as well as how the uncertainty in each parameter was established. As no information on fault dip was available, a distribution of steep dips was investigated as this is known to be an area of strike-slip faulting. Fig. 14.3c and 14.3d also show response surfaces in red, which use the most likely value in each distribution in Fig. 14.3 (indicated by the vertical dotted black lines in the other distributions) to show the required pore pressure to induce fault slip based on a fault's dip (Fig. 14.3c) or strike (Fig. 14.3d). The black horizontal line in Figs. 14.3c and 14.3d shows the 2 MPa expected pore pressure perturbation. This magnitude is based on the fact that pore pressure in the Arbuckle is about 2 MPa subhydrostatic (Nelson et al., 2015) and the observation that wellhead pressures remain subhydrostatic immediately after injection stops.

Walsh & Zoback presented the result as an empirical cumulative distribution function (CDF) showing the probability of slip on a known fault as a function of pore pressure

Managing the Risk of Injection-Induced Seismicity   411

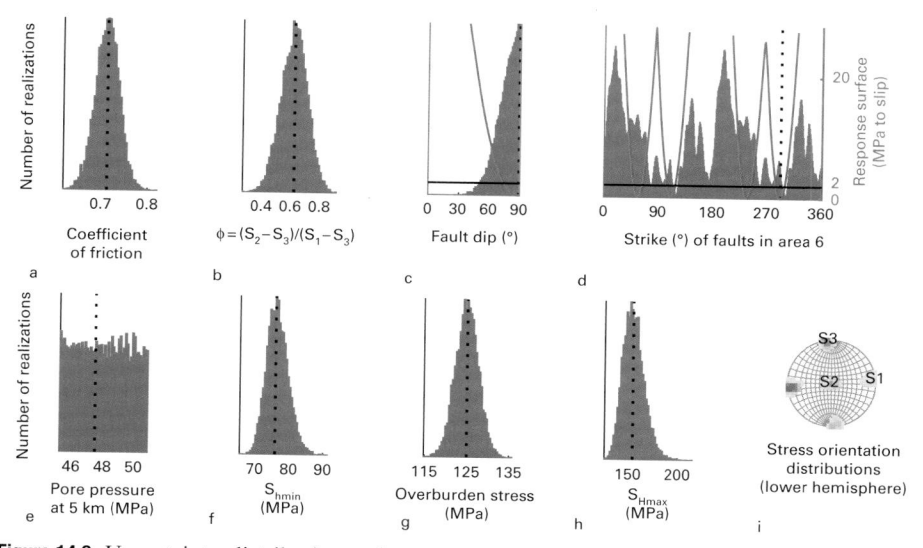

Figure 14.3 Uncertainty distributions of parameters used in 10,000 calculations of pore pressure required to cause slip on each fault segment in study area 6. Distributions of results from the bootstrapped moment tensor inversion are shown in b and i. Red (in c and d) show response surfaces of pore pressure to slip in the preferred geomechanical model described by vertical dotted black line in each parameter distribution. S_{hmin} and S_{Hmax}. From Walsh & Zoback (2016).

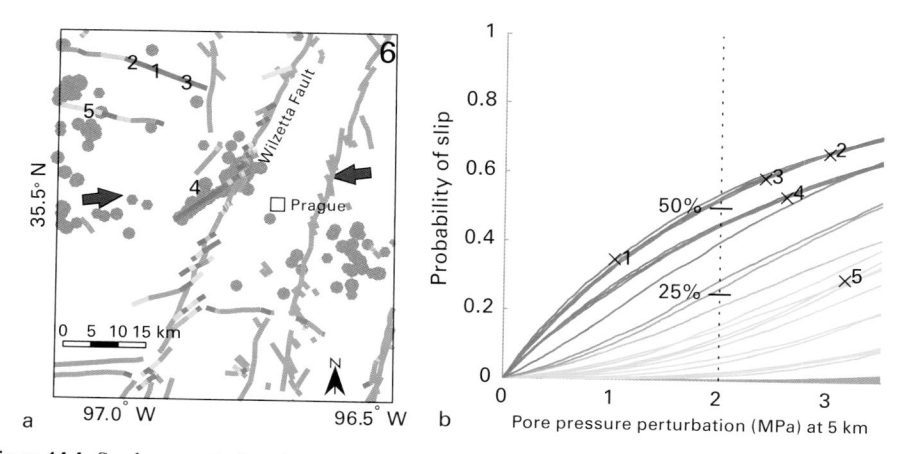

Figure 14.4 Study area 6 showing mapped faults and M ≥ 3 earthquakes (gray dots). Selected faults are colored and numbered to correspond to their respective cumulative distribution function (CDF). Colors of fault segments shown are based on corresponding CDF curves. Note that the CDF curves are related to pore pressure changes expected during saltwater disposal. 2 MPa was considered the largest possible pore pressure change occurring near the injection wells. From Walsh & Zoback (2016). *Reprinted with permission from Geological Science of America.*

increase. This is illustrated in Fig. 14.4. Each of the faults in the area are numbered. In cases in which the fault orientation changes, individual fault segments were numbered separately. A CDF for each fault segment identifies the potential for slip on the fault as a function of pore pressure. The highest possible pore pressure near individual injection wells was considered to be 2 MPa. The color scheme is arbitrary. If the probability was less than 1%, the fault was colored green. If greater than 33% it was colored red, with a gradation of colors for intermediate values. Note that most of the mapped faults are quite unlikely to slip in response to a 2 MPa pore pressure increase, including the large NNE trending faults like the Wilzetta fault. However, the splay of the Wilzetta that produced the M 5.6 Prague earthquake in 2011 was identifiable as a potentially active fault and could be seen on 3D seismic data from that area.

Figure 14.5 (from Walsh & Zoback, 2016) maps the probability of fault slip in response to a 2 MPa pore pressure change in all six study areas and an indication of the style of faulting (strike-slip, strike-slip–normal, or normal faulting) based on focal mechanism inversions. We observe that the majority of mapped faults are not likely to be activated by modest pore pressure changes. Also shown are $M \geq 3$ earthquakes and saltwater disposal wells that injected more than 300,000 barrels in any month from 2009 through 2014. The 13 February 2016 M 5.1 earthquake near Fairview, Oklahoma, is circled in the southwestern part of area 1. The focal mechanism of the earthquake indicates a steeply dipping, northeast-striking fault plane that aligns with similarly striking faults mapped to both the southwest and northeast of the epicenter, colored varying shades of yellow.

Retroactively Testing Fault Slip Potential in Oklahoma – The methodology described above has been applied to all $M \geq 5$ earthquakes in Oklahoma, indicated by the stars in Fig. 13.19a. In each case, had the fault on which the earthquake occurred been known, it would have been identified as being potentially active. This is illustrated in Fig. 14.6 for the M 5.8 Pawnee earthquake that occurred on September 3, 2016, the largest earthquake to have occurred in Oklahoma in historic time. The Pawnee mainshock occurred near the intersection of a north-east trending fault defined by foreshocks that is nearly coincident with a mapped northeast trending fault and an east-southeast trend of aftershocks where no fault had been mapped. The USGS focal plane mechanism defined two possible steeply dipping possible fault planes – one trending north-northeast and one trending east-southeast that is nearly parallel to the trend of aftershocks. Walsh (2017) analyzed the likelihood of induced slip on each of the two possible fault planes using the methodology outlined above. He demonstrated that the east-southeast striking plane corresponded to the likely fault because this plane was optimally oriented in the local stress field and could be triggered by a small increase in pore pressure (Fig. 14.6b) whereas the other plane would not be expected to slip unless pressures at depth were increased to close to the magnitude of S_{hmin}, or the frac gradient. Had this fault been known prior to the earthquake, the relative probability of earthquakes on the faults in the area would be as presented in Fig. 14.6c.

Managing the Risk of Injection-Induced Seismicity   413

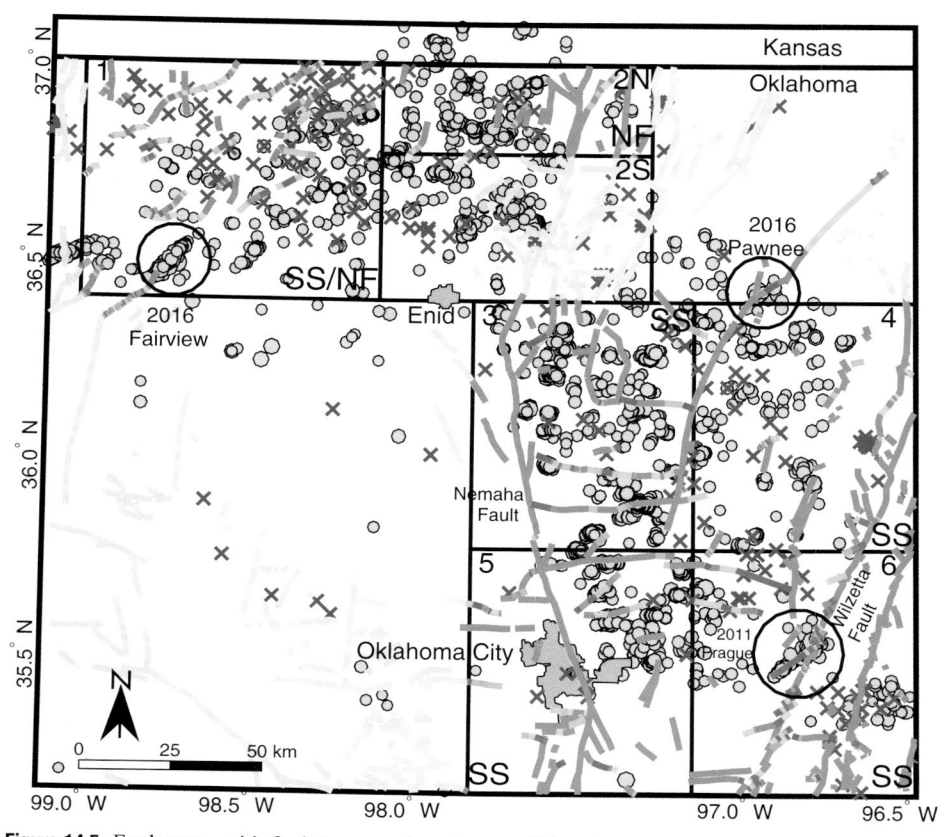

Figure 14.5 Fault map, with fault traces colored by conditional probability of slip in response to pore pressure perturbations triggered by injection. Fault segments colored red represent >33% probability of slip in response to 2 MPa pore pressure perturbation; those in green represent <1% probability of slip in response to the same pressure. Gray dots show M ≥ 3 earthquakes; gray lines show mapped faults that were not assessed. Disposal wells that injected >300,000 barrels in any month before 2015 are shown as blue crosses. In the corner of each study area, the principal style of faulting is shown: SS – strike-slip faulting observed; NF – normal faulting observed. Large black circles show earthquakes with M > 5. From Walsh & Zoback (2016).

Know Your Faults – To avoid injection into potentially active faults it is obviously necessary to know that a fault is present. In this regard it is quite clear that publicly available data is generally inadequate to identify all the faults of potential interest. As shown in Figs. 14.2 and 14.5, the great majority of earthquakes in north-central Oklahoma do not coincide with mapped faults, yet there is no doubt that the earthquakes did occur on faults. While it is well known that it is difficult to image faults in crystalline basement using seismic reflection data, the bigger faults (those of most interest) are often visible in seismic reflection data because they are also present in the overlaying sedimentary rocks. In order to use the methodology defined above, one must do everything possible to map all of the faults that are present in an area. Inevitably, faults will be missed but it would still be beneficial to avoid potentially problematic faults whenever they are identified.

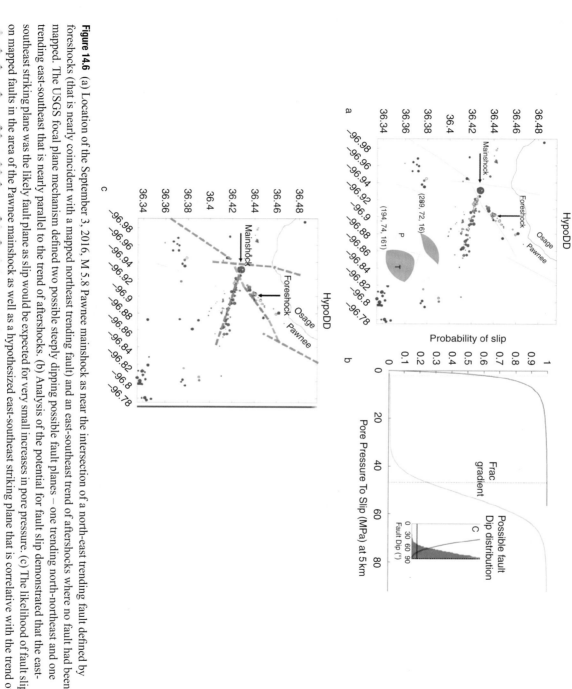

Figure 14.6 (a) Location of the September 3, 2016, M 5.8 Pawnee mainshock as near the intersection of a north-east trending fault defined by foreshocks (that is nearly coincident with a mapped northeast trending fault) and an east-southeast trend of aftershocks where no fault had been mapped. The USGS focal plane mechanism defined two possible steeply dipping possible fault planes – one trending north-northeast and one trending east-southeast that is nearly parallel to the trend of aftershocks. (b) Analysis of the potential for fault slip demonstrated that the east-southeast striking plane was the likely fault plane as slip would be expected for very small increases in pore pressure. (c) The likelihood of fault slip on mapped faults in the area of the Pawnee mainshock as well as a hypothesized east-southeast striking plane that is correlative with the trend of

Know Your Stress State – As shown by the maps presented in Chapter 7, there is an increasing number of areas where unconventional oil and gas is being developed where excellent data are available to constrain the orientation and relative magnitude of the horizontal principal stresses. This is certainly the case in north-central Oklahoma, the Fort Worth Basin and most of the Permian Basin. As discussed in Chapter 7, the methodologies to take advantage of already existing data (specifically image logs and cross-dipole shear velocity logs from in near-vertical wells) are well known, and many companies have willingly made such data available to enable the stress maps to be made. Only when reliable stress data are available, is it possible to identify potentially active faults.

FSP Online Tool – Following the methodology outlined above, Walsh et al. (2017) have made publicly available software for calculating the cumulative probability of a known fault exceeding the Mohr–Coulomb slip criteria from fluid injection. The program is entitled *FSP* to express the concept that the program is intended to quantify *fault slip potential*. This concept is important. For example, if an earthquake occurred on a fault in the not-distant past, the stress drop in that earthquake would make the probability of slip that might be induced by fluid pressure increases less likely. As this cannot be known, the intent of the software is to quantify the potential for slip, not to predict that slip will occur.

A graphical user interface is provided for the following input parameters: fault strike and dip, well locations and injection rates, hydrologic parameters and mechanical stress state parameters as well as the uncertainties in these parameters. Faults can be randomly generated or imported. The methodology works as follows: first, the Mohr–Coulomb pore pressure to slip on each fault is calculated using a deterministic approach. Next, a Monte-Carlo analysis of the same parameter is run on each fault, which yields the probability of each fault slipping as a function of pore pressure increase on it. Once the Monte-Carlo simulation has calculated probability of slip as a function of pore pressure, all that is needed to assess specific injection scenarios is a model that relates injection to pore pressure. By default, the code utilizes a simple radial flow model; more complex pressure models can be imported if available. The output of the hydrologic model is used as the pore pressure input to the probabilistic fault slip model, which yields an estimate of the cumulative probability of the fault slipping as a function of time (or pressure). The Monte-Carlo approach to uncertainty can optionally also be applied to the pressure increase on each fault. The program can only assess the probability of slip on known faults in response to a predictable pore pressure perturbation. The program is freely available through the website of the Stanford Center for Induced and Triggered Seismicity (scits.stanford.edu/software).

Estimation of Fault Slip Potential in the Permian and Fort Worth Basins

Two recent studies have applied the methodology described above to areas of timely interest. First is the Permian Basin of west Texas where seismicity apparently related to oil gas activities has increased in recent years and a great deal of drilling activity is likely to go on for the foreseeable future (note the ~1,000,000 potential drilling sites alluded to in Chapter 1). Second is the Dallas-Fort Worth metropolitan area in the Fort Worth Basin of north-central Texas. The rate of seismicity in that area increased dramatically from late 2008 through 2015, that coincided with injection of approximately 2 billion barrels of wastewater into deep aquifers. Although the rate of drilling and level of activity in the Barnett shale has slowed in recent years, over 6 million people live in the area affected by potential induced seismicity.

Figure 14.7a shows a generalized stress map of the Permian Basin from Lund Snee & Zoback (2018b). As shown in the detailed stress map presented in Fig. 7.3, the state of stress in the basin is quite variable. While smoothly varying stress orientations and relative magnitudes are seen in the Central Basin Platform and Midland Basin (approximately east–west compression in a strike-slip/normal to strike-slip faulting stress state), stress orientations rotate markedly from north to south in the Delaware Basin where the state of stress is characterized by normal faulting with less anisotropy between the two horizontal principal stresses. Consequently, Lund Snee & Zoback defined 16 study areas (shown in Fig. 14.7a) defined by fairly uniform A_ϕ values and S_{Hmax} orientations to minimize spatial variations of stress field in any given study area. They used fault traces in the public domain that generally do not specify fault dips. Thus, they made the reasonable assumption that potentially active faults dip in the range of 50° to 90°. This assumption implies that all fault segments could be ideally oriented for slip in either normal or strike-slip faulting environments at reasonable coefficients of friction, depending on the alignment of their strike with respect to S_{Hmax}.

Figure 14.7b shows the results of their fault slip potential analysis for all study areas across the Permian Basin. The color scale is such that dark green lines represent faults with ≤5% probability of being critically stressed at the specified pore-pressure increase; dark red indicates faults with ≥45% fault slip potential; and yellow, orange and light red represent intermediate values. The results shown in the figure indicate that high fault slip potential is expected for dramatically different fault orientations across the basin, reflecting the varying stress field. In the northern Delaware Basin and much of the Central Basin Platform, for example, faults striking ~east–west are the most likely to slip in response to a fluid-pressure increase. However, farther south in the southern Delaware Basin, faults striking northwest–southeast are the most likely to slip, and ~east–west-striking faults have relatively low slip potential. Notably, we find high slip potential for large fault traces mapped across the southern Delaware Basin and Central Basin Platform, and along the Matador Arch. Importantly, Fig. 14.7b also indicates the faults that are *unlikely* to slip in response to a modest fluid-pressure increase. We find that large groups of mostly north–south-striking faults, predominantly located along the Central Basin Platform, the western Delaware Basin and large parts of the Northwest Shelf have low fault slip potential at the modeled fluid-pressure perturbation.

Figure 14.8 shows a larger-scale view of area 10, an area of particularly dense faults. The figure clearly shows that even seemingly minor variations in fault strike can significantly change the fault slip potential. Figures 14.7 and 14.8 also show the locations of earthquakes that have been recorded since 1970 in relation to the mapped

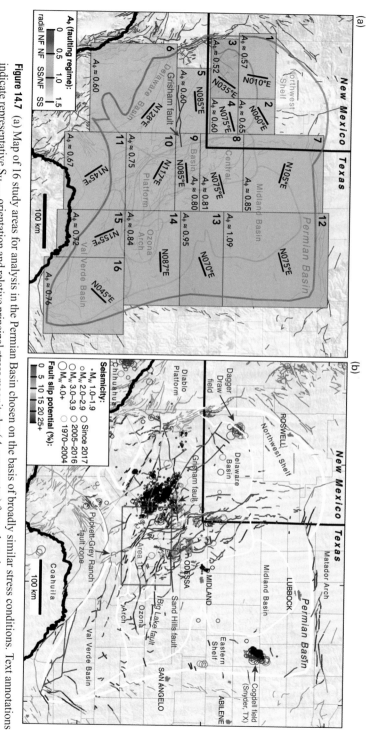

Figure 14.7 (a) Map of 16 study areas for analysis in the Permian Basin chosen on the basis of broadly similar stress conditions. Text annotations indicate representative S_{Hmax} orientation and relative principal stress magnitudes (A_ϕ parameter) for each study area. Gray lines in the background indicate fault traces in publicly available databases to which the FSP analysis will be applied. (b) Results of the FSP analysis. Earthquakes shown are from the USGS National Earthquake Information Center, the TexNet Seismic Monitoring Program. From Lund Snee & Zoback (2018).

faults. It is noteworthy that, like north-central Oklahoma, many earthquakes have occurred away from faults mapped at this regional scale. The most obvious examples are the groups of events near the Dagger Draw Field (southeast New Mexico); the Cogdell Field (near Snyder, Texas); the group of events near the town of Pecos, Texas; and a group of mostly M < 2 events located between the towns of Midland and Odessa, Texas. As the earthquakes undoubtedly occurred on faults, this observation again underscores the necessity of developing improved subsurface fault maps for use in areas that might experience injection-related pore-pressure increases. Nevertheless, Figs. 14.7 and 14.8 also show a number of earthquakes that appear to have occurred on mapped faults for which there is elevated fault slip potential. Of particular note are the earthquakes in southeastern Reeves and northwestern Pecos counties, Texas, of which an appreciable number occurred on or near yellow or orange faults. Potentially active faults are identified near some towns in the Permian Basin, including Odessa (Fig. 14.7b) and Fort Stockton, Texas (Fig. 14.8). In some areas, such as northern Brewster County, Texas, and parts of the northern Central Basin Platform, earthquakes occurred on or near orange or red faults that have relatively short along-strike lengths, making the faults appear fairly insignificant at this scale. In the area of active seismicity in Pecos and Reeves counties, we estimate relatively high slip potential for several significantly larger faults (>20 km along-strike length) on which few or no earthquakes have been recorded thus far. As discussed at length in Chapter 13, larger faults are of particular concern for seismic hazard because they are more likely to extend into basement and, therefore, to potentially be associated with larger magnitude earthquakes.

In the Fort Worth Basin, Hennings et al. (2019) carried out a detailed analysis of fault slip potential utilizing a new interpretation of faults in the basin that includes 250

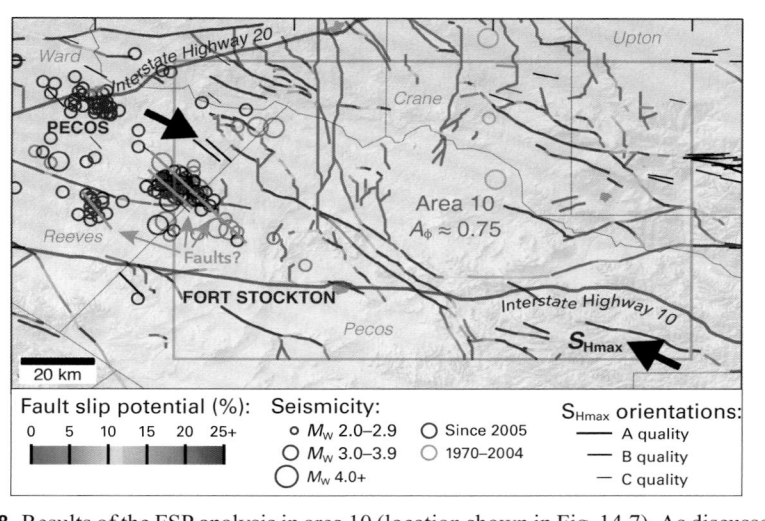

Figure 14.8 Results of the FSP analysis in area 10 (location shown in Fig. 14.7). As discussed in the text, most earthquakes in this area do not occur near mapped faults but there are both faults unlikely to slip under reasonably expected pressure changes as well as potentially problematic faults located throughout the area. From Lund Snee & Zoback (2018).

basement-rooted normal faults that strike dominantly to the NNE and updated stress information that shows that the basin has a relatively consistent NNE direction of maximum horizontal compression and a normal faulting (in the south) to normal/strike-slip faulting (in the north) state of stress (Fig. 7.4).

The fault map developed by Hennings et al. is shown in Fig. 14.9a and includes information from seismic reflection data provided by petroleum operators, new outcrop mapping and all publicly available sources of information including existing fault interpretations and thousands of well logs that were integrated into a 3D geological model that was used to interpret many of the faults that are shown. In addition, they interpreted faults or verified the existence of previously published faults from horizon mapping using control from 1,286 wells. Projecting from outcrop exposure of the southwest flank of the FWB in central Texas provided additional control. The faults shown offset the unconformity between Precambrian igneous and metamorphic rocks and overlying Phanerozoic sedimentary rocks. Like in Oklahoma, the great majority of earthquakes occurring in the Fort Worth Basin are in the crystalline basement rock underlaying the injection zone, which is typically the Ellenburger formation that, like the Arbuckle in Oklahoma and Ozark in Arkansas is a highly permeable carbonate sitting right on top of basement. Hennings et al. developed a confidence scale for the subsurface faults shown on the map such that the faults shown are listed as being either high confidence or moderate confidence. It is important to note that the authors concluded that the fault map was incomplete due to a lack of data from many areas. Earthquake hypocentral data from SMU's North Texas catalog, several local seismicity studies (Hornbach et al., 2015; Hornbach et al., 2016; Scales et al., 2017; Ogwari et al., 2018) and other published studies.

The stress map utilized by Hennings et al. is shown in Fig. 14.9b, along the outline of four areas with uniform stress orientations and magnitudes. As area 4 lacks wellbore stress data, stress conditions were interpolated from other data shown on the map and other data in Lund Snee & Zoback (2016). Focal plane mechanisms for events near Azle (A) and Venus (V) are also shown. In both cases the focal plane mechanisms principally indicate normal faulting with NE-striking nodal planes and there are mapped faults in both areas with strikes similar to the nodal planes of the focal plane mechanisms.

The result of the FSP analysis is shown in Fig. 14.9c. The Hennings et al. analysis shows that many faults in the basin are sensitive to only minor pressure increases, including faults located directly within the highly populated area. Identification of these potentially problematic faults should help both operators and regulators avoid siting injection wells in their vicinity.

Predicting Fault Slip and Casing Deformation with FSP – It is instructive to review an application of FSP to the problem of triggered fault slip shearing wells in the Sichuan Basin discussed in Chapter 13. As faulting induced by hydraulic fracturing is producing earthquakes large enough to be felt in this area, the methodology for avoiding causing damage to wells is basically the same as that to avoid inducing seismicity during hydraulic fracturing. Figure 14.10a shows an interpretation of faults from ant tracking applied to a 3D seismic reflection dataset in the area of the well shown in Fig. 13.21. After developing a geomechanical model for the area, Chen et al. (2018) applied FSP to

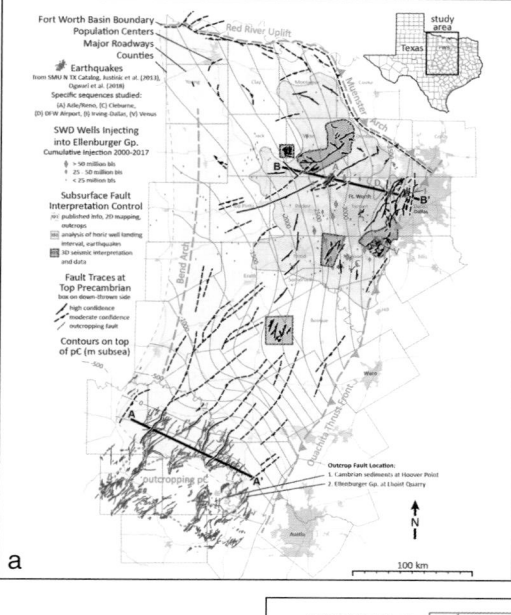

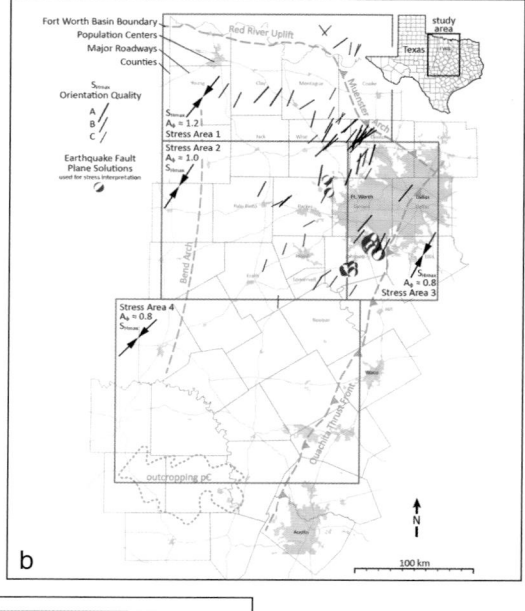

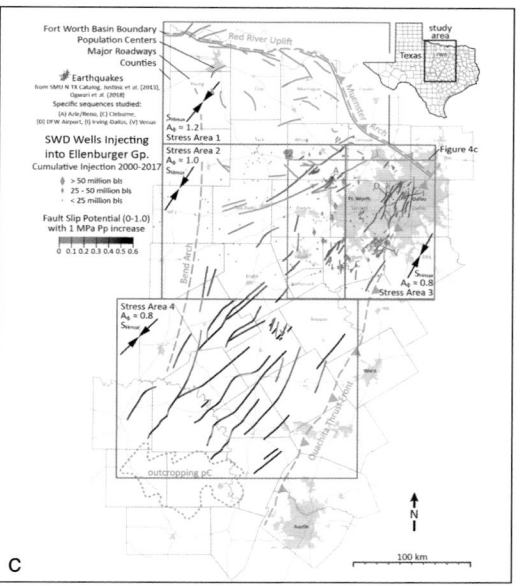

Figure 14.9 Study area in north-central Texas showing (a) Faults interpreted from a variety of sources (as explained in the text) that were used in the FSP analysis. (b) Stress map of the study area and the outlines of four areas where a constant stress state was assumed for the FSP analysis. Earthquake locations and focal plane mechanisms are also shown. (c) Results of the FSP analysis using the faults shown in (a) and the stress states shown in (b). In general, NE to NNE-striking faults have the highest likelihood of being activated by injection. Note that the scale for probability of fault slip is different than that used in the other figures in this chapter. From Hennings et al. (2019).

Managing the Risk of Injection-Induced Seismicity

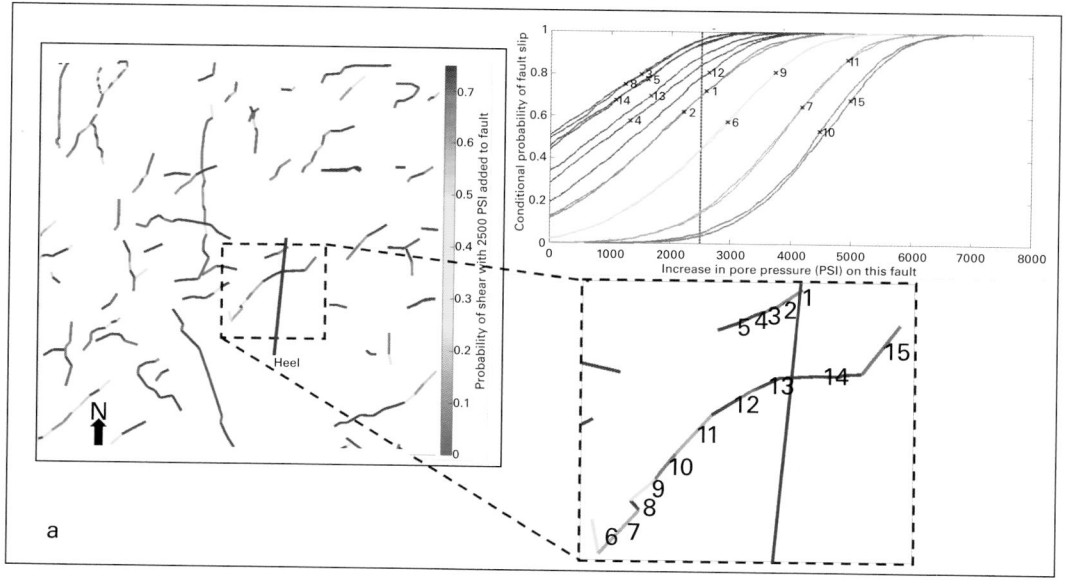

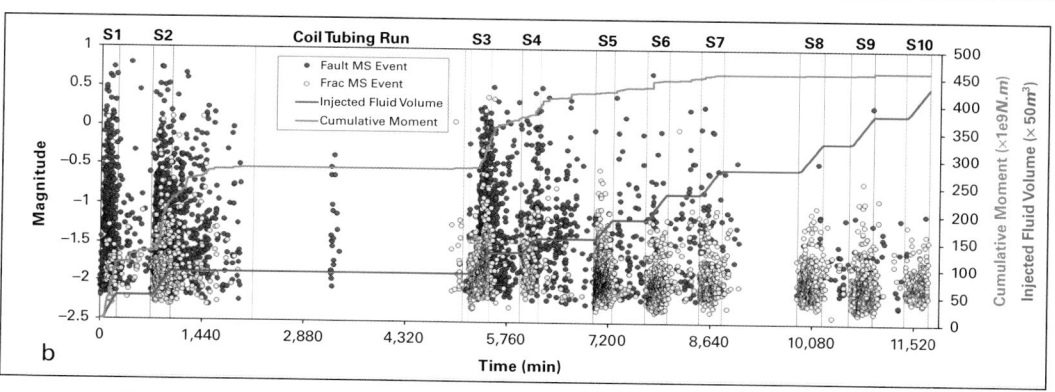

Figure 14.10 An image of faults derived from seismic reflection data in the vicinity of the well shown in Fig. 13.20 in the Sichuan Basin in China. The FSP analysis indicated a high likelihood of slip on both faults intersected by the well. From Chen et al. (2018).

the faults that were seismically mapped. The results are shown in Fig. 14.10, both for all the faults in the area as well as the two faults intersected by the well that were seismically imaged. As shown, both faults that were intersected had a high probability of slipping in response to the pressure perturbation associated with hydraulic fracturing. The fault near the toe (segments 1–5) was predicted to have a high probability for slipping by the FSP analysis and did, in fact, produced a great deal of seismicity as shown in Fig. 13.21. Earthquakes as large as M ~1 were associated with the different stages, implying that either there was a problem with zonal isolation – perhaps poorly cemented casing in the lateral – or a high degree of fracture and fault interconnectivity in the reservoir. Interestingly, the larger fault cutting through the well that was also predicted to slip (segments 12–14) did not produce seismic events (Fig. 13.21) but did offset the casing, apparently through aseismic creep.

Ignoring Poroelasticity – It is important to point out that a shortcoming of the FSP analysis utilized in the previous sections is that it ignores poroelasticity. As noted in Chapter 12, it is difficult to apply poroelastic theory in practical applications for a variety of reasons. For example, it is difficult to know the distribution of pore pressure in a fractured and faulted crystalline basement with very low matrix permeability. It is also difficult to know what poroelastic parameters should be used in the modeling as the majority of the pore pressure changes are occurring heterogeneously and concentrated along the fractures and faults. This said, Chang & Segall (2016) present a theoretical study of pore pressure and poroelastic effects considering a variety of scenarios in which injection into permeable sediments could trigger faulting in the basement. They show that due to poroelastic stress transfer, it is possible that triggering can occur without a direct hydraulic connection between the injection zone and the basement fault, although the tendency for triggering is much greater when the injection zone and basement fault are hydraulically connected. Barbour et al. (2017) present an analysis using a model of layered poroelastic half-space to investigate the processes associated with triggering the M 5.8 Pawnee earthquake discussed above. They found that direct pressure diffusion was the dominant mechanism for reducing the effective stress on the fault that triggered the earthquake; the magnitudes of shear and normal stresses induced by poroelastic coupling between elastic deformation of the solid matrix and pressure diffusion are comparable in magnitude to the pore-pressure changes.

Risk Management and Traffic Light Systems

Traffic light systems can be used in real time to define, in advance, how operating companies and regulators will respond to the occurrence of seismicity. In addition, they can be adapted to take into account local conditions such as the presence of population centers, critical infrastructure and overall risk tolerance. As reviewed by the US National Research Council (2013), traffic light systems have historically been used in enhanced geothermal settings and have been based on ground shaking or magnitude thresholds to signify whether the injection project should continue as planned (green), modify operations due to heightened risk (amber), or suspend operations due to severe risk (red). Such systems have been put in place by regulators to monitor hydraulic fracturing operations in the States of Ohio and Oklahoma in the US, in the Provinces of Alberta and British Columbia in Canada, in the United Kingdom and perhaps elsewhere.

Traffic Light Systems for Hydraulic Fracturing – Walters et al. (2015) proposed a generic traffic light system applicable to hydraulic fracturing (Fig. 14.11). The green, amber and red panels represent the levels of heightened concern because of specific observations. Recommended actions are also specified. Conceptually, hydraulic fracturing operations begin in the green zone where one would expect numerous microseismic

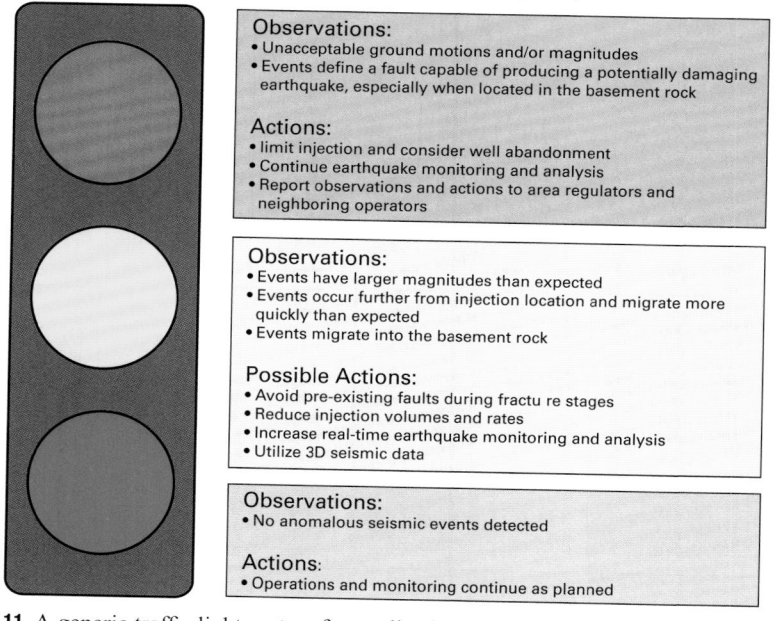

Figure 14.11 A generic traffic light system for application to real-time monitoring of hydraulic fracturing operations From Walters et al. (2015).

events, with M < −1. If *anomalously large* (see below) seismic events are detected, the project transitions to the amber zone. Anytime a project moves out of the green zone and into the amber or red zone, it is necessary to quickly evaluate to what extent operation practices might be adjusted or halted.

As might be expected, the term *anomalously large* is defined very differently in different areas. For example, as shown in Fig. 14.12, the magnitude threshold at which the amber warning light comes on (which usually implies notification of regulators) in Alberta is M 2.0, in Ohio it is M 1.5, in Oklahoma it is M 2.5 and in the UK it is M 0. Even more remarkable is the range of magnitudes for the red light to turn on (which involves stopping operations) in Alberta is M 4.0, in Ohio it is M 2.5, in Oklahoma it is M 3.5 and in the UK it is M 0.5. These number reflect the high degree of risk aversion sought by the respective regulatory agencies.

There are a number of important operational questions that must be addressed to utilize traffic light systems effectively. For example, when is real-time seismic monitoring required? As noted in Fig. 14.12, Ohio requires real-time monitoring on the basis of the proximity of a given site to basement faults and the occurrence of historic seismicity. Second, how will uncertainty in earthquake parameters be handled? Real-time determination of earthquake magnitudes can have a fair amount of uncertainty as can the locations of historical seismicity.

Traffic Light Systems for Flowback and Produced Water Disposal – Traffic light systems applicable to saltwater disposal (Fig. 14.13) are conceptually similar to what

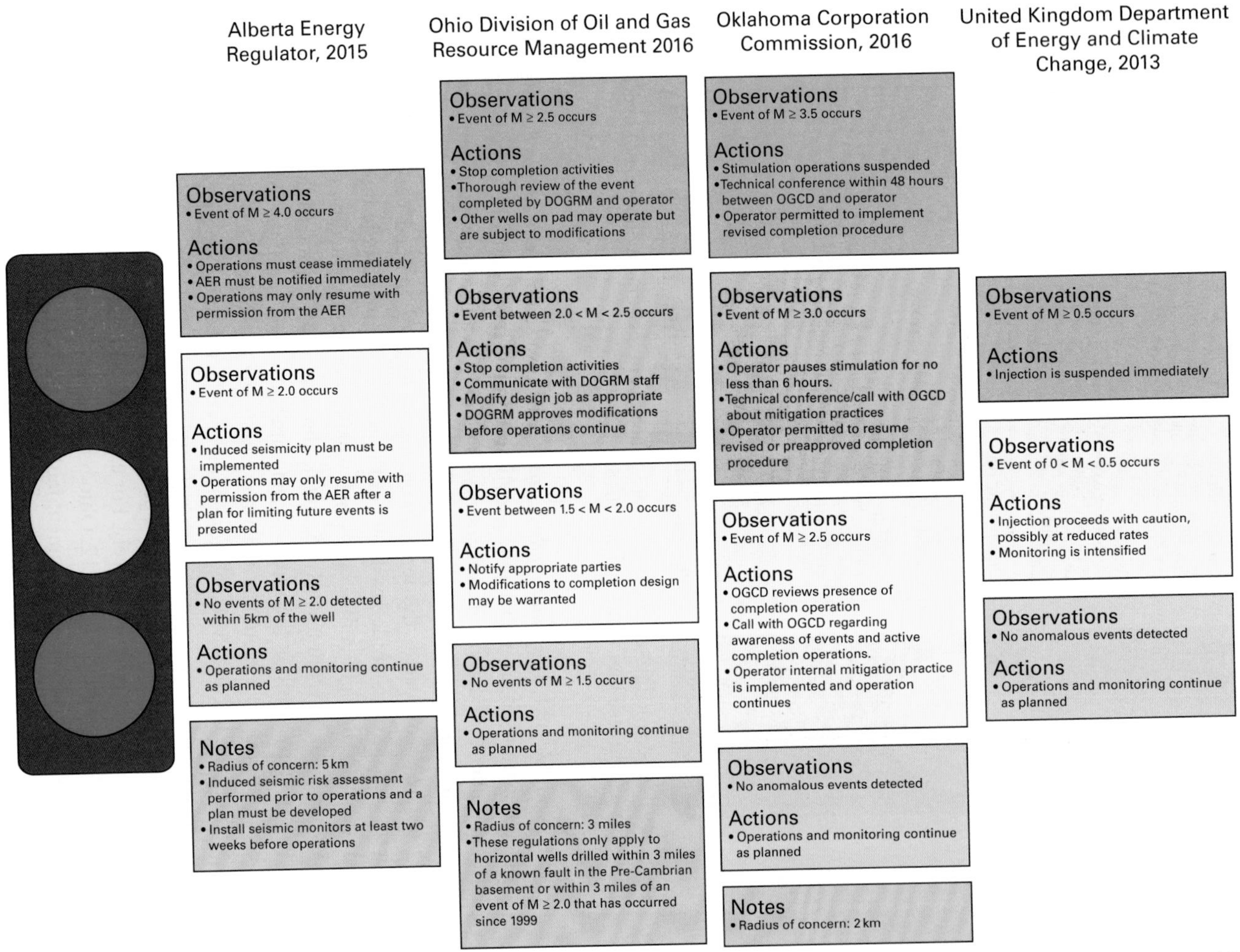

Alberta Energy Regulator, 2015

Observations
- Event of M ≥ 4.0 occurs

Actions
- Operations must cease immediately
- AER must be notified immediately
- Operations may only resume with permission from the AER

Observations
- Event of M ≥ 2.0 occurs

Actions
- Induced seismicity plan must be implemented
- Operations may only resume with permission from the AER after a plan for limiting future events is presented

Observations
- No events of M ≥ 2.0 detected within 5km of the well

Actions
- Operations and monitoring continue as planned

Notes
- Radius of concern: 5 km
- Induced seismic risk assessment performed prior to operations and a plan must be developed
- Install seismic monitors at least two weeks before operations

Ohio Division of Oil and Gas Resource Management 2016

Observations
- Event of M ≥ 2.5 occurs

Actions
- Stop completion activities
- Thorough review of the event completed by DOGRM and operator
- Other wells on pad may operate but are subject to modifications

Observations
- Event between 2.0 < M < 2.5 occurs

Actions
- Stop completion activities
- Communicate with DOGRM staff
- Modify design job as appropriate
- DOGRM approves modifications before operations continue

Observations
- Event between 1.5 < M < 2.0 occurs

Actions
- Notify appropriate parties
- Modifications to completion design may be warranted

Observations
- No events of M ≥ 1.5 occurs

Actions
- Operations and monitoring continue as planned

Notes
- Radius of concern: 3 miles
- These regulations only apply to horizontal wells drilled within 3 miles of a known fault in the Pre-Cambrian basement or within 3 miles of an event of M ≥ 2.0 that has occurred since 1999

Oklahoma Corporation Commission, 2016

Observations
- Event of M ≥ 3.5 occurs

Actions
- Stimulation operations suspended
- Technical conference within 48 hours between OGCD and operator
- Operator permitted to implement revised completion procedure

Observations
- Event of M ≥ 3.0 occurs

Actions
- Operator pauses stimulation for no less than 6 hours.
- Technical conference/call with OGCD about mitigation practices
- Operator permitted to resume revised or preapproved completion procedure

Observations
- Event of M ≥ 2.5 occurs

Actions
- OGCD reviews presence of completion operation
- Call with OGCD regarding awareness of events and active completion operations.
- Operator internal mitigation practice is implemented and operation continues

Observations
- No anomalous events detected

Actions
- Operations and monitoring continue as planned

United Kingdom Department of Energy and Climate Change, 2013

Observations
- Event of M ≥ 0.5 occurs

Actions
- Injection is suspended immediately

Observations
- Event of 0 < M < 0.5 occurs

Actions
- Injection proceeds with caution, possibly at reduced rates
- Monitoring is intensified

Observations
- No anomalous events detected

Actions
- Operations and monitoring continue as planned

Notes
- Radius of concern: 2 km

Figure 14.12 Comparison among different traffic light systems used in different areas. Note the very strong differences among the threshold magnitudes at which action is taken (see text).

Flowback Water Traffic Light System

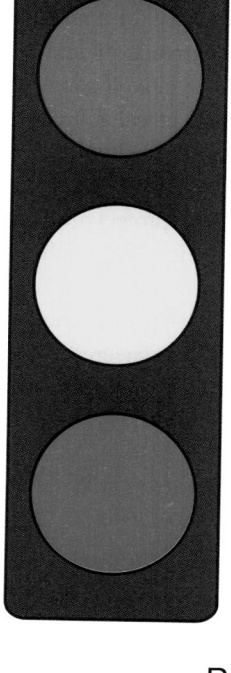

Observations:
- Unacceptable ground motions and/or magnitudes
- Events define a fault capable of producing a potentially damaging earthquake

Actions:
- Limit injection and consider well abandonment
- Continue earthquake monitoring and analysis
- Report observations and actions to area regulators and neighboring operators

Observations:
- Unexpected event(s) occurring

Actions:
- Increase real-time earthquake monitoring and analysis
- Decrease injection rates and volumes

Observations:
- No seismic events detected

Actions:
- Operations and monitoring continue as planned

Produced Water Traffic Light System

Observations:
- Unacceptable ground motions and/or magnitudes
- Events define a fault capable of producing a potentially damaging earthquake

Actions:
- Limit injection and consider well abandonment
- Continue earthquake monitoring and analysis
- Report observations and actions to area regulators and neighboring operators

Observations:
- Unexpected event(s) occurring
- Cumulative injection volumes between multiple wells becomes significant

Actions:
- Increase real-time earthquake monitoring and analysis
- Decrease injection rates and volumes

Observations:
- No seismic events detected

Actions:
- Operations and monitoring continue as planned

Figure 14.13 Generic traffic light system for application to monitoring of flowback or produced water disposal From Walters et al. (2015).

might be used for hydraulic fracturing, but are operationally different in several regards. What is common, of course, is that green, amber and red represent the levels of heightened awareness and required potential actions that might be taken. However, as fluids are injected into the subsurface and seismic events are monitored (potentially for a number of years), there are observations of potential risk other than the occurrence of earthquakes with magnitudes that exceed a certain threshold. For example, if relatively small earthquakes migrate farther from the injection site than expected it could indicate that fluid is potentially migrating through a permeable, potentially active fault. Similarly, the small earthquakes may illuminate a planar feature suggesting the presence of a potentially active fault.

It is important to note that with all traffic light systems, after a project moves to amber or red it may be possible to transition back to a lower risk level after a thorough evaluation of the hazard or a change in activities (such as reducing injection rates). This could include engaging engineers and subsurface geological and geophysical experts to review available subsurface data and, as necessary, to design and conduct engineered trials to adjust operating procedures, as indicated by the example discussed in the next section.

Utilizing Seismogenic Index Models to Manage Produced Water Injection

As discussed in Chapter 13, the extremely rapid increase in the number of earthquakes in the central and eastern US shown in Fig. 13.2 began around 2008 and accelerated markedly around 2012. While these earthquakes were associated with all three mechanisms – hydraulic fracturing, flowback water injection and produced water injection, the great majority of the events were associated with injection of water produced from formations such as the Mississippian limestone and the Hunton limestone in north-central Oklahoma and southernmost Kansas. This is shown clearly in Fig. 14.14 (updated from Langenbruch & Zoback, 2016). The blue line represents the monthly injection rate in north central Oklahoma (the area within the dashed lines in Fig. 13.19a, designated the areas of interest (AOI) by Oklahoma regulators) and the number of $M \geq 3$ earthquakes is shown by the green line. The spikes in the number of earthquakes at certain times are due to aftershock sequences of the larger earthquakes, as noted. There is a delay between the time of the rapid increase injection rates and the earthquakes that results from the time it takes fluid pressure to diffuse from the injection wells to the faults at depth. Nonetheless, the excellent correlation between injection rates and seismicity rates is clear – injection rates rapidly increase in 2012 and the earthquakes rapidly increased about a year later. When injection rates began to rapidly decrease in late 2015, the earthquake rate decreased rapidly as well after a short delay. The dramatic decrease in the number of earthquakes throughout the central and eastern US after 2015 seen in Fig. 13.2 is the direct reflection of the decrease in seismicity in north-central Oklahoma due to the reduction of produced water injection.

The decrease in injection rates in north-central Oklahoma came about as the result of two factors. First, a drop in the price of oil in 2015 led to a cut-back in activities in the Mississippian play. Second, recognition of the correlation between the rate of produced water injection and the rate at which earthquakes were occurring led Oklahoma regulators, in February and March 2016, to mandate a 40% reduction in the injection

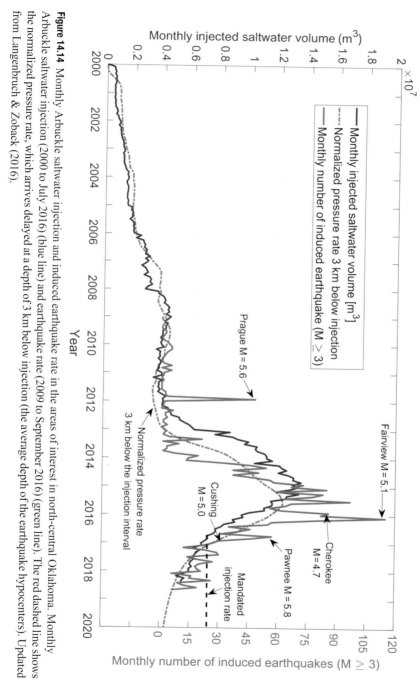

Figure 14.14 Monthly Arbuckle saltwater injection and induced earthquake rate in the areas of interest in north-central Oklahoma. Monthly Arbuckle saltwater injection (2000 to July 2016) (blue line) and earthquake rate (2009 to September 2016) (green line). The red dashed line shows the normalized pressure rate, which arrives delayed at a depth of 3 km below injection (the average depth of the earthquake hypocenters). Updated from Langenbruch & Zoback (2016).

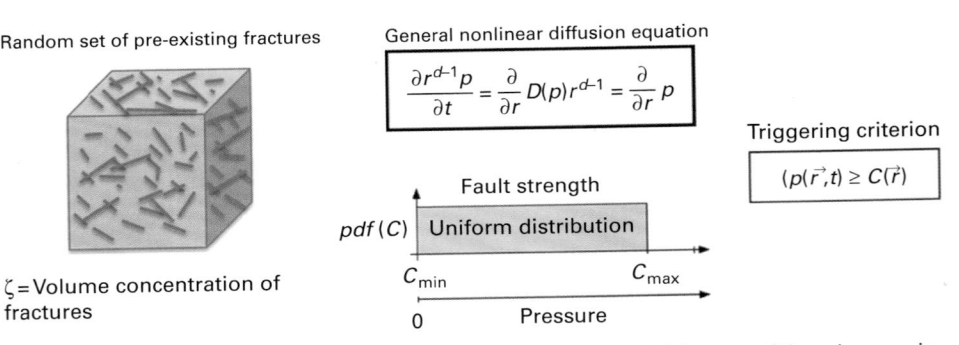

Figure 14.15 Schematic diagram of the physical assumptions underlying use of the seismogenic index model. Earthquake triggering occurs in response to pressure diffusion in the pore and fracture space of rocks. A random distribution of fractures are present with a broadly distributed fault strength sampled from a uniform distribution. The critical pressure corresponds to the pressure increase necessary to overcome the frictional strength of the individual pre-existing faults. (Figure courtesy of Cornelius Langenbruch.)

volume for 2016. This would essentially return the total volume of saltwater injected to the levels seen prior to the rapid increase in seismicity. Langenbruch & Zoback (2016) undertook a study that attempted to predict how future seismicity rates in the region would respond to the mandated decrease in injection rates.

Langenbruch & Zoback utilized a model based on the seismogenic index discussed in detail by Shapiro (2015). Physically, the ideas behind the seismogenic index model are shown in Fig. 14.15. It assumes there are pre-existing randomly distributed faults in the crust with a broad distribution of fault strengths, defined by the pressure increase necessary to overcome the frictional resistance to sliding. This is equivalent to the critically stressed crust model discussed in Chapter 13 – there are many faults with varied orientations in crystalline basement rocks, most of which are not potentially active in the current stress field. However, some of the pre-existing faults are well-oriented for slip in the current stress field and could slip in response to a small pressure perturbation (these could be thought of as having low strength). The process of earthquake triggering is controlled by diffusion of pressure in the pore fluid pressure and fractures of rocks. If the pressure perturbations exceed a critical level, an earthquake is triggered. The assumptions outlined above and uniform distribution of fault strength results in the earthquake rate being proportional to the injection rate (Shapiro et al., 2010).

Mathematically, the seismogenic index, Σ or SI, is defined by a modified Gutenberg–Richter relation for fluid-induced earthquakes

$$\log_{10}[N_{\geq M}(t)] = a(t) - bM = \log_{10}[V_I(t)] + \Sigma - bM \tag{14.1}$$

where $N_{\geq M}(t)$ is the number of earthquakes above magnitude M observed until the time t, $V_I(t)$ is the volume of injected fluid and a and b are the same a and b values of the Gutenberg–Richter law (Eqn. 9.15). Fundamentally, the SI incorporates the volume concentration of pre-existing faults and state of stress in an area. In contrast to the Gutenberg–Richter relation associated with natural earthquake sequences, Eqn. (14.1) does not assume stationarity of the *a* value (as typically used in probabilistic hazard analysis) but relates changes of earthquake activity in time to changes of the fluid injection rate in the crustal volume affected by the injection: $a(t) = \log_{10}[V_I(t)] + \Sigma$. Shapiro (2015) shows that the SI for different areas of induced seismicity remains constant through time. This supports the assumption that the SI is a characteristic of the *seismogenic state* of a crustal volume, independent of the anthropogenic perturbation.

Langenbruch & Zoback modified the SI model by introducing a critical level of injection which seems to be accommodated by the hydrogeologic system in Oklahoma without occurrence of seismicity. As can be seen in Fig. 14.14 this level had not been exceeded prior to 2008. After calibration of the modified SI model using the seismicity and injection rates in the first years of seismic activity, Langenbruch and Zoback predicted seismicity rates based on two assumptions (1) the produced water injection remain below the mandated levels, and (2) the decay of earthquakes will follow a well-known relation adopted from the decay rate of aftershock sequences following large tectonic earthquakes (Langenbruch & Shapiro, 2010). Figure 14.16a shows excellent correlation between the observed number of M ≥ 3 earthquakes in the AOI and the seismicity rate predicted by Langenbruch & Zoback (2016). This correlation has continued through the time of writing this book, two years after the prediction was made (Langenbruch & Zoback, 2017).

Based on the calibrated SI model Langenbruch & Zoback estimated the annual probability $P_E(M)$ to exceed a certain magnitude M, that is, the probability to observe one or more earthquakes above a specified magnitude in the complete AOI. If a Poisson process is used to describe the occurrence of induced earthquakes (following Shapiro et al., 2010 and Langenbruch et al., 2011) the magnitude exceedance probability can be computed based on the annual injected saltwater volume V_{IA} above the determined seismicity-triggering threshold

$$P_E(M) = 1 - P(0, M, V_{IA}) = 1 - \exp(-V_{IA}10^{\Sigma - bM}). \tag{14.2}$$

These probabilities are predicated based on the assumptions that the produced water injection remain constant at the mandated levels, and the decay of earthquakes will be related to that of aftershock sequences of induced earthquakes (Langenbruch & Shapiro, 2010). Figure 14.16b shows the resulting annual probability of earthquake occurrence to exceed a certain magnitude. One can see, as mentioned above, that the background probability exceeding an M 5 earthquake prior to 2009 was one per century. When the seismicity rate peaked in 2015, there was an 80% chance of exceeding an M 5 each year. The long-term response to the overall reduction of saltwater injection is for the probability curves to shift to the left, toward the background rate. It's important to note, however, that while the number of felt earthquakes of M ≥ 3 decreased rapidly between 2015 and 2017 (Fig. 14.16a), the probability of felt to potentially damaging earthquakes in 2017 was much higher than the background probability (Fig. 14.16b). For example, the model predicts that the earthquakes in Oklahoma will eventually return to tectonic background levels, but it will do so very slowly as large-scale injection in the area is still going on. Note that the annual probability of exceedance for moderate size earthquakes in 2025 is well above the background rate. Thus, the probability of occurrence of

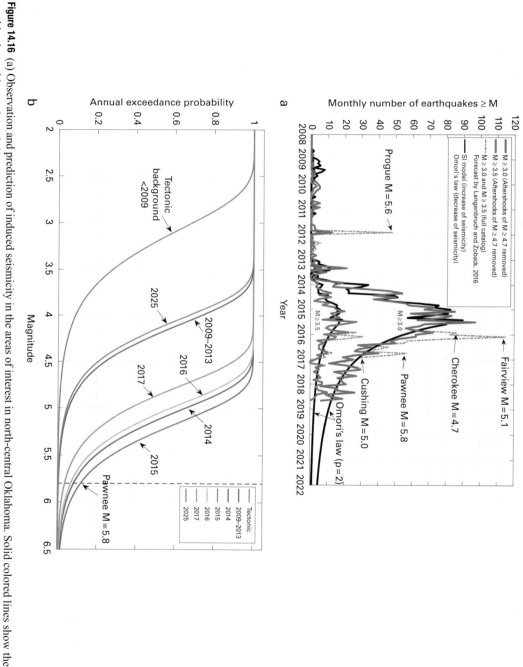

Figure 14.16 (a) Observation and prediction of induced seismicity in the areas of interest in north-central Oklahoma. Solid colored lines show the combined monthly number of observed earthquakes (green, M ≥ 3; red, M ≥ 3.5) with aftershocks of M ≥ 4.7 removed. The gray dashed lines present the complete earthquake catalog. The colored dotted lines present SI models calibrated through different times between June 2014 and December 2015. The black solid line shows a decay of M ≥ 3 earthquakes following Langenbruch and Shapiro (2010) and Omori's law (p = 2). (b) Combined annual probabilities of exceeding the magnitude shown on the abscissa. Updated from Langenbruch & Zoback (2016). Reprinted with permission from

potentially damaging earthquakes remains high for a number of years. The probability of a M $\geq$ 5 earthquake occurring in 2017 was about 37%, but had injection rates not been reduced, it would have been 80%.

The M5.8 Pawnee earthquake that occurred in September 2016 appears to have a higher magnitude than predicted by the Langenbruch and Zoback model because the annual maximum magnitude in 2016 is expected to be around M=5. However, the cumulative probability of exceeding M=5.8 by injection of the saltwater volume reported from 2009 through May 2017 was quite high, ~28% (Langenbruch & Zoback, 2017).

Improving the SI Model for Oklahoma Seismicity – Langenbruch et al. (2018) revisited the analysis presented above with the intention to address two shortcomings. Utilizing a hydrologic model of the area in which the seismicity was occurring (including southern Kansas), they were able to analyze spatial variability of the SI. As seen below, this translates directly into spatial variations of the likelihood of potentially damaging earthquakes. Second, they are able to calculate the expected rate at which the seismicity will decay, without assuming its functional form.

The hydrologic model used by Langenbruch et al. utilized all of the injection data available for the region. The numerical model of the hydrogeologic system of Oklahoma's deepest sediments and crystalline basement was aimed at predicting fluid injection-related pore pressure changes at depth in space and time. A three-dimensional hydrogeologic model was used (MODFLOW, Harbaugh et al., 2010) which assumes injected fluids are of constant density and dynamic viscosity. Injection occurs into the 2.1 km deep, 400 m thick Cambrian-aged, fractured dolomitic carbonate Arbuckle Group and is assumed to be uniform across the entire depth of the interface. The Arbuckle Group is underlain across the entire model domain by lower permeability crystalline basement to a depth of 20 km. The hydrogeologic model intentionally simplifies and idealizes the three-dimensional hydrogeologic medium by representing the system's large-scale bulk permeability with uniform layers. The initial model parameterizations are based on the reported range of Arbuckle Group permeability, 10^{-14} m^2 to 10^{-12} m^2, from core, outcrop and monitoring well tests. While lacking in direct measurements of crystalline basement permeability in Oklahoma and southern Kansas, they implemented bulk permeabilities of fractured basement rock from the literature (10^{-16}–10^{-14} m^2) and tested the sensitivity of model outputs to permeability by running several combinations of Arbuckle and basement permeability. A key feature in the model is the representation of the Nemaha fault as a regional-scale, low-permeability barrier to cross-fault flow. Two key baseline datasets constrain the large-scale Arbuckle Group permeability: (1) the observed hydraulic underpressure for the deepest hydrostratigraphy in Oklahoma and Kansas, and (2) large-scale trends in reported daily wellhead pressures for Arbuckle injection wells in Oklahoma. The observed hydraulic underpressure is well-known and discussed in Chapter 13. Large-scale trends in daily wellhead pressures show that more than half of all Arbuckle injection wells, even wells operating at rates greater than 10,000 m^3 day^{-1}, operate under gravity-feed injection requiring no wellhead pressure. Most of the remaining wells (~40%) operate at wellhead pressures between ~0.3 and 2 MPa, well within the pressure range expected from wellbore

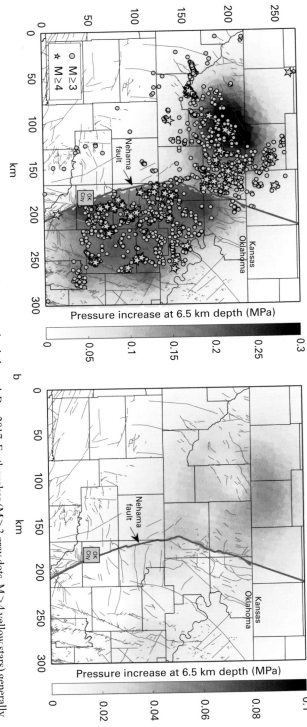

Figure 14.17 (a) Map of injection-induced pressure increase at depth through Dec 2017. Earthquakes (M≥3 gray dots, M≥4 yellow stars) generally occurred where injection increased pressure at depth. (b) Projected future pressure increases through Dec 2020. With the exception of a region east of Oklahoma City the pressure is expected to increase through 2020. Note the different range of the color schemes used in the figures. Mapped faults in the sedimentary cover are shown as gray lines. From Langenbruch et al. (2018).

Managing the Risk of Injection-Induced Seismicity   433

friction. These data indicate the large-scale, bulk permeability of the Arbuckle Group is likely towards the high end of the reported range.

Figure 14.17a shows the pressure increase at 6.5 km depth in the crystalline basement through December of 2017 resulting from the hydrogeologic modeling. The largest pressure increase was about 0.3 MPa, near the Oklahoma/Kansas border. Note the effect of assuming the Nemaha fault is a low-permeability barrier to cross-fault flow. It can be clearly seen that from 150 km to the south, pressure changes are essentially occurring only to the east of the Nemaha fault (due to the injection wells operating there), and so is the seismicity. Even though injection rates started to significantly decrease in 2015, Fig. 14.17b shows that, assuming injection rates remain at the March 2018 level, pressure will still be going up through the end of 2020. Peak pressure at seismogenic depth in the crystalline rock has not been reached yet. However, compared to past pressure increases (Fig. 14.7a) the projected maximum pressure increase from 2018 through 2020 is minor (less than 0.05 MPa).

With a hydrologic model available, Langenbruch et al. (2018) replaced injection rates (in the SI model Eqns. 14.1 and 14.2) by the modeled local rates of pressure increase and

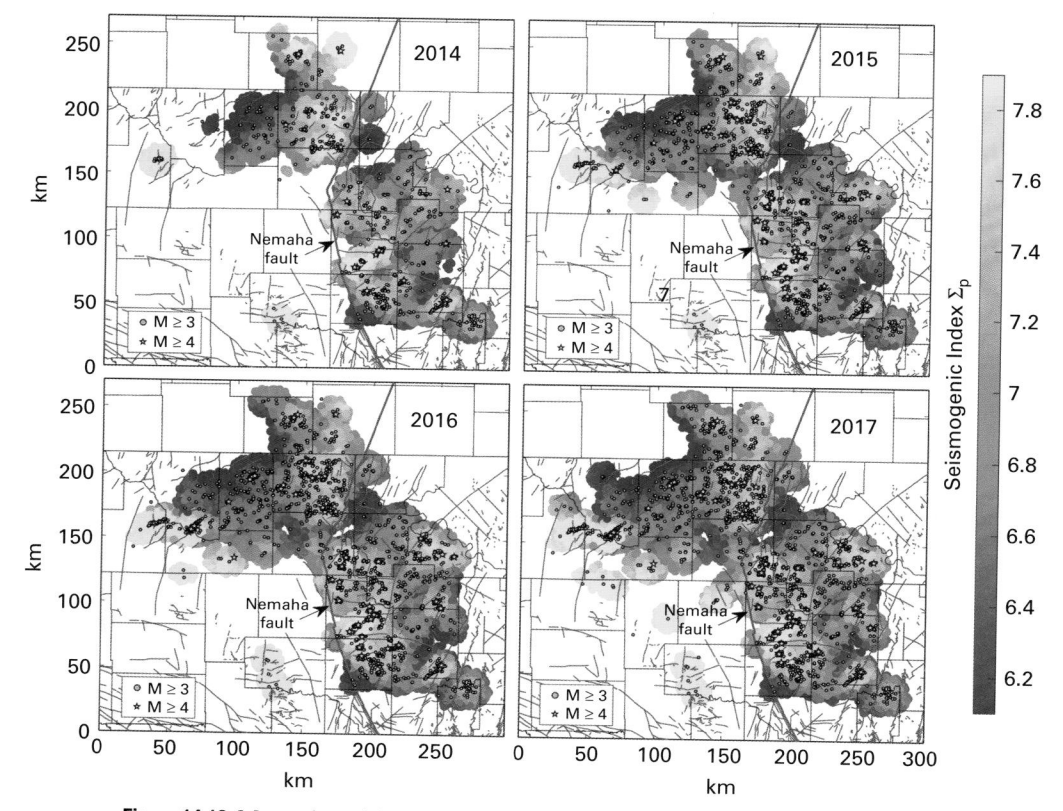

Figure 14.18 Mapped spatial variability of the SI in north-central Oklahoma and southern Kansas computed at 25,000 seed points with 10 km radius. Note that the variability of the SI is quite variable in space but stable through time. From Langenbruch et al. (2018).

computed the SI on a local scale. The SI was calibrated through different temporal endpoints between December 2014 and December 2017 in local-scale regions of 10 km radius around 25,000 seed points distributed throughout the study area. Note that the SI shows significant variability in space but it is stable in time (Fig. 14.18), another indication that it is characterizing the seismogenic state of the crust, independent of the hydrologic conditions. Regions of high and low SI can easily be differentiated.

Langenbruch et al. produced 1-year maps of the seismic hazard to assess the probability of potentially damaging induced earthquakes in Oklahoma and Kansas from 2015 through 2020, which are shown in Fig. 14.19. To produce forecasts, they only used observed earthquakes up to the forecast period and then used modeled pressure rates in the coming year to forecast expected seismicity throughout the study area. The expected annual rate ($N_{\geq M}$) of magnitude M or larger earthquakes is a Poisson process. The annual probability of one or more earthquakes above magnitude M within one year, can be computed according to

$$P_E(\mathrm{M}) = 1 - P_E(0, \mathrm{M}, N_{\geq M}) = 1 - \exp(-N_{\geq M}) \qquad (14.3)$$

They chose a magnitude of M = 4, because lower magnitudes are very unlikely to cause damage. The resulting prospective 1-year maps of the probability to exceed M = 4 (Fig. 14.19) show how the seismic hazard in time and space is changing in response to spatial and temporal variations of injection rates and spatial changes of the SI. In response to decreased saltwater injection rates, pressure increases are slowing down over a wide range of depths and the model forecasts a wide-spread reduction of the seismic hazard in 2017. In fact, the probability of damaging earthquakes through the region as a whole is essentially the same as shown in Fig. 14.16b and even without additional injection rate reductions after March 2018, the model predicts a further decrease of the seismic hazard in 2018 and 2020. East of the Nemaha fault, where injection rates have been reduced most significantly, the areas with strongest decreases were accurately predicted. In other parts of northern Oklahoma and southernmost Kansas, M ≥ 4 probabilities remain on a higher level. The maps identify three regions where M ≥ 4 exceedance probabilities in 2018 remain above 30%. Note that most earthquakes observed in 2018 occurred in or close to these regions.

Site Characterization Risk Frameworks

The Stanford Center for Induced and Triggered Seismicity (SCITS) developed an induced seismicity and risk assessment framework in an attempt to provide a unified approach to risk assessment that includes operational factors, exposure and vulnerability as well as risk tolerance. Walters et al. (2015) presents an overview of the proposed work flow and a summary of factors to consider. The end product of the workflow is to generate risk matrices, analogous to those proposed by Nygaard et al. (2013), that allow one to consider operational factors in the context of tolerance to different degrees of ground shaking.

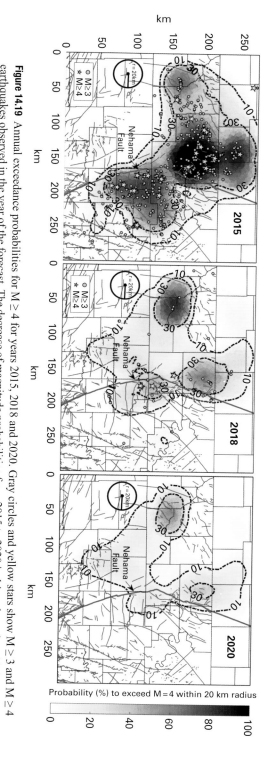

Figure 14.19 Annual exceedance probabilities for M ≥ 4 for years 2015, 2018 and 2020. Gray circles and yellow stars show M ≥ 3 and M ≥ 4 earthquakes observed in the year of the forecast. The decrease of magnitude probabilities from 2015 to 2020 is driven by reduced injection rates, which slow down the pressure increase at seismogenic depth. The strongest decrease of the seismic hazard is predicted east of the Nemaha fault, where injection rates were reduced most significantly. Probabilities for future years were computed assuming constant injection rates after March 2018. From Langenbruch et al. (2018).

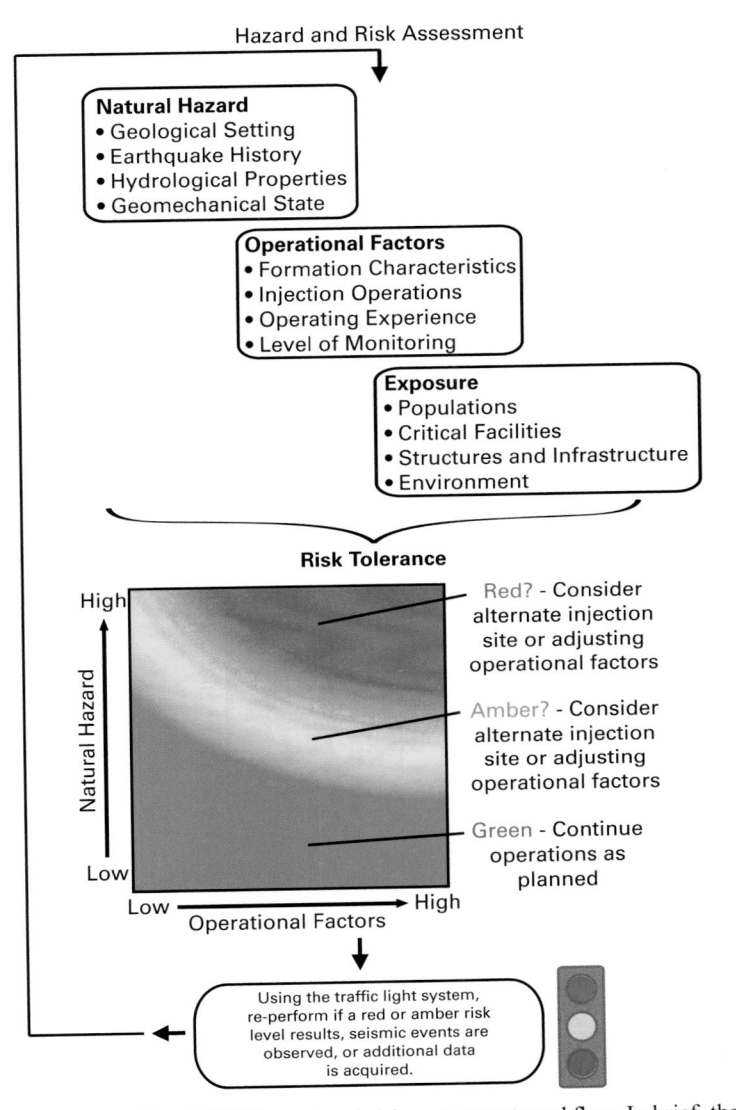

Figure 14.20 Overview of the SCITS hazard and risk-assessment workflow. In brief, the hazard, operational factors, exposure and tolerance for risk are evaluated prior to injection operations and reflected by shifting the green to red color spectrum in the risk-tolerance matrix. After injection begins, the occurrence of earthquakes in the region and additional site-characterization data could require additional iterations of the workflow. From Walters et al. (2015).

The proposed hazard and risk-assessment workflow (Fig. 14.20) for earthquakes triggered by both hydraulic fracturing and saltwater disposal is meant to be site specific and adaptable. It includes an analysis of the earthquake hazard at a site using the known geology, hydrology, earthquake history and geomechanics of the area. It can be used with a probabilistic seismic-hazard analysis (PSHA) (e.g., Kramer 1996) as the basis for determining the probable level of natural seismic hazard. In some cases there may be significant uncertainty in determining the level of hazard in an area due to the lack of historical seismicity and other factors. The determined natural hazard is then used in conjunction with operational factors that influence the potential for the occurrence of triggered earthquakes, including specific injection practices, the operating experience in the area and of the company responsible and the formation characteristics. Once probabilities of experiencing various levels of ground motions have been estimated based on possible triggered earthquake source locations and source sizes, they can be combined with the likely consequences to evaluate risk. Consequences depend upon the level of exposure of the site and surrounding area and the contributing operational factors. As such, risk assessment and planning need to occur jointly with planning of operations that might affect risk. Both the operational factors and exposure are described further below. The proposed workflow is intended to be implemented prior to injection operations and then used iteratively as new information related to the hazard and risk becomes available. In cases where the risk is non-negligible, mitigation can include additional monitoring and data collection (Nygaard et al., 2013). In severe cases, particular areas may be identified as having unreasonably high hazard and subsequent risk for fluid injection.

Operations – Along with the earthquake history and geologic, hydrologic and geomechanical characteristics of a site, a number of operational factors also contribute to the potential for triggered seismicity (Fig. 14.21a). Because operational factors are not included in PSHA procedures, we account for them separately in the formation of a project's risk-tolerance matrix. As it is not currently possible to link these operational factors in a quantitative or causative manner to earthquake occurrence, Walters et al. took the indirect approach presented below to consider operational factors as a separate metric to be used when assessing risk.

There are particular formation characteristics that may affect the risk at a site in addition to choosing injection well locations sufficiently far from potentially active faults. Also, examining whether the injection interval is in communication with crystalline basement is important as is whether the formation is initially underpressured. Obviously, the specific injection operations also have the potential to affect the level of risk associated with a project. The injection rates and volumes at single wells may be correlated with earthquake activity at a site. An increasingly significant operational consideration for saltwater disposal wells is the rate of injection of a group of wells in close proximity.

Exposure – The exposure associated with a particular site (Fig. 14.21b) depends on the number, proximity and condition of critical facilities, local structures and infrastructure, the size and density of the surrounding population, and protected sites that have the potential to experience ground shaking as a result of fluid injection. Specifically, population centers, hospitals, schools, power plants, dams, reservoirs, historical sites, hazardous materials storage, etc. If an injection project is proposed near one or more of these items, the risk for the project increases correspondingly. Because, as is the case in north-central Oklahoma, earthquakes can be triggered at some distance away from injection sites, and ground shaking from a moderate earthquake can be felt over a

Operational Factors	Formation Characteristics	Injection Operations		Operating Experience
		Wastewater Injection	Hydraulic Fracturing	
Significant	Injection horizon likely in communication with basement, underpressured injection interval	High cumulative injection volumes and rates	High fluid injection volumes and pressures near active faults, no flowback performed	Limited injection experience in region, past earthquakes clearly or ambiguously correlated with operations
Moderate	Injection horizon potentially in communication with basement, slightly underpressured injection interval	Moderate cumulative injection volumes and rates	Moderate fluid injection volumes and pressures near active faults, flowback possibly performed	Moderate injection experience in region with no surface felt ground shaking
Minor	Injection horizon not in communication with basement	Low cumulative injection volumes and rates	Low fluid injection volumes and pressures remote from active faults, flowback possibly performed	Extensive injection experience in region with no surface felt ground shaking

a

Exposure	Critical Facilities	Structures and Infrastructure	Environment	Populations
High	Facilities in the immediate vicinity with the potential to suffer damage	Few designed to withstand earthquakes based on current engineering practices	Many historical sites, protected species, and/or protected wildlands	High population density and/or total population
Moderate	Facilities in the nearby area	Many designed to withstand earthquakes based on current engineering practices	Few historical sites, protected species, and/or protected wildlands	Moderate population density and/or total population
Low	No facilities in the area	Most designed to withstand earthquakes based on current engineering practices	No historical sites, protected species, and/or protected wildlands	Low population density and/or total population

b

Figure 14.21 (a) Factors related to operations that contribute to the level of risk at an injection site. (b) Factors that contribute to the level of exposure at an injection site. From Walters et al. (2015).

wide region, determining this area of concern could be done in a way that incorporates the site-specific conditions of the geology, hydrology, geomechanical characterization, earthquake history and exposure to risk, as well as whether injection from neighboring operators may have a cumulative contribution to the risk in the area.

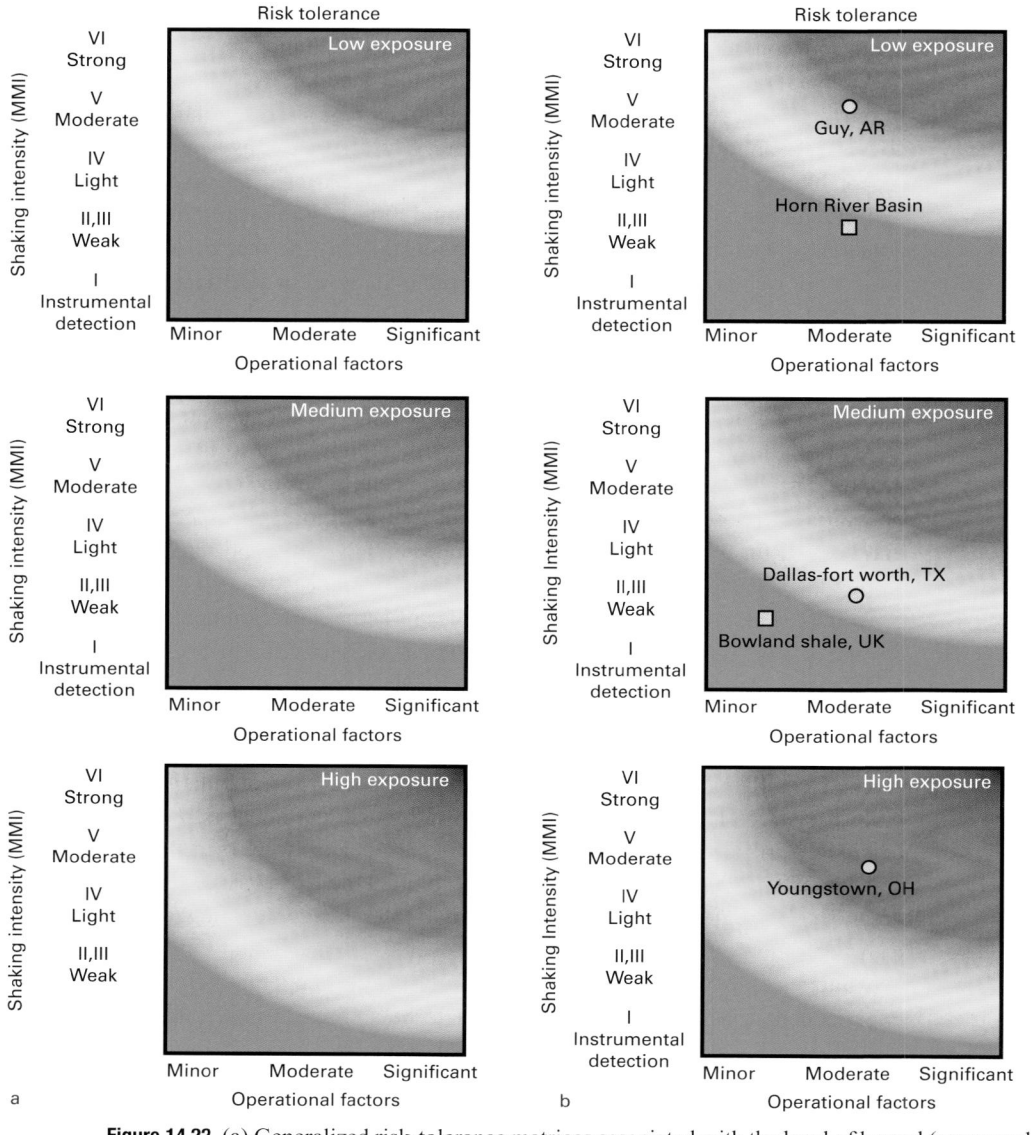

Figure 14.22 (a) Generalized risk-tolerance matrices associated with the level of hazard (expressed as probable shaking intensity determined using PSHA), as a function of operational factors for different levels of exposure. (b) Examples of projects being plotted on the risk-tolerance matrices in light of what we know after events have occurred. The squares represent hydraulic fracturing projects and the circles represent saltwater disposal projects. From Walters et al. (2015).

Risk Matrices – Once the seismic hazard, exposure and operational factors are determined for a given project, operators and regulators can aggregate the results using a risk-matrix method. Figure 14.22 shows how the results from the hazard assessment via PSHA (vertical axis), the operational factors (horizontal axis) and the exposure (top, middle, or bottom figure) can be aggregated to perform such an evaluation, as expanded upon from concepts proposed by Nygaard et al. (2013). Figure 14.22a shows generalized risk-tolerance matrices for areas of low exposure, medium exposure, or high exposure. In our proposed risk-tolerance matrices, the green regions would be considered favorable given appropriate operational practices; amber regions would be considered acceptable but may require enhanced monitoring, restricted operational practices and real-time data analysis; and red regions would require significant mitigating actions or potential site abandonment. An understanding of the risk that exists for a particular project allows the affected parties to determine the level of tolerance they have for the estimated risk. The tolerance for potential ground shaking will be shaped by the political, economic and emotional state of the populations involved, making it inherently site specific. In high-risk cases or for those who have a low tolerance for the determined risk, injection may not be allowed to proceed in certain locations. Alternatively, in other areas, the tolerance for risk may be sufficiently high so as to not interfere with the proposed injection project. Of course, how one determines the exact levels of exposure, operational factors, hazard and subsequent risk to inform the specific risk-tolerance matrix used for a particular project is somewhat subjective and requires collaboration among the stakeholders.

Walters et al. considered several examples of actual injection operations to illustrate the use of the risk-tolerance matrix shown in Fig. 14.22b. In each case, they only performed a rough analysis to provide context based on the current scientific literature. To reflect these differences in risk tolerance, the colored portions of the risk-tolerance matrices can shift either up or down to become more lenient or restrictive as appropriate.

The Horn River Basin is a remote location with very low population and little-to-no built infrastructure. It had experienced no significant earthquakes before development began in April 2009. By December 2011, 38 earthquakes were recorded between M 2.2 and 3.8. With all this in mind, Walters et al. plotted the Horn River basin example in the green-shaded region in the low-exposure risk-tolerance matrix, suggesting that significant mitigation efforts may not be needed even though seismicity was occurring.

Walters et al. considered the Bowland shale, a moderately populated area that experienced an M 2.3 earthquake in 2011 to be an area of moderate exposure. It was clear that the stakeholder involved in hydraulic fracturing operations in the United Kingdom had a very low tolerance for risk. Hence, while risk matrix analysis of Walters et al. might suggest that little to no action was appropriate, a highly restrictive traffic light system was adopted in the United Kingdom (Fig. 14.12) seemingly to prevent hydraulic fracturing from occurring.

The Dallas–Fort Worth airport saltwater disposal site was also considered to have medium exposure because of the airport facilities resulting in the project being located in the amber portion of the matrix. Undertaking activities such as enhance seismic monitoring, closer tracking of injected fluid and enhanced efforts to map faults in the area would seem appropriate.

Finally, the earthquakes that occurred in the Youngstown, Ohio saltwater disposal site (Kim, 2013) was in a populated area with nearby critical infrastructure and was thus considered a high exposure area and the moderate shaking experienced in the earthquakes placed this event in the red portion of the high-exposure risk-tolerance matrix. As appropriate, injection was terminated at that site.

Summary – Walters et al. presented a framework for risk assessment for triggered seismicity associated with saltwater disposal and hydraulic fracturing. This framework includes an assessment of the site characteristics, seismic hazard, operational factors, exposure and tolerance for risk. The process is intended to be site specific, adaptable and updated as new information becomes available and inclusive of the many issues that influence the decision-making process such as risk tolerance of operators, regulators and the public have when faced with understanding what to do about triggered earthquakes when they happen. The hazard and risk-assessment workflow can easily incorporate the use of a traffic-light systems that focuses on real-time seismological observations as a determining factor for whether a particular activity requires enhanced monitoring, decreased injection or a cessation of activities.

Perhaps the take-away message of this chapter is that when there is a possibility that injection-related activities could result in induced seismicity, it is important for operators and regulators to be pro-active. Such activities could involve siting injection wells based on detailed characterization, enhanced seismic monitoring, implementation of traffic light systems or the development risk matrices. Each of these steps would lead to a better understanding *what-and-why* earthquakes occur and more clearly define the steps operators and regulators should take to protect public and private interests.

References

Abercrombie, R. E. (1995). Earthquake source scaling relationships from -1 to 5 ML using seismograms recorded at 2.5-km depth. *Journal of Geophysical Research*, *100*(B12), 24015–24036.

Adachi, J., Siebrits, E., Peirce, A., & Desroches, J. (2007). Computer simulation of hydraulic fractures. *International Journal of Rock Mechanics and Mining Sciences*, *44*(5), 739–757. https://doi.org/10.1016/j.ijrmms.2006.11.006

Agarwal, K., Mayerhofer, M. J., & Warpinski, N. R. (2012). Impact of Geomechanics on Microseismicity. *SPE/EAGE European Unconventional Resources Conference and Exhibition*, (March), 20–22. https://doi.org/10.2118/152835-MS

Aguilera, R. (2016). Shale gas reservoirs: Theoretical, practical and research issues. *Petroleum Research*, *1*(1), 10–26. https://doi.org/10.1016/S2096-2495(17)30027–3

Ahmadov, R. (2011). Microtextural, elastic and transport properties of source rocks. Stanford University.

Ahmed, U. & Meehan, D. N. (eds.). (2016). *Unconventional Oil and Gas Resources: Exploitation and Development*. Boca Raton, FL: CRC Press.

Ajani, A. A. & Kelkar, M. G. (2012). Interference Study in Shale Plays. *SPE Hydraulic Fracturing Technology Conference*, (February), 6–8. https://doi.org/10.2118/151045-MS

Ajisafe, F., Solovyeva, I., Morales, A., Ejofodomi, E., & Marongiu-Porcu, M. (2017). Impact of Well Spacing and Interference on Production Performance in Unconventional Reservoirs, Permian Basin. *Unconventional Resources Technology Conference*, 24–26. https://doi.org/10.15530/urtec-2017–2690466

Aki, K., Fehler, M., & Das, S. (1977). Source mechanism of volcanic tremor: fluid-driven crack models and their application to the 1963 kilauea eruption. *Journal of Volcanology and Geothermal Research*, *2*(3), 259–287. https://doi.org/10.1016/0377–0273(77)90003–8

Aki, K. & Richards, P. G. (2002). *Quantitative Seismology* (2nd edn.). Sausalito, CA: University Science Books.

Akono, A. T. & Kabir, P. (2016). Microscopic fracture characterization of gas shale via scratch testing. *Mechanics Research Communications*, *78*, 86–92. https://doi.org/10.1016/j.mechrescom.2015.12.003

Al Alalli, A. (2018). Multiscale investigation of porosity and permeability relationship of shales with application towards hydraulic fracture fluid chemistry on shale matrix permeability. Stanford University.

Alalli, A. & Zoback, M. D. (2018). Microseismic evidence for horizontal hydraulic fractures in the Marcellus Shale, southeastern West Virginia. *Leading Edge*, *37*(5). https://doi.org/10.1190/tle37050356.1

Aljamaan, H. (2015). Multi-Component Physical Sorption Investigation of Gas Shales at the Core Level. *SPE Annual Technical Conference and Exhibition*. https://doi.org/10.2118/178736-STU

Aljamaan, H., Ross, C. M., & Kovscek, A. R. (2017). Multiscale Imaging of Gas Adsorption in Shales. In *SPE Canada Unconventional Resources Conference*. https://doi.org/SPE-185054-MS

Allmann, B. P. & Shearer, P. M. (2009). Global variations of stress drop for moderate to large earthquakes. *Journal of Geophysical Research: Solid Earth*, *114*(1), 1–22. https://doi.org/10.1029/2008JB005821

Alnoaimi, K. R., Duchateau, C., & Kovscek, A. R. (2015). Characterization and measurement of multiscale gas transport in shale core samples. *SPE Journal*, 1–16. https://doi.org/10.15530/urtec-2014–1920820

Alt, R. C. & Zoback, M. D. (2017). In situ stress and active faulting in Oklahoma. *Bulletin of the Seismological Society of America*, *107*(1), 1–13. https://doi.org/10.1785/0120160156

Alvarez, R. A., Alvarez, R. A., Zavala-Araiza, D., Lyon, D. R., Allen, D. T., Barkley, Z. R., . . . Pacala, S. W. (2018). Assessment of methane emissions from the U.S. oil and gas supply chain, Science, *361*(6398), 186–188. https://doi.org/10.1126/science.aar7204

Alzate, J. H. (2012). Integration of surface seismic, micro- seismic, and production logs for shale gas characterization: Methodology and field application. The University of Oklahoma.

Amer, A., Primio, R., & Ondrak, R. (2013). The impact of heat flow variations in shale gas evaluation: A Haynesville Shale case study. *Society of Petroleum Engineers*. https://doi.org/10.2118/164347-MS

An, C., Guo, X., & Killough, J. (2018). Impacts of Kerogen and Clay on Stress-Dependent Permeability Measurements of Shale Reservoirs. *SPE/AAPG/SEG Unconventional Resources Technology Conference*. https://doi.org/10.15530/urtec-2018–2902756

Anderson, E. M. (1951). *The Dynamics of Faulting and Dyke Formation with Applications to Britain*. Edinburgh: Oliver and Boyd.

Angelier, J. (1990). Inversion of field data in fault tectonics to obtain the regional stress – III. A new rapid direct inversion method by analytical means. *Geophysical Journal International*, *103*, 363–376.

Anovitz, L. M. & Cole, D. R. (2015). Characterization and Analysis of Porosity and Pore Structures. *Reviews in Mineralogy and Geochemistry*, *80*, 61–164.

Anovitz, L. M., Cole, D. R., Sheets, J. M., Swift, A., Elston, H. W., Welch, S., . . . Wasbrough, M. J. (2015). Effects of maturation on multiscale (nanometer to millimeter) porosity in the Eagle Ford Shale. *Interpretation*, *3*(3), SU59–SU70. https://doi.org/10.1190/INT-2014–0280.1

Aplin, A. C. & Macquaker, J. H. S. (2011). Mudstone diversity: Origin and implications for source, seal, and reservoir properties in petroleum systems. *AAPG Bulletin*, *95*(12), 2031–2059. https://doi.org/10.1306/03281110162

Araujo, O., Lopez-Bonetti, E., Garza, D., & Salinas, G. (2014). Successful Extended Injection Test for Obtaining Reservoir Data in a Gas-Oil Shale Formation in Mexico, SPE 169345-MS, Society of Petroleum Engineers. doi:10.2118/169345-MS

Arch, J. & Maltman, A. (1990). Anisotropic permeability and tormosity in deformed wet sediments structure in accrefionary tectonic features fluids flow from accrefionary within D6collement Project (DSDP)/ ODP drill holes which have penetrated major so the examination of microscopic s. *Journal of Geophysical Research*, *95*(89), 9035–9045. https://doi.org/10.1029/JB095iB06p09035

Arnold, R. & Townend, J. (2007). A Bayesian approach to estimating tectonic stress from seismological data. *Geophysical Journal International*, *170*(3), 1336–1356. https://doi.org/10.1111/j.1365-246X.2007.03485.x

Armitage, P. J., Faulkner, D. R., Worden, R. H., Aplin, A. C., Butcher, A. R., & Iliffe, J. (2011). Experimental measurement of, and controls on, permeability and permeability anisotropy of caprocks from the CO_2 storage project at the Krechba Field, Algeria. *Journal of Geophysical Research: Solid Earth*, *116*(12). https://doi.org/10.1029/2011JB008385

Ashby, M. F. & Sammis, C. G. (1990). The damage mechanics of brittle solids in compression. *Pure and Applied Geophysics*, *133*(3), 489–521. https://doi.org/10.1007/BF00878002

Atkinson, G. M., Eaton, D. W., Ghofrani, H., Walker, D., Cheadle, B., Schultz, R., . . . Kao, H. (2016). Hydraulic fracturing and seismicity in the Western Canada sedimentary basin. *Seismological Research Letters*, *87*(3), 631–647. https://doi.org/10.1785/0220150263

Aybar, U., Yu, W., Eshkalak, M. O., Sepehrnoori, K., & Patzek, T. (2015). Evaluation of production losses from unconventional shale reservoirs. *Journal of Natural Gas Science and Engineering*, *23*, 509–516. https://doi.org/10.1016/j.jngse.2015.02.030

Backeberg, N. R., Iacoviello, F., Rittner, M., Mitchell, T. M., Jones, A. P., Day, R., . . . Striolo, A. (2017). Quantifying the anisotropy and tortuosity of permeable pathways in clay-rich mudstones using models based on X-ray tomography. *Scientific Reports*, *7*(1), 1–12. https://doi.org/10.1038/s41598-017-14810-1

Baihly, J. D., Altman, R. M., Malpani, R., & Luo, F. (2010). Shale Gas Production Decline Trend Comparison over Time and Basins. *SPE Annual Technical Conference and Exhibition*. https://doi.org/10.2118/135555-MS

Bakulin, A., Grechka, V., & Tsvankin, I. (2000). Estimation of fracture parameters from reflection seismic data—Part I: HTI model due to a single fracture set. *Geophysics*, *65*(6), 1788. https://doi.org/10.1190/1.1444863

Bandyopadhyay, K. (2009). Seismic anisotropy: Geological causes and its implications to reservoir geophysics. Stanford University.

Bao, X. & Eaton, D. W. (2016). Fault activation by hydraulic fracturing in western Canada. *Science*, *354*(6318), 1406–1409. https://doi.org/10.1126/science.aag2583

Barbour, A. J., Norbeck, J. H., & Rubinstein, J. L. (2017). The effects of varying injection rates in Osage County, Oklahoma, on the 2016 Mw 5.8 Pawnee earthquake. *Seismological Research Letters*, *88*(4), 1040–1053. https://doi.org/10.1785/0220170003

Bardainne, T. (2011). Semi-Automatic Migration Location of Microseismic Events: A Good Trade-off between Efficiency and Reliability. In *Third Passive Seismic Workshop – Actively Passive!* (p. PSP09).

Barkley, Z. R., Lauvaux, T., Davis, K. J., Deng, A., Cao, Y., Sweeney, C., . . . Maasakkers, J. D. (2017). Quantifying methane emissions from natural gas production in northeastern Pennsylvania. *Atmospheric Chemistry and Physics Discussions*, 1–53. https://doi.org/10.5194/acp-2017-200

Baron, L. I., Longuntsov, B. M., & Pozin, E. Z. (1962). *Determination of Properties of Rocks*. Moscow: Gosgortekhizdat.

Barree, R. D., Barree, V. L., & Craig, D. P. (2007). Holistic Fracture Diagnostics. *SPE Rocky Oil & Gas Technology Symposium*. https://doi.org/10.2118/107877-PA

Barton, C. A., Zoback, M. D., & Moos, D. (1995). Fluid flow along potentially active faults in crystalline rock. *Geology*, *23*, 683–686.

Barton, C., Moos, D., & Tezuka, K. (2009). Geomechanical wellbore imaging: Implications for reservoir fracture permeability. *AAPG Bulletin*, *93*(11), 1551–1569. https://doi.org/10.1306/06180909030

Belyadi, H., Yuyi, J., Ahmad, M., Wyatt, J., & Energy, C. (2016). Deep Dry Utica Well Spacing Analysis with Case Study Dry Utica Rock Properties / Geological Overview. *SPE Eastern Regional Meeting*, (September), 13–15.

Bertier, P., Schweinar, K., Stanjek, H., Ghanizadeh, A., Clarkson, C. R., Busch, A., . . . Pipich, V. (2016). On the use and abuse of N2 physisorption for the characterization of the pore structure of shales. In *The Clay Minerals Society Workshop Lectures Series* (Vol. 21, pp. 151–161). https://doi.org/10.1346/CMS-WLS-21.12

Bhandari, A. R., Flemings, P. B., Polito, P. J., Cronin, M. B., & Bryant, S. L. (2015). Anisotropy and stress dependence of permeability in the Barnett Shale. *Transport in Porous Media, 108*(2), 393–411. https://doi.org/10.1007/s11242-015-0482-0

Biot, M. A. (1941). General theory of three-dimensional consolidation. *Journal of Applied Physics, 12*(2), 155–164.

Bishop, A. (1967). Progressive Failure with Special Reference to the Mechanism Causing It. In *Proc. Geotech. Conf.* (pp. 142–150). Oslo.

Bjørlykke, K. (1998). Clay mineral diagenesis in sedimentary basins – a key to the prediction of rock properties. Examples from the North Sea Basin. *Clay Minerals, 33*(1), 15–34. https://doi.org/10.1180/000985598545390

Blackwell, D. D., Richards, M. C., Frone, Z. S., Batir, J. F., Williams, M. A., Ruzo, A. A., & Dingwall, R. K. (2011). *SMU Geothermal Laboratory Heat Flow Map of the Conterminous United States, 2011.*

Boness, N. L. & Zoback, M. D. (2006). A multiscale study of the mechanisms controlling shear velocity anisotropy in the San Andreas Fault Observatory at Depth. *Geophysics, 71*(5). https://doi.org/10.1190/1.2231107

Bonnelye, A., Schubnel, A., David, C., Henry, P., Guglielmi, Y., Gout, C., . . . Dick, P. (2016a). Elastic wave velocity evolution of shales deformed under uppermost-crustal conditions. *Journal of Geophysical Research: Solid Earth, 122*(1), 1–45. https://doi.org/10.1002/2016JB013540

Bonnelye, A., Schubnel, A., David, C., Henry, P., Guglielmi, Y., Gout, C., . . . Dick, P. (2016b). Strength anisotropy of shales deformed under uppermost crustal conditions. *Journal of Geophysical Research: Solid Earth, 122*(1), 110–129. https://doi.org/10.1002/2016JB013040

Bott, M. H. P. (1959). The mechanics of oblique-slip faulting. *Geological Magazine, 96*, 109–117.

Bowker, K. (2007). Barnett Shale gas production, Fort Worth Basin: Issues and discussion, *AAPG Bulletin, 91*(4), 523–533.

Brace, W. F., Walsh, J. B., & Frangos, W. T. (1968). Permeability of granite under high pressure. *Journal of Geophysical Research, 73*(6), 2225–2236. https://doi.org/10.1029/JB073i006p02225

Brandt, A. R., Heath, G. A., & Cooley, D. (2016). Methane leaks from natural gas systems follow extreme distributions. *Environmental Science and Technology, 50*(22), 12512–12520. https://doi.org/10.1021/acs.est.6b04303

Brannon, H. D. & Bell, C. E. (2011). Eliminating Slickwater Compromises for Improved Shale Stimulation. In *SPE Annual Technical Conference and Exhibition* (Vol. 6, pp. 4578–4588). Retrieved from www.scopus.com/inward/record.url?eid=2-s2.0-84856719865&partnerID=40&md5=6fb0c8cbf9a6b1b0869baf1782970a12

Bratovich, M. W. & Walles, F. (2016). Formation evaluation and reservoir characterization of source rock reservoirs. In U. Ahmed & D. N. Meehan (eds.), *Unconventional Oil and Gas Resources: Exploitation and Development.* Boca Raton, FL: CRC Press.

Britt, L. K., Rock, B., Smith, M. B., & Klein, H. H. (2016). Production Benefits from Complexity – Effects of Rock Fabric, Managed Drawdown, and Propped Fracture Conductivity.

Burton, W. A. (2016). Multistage completion systems for unconventionals. In U. Ahmed & D. N. Meehan (eds.), *Unconventional Oil and Gas Resources: Exploitation and Development.* Boca Raton, FL: CRC Press.

Busch, A., Schweinar, K., Kampman, N., Coorn, A., Pipich, V., Feoktystov, A., . . . Bertier, P. (2017). Determining the porosity of mudrocks using methodological pluralism. *Geological Society, London, Special Publications*, *454*(1), 15–38. https://doi.org/10.1144/SP454.1

Byerlee, J. (1978). Friction of rocks. *Pure and Applied Geophysics*, *116*(4–5), 615–626. https://doi.org/10.1007/BF00876528

Caffagni, E., Eaton, D., van der Baan, M., & Jones, J. P. (2015). Regional seismicity: A potential pitfall for identification of long-period long-duration events. *Geophysics*, *80*(1), A1–A5. https://doi.org/10.1190/geo2014-0382.1

Caine, J. S., Evans, J. P., & Forster, C. B. (1996). Fault zone architecture and permeability structure, *Geology*, *24*(11), 1025–1028.

Candela, T., Renard, F., Klinger, Y., Mair, K., Schmittbuhl, J., & Brodsky, E. E. (2012). Roughness of fault surfaces over nine decades of length scales. *Journal of Geophysical Research: Solid Earth*, *117*(8), 1–30. https://doi.org/10.1029/2011JB009041

Cander, H. (2012). Sweet spots in shale gas and liquid plays: prediction of fluid composition and reservoir pressures. *Search and Discovery*, 40936, 29pp.

Cappa, F. & Rutqvist, J. (2011). Modeling of coupled deformation and permeability evolution during fault reactivation induced by deep underground injection of CO_2. *International Journal of Greenhouse Gas Control*, *5*(2), 336–346. https://doi.org/10.1016/j.ijggc.2010.08.005

Carey, J. W. (2013). Geochemistry of wellbore integrity in CO_2 sequestration: Portland cement-steel-brine-CO_2 interactions. *Reviews in Mineralogy and Geochemistry*, *77*(1), 505–539. https://doi.org/10.2138/rmg.2013.77.15

Chae, K. S. & Lee, J. W. (2015). Risk Analysis and Simulation for Geologic Storage of CO_2. In *Proceedings of the World Congress on Advances in Civil*, Environmental, *and Materials Research, Incheon*, Korea, *25–29 August 2015*.

Chalmers, G. R., Bustin, R. M., & Power, I. M. (2012a). Characterization of gas shale pore systems by porosimetry, pycnometry, surface area, and field emission scanning electron microscopy/transmission electron microscopy image analyses: Examples from the Barnett, Woodford, Haynesville, Marcellus, and Doig uni. *AAPG Bulletin*, *96*(6), 1099–1119. https://doi.org/10.1306/10171111052

Chalmers, G. R. L., Ross, D. J. K., & Bustin, R. M. (2012b). Geological controls on matrix permeability of Devonian Gas Shales in the Horn River and Liard basins, northeastern British Columbia, Canada. *International Journal of Coal Geology*, *103*, 120–131. https://doi.org/10.1016/j.coal.2012.05.006

Chambers, K., Kendall, J. M., Brandsberg-Dahl, S., & Rueda, J. (2010). Testing the ability of surface arrays to monitor microseismic activity. *Geophysical Prospecting*, *58*(5), 821–830. https://doi.org/10.1111/j.1365–2478.2010.00893.x

Chan, A. & Zoback, M. D. (2002). Deformation Analysis in Reservoir Space (DARS): A Simple Formalism for Prediction of Reservoir Deformation with Depletion – SPE 78174. In *SPE/ISRM Rock Mechanics Conference*. Irving, TX: Society of Petroleum Engineers.

Chandler, M. R., Meredith, P. G., Brantut, N., & Crawford, B. R. (2016). Fracture toughness anisotropy in shale. *Journal of Geophysical Research: Solid Earth*, *121*(3), 1706–1729. https://doi.org/10.1002/2015JB012756

Chang, C., Zoback, M. D., & Khaksar, A. (2006). Empirical relations between rock strength and physical properties in sedimentary rocks. *Journal of Petroleum Science and Engineering*, *51*(3–4), 223–237. https://doi.org/10.1016/j.petrol.2006.01.003

Chang, C. & Zoback, M. D. (2009). Viscous creep in room-dried unconsolidated Gulf of Mexico shale (I): Experimental results. *Journal of Petroleum Science and Engineering*, *69*(3–4). https://doi.org/10.1016/j.petrol.2009.08.018

Chang, K. W. & Segall, P. (2016). Injection-induced seismicity on basement faults including poroelastic stressing. *Journal of Geophysical Research: Solid Earth*, *121*(4), 2708–2726. https://doi.org/10.1002/2015JB012561

Charlez, P. & Baylocq, P. (2015). *The Shale Oil and Gas Debate*. Paris, France: Technip.

Chen, Z., Zhou, L., Walsh, R., & Zoback, M. (2018). Case study Fault slip and Casing Deformation Induced by Hydraulic Fracturing in Sichuan Basin – URTeC-2882313 20180501. In *URTeC 2882313*.

Cheng, Y. (2012). Impact of Water Dynamics in Fractures on the Performance of Hydraulically Fractured Wells in Gas Shale Reservoirs. *SPE International Symposium and Exhibition on Formation Damage Control*, (May 2011), 10–12. https://doi.org/10.2118/127863-MS

Chester, F. M., Evans, J. P., & Biegel, R. L. (1993). Internal structure and weakening mechanisms of the San Andreas fault. *Journal of Geophysical Research*, *98*, 771–786.

Chester, F. M. & Logan, J. M. (1986). Implications for mechanical properties of brittle faults from observations of the Punchbowl fault zone, California. *Pure and Applied Geophysics*, *124*, 79–106.

Chiaramonte, L., Zoback, M. D., Friedmann, J., & Stamp, V. (2008). Seal integrity and feasibility of CO_2 sequestration in the Teapot Dome EOR pilot: Geomechanical site characterization. *Environmental Geology*, *54*(8). https://doi.org/10.1007/s00254-007–0948-7

Cho, Y., Ozkan, E., & Apaydin, O. G. (2013). Pressure-dependent natural-fracture permeability in shale and its effect on shale-gas well production. *SPE Reservoir Evaluation & Engineering*, *16*(02), 216–228. https://doi.org/10.2118/159801-PA

Chong, K. K., Grieser, W. V., Passman, A., Tamayo, C. H., Modeland, N., & Burke, B. (2010). A Completions Roadmap to Shale-Play Development: A Review of Successful Approaches toward Shale-Play Stimulation in the Last Two Decades. *International Oil and Gas Conference and Exhibition in China*, 1–27. https://doi.org/10.2118/130369-MS

Ciezobka, J., Courtier, J., & Wicker, J. (2018). Results, 1–9. https://doi.org/10.15530/urtec-2018–2902355

Cipolla, C. & Wallace, J. (2014). Stimulated Reservoir Volume: A Misapplied Concept? *SPE Hydraulic Fracturing Technology Conference*, (February), 4–6. https://doi.org/10.2118/168596-MS

Cipolla, C., Motiee, M., Kechemir, A., & Corporation, H. (2018a). Integrating Microseismic, Geomechanics, Hydraulic Fracture Modeling, and Reservoir Simulation to Characterize Parent Well Depletion and Infill Well Performance in the Bakken, 1–12. https://doi.org/10.15530/urtec-2018–2899721

Cipolla, C., Gilbert, C., Sharma, A., & Lebas, J. (2018b). Case History of Completion Optimization in the Utica. *SPE Hydraulic Fracturing Technology Conference and Exhibition*, (January), 23–25.

Cipolla, C. L., Warpinski, N. R., Mayerhofer, M. J., Lolon, E. P., & Vincent, M. C. (2008). The Relationship between Fracture Complexity, Reservoir Properties, and Fracture Treatment Design. In *2008 SPE Annual Technical Conference and Exhibition, Denver, CO. 11–24 September 2008* (Vol. 1). Society of Petroleum Engineers. Retrieved from http://www.scopus.com/inward/record.url?eid=2-s2.0–79952850661&partnerID=40&md5=0139cfc6cf0b922340aca989490b6fe8

Clarke, H., Eisner, L., Styles, P., & Turner, P. (2014). Felt seismicity associated with shale gas hydraulic fracturing: The first documented example in Europe. *Geophysical Research Letters*, *41*(23), 8308–8314. https://doi.org/10.1002/2014GL062047

Clarkson, C. R., Solano, N., Bustin, R. M., Bustin, A. M. M., Chalmers, G. R. L., He, L., . . . Blach, T. P. (2013). Pore structure characterization of North American shale gas reservoirs using USANS/SANS, gas adsorption, and mercury intrusion. *Fuel*, *103*, 606–616. https://doi.org/10.1016/j.fuel.2012.06.119

Clerc, F., Harrington, R. M., Liu, Y., & Gu, Y. J. (2016). Stress drop estimates and hypocenter relocations of induced seismicity near Crooked Lake, Alberta. *Geophysical Research Letters*, *43*(13), 6942–6951. https://doi.org/10.1002/2016GL069800

Coates, D. F. & Parsons, R. C. (1966). Experimental criteria for classification of rock substances. *International Journal of Rock Mechanics and Mining Sciences*, *3*(3), 181–189. https://doi.org/10.1016/0148–9062(66)90022–2

Cocco, M., Tinti, E., & Cirella, A. (2016). On the scale dependence of earthquake stress drop. *Journal of Seismology*. https://doi.org/10.1007/s10950-016–9594-4

Council, N. R. (2013). *Induced Seismicity Potential in Energy Technologies*. Washington D.C.: National Academies Press.

Craig, D. P., Barree, R. D., Warpinski, N. R., & Blasingame, T. A. (2017). Fracture Closure Stress: Reexamining Field and Laboratory Experiments of Fracture Closure Using Modern Interpretation Methodologies, SPE187038-MS. In *SPE Annual Technical Conference and Exhibition, 9–11 October, San Antonio, Texas, USA*. Society of Petroleum Engineers. https://doi.org/10.2118/187038-MS

Crain, E. R. (2010). *Crain's Data Acquisition*.

Cramer, D. D. & Nguyen, D. H. (2013).Diagnostic Fracture Injection Testing Tactics in Unconventional Reservoirs, SPE 163863. Society of Petroleum Engineers. doi:10.2118/163863-MS

Crandall, D., Moore, J., Gill, M., & Stadelman, M. (2017). CT scanning and flow measurements of shale fractures after multiple shearing events. *International Journal of Rock Mechanics and Mining Sciences*, *100* (November 2016), 177–187. https://doi.org/10.1016/j.ijrmms.2017.10.016

Crawford, B. R., Faulkner, D. R., & Rutter, E. H. (2008). Strength, porosity, and permeability development during hydrostatic and shear loading of synthetic quartz-clay fault gouge. *Journal of Geophysical Research: Solid Earth*, *113*(3), 1–14. https://doi.org/10.1029/2006JB004634

Crone, A. J. & Luza, K. V. (1990). Style and timing of Holocene surface faulting on the Meers fault, southwestern Oklahoma. *Geological Society of America Bulletin*, *102*, 1–17.

Cui, X., Bustin, A. M. M., & Bustin, R. M. (2009). Measurements of gas permeability and diffusivity of tight reservoir rocks: Different approaches and their applications. *Geofluids*, *9* (3), 208–223. https://doi.org/10.1111/j.1468–8123.2009.00244.x

Curtis, C. D. (1980). Diagenetic alteration in black shales. *Journal of the Geological Society*, *137* (2), 189–194. https://doi.org/10.1144/gsjgs.137.2.0189

Curtis, J. B. (2002). Fractured shale-gas systems. *AAPG Bulletin*, *86*(11), 1921–1938. https://doi.org/10.1306/61EEDDBE-173E-11D7-8645000102C1865D

Curtis, M. E., Ambrose, R. J., Sondergeld, C. H., & Rai, C. S. (2011). Investigation of the Relationship between Organic Porosity and Thermal Maturity in the Marcellus Shale. *SPE Conference, SPE 114370*. https://doi.org/10.2118/144370-ms

Curtis, M. E., Cardott, B. J., Sondergeld, C. H., & Rai, C. S. (2012a). Development of organic porosity in the Woodford Shale with increasing thermal maturity. *International Journal of Coal Geology*, *103*, 26–31. https://doi.org/10.1016/j.coal.2012.08.004

Curtis, M. E., Sondergeld, C. H., Ambrose, R. J., & Rai, C. S. (2012b). Microstructural investigation of gas shales in two and three dimensions using nanometer-scale resolution imaging. *AAPG Bulletin*, *96*(4), 665–677. https://doi.org/10.1306/08151110188

Daigle, H. & Dugan, B. (2011). Permeability anisotropy and fabric development: A mechanistic explanation. *Water Resources Research*, *47*(12), 1–11. https://doi.org/10.1029/2011WR011110

Daniels, J., Waters, G., Le Calvez, J., Bentley, D., & Lassek, J. (2007). Contacting More of the Barnett Shale Through an Integration of Real-Time Microseismic Monitoring, Petrophysics, and Hydraulic Fracture Design. *Proceedings of SPE Annual Technical Conference and Exhibition*. https://doi.org/10.2118/110562-MS

Darold, A. P. & Holland, A. A. (2015). *Preliminary Oklahoma Optimal Fault Orientations. Oklahoma Open File Report oF 4–2015.*

Das, I. & Zoback, M. D. (2013a). Long-period, long-duration seismic events during hydraulic stimulation of shale and tight-gas reservoirs – Part 1: Waveform characteristics. *Geophysics*, *78*(6). https://doi.org/10.1190/GEO2013-0164.1

Das, I. & Zoback, M. D. (2013b). Long-period long-duration seismic events during hydraulic stimulation of shale and tight-gas reservoirs – Part 2: Location and mechanisms. *Geophysics*, *78*(6). https://doi.org/10.1190/GEO2013-0165.1

Day, S., Sakurovs, R., & Weir, S. (2008). Supercritical gas sorption on moist coals. *International Journal of Coal Geology*, *74*(3–4), 203–214. https://doi.org/10.1016/j.coal.2008.01.003

Detournay, E. (2004). Propagation regimes of fluid-driven fractures in impermeable rocks. *International Journal of Geomechanics*, *4*(1), 35–45. https://doi.org/10.1061/(ASCE)1532–3641(2004)4:1(35)

Detournay, E. (2016). Mechanics of hydraulic fractures. *Annual Review of Fluid Mechanics*, *48*(1), 311–339. https://doi.org/10.1146/annurev-fluid-010814–014736

Deutch, J., Holditch, S., Krupp, F., McGinty, K., Tierney, S., Yergin, D., & Zoback, M. D. (2011). *Shale Gas Production Subcommittee 90-Day Report.*

Dieterich, J. H. (1978). Time-dependent friction and the mechanics of stick-slip. In *Rock Friction and Earthquake Prediction* (pp. 790–806). Basel: Birkhäuser Basel. https://doi.org/10.1007/978–3-0348–7182-2_15

Dieterich, J. H. (1979). Modeling of rock friction 1. Experimental results and constitutive equations. *Journal of Geophysical Research*, *84*, 2161–2168.

Dieterich, J. H. & Kilgore, B. D. (1994). Direct observation of frictional contacts: New insights for state-dependent properties. *Pure and Applied Geophysics*, *143*(1–3), 283–302. https://doi.org/10.1007/BF00874332

Dohmen, T., Zhang, J., Barker, L., & Blangy, J. P. (2017). Microseismic magnitudes and b-values for delineating hydraulic fracturing and depletion. *SPE Journal*, *22*(5), 1–11. https://doi.org/10.2118/186096-PA

Dow, W. G. (1977). Kerogen studies and geological interpretations. *Journal of Geochemical Exploration*, *7*(C), 79–99. https://doi.org/10.1016/0375-6742(77)90078-4

Dunham, E. M., Belanger, D., Cong, L., & Kozdon, J. E. (2011). Earthquake ruptures with strongly rate-weakening friction and off-fault plasticity, part 2: Nonplanar faults. *Bulletin of the Seismological Society of America*, *101*(5), 2308–2322. https://doi.org/10.1785/0120100076

Eaton, D. W. (2018). *Passive Seismic Monitoring of Induced Seismicity*. Cambridge University Press.

Eaton, D. W. & Forouhideh, F. (2011). Solid angles and the impact of receiver-array geometry on microseismic moment-tensor inversion. *Geophysics*, *76*(6), WC77–WC85. https://doi.org/10.1190/geo2011-0077.1

Eaton, D. W. & Schultz, R. (2018). Increased likelihood of induced seismicity in highly over-pressured shale formations, *Geophysical Journal International*, *214*, 751–757.

Eaton, D. W., Igonin, N., Poulin, A., Weir, R., Zhang, H., Pellegrino, S., & Rodriquez, G. (2018). Tony Creek Dual Microseismic Experiment (ToC2ME). In *GeoConvention 2018*. Calgary, Canada.

Economides, M. J. & Nolte, K. (2000). *Reservoir Stimulation*. Wiley.

Edwards, R. W. J. & Celia, M. A. (2018). Shale gas well, hydraulic fracturing, and formation data to support modeling of gas and water flow in shale formations. *Water Resources Research*, 1–11. https://doi.org/10.1002/2017WR022130

EIA (2013). *Technically Recoverable Shale Oil and Shale Gas Resources: An Assessment of 137 Shale Formations in 41 Countries Outside the Unites States* (June).

Eisner, L., Fischer, T., & Rutledge, J. T. (2009). Determination of S-wave slowness from a linear array of borehole receivers. *Geophysical Journal International*, *176*(1), 31–39.

Eisner, L., Gei, D., Hallo, M., Opršal, I., & Ali, M. Y. (2013). The peak frequency of direct waves for microseismic events. *Geophysics*, *78*(6), A45–A49. https://doi.org/10.1190/geo2013-0197.1

Eisner, L., Hulsey, B. J., Duncan, P. M., Jurick, D., Werner, H., & Keller, W. (2010). Comparison of surface and borehole locations of induced seismicity. *Geophysical Prospecting*, *58*(5), 809–820. https://doi.org/10.1111/j.1365–2478.2010.00867.x https://doi.org/10.1111/j.1365-246X.2008.03939.x

Ellsworth, W. L. (2013). Injection-induced earthquakes. *Science*, *341* (1225942 12 July 2013). https://doi.org/10.1126/science.1225942

Engelder, T. & Fischer, M. P. (1994). Influence of poroelastic behavior on the magnitude of minimum horizontal stress, Sh, in overpressured parts of sedimentary basins. *Geology*, *22*(10), 949–952. https://doi.org/10.1130/0091–7613(1994)022<0949:IOPBOT>2.3.CO

Engelder, T., Lash, G. G., & Uzcátegui, R. S. (2009). Joint sets that enhance production from Middle and Upper Devonian gas shales of the Appalachian Basin. *AAPG Bulletin*, *93*(7), 857–889. https://doi.org/10.1306/03230908032

Engle, M. A., Reyes, F. R., Varonka, M. S., Orem, W. H., Ma, L., Ianno, A. J., . . . Carroll, K. C. (2016). Geochemistry of formation waters from the Wolfcamp and "Cline" shales: Insights into brine origin, reservoir connectivity, and fluid flow in the Permian Basin, USA. *Chemical Geology*, *425*, 76–92. https://doi.org/10.1016/j.chemgeo.2016.01.025

English, J. M., English, K. L., Corcoran, D. V. & Toussaint, F. (2016) Exhumation charge: The last gasp of a petroleum source rock and implications for unconventional shah resources. *AAPG Bulletin*, *100* (1), pp. 1–16. doi: 10.1306/07271514224.

Environmental Protection Agency (2015). *Assessment of the Potential Impacts of Hydraulic Fracturing for Oil and Gas on Drinking Water Resources: Executive Summary*, (June), 1–3. Retrieved from http://www2.epa.gov/sites/production/files/2015–07/documents/hf_es_erd_jun2015.pdf

EPA (2014). *Minimizing and Managing Potential Impacts of Injection-Induced Seismicity from Class II Disposal Wells: Practical Approaches*.

EPA (2016). *Hydraulic Fracturing for Oil and Gas: Impacts from the Hydraulic Fracturing Water Cycle on Drinking Water Resources in the United States*. Retrieved from www.epa.gov/hfstudy

Esaki, T., Du, S., Mitani, Y., Ikusada, K., & Jing, L. (1999). Development of a shear-flow test apparatus and determination of coupled properties for a single rock joint. *International Journal of Rock Mechanics and Mining Sciences*, *36*(5), 641–650. https://doi.org/10.1016/S0148-9062 (99)00044–3

Eyre, T. S. & van der Baan, M. (2017). The reliability of microseismic moment tensor solutions: Surface versus borehole monitoring. *Geophysics*, *82*(6), 1–46. https://doi.org/10.1190/geo2017-0056.1

Fan, L., Harris, B., & Jamaluddin, A. (2005). Understanding gas-condensate reservoirs. *Oilfield Review*, (Winter 2005/2006), 14–27. https://doi.org/http://dx.doi.org/10.4043/25710-MS

Fan, Z., Eichhubl, P., & Gale, J. F. W. (2016). Geomechanical analysis of fluid injection and seismic fault slip for the Mw4.8 Timpson, Texas, earthquake sequence. *Journal of Geophysical Research: Solid Earth*, *121*, 2798–2812. https://doi.org/10.1002/2016JB012821.Received

Fang, Y., Elsworth, D., Wang, C., Ishibashi, T., & Fitts, J. P. (2017). Frictional stability-permeability relationships for fractures in shales. *Journal of Geophysical Research: Solid Earth*, *122*(3), 1760–1776. https://doi.org/10.1002/2016JB013435

Farghal, N. (2018). Fault and fracture identification and characterization in 3D seismic data from unconventional reservoirs, PhD Thesis, Stanford University.

Farghal, N. S. & Zoback, M. D. (2015). Identification of slowly slipping faults in the Barnett Shale utilizing ant tracking; Identification of slowly slipping faults in the Barnett Shale utilizing ant tracking. In *SEG Technical Program Expanded Abstracts* (Vol. 34), 4919–4923. https://doi.org/10.1190/segam2015-5811224.1

Faulkner, D. R. & Rutter, E. H. (1998). The gas permeability of clay-bearing fault gouge at 20 C. *Geological Society, London, Special Publications*, *147*(1), 147–156. https://doi.org/10.1144/GSL.SP.1998.147.01.10

Faulkner, D. R., Lewis, A. C., & Rutter, E. H. (2003). On the internal structure and mechanics of large strike-slip fault zones: Field observations of the Carboneras fault in southeastern Spain. *Tectonophysics*, *367*(3–4), 235–251. https://doi.org/10.1016/S0040-1951(03)00134–3

Faulkner, D. R., Mitchell, T. M., Healy, D., & Heap, M. J. (2006). Slip on "weak" faults by the rotation of regional stress in the fracture damage zone. *Nature*, *444*(7121), 922–925. https://doi.org/10.1038/nature05353

Ferguson, W., Richards, G., Bere, A., Mutlu, U., & Paw, F. (2018). Modelling Near-Wellbore Hydraulic Fracture Branching, Complexity and Tortuosity: A Case Study Based on a Fully Coupled Geomechanical Modelling Approach. *SPE Hydraulic Fracturing Technology Conference and Exhibition*. https://doi.org/10.2118/189890-MS

Fisher, M. K. & Warpinski, N. R. (2012). Hydraulic-Fracture-Height Growth: Real Data. *SPE Production & Operations*. https://doi.org/10.2118/145949-PA

Fisher, M. K., Heinze, J. R., Harris, C. D., Davidson, B. M., Wright, C. A., & Dunn, K. P. (2004). Optimizing Horizontal Completion Techniques in the Barnett Shale Using Microseismic Fracture Mapping. *SPE Annual Technical Conference and Exhibition*. https://doi.org/10.2118/90051-MS

Fisher, M. K., Wright, C. A., Davidson, B. M., Goodwin, A. K., Fielder, E. O., Buckler, W. S., & Steinsberger, N. P. (2002). Integrating Fracture Mapping Technologies to Optimize Stimulations in the Barnett Shale. In *SPE Annual Technical Conference and Exhibition*. Soc. Petr. Engr. https://doi.org/10.2118/77441-MS

Fogden, A., Olson, T., Cheng, Q., Middleton, J., Kingston, A., Turner, M., . . . Armstrong, R. (2015). Dynamic Micro-CT Imaging of Diffusion in Unconventionals. *Proceedings of the Unconventional Resources Technology Conference*, 1–16. https://doi.org/10.15530/urtec-2015–2154822

Forand, D., Heesakkers, V., & Schwartz, K. (2017). Constraints on Natural Fracture and In-Situ Stress Trends of Unconventional Reservoirs in the Permian Basin, USA, 24–26. https://doi.org/10.15530/-urtec-2017–2669208

Foulger, G. R., Wilson, M. P., Gluyas, J. G., Julian, B. R., & Davies, R. J. (2018). Global review of human-induced earthquakes. *Earth-Science Reviews*, *178* (January 2017), 438–514. https://doi.org/10.1016/j.earscirev.2017.07.008

Friberg, P. A., Besana-Ostman, G. M., & Dricker, I. (2014). Characterization of an earthquake sequence triggered by hydraulic fracturing in Harrison County, Ohio. *Seismological Research Letters*, *85*(6), 1295–1307. https://doi.org/10.1785/0220140127

Friedrich, M. & Milliken, M. (2013). Determining the Contributing Reservoir Volume from Hydraulically Fractured Horizontal Wells in the Wolfcamp Formation in the Midland Basin. *Unconventional Resources Technology Conference, Denver, Colorado, 12–14 August 2013*, 1461–1468. https://doi.org/10.1190/urtec2013-149

Frohlich, C., Hayward, C., Stump, B., & Potter, E. (2011). The Dallas-Fort Worth earthquake sequence: October 2008 through May 2009. *Bulletin of the Seismological Society of America*, *101*(1), 327–340. https://doi.org/10.1785/0120100131

Frohlich, C., Ellsworth, W. L., Brown, W. A., Brunt, M., Luetgert, J., Macdonald, T., & Walter, S. (2014). The 17 May 2012M4.8 earthquake near Timpson, East Texas: An event possibly triggered by fluid injection. *Journal of Geophysical Research*, (August 1982), 581–593. https://doi.org/10.1002/2013JB010755.Received

Gaarenstroom, L., Tromp, R. A. J., Jong, M. C. de, & Brandenburg, A. M. (1993). Overpressures in the Central North Sea: Implications for trap integrity and drilling safety. In J. R. Parker (ed.), *Petroleum Geology of Northwest Europe: Proceedings of the 4th Conference* (pp. 1305–1313). London.

Gale, J. F. W., Elliott, S. J., & Laubach, S. E. (2018). Hydraulic Fractures in Core From Stimulated Reservoirs: Core Fracture Description of HFTS Slant Core, Midland Basin, West Texas, (1993). https://doi.org/10.15530/urtec-2018–2902624

Gale, J. F. W., Laubach, S. E., Olson, J. E., Eichhubl, P., & Fall, A. (2014). Natural fractures in shale: A review and new observations. *AAPG Bulletin*, *98*(11), 2165–2216.

Gale, J. F. W., Reed, R. M., & Holder, J. (2007). Natural fractures in the Barnett Shale and their importance for hydraulic fracture treatments. *AAPG Bulletin*, *91*(4), 603–622. https://doi.org/10.1306/11010606061

Gasparik, M., Ghanizadeh, A., Bertier, P., Gensterblum, Y., Bouw, S., & Krooss, B. M. (2012). High-pressure methane sorption isotherms of black shales from the Netherlands. *Energy and Fuels*, *26*(8), 4995–5004. https://doi.org/10.1021/ef300405g

Gasparik, M., Bertier, P., Gensterblum, Y., Ghanizadeh, A., Krooss, B. M., & Littke, R. (2014). Geological controls on the methane storage capacity in organic-rich shales. *International Journal of Coal Geology*, *123*, 34–51. https://doi.org/10.1016/j.coal.2013.06.010

Gdanski, R. D., Fulton, D. D., & Shen, C. (2009). Fracture-face-skin evolution during cleanup. *SPE Production & Operations*, *24*(1), 22–34. https://doi.org/10.2118/101083-pa

Geng, Z., Bonnelye, A., Chen, M., Jin, Y., Dick, P., David, C., . . . Schubnel, A. (2017). Elastic anisotropy reversal during brittle creep in shale. *Geophysical Research Letters*, *44*(21),10, 887–10, 895. https://doi.org/10.1002/2017GL074555

Gensterblum, Y., Merkel, A., Busch, A., & Krooss, B. M. (2013). High-pressure CH_4 and CO_2 sorption isotherms as a function of coal maturity and the influence of moisture. *International Journal of Coal Geology*, *118*, 45–57. https://doi.org/10.1016/j.coal.2013.07.024

Gensterblum, Y., Ghanizadeh, A., Cuss, R. J., Amann-Hildenbrand, A., Krooss, B. M., Clarkson, C. R., . . . Zoback, M. D. (2015). Gas transport and storage capacity in shale gas reservoirs – A review. Part A: Transport processes. *Journal of Unconventional Oil and Gas Resources*, *12*, 87–122. https://doi.org/10.1016/j.juogr.2015.08.001

Gephart, J. W. & Forsyth, D. W. (1984). An improved method for determining the regional stress tensor using earthquake focal mechanism data: application to the San Fernando earthquake sequence. *Journal of Geophysical Research*, *89*, 9305–9320.

Ghanbarian, B. & Javadpour, F. (2017). Upscaling pore pressure-dependent gas permeability in shales. *Journal of Geophysical Research: Solid Earth*, *122*(4), 2541–2552. https://doi.org/10.1002/2016JB013846

Ghanizadeh, A., Gasparik, M., Amann-Hildenbrand, A., Gensterblum, Y., & Krooss, B. M. (2013). Lithological controls on matrix permeability of organic-rich shales: An experimental study. *Energy Procedia*, *40*, 127–136. https://doi.org/10.1016/j.egypro.2013.08.016

Ghanizadeh, A., Amann-Hildenbrand, A., Gasparik, M., Gensterblum, Y., Krooss, B. M., & Littke, R. (2014b). Experimental study of fluid transport processes in the matrix system of the European organic-rich shales: II. Posidonia Shale (Lower Toarcian, northern Germany). *International Journal of Coal Geology*, *123*, 20–33. https://doi.org/10.1016/j.coal.2013.06.009

Gherabati, S. A., Hammes, U., Male, F., Browning, J., Smye, K., Ikonnikova, S. A., & McDaid, G. (2016). Assessment of Hydrocarbon-in-Place and Recovery Factors in the Eagle Ford Shale Play, URTeC: 2460252.

Godec, M., Koperna, G., Petrusak, R., & Oudinot, A. (2013). Potential for enhanced gas recovery and CO_2 storage in the Marcellus Shale in the Eastern United States. *International Journal of Coal Geology*, *118*, 95–104. https://doi.org/10.1016/j.coal.2013.05.007

Goertz-Allmann, B. P., Goertz, A., & Wiemer, S. (2011). Stress drop variations of induced earthquakes at the Basel geothermal site. *Geophysical Research Letters*, *38*(9), 1–5. https://doi.org/10.1029/2011GL047498

Goodway, B., Chen, T., & Downton, J. (1997). Improved AVO Fluid Detection and Lithology Discrimination Using Lame Petrophysical Parameters from P and S Inversions. *SEG Annual Meeting*. https://doi.org/10.1190/1.1885795

Goodway, B., Varsek, J., & Abaco, C. (2006). Practical applications of P-wave AVO for unconventional gas Resource Plays – Part 1: Seismic petrophysics. *CSEG Recorder, 31 Special* (March), 1–17.

Goodway, B., Perez, M., Varsek, J., & Abaco, C. (2010). Seismic petrophysics and itotropic-anisotropic AVO methods for unconventional gas exploration. *The Leading Edge, December*, 1500–1508. https://doi.org/10.1190/1.3525367

Gourjon, E. & Bertoncello, A. (2018). Impact of near Well-Bore Geology on Hydraulic Fractures Geometry and Well Productivity: A Statistical Look Back at the Utica Play. In *SPE Hydraulic Fracturing Technology Conference & Exhibition*.

Graham, S. A. & Williams, L. A. (1985). Tectonic, depositional, and diagenetic history of Monterey Formation (Miocene), Centra San Joaquin Basin, California. *American Association of Petroleum Geologists Bulletin*, *69*(3), 385–411. https://doi.org/10.1306/AD4624F7-16F7-11D7-8645000102C1865D

Grechka, V. & Heigl, W. M. (2017). *Microseismic Monitoring*. SEG.

Guo, B. (n.d.). Article in press.

Guo, Z., Chapman, M., & Li, X. (2012). Exploring the effect of fractures and microstructure on brittleness index in the Barnett Shale. In *SEG Technical Program Expanded Abstracts* (Vol. 2, pp. 1–5). https://doi.org/10.1190/segam2012-0771.1

Gupta, J. K., Zielonka, M. G., Albert, R. A., El-Rabaa, A. W., Burnham, H. A, & Choi, N. H. (2012). SPE 152224 Integrated methodology for optimizing development of unconventional gas resources. *Stress: The International Journal on the Biology of Stress*. https://doi.org/10.2118/152224-MS

Gutierrez, M. & Nyga, R. (2000). Stress-dependent permeability of a de-mineralised fracture in shale, *Marine and Petroleum Geology, 17*, 895–907.

GWPC & IOGCC (2017). *Potential Injection-Induced Seismicity Associated with Oil & Gas Development.*

Haege, M., Maxwell, S., Sonneland, L., Norton, M., & Resources, P. E. (2012). Integration of Passive Seismic and 3D Reflection Seismic in an Unconventional Shale Gas Play: Relationship between Rock Fabric and Seismic Moment of Microseismic Events. *2012 SEG Annual Meeting*, 1–5.

Hagin, P. N. & Zoback, M. D. (2004a). Viscous deformation of unconsolidated reservoir sands – Part 1: Time-dependent deformation, frequency dispersion, and attenuation. *Geophysics, 69*(3). https://doi.org/10.1190/1.1759459

Hagin, P. N. & Zoback, M. D. (2004b). Viscous deformation of unconsolidated reservoir sands – Part 2: Linear viscoelastic models. *Geophysics, 69*(3). https://doi.org/10.1190/1.1759460

Haimson, B. & C. Fairhurst (1967). Initiation and extension of hydraulic fractures in rocks. *Society of Petroleum Engineers Journal*, Sept.: 310–318.

Hajiabdolmajid, V. & Kaiser, P. (2003). Brittleness of rock and stability assessment in hard rock tunneling. *Tunnelling and Underground Space Technology, 18*(1), 35–48. https://doi.org/10.1016/S0886-7798(02)00100-1

Hakso, A. & Zoback, M. (2017). Utilizing multiplets as an independent assessment of relative microseismic location uncertainty. *The Leading Edge, 36*(10), 829–836. https://doi.org/10.1190/tle36100829.1

Hakso, A. & Zoback, M. D. (2019). The relation between stimulated shear fractures and production in the Barnett Shale: Implications for unconventional oil and gas reservoirs. *Geophysics*, in review.

Hallo, M., Oprsal, I., Eisner, L., & Ali, M. Y. (2014). Prediction of magnitude of the largest potentially induced seismic event. *Journal of Seismology, 18*(3), 421–431. https://doi.org/10.1007/s10950-014-9417-4

Harbaugh, A. W., Banta, E. R., Hill, M. C., & McDonald, M. G. (2010). *Modflow-2000, the U.S. Geological Survey Modular Groundwater Model – User Guide to Modularization Concepts and the Groundwater Flow Process.*

Healy, J. H., Rubey, W. W., Griggs, D. T., & Raleigh, C. B. (1968). The Denver earthquakes. *Science, 161*, 1301–1310.

Heller, R. & Zoback, M. (2014). Adsorption of methane and carbon dioxide on gas shale and pure mineral samples. *Journal of Unconventional Oil and Gas Resources, 8* (C). https://doi.org/10.1016/j.juogr.2014.06.001

Heller, R., Vermylen, J. P., & Zoback, M. D. (2014). Experimental investigation of matrix permeability of gas shales. *AAPG Bulletin, 98*(5), 975–995. https://doi.org/10.1306/09231313023

Hennings, P., Allwardt, P., Paul, P., Zahm, C., Reid, R., Alley, H., ... Hough, E. (2012). Relationship between fractures,faultzones, stress,andreservoir productivity in the Suban gas field, Sumatra, Indonesia. *AAPG Bulletin, 96*(4), 753–772. https://doi.org/10.1306/08161109084

Hennings, P., Lund Snee, J.-E., Osmond, J. L., Dommisse, R., DeShon, H. R., & Zoback, M. D. (2019). Slip potential of faults in the Fort Worth Basin of North-Central Texas, USA. *Science Advances*, in review.

Hill, R. (1963). Elastic properties of reinforced solids: Some theoretical principles. *Journal of the Mechanics and Physics of Solids, 11*(5), 357–372. https://doi.org/10.1016/0022-5096(63)90036-X

Ho, N. C., Peacor, D. R., & Van Der Pluijm, B. A. (1999). Preferred orientation of phyllosilicates in Gulf Coast mudstones and relation to the smectite-illite transition. *Clays and Clay Minerals*, *47*(4), 495–504. https://doi.org/10.1346/CCMN.1999.0470412

Holland, A. A. (2013). Earthquakes triggered by hydraulic fracturing in south-central Oklahoma. *Bulletin of the Seismological Society of America*, *103*(3), 1784–1792. https://doi.org/10.1785/0120120109

Hornbach, M. J., Deshon, H. R., Ellsworth, W. L., Stump, B. W., Hayward, C., Frohlich, C., . . . Luetgert, J. H. (2015). Causal factors for seismicity near Azle, Texas. *Nature Communications, 6*. https://doi.org/10.1038/ncomms7728

Hornbach, M. J., Jones, M., Scales, M., DeShon, H. R., Magnani, M. B., Frohlich, C., . . . Layton, M. (2016). Ellenburger wastewater injection and seismicity in North Texas. *Physics of the Earth and Planetary Interiors*, *261*, 54–68. https://doi.org/10.1016/j.pepi.2016.06.012

Horton, S. (2012). Disposal of hydrofracking waste fluid by injection into subsurface aquifers triggers earthquake swarm in Central Arkansas with potential for damaging earthquake. *Seismological Research Letters*, *83*(2), 250–260. https://doi.org/10.1785/gssrl.83.2.250

Hu, H., Li, A., & Zavala-Torres, R. (2017). Long-period long-duration seismic events during hydraulic fracturing: Implications for tensile fracture development. *Geophysical Research Letters*, *44*(10), 4814–4819. https://doi.org/10.1002/2017GL073582

Huang, J., Morris, J. P., Fu, P., Settgast, R. R., Sherman, C. S., & Ryerson, F. J. (2018). Hydraulic Fracture Height Growth Under the Combined Influence of Stress Barriers and Natural Fractures. *SPE Hydraulic Fracturing Technology Conference and Exhibition*. https://doi.org/10.2118/189861-MS

Huang, Y., Beroza, G. C., & Ellsworth, W. L. (2016). Stress drop estimates of potentially induced earthquakes in the Guy-Greenbrier sequence. *Journal of Geophysical Research: Solid Earth*, *121*(9), 6597–6607. https://doi.org/10.1002/2016JB013067

Hubbert, M. D. & Rubey, W. W. (1959). Role of fluid pressure in mechanics of overthrust faulting. *Geological Society of America Bulletin*, *70*, 115–205.

Hubbert, M. K. & Willis, D. G. (1957). Mechanics of hydraulic fracturing. *Petr. Trans. AIME*, *210*, 153–163.

Hucka, V. & Das, B. (1974). Brittleness determination of rocks by different methods. *International Journal of Rock Mechanics and Mining Sciences*, *11*(10), 389–392. https://doi.org/10.1016/0148–9062(74)91109–7

Hull, R. A., Leonard, P. A., & Maxwell, S. C. (2017a). Geomechanical Investigation of Microseismic Mechanisms Associated With Slip on Bed Parallel Fractures, 24–26. https://doi.org/10.15530/urtec-2017–26888667

Hull, R. A., Meek, R., Bello, H., Resources, P. N., & Miller, D. (2017b). Case History of DAS Fiber-Based Microseismic and Strain Data, Monitoring Horizontal Hydraulic Stimulations Using Various Tools to Highlight Physical Deformation Processes (Part A).

Hurd, O. (2012). Geomechanical analysis of intraplate earthquakes and earthquakes induced during stimulation of low permeability gas reservoirs. Stanford University.

Hurd, O. & Zoback, M. D. (2012a). Intraplate earthquakes, regional stress and fault mechanics in the central and eastern U.S. and Southeastern Canada. *Tectonophysics*, *581*. https://doi.org/10.1016/j.tecto.2012.04.002

Hurd, O. & Zoback, M. D. (2012b). Stimulated Shale Volume Characterization: Multiwell Case Study from the Horn River Shale: I. Geomechanics and Microseismicity. *SPE Annual Technical Conference and Exhibition, C*. https://doi.org/10.2118/159536-MS

Imanishi, K. & Ellsworth, W. L. (2006). Source scaling relationships of microearthquakes at Parkfield, CA, determined using the SAFOD pilot hole seismic array. In *Earthquakes: Radiated Energy and the Physics of Faulting* (pp. 81–90). American Geophysical Union. https://doi.org/10.1029/170GM10

Inamdar, A. A., Ogundare, T. M., Malpani, R., Atwood, W. K., Brook, K., Erwemi, A. M., & Purcell, D. (2010). Evaluation of Stimulation Techniques Using Microseismic Mapping in the Eagle Ford Shale. *Tight Gas Completions Conference*. https://doi.org/10.2118/136873-MS

Ingram, G. M. & Urai, J. L. (1999). Top-seal leakage through faults and fractures: the role of mudrock properties. *Geological Society, London, Special Publications*, *158*(1), 125–135. https://doi.org/10.1144/GSL.SP.1999.158.01.10

Ishibashi, T., Watanabe, N., Asanuma, H., & Tsuchiya, N. (2016). Linking microearthquakes to fracture permeability evolution. *Crustal Permeability*, 49–64. https://doi.org/10.1002/9781119166573.ch7

Al Ismail, M. I. (2016). Influence of nanopores on the transport of gas and gas-condensate in unconventional resources. Stanford University.

Al Ismail, M. I. & Zoback, M. D. (2016). Effects of rock mineralogy and pore structure on extremely low stress-dependent matrix permeability of unconventional shale gas and shale oil samples. *Royal Society Philosophical Transactions A*. https://doi.org/10.1098/rsta.2015.0428

Al Ismail, M. I., Hol, S., Reece, J. S., & Zoback, M. D. (2014). The Effect of CO_2 Adsorption on Permeability Anisotropy in the Eagle Ford Shale. *Unconventional Resources Technology Conference*, (1921520).

Ito, H. & Zoback, M. D. (2000). Fracture permeability and in situ stress to 7 km depth in the KTB scientific drillhole. *Geophysical Research Letters*, *27*, 1045–1048.

Jackson, R. B., Vengosh, A., Carey, J. W., Darrah, T. H., O'Sullivan, F., & Pétron, G. (2014). The environmental costs and benefits of fracking. *Ssrn*, (August), 1–36. https://doi.org/10.1146/annurev-environ-031113–144051

Jaeger, J. C. & Cook, N. G. W. (1971). *Fundamentals of Rock Mechanics* (3rd edn.). New York: Chapman and Hall.

Jarvie, D. M., Hill, R. J., Ruble, T. E., & Pollastro, R. M. (2007). Unconventional shale-gas systems: The Mississippian Barnett Shale of north-central Texas as one model for thermogenic shale-gas assessment. *AAPG Bulletin*, *91*(4), 475–499. https://doi.org/10.1306/12190606068

Jin, G. & Roy, B. (2017). Hydraulic-fracture geometry characterization using low-frequency DAS signal. *The Leading Edge*, *36*(12), 975–980. https://doi.org/10.1190/tle36120975.1

Jin, L. & Zoback, M. D. (2017). Fully coupled nonlinear fluid flow and poroelasticity in arbitrarily fractured porous media: A hybrid-dimensional computational model. *Journal of Geophysical Research*, *122*, 33. https://doi.org/10.1002/2017JB014892

Jin, L. & Zoback, M. D. (2018) Fully dynamic spontaneous rupture due to quasi-static pore pressure and poroelastic effects: An implicit nonlinear computational model of fluid-induced seismic events. *Journal of Geophysical Research: Solid Earth, 123*. https://doi.org/10.1029/2018JB015669

Jin, L. & Zoback, M. (2019). The Effects of Initial Stress State and Pore Pressure on Stress Changes Associated with Depletion. In *Hydraulic Fracturing Technology Conference*. The Woodlands, TX: Society of Petroleum Engineers.

Jin, X., Shah, S. N., Roegiers, J.-C., & Zhang, B. (2014a). Fracability evaluation in shale reservoirs – An integrated petrophysics and geomechanics approach. *SPE Journal*, (October 2015). https://doi.org/10.2118/168589-MS

Jin, X., Shah, S. N., Truax, J. A., & Roegiers, J.-C. (2014b). A Practical Petrophysical Approach for Brittleness Prediction from Porosity and Sonic Logging in Shale Reservoirs. *SPE Annual Technical Conference and Exhibition*, 18. https://doi.org/10.2118/170972-MS

Johnston, J. E. & Christensen, N. I. (1995). Seismic anisotropy of shales. *Journal of Geophysical Research*, *100*(B4), 5991–6003.

Johri, M., Dunham, E. M., Zoback, M. D., & Fang, Z. (2014a). Predicting fault damage zones by modeling dynamic rupture propagation and comparison with field observations. *Journal of Geophysical Research: Solid Earth*, *119*(2). https://doi.org/10.1002/2013JB010335

Johri, M., Zoback, M. D., & Hennings, P. (2014b). A scaling law to characterize fault-damage zones at reservoir depths. *AAPG Bulletin*, *98*(10), 2057–2079. https://doi.org/10.1306/05061413173

Jung, H., Sharma, Mukul M., Cramer, D., Oakes, S., & McClure, M. (2016). Re-examining interpretations of non-ideal behavior during diagnostic fracture injection tests, *Journal of Petroleum Science and Engineering*, *145*, 114–136, http://dx.doi.org/10.1016/j.petrol.2016.03.016

Jweda, J., Michael, E., Jokanola, O., Hofer, R., & Parisi, V. (2017). Optimizing Field Development Strategy Using Time-Lapse Geochemistry and Production Allocation in Eagle Ford – URTeC: 2671245 (pp. 24–26). Paper was prepared for presentation at the Unconventional Resources Technology Conference held in Austin, Texas, USA, 24–26 July 2017. https://doi.org/10.15530/urtec-2017–2671245

Kahn, D., Roberts, J., & Rich, J. (2017). Integrating Microseismic and Geomechanics to Interpret Hydraulic Fracture Growth. *Proceedings of the 5th Unconventional Resources Technology Conference*, (2016). https://doi.org/10.15530/urtec-2017–2697445

Kale, S., Rai, C., & Sondergeld, C. (2010). Rock Typing in Gas Shales. In *SPE Annual Technical Conference and Exhibition*. https://doi.org/10.2118/134539-MS

Kaluder, Z., Nikolaev, M., Davidenko, I., Leskin, F., Martynov, M., Shishmanidi, I., Platunov, A., Chong, K. K., Astafyev, V., Shnitiko, A., & Fedorenko, E. (2014) First High-Rate Hybrid Fracture in Em-Yoga Field, West Siberia, Russia. OTC 24712-MS, Offshore Technology Conference-Asia, http://www.onepetro.org/doi/10.4043/24712-MS

Kang, S. M., Fathi, E., Ambrose, R. J., Akkutlu, I. Y., & Sigal, R. F. (2011). Carbon dioxide storage capacity of organic-rich shales. *SPE Journal*, *16*(04), 842–855. https://doi.org/10.2118/134583-PA

Kanitpanyacharoen, W., Kets, F. B., Wenk, H. R., & Wirth, R. (2012). Mineral preferred orientation and microstructure in the Posidonia Shale in relation to different degrees of thermal maturity. *Clays and Clay Minerals*, *60*(3), 315–329. https://doi.org/10.1346/CCMN.2012.0600308

Kanitpanyacharoen, W., Wenk, H.-R., Kets, F., Lehr, C., & Wirth, R. (2011). Texture and anisotropy analysis of Qusaiba shales. *Geophysical Prospecting*, *59*(3), 536–556. https://doi.org/10.1111/j.1365–2478.2010.00942.x

Kassis, S. & Sondergeld, C. (2010). Fracture permeability of gas shale: Effects of roughness, fracture offset, proppant, and effective stress. *Society of Petroleum Engineers Journal*, (1), 1–17. https://doi.org/10.2118/131376-MS

Katz, D. (ed.). (1959). *Handbook of Natural Gas Engineering*. McGraw-Hill.

Keller, L. M. & Holzer, L. (2018). Image-based upscaling of permeability in Opalinus clay. *Journal of Geophysical Research: Solid Earth*, *123*(1), 285–295. https://doi.org/10.1002/2017JB014717

Kelly, S., El-Sobky, H., Torres-Verdín, C., & Balhoff, M. T. (2016). Assessing the utility of FIB-SEM images for shale digital rock physics. *Advances in Water Resources*, *95*, 302–316. https://doi.org/10.1016/j.advwatres.2015.06.010

Kennedy, R. L., Knecht, W. N., & Georgi, D. T. (2012). Comparisons and Contrasts of Shale Gas and Tight Gas Developments, North American Experience and Trends. *SPE Saudi Arabia Section Technical Symposium and Exhibition*, (August 2005). https://doi.org/10.2118/160855-MS

Kennedy, R., Luo, L., & Kusskra, V. (2016). The unconventional basins and plays – North America, the rest of the world and emerging basins. In U. Ahmed & D. N. Meehan (eds.), *Unconventional Oil and Gas Resources: Exploitation and Development*. Boca Raton, FL: CRC Press.

Keranen, K. M., Savage, H. M., Abers, G. A., & Cochran, E. S. (2013). Potentially induced earthquakes in Oklahoma, USA: Links between wastewater injection and the 2011 Mw5.7 earthquake sequence. *Geology*, *41*(6), 699–702. https://doi.org/10.1130/G34045.1

Kim, W. Y. (2013). Induced seismicity associated with fluid injection into a deep well in Youngstown, Ohio. *Journal of Geophysical Research: Solid Earth*, *118*(7), 3506–3518. https://doi.org/10.1002/jgrb.50247

King, G. C. P., Stein, R. S., & Lin, J. (1994). Static stress changes and the triggering of earthquakes. *Bulletin Of the Seismological Society of America*, *84*, 935–953.

King, G. E. (2012). Hydraulic Fracturing 101: What Every Representative, Environmentalist, Regulator, Reporter, Investor, University Researcher, Neighbor and Engineer Should Know About Estimating Frac Risk and Improving Frac Performance in Unconventional Gas and Oil Wells. *SPE Hydraulic Fracturing Technology Conference*, 1–80. https://doi.org/10.2118/152596-MS

King, G. E. (2014). Improving recovery factors in liquids-rich resource plays requires new approaches. *Editors Choice Magazine*. Retrieved from http://www.aogr.com/magazine/editors-choice/improving-recovery-factors-in-liquids-rich-resource-plays-requires-new-appr

King, G. E. & King, D. E. (2013). Environmental risk arising from well-construction failure–Differences between barrier and well failure, and estimates of failure frequency across common well types, locations, and well age. *SPE Production & Operations*, *28*(04), 323–344. https://doi.org/10.2118/166142-PA

King, G. E., Rainbolt, M. F., Swanson, C., & Corporation, A. (2017). SPE-187192-MS Frac Hit Induced Production Losses: Evaluating Root Causes, Damage Location, Possible Prevention Methods and Success of Remedial Treatments.

King, H. E., Eberle, A. P. R., Walters, C. C., Kliewer, C. E., Ertas, D., & Huynh, C. (2015). Pore architecture and connectivity in gas shale. *Energy and Fuels*, *29*(3), 1375–1390. https://doi.org/10.1021/ef502402e

Klaver, J., Desbois, G., Urai, J. L., & Littke, R. (2012). BIB-SEM study of the pore space morphology in early mature Posidonia Shale from the Hils area, Germany. *International Journal of Coal Geology*, *103*, 12–25. https://doi.org/10.1016/j.coal.2012.06.012

Klaver, J., Desbois, G., Littke, R., & Urai, J. L. (2015). BIB-SEM characterization of pore space morphology and distribution in postmature to overmature samples from the Haynesville and Bossier Shales. *Marine and Petroleum Geology*, *59*, 451–466. https://doi.org/10.1016/j.marpetgeo.2014.09.020

Klinkenberg, L. J. (1941). The permeability of porous media to liquids and gases. *API Drilling and Production Practice*, 200–2013. https://doi.org/10.5510/OGP20120200114

Kohli, A. H. & Zoback, M. D. (2013). Frictional properties of shale reservoir rocks. *Journal of Geophysical Research: Solid Earth*, *118*(9), 5109–5125. https://doi.org/10.1002/jgrb.50346

Kohli, A. H. & Zoback, M. D. (2019). Frictional properties of shale reservoir rocks II: Calcareous shales and thermal controls on frictional stability. *In Preparation*.

Kramer, S. L. (1996). *Geotechnical Earthquake Engineering*. Upper Saddle River, NJ: Prentice Hall.

Kranz, R. L., Saltzman, J. S., & Blacic, J. D. (1990). Hydraulic diffusivity measurements on laboratory rock samples using an oscillating pore pressure method. *International Journal of Rock Mechanics and Mining Sciences*, *27*(5), 345–352. https://doi.org/10.1016/0148–9062(90) 92709-N

Kronenberg, A. K., Kirby, S., & Pinkston, J. (1990). Basal slip and mechanical anisotropy of biotite. *Journal of Geophysical Research*, *95*, 19257–19278.

Krupnick, A., Gordon, H., & Olmstead, S. (2013). *What the Experts Say about the Environmental Risks of Shale Gas Development. Resources for the Future, Pathways to Dialogue* (Vol. 2). Retrieved from http://scholar.google.com/scholar?hl=en&q=What+the+Experts+Say+about+t he+Environmental+Risks+of+Shale+Gas+Development&btnG=≈sdt=1,5≈sdtp=#0

Kuang, W., Zoback, M., & Zhang, J. (2017). Estimating geomechanical parameters from micro-seismic plane focal mechanisms recorded during multistage hydraulic fracturing. *Geophysics*, *82*(1), KS1–KS11. https://doi.org/10.1190/geo2015-0691.1

Kubo, T. & Katayama, I. (2015). Effect of temperature on the frictional behavior of smectite and illite. *Journal of Mineralogical and Petrological Sciences*, *110*(6), 293–299. https://doi.org/ 10.2465/jmps.150421

Kuila, U., McCarty, D. K., Derkowski, A., Fischer, T. B., Topór, T., & Prasad, M. (2014). Nano-scale texture and porosity of organic matter and clay minerals in organic-rich mudrocks. *Fuel*, *135*, 359–373. https://doi.org/10.1016/j.fuel.2014.06.036

Kumar, A. & Hammack, R. (2016). Long period, long duration (LPLD) seismicity observed during hydraulic fracturing of the Marcellus Shale in Greene County, Pennsylvania, (Figure 1), 2684–2688. https://doi.org/10.1190/segam2016-13876878.1

Kumar, A., Chao, K., Hammack, R., & Harbert, W. (2018). Surface Seismic Monitoring of Hydraulic Fracturing Test Site (HFTS) in the Midland Basin, Texas National Energy Technology Laboratory, Department of Energy, Pittsburgh, PA, 1–11. https://doi.org/10.1553 0/urtec-2018–2902789

Kumar, A., Zorn, E., Hammack, R., & Harbert, W. (2017). Long-period, long-duration seismicity observed during hydraulic fracturing of the Marcellus Shale in Greene County, Pennsylvania, (July), 580–587. https://doi.org/10.1190/tle36070580.1

Kwon, O., Kronenberg, A. K., Gangi, A. F., Johnson, B., & Herbert, B. E. (2004). Permeability of illite-bearing shale: 1. Anisotropy and effects of clay content and loading. *Journal of Geophysical Research*, *109*(B10), 1–19. https://doi.org/10.1029/2004JB003052

Kwon, O., Kronenberg, A. K., Gangi, A., & Johnson, B. (2001). Permeability of Wilcox shale and its effective pressure law. *Journal of Geophysical Research*, *106*, 19339–19353.

LaFollette, R. F., Holcomb, W. D., & Aragon, J. (2012). Practical Data Mining: Analysis of Barnett Shale Production Results with Emphasis on Well Completion and Fracture Stimulation: SPE 152531. *Society of Petroleum Engineers*. https://doi.org/10.2118/152531-MS

Lakes, R. S. (1999). *Viscoelastic Solids*. CRC Mechanical Engineering Series.

Lalehrokh, F. & Bouma, J. (2014). Well Spacing Optimization in Eagle Ford. *SPE/CSUR Unconventional Resources Conference*. https://doi.org/10.2118/171640-MS

Lan, Q., Xu, M., Binazadeh, M., Dehghanpour, H., & Wood, J. M. (2015). A comparative investigation of shale wettability: The significance of pore connectivity. *Journal of Natural Gas Science and Engineering*, *27*, 1174–1188. https://doi.org/10.1016/j.jngse.2015.09.064

Langenbruch, C. & Shapiro, S. A. (2010). Decay rate of fluid-induced seismicity after termination of reservoir stimulations. *Geophysics*, *75*(6).

Langenbruch, C. & Zoback, M. D. (2016). How will induced seismicity in Oklahoma respond to decreased saltwater injection rates? *Science Advances*, *2*(11), 1–9. https://doi.org/10.1126/sciadv.1601542

Langenbruch, C. & Zoback, M. D. (2017). Response to comment on "How will induced seismicity in Oklahoma respond to decreased saltwater injection rates?" *Science Advances*, *3*(8). https://doi.org/10.1126/sciadv.aao2277

Langenbruch, C., Dinske, C., & Shapiro, S. A. (2011). Inter event times of fluid induced earthquakes suggest their Poisson nature. *Geophysical Research Letters*, *38*(21), 1–6. https://doi.org/10.1029/2011GL049474

Langenbruch, C., Weingarten, M., & Zoback, M. D. (2018). Physics-based forecasting 1 of manmade earthquake hazards in Oklahoma and Kansas. *Nature Communications*, *9*, 3946. DOI: 10.1038/s41467-018-06167-4 9

Langmuir, I. (1916). The constitution and fundamental properties of solids and liquids. Part I. Solids. *Journal of the American Chemical Society*, *38*(11), 2221–2295. https://doi.org/10.1021/ja02268a002

Laplace, P. S., Bowditch, N., & Bowditch, N. I. (1829). Mécanique céleste. *Meccanica*.

Laubach, S. E., Olson, J. E., & Gale, J. F. W. (2004). Are open fractures necessarily aligned with maximum horizontal stress? *Earth and Planetary Science Letters*, *222*(1), 191–195.

Lecampion, B., Bunger, A., & Zhang, X. (2017). Numerical methods for hydraulic fracture propagation: A review of recent trends. *Journal of Natural Gas Science and Engineering*. https://doi.org/10.1016/j.jngse.2017.10.012

Lee, D., Shrivastava, K., & Sharma, M. M. (2017). *Effect of Fluid Rheology on Proppant Transport in Hydraulic Fractures in Soft Sands*. American Rock Mechanics Association.

Lee, S., Fischer, T. B., Stokes, M. R., Klingler, R. J., Ilavsky, J., McCarty, D. K., . . . Winans, R. E. (2014). Dehydration effect on the pore size, porosity, and fractal parameters of shale rocks: Ultrasmall-angle X-ray scattering study. *Energy and Fuels*, *28*(11), 6772–6779. https://doi.org/10.1021/ef501427d

Letham, E. A. & Bustin, R. M. (2016). Klinkenberg gas slippage measurements as a means for shale pore structure characterization. *Geofluids*, *16*(2), 264–278. https://doi.org/10.1111/gfl.12147

Leu, L., Georgiadis, A., Blunt, M. J., Busch, A., Bertier, P., Schweinar, K., . . . Ott, H. (2016). Multiscale description of shale pore systems by scanning SAXS and WAXS microscopy. *Energy and Fuels*, *30*(12), 10282–10297. https://doi.org/10.1021/acs.energyfuels.6b02256

Li, J., Zhang, H., Sadi Kuleli, H., & Nafi Toksoz, M. (2011). Focal mechanism determination using high-frequency waveform matching and its application to small magnitude induced earthquakes. *Geophysical Journal International*, *184*(3), 1261–1274. https://doi.org/10.1111/j.1365-246X.2010.04903.x

Li, N., Lolon, E., Mayerhofer, M., Cordts, Y., White, R., & Childers, A. (2017). Optimizing Well Spacing and Well Performance in the Piceance Basin Niobrara Formation. *SPE Hydraulic Fracturing Technology Conference and Exhibition*. https://doi.org/10.2118/184848-MS

Li, S., Yuan, Y., Sun, W., Sun, D., & Jin, Z. (2016). Formation and destruction mechanism as well as major controlling factors of the Silurian shale gas overpressure in the Sichuan Basin, China, *Journal of Natural Gas Geoscience*, *1*(4), 287–294. doi: 10.1016/j.jnggs.2016.09.002.

Liang, B., Du, M., Goloway, C., Hammond, R., Yanez, P. P., & Richey, M. (2017). Subsurface Well Spacing Optimization in the Permian Basin. *Proceedings of the 5th Unconventional Resources Technology Conference*. https://doi.org/10.15530/urtec-2017–2671346

Lindsay, G. J., White, D. J., Miller, G. A., Baihly, J. D., & Sinosic, B. (2016). Understanding the Applicability and Economic Viability of Refracturing Horizontal Wells in Unconventional Plays. *SPE Hydraulic Fracturing Technology Conference*. https://doi.org/10.2118/179113-MS

Linker, M. F. & Dieterich, J. H. (1992). Effects of variable normal stress on rock friction: Observations and constitutive equations. *Journal of Geophysical Research*, *97*(92), 4923–4940.

Lisjak, A., Grasselli, G., & Vietor, T. (2014). Continuum-discontinuum analysis of failure mechanisms around unsupported circular excavations in anisotropic clay shales. *International Journal of Rock Mechanics and Mining Sciences*, *65*, 96–115. https://doi.org/10.1016/j.ijrmms.2013.10.006

Liu, F., Ellett, K., Xiao, Y., & Rupp, J. A. (2013). Assessing the feasibility of CO_2 storage in the New Albany Shale (Devonian-Mississippian) with potential enhanced gas recovery using reservoir simulation. *International Journal of Greenhouse Gas Control*, *17*, 111–126. https://doi.org/10.1016/j.ijggc.2013.04.018

Liu, S. & Valko, P. P. (2015). An Improved Equilibrium-Height Model for Predicting Hydraulic Fracture Height Migration in Multi-Layered Formations. *SPE Hydraulic Fracturing Technology Conference*, (1), 1–16. https://doi.org/10.2118/173335-MS

Lockner, D. A., Byerlee, J. D., Kuksenko, V., Ponomarev, A., & Sidorin, A. (1992). Observations of quasistatic fault growth from acoustic emissions. In *Fault Mechanics and Transport Properties of Rocks* (pp. 1–29). Academic Press.

Lockner, D. A., Tanaka, H., Ito, H., Ikeda, R., Omura, K., & Naka, H. (2009). Geometry of the Nojima fault at Nojima-Hirabayashi, Japan – I. A simple damage structure inferred from borehole core permeability. *Pure and Applied Geophysics*, *166*(10–11), 1649–1667. https://doi.org/10.1007/s00024-009–0515-0

Loucks, R. G. & Reed, R. M. (2014). Scanning-electron-microscope petrographic evidence for distinguishing organic-matter pores associated with depositional organic matter versus migrated organic matter in mudrocks. *GCAGS Journal*, *3*, 51–60.

Loucks, R. G. & Ruppel, S. C. (2007). Mississippian Barnett Shale: Lithofacies and depositional setting of a deep-water shale-gas succession in the Fort Worth Basin, Texas. *AAPG Bulletin*, *91*(4), 579–601. https://doi.org/10.1306/11020606059

Loucks, R. G., Reed, R. M., Ruppel, S. C., & Hammes, U. (2012). Spectrum of pore types and networks in mudrocks and a descriptive classification for matrix-related mudrock pores. *AAPG Bulletin*, *96*(6), 1071–1098. https://doi.org/10.1306/08171111061

Loucks, R. G., Reed, R. M., Ruppel, S. C., & Jarvie, D. M. (2009). Morphology, genesis, and distribution of nanometer-scale pores in siliceous mudstones of the Mississippian Barnett Shale. *Journal of Sedimentary Research*, *79*(12), 848–861. https://doi.org/10.2110/jsr.2009.092

Lund, B. & Townend, J. (2007). Calculating horizontal stress orientations with full or partial knowledge of the tectonic stress tensor. *Geophysical Journal International*, *170*(3), 1328–1335. https://doi.org/10.1111/j.1365-246X.2007.03468.x

Lund Snee, J. E. & Zoback, M. D. (2016). State of stress in Texas: Implications for induced seismicity. *Geophysical Research Letters*, *43*(19),10, 208–10,214. https://doi.org/10.1002/2016GL070974

Lund Snee, J.-E. & Zoback, M. D. (2018a). State of stress in the Central and Eastern U.S. Submitted to *Geology*.

Lund Snee, J.-E. & Zoback, M. D. (2018b). State of stress in the Permian Basin, Texas and New Mexico: Implications for induced seismicity. *The Leading Edge*, February. Retrieved from https://doi.org/10.1190/tle37020127.1.

Luo, M., Baker, M. R., & Lemone, D. V. (1994). Distribution and generation of the overpressure system, eastern Delaware Basin, western Texas and southern New Mexico. *American Association of Petroleum Geologists Bulletin*, *78*(9), 1386–1405. https://doi.org/10.1306/A25 FECB1-171B-11D7-8645000102C1865D

Lupini, J. F., Skinner, A. E., & Vaughan, P. R. (1981). The drained residual strength of cohesive soils. *Géotechnique*, *31*(2), 181–213. https://doi.org/10.1680/geot.1981.31.2.181

Lyman, S. N., Watkins, C., Jones, C. P., Mansfield, M. L., McKinley, M., Kenney, D., & Evans, J. (2017). Hydrocarbon and carbon dioxide fluxes from natural gas well pad soils and surrounding soils in Eastern Utah. *Environmental Science and Technology*, *51*(20), 11625–11633. https://doi.org/10.1021/acs.est.7b03408

Lynn, H. B. (2004). The winds of change: Anisotropic rocks – Their preferred direction of fluid flow and their associated seismic signatures – Part 2. *The Leading Edge*, *23*(12), 1258–1268. https://doi.org/10.1190/leedff.23.1258_1

Lyu, Q., Long, X., Ranjith, P. G., Tan, J., & Kang, Y. (2018). Experimental investigation on the mechanical behaviours of a low-clay shale under water-based fluids. *Engineering Geology*, *233* (July 2017), 124–138. https://doi.org/10.1016/j.enggeo.2017.12.002

Lyu, Q., Ranjith, P. G., Long, X., Kang, Y., & Huang, M. (2015). A review of shale swelling by water adsorption. *Journal of Natural Gas Science and Engineering*, *27*, 1421–1431. https://doi.org/10.1016/j.jngse.2015.10.004

Ma, L., Fauchille, A.-L., Dowey, P. J., Figueroa Pilz, F., Courtois, L., Taylor, K. G., & Lee, P. D. (2017). Correlative multi-scale imaging of shales: a review and future perspectives. *Geological Society, London, Special Publications*, *454*. https://doi.org/10.1144/SP454.11

Ma, L., Slater, T., Dowey, P. J., Yue, S., Rutter, E. H., Taylor, K. G., & Lee, P. D. (2018). Hierarchical integration of porosity in shales. *Scientific Reports*, *8*(1), 1–14. https://doi.org/10.1038/s41598-018–30153-x

Ma, L., Taylor, K. G., Lee, P. D., Dobson, K. J., Dowey, P. J., & Courtois, L. (2016). Novel 3D centimetre-to nano-scale quantification of an organic-rich mudstone: The Carboniferous Bowland Shale, Northern England. *Marine and Petroleum Geology*, *72*, 193–205. https://doi.org/10.1016/j.marpetgeo.2016.02.008

Ma, X. & Zoback, M. D. (2017a). Laboratory experiments simulating poroelastic stress changes associated with depletion and injection in low-porosity sedimentary rocks. *Journal of Geophysical Research: Solid Earth*, *122*(4), 1–26. https://doi.org/10.1002/2016JB013668

Ma, X. & Zoback, M. D. (2017b). Lithology-controlled stress variations and pad-scale faults: A case study of hydraulic fracturing in the Woodford Shale, Oklahoma. *Geophysics*, *82*(6). https://doi.org/10.1190/GEO2017-0044.1

Ma, X. & Zoback, M. D. (2018). In situ variations with heterogeneous lithology in the Woodford shale, Oklahoma and modeling through viscous stress relaxation. *Rock Mechanics and Rock Engineering*.

Mack, M. & Warpinski, N. R. (1989). Mechanics of hydraulic fracturing. In M. J. Economides & K. Nolte (eds.), *Reservoir Stimulation* (3rd edn., p. 807). Wiley.

Marone, C. & Cox, S. J. D. (1994). Scaling of rock friction constitutive parameters: The effects of surface roughness and cumulative offset on friction of gabbro. *Pure and Applied Geophysics*, *143*(1–3), 359–385. https://doi.org/10.1007/BF00874335

Marone, C. & Kilgore, B. (1993). Scaling of the critical slip distance for seismic faulting with shear strain in fault zones. *Nature, 362*(6421), 618–621. https://doi.org/10.1038/362618a0

Marone, C., Raleigh, C. B., & Scholz, C. H. (1990). Frictional behavior and constitutive modeling of simulated fault gouge. *Journal of Geophysical Research, 95,* 7007–7025.

Marongiu Porcu, M., Lee, D., Shan, D., & Morales, A. (2015). Advanced Modeling of Interwell Fracturing Interference: An Eagle Ford Shale Oil Study. *SPE Hydraulic Fracturing Technology Conference.* https://doi.org/10.2018/174902-MS

Martin, T., Kotov, S., Nelson, S. G., & Hughes, B. (2016). Stimulation of unconventional reservoirs. In U. Ahmed & D. N. Meehan (eds.), *Unconventional Oil and Gas Resources: Exploitation and Development.* Boca Raton, FL: CRC Press.

Martínez-Garzón, P., Ben-Zion, Y., Abolfathian, N., Kwiatek, G., & Bohnhoff, M. (2016). A refined methodology for stress inversions of earthquake focal mechanisms. *Journal of Geophysical Research: Solid Earth, 121*(12), 8666–8687. https://doi.org/10.1002/2016JB013493

Maury, J., Cornet, F. H., & Dorbath, L. (2013). A review of methods for determining stress fields from earthquakes focal mechanisms; Application to the Sierentz 1980 seismic crisis (Upper Rhine Graben). *Bulletin de La Societe Geologique de France, 184*(4–5), 319–334. https://doi.org/10.2113/gssgfbull.184.4–5.319

Mavko, G., Mukerji, T., & Dvorkin, J. (2009). *The Rock Physics Handbook, Second Edition: Tools for Seismic Analysis of Porous Media.* Cambridge University Press.

Maxwell, S. C. (2009). Assessing the Impact of Microseismic Location Uncertainties On Interpreted Fracture Geometries. *SPE Annual Technical Conference and Exhibition,* 13. https://doi.org/10.2118/125121-MS

Maxwell, S. C. (2012). Statistical Evaluation for Comparative Microseismic Interpretation. *74th EAGE Conference & Exhibition Incorporating SPE EUROPEC 2012,* (June 2012), D023.

Maxwell, S. C. (2014). *Microseismic Imaging of Hydraulic Fracturing: Improved Engineering of Unconventional Shale Reservoirs. Society of Exploration Geophysicists.* https://doi.org/10.1190/1.9781560803164

Maxwell, S. C. & Cipolla, C. (2011). SPE 146932 What Does Microseismicity Tell Us About Hydraulic Fracturing ? (November).

Maxwell, S. C., Raymer, D., Williams, M., & Primiero, P. (2012). Tracking microseismic signals fro the reservoir to the surface. *The Leading Edge,* (November).

Maxwell, S. C., Shemeta, J., Campbell, E., & Quirk, D. (2008). Microseismic Deformation Rate Monitoring. In *EAGE Passive Seismic Workshop – Exploration and Monitoring Applications* (p. SPE 116596). https://doi.org/10.2118/116596-MS

Maxwell, S. C., Urbancic, T. I., Steinsberger, N., & Zinno, R. (2002). Microseismic Imaging of Hydraulic Fracture Complexity in the Barnett shale, Paper 77440. In *Society Petroleum Engineering Annual Technical Conference.* San Antonio, TX: Society of Petroleum Engineers.

Mayerhofer, M. J., Lolon, E., Warpinski, N. R., Cipolla, C. L., Walser, D. W., & Rightmire, C. M. (2010). What is stimulated reservoir volume? *SPE Production & Operations, 25*(01), 89–98. https://doi.org/10.2118/119890-PA

McClure, M. W. (2017). The Spurious deflection of log-log superposition-time derivative plots of diagnostic fracture-injection tests. *SPE Reservoir Evaluation & Engineering, 20*(04) (November 2016), 12. https://doi.org/10.2118/186098-PA

McClure, M. W. & Horne, R. N. (2011). Investigation of injection-induced seismicity using a coupled fluid flow and rate/state friction model. *Geophysics, 76*(6), WC181–WC198. https://doi.org/10.1190/geo2011-0064.1

McClure, M. W. & Kang, C. A. (2017). A Three-Dimensional Reservoir, Wellbore, and Hydraulic Fracturing Simulator that is Compositional and Thermal, Tracks Proppant and Water Solute Transport, Includes Non-Darcy and Non-Newtonian Flow, and Handles Fracture Closure SPE-182593. In *SPE Reservoir Simulation Conference in Montgomery*, TX *20–22 February 2017*. Society of Petroleum Engineering. https://doi.org/10.2118/182593-MS

McClure, M. W., Babazadeh, M., Shiozawa, S., & Huang, J. (2016). Fully coupled hydromechanical simulation of hydraulic fracturing in 3D discrete-fracture networks. *SPE Journal*, *21*(04), 1302–1320. https://doi.org/10.2118/173354-PA

Mckernan, R., Mecklenburgh, J., Rutter, E., & Taylor, K. (2017). Microstructural controls on the pressure-dependent permeability of Whitby Mudstone. *Geological Society of London Special Publication, 454*. https://doi.org/10.1144/SP454.15

McNamara, D. E., Benz, H. M., Herrmann, R. B., Bergman, E. A., Earle, P., Holland, A., . . . Gassner, A. (2015). Earthquake hypocenters and focal mechanisms in central Oklahoma reveal a complex system of reactivated subsurface strike-slip faulting. *Geophysical Research Letters*, *42*(8), 2742–2749. https://doi.org/10.1002/2014GL062730

McNutt, S. R. (1992). Volcanic tremor. In *Encyclopedia of Earth System Science* (pp. 417–425). Academic Press Inc.

McPhee, C., Reed, J., & Zubizarreta, I. (2015). Best practice in coring and core analysis. *Developments in Petroleum Science*, *64*, 1–15. https://doi.org/10.1016/B978-0-444-63533-4 .00001-9

Meléndez-Martínez, J. & Schmitt, D. R. (2016). A comparative study of the anisotropic dynamic and static elastic moduli of unconventional reservoir shales: Implication for geomechanical investigations. *Geophysics*, *81*(3), D245–D261. https://doi.org/10.1190/geo2015-0427.1

Michael, A. J. (1984). Determination of stress from slip data: Faults and folds. *Journal of Geophysical Research*, *89*(B13), 11517–11526.

Middleton, R. S., Carey, J. W., Currier, R. P., Hyman, J. D., Kang, Q., Karra, S., . . . Viswanathan, H. S. (2015). Shale gas and non-aqueous fracturing fluids: Opportunities and challenges for supercritical CO_2. *Applied Energy*, *147*, 500–509. https://doi.org/10.1016/j .apenergy.2015.03.023

Miller, B., Paneitz, J., Mullen, M., Meijs, R., Tunstall, K., & Garcia, M. (2008). The Successful Application of a Compartmental Completion Technique Used to Isolate Multiple Hydraulic-Fracture Treatments in Horizontal Bakken Shale Wells in North Dakota. In *SPE Annual Technical Conf. and Exhibition*.

Miller, D. E., Daley, T. M., White, D., Freifeld, B. M., Robertson, M., Cocker, J., & Craven, M. (2012). Simultaneous Acquisition of Distributed Acoustic Sensing VSP with Multi-mode and Single-mode Fiber-optic Cables and 3C-Geophones at the Aquistore CO_2 Storage Site, *CSEG Recorder*, 28–33.

Miller, G., Lindsay, G., Baihly, J., & Xu, T. (2016). Parent well refracturing: Economic safety nets in an uneconomic market. *CC*, (May), 5–6. https://doi.org/10.2118/180200-MS

Milliken, K. L. & Day-Stirrat, R. J. (2013). Cementation in mudrocks: Brief review with examples from cratonic basin mudrocks. *AAPG Memoir*, *103*, 133–150. https://doi.org/10.1306/ 13401729H5252

Milliken, K. L., Rudnicki, M., Awwiller, D. N., & Zhang, T. (2013). Organic matter-hosted pore system, Marcellus Formation (Devonian), Pennsylvania. *AAPG Bulletin*, *97*(2), 177–200. https://doi.org/10.1306/07231212O48

Mitchell, J. K. & Soga, K. (2005). *Fundamentals of Soil Behavior*. New York: John Wiley & Sons.

Mitchell, T. M. & Faulkner, D. R. (2009). The nature and origin of off-fault damage surrounding strike-slip fault zones with a wide range of displacements: A field study from the Atacama fault system, northern Chile. *Journal of Structural Geology*, *31*(8), 802–816. https://doi.org/10.1016/j.jsg.2009.05.002

Mokhtari, M., Alqahtani, A. A., Tutuncu, A. N., & Yin, X. (2013). Stress-Dependent Permeability Anisotropy and Wettability of Shale Resources. In *Unconventional Resources Technology Conference* (pp. 2713–2728).

Montgomery, S. L., Jarvie, D. M., Bowker, K. A., & Pollastro, R. M. (2005). Mississippian Barnett Shale, Fort Worth basin, north-central Texas: Gas-shale play with multi-trillion cubic foot potential. *AAPG Bulletin*, *89*(2), 155–175. https://doi.org/10.1306/09170404042

Moore, D. E. & Lockner, D. A. (2004). Crystallographic controls on the frictional behavior of dry and water-saturated sheet structure minerals. *Journal of Geophysical Research*, *109*(B3), 1–16. https://doi.org/10.1029/2003JB002582

Moos, D. & Zoback, M. D. (1990). Utilization of observations of well bore failure to constrain the orientation and magnitude of crustal stresses: Application to continental deep sea drilling project and ocean drilling program boreholes. *Journal of Geophysical Research*, *95*, 9305–9325.

Moos, D. & Zoback, M. D. (1993). State of stress in the Long Valley caldera, California. *Geology*, *21*, 837. https://doi.org/10.1130/0091–7613(1993)021<0837:SOSITL>2.3.CO;2

Moos, D., Vassilellis, G., Cade, R., Franquet, J., Lacazette, A., Bourtembourg, E., & Daniel, G. (2011). Predicting Shale Reservoir Response to Stimulation in the Upper Devonian of West Virginia. *SPE Annual Technical Conference and Exhibition*, 16.

Morales, A., Zhang, K., Gekhar, K., Marongiu Porcu, M., Lee, D., Shan, D., . . . Acock, A. (2015). Advanced Modeling of Interwell Fracturing Interference: An Eagle Ford Shale Oil Study – Refracturing. *SPE Hydraulic Fracturing Technology Conference*. https://doi.org/10.2116/179177-MS

Morales, A., Zhang, K., Gakhar, K., Porcu, M., Lee, D. Shan, D. Malpani, R. Pope, T, Sobernheim, D, Acock, S. (2016). Advanced Modelling of Interwell Fracturing Interference: An Eagle Ford Shale Oil Study – Refracturing, SPE 179177-MS, SPE Hydraulic Fracturing Technology Conference, The Woodlands, Texas, 9–11 February 2016.

Morrow, C. A., Bo-Chong, Z., & Byerlee, J. D. (1986). Effective pressure law for permeability of Westerly granite under cycling loading. *Journal of Geophysical Research*, *91*(6), 3870–3876. https://doi.org/10.1029/JB091iB03p03870

National Academies Press (2017). *Flowback and Produced Waters*. National Academies Press. https://doi.org/10.17226/24620

National Research Council (2012). *Induced Seismicity Potential in Energy Technologies*. National Academies Press.

Nelson, P. H. (2003). A review of the multiwell experiment, Williams Fork and Iles Formations, Garfield County, Colorado. In *U.S. Geological Survey Digital Data Series DDS–69–B* (Chapter 15). U.S. Geological Survey.

Nelson, P. H. (2009). Pore-throat sizes in sandstones, tight sandstones, and shales. *AAPG Bulletin*, *93*(3), 329–340. https://doi.org/10.1306/10240808059

Nelson, P. H., Gianoutsos, N. J., & Drake, R. M. (2015). Underpressure in mesozoic and paleozoic rock units in the midcontinent of the United States. *AAPG Bulletin*, *99*(10), 1861–1892. https://doi.org/10.1306/04171514169

Nicholson, C. & Wesson, R. L. (1992). Triggered earthquakes and deep well activities. *Pure & Applied Geophysics*, *139*(3), 561–578.

Nicot, J. P., Scanlon, B. R., Reedy, R. C., & Costley, R. A. (2014). Source and fate of hydraulic fracturing water in the Barnett Shale: A historical perspective. *Environmental Science and Technology*, 2464–2471. https://doi.org/dx.doi.org/10.1021/es404050r

Nolte, K. (1979). Determination of Fracture Parameters From Fracturing Pressure Decline, SPE-8341-MS. *SPE Annual Technical Conference and Exhibition, Las Vegas, Nevada, USA, 23–26 September.* https://doi.org/10.2118/8341-MS.

Nolte, K., Maniere, J. L., & Owens, K. A. (1979). After-Closure Analysis of Fracture Calibration Tests. Paper SPE 38676, Society of Petroleum Engineers.

Nur, A. & Byerlee, J. D. (1971). An exact effective sress law for elastic deformation of rock with fluids. *Journal of Geophysical Research*, 6414–6419.

Nuttall, B. C., Drahovzal, J. A., Eble, C., & Bustin, R. M. (2005). CO_2 Sequestration in Gas Shales of Kentucky (Vol. 19).

Nygaard, K. J., Cardenas, J., Krishna, P. P., Ellison, T. K., & Templeton-Barrett, E. L. (2013). Technical Considerations Associated with Risk Management of Potential Induced Seismicity in Injection Operations. In *5to. Congreso de Producción y Desarrollo de Reservas*. Retrieved from https://pangea.stanford.edu/researchgroups/scits/sites/default/files/Argentina_Congress_May2013_TechConRiskManIndSeismicity_Final.pdf

Odling, N. E., Gillespie, P., Bourgine, B., Castaing, C., Chiles, J. P., Christensen, N. P., ... Watterson, J. (1999). Variations in fracture system geometry and their implications for fluid flow in fractures hydrocarbon reservoirs. *Petroleum Geoscience*, *5*(4), 373–384. https://doi.org/10.1144/petgeo.5.4.373

Ogwari, P. O., DeShon, H. R., & Hornbach, M. J. (2018). The Dallas-Fort Worth Airport earthquake sequence: Seismicity beyond injection period. *Journal of Geophysical Research: Solid Earth*, *123*(1), 553–563. https://doi.org/10.1002/2017JB015003

Ojha, S. P., Misra, S., Tinni, A., Sondergeld, C., & Rai, C. (2017). Pore connectivity and pore size distribution estimates for Wolfcamp and Eagle Ford shale samples from oil, gas and condensate windows using adsorption-desorption measurements. *Journal of Petroleum Science and Engineering*, *158*, 454–468. https://doi.org/10.1016/j.petrol.2017.08.070

Parshall, J. (2018). "Enormous" merge play resource rivals major world gas fields, largest discoveries. *Journal of Petroleum Technology*, 18 March.

Passey, Q. R., Bohacs, K. M., Esch, W. L., Klimentidis, R., & Sinha, S. (2010). From Oil-Prone Source Rock to Gas-Producing Shale Reservoir – Geologic and Petrophysical Characterization of Unconventional Shale-Gas Reservoirs. *CPS/SPE International Oil & Gas Conference and Exhibition in China, SPE145849*, 1707–1735. https://doi.org/10.2118/131350-MS

Patchen, D. G. & Carter, K. M., eds. (2015). *Utica Shale Appalachian Basin Exploration Consortium Final Report*. West Virginia University.

Patel, H., Johanning, J., & Fry, M. (2013). Borehole Microseismic, Completion and Production Data Analysis to Determine Future Wellbore Placement, Spacing and Vertical Connectivity, Eagle Ford Shale, South Texas. In *Unconventional Resources Technology Conference, Denver, Colorado*, 12–14 August 2013 (pp. 225–236). https://doi.org/10.1190/urtec2013-026

Paterson, M. S. & Wong, T. (2005). *Experimental Rock Deformation – The Brittle Field*. Berlin: Springer.

Pathi, V. S. M. (2008). *Factors Affecting the Permeability of Gas Shales*. University of British Columbia.

Patzek, T., Male, F., & Marder, M. (2013). Gas production in the Barnett Shale obeys a simple scaling theory. *Proceedings of the National Academy of Sciences*, *110*(49), 19731–19736. https://doi.org/10.1073/pnas.1313380110

Patzek, T., Male, F., & Marder, M. (2014). A simple model of gas production from hydrofractured horizontal wells in shales. *AAPG Bulletin*, *98*(12), 2507–2529. https://doi.org/10.1306/03241412125

Paul, P., Zoback, M., & Hennings, P. (2009). Fluid Flow in a Fractured Reservoir Using a Geomechanically Constrained Fault-Zone-Damage Model for Reservoir Simulation. *SPE Reservoir Evaluation and Engineering, August* (4).

Peltonen, C., Marcussen, Ø., Bjørlykke, K., & Jahren, J. (2009). Clay mineral diagenesis and quartz cementation in mudstones: The effects of smectite to illite reaction on rock properties. *Marine and Petroleum Geology*, *26*(6), 887–898. https://doi.org/10.1016/J.MARPETGEO.2008.01.021

Peng, S. & Xiao, X. (2017). Investigation of multiphase fluid imbibition in shale through synchrotron-based dynamic micro-CT imaging. *Journal of Geophysical Research: Solid Earth*, *122*(6), 4475–4491. https://doi.org/10.1002/2017JB014253

Peng, S., Yang, J., Xiao, X., Loucks, B., Ruppel, S. C., & Zhang, T. (2015). An integrated method for upscaling pore-network characterization and permeability estimation: Example from the Mississippian Barnett Shale. *Transport in Porous Media*, *109*(2), 359–376. https://doi.org/10.1007/s11242-015–0523-8

Perez Altamar, R. & Marfurt, K. (2014). Mineralogy-based brittleness prediction from surface seismic data: Application to the Barnett Shale. *Interpretation*, *2*(4), T255–T271. https://doi.org/10.1190/INT-2013–0161.1

Perez Altamar, R. & Marfurt, K. J. (2015). Identification of brittle/ductile areas in unconventional reservoirs using seismic and microseismic data: Application to the Barnett Shale. *Interpretation*, *3*(4), T233–T243. https://doi.org/10.1190/INT-2013–0021.1

Peters, K. E., Walters, C. C., & Moldowan, J. M. (2005). Origin and preservation of organic matter. *The Biomarker Guide*, *1*, 3–17.

Pettegrew, J. & Qiu, J. (2016). *Understanding Wolfcamp Well Performance – A Workflow to Describe the Relationship Between Well Spacing and EUR*, 1–3. https://doi.org/10.15530-urtec-2016–2464916

Poirier, J.-P. (1985). *Creep of Crystals:High-Temperature Deformation Processes in Metals, Ceramics and Minerals*. Cambridge University Press.

Pollastro, R. M. (2007). Total petroleum system assessment of undiscovered resources in the giant Barnett Shale continuous (unconventional) gas accumulation, Fort Worth Basin, Texas. *AAPG Bulletin*, *91*(4), 551–578. https://doi.org/10.1306/06200606007

Pollastro, R. M., Jarvie, D. M., Hill, R. J., & Adams, C. W. (2007). Geologic framework of the Mississippian Barnett Shale, Barnett-Paleozoic total petroleum system, Bend arch-Fort Worth Basin, Texas. *AAPG Bulletin*, *91*(4), 405–436. https://doi.org/10.1306/10300606008

Rahm, B. G. & Riha, S. J. (2012). Toward strategic management of shale gas development: Regional, collective impacts on water resources. *Environmental Science and Policy*, *17*, 12–23. https://doi.org/10.1016/j.envsci.2011.12.004

Rainbolt, M. F., Corporation, A., & Esco, J. (2018). SPE-189853-MS Paper Title: Frac Hit Induced Production Losses: Evaluating Root Causes, Damage Location, Possible Prevention Methods and Success of Remediation Treatments, Part II Case Study V, Wolfcamp, " PARENT " – " CHILD " Relationship, *187192* (partI).

Raleigh, C. B., Healy, J. H., & Bredehoeft, J. D. (1976). An experiment in earthquake control at Rangely, Colorado. *Science*, *191*, 1230–1237.

Ramanathan, V., Boskovic, D., Zhmodik, A., Li, Q., Ansarizadeh, M., Michi, O. P., & Garcia, G. (2015). A Simulation Approach to Modelling and Understanding Fracture Geometry with Respect to Well Spacing in Multi Well Pads in the Duvernay – A Case.

Randen, T., Pedersen, S., & Sønneland, L. (2001). Automatic extraction of fault surfaces from three dimensional seismic data. In *SEG Technical Program Expanded Abstracts* (pp. 551–554). https://doi.org/10.1190/1.1816675

Randolph, P. L., Soeder, D. J., & Chowdiah, P. (1984). Porosity and Permeability of Tight Sands. SPE Unconventional Gas Recovery Symposium. https://doi.org/10.2118/12836-MS

Rassouli, F. (2018). Laboratory study on the effects of carbonates and clay content on viscoplastic deformation of shales at reservoir conditions. Stanford University.

Rassouli, F. S. & Zoback, M. D. (2018). Comparison of short-term and long-term creep experiments in shales and carbonates from unconventional gas reservoirs. *Rock Mechanics and Rock Engineering*. https://doi.org/10.1007/s00603-018-1444-y

Rassouli, F. S., Ross, C. M., & Zoback, M. D. (2016). Shale Rock Characterization Using Multi-Scale Imaging. American Rock Mechanics Association.

Raterman, K. T., Farrell, H. E., Mora, O. S., Janssen, A. L., Gomez, G. A., Busetti, S., ... Warren, M. (2017). Sampling a Stimulated Rock Volume: An Eagle Ford Example (pp. 24–26). https://doi.org/10.15530/urtec-20172670034

Revil, A., Grauls, D., & Brevart, O. (2002). Mechanical compaction of sand/clay mixtures. *Journal of Geophysical Research*, *107*(B11), 2293. https://doi.org/10.1029/2001JB000318

Reynolds, A. C., Bonnie, R. J. M., Kelly, S., Krumm, R., & Group, P. O. (2018). Quantifying Nanoporosity: Insights Revealed by Parallel and Multiscale Analyses, (July), 1–12. https://doi.org/10.15530/urtec-2018-2898355

Rickman, R., Mullen, M. J., Petre, J. E., Grieser, W. V., & Kundert, D. (2008). A Practical Use of Shale Petrophysics for Stimulation Design Optimization: All Shale Plays Are Not Clones of the Barnett Shale. *SPE Annual Technical Conference and Exhibition*, (Wang), 1–11. https://doi.org/10.2118/115258-MS

Rittenhouse, S., Currie, J., & Blumstein, R. (2016). Using Mud Weights, DST, and DFIT Data to Generate a Regional Pore Pressure Model for the Delaware Basin, New Mexico and Texas, 1243–1252.

Rong, G., Yang, J., Cheng, L., & Zhou, C. (2016). Laboratory investigation of nonlinear flow characteristics in rough fractures during shear process. *Journal of Hydrology*, *541*, 1385–1394. https://doi.org/10.1016/j.jhydrol.2016.08.043

Ross, D. J. K. & Bustin, M. (2009). The importance of shale composition and pore structure upon gas storage potential of shale gas reservoirs. *Marine and Petroleum Geology*, *26*(6), 916–927. https://doi.org/10.1016/j.marpetgeo.2008.06.004

Roy, B., Hart, B., Mironova, A., Zhou, C., & Zimmer, U. (2014). Integrated characterization of hydraulic fracture treatments in the Barnett Shale: The Stocker geophysical experiment. *Interpretation*, *2*(2), T111–T127. https://doi.org/10.1190/INT-2013-0071.1

Rubinstein, J. L. & Mahani, A. B. (2015). Myths and facts on wastewater injection, hydraulic fracturing, enhanced oil recovery, and induced seismicity. *Seismological Research Letters*, *86* (4), 1060–1067. https://doi.org/10.1785/0220150067

Rucker, W. K., Oil, E., States, U., Bobich, J., Oil, E., States, U., ... States, U. (2016). Low Cost Field Application of Pressure Transient Communication for Rapid Determination of the Upper Limit of Horizontal Well Spacing, 2466–2474.

Rüger, A. & Gray, D. (2014). Analysis of anisotropic fractured reservoirs. *Encyclopedia of Exploration Geophysics*, N1–N14.

Ruina, A. (1983). Slip instability and state variable friction laws. *Journal of Geophysical Research*. https://doi.org/10.1029/JB088iB12p10359

Rutledge, J. & Phillips, W. S. (2003). Hydraulic stimulation of natural fractures as revealed by induced microearthquakes, Carthage Cotton Valley gas field, east Texas. *Geophysics, 68*(2), 441–452. https://doi.org/10.1190/1.1567214

Rutledge, J., Downie, R., Maxwell, S., & Drew, J. (2013). Geomechanics of Hydraulic Fracturing Inferred from Composite Radiation Patterns of Microseismicity. *Proceedings of SPE Annual Technical Conference and Exhibition*, SPE 166370. https://doi.org/10.2118/166370-MS

Rutledge, J., Phillips, W. S., & Mayerhofer, M. J. (2004). Faulting induced by forced fluid injection and fluid flow forced by faulting: An interpretation of hydraulic-fracture microseismicity, Carthage Cotton Valley gas field, Texas. *Bulletin of the Seismological Society of America, 94*(5), 1817–1830. https://doi.org/10.1785/012003257

Rutledge, J., Weng, X., Chapman, C., Yu, X., & Leaney, S. (2016). Bedding-Plane Slip as a Microseismic Source During Hydraulic Fracturing. *SEG International Exposition and 86th Annual Meeting*, 2555–2559. https://doi.org/10.1190/segam2016-13966680.1

Rutledge, J., Yu, X., Leaney, S., Bennett, L., & Maxwell, S. (2014). Microseismic shearing generated by fringe cracks and bedding-plane slip. *SEG Technical Program Expanded Abstracts 2014*, 2267–2272. https://doi.org/10.1190/segam2014-0896.1

Rutter, E. H. (1974). The influence of temperature, strain rate and interstitial water in the experimental deformation of calcite rocks. *Tectonophysics, 22*(3–4), 311–334. https://doi.org/10.1016/0040–1951(74)90089–4

Rutter, E. H. & Mecklenburgh, J. (2018). Influence of normal and shear stress on the hydraulic transmissivity of thin cracks in a tight quartz sandstone, a granite, and a shale. *Journal of Geophysical Research: Solid Earth, 123*(2), 1262–1285. https://doi.org/10.1002/2017JB014858

Rutter, E. H., Hackston, A. J., Yeatman, E., Brodie, K. H., Mecklenburgh, J., & May, S. E. (2013). Reduction of friction on geological faults by weak-phase smearing, *Journal of Structural Geology, 51*, 52–60.

Sahai, V., Jackson, G., & Rai, R. (2012). SPE 1557 Optimal Well W Spacing Configurations for Unconventional Gas Reservoirs. *Americas Unconventional Resources Conference, June 5–7, 2013*, (2010 0).

Saldungaray, P. & Palisch, T. (2012). Hydraulic fracture optimization in unconventional reservoirs. *World Oil*, (Spec. Suppl.), 7–13.

Sandrea, R. & Sandrea, I. (2014). New well-productivity data provide US shale potential insights. *Oil & Gas*, 66–77.

Sardinha, C., Petr, C., Lehmann, J., Pyecroft, J., Merkle, S., & Energy, N. (2014). Determining Interwell Connectivity and Reservoir Complexity Through Frac Introduction to the Horn River Shales, (Table 1), 1–15.

Saunois, M., Jackson, R. B., Bousquet, P., Poulter, B., & Candell, J. G. (2016). The growing role of methane in anthropogenic climate change. *Environmental Research Letters, 11*. https://doi.org/doi:10.1088/1748-9326/11/12/120207

Savage, H. M. & Brodsky, E. E. (2011). Collateral damage: Evolution with displacement of fracture distribution and secondary fault strands in fault damage zones. *Journal of Geophysical Research: Solid Earth, 116*(3). https://doi.org/10.1029/2010JB007665

Sayers, C. M. (1994). The elastic anisotrophy of shales. *Journal of Geophysical Research, 99*(B1), 767. https://doi.org/10.1029/93JB02579

Scales, M. M., DeShon, H. R., Magnani, M. B., Walter, J. I., Quinones, L., Pratt, T. L., & Hornbach, M. J. (2017). A decade of induced slip on the causative fault of the 2015 Mw4.0

Venus earthquake, Northeast Johnson County, Texas. *Journal of Geophysical Research: Solid Earth*, *122*(10), 7879–7894. https://doi.org/10.1002/2017JB014460

Scanlon, B. R., Reedy, R. C., & Nicot, J. P. (2014). Comparison of water use for hydraulic fracturing for unconventional oil and gas versus conventional oil. *Environmental Science & Technology*, *48*(20), 12386–93. https://doi.org/10.1021/es502506v

Scanlon, B. R., Reedy, R. C., Male, F., & Walsh, M. (2017). Water issues related to transitioning from conventional to unconventional oil production in the Permian Basin. *Environmental Science & Technology*, acs.est.7b02185. https://doi.org/10.1021/acs.est.7b02185

Schieber, J. (2010). Common Themes in the Formation and Preservation of Intrinsic Porosity in Shales and Mudstones – Illustrated with Examples Across the Phanerozoic. *SPE Unconventional Gas Conference*. https://doi.org/10.2118/132370-MS

Schoenball, M. & Ellsworth, W. L. (2017). Waveform-relocated earthquake catalog for Oklahoma and Southern Kansas illuminates the regional fault network. *Seismological Research Letters*, *88* (5), 1252–1258. https://doi.org/10.1785/0220170083

Schoenball, M., Walsh, F. R., Weingarten, M., & Ellsworth, W. L. (2018). How faults wake up: The Guthrie-Langston, Oklahoma earthquakes. *The Leading Edge*, *37*(2), 100–106. https://doi .org/10.1190/tle37020100.1

Scuderi, M. M. & Collettini, C. (2016). The role of fluid pressure in induced vs. triggered seismicity: insights from rock deformation experiments on carbonates. *Scientific Reports*, *6* (24852). https://doi.org/10.1038/srep24852

Scuderi, M. M., Collettini, C., & Marone, C. (2017a). Fluid-injection and the mechanics of frictional stability of shale-bearing faults. In *EGU General Assembly Conference Abstracts*.

Scuderi, M. M., Collettini, C., & Marone, C. (2017b). Frictional stability and earthquake triggering during fluid pressure stimulation of an experimental fault. *Earth and Planetary Science Letters*, *477*, 84–96. https://doi.org/10.1016/j.epsl.2017.08.009

Segall, P. (1995). A note on induced stress changes in hydrocarbon and geothermal reservoirs.

Segall, P., Grasso, J. R., & Mossop, A. (1994). Poroelastic stressing and induced seismicity near the Lacq Gas Field, Southwestern France. *Journal of Geophysical Research*, *99*(15), 423.

Segall, P., Rubin, A. M., Bradley, A. M., & Rice, J. R. (2010). Dilatant strengthening as a mechanism for slow slip events. *Journal of Geophysical Research: Solid Earth*, *115*(12). https://doi.org/10.1029/2010JB007449

Sen, V., Min, K. S., Ji, L., & Sullivan, R. (2018). Completions and Well Spacing Optimization by Dynamic SRV Modeling for Multi-Stage Hydraulic Fracturing SPE 191571. In *SPE ATCE*.

Shaffer, D. L., Chavez, L. H., Ben-Sasson, M., Castrillon, S., Yip, N., & Elimelech, M. (2013). Critical review desalination and reuse of high-salinity shale gas produced water: Drivers, technologies, and future directions. *Environmental Science & Technology*, *47*, 9569–9583. https://doi.org/dx.doi.org/10.1021/es401966e

Shaffner, J., Cheng, A., Simms, S., Keyser, E., & Yu, M. (2011). The Advantage of Incorporating Microseismic Data into Fracture Models. *Canadian Unconventional Resources Conference*, 1–12.

Shanley, K. W., Cluff, R. M., & Robinson, J. W. (2004). Factors controlling prolific gas production from low permeability sandstone reservoirs implications for resource assessment prospect development and risk analysis. *AAPG Bulletin*, *88*(8), 1083–1121.

Shapiro, S. A. (2015). *Fluid-Induced Seismicity*. Cambridge University Press.

Shapiro, S. A., Dinske, C., Langenbruch, C., & Wenzel, F. (2010). Seismogenic index and magnitude probability of earthquakes induced during reservoir fluid stimulations. *The Leading Edge, March*, 304–309.

Shemeta, J. & Anderson, P. (2010). It's a matter of size: Magnitude and moment estimates for microseismic data. *The Leading Edge*, (March).

Shen, Y., Ge, H., Meng, M., Jiang, Z., & Yang, X. (2017). Effect of water imbibition on shale permeability and its influence on gas production. *Energy and Fuels*, *31*(5), 4973–4980. https://doi.org/10.1021/acs.energyfuels.7b00338

Shrivastava, K., Hwang, J., & Sharma, M. M. (2018). Formation of Complex Fracture Networks in the Wolfcamp Shale: Calibrating Model Predictions with Core Measurements from the Hydraulic Fracturing Test Site SPE-191630. In *SPE Annual Technical Conference and Exhibition*.

Siddiqui, M. A. Q., Ali, S., Fei, H., & Roshan, H. (2018). Current understanding of shale wettability: A review on contact angle measurements. *Earth-Science Reviews*, *181* (October2017), 1–11. https://doi.org/10.1016/j.earscirev.2018.04.002

Siddiqui, S. & Kumar, A. (2016). Well Interference Effects for Multiwell Configurations in Unconventional Reservoirs. *SPE*.

Sigal, R. F. (2013). Mercury capillary pressure measurements on Barnett Core. *SPE Reservoir Evaluation & Engineering*, *16*(04), 432–442. https://doi.org/10.2118/167607-PA

Sileny, J., Hill, D. P., Eisner, L., & Cornet, F. H. (2009). Non-double-couple mechanisms of microearthquakes induced by hydraulic fracturing. *Journal of Geophysical Research: Solid Earth*, *114*(8), 1–15. https://doi.org/10.1029/2008JB005987

Simpson, R. W. (1997). Quantifying Anderson's fault types, *Journal of Geophysical Research*, *102*, 909–919.

Singh, H. (2016). A critical review of water uptake by shales. *Journal of Natural Gas Science and Engineering*, *34*, 751–766. https://doi.org/10.1016/j.jngse.2016.07.003

Sinha, S., Braun, E. M., Passey, Q. R., Leonardi, S. A., Wood, A. C., Zirkle, T., . . . Kudva, R. A. (2012). Advances in Measurement Standards and Flow Properties Measurements for Tight Rocks such as Shales. In *SPE/EAGE European Unconventional Resources Conference and Exhibition*. Society of Petroleum Engineers. https://doi.org/10.2118/152257-MS

Skempton, A. W. (1960). *Effective Stress in Soils, Concrete, and Rock, Pore Pressure and Suction in Soils*. London: Butterworths.

Skoumal, R. J., Brudzinski, M. R., & Currie, B. S. (2015). Earthquakes induced by hydraulic fracturing in Poland township, Ohio. *Bulletin of the Seismological Society of America*, *105*(1), 189–197. https://doi.org/10.1785/0120140168

Skoumal, R. J., Brudzinski, M. R., & Currie, B. S. (2018a). Proximity of Precambrian basement affects the likelihood of induced seismicity in the Appalachian, Illinois, and Williston Basins, central and eastern United States. *Geosphere*, *14*(3), 1365–1379. https://doi.org/10.1130/GES01542.1

Skoumal, R. J., Brudzinski, M. R., Barbour, A., Ries, R., & Currie, B. S. (2018b). Earthquakes Induced by Hydraulic Fracturing Are Pervasive in Oklahoma. In *2018 Banff International Induced Seismicity Workshop, 24–27 October 2018*.

Slatt, R. M. & Abousleiman, Y. (2011). Merging sequence stratigraphy and geomechanics for unconventional gas shales. *The Leading Edge*, *30*(3), 274. https://doi.org/10.1190/1.3567258

Smith, M. B. & Montgomery, C. (2015). *Hydraulic Fracturing: Emerging Trends and Technologies in Petroleum Engineering*. Boca Raton, FL: CRC Press.

Sondergeld, C. H., Ambrose, R. J., Rai, C. S., & Moncrieff, J. (2010a). Micro-Structural Studies of Gas Shales. *SPE Annual Technical Conference and Exhibition, SPE-131771*. https://doi.org/10.2118/131771-MS

Sondergeld, C., Newsham, K., Comisky, J., Rice, M., & Rai, C. (2010b). Petrophysical Considerations in Evaluating and Producing Shale Gas Resources. *SPE Unconventional Gas Conference*, 1–34. https://doi.org/10.2118/131768-MS

Sone, H. (2012). Mechanical properties of shale gas reservoir rocks and its relation to the in-situ stress variation observed in shale gas reservoirs. PhD Thesis, Stanford University. https://doi.org/10.1017/CBO9781107415324.004

Sone, H. & Zoback, M. D. (2013a). Mechanical properties of shale-gas reservoir rocks – Part 1: Static and dynamic elastic properties and anisotropy. *Geophysics*, *78*(5), D381–D392. https://doi.org/10.1190/geo2013-0050.1

Sone, H. & Zoback, M. D. (2013b). Mechanical properties of shale-gas reservoir rocks – Part 2: Ductile creep, brittle strength, and their relation to the elastic modulus. *Geophysics*, *78*(5), D393–D402. https://doi.org/10.1190/geo2013-0051.1

Sone, H. & Zoback, M. D. (2014a). Time-dependent deformation of shale gas reservoir rocks and its long-term effect on the in situ state of stress. *International Journal of Rock Mechanics and Mining Sciences*, *69*, 120–132. https://doi.org/10.1016/j.ijrmms.2014.04.002

Sone, H. & Zoback, M. D. (2014b). Viscous relaxation model for predicting least principal stress magnitudes in sedimentary rocks. *Journal of Petroleum Science and Engineering*, *124*, 416–431. https://doi.org/10.1016/j.petrol.2014.09.022

Song, F. & Toksöz, M. N. (2011). Full-waveform based complete moment tensor inversion and source parameter estimation from downhole microseismic data for hydrofracture monitoring. *Geophysics*, *76*(6), WC103–WC116. https://doi.org/10.1190/geo2011-0027.1

Stanek, F. & Eisner, L. (2013). New model explaining inverted source mechanisms of microseismic events induced by hydraulic fracturing, 2201–2205. https://doi.org/10.1190/segam2013-0554.1

Stanek, F. & Eisner, L. (2017). Seismicity induced by Hydraulic Fracturing in Shales: A Bedding Plane Slip Model. *Journal of Geophysical Research*, *122*. https://doi.org/10.1002/2017JB014213

Staněk, F., Eisner, L., & Vesnaver, A. (2017). Theoretical assessment of the full-moment-tensor resolvability for receiver arrays used in microseismic monitoring. *Acta Geodynamica et Geomaterialia*, *14*(2), 235–240. https://doi.org/10.13168/AGG.2017.0006

Stein, S. & Wysession, M. (2003). *An Introduction to Seismology, Earthquakes and Earth Structure*. Malden, MA: Blackwell Publishing.

Stephenson, B., Galan, E., Williams, W., Macdonald, J., Azad, A., Carduner, R., & Canada, S. (2018). SPE-189863-MS Geometry and Failure Mechanisms from Microseismic in the Duvernay Shale to Explain Changes in Well Performance with Drilling Azimuth, 1–20.

Stover, C. W. & Coffman, J. L. (1989). *Seismicity of the United States 1568–1989 (Revised) U.S. G.S. Professional Paper 1527* (Vol. 1989).

Suarez-Rivera, R. & Fjær, E. (2013). Evaluating the poroelastic effect on anisotropic, organic-rich, mudstone systems. In *Rock Mechanics and Rock Engineering* (Vol. 46, pp. 569–580). https://doi.org/10.1007/s00603-013–0374-y

Suarez-Rivera, R., Burghardt, J., Edelman, E., Stanchits, S., & Surdi, A. (2013). Geomechanics Considerations for Hydraulic Fracture Productivity. *47th US Rock Mechanics/Geomechanics Symposium*. Retrieved from https://www.onepetro.org/conference-paper/ARMA-2013–666

Suter, M. (1991). State of stress and active deformation in Mexico and western Central America. In *Neotectonics of North America* (Vol. 1, pp. 401–421). Geological Society of America Decade Map.

Takahashi, M., Mizoguchi, K., Kitamura, K., & Masuda, K. (2007). Effects of clay content on the frictional strength and fluid transport property of faults, *Journal of Geophysical Research, 112* (March). https://doi.org/10.1029/2006JB004678

Tarasov, B. & Potvin, Y. (2013). Universal criteria for rock brittleness estimation under triaxial compression. *International Journal of Rock Mechanics and Mining Sciences, 59*, 57–69. https://doi.org/10.1016/j.ijrmms.2012.12.011

Tembe, S., Lockner, D. A., & Wong, T.-F. (2010). Effect of clay content and mineralogy on frictional sliding behavior of simulated gouges: Binary and ternary mixtures of quartz, illite, and montmorillonite. *Journal of Geophysical Research, 115*(B3), 1–22. https://doi.org/10.1029/2009JB006383

Terzaghi, K. V. (1923). Die Berechnung der Durchassigkeitsziffer des Tones aus dem Verlauf der hydrodynamischen Spannungserscheinungen. *Sitzungsber. Akad. Wiss. Math. Naturwiss, 132*(105).

Teufel, L. W. & Logan, J. M. (1978). Effect of displacement rate on the real area of contact and temperatures generated during frictional sliding of Tennessee sandstone. *Pure and Applied Geophysics, 116*(4–5), 840–865. https://doi.org/10.1007/BF00876541

The Academy of Medicine, Engineering and Science of Texas (2017). *Environmental and Community Impacts of Shale Development in Texas*. The Academy of Medicine, Engineering and Science of Texas. https://doi.org/10.25238/TAMESTstf.6.2017

Thomsen, L. (1986). Weak elastic anisotropy. *Geophysics, 51*, 1954–1966.

Tien, C. (1994). *Adsorption Calculations and Modeling*. Butterworth-Heinemann.

Tinni, A., Fathi, E., Agarwal, R., Sondergeld, C., Akkutlu, Y., & Rai, C. (2012). Shale permeability measurements on plugs and crushed samples. In *SPE Canada Unconventional Resources Conference* (pp. 1–14). https://doi.org/10.2118/162235-MS

Tinni, A., Sondergeld, C., & Rai, C. (2017). Pore Connectivity Between Different Wettability Systems in Organic-Rich Shales. *SPE Reservoir Evaluation & Engineering, 20*(04), 1020–1027. https://doi.org/10.2118/185948-PA

Todd, T. & Simmons, G. (1972). Effect of pore pressure on the velocity of compressional waves in low-porosity rocks. *Journal of Geophysical Research, 77*(20), 3731–3743. https://doi.org/10.1029/JB077i020p03731

Torsch, W. C. (2012). Thermal and pore pressure history of the Haynesville Shale in north Louisiana: a numerical study of hydrocarbon generation, overpressure, and natural hydraulic fractures. Master's Thesis, Louisiana State University.

Townend, J. & Zoback, M. D. (2000). How faulting keeps the crust strong. *Geology, 28*(5), 399. https://doi.org/10.1130/0091–7613(2000)28<399:HFKTCS>2.0.CO;2

Ugueto, G., Huckabee, P., Reynolds, A., Somanchi, K., Wojtaszek, M., Nasse, D., . . . Ellis, D. (2018). SPE-189842-MS Hydraulic Fracture Placement Assessment in a Fiber Optic Compatible Coiled Tubing Activated Cemented Single Point Entry System.

Ulmer-Scholle, D. S., Scholle, P. A., Schieber, J., & Raine, R. J. (2015). *A Color Guide to the Petrography of Sandstones, Siltstones, Shales and Associated Rocks*. American Association of Petroleum Geologists. https://doi.org/10.1306/M1091304

US Energy Information Agency (2013). *Annual Energy Outlook 2013*. https://doi.org/DOE/EIA-0383(2015)

US Energy Information Agency (2017). *Annual Energy Outlook 2017*. Retrieved from http://www.eia.gov/outlooks/aeo/pdf/0383(2017).pdf

Vanorio, T., Prasad, M., & Nur, A. (2003). Elastic properties of dry clay mineral aggregates, suspensions and sandstones. *Geophysical Journal International, 155*(1), 319–326. https://doi.org/10.1046/j.1365-246X.2003.02046.x

Vanorio, T., Scotellaro, C., & Mavko, G. (2008). The effect of chemical and physical processes on the acoustic properties of carbonate rocks. *The Leading Edge*, *27*(8), 1040–1048. https://doi.org/10.1190/1.2967558

Vavryčuk, V. (2007). On the retrieval of moment tensors from borehole data. *Geophysical Prospecting*, *55*(3), 381–391. https://doi.org/10.1111/j.1365–2478.2007.00624.x

Vavryčuk, V. (2014). Iterative joint inversion for stress and fault orientations from focal mechanisms. *Geophysical Journal International*, *199*(1), 69–77. https://doi.org/10.1093/gji/ggu224

Vavrycuk, V., Bohnhoff, M., Jechumtálová, Z., Kolár, P., & Šileny, J. (2008). Non-double-couple mechanisms of microearthquakes induced during the 2000 injection experiment at the KTB site, Germany: A result of tensile faulting or anisotropy of a rock? *Tectonophysics*, *456*, 74–93. https://doi.org/10.1016/j.tecto.2007.08.019

Veatch, R. W. J. (1983). Overview of current hydraulic fracturing design and treatment technology – Part1. *Journal of Petroleum Technology*, *35*(4), 677–687. https://doi.org/10.2118/10039-PA

Vega, B., Dutta, A., & Kovscek, A. R. (2014). CT imaging of low-permeability, dual-porosity systems using high X-ray contrast gas. *Transport in Porous Media*, *101*(1), 81–97. https://doi.org/10.1007/s11242-013–0232-0

Vega, B., Ross, C. M., & Kovscek, A. R. (2015). Imaging-based characterization of calcite-filled fractures and porosity in shales. *SPE Journal*. https://doi.org/10.2118/2014–1922521-pa

Verberne, B. A., Spiers, C. J., Niemeijer, A. R., De Bresser, J. H. P., De Winter, D. A. M., & Plümper, O. (2013). Frictional properties and microstructure of calcite-rich fault gouges sheared at sub-seismic sliding velocities. *Pure and Applied Geophysics*, *171*(10), 2617–2640. https://doi.org/10.1007/s00024-013–0760-0

Vermylen, J. P. (2011). *Geomechanical Studies of the Barnett Shale, Texas, USA*. Stanford University.

Vermylen, J. P. & Zoback, M. D. (2011). Hydraulic Fracturing, Microseismic magnitudes, and Stress Evolution in the Barnett Shale, Texas, USA. In *Society of Petroleum Engineers – SPE Hydraulic Fracturing Technology Conference 2011*.

Vernik, L. (1994). Hydrocarbon-generation-induced microcracking of source rocks. *Geophysics*, *59*(4), 555. https://doi.org/10.1190/1.1443616

Vernik, L. & Nur, A. (1992). Ultrasonic velocity and anisotropy of hydrocarbon source rocks. *Geophysics*, *57*(5), 727–735. https://doi.org/10.1190/1.1443286

Vernik, L. & Liu, X. (1997). Velocity anisotropy in shales: A petrophysical study. *Geophysics*, *62*(2), 521. https://doi.org/10.1190/1.1444162

Waldhauser, F. & Ellsworth, W. L. (2000). A double-difference earthquake location algorithm: Method and application to the Northern Hayward Fault, California. *Bulletin of the Seismological Society of America*, *90*(6), 1353–1368. https://doi.org/10.1785/0120000006

Wallace, R. E. (1951). Geometry of shearing stress and relation to faulting. *Journal of Geology, 59*(2), 118–130. https://doi.org/10.1086/625831

Walls, J. & Nur, A. (1979). Pore Pressure and Confining Pressure Dependence of Permeability in Sandstone. In *7th Formation Evaluation Symposium*. Calgary, Canada: Canadian Well Logging Society.

Walsh, F. R. (2017). Seismotectonics of the central United States and probabilistic assessment of injection induced earthquakes. Stanford University.

Walsh, F. R. & Zoback, M. D. (2015). Oklahoma's recent earthquakes and saltwater disposal. *Science Advances*, *1*(5). https://doi.org/10.1126/sciadv.1500195

Walsh, F. R. & Zoback, M. D. (2016). Probabilistic assessment of potential fault slip related to injectioninduced earthquakes: Application to north-central Oklahoma, USA. *Geology*, *44*(12), 991–994. https://doi.org/10.1130/G38275.1

Walsh, F. R., Zoback, M. D., Pais, D., Weingarten, M., & Tyrrell, T. (2017). FSP1.0: A program for probabilistic estimation of fault slip potential resulting from fluid injection.

Walsh, J. B. (1981). Effect of pore pressure and confining pressure on fracture permeability. *International Journal of Rock Mechanics and Mining Sciences & Geomechanics Abstracts*, 18, 429–435.

Walters, R. J., Zoback, M. D., Baker, J. W., & Beroza, G. C. (2015). Characterizing and responding to seismic risk associated with earthquakes potentially triggered by fluid disposal and hydraulic fracturing. *Seismological Research Letters*, *86*(4). https://doi.org/10.1785/0220150048

Walton, I. & McLennan, J. (2013). The role of natural fractures in shale gas production. In *Effective and Sustainable Hydraulic Fracturing*. https://doi.org/10.5772/56404

Wang, F. P. & Gale, J. F. W. (2009). Screening criteria for shale-gas systems. *Gulf Coast Association of Geological Societies Transactions*, *59*, 779–793. Retrieved from http://archives.datapages .com/data/gcags_pdf/2009/WangGale.pdf

Wang, F. P. & Reed, R. M. (2009). Pore Networks and Fluid Flow in Gas Shales, SPE 124253. In *Proceedings of the SPE Annual Technical Conference and Exhibition*. New Orleans, LA: Society of Petroleum Engineers.

Wang, H. & Sharma, M. M. (2018). Estimating un-propped fracture conductivity and fracture compliance from diagnostic fracture injection tests. *SPE Journal*, *23*(05). Retrieved from http:// arxiv.org/abs/1802.05112

Wang, Z. (2002a). Seismic anisotropy in sedimentary rocks, part 1: A single-plug laboratory method. *Geophysics*, *67*(5), 1415. https://doi.org/10.1190/1.1512787

Wang, Z. (2002b). Seismic anisotropy in sedimentary rocks, part 2: A single-plug laboratory method. *Geophysics*, *67*(5), 1415. https://doi.org/10.1190/1.1512787

Warpinski, N. R. & Branagan, P. T. (1989). Altered-stress fracturing. *Journal of Petroleum Technology, SPE-17533-PA*, *41*(09), 990–997. https://doi.org/10.2118/17533-PA

Warpinski, N. R. & Teufel, L. W. (1986). A Viscoelastic Constitutive Model for Determining In-Situ Stress Magnitudes from Anelastic Strain Recovery of Core. In *SPE 61st ATCE New Orleans*.

Warpinski, N. R., Mayerhofer, M. J., & Agarwal, K. (2013). Hydraulic fracture geomechanics and microseismic source mechanisms. *SPE Journal*, *18*(4), 766–780. https://doi.org/10.2118/1589 35-PA

Warpinski, N. R., Moschovidis, Z. A., Parker, C. D., & Abou-Sayed, I. S. (1994). Discussion of comparison study of hydraulic fracturing models – test case: GRI staged field experiment no. 3. *SPE Production and Facilities (Society of Petroleum Engineers)*, *9* (SPE28158), 7–16.

Wenk, H. R., Voltolini, M., Mazurek, M., Van Loon, L. R., & Vinsot, A. (2008). Preferred orientations and anisotropy in shales: Callovo-oxfordian shale (France) and Opalinus clay (Switzerland). *Clays and Clay Minerals*, *56*(3), 285–306. https://doi.org/10.1346/ CCMN.2008.0560301

Williams, L. B. & Hervig, R. L. (2005). Lithium and boron isotopes in illite-smectite: The importance of crystal size. *Geochimica et Cosmochimica Acta*, *69*(24), 5705–5716. https:// doi.org/10.1016/j.gca.2005.08.005

World Stress Map. (n.d.). Retrieved from http://dc-app3-14.gfz-potsdam.de/pub/introduction/int roduction_frame.html

Wu, C.-H. & Sharma, M. M. (2016). Effect of Perforation Geometry and Orientation on Proppant Placement in Perforation Clusters in a Horizontal Well. *Hydraulic Fracturing Technology Conference*, 179117–MS. https://doi.org/10.2118/179117-MS

Wu, K. & Olson, J. (2013). Investigation of Critical In Situ and Injection Factors in Multi-Frac Treatments: Guidelines for Controlling Fracture Complexity. *Proceedings of 2013 SPE Hydraulic Fracturing Technology Conference*, (2000). https://doi.org/10.2118/163821-MS

Wu, W., Reece, J. S., Gensterblum, Y., & Zoback, M. D. (2017). Permeability evolution of slowly slipping faults in shale reservoirs. *Geophysical Research Letters*, 1–8. https://doi.org/10.1002/2017GL075506

Xu, S., Rassouli, F. S., & Zoback, M. D. (2017). Utilizing a Viscoplastic Stress Relaxation Model to Study Vertical Hydraulic Fracture Propagation in Permian Basin. *Unconventional Resources Technology Conference*. https://doi.org/10.15530/urtec-2017–2669793

Yang, F., Ning, Z., Zhang, R., Zhao, H., & Krooss, B. M. (2015). Investigations on the methane sorption capacity of marine shales from Sichuan Basin, China. *International Journal of Coal Geology*, 146, 104–117. https://doi.org/10.1016/j.coal.2015.05.009

Yang, F., Xie, C., Ning, Z., & Krooss, B. M. (2016). High-pressure methane sorption on dry and moisture-equilibrated shales. *Energy and Fuels*, 31(1), 482–492. https://doi.org/10.1021/acs.energyfuels.6b02999

Yang, Y. & Aplin, A. C. (2007). Permeability and petrophysical properties of 30 natural mudstones. *Journal of Geophysical Research: Solid Earth*, 112(3). https://doi.org/10.1029/2005JB004243

Yang, Y. & Mavko, G. (2018). Mathematical modeling of microcrack growth in source rock during kerogen thermal maturation. *AAPG Bulletin*. https://doi.org/10.1306/05111817062

Yang, Y. & Zoback, M. D. (2014). The role of preexisting fractures and faults during multistage hydraulic fracturing in the Bakken Formation. *Interpretation*, 2(3). https://doi.org/10.1190/INT-2013–0158.1

Yang, Y. & Zoback, M. D. (2016). Viscoplastic Deformation of the Bakken and Adjacent Formations and its Relation to Hydraulic Fracture Growth. *Rock Mechanics and Rock Engineering*, 49(2), 689–698. https://doi.org/10.1007/s00603-015-0866-z

Yang, Y., Sone, H., Hows, A., & Zoback, M. D. (2013). Comparison of Brittleness Indices in Organic-Rich Shale Formations. In *47th US Rock Mechanics / Geomechanics Symposium* (Vol. 13, pp. 1398–1404). Retrieved from http://www.scopus.com/inward/record.url?eid=2-s2.0–84892858564&partnerID=40&md5=f455e5d5e054a4c298d30166fdc0d351

Ye, Z., Janis, M., Ghassemi, A., & Riley, S. (2017). Laboratory Investigation of Fluid Flow and Permeability Evolution through Shale Fractures, 24–26. https://doi.org/10.15530/urtec-2017–2674846

Yoon, C. E., Huang, Y., Ellsworth, W. L., & Beroza, G. C. (2017). Seismicity during the initial stages of the Guy-Greenbrier, Arkansas, earthquake sequence. *Journal of Geophysical Research: Solid Earth*, 122(11), 9253–9274. https://doi.org/10.1002/2017JB014946

Yu, W. & Sepehrnoori, K. (2014). Optimization of Well Spacing for Bakken Tight Oil Reservoirs. *Proceedings of the 2nd Unconventional Resources Technology Conference*, (2013), 1981–1996. https://doi.org/10.15530/urtec-2014–1922108

Zecevic, M., Daniel, G., & Jurick, D. (2016). On the nature of long-period long-duration seismic events detected during hydraulic fracturing. *Geophysics*, 81(3), KS113–KS121. https://doi.org/10.1190/geo2015-0524.1

Zhang, D. & Yang, T. (2015). Environmental impacts of hydraulic fracturing in shale gas development in the United States. *Petroleum Exploration and Development*, 42(6), 876–883. https://doi.org/10.1016/S1876-3804(15)30085–9

Zhang, D., Ranjith, P. G., & Perera, M. S. A. (2016). The brittleness indices used in rock mechanics and their application in shale hydraulic fracturing: A review. *Journal of Petroleum Science and Engineering*, *143*(February), 158–170. https://doi.org/10.1016/j .petrol.2016.02.011

Zhang, J., Kamenov, A., Zhu, D., & Hill, A. D. (2014). Laboratory Measurement of Hydraulic Fracture Conductivities in the Barnett Shale. *SPE Hydraulic Fracturing Technology Conference*.

Zhang, J., Ouyang, L., Zhu, D., & Hill, A. D. (2015). Experimental and numerical studies of reduced fracture conductivity due to proppant embedment in the shale reservoir. *Journal of Petroleum Science and Engineering*, *130*, 37–45. https://doi.org/10.1016/j.petrol.2015.04.004

Zhang, X. & Sanderson, D. J. (1995). Anisotropic features of geometry and permeability in fractured rock masses. *Engineering Geology*, *40*(1–2), 65–75. https://doi.org/10.1016/0013–7 952(95)00040–2

Zhang, Y., Mostaghimi, P., Fogdon, A., Arena, A., Middleton, J., & Armstrong, R. T. (2017). Determination of Local Diffusion Coefficients and Directional Anisotropy in Shale From Dynamic Micro-CT Imaging and Microscopy. In *Proceedings of the 5th Unconventional Resources Technology Conference* (pp. 1–13). https://doi.org/10.15530/urtec-2017–2695407

Zhang, Y., Person, M., Rupp, J., Ellett, K., Celia, M. A., Gable, C. W., . . . Elliot, T. (2013). Hydrogeologic controls on induced seismicity in crystalline basement rocks due to fluid injection into basal reservoirs. *Groundwater*, *51*(4), 525–538. https://doi.org/10.1111/ gwat.12071

Zhang, Z. X. (2002). An empirical relation between mode I fracture toughness and the tensile strength of rock. *International Journal of Rock Mechanics and Mining Sciences*, *39* (3), 401–406. https://doi.org/10.1016/S1365-1609(02)00032–1

Zhu, H. & Tomson, R. (2013). Exploring Water Treatment, Reuse and Alternative Sources in Shale Production. Retrieved from http://shaleplay.loewy.com/2013/11/exploring-water-treat ment-reuse-and-alternative-sources-in-shale-production/

Ziarani, A. S. & Aguilera, R. (2012). Knudsen's permeability correction for tight porous media. *Transport in Porous Media*, *91*(1), 239–260. https://doi.org/10.1007/s11242-011–9842-6

Zoback, M. D. (2007). *Reservoir Geomechanics*. Cambridge University. https://doi.org/10.1017/ CBO9780511586477

Zoback, M. D. (2012). Managing the seismic risk posed by wastewater disposal. *Earth, (April)*.

Zoback, M. D. & Arent, D. J. (2014). The opportunities and challenges of sustainable shale gas development. *Elements*, *10*(4). https://doi.org/10.2113/gselements.10.4.251

Zoback, M. D. & Byerlee, J. D. (1975). The effect of cyclic differential stress on dilatancy in Westerly granite under uniaxial and triaxial conditions. *Journal of Geophysical Research*, *80*, 1526–1530.

Zoback, M. D. & Harjes, H. P. (1997). Injection induced earthquakes and crustal stress at 9 km depth at the KTB deep drilling site, Germany. *Journal of Geophysical Research*, *102*, 18477–18491.

Zoback, M. D. & Lund Snee, J.-E. (2018). Predicted and observed shear on pre-existing faults during hydraulic fracture stimulation. In *SEG Technical Program Expanded Abstracts*. Society of Exploration Geophysicists.

Zoback, M. D. & Townend, J. (2001). Implications of hydrostatic pore pressures and high crustal strength for the deformation of intraplate lithosphere. *Tectonophysics*, *336*, 19–30.

Zoback, M. D. & Zoback, M. L. (1991). Tectonic stress field of North America and relative plate motions. In D. B. Slemmons, E. R. Engdahl, M. D. Zoback, & D. D. Blackwell (eds.),

Neotectonics of North America. Geol. Soc. Amer., Decade Map (Vol. 1). Boulder, Co.: Geological Society of America.

Zoback, M. D., Mastin, L., & Barton, C. (1987). In situ stress measurements in deep boreholes using hydraulic fracturing, wellbore breakouts and Stonely wave polarization. In O. Stefansson (ed.), *Rock Stress and Rock Stress Measurements*, (pp. 289–299). Stockholm, Sweden: Centrek Publ., Lulea.

Zoback, M. D., Kohli, A., Das, I., & McClure, M. (2012). The Importance of Slow Slip on Faults During Hydraulic Fracturing Stimulation of Shale Gas Reservoirs. In *Society of Petroleum Engineers – SPE Americas Unconventional Resources Conference 2012*.

Zoback, M. D., Townend, J., & Grollimund, B. (2002). Steady-state failure equilibrium and deformation of intraplate lithosphere. *International Geology Review, 44*(5). https://doi.org/10.2747/0020–6814.44.5.383

Zoback, M. L. (1992a). First and second order patterns of tectonic stress: The World Stress Map Project. *Journal of Geophysical Research, 97*, 11, 703–711, 728.

Zoback, M. L. (1992b). Stress field constraints on intraplate seismicity in eastern North America. *Journal of Geophysical Research, 97*(B8), 11761–11782.

Zoback, M. L. & Zoback, M. D. (1980). State of stress in the conterminous United States. *Journal of Geophysical. Research, 85*, 6113–6156.

Zoback, M. L. & Zoback, M. D. (1989). Tectonic stress field of the continental U.S. in geophysical framework of the continental United States. *GSA Memoir, 172*, 523–539.

Zuo, J. Y., Guo, X., Liu, Y., Pan, S., Canas, J., & Mullins, O. C. (2018). Impact of capillary pressure and nanopore confinement on phase behaviors of shale gas and oil. *Energy and Fuels, 32*(4), 4705–4714. https://doi.org/10.1021/acs.energyfuels.7b03975.

Index

air pollution, 7, 378
Anadarko Basin, 8, 206
Andersonian faulting theory, 182
 A_ϕ and relative stress magnitudes, 182
 Extended Andersonian faulting theory, 182
 normal faulting, 182
 normal/strike-slip faulting, 182
 reverse faulting, 182
 strike-slip faulting, 182
 strike-slip/reverse faulting, 182
anisotropy, 37, 41
 creep compliance, 73, 79, 90
 elastic moduli, 43, 45, 50, 51, 52, 68, 186, 198
 HTI, 27, 228
 mineral, 32, 34
 permeability, 137, 167, 174
 porosity, 35, 125, 129
 rock strength, 67
 seismic, 181, 228, 229, 265, 269, 292
 stress, 27, 60, 88, 111, 182, 202, 209, 313, 342, 361, 369, 372
 structural, 27, 39, 48, 69, 73, 116, 132, 198
 VTI, 27, 41
ant tracking, 219, 220, 224, 225, 405, 419
Appalachian Basin
 stress map, 190
aquifers
 potable, 233, 254, 378, 384, 385, 393
 saline, 381, 416
aseismic (slow) slip, 68, 101, 113, 317, 350, 358, 359, 405, 422
attenuation, 264, 277, 283, 285
azimuthal AVO, 228, 229, 230
A_ϕ, 182, 184, 193, 204, 210, 417

bacteria, 11, 255, 383
Bakken formation, 21, 22, 88, 274, 318, 329, 365, 397
 map, 17
 pore pressure, 206
 poroelasticity, 56, 366
 produced water, 384
 production, 7, 18, 19
 samples, 61

 stratigraphy, 16
 stress state, 369
 water use, 379
Barnett shale, 10, 15, 21, 88
 3D seismic reflection, 225
 capillary pressure, 143
 deposition, 10
 fractures, 214, 305, 306, 318
 geothermal gradient, 208
 hydraulic fracturing, 239, 241, 250
 map, 13
 microseismicity, 218, 242, 254, 265, 277, 303, 304, 314
 pore pressure, 206
 produced water, 382, 384
 production, 5, 7, 12, 18, 233, 302, 362
 samples, 33, 61
 seismic anisotropy, 227
 stratigraphy, 11, 57, 58
 stress state, 111
 wells, 12, 219, 377
bedding, 10, 34, 39, 41, 116
 creep compliance anisotropy, 72
 definition, 27
 permeability anisotropy, 167
 slip during hydraulic stimulation, 310, 311
 strength anisotropy, 67
 velocity anisotropy, 27, 44, 228
Biot coefficient, 153, 354
 definition, 53
 laboratory data, 54, 56
brittleness
 definition, 84
 emprical relationships, 85
 rethinking brittleness, 87
bulk modulus, 47, 53, 55
Byerlee's law, 104

carbonates, 32, 35, 39, 41, 104, 120
 calcite, 32, 47, 48, 105, 107, 132, 212
 dolomite, 32, 48, 121, 217, 431
cementation, 14, 120, 121
Central Basin Platform, 186, 312, 313, 384, 416
chert, 31

clay
  content, 14, 32
  elastic properties, 47
  fabric, 27, 32, 34, 39, 116
  friction, 104, 105, 107
  illite, 32, 105, 172
  interlayer illite–smectite, 32
  porosity, 34, 35, 38, 120
  smectite, 173
  smectite to illite transition, 14, 32
  smectite–illite interlayer structure, 120, 173
  stabilizers, 255, 383
  swelling, 34, 126, 172, 173, 176
CO_2
  adsorption, 126, 132, 171, 172
  emissions, 5, 378, 387
  fracturing fluid, 256
  injection, 21, 173
  molecular size, 118
  permeability, 173
  storage, 6, 171, 409
coal, 5, 11, 172, 378
compaction, 14, 39, 102, 110, 120, 211
  pore, 55, 69, 123, 151, 159, 165, 167, 173
compressibility
  gas, 156, 346
  pore, 153, 163, 165, 354
compressional wave (P-wave) velocity, 44, 45, 46, 47, 228
compressive strength, 66, 175
corner frequency, 283, 284, 349
crack growth, 67, 69, 72
creep
  anisotropy, 73, 79
  compliance, 70, 165
  constitutive laws, 77
  definition, 69
  measurement, 43, 66
  on faults, 110, 112, 422
  stress and strain partitioning, 73
  stress dependence, 70
critical stress intensity factor, 67, 333
critically stressed faults, 112
  permeability, 397
  theory, 93

Dallas Fort Worth airport earthquakes, 403
damage zone, 68, 217
  effects on hydraulic fracturing, 219, 242
  modeling, 223
  observations, 219, 224
  permeability, 205, 217, 223
Delaware Basin, 364
  fault slip potential (FSP), 416
  pore pressure, 206, 344
  stress map, 187, 313, 416

  water disposal, 382
  well spacing, 330
depletion, 139, 168
  effects on permeability, 151, 159, 179, 256, 315
  heterogeneity, 230, 362, 372
  induced stress changes, 360, 366, 370
  poroelastic response, 54
DFIT test, 199, 206, 251, 369
diagenesis, 14, 34, 39, 120
distributed temperature sensing (DTS), 219, 242, 269, 303, 350
drilling mud, 235
  chemistry, 236
  weight, 202, 203, 206
drilling-induced tensile fractures (DITFs), 186
ductility, 60, 66, 69, 72, 111
Duvernay shale, 235, 396, 400
dynamic rupture modeling, 223, 358
dynamic stimulated rock volume (DSRV), 330

Eagle Ford
  composition, 37, 62, 119, 121, 133, 135
  core through/drill through experiment, 245, 246
  creep, 71, 72, 88, 166
  fluid flow, 121, 127, 132, 133, 135, 144, 157, 158, 159, 161, 166, 174
  friction, 103, 105
  general properties, 15, 17, 21, 206, 208, 239
  hydraulic fracturing, 324, 329, 333, 365
  map, 18
  physical properties, 46, 48, 51, 52, 67, 68, 69, 71, 317, 321
  production, 18, 19, 20
  stratigraphy, 18, 27, 28
  stress, 198, 246
  transport properties, 162
  water, 379
earthquake
  moment tensor, 271
  scaling relations, 286
  source dimension, 284, 285, 286
  stress drop, 285
earthquake focal plane mechanisms, 192, 273, 274, 278, 282
  CLVD, 272
  definitions of components, 273
  normal, strike-slip, thrust mechanisms, 193
  stress inversion, 194, 196
earthquakes, 95
  casing deformation, 404, 405
  double-difference location method, 297
  Gutenberg–Richter relation, 287
  induced by flow back water injection, 189
  induced by hydraulic fracturing, 394, 403
  induced by produced water injection, 189
  induced by water injection, 393, 399, 400

intraplate, 379, 390
location using Geiger's method, 287
location using migration, 292
magnitude, 283
moment tensor, 270
moment tensor inversion, 271
multiplets and relative location accuracy, 294
proximity to basement, 397
risk management, 406
scalar moment, 271
scaling, 285
scaling relationships, 285
schematic earthquake machine, 392
seismic wave radiation, 270
slip, 223, 281
stress drops, 286
effective stress
exact, 153, 154
simple (Terzaghi), 53
Terzaghi (simple), 53, 54, 55, 56, 92, 108, 154, 157, 159, 209, 358
elastic moduli
anisotropy, 198
definitions, 43, 47, 49
dynamic, 43, 46, 48, 49
static, 43, 46, 48, 49, 51, 73
elastic properties, 47, 56, 58, 66, 72, 79
environmental impacts, 5, 22, 378, 379, 384, 387
drilling and production, 382, 386
hydraulic fracturing, 382, 383, 385
methane leakage, 387
well construction, 386

fault slip potential (FSP), 409, 415
casing deformation, 421
Fort Worth Basin, 418, 420
Oklahoma, 96, 411, 413, 414
Permian Basin, 415, 417, 418
faults
basement faults, 393, 395, 397, 423
damage zones, 68
identification of faults in 3D seismic data, 225
identification of potentially active faults, 409
Mohr circles, 94, 304, 408
pad-scale faults, 181, 365
roughness, 67, 99, 223, 318
stereonets, 213, 214, 273, 304
feldspar, 32
flowback water, 239, 383, 384, 389
composition, 378, 384
disposal, 377, 381, 382, 385, 392, 396, 403
Fort Worth Basin, 12
earthquakes, 383, 419
fractures, 215
potentially active faults, 416
stress, 185

fossils, 37, 121, 132
frac gradient, 88, 89, 209
variation with depth, 237, 251, 307, 314, 326, 327, 335, 336, 342
frac hit, 300, 344, 362, 367, 372
fracture toughness, 67, 243, 333
fractures (pre-existing), 28, 179, 198, 211, 217, 223, 225, 229, 312
microfractures, 37, 39, 116, 118, 119
model, 25, 360, 361, 372
opening mode, 212, 213, 281, 333
orientation, 182, 211, 213, 215
permeability and normal stress, 318
permeability and shear slip, 315, 316, 319, 345, 352, 361, 396
representation in stereonets, 213
representation on Mohr circles, 94
shear slip, 25, 114, 181, 211, 255, 301, 304, 305, 307, 312, 313
friction
compositional controls, 103, 105, 107
Coulomb faulting, 92, 93, 310
frictional strength, 66, 67, 68, 69, 104, 410
internal friction, 66, 68, 69
pore fluid effects, 108, 109, 209, 391
rate-state, 98, 99, 100, 103
static and dynamic, 95
stress magnitudes, 93, 203
thermal controls, 103, 105

geologic stress indicators, 185
geomechanical model, 335, 409, 411, 419
geothermal gradient, 11
Green's function, 276
Guy-Greenbrier earthquakes, 402, 403

Hashin–Shtrikman bounds, 46
Haynesville, 15, 20, 21
creep, 88, 89
friction, 105
stress, 95
temperature, 208
heat flow, 206
HFTS-1, 246
Hill's (arithmetic) average, 47
Hooke's law, 43
horizontal drilling, 21
general characteristics, 21
landing zone, 236, 237, 323, 325, 327
optimal direction, 235
pad drilling, 23
well construction, 388
Horizontal Transverse Isotropy (HTI), 27, 228
hydraulic fracturing, 238
controlled entry point, 241
determination of S_{hmin}, 199

hydraulic fracturing (cont.)
environmental impacts, 385
fluid viscosity and proppant transport, 258
frac-hit phenomenon, 345, 362, 367
fracturing fluid composition, 235, 238, 255, 258, 383, 384
in isotropic stress fields, 88, 328
induced seismicity, 392, 396, 397, 403, 404, 423, 439
instantaneous shut-in pressure (ISIP), 199
interaction during simultaneous propagation, 248
leak-off tests, 199
modeling growth, 242, 248, 335, 354
multi-stage, 21, 23
net pressure, 201
observed in core, 245
observed in image logs, 213, 245
optimization, 344
overview of parameters in unconventional basins, 239
perforations, 335
plug-and-perf, 238, 240
propagation into depleted areas, 366
proppant, 255, 258, 337
slickwater, 239, 255, 260, 379
sliding sleeve, 240
stress orientation, 199
stress shadow, 247
vertical growth, 60, 253
zipper fracturing, 249
hydrocarbon maturation, 12, 15, 16, 120, 138, 206, 208
hydrocarbon migration, 120
hydrogen index (HI), 12

imaging, 118, 130
focused ion beam (FIB-SEM), 123, 136
helium ion microscopy (HIM), 124
optical petrography, 123
porosity, 130, 131
scanning electron micrograph (SEM), 163
scanning electron microscopy (SEM), 123
thin section photomicrographs, 28
transmission electron microscopy (TEM), 124
wellbore, 237
X-ray computed tomography (CT), 123, 136
in-fill drilling, 346, 364

kerogen, 11, 14, 37
KGD hydraulic fracture model, 242
Klinkenberg effect, 152, 163
Knudsen diffusion, 150, 151, 178

limestone, 31, 57
logging while drilling (LWD), 236, 237

Mancos shale, 8, 42
Marcellus, 29
general properties, 15
map, 191, 398
microseismicity, 254, 326, 327
production, 18, 19, 20, 140
water budget, 385
mean free path, 150, 152, 173
methane
adsorption, 170, 177
coal bed, 11, 389
diffusion, 348
injection, 21
leakage, 378, 387
microseismic events, 25
alignments, 199
defining an SRV, 302, 361
defining faults, 219, 220, 254
focal plane mechanisms, 273
fracture networks, 307, 308, 349
location techniques, 289
magnitude, 25, 283
mechanism, 248, 278
migration, 281, 354
monitoring with downhole arrays, 264
monitoring with surface arrays, 265
relative location accuracy, 294
zonal isolation, 242
Midland basin, 246
map, 312, 382
water production and disposal, 384
Mohr diagram, 92, 93, 304, 359, 408
Mohr–Coulomb failure criterion, 66, 92, 409, 415
Coulomb Failure Function (CFF), 92, 306, 310, 358
moment tensor
moment tensor inversion, 276
Monterey formation, 16
mudstone, 157, 162, 209
multi-stage hydraulic fracturing, 238, 240, 301

natural gas, 6, 7, 9, 12, 13, 139
natural gas liquids, 11, 12, 13, 138
New Mexico
induced seismicity, 389, 418
pore pressure, 207
stress state, 187
unconventional plays, 8
normal faulting, 88, 90, 94, 183, 193, 204, 209, 210, 313, 413

Ohio
earthquakes, 389, 397, 398, 403, 422
oil production, 7, 12, 19

Oklahoma
  induced seismicity, 96, 389, 395, 399, 400, 401, 427
  managing seismic risk, 413, 422, 430, 432, 433
  stress map, 184, 187, 189, 195, 410
  unconventional plays, 8, 29
Opalinus clay, 67, 125, 131, 138
optical fibers and distributed acoustic sensing, 269

permeability, 162
  anisotropy, 161, 162, 167, 174
  changes with depletion, 153, 157, 158, 164, 174, 320
  changes with shear slip, 315, 317, 319, 321, 331, 355, 357
  measurement issues, 151, 152, 155, 156
  relative, 141, 144, 147
Permian Basin
  earthquakes, 417
  faults and fractures, 215, 312
  potentially active faults, 418
  stress map, 184, 187
  water cycle, 382
PKN hydraulic fracture model, 242
plug-and-perf, 240, 302
Poisson's ratio, 43, 46, 58
pore pressure, 89
  depletion, 360, 368, 369, 372
  hydrostatic, 89, 209
  in unconventional formations, 205, 206, 344
  measurement in unconventional formations, 201
  overpressure, 95, 209, 230
  stimulating shear, 92, 108, 203, 209, 210, 305, 355, 359, 392, 401, 408, 409, 411, 428, 432
pores
  interparticle, 35, 39, 120
  intraparticle, 54, 121
  organic matter, 122, 133, 142
  size distribution, 124, 125, 127, 135
  sources of porosity, 118
poroelasticity
  laboratory data, 56, 57
  modeling, 353, 355, 356
  stress changes, 369, 370, 371, 372, 373
  theory, 53
porosity
  measurement methods, 117, 123, 128, 130, 131
  unconventional reservoirs, 15, 38, 51, 116, 119
principal stresses, 183
production from unconventional reservoirs, 176
  correlation with production, 205
  global, 9
  simulation, 178, 340, 346
  US, 7, 19, 20
pyrite in unconventional reservoirs, 61, 121

QFP (quartz, feldspar, pyrite), 41, 105
quantitative risk assessment (QRA), 409
quartz in unconventional reservoirs, 61

recovery factors, 20, 361
representative elementary volume (REV), 41
risk framework for induced seismicity, 406, 434
risk tolerance assessment, 440
rock fabric, 31, 39, 123

screen-out, 30
seismic velocity anisotropy, 44, 226, 227, 228, 268
Seismogenic Index model, 426, 428
  Oklahoma, 430, 433
S_{Hmax}
  definition, 182
  magnitude, 203, 204, 209
  orientation, 184
S_{hmin}
  control on hydraulic fracture orientation, 24, 245
  definition, 23
  drilling direction, 236
  magnitude, 94, 95, 204, 209
stereographic projection, 193, 215, 273, 305
stimulated rock volume (SRV), 4, 302, 330, 361
stress
  affect of uncertainty on fault slip prediction, 411
  changes due to depletion, 346, 370, 371, 372, 373
  concentration around vertical well, 191
  dynamic stress changes, 223
  earthquake stress drop, 284, 285, 286
  effective normal stress, 92, 315
  from focal plane mechanisms, 193, 195, 197, 275
  horizontal stress anisotropy, 28, 313
  magnitudes and pore pressure, 207, 209, 210
  map of S_{Hmax} and A_ϕ, 96, 184, 187, 188, 189, 190
  measurement techniques, 185
  polygon, 204
  quality criterion, 186
  ratio of maximum principal effective stresses, 93
  relative stress magnitudes, 183
  shadows, 238, 248, 249, 250, 261
  shear stress, 92
  variations with depth, 23, 233, 251, 325, 327, 337
  viscoplastic stress relaxation, 60, 80, 88, 341, 342, 343

temperature at depth, 208
tensile fractures, 114
Texas
  seismicity, 403, 408
  stress map, 184
Thomsen parameters, 228

total organic content (TOC), 15, 32, 36, 51, 61, 113, 124, 139
Tournemire shale, 67, 68
traffic light systems, 422, 423, 424, 425

unconfined compressive strength (UCS), 204
unconventional reservoirs
    compositional data, 61
    general properties, 15, 21, 33, 329, 365
    pore pressure, 206
Utica
    pore pressure, 205
    stress map, 191

Vertical Transverse Isotropy (VTI), 228
viscoelastic
    properties, 79
    theory, 65, 76, 77, 78, 81
viscoplastic
    stress relaxation, 89, 112
vitrinite reflectance (R_o), 11, 13
volumetric strain, 54, 153

water
    depth of aquifers, 253
    disposal, 94, 189, 382, 385, 393
    imbibition, 175
    pore, 115, 140

production, 147, 364, 381, 384
re-use, 138, 385
saline aquifers, 381
water use in hydraulic fracturing, 239, 380, 383
water-wet systems, 141, 144
wellbore breakouts, 186, 191
wells (unconventional)
    construction, 386
    general information, 21, 234
    integrity, 387
    leakage, 387
    path, 29
    potential in US, 8
    schematic, 23, 25, 234, 388
well-to-well communication, 362
Wolfcamp formation, 247, 251, 363
Woodford formation
    stimulation, 29, 220
    variation of stress with lithology, 323, 325, 343
World Stress Map, 185, 472, 475

X-ray goniometry, 34, 36

Young's modulus, 43, 46, 51, 58, 70, 72, 75, 79, 166, 343

zonal isolation, 241, 300